25th International Compressor Engineering Conference at Purdue 2021

AA005958

West Lafayette, Indiana, USA
24-28 May 2021

ISBN: 978-1-7138-3070-2

Printed from e-media with permission by:

Curran Associates, Inc.
57 Morehouse Lane
Red Hook, NY 12571

Ray W. Herrick Laboratories
School of Mechanical Engineering
Purdue University
177 S. Russell Street
West Lafayette, IN 47907-2099, USA.

Phone: +1 (765) 494-2132
Fax: +1 (765) 494-0787

rhlab@ecn.purdue.edu

Additional copies of this publication are available from:

Curran Associates, Inc.
57 Morehouse Lane
Red Hook, NY 12571 USA
Phone: 845-758-0400
Fax: 845-758-2633
Email: curran@proceedings.com
Web: www.proceedings.com

TABLE OF CONTENTS

CENTRIFUGAL COMPRESSORS

MODELING AND MEASUREMENT TECHNIQUES

ROTARY COMPRESSORS

SCREW COMPRESSORS

SCROLL COMPRESSORS

TRIBOLOGY, LUBRICATION AND NVH

ADDITIONAL PAPERS

Optimization of Combined Casing Treatment Structure Applied in a Transonic Axial Compressor Based on Surrogate Model

Jiaxuan Liu, Min Dai, Fusheng Yang*, Jianqiang Deng, Zaoxiao Zhang

Department of Process Equipment and Control Engineering, Xi'an Jiaotong University,
Xi'an, Shanxi Province, China
yang.fs@mail.xjtu.edu.cn

* Corresponding Author

ABSTRACT

For modern high load compressors, an excellent stability-enhancing capability by casing treatment (CT) is desirable. However, it is very time consuming to accomplish effective CT design. In this study, a new combined CT structure composed of axial skewed slots and end-wall injection, was proposed to be installed in transonic axial compressors to improve the overall performance. Considering the high computation cost for CFD simulation of the flow field in transonic compressor, a Gaussian Process Regression (GPR) surrogate model combined with Latin hypercube sampling, was utilized to predict compressor performance. For optimization process, a multi-objective evolutionary algorithm (NSGA-II) was adopted to obtain the Pareto-optimal front. The main geometric parameters of the slot and the mass-flow rate of injection were selected as design parameters, with the peak efficiency and pressure ratio being two objectives. The results indicated that the surrogate model works well in capturing the key features of the concerning target and accelerating the optimization process. The optimal scheme of the combined CT was found able to increase stall margin (SM) by 19.5% with low efficiency penalty, showing a better performance than the reference combined casing treatment (CCT) scheme. What's more, the analysis results of entropy generation showed that the superior effect of optimized scheme (OPT) can be attributed to the improvement of exchange flow in slots and the decreased loss in the whole passage.

1. INTRODUCTION

With the development of aerospace, energy and chemical industries, increasingly high requirements for axial compressors in practical application scenarios are posed, e.g. operating stability and high load. However, under transonic working condition, flow separation tends to be intensified and even blocks the flow passage, which may cause the rotating stall and surge (Wennerstrom, 1990; Han, 1999). Therefore, researchers have put forward numerous control techniques to enhance the operating stability of axial compressor over the last 50 years (Hergt et al., 2013; Tang et al., 2017; Houghton and Day, 2012), which can be divided into two categories: passive and active. The passive control method features simple structure and easiness to implement, but may not work well under varied operating conditions and produce additional loss (Wang et al.,2019). The active control method, on the contrary, involves relatively complex structure, yet its advantage lies in the flexibility of control effect by adjusting the parameters according to the state.

The optimal design of CTs has always been critical issue, which is helpful for improving the performance of the compressor. However, it is very time consuming to realize effective CT due to the difficulty to determine its effect on SM improvement before test. Hence, in order to design an excellent stability-enhancing CT with little efficiency penalty, different optimization strategies have been put forward. Abdus et al. (2008) chose the angles of sweep, lean, and skew as design variables, while the polynomial response surface approximation, Kriging, and radial basis neural network were utilized respectively. A 1.4% relative increase in peak efficiency was obtained using radial basis neural network. Kim et al. (2013) optimized a transonic axial compressor with circumferential casing grooves by adopting a weighted-average surrogate model. Three design variables including the tip clearance, blade tip angle, and depth of the grooves, were chosen during the optimization process. The results of steady state simulation indicated that the optimized CT increases the stall margin by 5.13% with the peak efficiency decreased by only 0.18%. Jiang et al. (2019) studied a novel CT based on the bimodal Gaussian function curves, which is optimized by a genetic algorithm with the Kriging model and orthogonal design. Three different configurations were simulated and the result showed that

CT influenced the vortex topology significantly, the casing and blade tip treatments can reduce leakage flow rate by 13.73% and lead to a 5.39% decrease of total pressure loss. Ba et al. (2020) introduced a novel axial slot casing treatment, and a series of new parameters controlling the meridional profile were put forward besides the conventional design parameters like the height, length and position. The optimization process was carried out based on NSGA-II and Kriging surrogate model, and the Pareto front was obtained considering both stability and efficiency. The optimal result shows improvement of the stall margin by 11.23% with the peak efficiency reduced by 0.21%.
On the other hand, scholars have done much research on combining both passive and active CTs due to the limitation of individual method. Kim et al. (2013) combined the injector with grooves. The simulation results showed that the combined CT increased the SM and the peak efficiency by 6.08% and 0.6%, respectively. The combination of two structures also reduced the entropy near the shroud and decreased the low-velocity region remarkably. In a follow-up study, Kim et al. (2013) discussed the influences of the length and height of the grooves and the mass flow rate of the injection on the performance of compressor, using a hybrid multi-objective evolutionary algorithm based on NSGA-II. The optimized combined CT improved the SM and peak efficiency by 8.585% and 0.96%, respectively. An obvious correlation was also found between the decrease in the peak adiabatic efficiency and the expanding SM. Dinh et al. (2015) added an ejection structure in the grooves on the basis of Kim's work, the influences of the length and the height of the grooves, the mass flow rate of injection/ ejection on the total pressure ratio were studied. They found that the loss of the total pressure ratio of the transonic axial compressor decreases compared with the case in which the groove combined with injection only. The influence of mass-flow rate on stall margin was found smaller than that of geometric parameters. However, current studies on combined CTs mainly focus on the leakage flow on the blade tip, while the reduced mass flow in the lower part of the blade caused by the additional structure is also worth noting according to Müller et al. (2011).
In this study, a new combined CT composed of axial skewed slots and end-wall injection is introduced for an axial transonic compressor in order to enhance the stability of compressor and reduce the efficiency penalty. A surrogate model was presented using Gaussian Process Regression (GPR) model and then a multi-objective evolutionary algorithm (NSGA-II) was adopted to obtain a global Pareto-optimal front for the design of such combined CT. The width and depth of the slots as well as the mass-flow rate of injection, were selected as the design variables. The performance and flow characteristic are analyzed through steady-state simulation. What's more, the analysis of entropy generation in the flow passage was also conducted to explore the working mechanism of the newly proposed CT for superior effect on flow control.

2. MODEL DISCRIPTION

NASA rotor 67 known for typical stall phenomena at the blade tip when operated close to surge, is taken as the research object in this paper. The number of blades is 22, the design speed is 16043 r/min, and the design pressure ratio is 1.63. The rotor operates at a design mass-flow rate of 33.25 kg/s. Detailed geometric data are referred to the literature (Chima and Strazisar, 1982), and the schematic is shown in Figure 1.

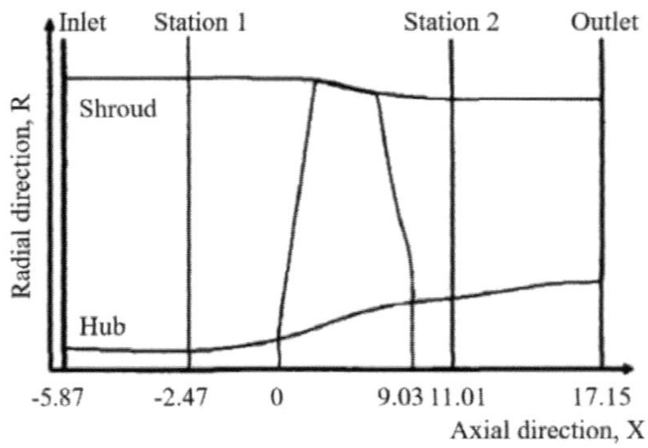

Figure 1: Geometric design of NASA rotor67.

Liu et al. (1987) showed that the axial skewed slots perform best with regard to stability enhancement of rotor under the same area ratio and length of slot, hence such structure is adopted as shown in Figure 2. The initial main structural parameters are shown in Table 1. The starting point of the slot is located at the leading edge of the rotor tip. In addition, the end-wall injection is introduced by groove at the hub. The groove is placed at 90% of the chord length with a width of 1mm, as shown in Figure 3, and the original mass-flow rate of injection is 0.01kg/s.

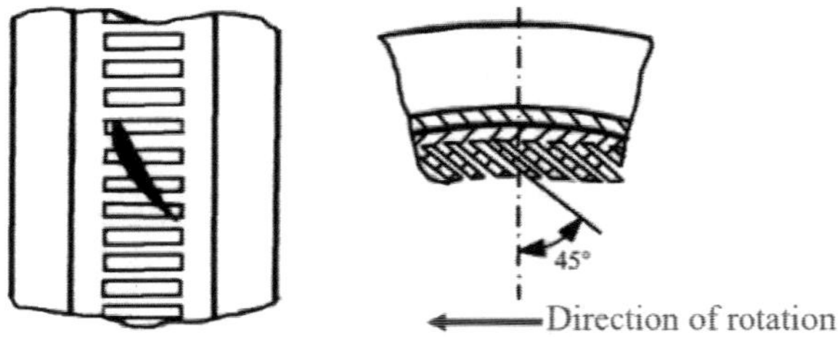

Figure 2: Geometry of axial skewed slots.

Table 1: Basic parameters of axial skewed slots.

Parameters	Value
Depth	11mm
Width	3mm
Number of slots	6
percent of slot area	50%
Skewed angle	45°

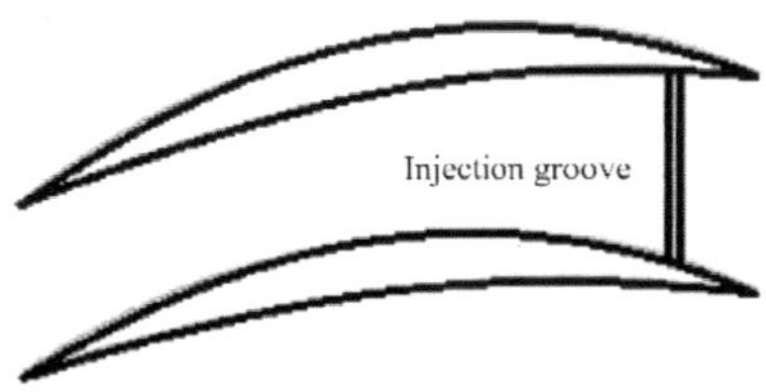

Figure 3: Geometry of injection groove.

The computational domain is discretized into multi-blocks of structured grids, as shown in Fig. 4(a). The mesh of blades is covered by the O-blocks, which are used to accurately predict the boundary layer flows. The rest passage domain is filled up with the H/J/C/L-blocks generated using the software TurboGrid in ANSYS Workbench 17.0. The grids in the slots and groove form an individual block created by the meshing tool ICEM. The dimensionless parameter, y+, is kept below 3. The total number of grids inside the passage is 1.5×10^6, with each single axial skewed slot and injection groove being resolved with 2×10^4 and 1.5×10^4 grids, respectively.

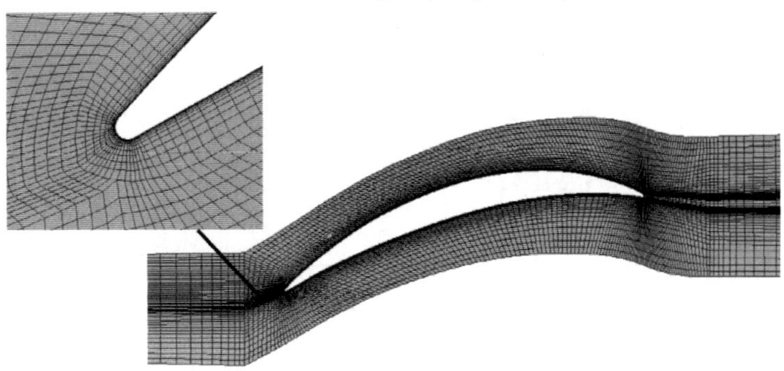

(a)

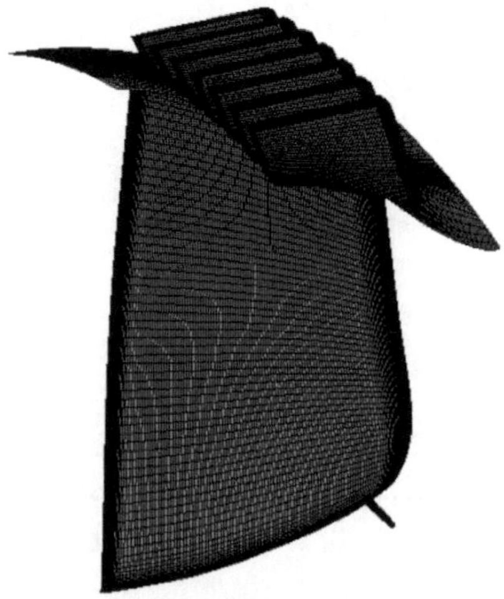

(b)

Figure 4: Local computational domain and grids. (a) passage grid;(b) three-dimensional view.

The simulations are carried out in the CFX solver using RANS model with the energy equation considered. The SST k-omega turbulence model is used to the discrete equations. The physical time step is 5×10^{-4}s with the residual target of 1×10^{-6}. A single passage computation is performed considering the periodicity. The total pressure and temperature of inlet and the mass flow rate of outlet are given as 1.0 atm, 288.15 K and 33.25kg/s, respectively. No-slip wall condition is specified on walls, and the "frozen rotor" is used for the rotor-to-stator interface. As reported by Prince et al. (1974), the steady flow distribution is what matters for the effect of casing treatment, hence the study was carried out for steady state simulation. The numerical model for the single rotor was validated against the experimental data previously (Liu et al., 2020), and the results showed a satisfactory prediction of performance curve with a relative error within 3%. The grid independence of the computations was also tested.

It should be noted that the mass-flow rate at the inlet and outlet of the flow passage is no longer conserved with the injection, so the general formula of isentropic efficiency is not applicable. According to the method adopted by Wang et al. (2009), the modified efficiency formula considering the injection could be obtained as follows:

$$\eta_{in} = \frac{W_a}{W} = \frac{\left(\frac{P_{out}}{P_{in}}\right)^{\frac{\gamma-1}{\gamma}} - 1}{\left(\frac{T_{out}}{T_{in}}\right) - \left(\frac{m_i}{m_{out}}\right) * \left(\frac{T_i}{T_{in}} + m_{in}/m_i\right)} \tag{1}$$

Where W_a is the required isentropic compression work , W is the actual work done, P_{out} is the outlet pressure, P_{in} is the inlet pressure, γ is the specific heat ratio, T_{in} is the inlet temperature,T_{out} is the outlet temperature, T_i is the temperature of the injection, m_{in} is the mass-flow rate of inlet, m_{out} is the mass-flow rate of outlet and m_i is the mass-flow rate of injection.

3. OPTIMIZATION PROCEDURE

The efficiency and pressure ratio were determined as the optimization objectives since Ba et al. (2020) found that a trade-off correlation between efficiency and SM can be obtained from the optimization results, meanwhile the computational expense can be decreased. The depth of slot, number of slots, skewed angle of slot and mass-flow rate of injection were selected as the design variables through parametric studies. The range of variables are shown in Table 2, and the design points, were selected by the screening method. The values of the objectives at these design points were calculated and the results were used to construct the surrogate model.

Table 2: Design space

Variables	Low bound	Upper bound
Depth of slot	9.9mm	18.4mm
Number of slots	4	8
Skewed angle	30°	75°
Mass-flow rate of injection	0.001kg/s	0.03 kg/s

Then the Gaussian Process Regression (KRG) is chosen to fit and forecast the data, since it is suitable for the regression problem of low dimension and small sample compared with other surrogate models e.g. response surface. The error between prediction results by the surrogated model and those by RANS is within 0.1%, which indicates that the model can meet the accuracy requirement. To achieve an ideal compromise between the two objectives, a multi-objective evolutionary algorithm NSGA-II is applied to obtain a global Pareto-optimal front. The maximum number of evolutionary generation, population size, crossover rate and mutation rate are set as 50, 200, 0.8 and 0.005, respectively. In addition, to ensure the application of compressor, the optimization procedure is executed with a mass-flow rate near the original design point. The optimization results are shown in Figure 4. It's obvious that the Pareto-optimal solution resembles a convex front, indicating a trade-off correlation that higher peak adiabatic efficiency is always accompanied by lower SM and vice versa. One point on the Pareto front is chosen as the optimal scheme (OPT), the parameters of which are shown in Table 3.

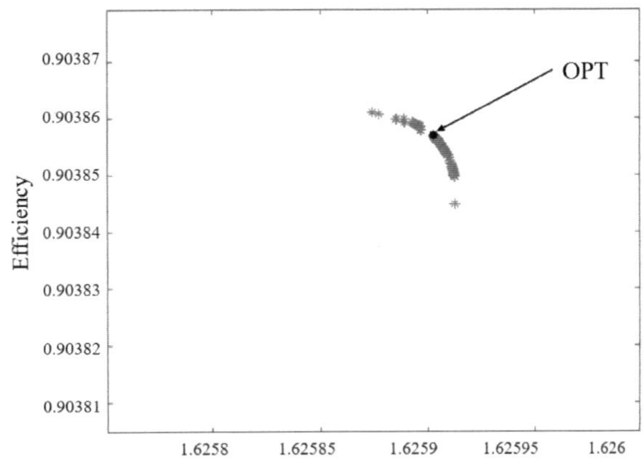

Figure 4: Optimization results.

Table 3: Parameters of OPT scheme.

Variables	Value
Skewed angle	60°
Depth of slot	11.28mm
Number of slots	6
Mass-flow rate of injection	0.0076kg/s
Isentropic efficiency	90.46%
Pressure ratio	1.6114

4. RESULT AND DISCUSSION

The definition of the SM of compressor is:

$$SM=\left(\frac{\pi_{NS}^{*}m_{NPE}^{*}}{\pi_{NPE}^{*}m_{NS}^{*}}-1\right)\times 100\% \quad (2)$$

Where π_{NS}^{*} and m_{NS}^{*} represent the pressure ratio and mass-flow rate near the stall point, π_{NPE}^{*} and m_{NPE}^{*} represent the pressure ratio and mass-flow rate near the peak efficiency point.

3.1 Performance Analysis

Figure 5 shows a comparison of performance curves for different schemes, in which SC, CCT, OPT represent the smooth casing scheme, the combined casing treatment schemes before and after optimization, respectively. The simulations were carried out at the design rotating speed with various back pressures. It is obvious that the application of combined CT expands the stable operating range with relatively large efficiency penalty in CCT scheme, while in OPT scheme the surge is further postponed with the efficiency penalty decreased compared to CCT. The SM of SC, CCT and OPT schemes are 10.8%, 25.53%, 30.55% respectively, indicating the stability of compressor is improved significantly after optimization.

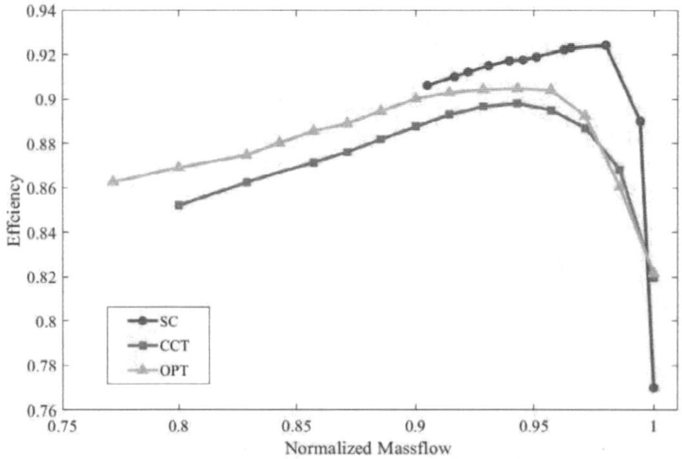

Figure 5: Comparison of performance curves.

3.2 Analysis of Leakage Flow

To study the effect of stall postponing of different schemes, a comparison of leakage flow at stall point of SC is shown in Figure 6. It's obvious that there are three vortices in Figure 6 (a), i.e. leakage vortex, induced vortex and separation vortex (Hu et al. 2015). The leakage flow at the leading edge of the blade presents a nearly circumferential pattern and blocks the passage. The induced vortex is nearly perpendicular to the main flow and intersects with the leakage vortex. Besides, the low-velocity fluid separates under the shear action of the wall boundary layer and forms a separation vortex. The flow loss near the tip increases and compressor tends to get unstable with significant reduction in efficiency.

In Figure 6 (b), there is no apparent blockage and the development of leakage vortex is suppressed because of the introduction of slots. The growth of the separation vortex at the trailing edge is also restrained since the injection increases the momentum of low-velocity fluid, thus the range of stable operation is expanded. However, the phenomenon of the flow crossing into an adjacent passage still exists. In Figure 6 (c), the leakage flow is significantly improved. There is no obvious expansion when the leakage flow moves downstream, further reducing its influence area. In addition, the fluid close to the blade flows along the surface without boundary layer separation. It is further demonstrated that a well-designed CT can improve the stability of the compressor remarkably by eliminating the vortexes.

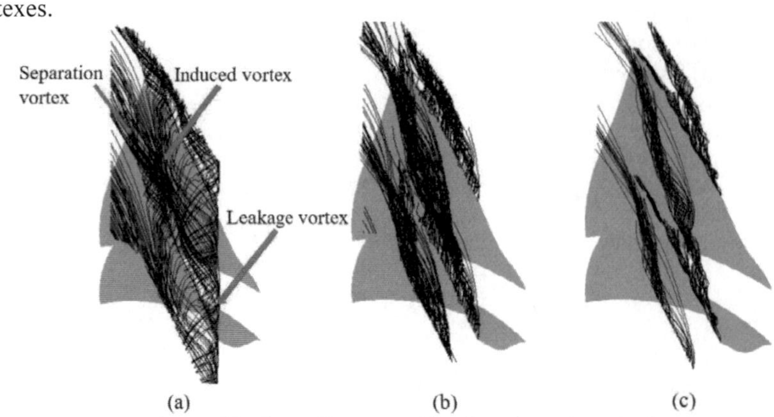

Figure 6: Comparison of leakage flow at stall point of SC (a) SC;(b) CCT;(c) OPT.

3.3 Analysis of Flow Field

Figure 7 shows a comparison of Mach number (Ma) contour at 80% span of different schemes at near design point. In Figure 7 (a), the shock wave occurs and forms a shape of "λ" with main flow. A high Ma region (HMR) occurs and the boundary layer separates at the trailing edge, resulting in a low-velocity region near the outlet of the passage. In Figure 7 (b), the casing wall and the flow passage are both changed partially by slots, the location of the shock wave moves forward to the leading edge, the intensity of shock wave is also decreased. The thickness of the boundary layer as well as the area of low-velocity region are significantly reduced. Brandstetter et al. (2016) concluded that the effect of slots could be accounted for by the so-called exchange flow, i.e. flows with higher static pressure in the downstream blocking area of shock waves are pumped into slots, accelerated, and then injected upstream to the leading edge. In this regard, in CCT scheme, the injection replenishes the fluid at lower part of the blade and increases the radial momentum in the passage at the same time, improving the suction effect of exchange flow and hence the stability. However, when the exchange flow in the slots reach its maximum, excess flow near the slots will increase mixing loss and even block the passage (Fan, 2015). In Figure 7 (c), the excess flow is prevented by being pumped timely with enhanced exchange flow and weakened injection. The value of Ma in HMR reduce, hence the turbulence intensity and mixing loss. Accordingly, the improvement on stability enhancement and efficiency is a mutual balance of slot structure and injection parameters.

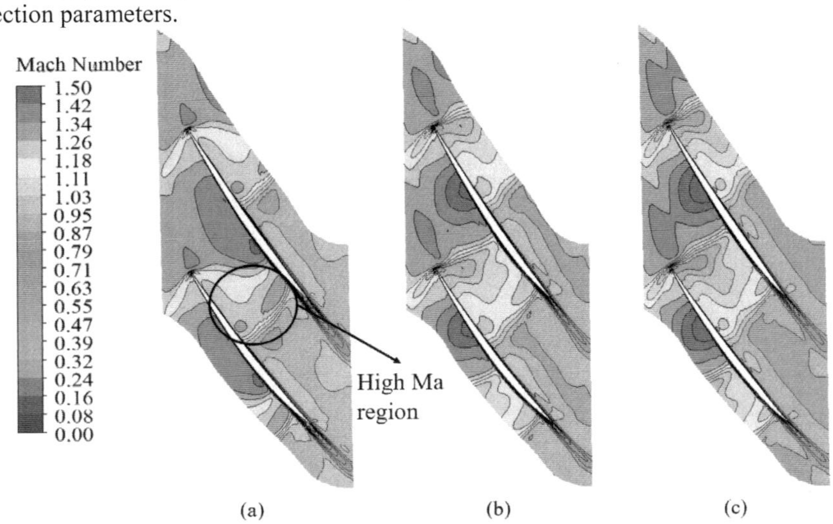

Figure 7: Comparison of Ma contour at 80% span. (a) SC;(b) CCT;(c) OPT.

3.4 Entropy Analysis

In the last several years, a new method based on local entropy generation (Kock and Herwig, 2004; Nan et al., 2016) was developed for analyzing loss mechanisms and identifying entropy quantitatively. In this study, in order to further clarify the mechanism of the effect of combined CT on stability enhancement, the entropy generation in the concerning flow passages is analyzed. Since steady state simulation was conducted for the adiabatic compressor system, the entropy generation in the compressor can be calculated from numerical RANS results as follows (Zhang et al., 2019):

$$\dot{S}_{g,local} = \frac{k}{T^2}\frac{\partial \bar{T}}{\partial z}\frac{\partial \bar{T}}{\partial z} + \frac{\alpha_t}{\alpha}\frac{k}{T^2}\frac{\partial \bar{T}}{\partial z}\frac{\partial \bar{T}}{\partial z} + \frac{2\mu \overline{s_{ij} s_{ij}}}{\bar{T}} + \frac{\varepsilon \bar{\rho}}{\bar{T}} \quad (3)$$

Where α_t is the turbulent thermal conductivity, " ¯" means the numerical Reynolds averaged parameters, k is thermal conductivity; T is local temperature of fluid; μ is fluid dynamic viscosity, s_{ij} is shear strain rate, ρ is density and ε is turbulent dissipation rate.

Figure 8 shows a comparison of contour of $\dot{S}_{g,local}$ and backward streamline in the slot. The exchange flow is obvious in Figure 8(a). However, there is a flow blockage near the wall in CCT scheme, in addition, the strong shear force increases entropy generation rate significantly and hence flow loss. In Figure 8(b), as the depth increases and the injection weakens, the excess flow near the tip decreases, hence the exchange flow is enhanced with lower entropy generation rate after optimization, and the SM is further increased with lower efficiency penalty.

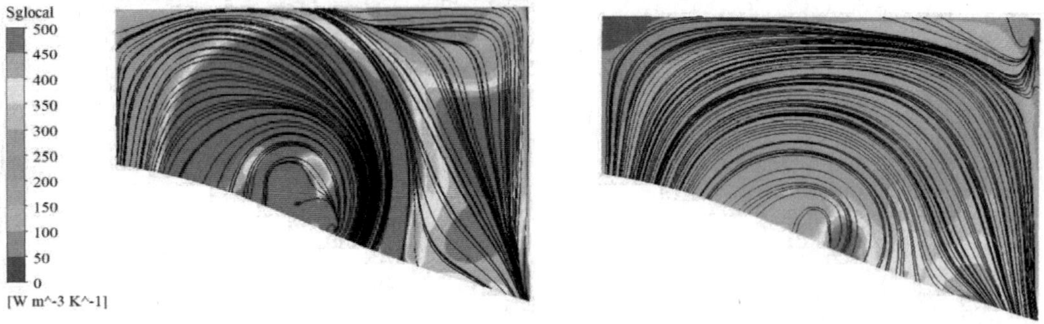

(a) (b)

Figure 8: comparison of contour of $\dot{S}_{g,local}$ and backward streamline. (a) CCT;(b) OPT.

Furthermore, to determine the region which brings about the most entropy generation and compare the influences between CCT and OPT, the whole passage is divided into three parts, including upper section, lower section and main flow section, as shown in Figure 9. The total entropy generation of each part is calculated according to Eq. (3), and the relative variation of loss is defined as follows:

$$\Delta L = \frac{L_{ORI} - L}{L_{ORI}} \tag{4}$$

Where L_{ORI} represents the loss of SC scheme, L represents the loss of CCT or OPT scheme.

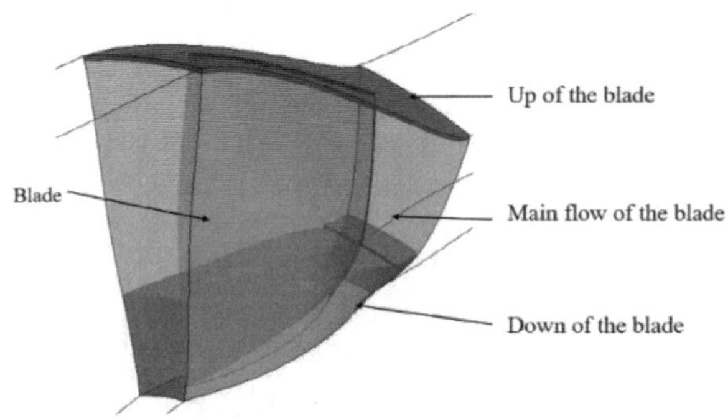

Figure 9: Schematic drawing of control volumes.

The relative variation of loss is shown in Table 4. Apparently, the effect of loss reduction in CCT scheme mainly concentrates in the lower region since the mass flow in the lower part of the blade is replenished and the flow separation is inhibited, while the loss of main flow and upper region both increase because of strong shear action of flow. The total loss in the whole passage, however, decreases by 2.68%. On the other hand, the total loss in OPT scheme is 6.21% lower than the original scheme. The main effect still lies in the lower region, in addition, the loss of upper region and mainstream region both decrease just as entropy analysis indicated.

Table 4: Relative variation of loss.

	Up	Down	Main	Sum
CCT	-0.03%	18.49%	-2.07%	2.68%
OPT	6.04%	21.21%	1.47%	6.21%

5. CONCLUSIONS

An optimization process based on NSGA-II and GPR of a novel CT structure combining axial skewed slots and end-wall injection, is proposed in this paper. The analysis of performance curve, leakage vortices, flow field and entropy generation in three schemes are studied by numerical analysis. It was found that the selected optimal representative design, OPT, were superior to the reference design, the SM is increased by 19.5% with little efficiency penalty, greatly expanding the stable operating range. The optimized CT structure embodies a mutual balance of slot geometric parameters and injection rate. The superior flow control performance by the optimized structure can be attributed to the improvement of exchange flow, and the reduction of flow loss throughout the whole passage is also confirmed via entropy generation analysis.

NOMENCLATURE

η_{in}	isentropic efficiency	-
T	temperature	K
P	pressure	Pa
γ	specific heat ratio	-
m	mass-flow rate	kg/s
π	pressure ratio	-
$\dot{S}_{g,local}$	entropy generation rate	$W/K \cdot m^3$
k	turbulence kinetic energy	J
α	thermal conductivity	-
μ	fluid dynamic viscosity	$Pa \cdot s$
ρ	density	kg/m^3
ε	turbulent dissipation rate	-
s_{ij}	shear strain rate	1/s
ΔL	relative variation of loss	-
L	loss	W

Subscript

in	inlet
out	outlet
i	injection
NS	near stall
NPE	near peak efficiency
t	turbulent
SC	SC scheme
c	CCT or OPT scheme

REFERENCES

Ba, D., Zhang, Q. F., Du, J., Li Z. H., Zhang, H. W., Nie, C. Q. (2020). Design optimization of axial slot casing treatment in a highly-loaded mixed-flow compressor. *Aerospace ence and Technology*, 107,1270-9638.

Brandstetter, C., Wartzek, F., Werner, J., Schiffer, H. P., Heinichen, F. (2016). Unsteady measurements of periodic effects in a transonic compressor with casing treatments. *Journal of Turbomachinery*, 138(5), 051007.1-051007.9.

Chima, R. V., Strazisar, A. J. (1982). Comparison of two- and three-dimensional flow computations with laser anemometer measurements in a transonic compressor rotor. *Journal of Engineering for Gas Turbines & Power*, 105(3), 596-606.

Dinh, C. T., Heo, M. W., Kim, K. Y. (2015). Aerodynamic performance of transonic axial compressor with a casing groove combined with blade tip injection and ejection. *Aerospace Science and Technology*, 46, 176-187.

Fan, L., (2015). Stall of an axial-compressor rotor with casing treatment. *Journal of Aerospace Power*, 30(11), 2647-2657.

Hah, C., Loellbach, J. (1999). Development of Hub Corner Stall and Its Influence on the Performance of Axial Compressor Blade Rows. *ASME Journal of Turbomachinery*, 121(1), 67-77.

Hergt, A., Meyer, R., Engel, K. (2013). Effects of Vortex Generator Application on the Performance of a Compressor Cascade. *ASME Journal of Turbomachinery*, 135(2), 021026.1-021026.10.

Houghton, T., Day, I. (2012). Stability Enhancement by Casing Grooves: The Importance of Stall Inception Mechanism and Solidity. *ASME Journal of turbomachinery*, 134(2), 021003.1-021003.8.

Jiang, S., Chen, F., Yu, Y. J., Chen, S. W., Song, Y. P. (2019). Treatment and optimization of casing and blade tip for aerodynamic control of tip leakage flow in a turbine cascade. *Aerospace ence and Technology*, 86, 704-713.

Kim, D. W., Kim, J. H., Kim, K. Y. (2013). Parametric Study on Aerodynamic Performance of a Transonic Axial Compressor with a Casing Groove and Tip Injection. *Applied Mechanics and Materials*, 284-287, 872-877.
Kim, J. H., Choi, K. J., Kim, K.Y. (2013). Aerodynamic analysis and optimization of a transonic axial compressor with casing grooves to improve operating stability. *Aerospace ence and Technology*, 29(1), 81-91.
Kim, J. H., Kim, D. W., Kim, K. Y. (2013). Aerodynamic optimization of a transonic axial compressor with a casing groove combined with tip injection. *Journal of Power and Energy*, 227(8), 869-884.
Kock, F., Herwig, H. (2004). Local entropy production in turbulent shear flows: a high-Reynolds number model with wall functions. *International Journal of Heat and Mass Transfer*, 47(10-11), 2205 2215.
Liu, J. X., Yang, F. S., Deng, J. Q., Zhang, Z. X., (2020). Influence of Combined Casing Treatment on Flow Characteristics of Transonic Compressor. *Journal of Xi'an Jiaotong University*, 54(04), 173-178.
Liu, Z. W., Zhang, C. S., Shi, J. X., Wang, Z. S., Huang, J. G. (1987). A study on effects of axial positions of skewed slots CT on compressor performance. *Journal of Engineering Thermophysics*, 8(1).
Mu¨ller, M. W., Schiffer, H. P., Voges, M., Hah, C. (2011). Investigation of Passage Flow Features in a Transonic Compressor Rotor With Casing Treatments. *Proceedings of the ASME 2011 Turbo Expo: Turbine Technical Conference and Exposition. Volume 7: Turbomachinery, Parts A, B, and C* (65-75). Vancouver, British Columbia, Canada: ASME.
Nan, X., Liu, L., Ma, N., Lu, Q., Lin, F. (2016). Numerical investigation of entropy generation distribution in a transonic compressor. *Proceedings of the ASME Turbo Expo 2016: Turbomachinery Technical Conference and Exposition. Volume 2A: Turbomachinery* (GT2016-56519), Seoul, South Korea: ASME.
Prince, D. C., Wisler, D. D., Hilvers, D. E. (1975). Study of Casing Treatment Stall Margin Improvement Phenomena. *Proceedings of the ASME 1975 International Gas Turbine Conference and Products Show. Volume 1A: General* (75-GT-60). Houston, Texas, USA: ASME.
Samad, A., Kim, K. Y., Goel, T., Haftka, R. T., Shyy, W. (2008). Multiple Surrogate Modeling for Axial Compressor Blade Shape Optimization. *Journal of Propulsion & Power*, 24, 302-310.
Tang, M. Z., Jin, D. H., Gui, X. M. (2017). Modeling and Numerical Investigation of the Inlet Circumferential Fluctuations of Swept and Bowed Blades. *ASME Journal of Thermal Science*, 26(1), 1-10.
Wang, R. G., Hu, J. G., Yu, C., Li, K. (2015). Research on secondary flow vortex structure in transonic compressor rotor. *Journal of Propulsion Technology*, 36(04), 504-512.
Wang, S. T., Qiang, X. Q., Feng, G. T., Lin, W. C., Wang, Z. Q. (2009). Study of low-reaction boundary layer suction axial compressor and it's internal flow control. *Journal of Engineering Thermophysics*, 30(1), 35-40.
Wang, W., Lu, J. L., Luo, X. Q., Chu, W. L. (2019). Coupling method of stability enhancement based on casing treatments in an axial compressor. *Aerospace Science and Technology*, 95, 1270-9638.
Wennerstrom, A. J. (1990). Highly Loaded Axial Flow Compressors: History and Current Developments. *ASME Journal of Turbomachinery*, 112, 567–578.
Zhang, Q., Du, J., Li, Z., Li, J., Zhang, H. (2019). Entropy generation analysis in a mixed-flow compressor with casing treatment. *Journal of Thermal Science*, 28(5), 915-928.

ACKNOWLEDGEMENT

This work is financially supported by National Key R&D Program of China (NO.2017YFB0903600)

1D Design and Optimization of a Micro-Centrifugal Compressor Design for Air Conditioning Applications

Zhenyuan Mei[1], Tao Cao[1*], Yunho Hwang[1]

[1]Center for Environmental Energy Engineering
Department of Mechanical Engineering, University of Maryland,
4164 Glenn Martin Hall Bldg., College Park, MD 20742, United States

*Corresponding author: Tel: (301) 405-8672, email: taocao@umd.edu

ABSTRACT

Miniature compressors play an important role in compact air conditioning systems. Its small size means smaller weight and less material cost. Centrifugal compressors do not have displacement volume and has only one major moving part, which makes it a good candidate for miniature compressor. Typical residential air conditioning systems use R410A as refrigerant due to its good thermodynamic properties and no ozone depletion potential. However, its global warming potential is even higher than that of R22. In addition, it requires high operating speed for a centrifugal compressor. One alternative for R410A is R600a. It has good thermodynamic performance and low impact on environment. Also, its operating pressure is much lower than that of R410A and it requires lower rotational speed. In this paper, a one-dimensional design was performed for a micro-centrifugal compressor for an air conditioning system which has 7 kW cooling capacity and uses R600a as the refrigerant. Furthermore, a single-objective optimization was performed to improve its efficiency using the genetic algorithm. The simulation result shows that the obtained design has 69.7% isentropic efficiency and 2.74 pressure ratio at 250 KRPM.

Keyword: 1D design and optimization, Micro-centrifugal compressor, R600a

1. Introduction

Miniature compressors play an important role in compact air conditioning systems. Its small size means smaller weight and less material cost. Centrifugal compressors do not have displacement volume and has only one major moving part, which makes it a good candidate for miniature compressor. R410A is a refrigerant that has been widely used in residential air conditioning applications due to its good thermodynamic performance and zero ozone depletion potential (ODP). However, its 100-year global warming potential (GWP) is 2,088, which is 15% higher than the GWP of R22 (*IPCC AR4*, 2007; Tian et al., 2015). This means its effect on global warming is about 2,000 times higher than that of carbon dioxide. In addition, it requires high operating speeds for a centrifugal compressor. One alternative to R410A is R600a (isobutane). As a hydrocarbon refrigerant, it is environmentally friendly and has a good thermodynamic performance. Its ODP and GWP are 0 and 5, respectively. And it has large latent heat, which is beneficial for the thermodynamic cycle. Moreover, its operating pressure is relatively lower than that of R410A and it requires lower rotational speed. However, it is highly flammable, which limits its application. Recently, it is gradually being allowed to be used in the refrigerator system with a charge limitation of 150 g. In the future, a compact air conditioner with next generation heat exchangers and compressors may enable its application for residential air conditioning. There have been significant work done on small-scaled centrifugal compressors. Schiffmann et al. (2009) designed a small-scaled centrifugal compressor for a two-stage R134a domestic heat pump system. Its rotational speed was 210,000 RPM, and its peak efficiency was 79%. Javed et al. (2016) designed a two-stage centrifugal compressor for a heat pump system using R134a. Its efficiency at the design point was around 79.5%. Casey et al. (2013) developed a small-scaled centrifugal compressor for a single-stage heat pump using R600. Its rotational speed was 250,000 RPM, and its diameter was around 21 mm. In this paper, preliminary one-dimensional (1D) design and optimization were performed for an air conditioning system using R600a as the refrigerant.

2. Design boundaries

Before performing the preliminary design, it is very important to define the design boundaries. In this paper, the small-scaled centrifugal compressor was designed for a basic four-component vapor compression cycle. The evaporating temperature and condensing temperatures were set to be 10°C and 45°C, respectively. And the cooling capacity was set to be 7 kW. A thermodynamic model was developed in Engineering Equation Solver (EES) to calculate the corresponding refrigerant mass flow rate, suction pressure, and discharge pressure. In this thermodynamic model, it was assumed that there is no pressure drop inside any component. And the maximum isentropic efficiency (85%) from the $n_s d_s$ diagram for single stage compressors from Balje (1981) was used. The suction superheat and condenser subcooling were fixed at 5 K. The calculation result shows that the designed compressor can deliver the refrigerant mass flow rate of 0.025 kg/s at the pressure ratio of 2.74. These values were used as the design boundaries for the centrifugal compressor.

3. 1D centrifugal compressor model and design variables

3.1 1D Compressor Model

In general, there are two types of 1D compressor models (Japikse and Baines, 1994): single zone model and two-zone model. The single zone model assumes the flow inside centrifugal compressor is one-dimensional, and uses a mid-streamline to represent the entire flow passage. And compressor losses are calculated by empirical correlations. Another modeling approach is the two-zone model, which models the flow passage as two zones: the primary zone and the secondary zone. The flow inside the primary zone is considered as an isentropic flow, and the secondary zone consists of low velocity regions such as near wall boundary layer and viscous region which contains substantial losses. The fluid properties of each zone are calculated separately. Eventually, they are mixed at the exit of the impeller, where mixing loss is applied. Japikse (1985) compared the single zone and two-zone models by using the Eckardt compressor. The results show that the deviation of the two-zone model is much less than that of the single zone model. In this paper, the compressor was designed by using the two-zone model in COMPAL®, which is a commercial software developed by NREC. And in our calculation, specific heat polynomial fluid model was used. Where the specific heat of the refrigerant is a function of fluid temperature.

3.2 Rotational Speed

It is important to determine the rotational speed of the centrifugal compressor before designing the compressor. The $n_s d_s$ diagram for single stage compressors from Balje (1981) was used to calculate the estimated rotational speed. Where n_s is the specific speed and d_s is the specific diameter. Note that this diagram was generated for $Re = 2 \times 10^6$ and $s/b_2 = 0.02$, which could be different from the actual design. Therefore, the actual efficiency could be different. According to the $n_s d_s$ diagram, the minimum specific speed required for the maximum efficiency for radial machine was around 0.64, and the corresponding specific diameter was around 4. The calculated rotational speed and impeller diameter were 247 KRPM and 20 mm. In the base design, 240 KRPM was used.

3.3. Incidence Angle

Incidence angle is the difference between the inlet relative flow angle and the inlet blade angle. And it has a positive correlation with the incidence loss (Conrad et al., 1980). In the base design, 1° was used to delay inducer stall as suggested by Javed et al. (2016).

3.4 Hub-to-tip Ratio at Impeller Inlet

The hub-to-tip ratio is defined as the impeller inlet hub radius over the impeller inlet tip radius, which defines the fraction of flow area over the inlet cross-section. This value typically ranges from 0.3 to 0.5 (Javed et al., 2016). Javed et al. (2016) did a parametric study for a similar-sized compressor with the same refrigerant. They found that 0.35 was the optimum value for their compressor. Therefore, in the base design, 0.35 was used.

3.5 Hub Diameter

Many factors affect the minimum hub diameter. For example, the hub diameter needs to be large enough to hold enough blades. Other limiting factors are the minimum shaft diameter required to transport torque and inducer relative Mach number (Javed et al., 2016). Here, the calculation of the minimum shaft diameter required is explained.

Many researchers used aluminum as their impeller material (Javed et al., 2016; Schiffmann and Favrat, 2010) in the past. And in this preliminary design, aluminum 6061 is selected as the material. This minimum diameter is then calculated by the equation derived from Soderberg's failure criteria (Budynas et al., 2011):

$$d = \left(\frac{16n}{\pi}\left\{\frac{1}{S_e}\left[4\left(K_f M_a\right)^2 + 3\left(K_{fs} T_a\right)^2\right]^{0.5} + \frac{1}{S_{yt}}\left[4\left(K_f M_m\right)^2 + 3\left(K_{fs} T_m\right)^2\right]^{0.5}\right\}\right)^{1/3} \quad (1)$$

Where n is the safety factor, and 1.5 was used as the safety factor calculation. $\boldsymbol{S_e}$ is the endurance limit, S_{yt} is the yielding strength, M_a and T_a are the alternating bending moment and alternating torque, M_m and T_m are the midrange bending moment and midrange torque. The calculated minimum diameter was 1.77 mm. Therefore, if the hub inlet diameter was larger than 1.77 mm, it should be able to transport the required torque.

3.6 Number of Blades

Meroni et al. (2018) used an empirical equation from Xu and Amano (2012) to calculate the number of blades required for the impeller as shown in Equation (2). Both equations results were rounded to the closest integer. In this design, the calculated number of blades is 23, which is hard to be fabricated in such a small dimension. Eventually, 14 blades at the impeller exit were used in the base design.

$$N_b = \begin{cases} round(12.03 + 2.544 \times PR_{tt}) & without\ splitter \\ round(-4.527 \times e^{\frac{1.865}{PR_{tt}}} + 32.22) & with\ splitter \end{cases} \quad (2)$$

3.7 Backswept Angle

Nowadays, backswept impeller is widely used since it provides wide surge margin (Boyce, 2003). In the base design, the backswept angle is set to be -45° (Schiffmann and Favrat, 2010).

3.8 Diffuser

There are many types of diffusers. Since in the base design, the impeller is expected to be very small, it is hard to implement a vaned diffuser. Also, the vaneless diffuser usually has a wider operating range. Therefore, the vaneless diffuser was used in this design. The width of the vaneless diffuser was constant and equaled to the impeller exit width, and the ratio of its exit radius over the impeller exit radius was 1.65. After all the design inputs were provided, COMPAL® generated the base compressor design, which has a 2.74 total-to-static pressure ratio and 66% total-to-static efficiency. And the compressor power at the design point was 1.6 kW.

4. DESIGN Optimization

Design optimization was performed to maximize the compressor's efficiency at the design point. Effects on the operating range was not considered during this optimization. Pymoo (Blank and Deb, 2020), which is a genetic algorithm optimization package available in Python, was used. And since the current problem was a single objective optimization, the basic genetic algorithm was used. The problem formulation is shown below:

$$\begin{aligned} & \mathrm{f} = -\eta_{is,ts} \\ s.t. & \\ & no\ stall\ or\ choke \\ & 0.9 - r_{1h} \leq 0 \end{aligned} \quad (3)$$

A Python script was written to bridge Pymoo and COMPAL®. The objective function is essentially the compressor design software whose outputs are η , and other important design parameters. Because the solver solves the minimization problem, a minus sign is added to the calculated efficiency to convert it into a minimization problem. There are five design variables, and their ranges are listed in Table 1.

Table 1: Range of design variables.

Design variables	I_{1t}	r_{1h}/r_{1t}	β_{2b}	N	r_5/r_{2t}
Unit	degree	-	degree	KRPM	-
Lower bound	-2.0	0.3	-50	200	1.05
Upper bound	2.0	0.5	0	250	2.00

The population size was 50. Objective space tolerance was used as the stopping criteria. The average change in the objective space should be less than 10^{-5} to stop the solver. For better robustness, the maximum value in the last 50 generations was picked. Note that this stopping criterion was evaluated every 10th generation.

Three optimization runs were performed. The initial values were selected randomly. The best fitness values of each generation for each run are shown in Figure 1. Because of the stopping criteria, the number of generations for each run is different. Eventually, all three runs yield a similar result.

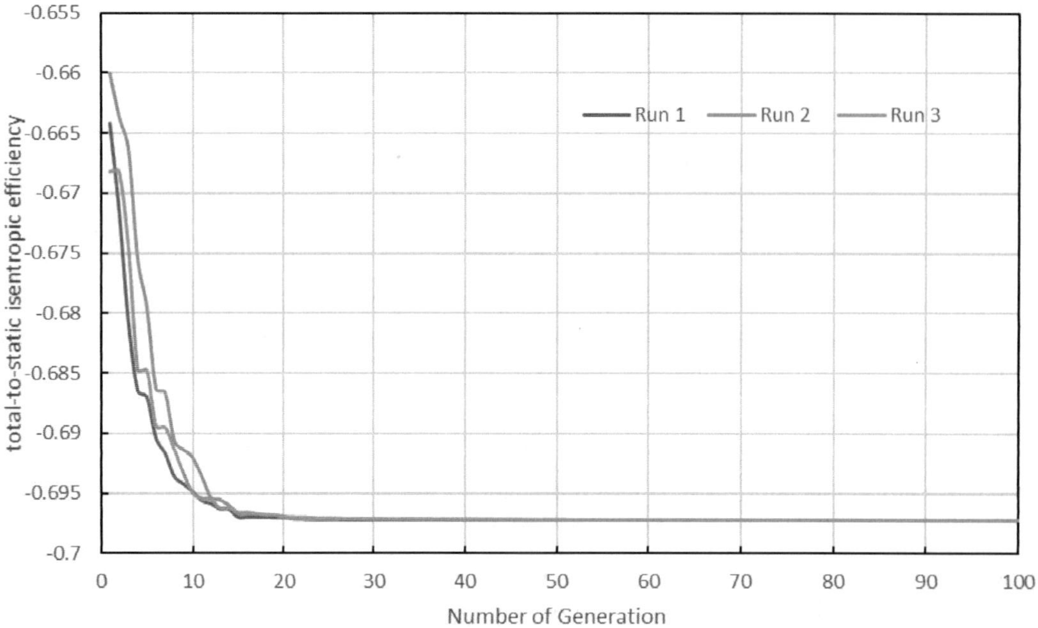

Figure 1: Optimization calculation history.

The comparison between the base design and optimized design is shown in Table 2. The optimized design had a smaller inlet hub-to-tip ratio, and the incidence angle was 2°. Moreover, a higher rotational speed was required for the optimized design. The total-to-static isentropic efficiency was improved by 3.5, from 66.2% to 69.7%.

Table 2: Comparison between the base design and the optimized design.

Parameters	**Base design**	**Optimized design**
Number of main blades	7	7
Number of splitter blades	7	7
Inlet tip blade angle	-52.3°	-50.6°
Inlet hub-to-tip ratio	0.35	0.30
Impeller outlet blade angle	-45°	-50°
Impeller outlet radius	12 mm	11 mm

Impeller exit width	0.6 mm	0.7 mm
Impeller diffuser radius	19 mm	22 mm
Design speed	240 KRPM	250 KRPM
Total-to-static efficiency	66.2%	69.7%

Figure 2 and Figure 3 show the compressor maps of the base and optimized design. The compressor performance maps were generated by varying the fraction of compressor speed from 0.6 to 1.0 with an interval of 0.2. Furthermore, 1.1 times design speed was also added to check its off-design performance at a higher speed.

In general, the optimized design has a higher efficiency than the base design. Especially when the refrigerant mass flow rate and compressor speed are low, which could be the part-load condition. Also, the base design shows a sharp efficiency drops when the refrigerant mass flow rate increases, but the optimized design shows a more flattened shape. In addition, it may be noticed that the optimized design has a narrow operating range as compared to the base design. Overall, the optimized design has a narrower operating range when its speed is high but has higher efficiency, especially in the low-speed region.

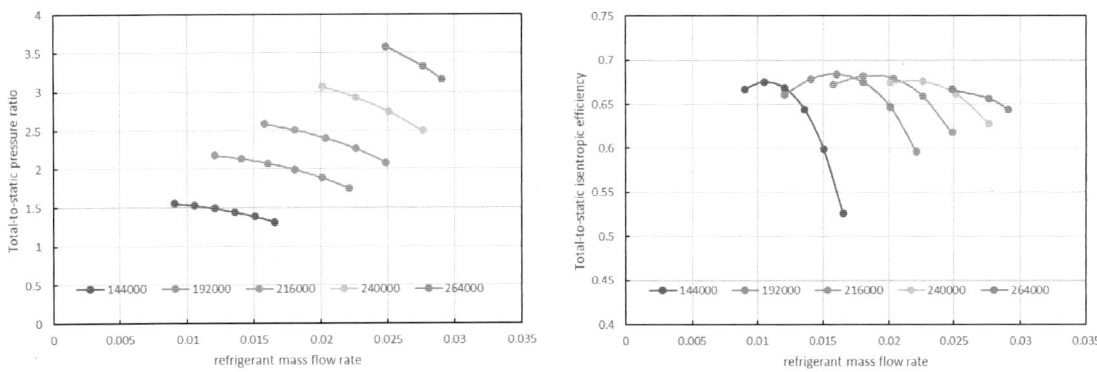

Figure 2: Compressor performance map - base design: (left) pressure ratio, (right) efficiency.

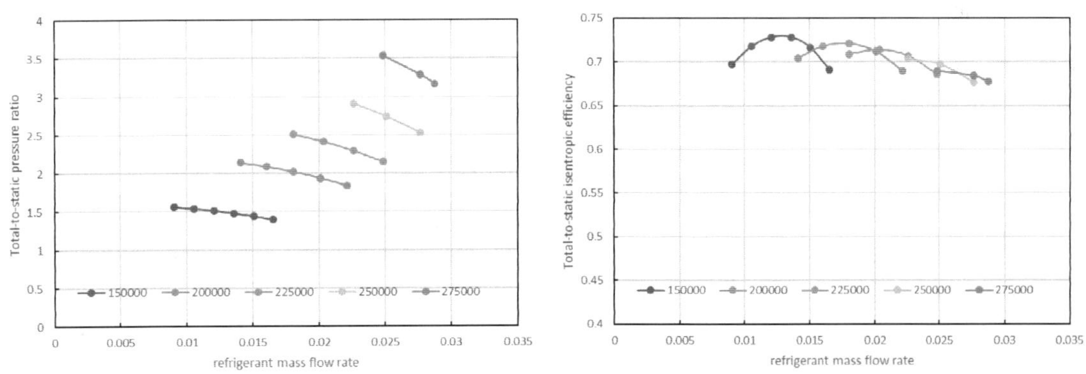

Figure 3: Compressor performance map - optimized design: (left) pressure ratio, (right) efficiency.

5. Conclusions

In conclusion, a 1D design was performed for a small-scale centrifugal compressor for air conditioning systems. A genetic algorithm was used to optimize the centrifugal compressor design. The final 1D design has an 11 mm radius. It reaches 69.7% total-to-static isentropic efficiency at design speed of 250 KRPM. Its total-to-static pressure ratio is 2.74. The low isentropic efficiency is probably due to limited compressor rotational speed and relatively large relative tip clearance due to a small dimension. For future work, it is necessary to conduct 3D CFD simulations to evaluate the compressor's performance in depth and followed by experimental evaluations.

Nomenclature

d_s	specific diameter, $\frac{D\Delta h^{0.25}}{\sqrt{\dot{Q}_{in}}}$	(-)
Δh	adiabatic head	(kJ/kg)
GWP	Global Warming Potential	(-)
I	incidence angle	(°)
KRPM	thousand revolution per minute	(-)
ODP	Ozone Depletion Potential	(-)
PR	pressure ratio	(-)
$\dot{Q}_{in}$	impeller inlet volumetric flow rate	(m^3/s)
r	radius	(mm)
Re	Reynold number	(-)
s	tip clearance	(mm)
S_e	endurance limit	(MPa)
S_{yt}	yielding strength	(MPa)
T	torque	(Nm)
M	bending moment	(Nm)
N_b	number of blades	(-)
n	safety factor	(-)
n_s	specific speed, $\frac{\omega\sqrt{\dot{Q}_{in}}}{\Delta h^{0.75}}$	(-)
β	blade angle	(°)
η	efficiency	(-)
ω	impeller rotational speed	(rad/s)

Subscript

$1h$	impeller inlet hub
$1t$	impeller inlet tip
a	alternating
m	midrange
ts	total-to-static
Tt	total-to-total
is	isentropic

Reference

AR4 Climate Change 2007. (2007). IPCC. https://www.ipcc.ch/report/ar4/wg1/

Balje, O. E. (1981). *Turbomachines-A guide to design, selection, and theory*. John Wiley & Sons.

Blank, J., and Deb, K. (2020). pymoo: Multi-objective Optimization in Python. *ArXiv:2002.04504 [Cs]*. http://arxiv.org/abs/2002.04504

Boyce, M. P. (2003). *Centrifugal compressors: A basic guide*. PennWell.

Budynas, R. G., Nisbett, J. Keith., and Shigley, J. Edward. (2011). *Shigley's mechanical engineering design* (9th ed.). McGraw-Hill; WorldCat.org.

Casey, M. V., Krähenbuhl, D., and Zwyssig, C. (2013). The Design of Ultra-High-Speed Miniature Centrifugal Compressors. *European Conference on Turbomachinery Fluid Dynamics and Thermodynamics ETC*, *10*, 1–13.

Conrad, O., Raif, K., Wessels, M., and Aktiengesellschaft, D.-B. (1980). THE CALCULATION OF PERFORMANCE MAPS FOR CENTRIFUGAL COMPRESSORS WITH VANE-ISLAND DIFFUSERS. *Proceedings of the Twenty-Fifyh Annual International Gas Turbine Conference and Exhibit*, 14.

Japikse, D. (1985). Assessment of Single- and Two-Zone Modeling of Centrifugal Compressors, Studies in Component Performance: Part 3. *Volume 1: Aircraft Engine; Marine; Turbomachinery; Microturbines and Small Turbomachinery*, V001T03A023. https://doi.org/10.1115/85-GT-73

Japikse, D., and Baines, N. C. (1994). *Introduction to Turbomachinery*. Oxford: Oxford University Press.

Javed, A., Arpagaus, C., Bertsch, S., and Schiffmann, J. (2016). Small-scale turbocompressors for wide-range operation with large tip-clearances for a two-stage heat pump concept. *International Journal of Refrigeration*, *69*, 285–302. https://doi.org/10.1016/j.ijrefrig.2016.06.015

Meroni, A., Zühlsdorf, B., Elmegaard, B., and Haglind, F. (2018). Design of centrifugal compressors for heat pump systems. *Applied Energy*, *232*, 139–156. https://doi.org/10.1016/j.apenergy.2018.09.210

Schiffmann, J., and Favrat, D. (2009). Experimental investigation of a direct driven radial compressor for domestic heat pumps. *International Journal of Refrigeration*, *32*(8), 1918–1928. https://doi.org/10.1016/j.ijrefrig.2009.07.006

Schiffmann, J., and Favrat, D. (2010). Design, experimental investigation and multi-objective optimization of a small-scale radial compressor for heat pump applications. *Energy*, *35*(1), 436–450. https://doi.org/10.1016/j.energy.2009.10.010

Tian, Q., Cai, D., Ren, L., Tang, W., Xie, Y., He, G., and Liu, F. (2015). An experimental investigation of refrigerant mixture R32/R290 as drop-in replacement for HFC410A in household air conditioners. *International Journal of Refrigeration*, *57*. https://doi.org/10.1016/j.ijrefrig.2015.05.005

Xu, C., and Amano, R. S. (2012). Empirical Design Considerations for Industrial Centrifugal Compressors. *International Journal of Rotating Machinery*, *2012*, 15. https://doi.org/10.1155/2012/184061

ACKNOWLEDGEMENT

We gratefully acknowledge the support of the Center for Environmental Energy Engineering (CEEE) at the University of Maryland.

Regenerative Gas Turbine Power Plant: Performance & Evaluation

Abdalla, Momin Elhadi*[1]; Siddharth Pannir [2]; Hussein Ahmed, Amina[3]; Mahjob, Aya[4]; Abdalla, Salah Ahmed[5]

Organization(s): 1,3,4: University of Khartoum, Chemical Engineering Department, Khartoum, Sudan
2: GenH, Charlestown, Massachusetts, United States
5: University of Khartoum, Energy Research Centre, Khartoum, Sudan

[1] University of Khartoum, Chemical Engineering Department,Khartoum, Sudan, mominhadi@yahoo.com
[2]GenH Inc, Charlestown, Massachusetts, United States, Siddharth.pannir@genh.co

ABSTRACT

In this work, comprehensive operational and conceptual design basics of the Regenerative Gas Turbine were studied and applied to the Khartoum North Thermal Power Station, Sudan, which has a total power of 187MW. The analysis and results of this work were executed using the Engineering Equation Solver.The results show that, the increasing the effectiveness of the regenerative cycle increased the thermal efficiency. However, there is a turning point of compressor inlet temperature, after which the further increase of temperature and regenerator effectiveness will lead to decline in the thermal efficiency of the cycle. At lower regeneration and moderate regenerator effectiveness, the increase in compression ratio leads to an increase in thermal efficiency of the cycle. At the highest values of regeneration effectiveness, the increase in compression ratios reduced the thermal efficiency of the cycle. The results revealed that regeneration is more effective at lower pressure ratios, ambient temperatures, and low minimum (compressor) to maximum (combustor) temperature ratios. An increase in regeneration effectiveness decreases the specific fuel consumption for lower and moderate compression ratios. At higher compression ratios, increasing regenerator effectiveness leads to an increase in the specific fuel consumption (SFC) of the cycle. At low and moderate compressor inlet temperature, increasing the regenerator effectiveness decreases fuel demand in the combustor, which reflects in decreasing the heat rate to the combustor especially at higher regenerative effectiveness (e=95%). As the effectiveness varies between 10-75%, the compressor inlet temperature varies from 200K to 350K and the regenerator exhaust temperature exhibited different profiles according to the conditions of inlet temperature. It was found that power curve declines smoothly due to the increase in irreversibility of regeneration cycle and remains high at higher turbine inlet temperatures. Compressor inlet temperatures between 100-330K increase the regeneration effectiveness varying between 10-95%, resulting a in different profile of the combustor inlet temperature. The mass flow rate of the fuel in the combustor decreases with increasing regeneration effectiveness at lower compressor inlet temperatures. At higher inlet temperatures, the fuel flow rate will gradually increase with the regeneration effectiveness due increasing irreversibilities of the regenerator. For a compression ratio of 15, the fuel mass flow rate reaches the lowest value of (6.30 kg/sec) at the lowest ambient temperature of 200 K and a regenerative effectiveness of 95%. The increase of the lower heating value (LHV) leads to a gradual increase in the thermal efficiency of the regenerative gas turbine (RGT), due to increasing cycle power and combustor capacity. The results concluded that the regeneration effectiveness is higher at low and moderate compressor inlet temperatures and compression ratios, through which, avoiding the regenerator's irreversibility is possible.

1. INTRODUCTION

In light of the detrimental effects that fossil fuels have on changing the climate system, multiple attempts have been made to find ways for a safe and environmentally benign thermal power source. This has led to extensive research on lowering the gas emissions as well as increasing power plant efficiency based on complex [3] [24] gas turbine (GT) setups. Reducing long-term greenhouse gas (GHG) emissions of the energy industry [25] is one of the toughest challenges of the energy transition. These emissions result from the combustion of fossil fuels for energy purposes and from other process emissions.

Continuous population growth and increasing economic activity inSudan, has led to an increased need to demand a large build power supplier of the GT Plants [16]. These plants generate maximum output at summer ambient temperature ratings. The use of GT to generate electricity has become an attractive endeavor due to the comparatively low initial capital cost as well as its stability of supply under varying circumstances. Another outstanding feature of this equipment is its capability of quick starting using a wide variety of Fuels from natural gas to residual oil, biomass, and powdered coal[1][12][22][23]. Additionally,GT equipment benefit from better construction materials and the use of adequate blade cooling systems [15][17] to counter the inlet gas temperature which can often exceed 1200°C [1][5][8][9]. As a result of this, the overall thermal efficiency of a GT plant can be about 35%, which is almost the same as that of a conventional steam power plant. Moreover, the GT normally characterized with its low weight per unit power, is used to drive aviation systems on all kinds on aircrafts. It is also being increasingly used in land vehicles like buses and trucks and to drive locomotives and marine ships. In oil and gas industries, the GT is widely employed to drive auxiliaries like compressors, blowers, and pumps [1] [11].

Researchers have conducted research into different methods to increase thermal efficiency of regenerative GTcycles [13][19], one of which is the reheating process used to increase thermal efficiency of gas and steam turbine cycles. Similarly, regeneration is utilized to increase thermal efficiencies of both the simple GT and steam turbine cycles. Another important procedure to increase the thermal efficiency of the power plant cycle is the combined cycle, which consists of a GT and a steam turbine cycles [1] [2] [6].

This work aims to reinforce the understanding of a regenerative GT as a thermal process utilizing the regeneratingenergy of the exhaust gases departing the turbine unit as well as applying similar design parameters to theKhartoum North Thermal Power Station (GT, 187 MW) [16] in Sudan. The code of the performance model for the regenerative GTwas developed on theEES.The effects of the operating parameters were analyzed on the RGT power plant. These operating parameters include compression ratio, ambient temperature, turbine inlet temperature, and regenerator effectiveness. This study pushes for establishing a qualified operational and conceptual design procedure for the regenerative GTunit. The work also presents a preliminary strategy toidentify the performance and evaluation criterion of the regenerative GT power plant utilizing the effect of various operating conditions.

2. MODELING OF COMPONENTS

Most of the properties of air and combustion gas products were predicted by the variation of specific heat and thermodynamic functions.

The GT power plants consist of four components including the compressor, combustion chamber (CC), turbine, and regenerator. The combined cycle arrangement considered in Fig.1 is a clear presentation on how to utilize the hot turbine exhaust gas. Fresh atmospheric air is filtered and drawn continuously into the compressor, and thenthe energy is added by the combustion of the fuel in the combustion chamber unit. The products of combustion are expanded through the turbine [7]and consequently produce electrical work while the rest of the exhaust gases aredischarged into theregenerator. The counter current *regenerator* allows the air exiting the compressor to be *preheated* before entering the combustor [4], thereby reducing the amount of the fuel that must be supplied to the combustor itself.

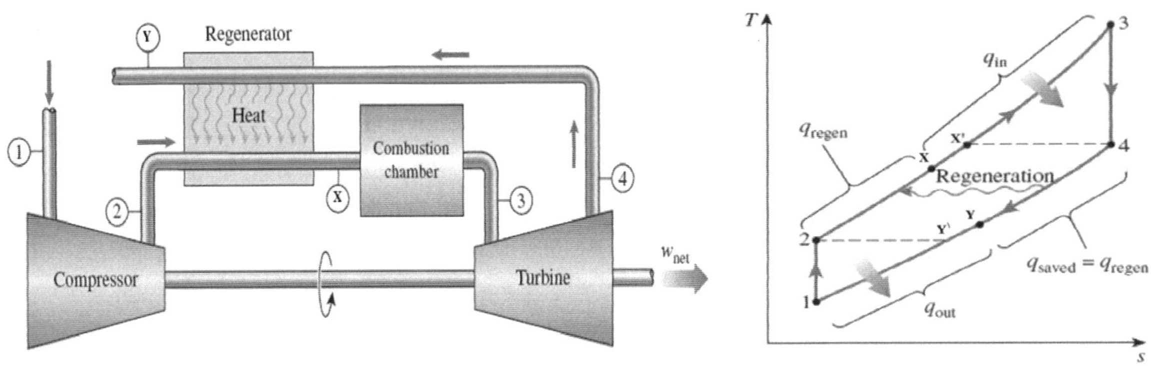

Fig.1: The regenerative GT cycle, T-S diagram.

The intake pressure at the compressor inlet was modeled with the following equation [13]:

$$P_1 = P_{atm} - \Delta P_{intake} \tag{1}$$

Where the intake pressure drop (ΔP_{intake}) was taken to be 0.005 bar, and the intake temperature was modeled as the ambient temperature. Theprocess on the temperature-entropy diagram is represented in Fig.1.The compressor compression ratio (r_P) can be defined as [2]:

$$r_P = \frac{P_2}{P_1} \tag{2}$$

where $P_1$and $P_2$are compressor inlet and outlet air pressure, respectively. Accordingly, the isentropic outlet temperature leaving the compressor is modeled by the equation[1][14][18]:

$$\frac{T_1}{T_{2s}} = \left(\frac{P_2}{P_1}\right)^{\frac{\gamma_a - 1}{\gamma_a}} \tag{3}$$

The specific heat ratio for air $\gamma_a$was taken as 1.4 and was predicted at $\gamma_g = 1.3$ for the gas.The isentropic efficiency of the compressor and turbine was taken to be in the range of 85% to 90%. The isentropic compressor efficiencyis expressed by the equation [4] [21]:

$$\eta_c = \frac{T_{2s} - T_1}{T_2 - T_1} \tag{4}$$

Where, T_1 and T_2 are the compressor inlet and outlet air temperatures respectively and T_{2s} is the compressor isentropic outlettemperature. Thespecific work required to run the compressor work (W_C) is modeled with the following equation [21]:

$$\dot{W_c} = \dot{m}_a C_{P_a}(T_2 - T_1) = \dot{m}_a C_{P_a} T_1 \left[\frac{r_p^{\frac{\gamma_a - 1}{\gamma_a}} - 1}{\eta_c}\right] \tag{5}$$

With the specific heat of air taken as $C_{P_{air}} = 1.005 \frac{kJ}{kgK}$, which can be substituted into Equations (6) and (7) for the range of [21]:

If($T_1 \leq 800K$)

$$C_{P_{air}} = 1018.9 - 0.1378 \times T_1 + 1.9843 \times 10^{-4} \times T_1^2 + 4.2399 \times 10^{-7} \times T_1^3 - 3.7632 \times 10^{-10} \times T_1^4 \tag{6}$$

If ($T_1 > 800K$)

$$C_{P_{air}} = 7.9865 \times 10^2 - 0.5339 \times T_1 - 2.2882 \times 10^{-4} \times T_1^2 + 3.7421 \times 10^{-8} \times T_1^3 \tag{7}$$

T
he specific heat of the flue gas (C_{pg}) is given by Naradasuetal.(2007) [21]:

$$C_{P_g} = 1.8083 - 2.3127 \times 10^{-3} \times T + 4.045 \times 10^{-6} \times T^2 - 1.7363 \times 10^{-9} \times T^3 \tag{8}$$

From the energy balance in the combustion chamber [1]:

$$\dot{m}_a C_{P_a} T_x + \dot{m}_f LHV + \dot{m}_f C_{P_f} T_f = (\dot{m}_a + \dot{m}_f) C_{P_g} T_{IT} \tag{9}$$

Where $\dot{m}_f$is the fuel mass flow rate in (kg/sec), $\dot{m}_a$is the air mass flow rate (kg/sec), LHV is the fuel'slower heating value (the fuel used has a value of 48 MJ/kg),$T_{IT}$is the turbine inlet temperature, $C_{P_f}$is the specific heat of fuel, andT_f is the temperature of the fuel. The specific heat of the flue gas was modeled with $C_{P_g} = 1.07\ kJ/kg.K$; efficiency was set at 95%, and a pressure drop of$\Delta P_{C,C} = 0.4785\ bar$ in the combustor. Accordingly, the efficiency of the combustor was modeled as [1]:

$$\eta_{C,C} = \frac{\dot{m}_g C_{P_g} T_{IT} - \dot{m}_a C_{P_a} T_x}{\dot{m}_f LHV} \tag{10}$$

The air fuel ratio at the combustor was modeled according to the following equation:

$$AFR = \frac{A}{F} = \frac{\dot{m}_a}{\dot{m}_f} \tag{11}$$

Where the total mass flow rate is given by:

$$\dot{m}_g = \dot{m}_a + \dot{m}_f \tag{12}$$

The discharge gas of the turbine was predicted according to the equation:

$$\frac{T_3}{T_{4s}} = \left(\frac{P_3}{P_4}\right)^{\frac{\gamma_g - 1}{\gamma_g}} \tag{13}$$

Where the actual outlet temperature leaving the turbine at the isentropic conditions was modeled according to:

$$\eta_t = \frac{T_3 - T_4}{T_3 - T_{4s}} \tag{14}$$

The regenerator effectiveness ε was modeled according to the equation [4]:

$$\varepsilon = \frac{T_x - T_2}{T_4 - T_2} \tag{15}$$

where T_x is the combustor inlet temperature.The shaft work produced from the turbine is determined by the equation [20]:

$$\dot{W}_{RGT} = \dot{m}_g C_{P_g}(T_4 - TIT) = \dot{m}_g C_{P_g} \eta_t T_{IT}\left[1 - \frac{1}{r_p^{\frac{\gamma_a - 1}{\gamma_a}}}\right] \tag{16}$$

The network from the GT unit was expressed by the equation:

$$\dot{W}_{RGT,Net} = \dot{W}_{RGT,Net} - \dot{W}_C = \dot{m}_g C_{P_g} \eta_t T_{IT}\left[1 - \frac{1}{r_p^{\frac{\gamma_g - 1}{\gamma_g}}}\right] - \dot{m}_a C_{P_a} T_1 \left[\frac{r_p^{\frac{\gamma_a - 1}{\gamma_a}} - 1}{\eta_c}\right] \tag{17}$$

The output power from theGT is expressed with the equation [1] [20]:

$$P_{RGT} = [\dot{W}_{RGT,Net} - \dot{W}_C] \times \eta_{Mech} \eta_{Gen} \tag{18}$$

The mechanical (η_{Mech}) and generator (η_{Gen}) efficiencies were taken to be 92% and 95% respectively. The heat supplied was expressed with the equation [1]:

$$\dot{Q}_{add} = \dot{m}_g C_{P_g} T_{IT} - \dot{m}_a C_{P_a} T_x \tag{19}$$

The heat supplied (per kg. air) to the combustor was modeled according to the equation [1]:

$$Q_{add} = \frac{\dot{m}_f \times \eta_{C,C} \times LHV}{\dot{m}_{air}} = \frac{\eta_{C,C} \times LHV}{AFR} \tag{20}$$

The GT efficiency was determined by the equation [1]:

$$\eta_{over,RGT} = \frac{\dot{W}_{RGT,Net}}{\dot{Q}_{add}} \tag{21}$$

Accordingly, the heat rate (HR) which is defined as the consumed heat to generate unit energy of electricity was determined by the equation [1] [10]:

$$HR = \frac{3600 * \dot{m}_f * LHV}{\eta_{over,RGT}} \tag{22}$$

The specific fuel consumption (SFC) is determined by the equation [1]:

$$SFC = \frac{3600 * \dot{m}_f}{\eta_{over,RGT}} \tag{23}$$

3. RESULTS AND DISCUSSIONS

The analysis and results of this work was executed using thermodynamic EES codes. The simulation results display the effectiveness of regeneration and other important parameters on the performance of the RGT.

As can be observed in Fig.2, the increaseof the compressor inlet temperature leads to a decrease in thermal efficiency of the cycle due to change in air density, increase in compressor work, and fuel demand.It was also observed that at constant compressor inlet temperature, the increases in effectiveness of the regenerative cycle increased the thermal efficiency (Fig.2) of the GTcycle. The results show that, there is a turning point of compressor inlet temperature (280K), through which the further increase of the temperature and the regenerator effectiveness will lead to declining thermal efficiency of the cycle. Regenerative effectiveness peaks and then declines due to friction, mechanical losses, and shifting of pressure drops during the heat exchange process between the regenerator and the combustor.

To a certain extent in RGT power plants, increasing the compression ratio results in an optimum thermal efficiency at varying regenerator effectiveness. As indicated in Fig.3, at lower and moderate regeneration effectiveness, increase in compression ratio leads to an increase in thermal efficiency of the cycle. However, at the highest values of regeneration effectiveness, increase in compression ratios will lead to a decline in thermal efficiency of the cycle i.e., for each degree of regeneration there is an optimum compression ratio for maximum RGT thermal efficiency. Generally, thermal efficiency reaches a maximum value at optimum compression ratio through which maximum real work occurs. Thereafter, work will decrease and increasing the compression ratio will reduce the thermal efficiency of the cycle, as shown in Fig.4.
With an air flowrate of 500 kg/sec, the thermal efficiency increases sharply (Fig.4), especially between compression

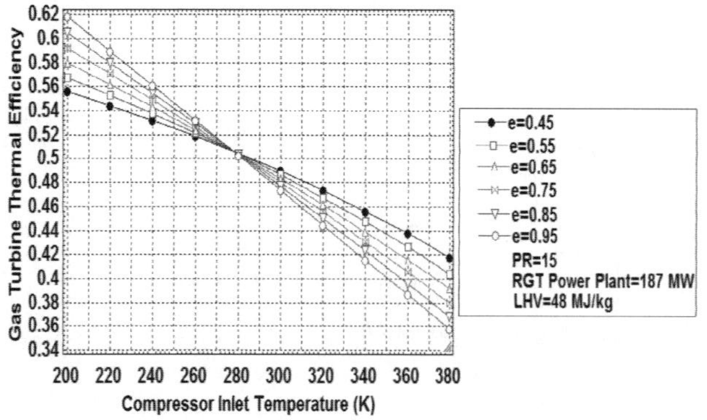

Fig.2: Thermal efficiency versus air (ambient) temperature for different regenerative effectiveness (e).

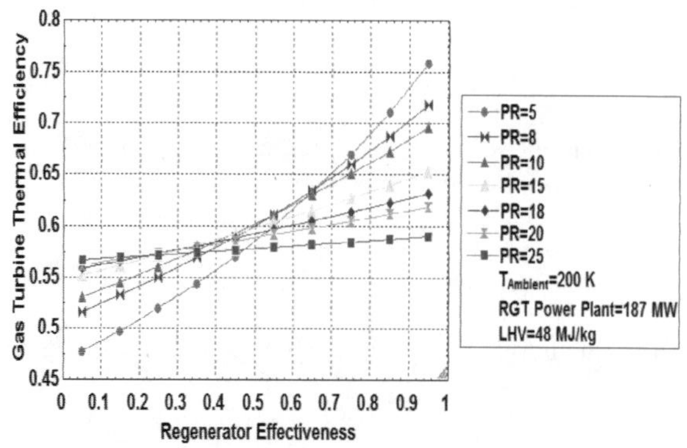

Fig.3: Variation of regenerator effectiveness with GT thermal efficiency at different compression ratios.

ratios of 4-10 where the efficiency isbetween 25-36%. Regeneration is more effective at lower pressure ratios, ambient temperaturesandlow minimum (compressor) to maximum (combustor) temperature ratios(10-17%).

The effect of different compression ratios, from 5 to 30, on specific fuel consumption (SFC) and ambient temperatures, between 200K and 300K,isshown in Fig.5. At constant compression ratio the increase of ambient temperature will lead to increased specific fuel consumption due to change in air mass and compressor work. The results indicate that the decrease in ambient temperature leads to decrease in specific fuel consumption for the whole cycle's conditions. As is evident from Fig.5, the decrease in ambient temperature led to the lowest specific fuel consumption (0.132 kg/kW.h), at optimum lowest compression ratio (rp=8) at 200K.Thereafter, the increase in compression ratio will lead to increase in fuel demand in the cycle at higher ambient temperatures.

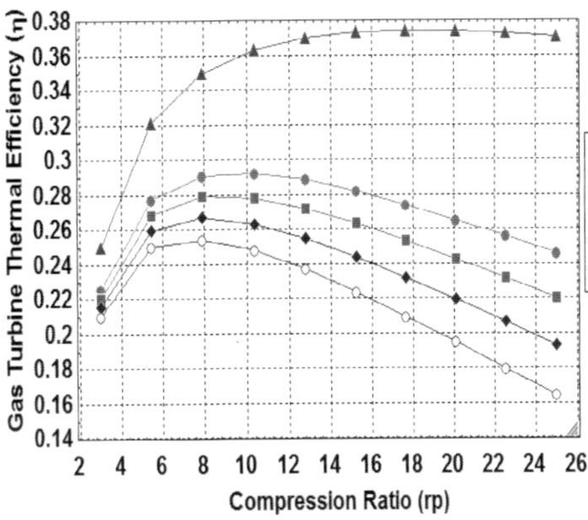

Fig.4: Thermal efficiency versus compression ratio for different inlet air temperatures.

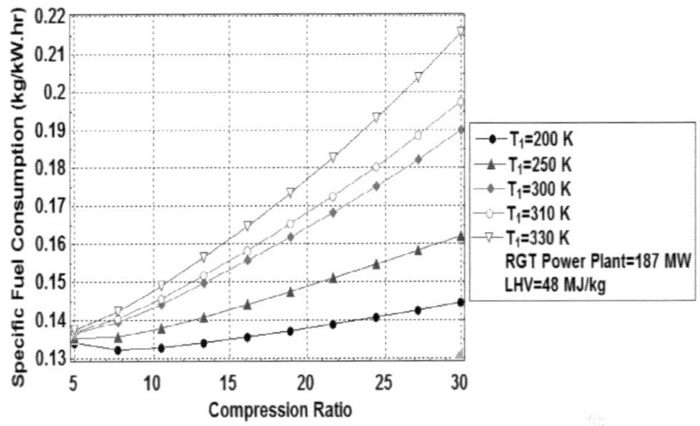

Fig.5: Effect of the compression ratio on the SFC for different ambient temperatures.

Fig.6, plots variationin compression ratio, ranging from4 through 30, with specific fuel consumption at different regeneration effectiveness,rangingfrom45% through 95%, at a constant inlet air temperature (T=200 K), and regenerative power of 187 MW.The results indicate that increasing the compression ratio leads to an increase in SFC. The general trends exhibit an increase in regeneration effectiveness which will decrease the specific fuel consumption for a lower and moderate compression ratio. At higher compression ratios the increase of regenerator effectiveness led to increase the SFC of the cycle.

Consequently, Fig.7 represents the effect of compression ratio values,rangingfrom 3 to 25, and regeneration effectiveness,varying between 45-95%,on thermal efficiency of the cycle. The addition of regenerationhas increased the thermal efficiency at constant ambient temperature, for low and moderate compression ratio's conditions. As evident in Fig.7, for the lowest ambient temperature (T=250 K), there is an optimum compression ratio (r_p=5) for which the maximum thermal efficiency (η=39%) varied with the maximum regeneration effectiveness (e=95%).

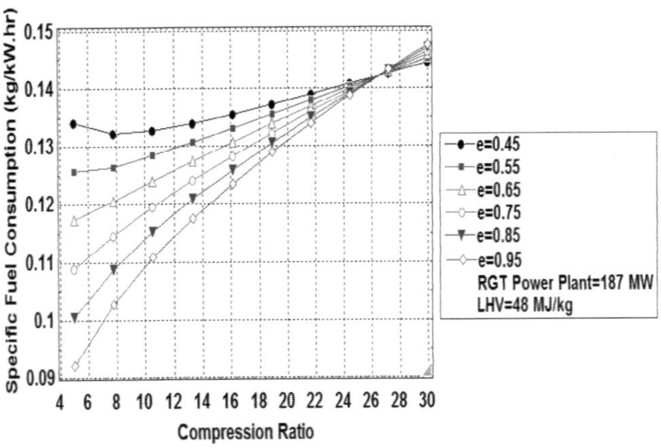

Fig.6: Influence of compression ratio on specific fuel consumption at different regeneration effectiveness(e.)

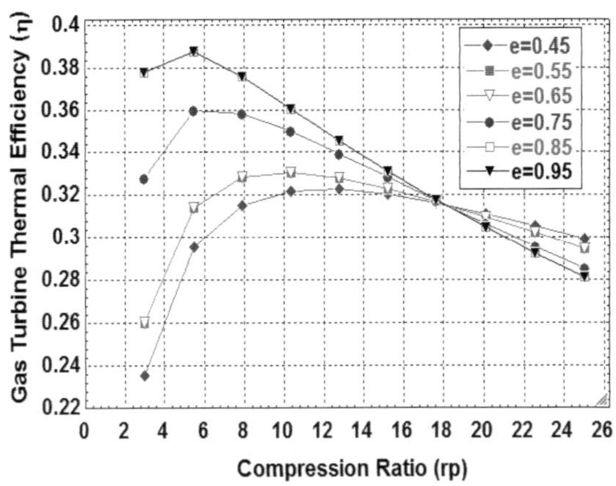

Fig.7: Effect of compression ratio and regeneration effectiveness (e) on thermal efficiency.

Fig.8, displays the influence of compressor inlet temperatures between 200K to 340K, on the combustor heat rate at different regeneration effectiveness varying from 45-95%.At constant overall compression ratios of (PR=15) and overall cycle power of 187 MW, the increase of compressor inlet temperature leads to increase the combustor heat rate. At low and moderate compressor inlet temperature, the addition of the regenerator effectiveness decreases fueldemand in the combustor, which reflects inthe decreasing heat rate to the combustor. At higher compressor inlet temperature, the demand of fuel by combustor will increase and thus increasing theregenerator effectiveness will lead to an increase in the combustor heat rate.

At constant overall cycle power of 187 MWand regeneration effectiveness of 95%, Fig.9 plots the variation of different compression ratio, varying between, 5-40, with the compressor work at different ambient temperatures ranging from 200-300K. Increasing the compression ratio led toan increase in the compressor's work and the highest values of work were reached at higher ambient temperatures.

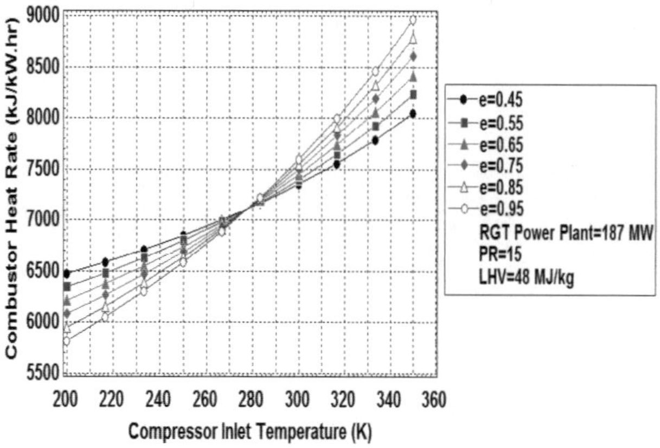

Fig.8: Variation of compressor inlet temperature with the combustor heat rate including different regeneration effectiveness (e).

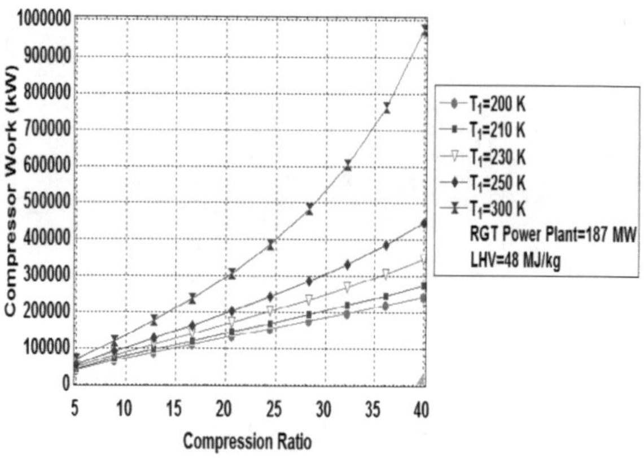

Fig.9: Variation of compression ratios with compressor work at different ambient temperatures.

The relationship betweenregenerator effectiveness, varying from5-95%, andcombustor fuel mass flow rate at different compression ratios (PR=5-25), is plotted in Fig.10. As can be observed from Fig.10, at lower and moderate regenerator effectiveness, anincrease in compressionratios decreases the combustor fuel mass flow rate of the cycle. At higher regenerator effectiveness, the increase of the compression ratio leads to increase in fuel demand due to irreversibilities at the regenerator and the combustor.

In Fig.11, the relationship of regenerator effectiveness to the regenerator exhaust temperature at varying compressor inlet temperatures is plotted. As the effectiveness varies between 10-75%,the compressor inlet temperature varies from 200K to 350K, the regenerator exhaust temperature exhibited different profiles according to the conditions of the inlet temperature. The regenerator exhaust temperature revealed decreasing value at lower compressor temperatures and increasing values at higher compressor inlet temperatures due to the increase in the irreversibilities at the compressor and regenerator.

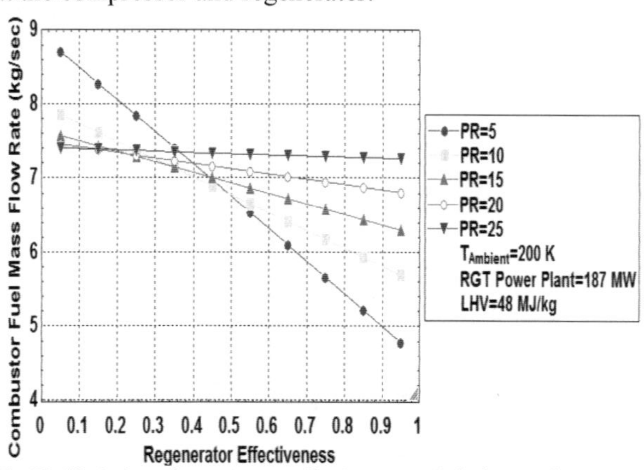

Fig.10: Variation of regenerator effectiveness with fuel mass flow rate in the combustor at different compression ratios.

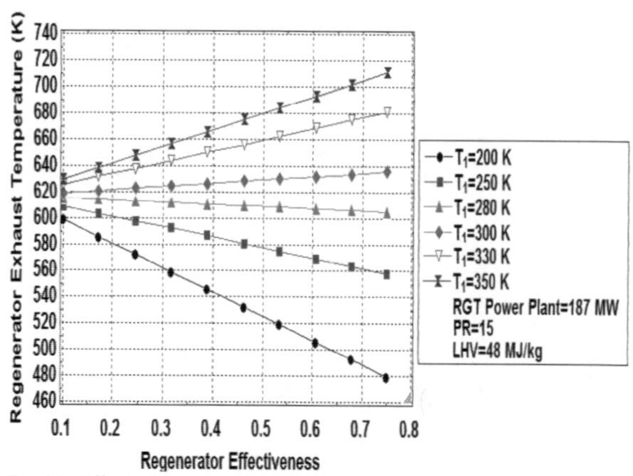

Fig.11: Effect of regenerator effectiveness on the regenerator exhaust temperature at different compressor inlet temperatures.

Fig.12,shows that increasing the turbine inlet temperature (T_{IT}) with a low regenerator effectiveness will result in an increased regenerator exhaust temperature due to gradual increase in cycle power and turbine outlet temperature. Although regenerator heat exchanger factor "effectiveness" promotes higher turbine inlet temperatures rather than the exhaust temperatures, the irreversibilities friction, mechanical losses, and fluctuation of average mean temperature will lead to depreciating efficiency.

Fig.13, plots the effect of turbine inlet temperature, between 1000-1800K, on RGT thermal efficiency for regenerationeffectiveness, ranging from45-95%. As can be noted from Fig.13, the thermal efficiency of the cycle increasesgraduallywith increasingturbine inlet temperature, as there is a further increase of the cycle's power. Thermal efficiency remains high at higher regeneration effectiveness.

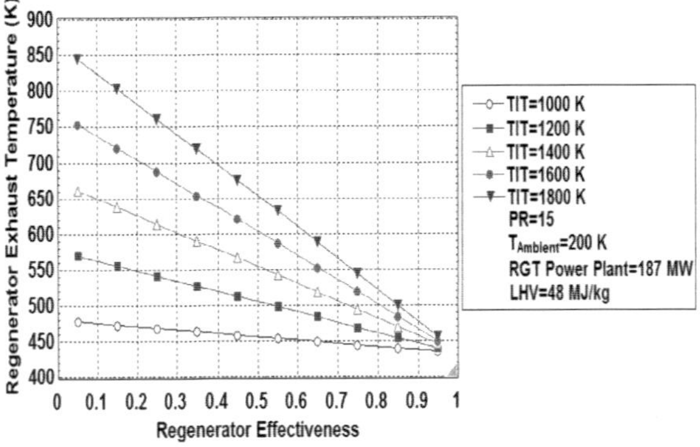

Fig.12: Effect of regenerator effectiveness(e) on regenerator exhaust temperature at different turbine inlet temperatures.

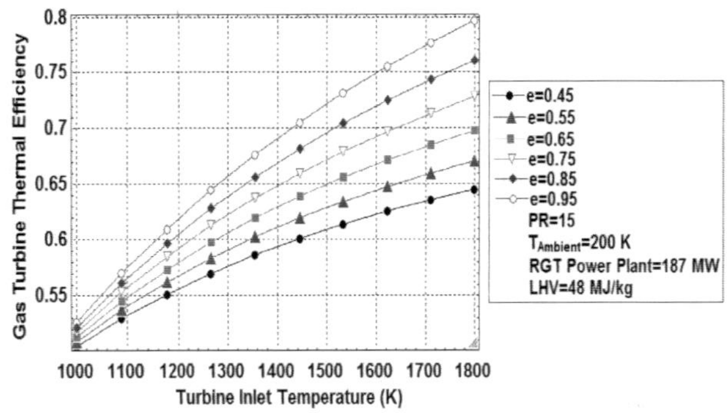

Fig.13: Effect of turbine inlet temperatures (TIT) on RGT thermal efficiency at different regenerator effectiveness.

With compression ratio held constant at 10,air flow rate at 500 kg/sec, and compressor inlet temperature of 200K,Fig.14presents the effect of regenerator effectiveness on the GT power at different turbine inlet temperatures. The power curve smoothly declines due to increase inthe regenerator's irreversibility. The power remains high at higher turbine inlet temperatures.

At compressor inlet temperaturesbetween 100-330K and regeneration effectivenessbetween 10-95%,there aredifferent profilesof the combustor inlet temperature as shown in Fig.15. Increasing the combustor inlet temperature sharply reduces the amount of specific fuel consumption, particularly at lower ambient temperatures.

The combustor inlet temperature increases value at lower compressor temperatures, while decreasing at higher compressor inlet temperatures from the increase in irreversibilities at the regenerator and combustor. At the highest regeneration effectiveness, the combustor inlet temperature stabilizes.

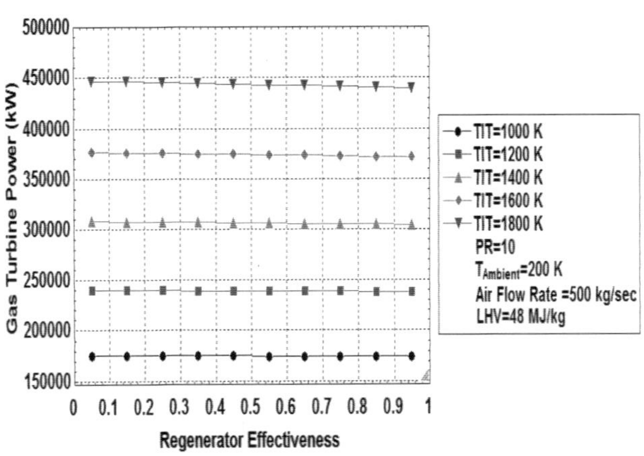

Fig.14: Effect of regenerator effectiveness on RGT thermal efficiency at different turbine inlet temperatures (TIT).

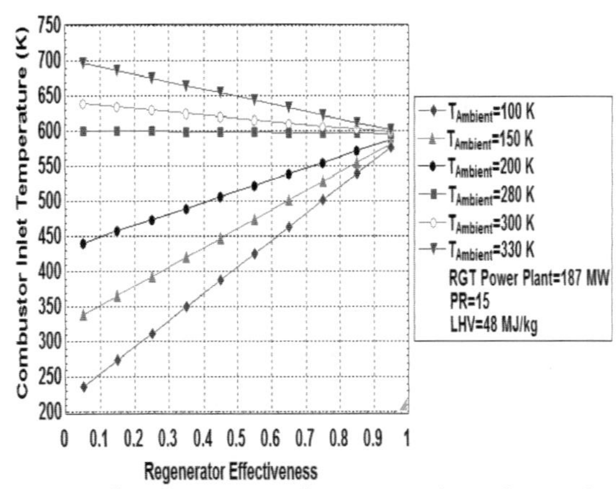

Fig.15: Effect of regenerator effectiveness on the combustor inlet temperature at different compressor inlet temperatures.

Fig.16, plots the variation of air mass flow rate, between 200-500 kg/s,toGT power at different compressor inlet temperatures rangingfrom 200-330K. The increase in mass flow rate of air directly increasesthe power of the plant, reaching a maximum value at the lowest ambient temperatures. The flow rate of the air is the major controlling parameter of increasing the power for the GTcycle. However,increasing the airflow rate will require more fuel inside the combustor, gradually increasing the specific fuel consumption of the cycle.

Fig.17, indicates the influence of regenerator effectiveness on the combustor's fuel mass flow rate at different compressor inlet temperatures varying between 200-350K.As the regenerator effectiveness varies from 5% to 95%,the compressor inlet temperature ranges from 200K to 350K, the mass flow rate of the fuel in the combustor exhibited different profiles according to the conditions of the inlet temperature, as shown in Fig.17.

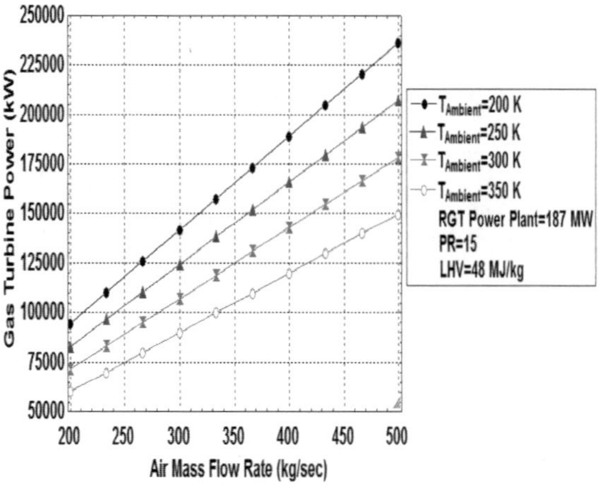

Fig.16: Variation of air mass flow rate with RGT power at different compressor inlet temperatures.

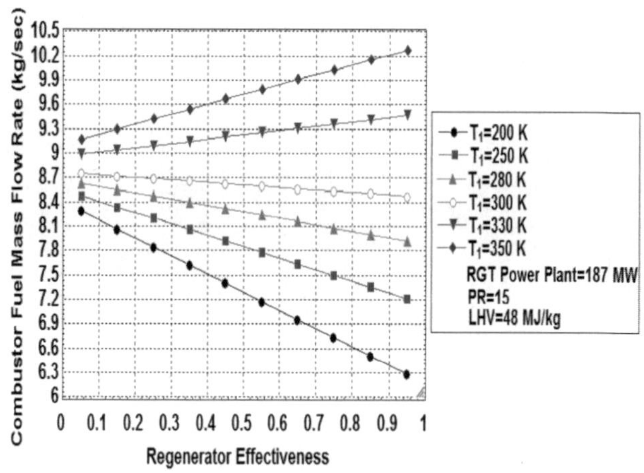

Fig.17: Influence of regenerator effectiveness on the fuel mass flow rate in the combustor at different compressor inlet temperatures.

Fig.18, shows that the fuel lower heating value (LHV) has great influence on the cycle's efficiency. The increase in LHV leads to a gradual increase in the thermal efficiency of the RGT, because of an increased in cycle power and the combustor capacity. At higher LHV of 50 MJ/kg, inlet temperature of 200 K, and power output of 187MW, the regenerative effectiveness increases the RGT thermal efficiency gradually, reaching a lower value of 59.40% at 45% regenerator effectiveness and a higher value of 65.40% at 95% regenerator effectiveness.The results show that the regeneration effectiveness is more effective at low inlet temperatures through which the regenerator's irreversibility can be avoided.

The mass flow rate of the fuel in the combustor decreases with increasing regeneration effectiveness at lower compressor inlet temperatures as shown in Fig. 19. However, at ~318K, the regenerator effectiveness does not affect the relationship between combustor fuel mass flow rate and compressor inlet temperature. Following this point, higher compressor inlet temperature and regenerator effectiveness increase the fuel flow rate, from increasingirreversibilities in the combustor and regenerator. For a regenerative power of 187MW, and compression

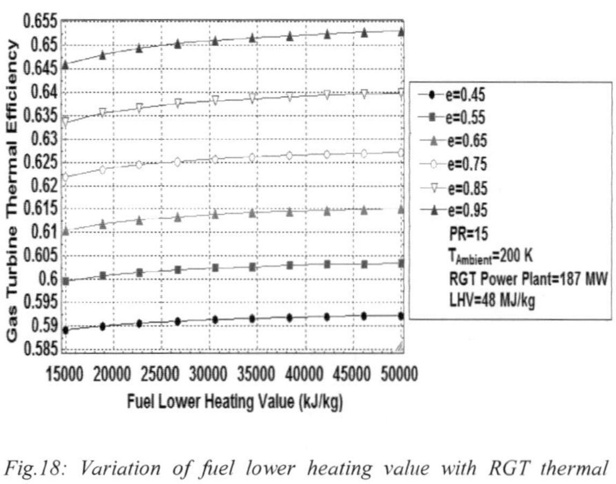

Fig.18: Variation of fuel lower heating value with RGT thermal efficiency at different regenerator effectiveness.

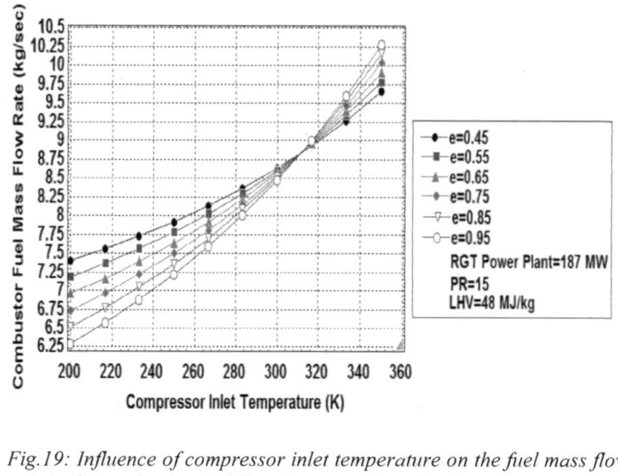

Fig.19: Influence of compressor inlet temperature on the fuel mass flow rate in the combustor at different regenerator effectiveness.

ratio of 15, the fuel mass flow rate reaches the lowest value of (6.30 kg/sec) at the lowest ambient temperature of 200 K and a regenerative effectiveness of 95%.The fuel mass flow rate reaches the highest value of (10.25 kg/sec) at the highest ambient temperature of 350 K and a regenerative effectiveness of 95%.

CONCLUSIONS

This work discussed the performance and evaluations of the RGT power plants including the effect of the regeneration. The results show various reasons and justifications for using the regenerative unit including different aspects of the fuel demand and thermal efficiency. A rigorous parametric study was introduced and executed for each unit of the plant's cycle. In addition, the work delivered various results and investigations with different variables such as compressor parameters and regeneration effects on the output power and the thermal efficiency of the RGT power plant applied to the Khartoum North Thermal Power Station (GT,187 MW).The variation in operating conditions (regenerative effectiveness, compression ratio, turbine inlet and exhaust temperature, combustor inlet temperature, combustor fuel mass flow rate,fuel'scaloric value, and ambient temperature) on the performance of GT (thermal efficiency, compressor work, power, specific fuel consumption, heat rate) were successfullyinvestigated. The parametric study revealed thatthe regenerative effectiveness, compression ratio, inlet air temperaturehad a significant effect on the thermal efficiency and power output of a RGTpower plant. The major suggestions to enhance the thermal efficiency of the regenerative cycle is the development of multistage turbine expansions with reheat units to increase the turbine inlet temperature beside multistage compressions with intercoolingunits which demands lower compressor inlet temperatures and fuel consumptions.

NOMENCLATURE

Symbols

T	Temperature	(K)	T_x	Combustor Inlet Temperature	(K)
S	Entropy	(kJ/kg.K)	$\Delta P_{C,C}$	Combustor Pressure Drop	(bar)
P	Pressure	(kPa)	$\eta_{C,C}$	Combustor Efficiency	-
r_P	Compression Ratio	-	ε	Regenerator Effectiveness	-
γ	Specific Heat Ratio	-	$\dot{W}_{GT}$	Turbine Shaft Work	(MW)
η_C	Isentropic Compressor Efficiency	-	η_T	Turbine Efficiency	-
T_S	Compressor Isentropic Temperature	(K)	P_{GT}	GT Power	(MW)
$\dot{W}_C$	Specific Compressor Work	(MW)	$\dot{Q}_{add}$	Heat Supplied	(kW)
$\dot{m}_a$	Air Mass	(kg.air)	C_{P_a}	Heat Capacity of Air	(kJ/kg.K)
$\dot{m}_f$	Fuel Mass	(kg.fuel)	C_{P_f}	Heat Capacity of Fuel	(kJ/kg.K)
$\dot{m}_g$	Gas Mass	(kg.gas)	C_{Pg}	Heat Capacity of Flue Gas	(kJ/kg.K)

Subscripts

PR	Pressure ratio
RGT	Regenerative Gas Turbine
Mech	Mechanical
Gen	Generator
AFR	Air Fuel Ratio
GHG	Greenhouse Gas
EES	Engineering Equation Solver
CC	Combustion Chamber
ATM	Atmospheric
LHV	Caloric Value
TIT	Turbine Inlet Temperature
HR	Heat Rate

SFC Specific Fuel Consumption

REFERENCES

1. Nag, P.K. (2008) Power Plant Engineering. Tata McGraw-Hill Publishing Company Limited, New Delhi.
2. Al-Sayed AF (2008). Aircraft Propulsion and GTEngines, Taylor and Francis, ISBN 978-0-8493-9196-5.
3. GT world, vol. 40. Southport, CT: Pequot Publishing Inc.; 2010. No. 5,pp. 10e14.
4. Moran, M.J. and Shapiro, H.N. (2008): Fundamentals of Engineering Thermodynamics. John Wiley & Sons, INC, New York.
5. P.W. Gill, J.H. Smith, Jr., E.J.Ziurys, Fundamentals of Internal Combustion Engines, Fourth Edition, Oxford &IBH Publishing Co.,1959.
6. T.D. Eastop and A.McKonkey, Applied Thermodynamics for Engineering Technologists, Fifth Edition, Addison-Wesley Longman Ltd., U.K.,1993.
7. R.K.Rajput , A. Text Book of Power Plant Engineering, Laxmi Publications, New Delhi, 1995.
8. V. Gansesan, Internal Combustion Engines, Tata McGraw-Hill, New Delhi,1992.
9. C.F. Taylor, The Internal Combustion Engine in Theory and Practice,Vol. I and II, MIT Press, Cambridge, Mass.,1985.
10. SaravanaMuttoo, H., Rogers, G., Cohen, H. and Straznicky. GT Theory. New York: Prentice Hall, 2009.
11. Meherwan P. Boyce (2002). GT Engineering Handbook, Gulf Professional Publishing, New Delhi.
12. Farouk N, Sheng L. Effect of Fuel Types on the Performance of GTs. International Journal of Computer Science Issues (IJCSI). 2013; **10**(1):436.
13. HossinOmar,AlyKamel,MohammedAlsanousi ,Performance of Regenerative GT Power Plant, Energy and Power Engineering, 2017,9,136-146, https://doi.org/10.4236/epe.2017.92011
14. Konstantin Volkov (2012): Efficiency, Performance and Robustness of GTs, Published by InTech, ISBN 978-953-51-0464-3, Croatia.
15. A.K. Mohapatra, Sanjay:Thermodynamic assessment of impact of inlet air cooling techniques on GT and combined cycle performance, Energy 68 (2014) 191-203
16. Johnke T, Mast M (2002). GT Power Boosters to enhance power output. Siemens Power for generation, Siemens Power J.
17. Rahman MM, Ibrahim TK, Kadirgama K, Mamat R, Bakar RA (2011).Influence of operation conditions and ambient temperature on performance of GT power plant. Adv. Mater. Res., 189-193:3007-3013.
18. Mohapatra, A.K. and Prasad, L. (2012) Parametric Analysis of Cooled GT Cycle with Evaporative Inlet Air Cooling. International Journal of Scientific & Engineering Research, 3, Issue 3.
19. Mahmood, F.G. and Mahdi, D.D. (2009) A New Approach for Enhancing Performance of a GT (Case Study: Khangiran Refinery). Applied Energy , 86, 2750-2759. http://dx.doi.org/10.1016/j.apenergy.2009.04.017.
20. Omar Shakir Mahmood, Mohammad Tariq (2014), Analysis of a Regenerative GT Cycle for Power Plant, International Journal of Scientific Engineering and Technology Research, ISSN 2319-8885, Vol.03,Issue 4, pp. 611-616.
21. M.M. Rahman, Thamir K.Ibrahim, Ahmed N. Abdalla, Thermodynamic Performance Analysis of GT Power Plant, International Journal of the Physical Sciences Vol.6(14), pp. 3539-3550,2011.
22. ASME GT Fuels B 133.7M Published: 1985 (Reaffirmed year:1992).
23. ISO, Natural Gas-Calculation of Calorific Value, Density and Relative Density, International Organization for Standardization ISO 6976-1983(E).
24. ASME GT Installation Sound Emissions B133.8 Published:1977 (Reaffirmed:1989).
25. ASME Measurement of Exhaust Emissions from Stationary GT Engines B133.9 Published:1994.

ACKNOWLEDGMENTS

The authors greatly acknowledge the technical support ofChemical Engineering Department and the Energy Research Centre of the University of Khartoum, Faculty of Engineering for providinglaboratory and facilitating thefield of works.

Novel Refrigeration Cycle with Continuous Cooling Turbo Compressor and Condensing Ejector Using Water as Refrigerant

Tadayoshi SHOYAMA [1,2]*, Bunki KAWANO [1], Iori MARUHASHI [1], Michiyoshi KUSAKA [1],
Hongzhi SUN [1], Junki YOSHIMOTO [1], and Masaru MATSUI [1]

[1]Panasonic Corporation, Appliances Company,
Moriguchi, Osaka 570-8501, Japan
[2]E-mail: shoyama.tadayoshi@jp.panasonic.com

* Corresponding Author

ABSTRACT

Turbo chiller using water as a refrigerant was developed. A refrigeration cycle using water (R718) as a refrigerant has a high theoretical efficiency, but the cycle pressure ratio is high, and the compressor discharge temperature is elevated. Although the conversion of the compressor to a multistage compressor and the addition of an intercooler between each stage are known to lower the discharge temperature and improve performance, the resulting increased size is problematic. To resolve this issue, this study develops a refrigeration cycle that lowers discharge temperature during the compression process without enlarging the device. Using a continuous cooling compressor that compresses refrigerant vapor while continuously cooling it in process and a condensing ejector that condenses the vapor while compressing it, we experimentally verify the performance of each element. Based on the results, we verify the basic principles of a continuous cooling compressor for continuously cooling compressed vapor with refrigerant droplets sprayed by nozzles installed on the impeller. In addition, the condensing ejector uses a high-speed fine refrigerant droplets to transfer momentum to the refrigerant vapor. And thus raises its pressure within a two-phase state and simultaneously condensing refrigerant vapor.

1. INTRODUCTION

In recent years, reduced global warming potential (GWP) has been sought after the Kigali amendment to the Montreal Protocol and the requirements of F-gas regulations. Thus, there has been a movement to switch to low-GWP refrigerants, and a lot of attention has been given to the development of natural refrigerants. Natural refrigerants include propane (R290) and ammonia (R717); however, their flammability and toxicity are a problem. On the other hand, water (R718) may be the ultimate refrigerant as it is nontoxic, inflammable, and highly stable. It is also zero rated in terms of its GWP and ozone depletion potential (ODP). Furthermore, its theoretical coefficient of performance (COP) is high (Kilicarslan and Müller, 2005). A typical chiller for a water refrigerant is an absorption chiller heater, which uses a gaseous vapor absorption cycle. Several electrical vapor compression cycle-type products are available, but these consist only of the eChiller by Efficient Energy and MizTurbo, which was developed by Kawasaki Heavy Industries. Our company has developed a water refrigerant system equipped with a two-stage turbo compressor using water both as a refrigerant and a compressor bearing lubricant, and it has achieved a COP of 5.0, which corresponds to the same efficiency as conventional refrigerants (Shoyama et al., 2020). However, due to the high cycle pressure ratio, the compressor discharge superheat is high. Therefore, we mixed refrigerant droplets into compression process refrigerant vapor to develop a novel continuous cooling turbo compressor that compresses with no superheat. We also develop a condensing ejector that compresses and condenses refrigerant vapor with no superheat using fine and high-speed refrigerant droplet spray. This paper reports on our construction of a refrigeration cycle using the above new elemental technology, and we demonstrate its principle of operation.

2. THEORY AND DESIGN

2.1 Refrigeration Cycle

Figure 1 shows the refrigeration cycle diagram of a conventional water refrigerant chiller, and Fig. 2 shows the developed refrigeration cycle. A conventional refrigeration cycle cools compressor discharge and prevents superheat using an intercooler. The continuous cooling turbo compressor constructed in this study has a three-stage compression configuration, and it achieves both miniaturization (high specific speed) and high efficiency (inlet Mach number $\leq$ 0.8). The impeller at each stage is equipped with a nozzle for spraying water using the centrifugal force of the impeller rotations, and it compresses the refrigerant vapor while cooling it with sprayed refrigerant droplets. It compresses the refrigerant vapor while removing excess heat from the compression process, thus reducing the compression power. In addition, refrigerant vapor discharged from the compressor is condensed by a heat exchanger in a conventional refrigeration cycle. In contrast, the condensing ejector consists of a nozzle and a mixing chamber, and it sprays refrigerant liquid that has had its heat dissipated with a cooling tower via a plate heat exchanger at high-speed from a nozzle within the mixing chamber. Then, it compresses refrigerant vapor discharged from the continuous cooling turbo compressor with sprayed refrigerant droplets. While in a gas–liquid two-phase state, it is then pressurized, and any excess heat in the compression process is continuously cooled, such that refrigerant vapor condenses from gas–liquid nonequilibrium (Marble, 1968). Refrigerant liquid condensed by the condensing ejector is returned to an evaporator via a two-stage economizer system. The design values of the water refrigerant chiller are a cycle pressure ratio of 7.0, continuous cooling turbo compressor pressure ratio of 5.3, condensing ejector pressure ratio of 1.3, and cycle COP of 6.0.

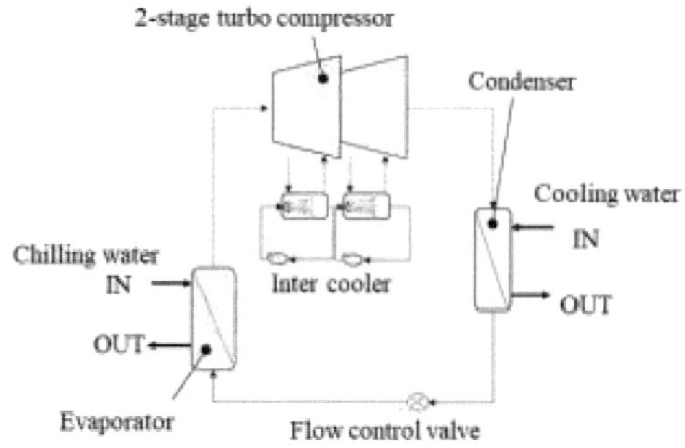

Figure 1: Conventional refrigeration chiller.

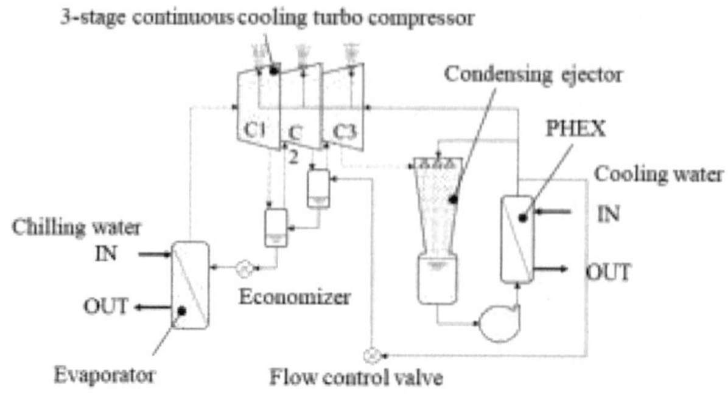

Figure 2: Developed refrigeration chiller.

2.2 Continuous Cooling Compressor

The spraying of refrigerant droplets from the compressor inlet is known as a method of cooling and simultaneous compression using the latent heat of evaporation, and it also contributes a decrease in theoretical power (Wu, et al. 2019). However, in a water refrigerant turbo compressor in which the impeller and refrigerant droplets have a high relative velocity, blades erosion poses a problem. In the continuous cooling compression employed in this study, centrifugally pressurized spray is used. Centrifugally pressurized spray involves spraying with a nozzle embedded within a rotating impeller, as shown in Fig. 3. Using the centrifugal force of the impeller, refrigerant droplets that are fine enough to be carried by the compressed vapor flow are generated and sprayed inside the impeller. Water is supplied to the hub surface from the axial channel passing through the shaft center. A nozzle is attached to the end of the radial channel, thus pressurized water is supplied by the centrifugal action of the rotations, and is then sprayed from the nozzle. This axial channel also provides a centrifugally pressurized supply to water-lubricated bearings. The benefits of this approach are that the relative velocity of the refrigerant droplets and impeller is low, thus the erosion is suppressed, and a sufficiently high pressure is achieved from the high-speed rotations of the impeller. Providing a channel and nozzle on the impeller results in vibration problems due to increased unbalance, but these can be overcome by adjusting the nozzle placement; thus, we adopted a centrifugally pressurized spray. Below, we describe the design of the compressor.

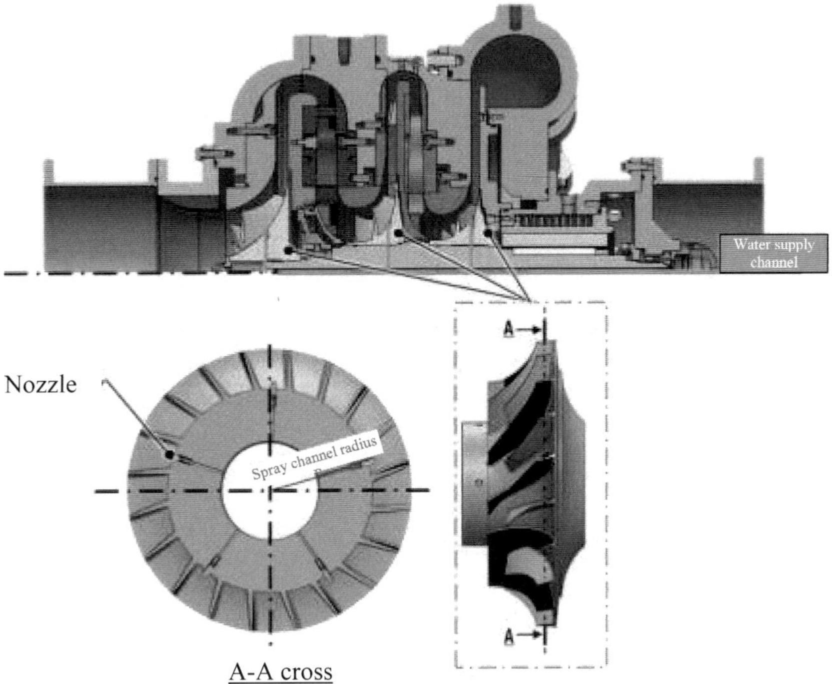

Figure 3: Cross section of the turbo compressor and its impeller.

In a previous study (Shoyama et al., 2020), a turbo compressor was developed for a water refrigerant heat pump with a 30RT refrigeration capacity, and a centrifugal impeller with a specific speed of 0.7 and an outer diameter of φ360 mm was designed. In the aerodynamic design of the impellers, the specific speed was increased to 1.4 with three stage impellers. The inlet Mach number increases, but by keeping it below 0.8, it is possible to prevent reduction of aerodynamic performance due to shock waves that may be generated at the leading edge.

Three impellers (I), one motor (M), and two bearings (B) are arranged on the main shaft. This arrangement was necessary considering multiple factors, including the rotor dynamics of the high-speed rotor, configuration of the refrigerant channel, and assembly and disassembly work. Table 1 shows a comparison of each shaft arrangement. The BIIIMB arrangement in which bearings are arranged at both shaft ends is the same as that of the 30RT compressor that was developed. There is the benefit of being able to replace the bearing without removing the impeller; mechanical loss is also minimized because the bearing size can be reduced. However, because the first-stage impeller inlet has a bearing, the refrigerant suction area is narrow, which is a major disadvantage of the impeller miniaturization. In the IBIMIB arrangement, the thrust force can be canceled by a pair of back-to-back impellers. In addition, this configuration can be regarded as a configuration of the 30RT compressor (BIMIB arrangement) where a first-stage impeller has been added to one shaft end. Therefore, one advantage is that developing a rotation system is associated with low risk. However, because the discharge side of the impeller and the suction side of the next stage are separated, a piping is difficult to connect the discharge surface to the suction surface of the next stage. It also increases the size of the entire system.

In the same IBIMIB arrangement, when the impellers are arranged in the same direction, no piping problems exist. However, when the motor is arranged close to the middle, the middle part assumes a larger diameter for the shaft to have a higher bending natural frequency, which enlarges the motor diameter. Due to a large refrigerant vapor specific volume, the size of a motor is small relative to a conventional fluorocarbon refrigerant turbo compressor. When a large motor is arranged in the middle of the shaft, an over-engineered motor is required, which is not only high cost but has a low maximum rotation speed owing to centrifugal force limitations, making it unsuitable for the operating frequency. Consequently, it is better to arrange a motor near the shaft ends where small diameter motor can be applied.

The IBIIMB arrangement that was finally adopted satisfies the required conditions of the first-stage inlet Mach number, motor size, and pipe arrangement mentioned above. Although it poses disadvantages in terms of the bearing replacement and large thrust force, these are acceptable. The advantage is that the bending natural frequency of the shaft is increased, thus it is possible to sufficiently separate the operating frequency range and bending resonance frequency.

Table 1: Ground design.

	BIIIMB	IBIMIB		IBIIMB
	Outboard bearing	Back-to-back impeller motor center	Same direction impeller motor center	Same direction impeller motor shaft ends
	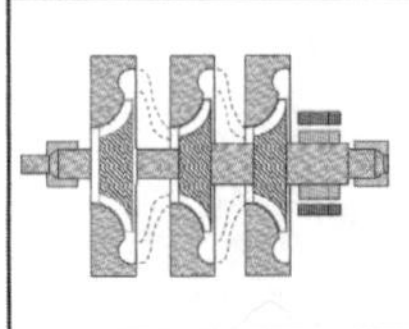	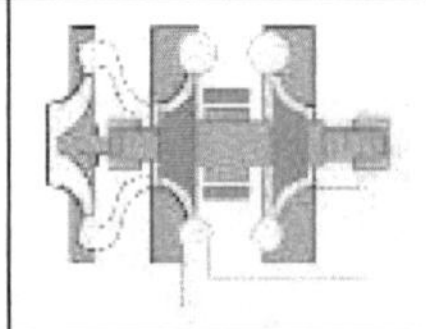	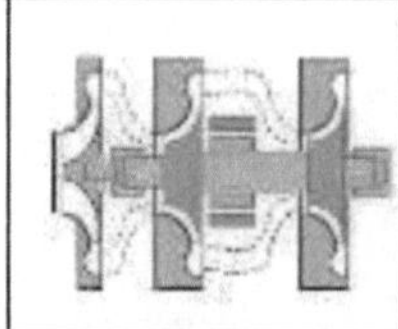	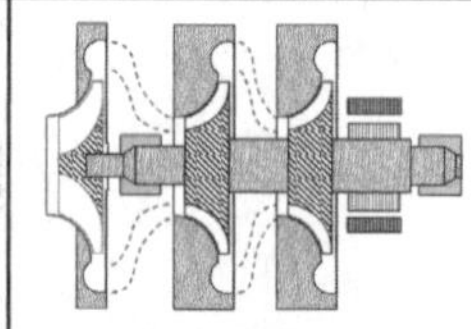
1st stage inlet area	×	○	○	○
Bearing replacement work	◎	○	○	○
Thrust force	○	◎	○	○
Motor size	○	×	×	○
Pipe arrangement	○	×	○	○
Total	×	×	×	○

For the bearing, we used a saturated water journal bearing, which has been used with a turbo compressor for the previous water refrigerant heat pump. However, because the three impellers are all facing the same direction and the thrust force is not cancel, we used a bearing in which the thrust support part and radial support part were integrated for the 1st-stage side bearing. Figure 4 shows a cross-sectional diagram of a 1st-stage side bearing. Because the thrust force of the impeller acts toward the 1st-stage side bearing, the 3rd-stage side bearing does not have to have a thrust support part. However, considering the possibility of the shaft movement to the rear-stage side due to disturbances at the start and stop, we used a tapered radial bearing. The cross-section of the bearing surface has 3-lobe shape with a preload factor of 0.6 to reduce self-excited vibration, such as whirl and whip.

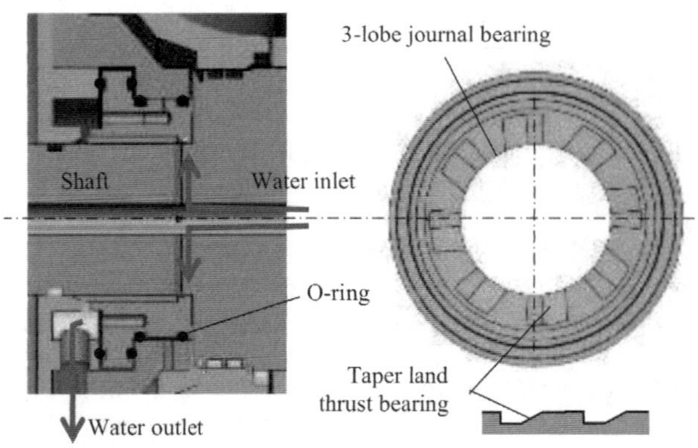

Figure 4: Water-lubricated integrated bearing.

To achieve both atomization and reduced erosion risk, we did not adopt a high-pressure atomization nozzle using an external pump for the compressor. instead, we used a centrifugal pressure nozzle built into the compressor. The centrifugal pressure nozzle rotates at the same time as the impeller, so the relative velocity of the impeller and refrigerant droplets is small, and there is little risk of erosion of the impeller from droplet collisions. The pressure applied to the nozzle during high-speed rotations is determined by the rotational velocity and rotational radius. The pressure P applied to the nozzle is expressed as follows:

$$P = \frac{1}{2}\rho(\omega r)^2 \tag{1}$$

Where ρ is the water density, ω is the angular rotation speed, and r is the radius with nozzle attached. In the compressor operating conditions (design values) shown in Table 2, 4.2 g/s was the calculation result obtained for the spray flow rate that is required to achieve continuous cooling compression process, under the assumption that compression process with an adiabatic efficiency of 80 %.

Table 2: Nozzle spray flow rate.

	Rotation speed, rpm	Spray radius, mm	Pressure, MPa	Spray flow rate, g/s
1st stage	24500	73.6	17.8	1.3
2nd stage	24500	85.6	24.1	1.4
3rd stage	24500	92.6	28.2	1.5

Figure 5 shows a cross-sectional diagram of a centrifugal pressure nozzle embedded in an impeller. The centrifugal pressure nozzle consists mainly of an internal nozzle channel, a filter to prevent orifice blocking (sintered metallic filter element, 10 μm), an O-ring to prevent water refrigerant from leaking outside the impeller, a swirl chamber for generating rotational flow, and an orifice for generating a thin liquid film. Water refrigerant pressurized by the centrifugal force of the impeller flows from the top part of the figure into the centrifugal pressure nozzle tangential to the swirl chamber, and then a thin liquid film is generated within the orifice by a turning force and sprayed. The sprayed liquid film is further ruptured, and fine particles are generated. The relationship between the pressure, spray droplet size and the flow rate was acquired from computational fluid dynamics (CFD) and nozzle unit tests using an external pressurization pump, realizing SMD 5.8 μm.

A 4.2 g/s spray requires 5 nozzles for each impeller. To suppress rotor unbalance, the nozzle arrangement on the shaft needs to be symmetrical. Because the nozzles are mounted between the blades, the number of nozzles that can be selected is the divisor of the number of impeller blades or the number obtained by subtracting the divisor from the number of blades. Because this needs to be considered when selecting the number of blades, prime numbers should be avoided. For example, if the first-stage impeller of this compressor has 15 blades, then the allowable number of nozzles is 3, 5, 10, 12, or 15.

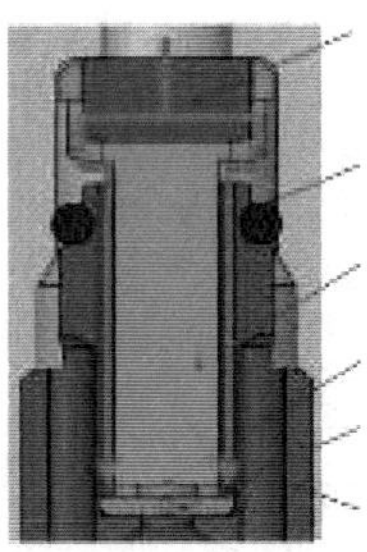

Figure 5: Centrifugal pressure nozzle configuration.

2.3 Condensing Ejector

The condensing ejector uses a high-speed fine refrigerant droplets to transfer momentum to the refrigerant vapor. And thus raises its pressure within a two-phase state. Figure 6 shows the exterior of the constructed 30RT condensing ejector. Because the spraying of refrigerant droplets using nozzles is necessary to limit the collision of refrigerant droplets with the wall surface of the mixing chamber, we require a spray form that conforms with the shape of the mixing chamber. Hence, we adopted a wall-impingement spray nozzle (Inamura et al., 2012). A wall-impingement spray nozzle consists of an impingement wall configured on concentric circles and multiple orifices for spraying water jets. The water jets impact the impingement wall, and thus form a liquid film. A fine droplet spray, which flies into space, is then formed by liquid film instability. Because each spray flies out in a direction conforming to the impingement wall, it is possible to form a spray in any shape conforming to the shape of the mixing chamber. In the mixing chamber, momentum is transferred to refrigerant vapor from the high-speed fine refrigerant droplet spray, where vapor pressure is raised, and a differential is produced between the refrigerant droplet spray saturation pressure and the refrigerant vapor pressure. Thus, refrigerant vapor condenses into refrigerant droplets, and the vapor sucked into the condensing ejector is completely condensed at the outlet. That is, the vapor volume compressed by the condensing ejector is normally reduced compared to the compressor, and excess heat from the refrigerant vapor is constantly transferred to refrigerant droplets in the pressurization process, thereby significantly reducing the theoretical power.

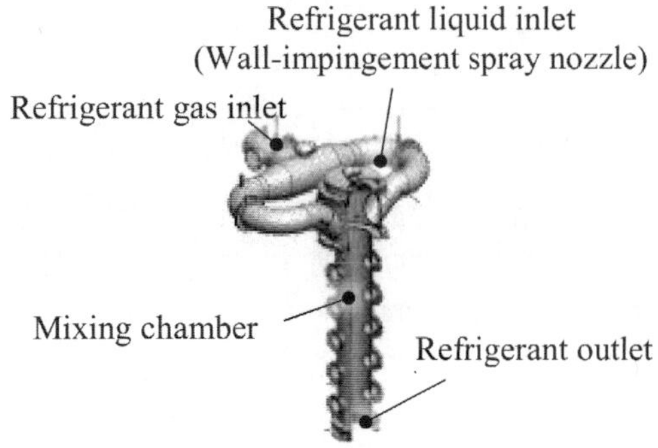

Figure 6: 30RT Condensing Ejector.

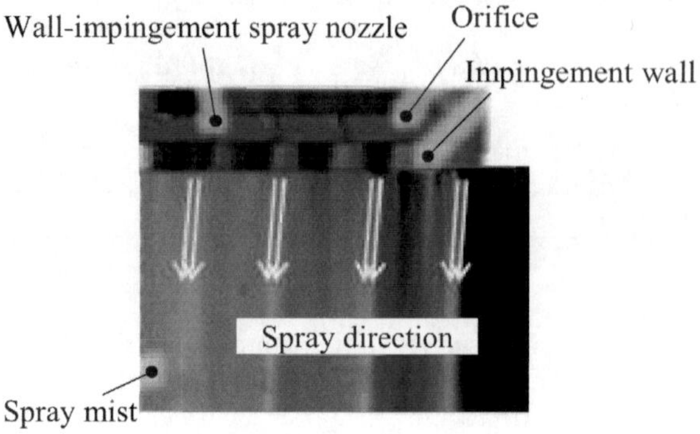

Figure 7: Wall-impingement spray nozzle.

Figure 8 shows a schematic of a condensing ejector, and Fig. 9 shows a Mollier diagram. The condensing ejector consists of a wall-impingement spray nozzle (5) and a mixing chamber (2). Liquid refrigerant cooled by the plate heat exchanger (PHE) (4) is supplied to the wall-impingement spray nozzle (5) and then sprayed as fine, high-speed refrigerant droplets. In the mixing chamber (2), refrigerant vapor discharged by the continuous cooling compressor (1), and fine, high-speed refrigerant droplets sprayed from the wall-impingement spray nozzle (5) flow inside. Then, momentum is transferred to the refrigerant vapor from refrigerant droplets, where the refrigerant is compressed in a two-phase state. When this occurs, pressure rises in the mixing chamber with the refrigerant in a gas–liquid two-phase state, thus a pressure difference is produced between the saturation pressure of the sprayed liquid refrigerant and the pressure of the refrigerant vapor, and condensation occurs owing to the gas–liquid nonequilibrium. Consequently, the refrigerant vapor discharged from the continuous cooling compressor (1) completely condenses in the process of being compressed by the condensing ejector, and is discharged to a buffer tank (3) as liquid refrigerant. The PHE (4) consists of a heat dissipating circuit via a cooling tower, and it dissipates the heat condensed by the condensing ejector. The pressure ratio of the condensing ejector is determined by the temperature difference cooled by the PHE (4), and vapor is completely condensed in the process of being compressed by the condensing ejector at the 33.5 °C spray temperature of the wall-impingement spray nozzle (5) and the 38.5 °C condensing ejector discharge liquid refrigerant temperature; it is then discharged to the buffer tank Oas liquid refrigerant. If the PHE (4) temperature difference is 5 °C, then the pressure ratio reaches a maximum of 1.3. Table 3 shows the main specifications.

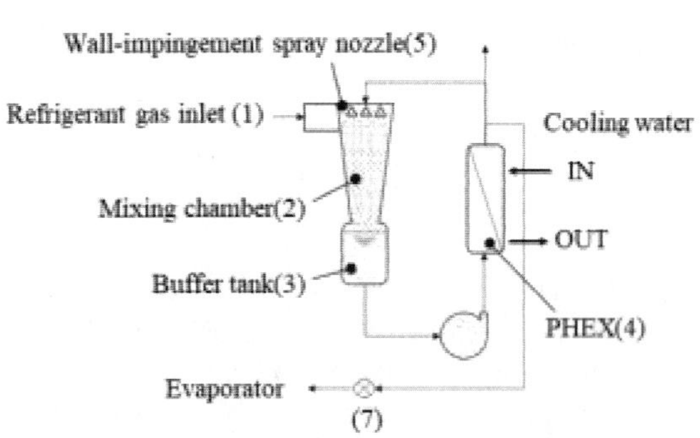

Figure 8: Schematic of condensing ejector.

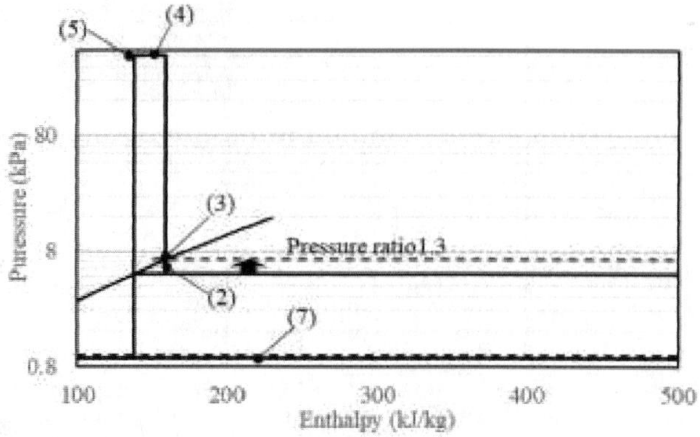

Figure 9: Mollier diagram of condensing process.

Table 3: Specification condensing ejector specification.

Ejector pressure ratio	1.30
Flow rate	300 [L/min]
External size (W × D × H)	Φ300 × 1700 [mm]
Spray pressure	200 [kPa]
Condensing capacity	120 [kW]

3. EXPERIMENTS

3.1 Refrigeration Cycle

To demonstrate the effects of a continuous cooling compressor and condensing ejector, we constructed a 30RT continuous cooling turbo chiller. Figure 10 shows a photograph of a continuous cooling turbo chiller. We demonstrated that a compression process that limits the degree of superheat can be realized using a continuous cooling compressor that sprays refrigerant droplets with a nozzles installed on impellers, and simultaneous compression and condensing can be realized with a condensing ejector that compressing vapor by refrigerant droplets. However, it was not possible to perform an evaluation using the rated pressure ratio owing to the insufficient pressure ratio of the condensing ejector. Realizing the target cycle performance requires the realization of a 1.3 pressure ratio by promoting atomization using the condensing ejector, and achieving power savings requires atomization using the centrifugal pressure nozzle of the continuous cooling turbo compressor. A COP of 4.86 can be expected with the former improvement, and a COP of 6.0 can be expected with the latter.

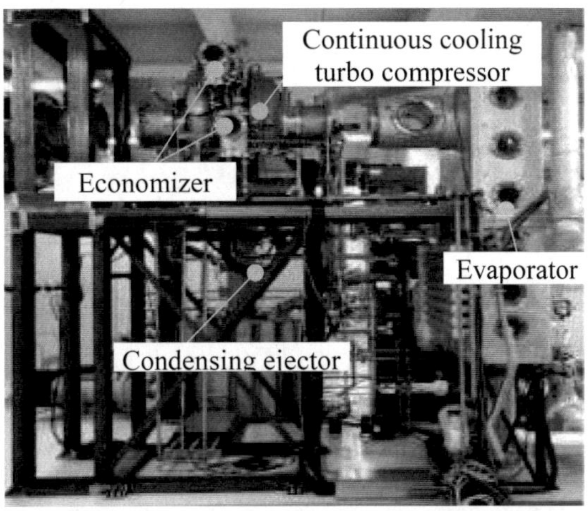

Figure 10: Continuous cooling turbo chiller.

3.2 Continuous Cooling Compressor

Here, we describe the experimental results for the spray flow rate and superheat of discharged refrigerant in continuous cooling compression. Cooling effects were insufficient and discharge superheat was 50 K with the 4.2 g/s spray flow rate of the 5 nozzles installed at each impeller of the initial design. To increase the cooling effect, nozzle installation holes were added between all of the blades. Rotor balancing was performed with plugs inserted with the same mass as the actual nozzles into the unused holes. Figure 11 shows the relationship between the total spray flow rate and the discharge superheat. When we mounted nozzles between all of the blades and increased the spray flow rate to 15.5 g/s, superheat of the discharged refrigerant vanished. The COP was less than when discharge superheat existed, but continuous cooling compression was demonstrated. The cause of the decrease in COP is assumed to be the reduced aerodynamic efficiency resulting from the increased spray flow rate. Furthermore, excess sprayed water flew out in large quantities from drain holes at the bottom of the volute. Therefore, we used 10, 8, and 7 nozzles for stages 1 to 3, respectively, to achieve a total spray flow rate of 6.9 g/s. Then the discharge superheat was 35 K, but shaft vibrations during compressor operation were excessively large, and abnormal bearing wear was observed. The number of nozzles for the 3-stage impeller was not the divisor of 20 blades, and they were asymmetrically arranged. However, as described above, rotor balancing was performed by inserting plugs of equal mass. The cause of the unbalance is believed to have been sprayed water absent during balancing. This was present only on the asymmetrically arranged nozzles, which resulted in unbalance only during operation.

The spray flow rate required for continuous cooling compression with 0 K discharge superheat was 9.5 g/s. The cause of the increase from design values is considered to be an increased compressed vapor temperature due to decreased impeller adiabatic efficiency. The flow between the blades was analyzed and designed with CFD, accounting for the three dimensional blade shape. However, blockage due to droplets and the effects of continuous cooling were not considered in the analysis. The latter is considered to have little influence on the efficiency resulting from changes in the volumetric flow rate. With respect to the former, it is possible that the spray flow rate is increased due to coarse particles that does not contribute to the continuous cooling, which is inevitably included in particle diameter distribution. The channel area is blocked by the particles. Furthermore, the momentum transfer between the spray and vapor causes rapid deceleration of the compressed vapor flow, creating a large low-energy region. This results in the reduction of the actual flow area, causing a choke phenomenon. In addition, because water refrigerant has large viscosity, then a thick boundary layer is formed on the hub face of the impeller. The thickness would be increased by the spray droplets and form a large low-energy region. When designing a continuous cooling turbo impeller, such effects must be taken into consideration.

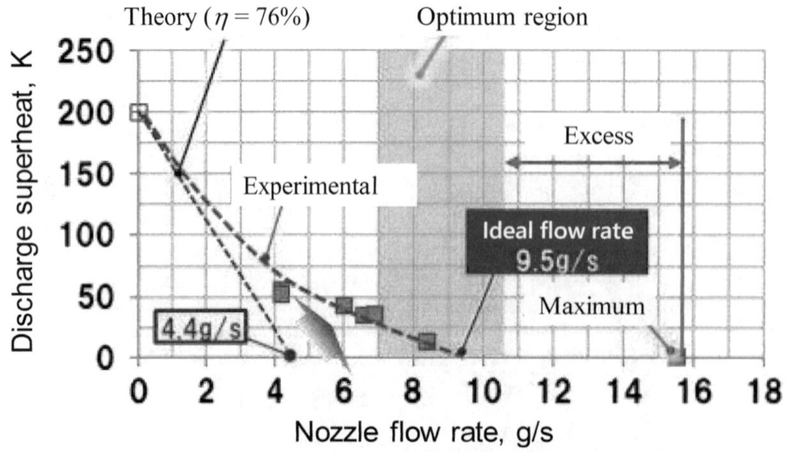

Figure 11: Discharge superheat and spray flow rate.

3.3 Condensing Ejector

Next, we describe the experimental results of the condensing ejector. Table 4 shows the pressurizing effects of the condensing ejector. The vapor that is sucked into the condensing ejector is pressurized at a pressure ratio of 1.11. The saturation temperature at a discharge vapor pressure of 3.16 kPa is 24.9 °C, which is in good agreement with the discharged refrigerant liquid temperature. Therefore, we demonstrated the basic principle of a condensing ejector, which compresses refrigerant vapor with the transfer of momentum from fine, high-speed droplets, allowing it to be pressurized in a two-phase state so that the vapor is completely condensed. However, the degree of pressurization was insufficient relative to the 1.3 design pressure ratio.

Table 4: Experimental result of condensing ejector.

Suction Vapor Pressure	2.84 kPa
Discharge Vapor pressure	3.16 kPa
Spray water Temperature	20.6 °C
Discharge water temperature	24.8 °C
Pressure ratio	1.11

Fluidity within the condensing ejector involves refrigerant droplets hovering about the refrigerant vapor at high speed. If the velocity of refrigerant droplets is faster than refrigerant vapor, then momentum is transferred to the refrigerant vapor. Thus, the refrigerant vapor pressure is increased, and condensing occurs owing to the gas–liquid nonequilibrium with refrigerant droplets. Formula (2) shows the amount of refrigerant vapor condensation S_{mas}, and Formula (3) shows the amount of momentum transferred to refrigerant vapor from refrigerant droplets per unit volume and unit time S_{mas}. Formula (4) shows the density of refrigerant droplets per unit volume.

$$S_{mas} = 0.23\rho_d 4\pi r^2 \rho_G a \left(\frac{P_G - P_{sat}(T_L)}{P_G}\right) \tag{2}$$

$$S_{mom} = -S_{mas} u_L + \rho_d 6\pi r \mu_G (u_L - u_G) \tag{3}$$

$$\rho_d = \frac{3M_L}{4\pi r^3} \frac{1}{A u_L} \tag{4}$$

The condensing of refrigerant vapor is produced by the difference between the saturation pressure of droplet $P_{sat}(T_L)$ and refrigerant vapor pressure P_G at the refrigerant droplet temperature. Momentum transfer is produced by the transfer owing to condensation, as well as the difference between the refrigerant droplet velocity u_L and refrigerant vapor flow rate u_G. This transfer increases as the refrigerant droplet radius r decreases. Consequently, the large particle size of the refrigerant droplets causes an insufficient momentum transfer, resulting in an insufficient pressure increase compared to the design pressure ratio of 1.3.

4. CONCLUSIONS

- The continuous cooling turbo chiller equipped with a continuous cooling compressor and condensing ejector was constructed to demonstrate the basic principle of each element.
- Realizing the target COP requires the further atomization of refrigerant droplets sprayed by the centrifugal pressure nozzle of the continuous cooling compressor and wall-impingement spray nozzle of the condensing ejector. A COP of 4.86 could be expected by promoting atomization by the wall-impingement spray nozzle, and a COP of 6.0 could be expected by promoting atomization by the centrifugal pressure nozzle.
- Suppression of discharge superheat was demonstrated by the continuous cooling compression technology that sprays centrifugally pressurized water refrigerant from a impeller
- The impeller efficiency is diminished by the blockage effect of unevaporated residual droplets, which decreases the COP. Designing the aerodynamics of impellers requires some consideration of the effects of spray droplets as well.
- The basic principle of a condensing ejector was demonstrated. The condensing ejector uses a high-speed fine refrigerant droplets to transfer momentum to the refrigerant vapor. And thus raises its pressure within a two-phase state and simultaneously condensing refrigerant vapor.
- Momentum transfer is achieved by the difference between the refrigerant droplet velocity u_L and the refrigerant vapor velocity u_G, in addition to the volume of refrigerant transferred by condensing, and it increases as the refrigerant droplet radius r becomes smaller. Realizing the design pressure ratio requires the construction of a nozzle that sprays even finer refrigerant droplets.

NOMENCLATURE

ρ_d	droplet density	kg/m^3
r	radius of droplet	m
ρ	density	kg/m^3
a	sonic speed	m/s
P	pressure	Pa
u	velocity	m/s
μ	viscosity	Pa・s
M	volume flow rate	m^3/s
A	cross section area	m^2

Subscript

G	gas
L	liquid
sat	saturation

REFERENCES

Kilicarslan A, & Müller N. (2005). A Comparative Study of Water as a Refrigerant with Some Current Refrigerants. Int. J. Energy Res, *29*(11), 947–959.

Shoyama, T., Kawano, B., Ogata, T., Matsui, M., Furukawa, M., & Dousti, S. (2020). Centrifugal Turbo Chiller Using Water as Refrigerant and Lubricant. *J. Proc. Mech. Eng.*, 0954408920938197.

Wu, H., Yin, H., Li, Y., & Xu, X. (2019). Effect of Droplets on Water Vapor Compression Performance. *Desalin.*, *464*, 33–47.

Inamura, T., Yanaoka, H., & Tomoda, T. (2012). Prediction of Mean Droplet Size of Sprays Issued from Wall Impingement Injector. *AIAA Journal*, *42*(3), 614–621.

Marble, F. E. (1968). Some Gasdynamic Problems in the Flow of Condensing Vapors. *Acta Astronautica*, *14*, 585–614.

ACKNOWLEDGEMENT

This study was conducted as part of the project "Technological Development and Demonstration of a Chiller for Natural Refrigerant Central Air Conditioning". We hereby express our sincerest gratitude to the Ministry of the Environment, Japan.

Computational Fluid Dynamics Study on Transonic Axial Compressors using Cartesian Cut-Cell Based Method with Adaptive Mesh Refinement and Boundary Layer Mesh

Yanheng LI*, David Henry ROWINSKI*, Karthik RUDRA REDD, Karan BANSAL,
Convergent Science Inc.,
Madison, Wisconsin, United States

Contact Information (+1-608-230-1552, yanheng.li@convergecfd.com,
david.rowinski@convergecfd.com)

* Corresponding Author

ABSTRACT

Axial compressors are used extensively in the energy, power, and transportation industries. Computational Fluid Dynamics (CFD) has been widely used in research and development of dynamic compressors. CFD modeling for such designs often presents great challenges in terms of meshing and computational cost due to moving parts with complex shapes, tiny gaps, and a large range of length scales and time scales to resolve. In this work, a Cartesian cut-cell based method with adaptive mesh refinement (AMR) is used to study Rotor 67, a transonic axial compressor design from NASA. The adopted method is demonstrated to be easily implemented and copmutationally efficient through a mesh convergence study, largely due to the advantage of an autonomously generated Cartesian cut-cell grid and AMR. Additionally, a boundary layer mesh can be used in conjunction with the Cartesian cut-cell mesh in order to resolve the near-wall flow more efficiently. Both the frozen-rotor approach with a single non-inertial reference frame (SRF) and a moving-rotor approach in a single inertial reference frame are used for the computation of the global pressure ratio and the isentropic efficiency as well as the local flow velocity, pressure, and temperature. Results show great grid convergence and good agreement with previously published experimental data for multiple operating conditions in terms of both global and local flow quantities.

1. INTRODUCTION

Axial compressors are an important type of dynamic compressor and have a wide variety of applications in the energy, power generation, and transportation industries due to their advantages of being able to accommodate a high volumetric flow rate, high efficiency, smoothness, and ease in multi-stage operation. Computational Fluid Dynamics (CFD) has been used extensively in the modeling and design of axial compressors, from global performance evaluations to studies of local flow details. However, complex curvature of the geometries, high rotational speeds, and tiny tip clearances often create great challenges for CFD meshing and often incurs a conservation/stability issue for the numerical solution: the cells are often skewed with poor aspect ratios near the complex locations and the conservation of volume is difficult. The meshing for such designs is often time-consuming with difficulties in assuring the mesh quality, which consequently affects the numerical performance of a CFD solution.

There are many existing efforts on computational modeling of the flow dynamics of an axial compressor. Grosvenor (2008) studied the NASA Rotor 67 transonic axial compressor using the Jameson-Schmidt-Turkel (JST) scheme and the Spallart-Allmaras turbulence modeling with a body-fitted mesh and compared against the original test data. Conelius et. al. (2014) employed both a transient and a steady-state solver to study multi-stage axial compressors with an unstructured mesh. Charalambous et. al. (2004) used a similar methodology and tools to study the effect of the inflow distortion on compressor performance. Most of the exiting studies require a dedicated user-created mesh with professional or commercial mesh generation tools, which typically involves long time and a recursiverevision process without precise knowledge beforehand on where the mesh should be further refined to capture the flow physics more accurately. Moreover, for different operating conditions, the flow could vary considerably, and the mesh may need appropriate adjustments which further increase the time cost in creating a mesh suitable for a wide variety of operating conditions of an axial compressor. A grid convergence study could require additional meshing time to investigate the uncertainty due to different grid resolution, which is an essential part of any detailed CFD study.

In this study, a Cartesian cut-cell method is used to investigate the NASA Rotor 67 transonic axial compressor characterized in Strazisar (1989). A single non-inertial reference frame (SRF) is used for most of the design performance studies and comparison with test data. This study features a novel autonomous meshing method which automatically generates a Cartesian mesh on-the-fly to fill the original triangulated surface geometry without requiring any user meshing time. At the locations where a Cartesian cell meets the boundary, the cell is cut to fit the boundary geometry exactly, without the need to distort the original boundary geometry, hence perfectly conserving the domain volume. Based on the gradient or curvature of the velocity and temperature fields, the grid is dynamically refined through adaptive mesh refinement (AMR), without requiring any a-priori knowledge of the flow field. The total pressure rise, adiabatic efficiency, and local flow details at the impeller exit have been studied and compared with the experiment data, with a reasonable match and good grid convergence. It is demonstrated that, although most of these studies are performed using a single blade passage sector geometry in a SRF and with a steady-state solver for computational efficiency, the same problem is also solved using the moving full geometry to conduct transient behavior studies, which renders this methodology tractable for industrial problems in which the assumptions of such geometry simplifications may not be practical. In addition, the adopted meshing technique can be optionally combined with body-fitted boundary layer mesh to resolve the near-wall and gap flow details with higher accuracy. Similar approaches have been successfully applied for several similar pump and compressor studies, including screw compressors, reciprocating compressors [Rowinski (2016)], and centrifugal fans [Li(2018)].

2. TEST CASE AND METHODOLOGY

The NASA Rotor 67 model is a single stage transonic axial compressor, originally built and experimentally tested in 1989 at the NASA Glenn Research Center [Strazisar (1989)]. It has 22 back-swept blades, with a converging cross-sectional area from the inlet to the outlet (Figure 1). The test rig has a standard test condition of the inlet total air temperature at 288.15K and total pressure of 1 atm (101325Pa). At the tested speed of 16,043 revolutions per minute, the impeller tip speed is 429 m/s with a Mach number of 1.38. A single self-nulling combination probe wase used for measuring total pressure, total temperature, and flow angle, while single self-nulling 18-degree wedge-angle probes were used for static pressure measurement. The velocity was measured by laser anemometer. Most of the presented studies focus on the steady state result, hence a sector geometry is used for computational efficiency, and both a standard model with a tip clearance of 0.61mm and a model with zero tip clearance are investigated. The full moving geometry is demonstrated to be useful for further transient studies and yields steady-state results consistent with the sector model. Both geometries are based on the study from [Doi (2002)].

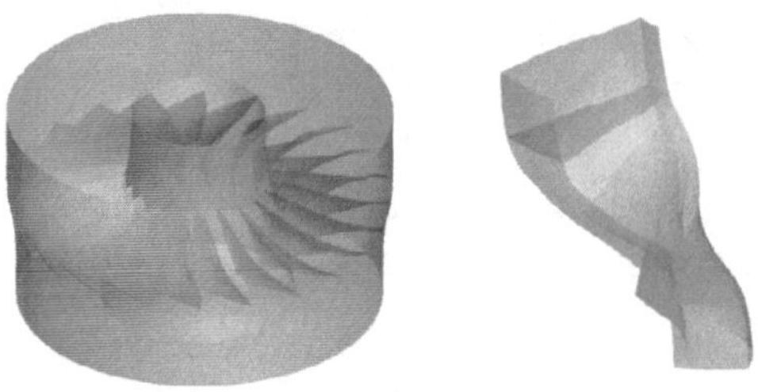

Figure 1: Rotor 67 full geometry (left) and sector geometry with a single blade passage (right).

The commercial code CONVERGE [(Richards et. al. (2017)] is used for the study, which adopts a second-order finite volume scheme with the aforementioned Cartesian cut-cell grid, originally proposed by Senecal et. al. (2007). The governing momentum, energy and turbulence transport equations are solved on a finite-volume grid with all quantities collocated at the cell center. The Pressure Implicit with Splitting of Operators (PISO) method [Issa (1985)] is used for pressure-velocity coupling. Since the Cartesian cells may be cut by the wall boundary, to ensure a good aspect ratio and sufficient cell volume, small cut-cells will be merged with nearby full cells in a technique termed cell-pairing, the degree to which can be controlled by a single parameter termed the minimum volume ratio. In this study, a value of 0.3 is used for this parameter, which means that if a ratio of a cut-cell's volume to its volume before being cut is less than 0.3, it will be paired to the neighboring cell with which it shares the most volume. The cell-pairing scheme can further improve the numerical stability on the basis of Cartesian cut-cell grid's overall good orthogonality and perfect aspect ratio. Automatic and variable time step control is used based on Courant-Friedrichs-Lewy (CFL) numbers. Message Passing Interface (MPI) and shared memory are used for parallelization. Automatic Mesh Refinement (AMR) is employed to adaptively refine the mesh based on local variations of velocity and temperature to efficiently refine the mesh around the locations where the flow is most sensitive (for example, at the shock wave across the blade tip) for a highly-efficient mesh for a wide range of operating conditions as demonstrated in Pomraning(2014). A typical overview of the Cartesian cut-cell grid system is shown in Figure 2.

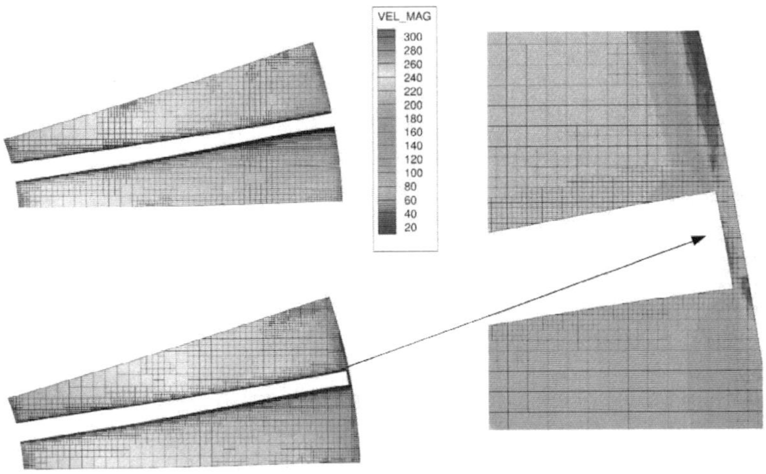

Figure 2: The Cartesian cut-cell grid system for zero tip clearance model (left upper) and for the model with a 0.61 mm tip gap (left lower and right).

A compressible gas solver is used with the ideal gas as equation of state. To avoid the common checker-boarding issue for collocated cells, the scheme from Rhie and Chow (1983) is used with the pressure-velocity coupling methodology. Due to the high velocities and prominent turbulent flow features in these cases, the Reynolds Averaged Navier-Stokes (RANS) equations are used for turbulence modeling. Most cases in this work are run with the Reynold-Stress Model (RSM) RANS turbulence modeling [Liu et. al. (2020)] except in cases specifically mentioned. The RSM model, involves seven additional transport equations for turbulent kinetic energy, dissipation rate, and the individual components of the Reynolds stress tensor, and has shown good applicability in rotating flows. For meshes with finer resolution (y+<30), the Shear Sress Transport (SST-k-Omega) RANS model from Menter (1994) is used instead. The transport equations are solved by the Successive Over-Relaxation (SOR) algorithm. At the wall boundaries, standard law-of-the-wall models are used for velocity for RSM turbulence model and automatic wall functions are used for SST turbulence model. The inlet is set at 1 atm of total pressure according to the test condition, and different outflow static pressures are prescribed according to the multiple test conditions. A steady state solver is used for most of the studies, with the exception of the moving full geometry case.

3. Results and Discussion

3.1 Global Performance Predictions

For the baseline runs, a 5 mm base grid is used with maximum level of refinement of 4, corresponding with the minimum grid size of $5/2^4=0.3125$ mm and an average y+ around 30 near the pressure side of the blade and a maximum cell count of 1.62M cells. The maximum CFL number is set at 10, but it will reduce to 1 at the final stage of each run, which corresponds to a time step range within 2e-7s to 2e-6s. A single condition can be run to a steady state within 12 hours with 16 CPUs. According to the original test, the pressure ratios are based on the total pressure measured at station 1 (impeller inlet at z=2.743cm) and station 2 (impeller exit at z=-11.011cm). The adiabatic efficiency is calculated based on equation (1).

$$\eta = \frac{\left(\frac{p_{t2}}{p_{t1}}\right)^{\frac{\gamma-1}{\gamma}} - 1}{\frac{T_{t2}}{T_{t1}} - 1}, \qquad (1)$$

Where p_t is the total pressure, T_t is the total temperature, γ is the heat capacity ratio (1.4 for dry air). Subscripts 1 and 2 denote station 1 and station 2, respectively. With the baseline setup and RSM RANS turbulence model, the model performance is compared against test data in Figure 3.

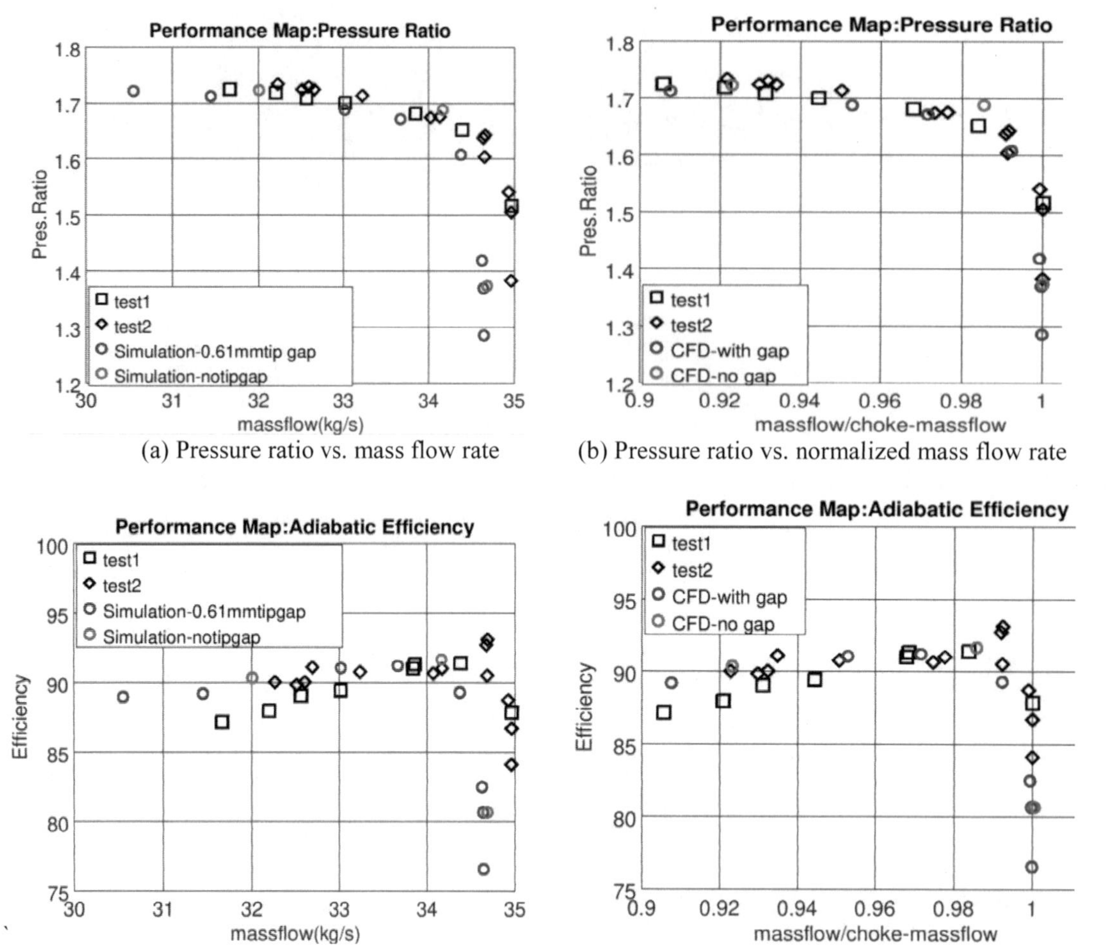

(a) Pressure ratio vs. mass flow rate (b) Pressure ratio vs. normalized mass flow rate

(c) Adiabatic Efficiency vs. mass flow rate (d) Adiabatic efficiency vs. normalized mass flow rate

Figure 3: Pressure ratio and Adiabatic Efficiency vs. mass flow rate.

From figures 3(a) and 3(c) illustrate that the overall trends of both pressure ratio and the efficiency curve are well matched against the measured test data, with the predicted choke mass flow rate of 34.70 kg/s within 1% of the 34.96 kg/s test data. For most of existing studies, the dimensionless mass flow normalized by the predicted choke mass flow is often used as a metric for comparison [Grosvenor (2008)]. Therefore, the normalized pressure ratio and efficiency curve are also shown in Figures 3(b) and 3(d), respectively. The normalized curve indicates that the peak efficiency location and the near stall condition match well with the test data. The leakage effect of the tip clearance can be observed when comparing the previous results to the results from the cases without tip gap. The tip gap results in overall slight lower performance due to the additional leakage loss.

In CFD studies on dynamic compressors, numerical error from the grid resolution is a common issue that needs to be carefully addressed. For this work, the choke mass flow rate is an important quantity to be used in normalization, and it ought to demonstrate good grid convergence or low grid uncertainty. Figure 4 shows the grid dependency of the choke mass flow rate on five grids of varying resolution, whose y+ values range from 100 on grid 0 to 30 on grid 4, considering the average on the pressure side of the blade. Good grid convergence can be observed, and most of the studies in this work utilize grid 4. The adopted Cartesian cut-cell meshing approach allows a global grid convergence study to be performed easily and robustly, simply by changing the base grid size and optionally the maximum level of refinement in AMR.

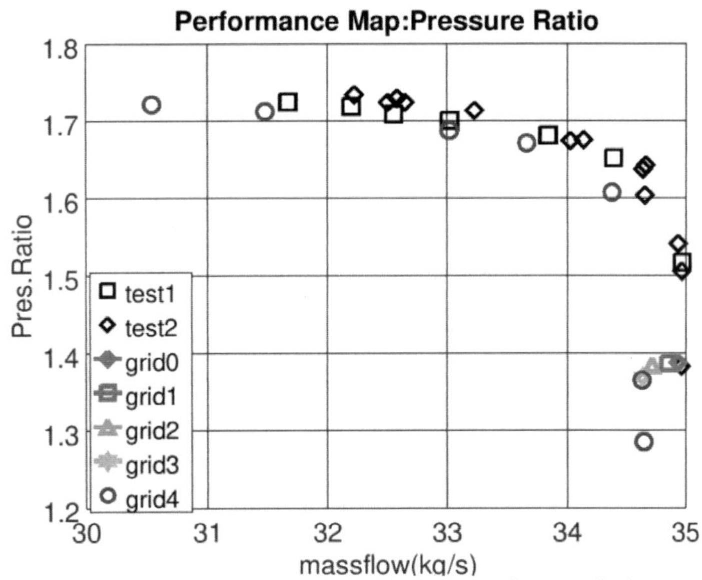

Figure 4: Grid Convergence on choke mass flow prediction.

3.2 Predictions of Local Flow Details

To further validate the proposed method, the local flow details for the near-peak and near-stall condition are also assessed in this work for the model including the gap clearance. In the original experiment, the radial distribution of total pressure, total temperature, static pressure and flow angle are presented for both a near-stall condition (corresponding to a total pressure ratio of 1.728) and a near-peak condition (corresponding to a total pressure ratio of 1.642) at station 2. Since in this work the static pressure boundary condition is used at the outlet, the results from two conditions closest to the test condition (with less than 1% difference in total pressure ratio) are averaged. The comparisons are shown in Figure 5.

Figure 5 shows that even at the relatively coarse resolutions (where the y+ is around 30) used as the baseline setup, the prediction of pressure and temperature distribution at impeller exit matches reasonably well with the test data for both the near-peak and near-stall conditions. The results also suggest that the ideal gas law used for the equation of state is sufficiently accurate for the dry air at the given condition, and that there is no need to choose a more advanced equation of state for these conditions (which would otherwise increase the computational cost slightly and make the adiabatic efficiency used in Eq. (1) less precise). As for the flow angle (defined as the angle between the flow direction and velocity component normal to the flow), there is overall around 4 degrees of difference compared with test data, but the minimum angle radial positions are accurately captured compared with the test data. The overall validation shown in Figures 3, 4 and 5 demonstrate the validity of the proposed methodology. The static temperature, static pressure and velocity contours at the middle axial chord (z=-4.62cm) are shown in Figure 6 for near-stall, near-peak, and near-choke conditions using the same color scale for each quantity. The shock wave which starts forming from choke to peak at the pressure side, and then shifting to the suction side, is clearly evident in these contour plots.

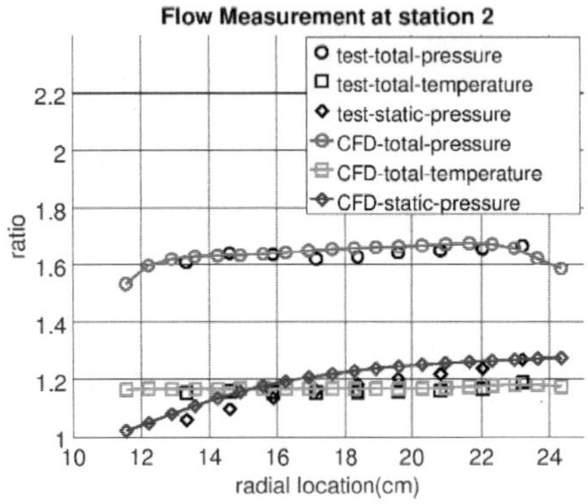

(a) pressure/temperature-near peak

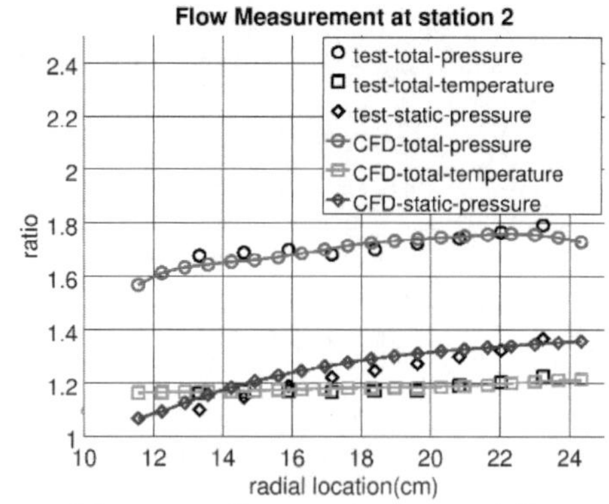

(b) pressure/temperature-near stall

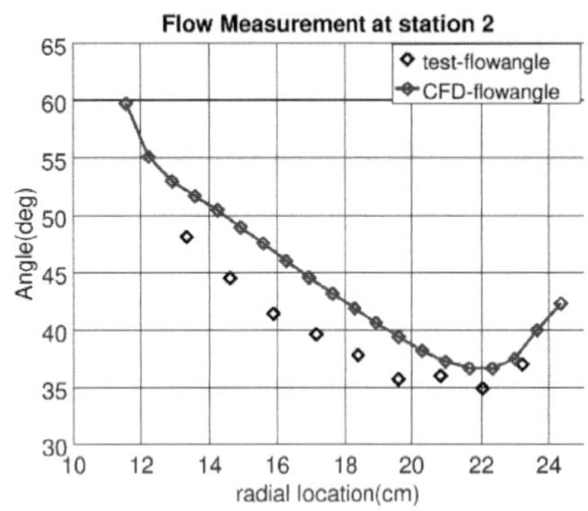

(c) flow angle-near peak

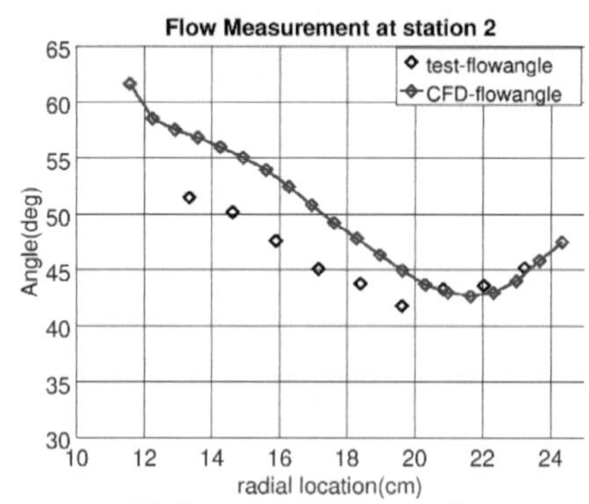

(d) flow angle-near stall

Figure 5: Radial distriubtion of flow measurement compared with test.

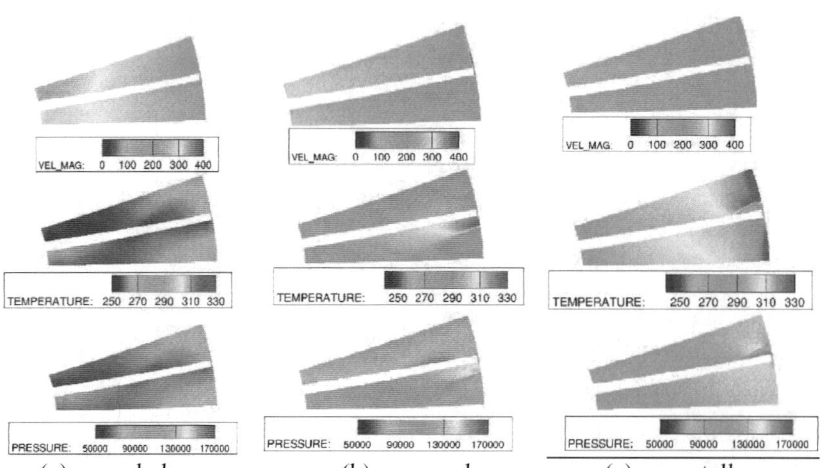

(a) near-choke (b) near-peak (c) near-stall

Figure 6: Flow field comparison at different conditions at middle axial chord.

3.3 Other Investigations and Future Work

For the above studies of the global and local flow fields, the sector geometry is used for computational efficiency. It is necessary to verify that the sector geometry and non-inertial reference frame are an accurate compared to the actual full moving geometry. Two coarse grid runs (with y+ values around 100, base grid sizes of 8 mm, and minimum grid sizes of 1 mm), one with a sector geometry and one with full moving geometry, are set up in exactly the same way and run side-by-side for the near-peak condition . The full moving geometry has a total cell count of 12M cells while the sector geometry has a total cell count of 0.5M cells. The pressure and velocity contours at the mid chord (z=-

4.5cm) location are presented as Figure 7. From the comparison, it is evident that the axisymmetry of the flow field is sufficient to justify the choice of a sector geometry, and for these models and conditions, there is very little blade to blade variation in flow quantities. Although with all Cartesian grids, there may be slight numerical effects due to the sector geometry alignment angle with the orthogonal direction of the grid. Different grid alignments may cause slight resolution change near the wall, while the full moving geometry does not have such an issue.

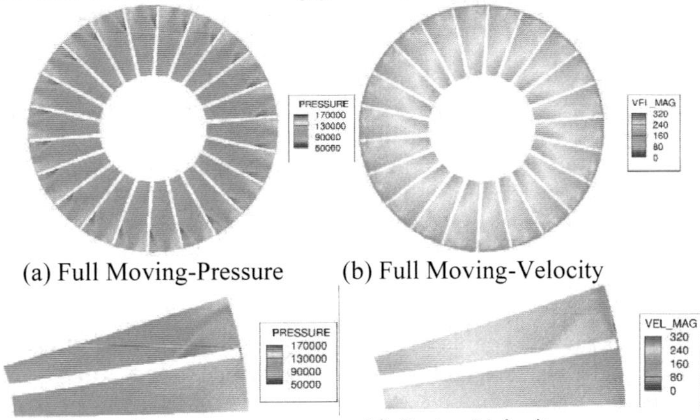

(a) Full Moving-Pressure (b) Full Moving-Velocity

(c) Sector-Pressure (d) Sector-Velocity

Figure 7: Comparison between full moving geometry and sector geometry run at near peak condition.

Finally, if detailed vortex structures near the wall and the gap are of specific interest, further grid refinement is needed than the current y+ levels of 30, at which point the structured Cartesian grid may become less efficient as the body-fitted boundary layer mesh for the curved surface of an axial compressor. The proposed numerical scheme allows a hybrid scheme: the user can create a boundary layer mesh for the blade before the simulation, and for the remaining majority of the domain, a Cartesian cut-cell will still be automatically created on-the-fly. The interface between the Cartesian cut-cell and the boundary layer mesh will be handled automatically by the proposed methodology during run time as shown in Figure 8. With the help of this hybrid meshing strategy, a fine resolution of 6E-6m and y+ values less than 2 (compared with the 0.625mm finest resolution for the Cartesian cut-cell baseline setup) can be achieved while still maintaining similar base Cartesian cut-cell resolution level and allowing the AMR to resolve the shock. The total cell count only increased from 1.6M cells to 3M cells, making the overall run time still within an acceptable level. It can be seen from the pressure contour in Figure 8 that more small scale flow details are captured with the help of the finer resolution.

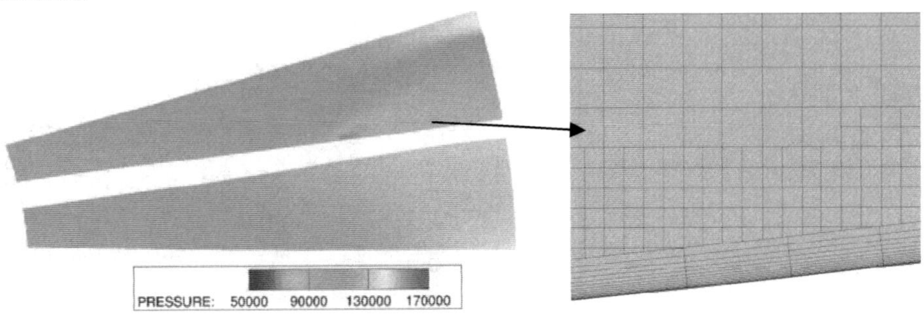

Figure 8: Hybrid mesh near blade and the pressure field.

4. CONCLUSIONS

This work presents a CFD modeling study of an axial compressor design (NASA Rotor 67) using an automated Cartesian cut-cell based finite-volume method with optional hybrid boundary-layer mesh. The employed method largely eliminates mesh-related user time and instability/numerical diffusion due to poor mesh quality. It offers the advantage of easily performed grid convergence studies and multi-condition study of the design. With the help of adaptive mesh refinement, the computational efficiency is further enhanced without the need to know the flow field before the simulation for a grid strategy that performs efficiently across a variety of operating conditions.

Global flow quantities are compared against the original test data for total pressure ratio and adiabatic efficiency as functions of mass flow rate, and demonstrated good agreement across the full operation range. This includes the predictions for choke mass flow rate, the peak efficiency location, and overall near stall performance. Grid convergence trends have been clearly observed with the chosen grid resolutions. For the local flow quantities, the radial distribution of total pressure, static pressure, total temperature and flow angle at the impeller exit are studied and validated against test data, with reasonable match in the average value, extreme value locations, and overall trends at both near-stall and near-peak conditions. These validations justify the choice of the grid resolution, the RSM RANS turbulence model, and the overall solution scheme. The mid-chord velocity, pressure, and temperature contours are shown to illustrate how the shock wave is generated and shifts from the pressure side to the suction side when the condition changes from near-choke to near-peak and to near-stall. Moreover, a comparison between the sector geometry and the full moving geometry result at the mid-chord proves the validity of using a sector geometry for the steady state study since the observed flow field is quite axisymmetric for the studied model and operation conditions.

Finally, although the all Cartesian cut-cell grid strategy can theoretically handle any type of geometry, it is not as efficient in fully resolving the boundary layer of an axial compressor blade surface which is curved and not necessarily aligned with the grid line direction. A hybrid meshing strategy can be used to solve this problem, which requires the user provide the boundary-layer mesh before the simulation while the Cartesian cut-cell grid will be created on-the-fly for remaining part of the domain. The two mesh interfaces will be automatically handled by the proposed method. With this hybrid meshing approach, a similar grid resolution for the location far from the boundary is maintained, while utilizing a boundary-layer mesh to fully and efficiently resolve the near wall effect and allowing AMR to efficiently resolve the shock wave.

REFERENCES

Charalambous, Nikolaos, et al. "Axial compressor response to inlet flow distortions by a CFD analysis." *Turbo Expo: Power for Land, Sea, and Air*. Vol. 41707. 2004.

Cornelius, Christian, et al. "Experimental and computational analysis of a multistage axial compressor including stall prediction by steady and transient CFD methods." *Journal of Turbomachinery* 136.6 (2014).

Doi, H. and Alonso, J.J., "Fluid/Structure Coupled Aeroelastic Computations for Transonic Flows in Turbomachinery", ASME Turbo Expo 2002, GT-2002-30313, Amsterdam, the Netherlands, June 3-6, 2002.

Grosvenor, A. D. 2008. Numerical studies toward prediction, analysis and treatment of SWBLI in transonic compressors. 14th Conference (International) on the Methods of Aerophysical Research Proceedings. Section 2. Novosibirsk, Russia. CD. ISBN 978-5-98901-040-0. 10 p.

Issa, R. I. (1985). Solution of the Implicitly Discretised Fluid Flow Equations by Operator-Splitting. *Journal of Computational Physics*, 62.

Li, Y., Rowinski, H.D., Bansal, K., & Reddy, R.K., CFD modeling and performance evaluation of a centrifugal fan using a cut-cell method with automatic mesh generation and adaptive mesh refinement, International Compressor Engineering Conference, 2018.

Liu, Z., Wijeyakulasuriya, S., Mashayekh, A., and Chai, X., "Investigation of Reynolds Stress Model for Complex Flow Using CONVERGE," SAE Technical Paper 2020-01-1104, 2020, https://doi.org/10.4271/2020-01-1104.

Menter, F.R., (1994), Two-equation eddy-viscosity turbulence models for engineering applications. AIAA-Journal, 32(8), pp. 269-289, 1994.

Pomraning, E., Richards, K., and Senecal, P. (2014). Modeling Turbulent Combustion Using a RANS Model, Detailed Chemistry, and Adaptive Mesh Refinement. *SAE Technical Paper 2014-01-1116.*

Rhie, C. M. and Chow, W. L. (1983). Numerical Study of the Turbulent Flow Past an Airfoil with Trailing Edge Separation.AIAA J., 21, 1525-1532.

Richards, K. J., Senecal, P. K., Pomraning, E. (2017). CONVERGE 2.4 Manual, Convergent Science Inc., Madison, WI.

Rowinski, D. H. and Davis, K. E. (2016). Modleing Reciprocating Compressors Using A Cartesian Cut-Cell Method With Automatic Mesh Generation. International Compressor Engineering Conference (2016).

Senecal, P. K., Pomraning, E., Richards, K. J., Briggs, T. E., Choi, C. Y., McDavid, R. M., Patterson, M. A., Hou, S., & Shethaji, T. (2007). A New Parallel Cut-Cell Cartesian CFD Code for Rapid Grid Generation Applied to In-Cylinder Diesel Engine Simulations. SAE Technical Paper, #2007-01-0159.

Strazisar, A.J., Wood, J.R., Hathaway, M.D., Suder, K.L., 1989, "Laser Anemometer Measurements in a Transonic Axial-Flow Fan Rotor," NASA TP-2879.

Compressor Performance Comparison of R454A and R454C for Refrigeration

Lars SJOHOLM[1], Young Chan MA[2*], Gurudath NAYAK[3], Matt STINSON[4]

[1] Trane Technologies — Thermo King,
Minneapolis, Minnesota, USA
Lars_Sjoholm@tranetechnologies.com

[2*] Trane Technologies — Thermo King,
Minneapolis, Minnesota, USA
Youngchan_Ma@tranetechnologies.com

[3] Trane Technologies, Modeling and Simulation,
Bengaluru, India
Gurudath.Nayak@tranetechnologies.com

[4] Trane Technologies, Modeling and Simulation,
Minneapolis, Minnesota, USA
Matthew.Stinson@tranetechnologies.com

* Corresponding Author

ABSTRACT

The intent of this paper is to compare compressor performance of two low GWP refrigerants that are mildly flammable. Since there are limited test capabilities for compressors with flammable refrigerants, the compressor tests in this paper are based on a non-flammable refrigerant, in this case R513A. The tested compressor is an open type screw compressor that is driven by an external motor and is vapor injected. A physical compressor model is created from test data. This compressor model is not based on maps for efficiencies, capacity, or power consumption, but rather a basis of physical bulk elements, such as pressure drop from orifices (at suction, discharge and vapor injection), leakage orifices, volume ratio between the suction port and the vapor injection port as well as volume ratio between the suction port and the discharge port. The compressor displacement and speeds are the most fundamental inputs for both the model and the tests. For mechanical losses a friction coefficient and friction exponent, relative to the speed, are obtained from the developed compressor model. This model can predict the actual compressor operation with different refrigerants. This compressor model developed based on R513A testing is applied to predict the performance of R454A and R454C refrigerants that are potential candidates for low GWP transport refrigeration. R454A always gives higher capacity than R454C. For COP, R454A outperforms R454C in some conditions while R454C outperforms R454A in others.

1. INTRODUCTION

When looking at many different refrigerants, especially A2L or mildly flammable, the use of compressor modelling and simulations becomes very important in early development because there are limited testing capability available for flammable refrigerants. Standard compressor rating models (AHRI, 2020), where refrigeration capacity and power consumption are curve fitted, would not work. The traditional approach would be to use some kind of compressor efficiency model (Cambio, 2016; Erickson, 1998; Sjoholm & Ma, 2018). Furthermore, the isentropic efficiency is hard to define for a vapor injected compressor due to the assumed mixing behavior between the main suction flow and the vapor injection flow. Another approach would be to use a one dimensional physical model, where the model is driven by the geometry of the compressor. The problem with such a model is that it still needs some correlation between simulation and test. The model used in this paper, is a physical model based on bulk elements that is test data driven directly. This way no special correlation procedure is needed. To use this physical model, at least in theory, it should be very good for different refrigerants assuming that the model is a good representation of the compressor. CFD models are not efficient because of very different grid representation in flow zones and tight clearance zones. Besides this, the compressor model has to handle economizer flow or vapor injection. To model this with good results is challenging because the vapor injection port is open to a relative large part of the volume curve. This makes it hard to define the vapor injection port as a positive displacement port. The larger the operating range is, regarding pressures, temperatures, and speed, the harder it is to predict this flow.

The objective of this paper is not to show what refrigerant is better between R454A and R454C. Rather, it is to show some thermodynamic behavior differences. To figure out what refrigerant best suits a certain application, the whole refrigeration system and the user profiles have to be taken into account.

2. LOW GWP REFRIGERANTS

The target refrigerants for this study are R454A and R454C. The other refrigerants R32, R452B, R454B and R466A are excluded due to too high discharge temperature. R1234yf, R1234ze, R513A, and other R134a like refrigerants are excluded due to too low suction pressure. Any refrigerant blends with R1234ze, R13I1 and R744 as components are not considered due to their different or unknown behaviors. No components based on A2 or A3 were considered. The desired GWP should be below 700 or 150. Some basic data for R454A and R454C are shown on Table 1.

Table 1: The Basic Data for R454A and R454C

ASHRAE Designation	ASHRAE Class	GWP (AR4)	R32	R1234yf	Molecular Weight	Normal Bubble Point	Normal Dew Point	Critical Temperature	Critical Pressure
			Mass Fraction		g/mol	°C	°C	°C	kPa
R454A	A2L	239	0.350	0.650	80.468	−48.33	−41.61	85.72	4900.8
R454C	A2L	148	0.215	0.785	90.776	−45.89	−37.72	88.50	4474.0

Two temperature reference points are selected for this study. For low temperature refrigeration, −35/40 °C (suction dew temperature/discharge dew temperature) is chosen and medium temperature refrigeration −10/45 °C is chosen (AHRI, 2020). Table 2 shows the temperature glide at the selected temperatures and the two refrigerants. Hundy & Vittal (2000) discusses handling the glide effects on compressor performance definitions.

Table 2: The Temperature Glide Comparison of R454A and R454C

	R454A		R454C	
Dew Temperature	Pressure	Temperature Glide	Pressure	Temperature Glide
°C	kPa	K	kPa	K
−35	141.53	5.71	115.11	7.82
10	387.09	5.72	319.56	7.85
40	1729.58	5.04	1458.82	6.95
45	1958.55	4.87	1645.42	6.74

3. THE EXPERIMENTAL COMPRESSOR

An open type screw compressor was chosen as an experimental test compressor to simplify the modelling without hermetic motor. This compressor is a screw compressor with only rolling element bearings to manage a wide range of lubrication conditions. It is also equipped with a vapor injection port (economizer port). However, the final solution of compressor for A2L refrigerants would be a hermetic or semi-hermetic compressor. The basic compressor data are shown on Table 3 along with Figure 1.

4. TEST SET UP AND TEST CONDITIONS

The test set up is displayed in Figure 2. The compressor installed is screw type with no internal oil separator. The external oil separator is added. Water cooled sub-cooler is utilized to keep the sub-cooling to be equal or larger than 10 °F. Two sight glasses are also added after a sub-cooler and before the compressor oil return port. The economizer

Table 3: The basic experimental compressor data

Type	Displacement	Suction to Discharge Volume Ratio	Vapor Injection Volume Ratio*	Speed Range	Lubricant	Oil Separator
	cc/rev	1	1	RPM		
Open Screw	384.76	3.5	1.07	1000–4000	POE ISO 120	External

* The suction volume relative to the volume of the flute when the injection port is in the middle of the flute.

Figure 1: The experimental screw compressor showing suction, discharge, oil injection, and vapor injection

heat exchanger and three mass flow meters—auxiliary (economizer) refrigerant flow meter, main refrigerant flow meter, and oil flow meter—are included. For more accurate data acquisition, the pressures and temperatures are measured at each points in Figure 2. Test conditions are chosen based on the speeds (1000, 2000, 3000 and 4000 rpm) without economizer in Figure 3a and with economizer in Figure 3b. The compressor test rig is a full condensing refrigerant loop, including a brazed plate economizer heat exchanger. The size of the economizer heat exchanger influences the vapor injection pressure to some extent.

5. NUMERICAL COMPRESSOR MODEL

Compressor models can take many forms: from basic empirical map-based models to general-purpose high-fidelity 3D models, with various levels of fidelity in between. The choice of modelling approach should normally be driven by what the model is being used for or what questions it is helping to answer. For example, for system simulations with a given compressor and refrigerant, a map based model may be ideal to help predict system power and capacity. Meanwhile, a compressor manufacturer designing the internals of the compressor may need a more fundamental model with 3D elements to help optimize the detailed compressor design.

In this work, it was desirable for the compressor model to have strong physical basis to enable predictability across refrigerants and over a wide operating envelope, while still having the ability to fit the model using a black box data driven approach. Neither an empirical map based approach nor a full 3D model met both of those criteria. Instead, a modeling approach comprised of lumped physical elements assembled to represent key behaviors was used (Stulgies et al., 2009; Cavalcante et al., 2008). A model was assembled to capture the following key physical phenomena for a suction gas cooled hermetic motor or open scroll/screw compressor with an injection port: pressure drop at suction, discharge, and injection ports; leakage from discharge to suction volumes; mechanical friction; suction gas cooling of a motor; and heat loss to the ambient. A schematic detailing the elements describing the hermetic scroll compressor is shown in Figure 4. For an open compressor, the motor is removed along with the motor loss that is heating the suction gas. The model is built in the Modelica language using the TIL-Suite modeling library (TLK-Thermo GmbH, 2020).

The high level mathematical model will now be described element by element, leaving room for flexibility in the

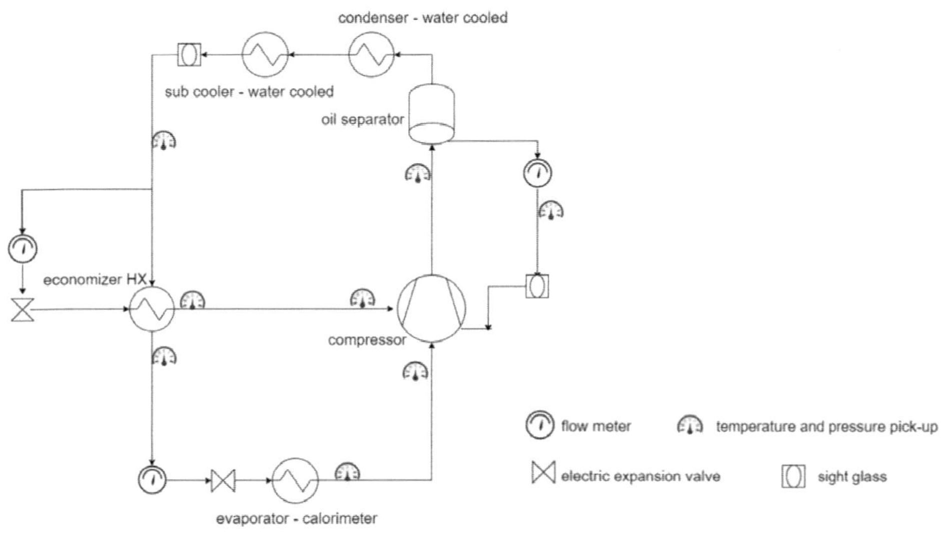

Figure 2: Test Stand Configuration for Economizer Compressor

detailed implementation:

Suction and Discharge Volumes The suction and discharge volume elements are described by constant pressure mixing volumes, for which the 1st law of thermodynamics can be applied:

$$\frac{dU_{CV}}{dt} = \sum Q_{in} - \sum Q_{out} + \sum \dot{m}h_{in} - \sum \dot{m}h_{out} \tag{1}$$

Motor The most straightforward approach for modeling the motor is with a constant efficiency. Alternatively, the efficiency could be described as a function of the mechanical torque and speed and other electrical boundary conditions. Given the motor efficiency, the motor can be described by:

$$\eta_{motor} = \frac{P_{shaft}}{P_{electric}} \tag{2}$$

$$Q_{motor,loss} = (1 - \eta_{motor})\, P_{electric} \tag{3}$$

Pressure Drop Elements A semi-empirical orifice model derived from Bernoulli's principle can be used to describe the four pressure drop elements in the compressor model:

$$\dot{m} = A_{eff}\sqrt{2\rho\Delta p} \tag{4}$$

While the above form is an incompressible formulation, this equation could be further enhanced to consider compressibility effects.

Mechanical Friction In the context of the compressor model, the mechanical friction refers primarily to the fluid viscous dissipation effects internal to the compressor. It is expected to be a strong function of the compressor RPM and should behave within the laminar and turbulent limits. Given these considerations, a reasonable model for the viscous dissipation in terms of the heat generated can be stated as:

$$Q_{friction} = a \cdot RPM^{n} \tag{5}$$

where a is the friction coefficient and n is expected to be within 1.0 (laminar) and 2.0 (turbulent).

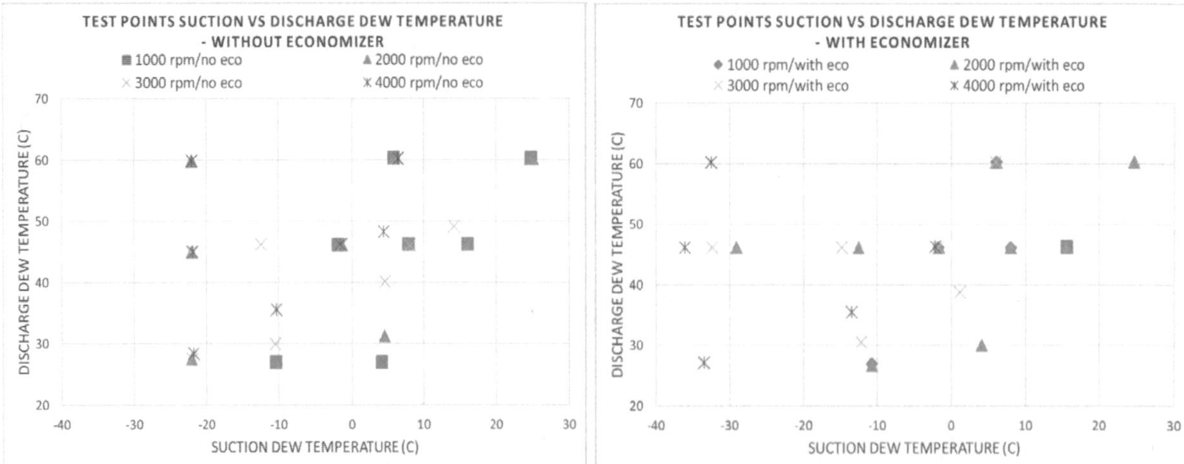

(a) Without Economizer　　　　**(b)** With Economizer

Figure 3: Test Points Suction Dew Temperature vs. Discharge Dew Temperature

Shell Heat Transfer For an adiabatic compressor, the discharge temperature will be resolved automatically given the upstream boundary conditions, the mass flow, and the power. However, if more precision is required, then the heat transfer loss through the shell can be considered, which will normally result in a lower discharge temperature. A simple model for the shell heat transfer is to use an overall driving temperature difference with a *UA* conductance coefficient (overall heat transfer coefficient):

$$Q_{\text{shell,loss}} = UA\,(T_{\text{dis}} - T_{\text{amb}}) \tag{6}$$

Ideal Compressor The ideal compressor element internal to the complete compressor model can be considered as lossless except for the losses that the compressor has no practical means to avoid, such as losses associated with mixing (due to the mainstream mixing with the injected stream) and the over/under-compression. Depending on the effects considered and where and how the mixing effect is implemented, the specifics can take various forms with up to three parameters: the displacement, the displacement ratio (effective displacement ratio between suction and injection locations), and the built-in volume ratio (or compression ratio, the volume ratio between suction and discharge locations). As an example, if the mixing effect occurs at the injection pressure level in a constant pressure mixing volume with no over/under-compression effects considered, then the following describes the ideal compressor model:

$$\dot{m}_{\text{suc,int}} = \rho_{\text{suc,int}} \frac{RPM \cdot displacement}{60} \tag{7}$$

$$\dot{m}_{\text{dis,int}} = \rho_{\text{mix,int}} \frac{RPM \cdot displacement/displacementRatio}{60} \tag{8}$$

$$\dot{m}_{\text{dis,int}} = \dot{m}_{\text{suc,int}} + \dot{m}_{\text{inj,int}} \tag{9}$$

$$\rho_{\text{mix,int}} = \rho\left(p_{\text{inj,int}}, h_{\text{mix,int}}\right) \tag{10}$$

$$\dot{m}_{\text{dis,int}} h_{\text{mix,int}} = \dot{m}_{\text{suc,int}} h\left(p_{\text{inj,int}}, s_{\text{suc,int}}\right) + \dot{m}_{\text{inj,int}} h\left(p_{\text{inj,int}}, T_{\text{inj,int}}\right) \tag{11}$$

$$P_{\text{shaft,ideal}} = \dot{m}_{\text{suc,int}}\left(h\left(p_{\text{inj,int}}, s_{\text{suc,int}}\right) - h\left(p_{\text{suc,int}}, T_{\text{suc,int}}\right)\right) + \dot{m}_{\text{dis,int}}\left(h\left(p_{\text{dis,int}}, s_{\text{mix,int}}\right) - h\left(p_{\text{inj,int}}, h_{\text{mix,int}}\right)\right) \tag{12}$$

The complete compressor model used in this work used 10 parameters in total that can affect the steady state result (the four A_{eff} values, *UA*, *displacement*, *displacementRatio*, *compressionRatio*, *a*, and *n*). While more details could be added element by element to increase the number of parameters, care must be taken since the model becomes more difficult for an optimizer to fit robustly. By focusing on just the key physical elements with simple physical descriptions, the model has the ability to predict the compressor behavior robustly.

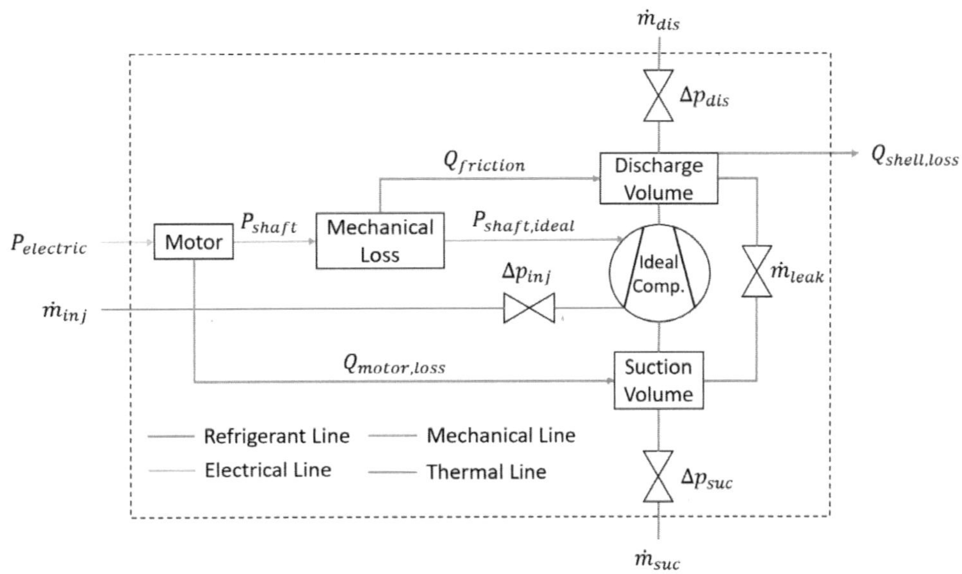

Figure 4: Schematic Model of a Hermetic or Open Scroll/Screw Compressor with Vapor Injection

Given the complete model and test data, a nonlinear optimization tool can be used to fit the parameters. The minimization function must first be defined. For many map based models, there are effectively multiple fits: a fit describing the power and a fit describing the mass flow as an example. In contrast, here there is only one fit, which is conceptually correct since in an actual physical compressor, any elemental physical changes will generally affect mass flows, power, and discharge temperature simultaneously. Therefore the objective function (minimizing the residual sum of squares (RSS) for the test conditions detailed in Figure 3) can be defined as follows:

$$RSS = \sum_{i=1}^{n} \left[\frac{\left(\dot{m}_{\mathrm{suc},i} - \hat{\dot{m}}_{\mathrm{suc},i}\right)^2}{\sigma^2_{\dot{m}_{\mathrm{suc}},i}} + \frac{\left(\dot{m}_{\mathrm{inj},i} - \hat{\dot{m}}_{\mathrm{inj},i}\right)^2}{\sigma^2_{\dot{m}_{\mathrm{inj}},i}} + \frac{\left(P_i - \hat{P}_i\right)^2}{\sigma^2_{P,i}} + \frac{\left(T_{\mathrm{dis},i} - \hat{T}_{\mathrm{dis},i}\right)^2}{\sigma^2_{T_{\mathrm{dis}},i}} \right] \tag{13}$$

Note that the hat symbol (as in $\hat{x}$) denotes the model prediction. It is important for the user to give reasonable estimates for the variance terms, especially since the measurement data have different units and uncertainty levels. Finally, starting guesses are given for each parameter, and the best fit can now be performed. In this work, the Levenberg-Marquardt optimizer was used. After the fit process is completed, various sanity checks are made and the goodness of fit is assessed. The process can be sensitive to the assumed variances and the initial guesses. Often, multi starting guesses are needed to find the minimum. Additionally, it is critical to have a good sized dataset over a large portion of the operating map with many levels in each of the independent variables (typically pressures and speeds on a compressor test stand).

6. NUMERICAL COMPRESSOR MODEL VALIDATION

The compressor test data with R513A refrigerant is used to fit different compressor physical parameters. For the fixed compressor displacement, the independent variables such as compressor suction pressure and temperature, vapor injection port inlet pressure and temperature, and discharge pressure are the required inputs for the fitting process along with the dependents compressor suction mass flow rate, vapor injection mass flow rate, compressor shaft power and discharge temperature. Compressor physical parameters such as areas of the suction valve, the discharge valve, the economizer port, leakages, the displacement ratio and the built-in volume ratio predict the refrigerant mass flow rates through the suction and vapor injection port, combined with additional frictional parameters which predict the compressor shaft power. *UA* captures the heat loss from the compressor shell which directly relates to the compressor discharge temperature prediction.

Table 4 provides the details of the compressor individual fitting targets for the dependents which are based on the compressor test data. The finished compressor fitting process with the individual fitting targets (σ, standard deviation) has resulted in the following R2 (R, Residual) and RMSD (Root Mean Square Deviation) values, which indicates that there is a very good match with that of the test data. Figure 5a provides the comparison of the simulation vs test data for the compressor suction mass flow rate. Figure 5b provides the comparison of the simulation vs test data for the vapor injection mass flow rate. It can be seen that the injection mass flowrate prediction is not very smooth as compared to the suction mass flow rate, which indicates many factors may have influence on the same. Figure 5c compares the compressor shaft power where the predictions are in very good agreement for the entire operating map. Figure 5d compares the compressor discharge temperature, which is within ±5 K over entire data set excluding a few points.

Table 4: Compressor Fitting Targets and Statistics

Dependent	σ	R^2	RMSD
Suction Mass Flow Rate	0.0009 kg/s	0.9978	5.320×10^{-3} kg/s
Vapor Injection Mass Flow Rate	0.003 kg/s	0.9653	3.843×10^{-3} kg/s
Compressor Shaft Power	99 W	0.9961	2.702×10^{2} W
Discharge Temperature	0.97 K	0.9533	3.476×10^{0} K

Once the compressor physical parameters are finalized based on the fitting process, the parameter based compressor model can now be used to predict the performance when integrated with a refrigeration system with any other refrigerants. In this paper, the original low pressure refrigerant R513A can be replaced with low GWP medium pressure refrigerants R454A or R454C, which are R452A refrigerant alternatives.

7. RESULTS

A system model is set up to run the simulations at AHRI 540-2019 standard rating conditions stated in the Table 5, with R454A and R454C refrigerants. For the medium rating condition where the suction dew temperature is −10 °C and the discharge dew temperature is 45 °C, simulations are performed both with and without economizer (vapor injection). For the low rating condition where the suction dew temperature is −35 °C and the discharge dew temperature is 40 °C, simulations are performed only with the vapor injection because the dew temperature difference between the suction and the discharge is large. The effect of the temperature glide on the thermal performance in the evaporator and the condenser are not considered since there are no physical evaporator and condenser in the simulations.

Table 5: AHRI SI Rating Conditions

Dependent	Low	Medium
Suction Dew Temperature (°C)	−35	−10
Suction Superheat (K)	10	10
Discharge Dew Temperature (°C)	40	45
Condenser Subcooling (K)	0	0
Ambient Temperature Surrounding Compressor (°C)	20	20
Economizer Superheat (K)	10	10
Compressor Speed (rpm)	1000, 2000, 3000, 4000	1000, 2000, 3000, 4000
Refrigerant Name	R454A, R454C	R454A, R454C

Figure 6a shows the capacity ratio between R454C and R454A (i.e. $\text{Cooling Capacity}_{R454C}/\text{Cooling Capacity}_{R454A}$) at both the rated conditions at different compressor operating speeds. It can be seen that the capacity ratio is typically in the range 0.82–0.86 indicating that the refrigerant R454C requires a larger compressor displacement or a higher compressor speed to match the capacity. It could also be seen that the capacity ratio decreases slightly as the compressor speed is increased. Figure 6b shows the COP ratio of R454C and R454A (i.e. $\text{COP}_{R454C}/\text{COP}_{R454A}$) with the increase in the compressor speed. It can be seen that the refrigerant R454C has slight advantage over R454A at lower speeds at low rating condition and it decreases as the speed is increases. At the medium rating condition, the refrigerant R454C has slight advantage over R454A, and the difference is reduced as the compressor speed is increased. Figure 6c shows the difference between the compressor discharge temperatures between R454A and R454C (i.e. $T_{\text{dis,R454C}} - T_{\text{dis,R454A}}$).

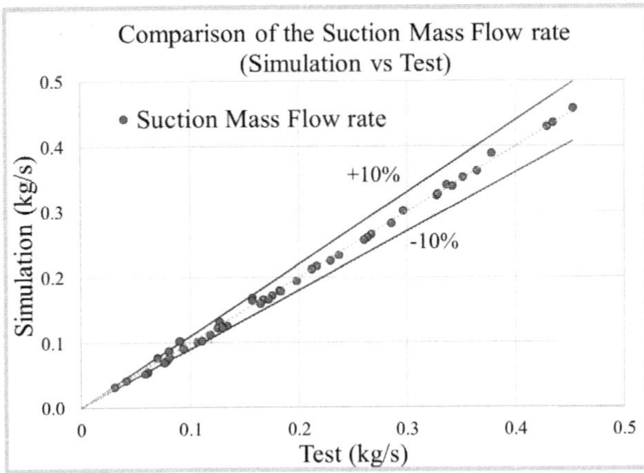

(a) Suction Mass Flow Rate

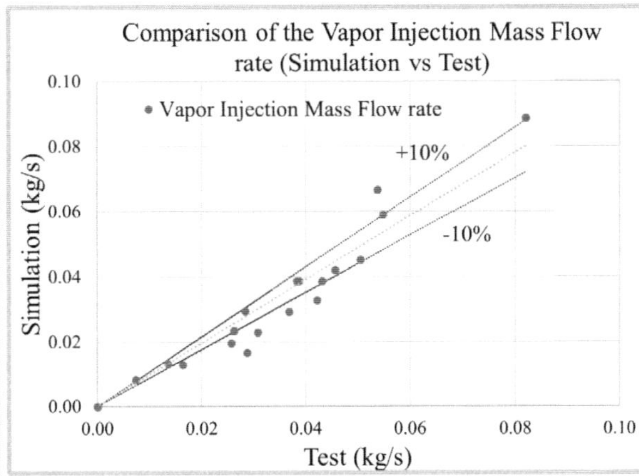

(b) Vapor Injection Mass Flow Rate

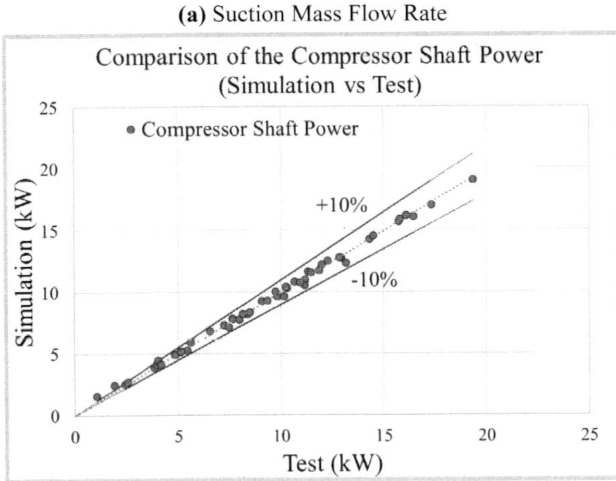

(c) Shaft Power

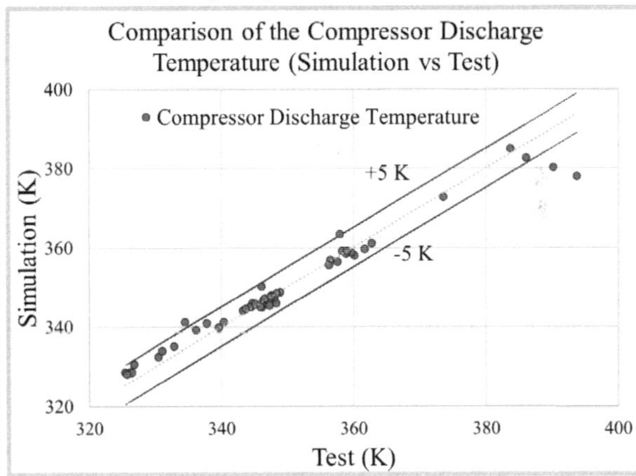

(d) Discharge Temperature

Figure 5: Simulation vs. Test Comparison

Since the refrigerant R454A has higher mass fraction of R32 in the mixture composition compared to R454C, it always has significantly higher compressor discharge temperature which may require liquid refrigerant injection as a mitigation option. The difference is observed to be larger for −35/40 °C (suction dew temperature/discharge dew temperature) condition compared to −10/45 °C condition. Figure 6d shows the COP ratio with economizer and without economizer (i.e. $COP_{\text{with eco}}/COP_{\text{without eco}}$) for both refrigerants at low and medium temperature conditions. It can be seen that there is an improvement of the COP by about 15 % with an economizer. The COP ratio also decreases as the compressor speed is increased.

8. CONCLUSION

The compressor performance prediction utilizing a numerical physical bulk model with R454A and R454C is conducted using vapor injection with economizer heat exchanger based on compressor test data with R513A. Following findings are observed:

- R454A always gives higher capacity than R454C at the same compressor displacement and speed (Figure 6a).
- R454A gives higher COP than R454C at frozen conditions and higher speeds while R454C gives higher COP at lower speeds for the same frozen conditions (Figure 6b).
- R454C gives always higher COP at fresh conditions independent of speed (Figure 6b).
- R454C gives always lower discharge temperature than R454A (Figure 6c).
- Both refrigerants are expected to need liquid injection to cover the entire application. The proportion of the

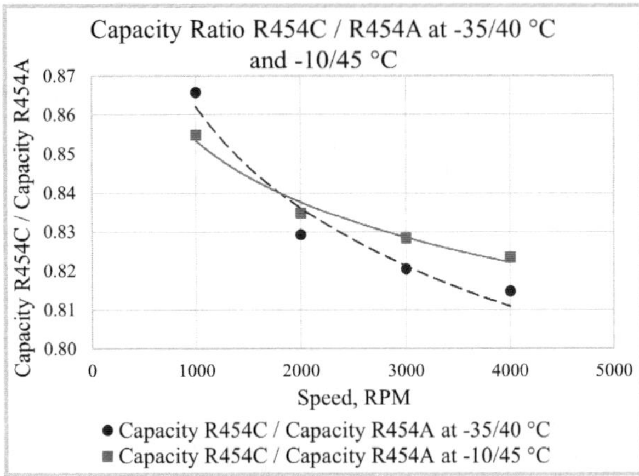

(a) Capacity Ratio (R454C/R454A)

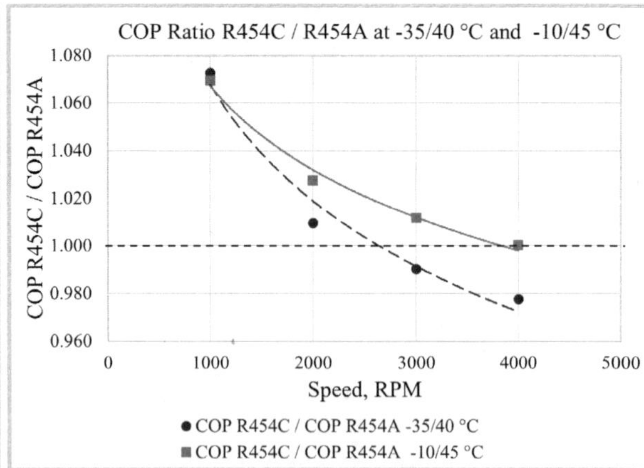

(b) COP Ratio (R454C/R454A)

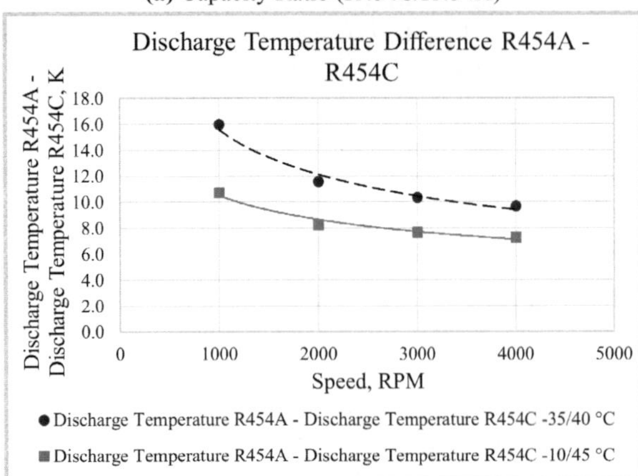

(c) Discharge Temperature Difference (R454A-R454C)

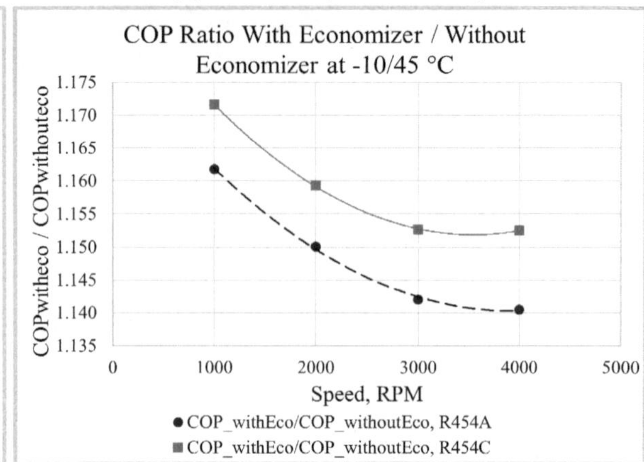

(d) $COP_{withEcon}/COP_{withoutEcon}$

Figure 6: Comparison Between R454C and R454A

entire application requiring liquid injection will be larger with R454A compared to R454C.

- The benefit of using vapor injection with R454C is greater than R454A (Figure 6d).
- Additional system performance studies will be required for better understanding of R454A vs. R454C applications.

NOMENCLATURE

A_{eff}	effective area	(m^2)	Δp	pressure difference	(Pa)
$\dot{m}$	mass flow rate	(kg/s)	η	efficiency	(1)
h	specific enthalpy	(J/kg)	ρ	density	(kg/m^3)
p	pressure	(Pa)	σ_x^2	variance	($[x^2]$)
P	power	(W)			
Q	heat flow rate	(W)			
s	entropy	(J/(kg K))			
t	time	(s)			
T	temperature	(K)			
U	internal energy	(J)			
UA	overall heat conductance	(W/K)			

Subscript

amb	ambient	int	internal ideal compressor
CV	control volume	leak	leakage
dis	discharge	mix	mixed state
in	inflow	out	outflow
inj	injection	suc	suction

REFERENCES

AHRI. (2020). *Ahri standard 540 for performance rating of positive displacement refrigerant compressors and compressor units* (Standard). Arlington, VA, USA: Air-Conditioning, Heating, and Refrigeration Institute.

Cambio, M. (2016). Representation of a positive displacement compressor map with vapor injection. In *Ashrae annual conference seminar 13: Advances in compressor design, testing, and performance modeling for new efficiency standards and alternate refrigerants.* Orlando, FL, USA.

Cavalcante, P., Farsterling, S., Tegethoff, W., Stulgies, N., & Köhler, J. (2008). Transient modeling and sensitivity analysis of a controlled r744 swash plate compressor. In *International compressor engineering conference.* West Lafayette, IN, USA.

Erickson, L. (1998). Rating equation for positive displacement compressors with auxiliary suction ports. In *International compressor engineering conference.* West Lafayette, IN, USA.

Hundy, G., & Vittal, R. (2000). Compressor performance definitions for refrigerants with glide. In *International compressor engineering conference.* West Lafayette, IN, USA.

Sjoholm, L., & Ma, Y. (2018). A study of low gwp refrigerants for transport refrigeration based on hermetic scroll compressors with an economizer. In *International compressor engineering conference.* West Lafayette, IN, USA.

Stulgies, N., Gräber, M., Tegethoff, W., & Försterling, S. (2009). Evaluation of different compressor control concepts for a swash plate compressor. In *Proceedings of the 7th international modelica conference.* Como, Italy.

TLK-Thermo GmbH. (2020). TIL Suite (Version 3.8.0) [Computer software]. Braunschweig, Germany.

ACKNOWLEDGMENT

The authors wish to thank the management of Trane Technologies and Thermo King for permission to publish this paper. The authors also wish to thank Bruce Wynn, Ken Schultz and Robert Isaacs for supporting this work.

Quantitative and Qualitative Evaluation of Various Positive-Displacement Compressor Modeling Platforms

M. Mohsin Tanveer[1*], Craig R. Bradshaw[1]

[1] Center for Integrated Building Systems, Oklahoma State University, Stillwater, OK 74078
Contact Information (mohsin.tanveer@okstate.edu)

* Corresponding Author

ABSTRACT

Several regulatory measures aimed to mitigate climate change are forcing compressor manufacturers to replace current refrigerants to those with low Global Warming Potential (GWP). New refrigerants need to be evaluated to ensure adequate efficiency for use in modern products. Evaluation can be done heuristically, which is expensive and time-consuming, while a carefully designed simulation model can provide similar outcomes for a significantly reduced cost. This paper presents a comparison between various user-developed and existing reciprocating compressor models to assist in the selection of a suitable modeling platform for a wide-ranging study. The reciprocating compressor is selected because of the simplicity of the model to ensure consistency across different platforms. The user-developed models are developed in MATLAB™ and Modelica™ for the reciprocating compressor. The same compressor is also modeled using existing compressor modeling platforms, PDSim and GT-Suite™. The compressor model includes three main components; geometry, compression process and frictional losses. Other sub-models, like valve model and heat transfer model, are also part of the compression process. These platforms are evaluated based on both quantitative and qualitative criteria. Modelica™ is found to be computationally efficient while GT-Suite™ took maximum time for simulation among the compared platforms. On a qualitative basis, PDSim is potentially a better platform for compressor optimization; which is also readily available to end user due to its open-source nature and prospects for future model development.

1. INTRODUCTION

A 2016 amendment to Montreal protocol directs a shift towards the low-GWP refrigerants, which will result in phasing down the use of HFC refrigerants in HVAC&R systems Consequently, HFC refrigerants will need to be replaced with low-GWP alternatives or natural refrigerants. The effect of this transition on the performance of existing compressor technology is a critical evaluation step in developing next-generation products.

The development of prolific and powerful computer technology has enabled detailed compressor models to become a more viable substitute for heuristic compressor design. The statistical correlations are the simplest among various compressor models and can only be applied within the ranges of the training data for the model. The extrapolation of these models often entails very high uncertainties and can not be used as predictive tools.

The fidelity of the models increases as they incorporate the physics involved in the compression process, and their reliability depends on the accuracy of the individual sub-models. The mechanistic chamber model extends the fidelity beyond a statistical model and is based on a control volume (or chamber). It can be implemented for various positive displacement compressors and can also be used as a predictive tool. Chen et al. (2002) suggested that a comprehensive simulation model for the scroll compressor can predict the real behavior of the compressor and can be used for compressor design and optimization purposes.

Fully distributed (or 3D) models typically manifest as CFD-based. Though CFD provides a detailed analysis of fluid behavior in the compressor, often such detail is not required for performance evaluation. Moreover, computer technology developed and readily available in the past decade but still, CFD is computationally impracticable for holistic performance optimization of compressors. The same outcomes can be accomplished in significantly reduced time and cost by utilizing properly tuned and validated 0D or 1D mechanistic model (Pereira et al., 2008).

While the physics of the model is important for reliable results, numerical solution techniques are also vital for dynamic models. Dynamic behavior might not be modeled properly, if proper solution technique is not selected, due to the stiff nature of compression process equations. The selection of suitable solver technique is important to reduce the computational time and fulfill the model requirement. Moreover, several proprietary, closed-source and open-source compressor modeling platforms are available. A qualitative comparison of these available platforms is required to explore their capabilities for compressor modeling. Additionally, simulation time is very important for complex physics-based models and a comparison of available platforms should be conducted to help in platform selection for future model development.

In summary, a mechanistic chamber model can be used to predict the compressor performance at conditions different than for which it had been validated. This makes this type of model well-suited to explore the future of compressor technologies amid the transition of working fluids. This work presents the first step in the process of model development by the selection of the modeling platform that will allow the most consistent, and rapid, development of a plurality of compressor models to support this effort. To this end, four different compressor modeling platforms are evaluated based on qualitative and quantitative parameters applied to the modeling of a reciprocating compressor. The first two platforms are user-developed mechanistic chamber models, developed in MATLAB™ (MATLAB, 2019) and Modelica™. The third platform is PDSim (Bell et al., 2020a), which is a Python-based package and utilizes the same mechanistic chamber modeling approach, developed to model positive displacement compressors. The fourth platform is GT-Suite™ (GT-Suite™, 2019) which is based on 1-D CFD. Model templates for certain compressors (i.e. reciprocating compressors, scroll compressors) are available in GT-Suite™ and has been employed for various applications by number of authors.

2. MECHANISTIC CHAMBER MODEL OF A RECIPROCATING COMPRESSOR

This section presents the physics of the model developed in all four modeling platforms (MATLAB™, Modelica™, PDSim, and GT-Suite™). This model is constructed using the mechanistic chamber model approach and is similar to the models presented by Bell et al. (2020a). This approach models the compressor by applying the mass and energy conservation to the control volume (defined inside the compression chamber) and predicting the refrigerant properties on each step of the crankshaft rotation. The general approach of the model is presented in Figure 1. The modeling process starts by defining the guess values for the initial thermodynamic state in the control volume. The volume model is used to calculate the instantaneous volume of the control volume. Then, mass and energy conservation equations are used to calculate the next thermodynamic state which requires inputs from sub-models of volume, mass flow and heat transfer. The process is assumed to be quasi-steady and thermophysical properties are considered constant for each time step. The process is repeated for one crank rotation. The compressor wall temperature is considered constant for each cycle and re-evaluated at the end of each iteration. At the end of the cycle, residuals are calculated for instantaneous temperature, density and time-averaged wall temperature. If the residuals are greater than tolerance than final states are used as an initial guess and the process will be repeated until convergence is achieved. The details of each sub-model is described in the following sections.

2.1 Compression Process Equations

The compression process is modeled by applying the mass-energy balance to a general control volume. The general mass conservation equation of a control volume can be re-cast to calculate the instantaneous density of a generic control volume, written as,

$$\frac{d\rho}{dt} = \frac{1}{V}\left[-\rho\frac{dV}{dt} + \left(\sum \dot{m}_{in} - \sum \dot{m}_{out}\right)\right]. \tag{1}$$

Similarly, by ignoring kinetic and potential energy changes, the conservation of energy can be represented as,

$$\frac{dE_{CV}}{dt} = \frac{du}{dt} = \frac{\partial u}{\partial \rho}\frac{d\rho}{dt} + \frac{\partial u}{\partial T}\frac{dT}{dt}. \tag{2}$$

These expressions can be combined and in then applying the definition of enthalpy and boundary work to generate the

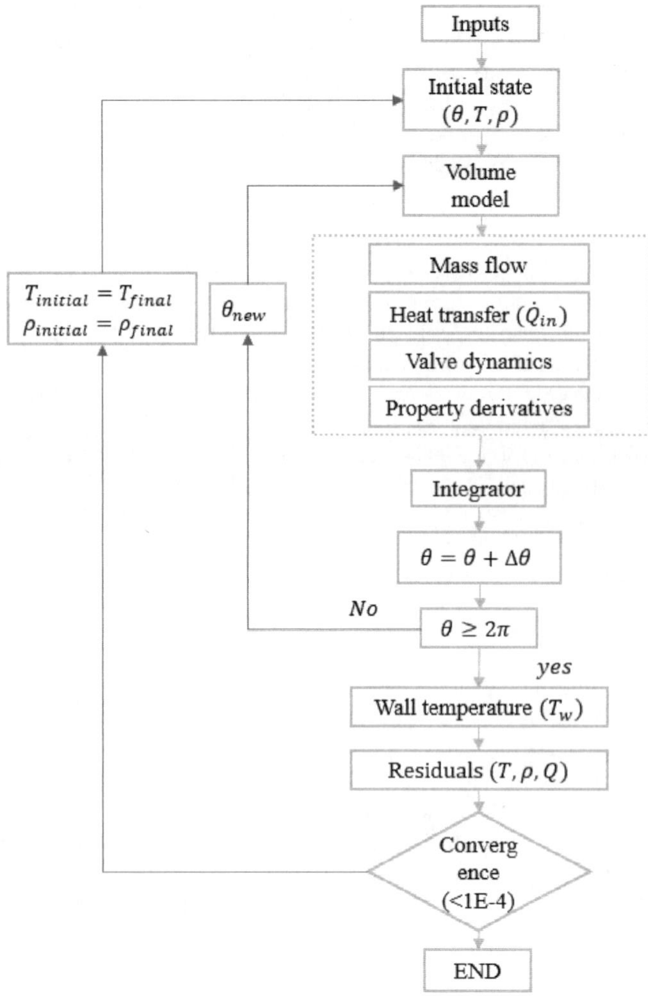

Figure 1: Flow chart for mechanistic chamber model.

final form of the energy balance equation,

$$\frac{dT}{dt} = \frac{-1}{\rho V \frac{\partial u}{\partial T}} \left[\left(\rho V \frac{\partial u}{\partial \rho} + V u \right) \frac{d\rho}{dt} - (P + \rho u) \frac{dV}{dt} \right.$$
$$\left. + \sum \dot{m}_{in} h_{in} - \sum \dot{m}_{out} h_{out} + \dot{Q}_{in} \right]. \tag{3}$$

This general form of the compression process equations is generic for all positive displacement types and requires, as inputs, information for the various terms in the expressions that define the specific type of compressor being analyzed. In this case, the volume, heat transfer, and mass flow models developed are specific for a reciprocating compressor.

2.2 Volume Modeling

The mass and energy balance equations require as an input, the instantaneous volume in the cylinder to define the behavior of the control volume. For a reciprocating compressor, this can be calculated by using two different approaches, either kinematic expressions or assuming a sinusoidal volume change. The model selected depends on the availability of information. As current work is focused on the capabilities of the platform, a hypothetical compressor is modeled by assuming displaced volume because detailed compressor geometry is not available. Therefore, for the current application, the second approach is more suitable that assumes a sinusoidal piston movement,

$$V = V_0 + \frac{V_{disp}}{2(1 - cos(\theta))}. \tag{4}$$

The clearance volume is assumed to be 1% of the displaced volume.

2.3 Thermo-Physical Properties

Thermophysical properties are a vital part of the model and are required in almost all the sub-models. Consequently, model accuracy is strongly dependent on the accuracy of the method used for property calculation. Moreover, these property calculations take around 90% of the total simulation time. It is important to discuss different available methods for property calculations. The thermophysical properties are calculated using pre-coded Equations of State (EOS) from various libraries. The REFPROP is used in the model developed in MATLAB™ and GT-Suite™. TIL media library is used for the model developed in Modelica ™ which is developed by TLK-Thermo GmbH and allows a user to access external database i.e. REFPROP. Being an open-source platform, PDSim utilizes an open-source library (CoolProp) for property calculation (Bell et al., 2014). Density and temperature are selected as independent properties to fix the thermodynamic state of the refrigerant at each each step.

2.4 Instantaneous Heat Transfer in Cylinder

Heat transfer from the refrigerant to the cylinder wall is calculated initially by assuming a constant cylinder wall temperature over a cycle because instantaneous cylinder wall temperature changes very little as compared to the time-averaged wall temperature over a cycle. The new wall temperature is computed at the end of the cycle which is used for the next iteration, convergence is achieved. Instantaneous heat transfer in the cylinder is calculated using Newton's law of cooling as,

$$\dot{Q}_{in} = h_{ref} A_p (T_w - T), \tag{5}$$

where convection heat transfer coefficient is calculated as given below,

$$h_{ref} = 0.053 \frac{k}{d_p} \mathrm{Pr}^{0.6} \mathrm{Re}^{0.8}. \tag{6}$$

For Reynolds number calculation, the velocity of refrigerant is assumed to be equal to the velocity of the piston. The net heat transfer between the refrigerant and the wall is then calculated for each compression cycle by taking the average of the instantaneous heat transfer over the entire cycle.

2.4.1 Ambient Heat Balance The average cylinder heat transfer is coupled to the ambient through an ambient heat balance. The cylinder wall temperature is calculated at the end of each iteration by setting up heat transfer from the cylinder wall to the ambient as given below,

$$T_w = \frac{\dot{Q}_{amb}}{h_a A_{cyl}} + T_{amb} \tag{7}$$

The ambient heat transfer $\dot{Q}_{amb}$ is computed as,

$$\dot{Q}_{amb} = h_a A_{cyl} (T_w - T_{amb}), \tag{8}$$

and heat transfer coefficient for ambient air is calculated by following correlation

$$\mathrm{Nu} = 0.683 \mathrm{Re}^{0.466} \mathrm{Pr}^{\frac{1}{3}}, \tag{9}$$

where, Reynolds number is calculated by assuming fixed wind velocity.

2.5 Mass Flow and Valve Model

Mass flow through suction and the discharge port is calculated by assuming an isentropic compressible flow assumption and calculated using the following expression,

$$\dot{m} = \rho\sqrt{\gamma RT}A_v\sqrt{\left(\frac{P_{high}}{P_{low}}\right)^{\left(\frac{\gamma-1}{\gamma}\right)}\frac{2}{\gamma-1}}. \tag{10}$$

The model is designed to work with or without valve dynamics to observe the effect of the valve model on the stiffness of compression process equations and computational time. Valves in reciprocating compressors are typically a reed valve, which is a flat, constrained, beam that behaves like a spring when moved in response to the pressure differences. The flow area available to the mass flow model requires the valve opening, which is used to calculate of flow area. This information is collected by modeling the dynamics of the reeds assuming they behave as one degree of freedom mass-spring system. Valve dynamics is modeled using the method presented by Kim & Groll (2007).

3. NUMERICAL INTEGRATORS

Three numerical integration techniques are compared in this study using the various software platforms. This comparison will add an additional dimension to the platform comparison as some platforms allow precise control over numerical integration (*e.g.* MATLAB™, PDSim, Python) while others offer very little (*e.g.* Modelica™, GTSuite™). As a basis of comparison, this study will compare the most basic numerical techniques, Euler and Modified Euler, against more advanced solvers like the Adaptive-RK and the built-in solvers in the commercial packages. Additional information on each numerical integrator can be found in (Bell et al., 2020b).

4. MODELING PLATFORMS

Four different platforms are selected for quantitative and qualitative evaluation to help the selection of a suitable platform for future compressor performance evaluation. A user-developed program is modeled in two of these platforms. The rest of the platforms have already developed model components or libraries which are utilized for compressor modeling. A detailed description of the model developed in each platform is presented below. The code for each of the four models is also provided as an electronic annex (available at doi:10.5281/zenodo.3923415).

4.1 User Developed Code

4.1.1 MATLAB ™, Code is developed in MATLAB™ and program flow is presented in Figure 1. The model physics is described in Section 2. The sub-models are developed in the form of functions and integrated for compressor modeling. Three different integration approaches, as discussed in Section 3, are used to demonstrate their capability for compressor performance and their efficiency in terms of computational time.

4.1.2 Modelica ™, The same model is replicated in Modelica™ using available Euler and RK45 solvers. Only suction and discharge mass flow rate models are provided in the form of functions while rest of the program is structured in a single Modelica™ model. Contrary to most programming software that interprets the program as a set of instructions, Modelica™ tries to simulate the program w.r.t time so there is no required flow for programming. Besides, equations are not required to be manipulated for the calculated variable to be on the left side. The only criteria that need to be fulfilled, is the number of provided equations should be equal to the number of variables. Also, variables used in the model need to be declared at the top along with their type.

4.2 Pre-Developed Model Bases

4.2.1 PDSim, PDSim is a Python-based, object-oriented, positive-displacement compressor and expander modeling platform described in (Bell et al., 2020a) and (Ziviani et al., 2020). It utilizes the high-level features of python and incorporates low-level code using Cython for performance optimization purposes. It has some libraries, developed in an object-oriented fashion, to simplify the modeling of positive displacement compressors. It allows users to use a text-based programming approach to model the compressor which provides adequate flexibility for modification or its Graphical User Interface (GUI) can be used by simply providing the compressor parameters. Different pre-developed example compressor models are provided in the package, including rolling piston compressor, scroll compressor and reciprocating compressor.

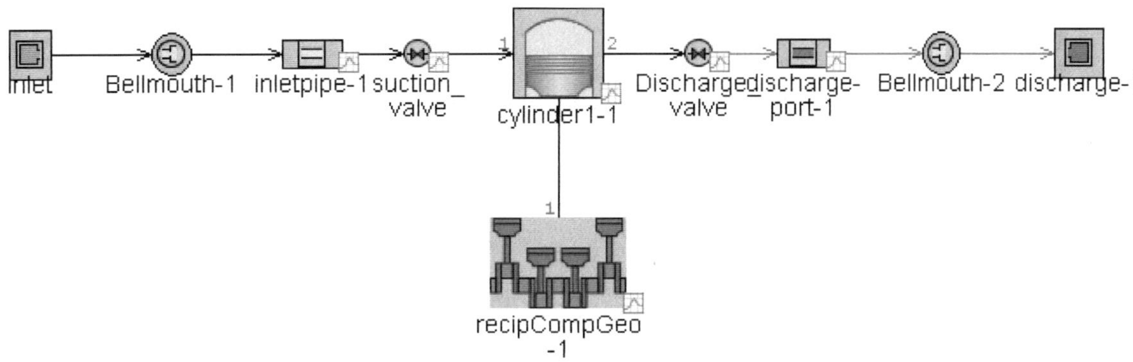

Figure 2: Schematic of compressor model in GT-Suite™.

4.2.2 GT-Suite™, GT-suite™ is a software package that utilizes a 1-D computational fluid dynamics (CFD) engine to do the dynamic analysis of a system of components. It has built-in physics-based models of components that can be connected to model a system, including compressors. A schematic of the reciprocating compressor model is shown in figure 2. The cylinder is connected to the inlet and discharge side by pressure-driven valves and ports. Cylinder volume is calculated using crank kinematics and is capable of modeling a multi-cylinder compressor. Detailed list of parameters used for setting up the compressor model in GT-Suite is presented in (Tanveer & Bradshaw, 2020) and can also found in the electronic annex.

5. RESULTS AND QUANTITATIVE COMPARISON

5.1 Model Prediction Comparison

The compressor with the same dimensions and parameters is modeled across four platforms to compare the simulation time. Table 1 and Figure 3 shows that models developed across platforms are similar enough to be considered identical for this study. The calculated mass flow rate, power and volumetric efficiency are reasonably close to each other for four developed models. The pressure-volume curve indicates that the results for all four platforms present very similar thermodynamic behavior as well. The difference in the indicator curve of GT-Suite™ can be explained by the dynamics of the valve component. The valve component in GT-Suite™ is very detailed which considers the non-linear behavior of the valve (i.e. damping etc.) while the rest of the models use a 1-D spring-mass-damper model. This difference in the valve modeling approach results in mass flow fluctuations on the discharge side. Additionally, the Modelica™ model does not have a valve sub-model because Modelica™ requires variables to be a continuous function of time. In the valve model, the integration of maximum valve lift in terms of stopper height makes it a piece-wise function. The valve model can be integrated by using the "$Algorithm$" feature in Modelica™ but it will limit the use of Modelica's other features. Although there are minor differences in modeling approaches across different platforms, overall the compressor outputs are identical as evident from Table 1. Small differences are found in the indicated power which is caused by slight differences in valve behavior and discharge temperature. For MATLAB™, it is also observed that reducing the solver tolerance gives better agreement with PDSim and GT-Suite™ results. The results, presented in Table 1, are obtained at same solver tolerance to make a comparison between platforms at consistent settings. Similarly, as mentioned earlier, the model in Modelica™ does not account for valve losses and is slightly under-predicting the indicated power as a result. Despite the small differences in predicted outputs, these results are considered sufficiently close to compare the qualitative and quantitative metrics associated with the model development process.

Table 1: Compressor performance parameters calculated from four different platforms

parameter	MATLAB™	PDSim	Modelica™	GT-Suite™
Mass flow rate $[g/s]$	11.12	11.15	11.15	11.20
Indicated Power $[W]$	261.16	247.7	235.05	244.59
Volumetric efficiency [%]	97.63	97.83	97.85	98.42

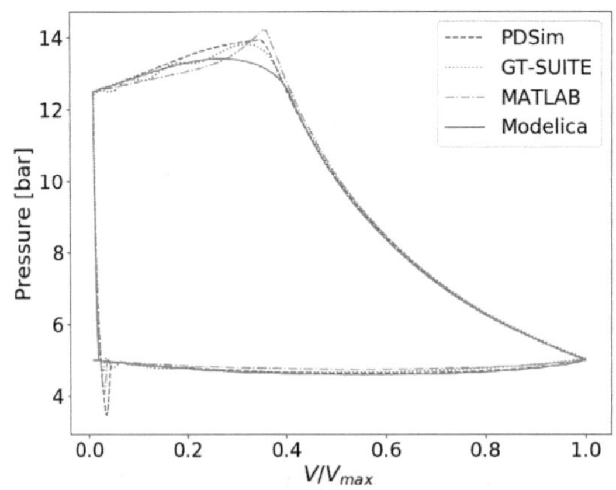

Figure 3: Indicator diagram across different platforms

5.2 Temporal Results for Four Platforms and Three Integrators

Figure 4a [1] shows the simulation time required by the four platforms for compressor modeling using the RK45 method at various solver error tolerances. The simulation time increases with reduced error tolerance due to the reduced step size requirement necessitating smaller steps. It was found that a significantly reduced minimum step size limit is required for large error tolerances. Often, choosing large error tolerance cause stability issues with the model.

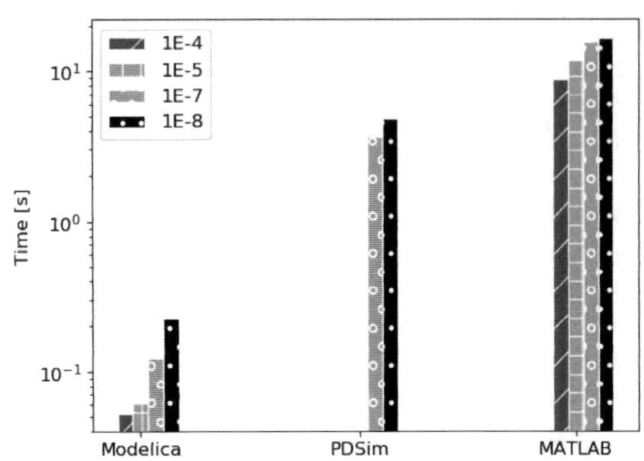

(a) Effect of error tolerance in RK45

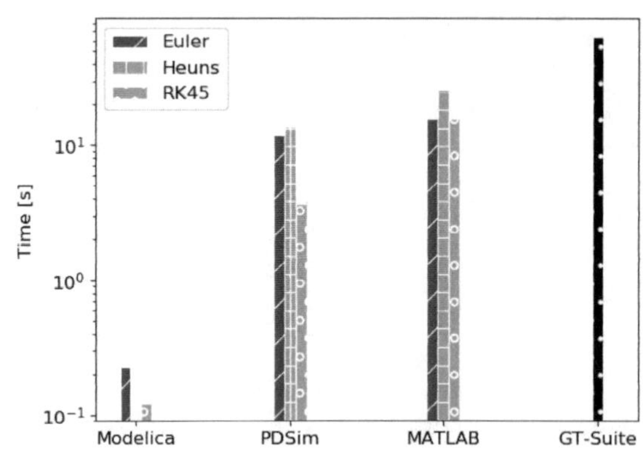

(b) Simulation time for different numerical integrators

Figure 4: Simulation time comparison

Figure 4a also provides a comparison of solution time for the compared platforms. Modelica™ is found to be significantly faster than other platforms compared in this article. One possible reason is the utilization of its built-in solver in the model which is optimized computationally for dynamic modeling. PDSim is found to be slightly faster than MATLAB™ with a couple of possible explanations being the use of Cython in PDSim for a reduction in interpreted code steps or the use of REFPROP in the MATLAB™ model and CoolProp in PDSim. Although there is a difference in simulation time for different platforms, It is not significant enough for a user to change modeling platform.

[1] The simulation time results for Modelica™, presented in Figure 4a and 4b, are CPU time instead of actual simulation time. A detailed explanation is presented in section 5.2.

Figure 4b presents the performance of different numerical integrators for compressor modeling, showing the integrators used on each platform and the impact on simulation speed. Although the RK45 is the most complex of the compared solvers, it takes much less simulation time as compared to Euler and Heun's method. For simulation time comparison across different platforms(excluding GT-Suite™), the compressor is simulated using the Euler method with a fixed number of steps. The Euler method is used with 7000 steps due to computational limitations and it should be noted that it would take a significantly higher than 7000 steps to match the same level of error tolerance for a direct comparison to RK45. Despite this, RK45 was still significantly faster than the Euler and Heun's integrator. It was found that Modelica™ takes the least amount of time for a solution while PDSim is slightly faster than the MATLAB™ based model. Contrary to the other platforms, the simulation time for Modelica™ is the CPU time as it is the only profiling variable that can be measured in Modelica™ with reasonable accuracy. It is observed that there is some overhead time (few seconds) associated with some simulation initialization that can only be recorded using a stopwatch. A stopwatch analysis shows that the actual simulation time for Modelica™is higher than CPU time presented in Figure 4a and 4b but with high levels of variability. The actual simulation time are similar orders of magnitudes as PDSim and MATLAB ™ but due to uncertainties in time measurement, these results are not reported in the text.

6. QUALITATIVE ANALYSIS OF FOUR PLATFORMS

All the aforementioned platforms are found to be capable of compressor modeling, but each one of them has some pros and cons. These platforms are evaluated on the basis of four criteria 1) User friendliness, 2)Ease for model development and 3) Future prospects.

6.1 User-Friendliness

A user-friendly simulation model should require minimum knowledge about the model from the user; but, at the same time, should be flexible enough to cater to the diverse needs of the different users. A GUI based platform is often very easy to use, with minimal knowledge requirement from a user, as it shadows most of the strenuous physics and programming. Among different platforms, compared in the text, GT-Suite™ is completely GUI based while others have some kind of GUI development tools available. GT-Suite™ has a drag and drop configuration, where components can be connected to model a system. It also allows users to define inputs in the form of parametric equations but does not provide a user the option to customize or change any background physics for a component.

Modelica™ also provides component development tools in the form of blocks. A block can be developed for each sub-model and, later on, these blocks can be connected in different orientations for system modeling. In Modelica™, the model should be a continuous function of time for successful simulation. The number of equations and variables should be equal because Modelica™ solves the equations simultaneously. These features are very useful for dynamic simulation and reduce the number of lines required for programming. It is very convenient for systems that involve less complex components but as the component complexity increases, its simultaneous solution structure provides less control and often causes instability to the simulations.

PDSim also has a GUI interface available, where a user can enter compressor parameter (*i.e.* dimensions, flow parameters, etc.) to get results and it requires minimum compressor knowledge. It also provides the user with an option to modify the individual models by adding Python functions and/or model the entire compressor using a script to call pre-coded software objects. The GUI interface of PDSim is not as robust as it is in GT-Suite ™. It does not allow flexibility for connecting components in different orientations for example modeling of multi-cylinder compressor. But its object oriented model provides full control to the user for model modification and development activities.

6.2 Model Troubleshooting

Another most important feature specific to a platform is easiness for troubleshooting. Most of the effort goes into troubleshooting by a user when using previously developed models. A platform must be able to provide sufficient information for troubleshooting. Apart from syntax or platform-specific errors, errors could be related to the model and can be anticipated by the developer. Flags should be added to the program to raise those errors to ease up the troubleshooting process. In GT-Suite™, if a simulation fails, it most often refers to the individual component and in some cases to the parameter which is causing a problem. It often proves useful in identifying the problem. In Modelica™, as variables can be assigned types and units, it makes it easy to find unit errors but sometimes, due to its simulation equation solution structure, troubleshooting a code becomes very difficult. For example, if a model is over-defined or under-defined, then it will only raise an error mentioning the model is structurally singular. Often, it becomes very hard to find which variable is missing or which equation needs to be added to resolve the issue. Contrary, as PDsim is based on Python, it treats the code as a set of instructions to execute. If something causes a simulation error, it returns the last instance which causes the error. Additionally, models related problems, for example specifying the wrong values to a parameter, are anticipated and flags are added to point out the problem.

6.3 Future Prospects

Finally, the last comparison is for prospects of compressor modeling across each platform. For GT-Suite™, being commercial software, it is not individual discretion to improve or add new compressor models to the software. Modelica™ has a fast computational time and easy object-oriented based syntax but its simulation focused solution structure is not convenient for complex compressor model development. Components in Modelica™ can be developed easily but it could be very time consuming for a user to troubleshoot it. As the purpose of such model development is to ease the efforts at a later stage, Modelica™ might not be a suitable option for compressor modeling. MATLAB™ is a powerful tool for model development and Simulink could be very helpful in defining drag and drop configuration but it is a closed source platform and may limit the researchers from contributing towards model development. Additionally, the availability of the model to the user will be limited. In contrast, PDSim is an open-source package. So, anyone can contribute to its development. Moreover, it is based on Python which also an open-source high-level programming language and is widely used among the research community. While sufficiently good results can be obtained from PDSim's GUI interface, it involves a steep learning curve associated with its text-based model development.

7. CONCLUSION

Platforms for compressor modeling are compared by modeling the same reciprocating compressor on both quantitative and qualitative bases. The model developed on each platform was found to be similar enough for direct comparison. Each platform was found to have pros and cons associated with it and depending on the modeling application, some platforms may be better suited to specific circumstances for compressor modeling. For performance evaluation of low-GWP refrigerants, PDSim possesses the better combinations of features. Although its simulation time is higher than Modelica™, the simulation time difference is not significant. Additionally, the availability of pre-coded libraries for mechanistic chamber model can be helpful in reducing the time associated with model development. While GT-Suite is more user friendly than PDSim, PDSim offers more flexibility for model improvement and customization. Moreover, it is an open-source platform that allows researchers to contribute to its development.

NOMENCLATURE

$\dot{m}$	Mass flow rate	[kg/s]
$\dot{m}_{out}$	Total mass leaving the control volume	[kg/s]
$\dot{Q}_{amb}$	Average heat transfer rate from cylinder wall to ambient	[W]
$\dot{Q}_{in}$	Instantaneous heat transfer rate from cylinder wall to the refrigerant	[W]
$\dot{W}$	Rate of work done	[W]
γ	Heat capacity ratio	[-]
ρ	Density	[kg/m^3]
Re	Reynolds number	[-]
θ	Crank angle	[rad]
A_p	Cylinder cross-sectional area	[m^2]
A_v	Area of the valve	[m^2]
A_{cyl}	Outer surface area of the cylinder	[m^2]
E_{CV}	Energy inside the control volume	[J]
h_a	Ambient air heat transfer coefficient	[J/kg]
h_{in}	Specific enthalpy of the refrigerant at suction	[J/kg]
h_{out}	Specific enthalpy of refrigerant at discharge	[J/kg]
h_{ref}	Refrigerant convection heat transfer	[J/kg]
P	Pressure	[kPa]
P_{high}	Pressure on high pressure side of the valve	[kPa]
P_{low}	Pressure on low pressure side of the valve	[kPa]
R	Specific gas constant	[J/kg.K]
T	Instantaneous temperature in control volume	[K]
t	Time	[s]
T_w	Cylinder wall temperature	[K]
T_{amb}	Ambient air temperature	[K]
u	Specific internal energy	[J/kg]

V	Instantaneous volume of the control volume	$[m^3]$	Nu	Nusselt number	[-]
V_0	Clearance volume	$[m^3]$			
V_{disp}	Total displaced volume	$[m^3]$	Pr	Prandtl number	[-]

REFERENCES

Bell, I. H., Wronski, J., Quoilin, S., & Lemort, V. (2014). Pure and Pseudo-Pure Fluid Thermophysical Property Evaluation and the Open-Source Thermophysical Property Library CoolProp. *Industrial & Engineering Chemistry Research*, *53*, 2498–2508.

Bell, I. H., Ziviani, D., Lemort, V., Bradshaw, C. R., Mathison, M., Horton, W. T., ... Groll, E. A. (2020a). PDSim: A General Quasi-Steady Modeling Approach for Positive Displacement Compressors and Expanders. *International Journal of Refrigeration*, *110*, 310–322. doi: 10.1016/j.ijrefrig.2019.09.002

Bell, I. H., Ziviani, D., Lemort, V., Bradshaw, C. R., Mathison, M., Horton, W. T., ... Groll, E. A. (2020b). Pdsim: A general quasi-steady modeling approach for positive displacement compressors and expanders. *International Journal of Refrigeration*, *110*, 310–322.

Chen, Y., Halm, N. P., Braun, J. E., & Groll, E. A. (2002). Mathematical Modeling of Scroll Compressors- part II: Overall Scroll Compressor Modeling. *International Journal of Refrigeration*, *25*(6), 751–764. doi: 10.1016/S0140-7007(01)00072-X

GT-Suite™. (2019). *v2019*. Westmont, Illinois: Gamma Technologies.

Kim, J. H., & Groll, E. A. (2007). Feasibility Study of a Bowtie Compressor with Novel Capacity Modulation. *International Journal of Refrigeration*, *30*(8), 1427–1438.

MATLAB. (2019). *version 9.7.0 (R2019b)*. Natick, Massachusetts: The Mathworks Inc.

Pereira, E. L. L., Deschamps, C. J., & Ribas, F. A. (2008). A Comparative Analysis of Numerical Simulation Approaches for Ring Valve Dynamics. In *International compressor engineering conference.*

Tanveer, M. M., & Bradshaw, C. R. (2020). Quantitative and qualitative evaluation of various positive-displacement compressor modeling platforms. *International Journal of Refrigeration*, *119*, 48–63.

Ziviani, D., Bell, I. H., Zhang, X., Lemort, V., Paepe, M. D., Braun, J. E., & Groll, E. A. (2020). PDSim: Demonstrating the Capabilities of an Open-Source Simulation Framework for Positive Displacement Compressors and Expanders. *International Journal of Refrigeration*, *110*, 323–339. doi: 10.1016/j.ijrefrig.2019.10.015

Experimental Study of the Heat Pump with Variable Speed Compressor for Domestic Heat Load Applications

Muhammad ABID [1]*, Neil HEWITT[1], Ming Jun HUANG[1], Christopher WILSON[1], Donal COTTER[1]

[1] Belfast School of Architecture and the Built Environment, Faculty of Computing, Engineering and the Built Environment, Ulster University, Jordanstown Campus, United Kingdom
E-mails: abid-m@ulster.ac.uk, nj.hewitt@ulster.ac.uk, m.huang@ulster.ac.uk, c.wilson@ulster.ac.uk, d.cotter@ulster.ac.uk
Phone: 00442890366413

* Corresponding Author

ABSTRACT

Capacity control for heat pump (HP) systems is essential to efficiently meet variable heating loads in domestic buildings. A low-to-medium temperature hot water supply within a 30 °C-55 °C range is commonly required for domestic heating load applications. In this study, the performance of a variable speed air source heat pump (ASHP) system was studied in the laboratory under a series of ambient conditions at -2 °C, 2 °C, 7 °C and 15°C, examining performance at different water supplying temperatures (WST) of 30 °C, 35 °C, 40 °C, 45 °C, 50 °C, and 55 °C. The performance of the variable speed HP system was evaluated in a conditioning chamber at the above-mentioned ambient temperatures and corresponding humidity levels as per BS EN 14511 standards. The HP performance at six different heating capacities of 18kW, 15kW,12kW, 9kW, 6kW, and 3kW was measured and evaluated via co-efficient of performance (CoP) values. The CoP values changes according to the load, water supply temperature (WST) and ambient temperature conditions. The CoP values for fixed ambient temperature conditions and WST i.e., 7 °C and WST 30 °C (7A30W) at 15kW, 12kW, 9kW, 6kW, and 3kW were found to be 3.56, 3.83,4.35, 4.19, and 3.39, respectively. The corresponding frequency of operation was 112.25Hz, 105.87Hz, 81.57Hz, 61.05Hz, 37.64Hz, 18.75 Hz with an absolute percentage difference of 87.01%, 76.45%, 35.95%, 1.75%, 37.26%, and 68.75% to that of the nominal frequency of 60Hz. The trend was similar by changing WST to 35 °C, 40 °C, 45 °C, 50 °C and 55 °C, but with lower CoP values and a smaller variation range. The difference between water inlet and supply temperature (delta T) is another important parameter in domestic heat load applications, and this study also presents results of comparative testing of delta T of 5, 10 and 20 at ambient conditions of 15 °C for WST of 40 °C, 45 °C, 50 °C, and 55 °C. The experimental results indicate that this variable speed-based HP system has the potential to meet different domestic heating demands within a range of WST and ambient conditions.

1. INTRODUCTION

Domestic building sector in United Kingdom (UK) consumes 29% of the total final energy consumption (Energy Consumption report, 2017). A major part of the domestic building energy consumption (79%) is because of space heating (SH) and domestic hot water (DHW) production, while cooking, lighting and other appliances consume the rest (Druckman and Jackson, 2008). UK target as committed in Climate Change Act, 2008 is to reduce carbon emissions by 80% to that of 1990 levels by 2050 (Climate Change Act, 2008). Hence, there is a great potential of CO_2 emission reduction if the domestic building heating demand could be meet by renewable energy-based energy devices like heat pump instead of fossil fuel-based gas/oil boilers.

The ASHP heating device utilizes environmental thermal heat energy from the air, with proportion of electrical energy mainly for the compressor operation. In seeking a system which can modulate its capacity according to the load requirements and operate over a wide range, a variable speed compressor-based HP system is a viable option. The performance of ASHP system depends on the ambient temperature conditions, WST, and the heating capacity. Energy efficiency is improved using variable speed compressor due to capacity match at different part loads, resulting in reduced cycling losses. The current standard BS EN14511 classify the HP system depending on the WST, as low temperature (35 °C), medium temperature (45 °C), high temperature (55 °C), and very high temperature (above 65 °C) (BS EN14511, 2004). The performance assessment of the variable speed compressor-based HP system at different heating capacities, heat supply temperature, ambient conditions require investigation for potential improvement in performance. Hence the HP development, testing under the controlled laboratory conditions through experimental analysis becomes very crucial to see the system actual performance.

1.1 Literature Review

The control of the system at part loads conditions basically differentiates variable speed drive from the conventual fixed speed HP system. Higher CoP values, faster temperature control, low starting current, noise, vibrations, and more control were found well-established advantages of variable speed HP (Qureshi and Tassou, 1996). However, there were certain issues reported, i.e., impact of variable speed operation on the isentropic and volumetric efficiency of the compressor, pressure ratio (Pr) and inverter losses. The CoP of ASHP has been reported to be in the range of 3.2-4.5, according to lab scale tests as per EN14511 test standards at 2A35W (Bundesamt *et al.*, 2016). The earlier studies on the variable speed compressor HP system for performance improvement, associated issues, volumetric and isentropic efficiency is discussed with research gaps in the following.

The inverter losses due to change in compressor supplied frequency was analysed experimentally for ground source heat pump (GSHP) while keeping constant load/source side temperature of 26 °C/4.5 °C by (Madani *et al.*, 2010). The results showed that overall, 30% reduction in CoP values was noticed according to the frequency of operation. The experiments were conducted at single WST with frequency variation in the range of 30Hz-90Hz. The isentropic efficiency was highest when the compressor frequency was near to the nominal value of 50 Hz. ASHP system with nominal capacity of 9.8kW at -5 °C was developed for retrofitting applications and tested the performance at a single WST of 60 °C, different ambient conditions as per test standard BS EN14511 (Quinn, 2012). The frequency modulation range was established experimentally in between 37.5 Hz-75 Hz for the safe compressor operation. The HP was tested at four fixed frequencies of 37.5 Hz, 50Hz, 60Hz, and 75 Hz using off the shelf inverter. The heating capacity variation range was 3.14-4.87kW, 3.4-5.38kW, 3.43-5.57kW, 3.49-5.86kW, 3.59-6.03kW, 3.73-6.597kW due to frequency modulation range (37.5-75 Hz) for fixed ambient temperature conditions of -15 °C, -7 °C, -2 °C, 2 °C, 7 °C, and 15 °C, respectively. The highest performance was recorded at the system nominal frequency of 60Hz, while keeping other conditions constant. However, the speed modulation range was very limited with single WST. Variable speed ASHP system was experimentally studied for the performance evaluation in the conditioning chamber for various WST of 35 °C, 45 °C, 50 °C at ambient temperature conditions range of -15 °C to 7 °C without controlling the heating capacity at constant value (Hewitt *et al.*, 2011). The water mass flow rate was fixed with no control over delta T. The CoP value increased from 2.43 to 4.26 depending on the operating frequency and the ambient temperature variation was in between -15 °C and 7 °C at fixed WST of 35 °C. At higher WST of 50 °C the same trend of increase in performance has been shown with increase in ambient temperature but with the CoP value of 1.73 & 3.15 at -15 °C and 7 °C respectively, which is 40.46%, 35.23% lower than that of 35 °C. Capacity control ground source heat pump (GSHP) was investigated with an aim to increase the system efficiency with the variable speed compressor, fan, water circulating pumps, where two different prototypes were tested and investigated under the laboratory conditions (Karlsson, 2007). It was reported that the thermal inertia of various hydronic heating distribution system (underfloor, hydronic fan coil, and radiator) influences the cyclic properties of heat pump for space heating. The cycling characteristics has less effect while using variable speed heat pump system due to load matching and at same time the temperature of water inside the system was maintained at set-point more precisely. It was found that the response time at start-up was also affected by thermal inertia of heat distribution type, in addition to the system configuration, the way it was linked to hot water supply temperature (Karlsson and Fahlen, 2008). It was concluded that the water supply and entering water temperature in addition to the difference between these two value effects the seasonal performances significantly. The reduction in water supply/return temperature values from 55/45 °C to 35/28 °C results in increasing seasonal performance by 30-35%. The performance improvement was because of the reduction in WST and the impact of delta T value variation was not considered. Another experimental study on the variable speed compressor with application for industrial plant cold storage was conducted to investigate the energy saving potential (Aprea *et al.*, 2009). An increase in energy saving of up to 20% was reported for various conditions while using the variable speed control in comparison to the thermostatic control. The study conducted for finding the optimal frequency while maintaining the required cold storage temperature, by keeping evaporating, condensing temperature, and heat load constant. The system frequency was varied for power saving by matching the load demand and comparing the results with that of nominal frequency of 50Hz. The HP system was evaluated at WST of 34°C & 42°C and approximately 30% of energy saving could be achieved at frequency of 24 Hz and 36 Hz satisfying the loads at set conditions, compared to that nominal frequency.

The study investigating the variable speed compressor isentropic, & volumetric efficiencies, inverter losses, pressure ratio (Pr) at different frequency of 35 Hz, 40 Hz, 50 Hz, 60 Hz and 75Hz was conducted (Cuevas and Lebrun, 2009). It was concluded that the compressor efficiencies were mainly dependent on the pressor ratio (Pr) and effected very slightly with the compressor supplied frequency except for the tests of 40 Hz and 75Hz and the reason was linked to lubrication issue and other electromechanical losses at low and high-speed operation. The maximum isentropic efficiency was 65% when the pressure ratio was 2.2, while the volumetric efficiency showed linear decrease from 98% to 83% with increase of pressure ratio from 1.5 to 5.6.

1.2 Present Work Contribution

The above literature review on the variable speed HP sytem shows that the focus of the studies were on individual component(i.e compressor isentropic and volumetric efficiencies, inverter losses), single water supply and return temperature, uncontrolled delta T values. The variable speed heat pump system for the of fixed heating capcities simulating the real domestic heat load demand, WST, and ambient conditions under the lab conditions is missing from the literature, to the best of author's knowledge. Therfore in this study variable speed ASHP prototype developed and tested as per EN14511-3 (EN 14511–3, 2004) method at various steady state heating capacity, WST, and ambient conditions to evaluate the performance of the system for the domestic heat load applications. The establishment of the optimal operating point and the possibility for potential energy saving, needs detailed experimental results analysis, which could provides information for further performance predictions.

2. METHODOLOGY

2.1 Test Facility

The test facility mainly includes the developed ASHP system, conditioning chamber, heat source, and load side arrangements. The experimental setup/ test facility with schematic diagram is shown in Fig. 1(a, b). It could provide range of ambient conditions on source side and different water supply/entering temperature using PID controlled valves in combination to the water flow rates adjustment via variable speed water circulation pump. The delta T value, difference between WST and entering water temperature (EWT), is controlled by PID valve on the load side. As can be seen from schematic diagram the heat load is dumped to the water tank and/or heat exchanger via fresh water supply. The PID valve, which open and closes according to the temperature set points for managing delta T.

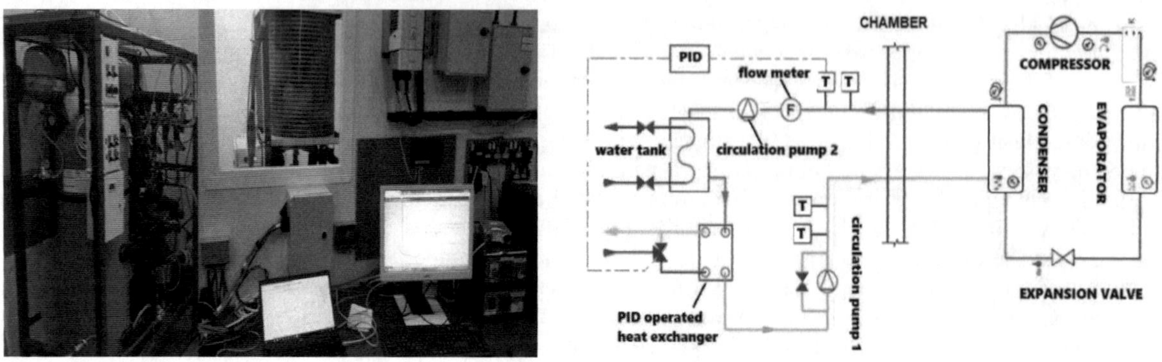

(a) (b)

Figure 1: a) Experimental setup, b) Simplified schematic diagram for the HP inside the conditioning chamber and heat load dumping rig

Proper measurement, data loggers and the control mechanism setup were established. The uncertainty associated with the measurement devices is shown in Table 1. Variable speed scroll compressor model number XPV0302E-4X9 and controller was provided by Emerson Climate Technologies (Copeland Ltd). The controller (SEC) developed for XPV/XHV Copeland variable speed scroll families for heating and cooling applications, provide the speed control for compressor operation.

Table 1: Uncertainties ranges for measurement instruments

Measured quantity	Measurement device	Units	Uncertainties
Temperature	Inline and Surface PT100, Eltek GD 24	°C	±0.3
Liquid mass flow rate	Electromagnetic, Eltek, GC 62	kg/s	±1.5 %
Electric power consumption meter	Landis and Gr P350	kW	± 1 %
Compressor speed	XPV0302E-4X9	rpm	± 0.5 %
Refrigerant Pressure	PT5 Pressure transmitters	kPa	± 1 %
Refrigerant Temperature	NTC (ECN-EG30)	°C	± 0.5

2.2 Testing Method

The testing procedure for ASHP system mentioned in EN14511-3, 2004 have been followed in terms of methodology, sensor, quantity, & measurement accuracy requirements. The ambient conditions in the conditioning chamber for the testing was maintained with PID controlled heater, cooler & humidifier at test conditions of -2 °C, 2 °C, 7 °C & 15 °C. The PID humidifiers was able to meet the requirements mentioned in the standard for the relative humidity (RH) upper limits and lower limits. On heat source side inside the conditioning chamber the ambient temperature and humidity was maintained as per standard requirements with heater, cooler and steamers controlled by the PIDs. Two separate PIDS controlling the heater, cooler independently was keeping the set point temperature inside the conditioning chamber. Similarly, two steamers controlled by the PIDs were operating to maintain the required humidity requirements. The set point values for water supply temperature, ambient temperature, humidity was controlled with high accuracy. Four different ambient conditions with varying water supply temperature (WST), and various delta T values have been chosen to investigate the system for low to medium WST applications of 35 °C to 55 °C. The tested conditions ambient temperature, humidity, and WST values is shown in Table 2. Although, delta T values mentioned in the standard was 5, which has been followed and extended to 10 and 20 to see the impact on the HP performance. The delta T values have application for system performance improvement and radiators are designed in the range of 10 °C - 20 °C value.

Table 2: HP lab testing conditions

HC (kW)	T_a (° C)	RH (%)	RH (%) - upper limit	RH (%)- lower limit	WST (°C)	delta T
12,9,6,3	-2	79.3	85.5	73.2	35, 45	10
12,9,6,3	2	83.9	88.7	79.1	35, 45, 55	10
15,12,9,6,3	7	86.9	90.8	83	30, 35, 40,45, 50,55	10
18,15,12,9,6,3	15	90	93	87.1	30, 35, 40,45, 50, 55	10
18,15,12,9,6,3	15	90	93	87.1	25, 30, 35, 40,45, 50,55	5
18,15,12,9,6,3	15	90	93	87.1	40, 45, 50,55	20

The superheat set point was 10 K during all the tests and it was maintained by using electronic expansion valve model number EXL-BF1-Unipolar stepper motor valve. The controller was controlling the superheat value according to the set point and whenever there was an abrupt change on the heat supply load side due to this PID controlled valve opening or closing then the controller responded mainly by changing the superheat to the new load conditions. The heating capacity on the water side of condenser was balanced to that of the refrigerant side. The energy balance analysis was performed to see the reliability of experimental results as shown in Fig. 2. On water side of the condenser the heating capacity was calculated with equation (1) using measured water inlet and outlet temperature difference, the water specific heat and mass flow rate. The refrigerant mass flow rate through the condenser and evaporator was measured through the flow meter installed in the liquid refrigerant line. The heating capacity on refrigerant side was calculated by equation (2) using the measured values of the refrigerant mass flow rate and the enthalpy difference between the inlet and outlet of the condenser.

$$Q_w = \dot{m}_w \; C_P \Delta T \tag{1}$$

$$Q_r = \dot{m}_r \Delta h \tag{2}$$

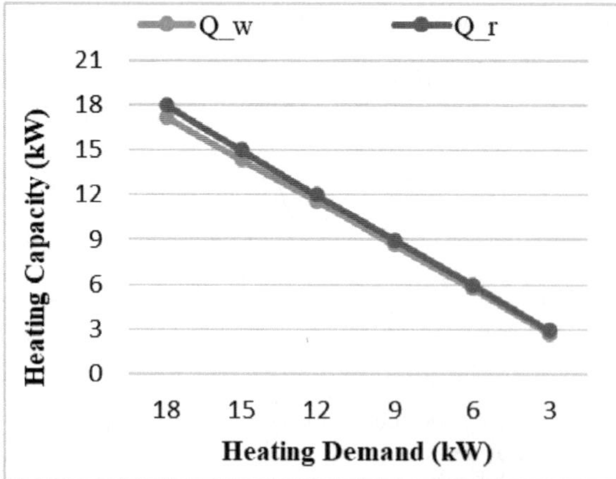

Figure 2: Energy balance between refrigerant and water side of condenser

Following the energy balance analysis, all other tests were completed, and the results are presented in the following section. The tests results were repeated to maintain the accuracy and errors limits. The CoP, ratio between the heat output and total electric power consumption was calculated by equation (3);

$$CoP = \frac{Q}{P_{total}} = \frac{C_p * \dot{m}_w * \Delta T}{p_{comp} + p_{aux}} \tag{3}$$

whereas Q is heating capacity, P is the electric power consumption, c_p the water specific heat, $\dot{m}$ is water mass flow rate (kg/s), and ΔT is the difference between the water supply temperature (WST) and entering water temperature (EWT).

3. RESULTS

The modulation range for the compressor was in between 15-120Hz. The summary results for HP performance under the conditions of ambient temperature at 15 °C, 7 °C, 2 °C and -2 °C, with (WST) of 35 °C, 45 °C, 55 °C for different heating capacity is presented in Table 3. The operation of the system at nominal frequency of 60Hz shows good performance compared to its operation above/ below to that of the nominal value while keeping other conditions fixed. The nominal frequency of 60Hz for a fixed WST and ambient conditions results in higher CoP values, was mainly because of higher compressor isentropic efficiencies, lower-pressure ratio. Notation used in the table, i.e. 15A35W represent the ambient temperature being 15 °C and WST of 35 °C. The HP works at a wide range of changing pressure ratio and only at single nominal operating point, external pressure ratio is equal to that of compressor internal pressure ratio. At all other points under/ over compression results in smaller efficiency.
Fig. 3 (a, b, c, d) shows CoP trends against the constant heating capacities at ambient conditions of 15 °C, 7 °C, 2 °C & -2 °C, respectively. At ambient conditions of 7 °C, test for 18kW heating demand was conducted but the heating capacities achieved were slightly lower than 18kW and therefore not presented. At 2 °C & -2 °C, 15kW tests were not possible due to conditioning chamber limitations and a maximum heating capacity achieved were of the order of 12kW. The CoP values for the fixed heating capacities depends on the ambient conditions, and WST. The CoP values for 18kW,15kW, 12kW, 9kW, 6kW, 3kW varied in the range of 3.91-2.63, 4.72-2.78, 5.34-2.87, 5.31-2.94, 5.23-2.83, 4.01-2.18 for a WST variation between 30 °C & 55 °C at ambient conditions of 15 °C (Fig. 3, a). At 7 °C (Fig. 3b), the CoP values for each heating capacities of 15kW, 12kW, 9kW, 6kW, 3kW variation range is 3.56-2.43, 3.83-2.43, 4.35-2.41, 4.19-2.43, 3.39-2.28 for a WST variation between 30 to 55 °C. The CoP values with 12kW, 9kW,6kW, 3kW with WST of 35 °C- 55 °C are in the range of 3.04-2.21, 3.43-2.21, 3.29-2.22, 2.79-2.14 respectively at 2 °C (Fig. 3., c). At low ambient conditions of -2 °C, tests for 55 °C WST was not possible with 12kW, and 3kW due to higher discharge line temperature (DLT) limitations of 120 °C. Fig. 4 (a, b, c, d) depicts the frequency variation against the fixed heat heating capacities for the range of WST at four ambient temperature conditions of 15 °C, 7 °C, 2 °C & -2 °C, respectively. The frequency requirements increase with increase in heating capacity while keeping the WST and ambient temperature conditions fixed.

Table 3: Summary results for the heat pump (HP) test

EWT/WST	HD (kW)	HC (kW)	$\dot{m}_{\_w}$ (kg/s)	RH (%)	T_a (°C)	Φ (Hz)	Pr	P(kW)	COP
				15A35W					
25/35	18	17.97	0.43	90.25	14.91	101.35	3.62	4.94	3.64
25/35	15	14.98	0.36	89.70	15.16	93.10	3.48	3.65	4.11
25/35	12	11.98	0.29	91.18	15.19	62.87	2.71	2.42	4.96
25/35	9	8.98	0.22	91.87	14.95	45.60	2.50	1.88	4.78
25/35	6	5.98	0.14	89.34	15.11	30.04	2.39	1.29	4.64
25/35	3	2.98	0.07	91.47	14.92	15.21	2.33	0.85	3.50
				15A45W					
35/45	18	17.98	0.43	90.99	15.25	112.58	4.16	5.72	3.14
35/45	15	14.98	0.36	90.12	14.87	97.02	3.87	4.37	3.43
35/45	12	11.98	0.29	91.54	14.87	65.27	3.46	3.13	3.83
35/45	9	8.98	0.22	90.56	14.88	47.46	3.26	2.40	3.75
35/45	6	5.98	0.14	89.27	14.88	30.99	3.06	1.64	3.65
35/45	3	2.98	0.07	91.54	14.87	15.52	3.08	1.11	2.68
				15A55W					
45/55	18	17.97	0.43	92.61	14.92	114.23	4.95	6.84	2.63
45/55	15	14.98	0.36	91.23	14.81	97.42	4.73	5.39	2.78
45/55	12	11.98	0.29	89.82	14.43	68.64	4.50	4.18	2.87
45/55	9	8.98	0.22	91.10	14.61	50.52	4.26	3.06	2.94
45/55	6	5.98	0.14	89.92	14.80	32.49	3.95	2.11	2.83
45/55	3	3.00	0.07	91.78	14.89	15.97	3.93	1.38	2.18
				7A35W					
25/35	15	14.95	0.36	89.14	6.90	107.69	4.53	4.80	3.12
25/35	12	11.98	0.29	88.37	6.82	85.27	4.14	3.57	3.35
25/35	9	8.98	0.22	87.85	6.76	58.77	3.47	2.34	3.84
25/35	6	5.98	0.14	87.91	6.79	37.78	3.19	1.57	3.80
25/35	3	2.98	0.07	87.77	7.32	21.73	3.69	1.06	2.81
				7A45W					
35/45	15	14.98	0.36	88.07	6.82	108.33	5.42	5.56	2.64
35/45	12	11.99	0.29	88.22	7.71	88.50	5.32	4.34	2.76
35/45	9	8.99	0.22	88.42	6.86	59.69	4.43	2.89	3.11
35/45	6	5.98	0.14	88.68	6.83	38.95	4.15	1.96	3.05
35/45	3	2.98	0.07	88.81	6.95	19.15	4.02	1.23	2.43
				7A55W					
45/55	15	14.97	0.36	87.97	7.62	110.90	5.93	6.17	2.43
45/55	12	11.98	0.29	87.32	7.08	89.69	5.90	4.93	2.43
45/55	9	8.96	0.22	89.13	6.20	63.22	5.86	3.72	2.41
45/55	6	5.98	0.14	87.87	6.36	40.00	5.30	2.46	2.43
45/55	3	2.98	0.072	88.21	6.59	19.85	5.50	1.31	2.28
				2A35W					
25/35	12	11.99	0.29	85.67	2.12	92.25	5.10	3.94	3.04
25/35	9	9.00	0.22	87.04	2.13	67.19	4.10	2.63	3.43
25/35	6	5.99	0.14	84.47	1.92	43.48	4.01	1.82	3.29
25/35	3	2.98	0.07	86.10	2.19	21.10	3.72	1.07	2.79
				2A45W					
35/45	12	11.99	0.29	86.46	2.10	93.84	5.78	4.60	2.61
35/45	9	8.99	0.22	88.78	2.18	67.64	5.42	3.29	2.73
35/45	6	5.99	0.14	85.08	2.04	44.39	5.67	2.36	2.54
35/45	3	2.99	0.07	85.97	2.08	20.79	5.12	1.36	2.20
35/45	3	3.06	0.07	86.06	2.15	21.95	5.78	1.39	2.20
				2A55W					
45/55	12	11.98	0.29	86.85	2.21	93.52	7.05	5.43	2.21
45/55	9	8.99	0.22	87.93	2.02	69.37	6.88	4.07	2.21
45/55	6	5.99	0.14	85.45	1.88	44.83	6.39	2.70	2.22
45/55	3	2.92	0.07	85.19	2.05	24.34	5.99	1.37	2.14
				-2A35W					
25/35	12	11.97	0.29	83.25	-2.33	102.84	5.54	4.26	2.81
25/35	9	8.98	0.22	82.23	-2.23	74.32	4.86	2.95	3.05
25/35	6	5.99	0.14	82.52	-2.22	48.64	4.60	1.96	3.05
25/35	3	2.99	0.07	81.93	-1.85	23.80	4.42	1.14	2.62
				-2A45W					
35/45	12	11.93	0.29	82.29	-2.28	105.17	6.61	4.94	2.41
35/45	9	8.98	0.22	83.25	-2.25	75.63	5.91	3.43	2.62
35/45	6	5.99	0.14	80.43	-2.23	48.83	5.87	2.39	2.51
35/45	3	2.98	0.07	81.82	-1.79	23.07	5.46	1.39	2.15

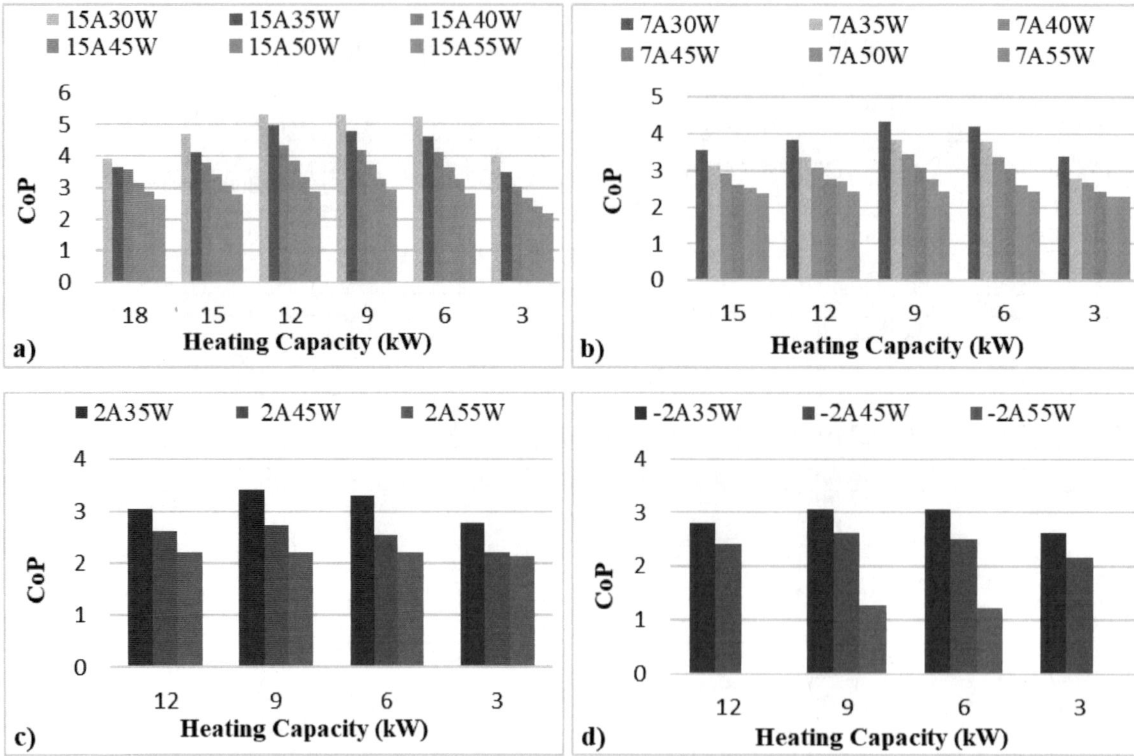

Figure 3: Heating Capacity vs. CoP at ambient temperature of a) 15 °C, b) 7 °C, c) 2 °C, d) -2 °C

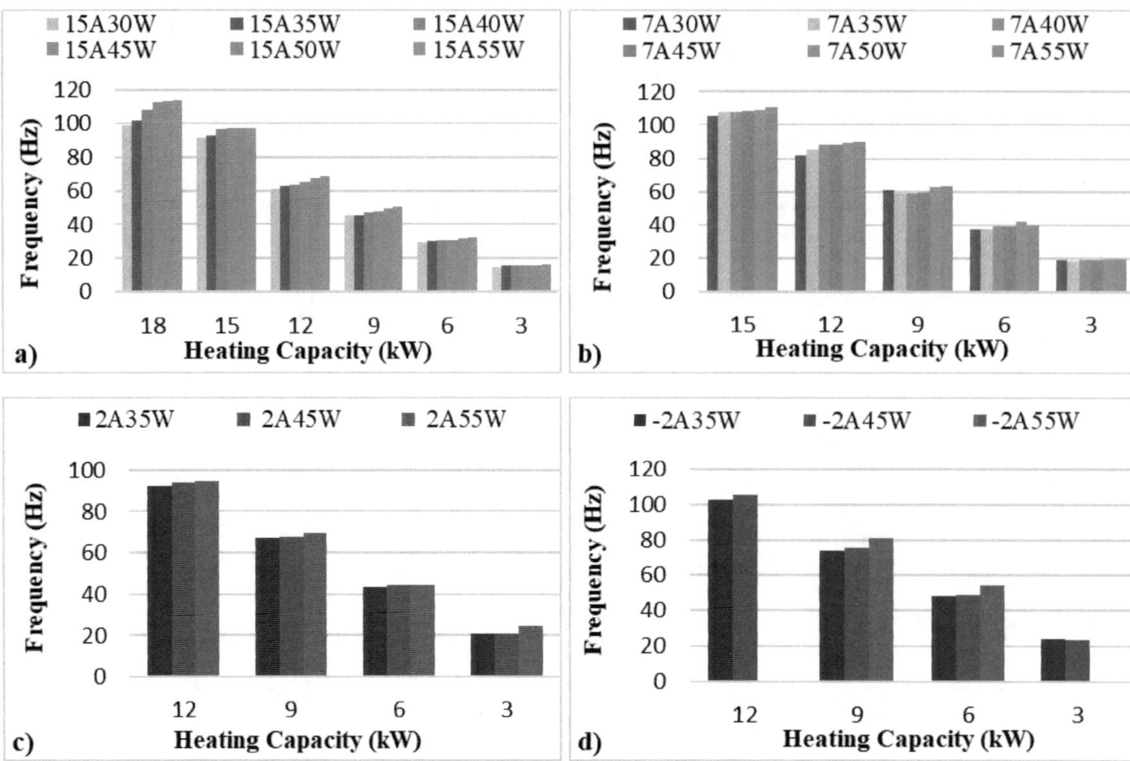

Figure 4: Heating Capacity vs. frequency at ambient temperature of a) 15 °C, b) 7 °C, c) 2 °C d) -2 °C

Table 4 shows the impact of changing delta T on the CoP and pressure ratio values for constant WST of 40 °C, 45 °C, 50 °C, and 55 °C at ambient conditions of 15 °C. The CoP value for 18kW test with WST 40 °C, 45 °C, 50 °C, and 55 °C are 3.66 &3.55, 3.29 & 3.14, 3.05& 2.86, 2.88 & 2.62 for delta T of 20 and 10, respectively. While the pressure ratio shows the reverse pattern, where the pressure ratio is 3.61& 3.72, 3.96 &4.16, 4.31&4.53, 4.54 &4.94 for delta T of 20 and 10 respectively. At the same heating capacity, the CoP value changes according to the delta T value. Keeping delta T value smaller results in lower CoP values for the constant heating capacity and WST. The reasons for this CoP drop due to smaller delta T was higher pressure ratio (Pr) and lower isentropic efficiencies.

Table 4: Summary of test results for changing delta T values at 15 °C

	delta T=20					delta T=10					delta T=5				
HC (kW)	**P_suc(bar)**	**P_dis (bar**	**Pr**	**CoP**	**Isen. eff.**	**P_suc**	**P_dis(bar)**	**Pr**	**CoP**	**Isen eff.**	**P_suc(bar)**	**P_dis (bar**	**Pr**	**Isen eff.**	**CoP**
WST=40 °C															
18	6.43	23.23	3.62	3.66	65.95	6.45	24.00	3.72	3.56	65.65	-	-	-	-	-
15	6.98	21.60	3.10	4.22	69.75	6.93	24.03	3.47	3.80	68.81	-	-	-	-	-
12	7.47	21.17	2.83	4.60	72.45	7.69	23.68	3.08	4.36	71.83	-	-	-	-	-
9	7.79	21.51	2.76	4.57	72.55	8.01	22.96	2.87	4.21	72.55	8.32	25.07	3.01	72.60	4.12
6	8.15	20.73	2.54	4.41	68.80	8.17	22.21	2.72	4.12	67.23	8.30	24.24	2.92	68.25	3.84
3	8.45	20.25	2.40	3.36	57.78	8.04	21.89	2.72	3.02	54.16	8.44	23.28	2.76	54.21	2.94
WST=45 °C															
18	6.71	26.65	3.97	3.29	65.18	6.67	27.78	4.16	3.14	65.02	-	-	-	-	-
15	6.87	24.42	3.55	3.72	68.56	7.03	27.17	3.87	3.43	70.76	-	-	-	-	-
12	7.55	24.19	3.20	4.06	71.95	7.67	26.57	3.46	3.83	71.36	-	-	-	-	-
9	7.72	23.86	3.09	4.10	71.38	8.08	26.31	3.26	3.75	70.98	7.98	28.27	3.54	70.37	3.55
6	8.05	23.41	2.91	3.89	68.25	8.28	25.32	3.06	3.65	68.14	8.29	27.50	3.32	68.23	3.37
3	8.40	21.78	2.59	3.09	56.03	8.06	24.80	3.08	2.68	52.02	7.80	26.50	3.40	49.66	2.49
WST=50 °C															
18	6.73	29.03	4.32	3.06	65.18	6.94	31.44	4.53	2.87	63.57	-	-	-	-	-
15	7.12	27.50	3.86	3.41	67.92	7.14	30.72	4.30	3.06	69.73	-	-	-	-	-
12	7.25	27.65	3.81	3.48	71.38	7.59	30.22	3.98	3.34	70.06	-	-	-	-	-
9	7.79	27.32	3.51	3.62	71.37	8.02	29.84	3.72	3.25	69.91	7.84	31.79	4.06	69.03	3.10
6	8.01	26.84	3.35	3.40	68.79	8.31	28.45	3.43	3.26	68.23	8.23	30.92	3.76	66.09	2.99
3	8.21	23.65	2.88	2.82	52.02	8.27	28.06	3.39	2.41	49.66	8.07	30.66	3.80	46.32	2.22
WST=55 °C															
18	6.88	31.20	4.54	2.88	63.50	7.04	34.81	4.95	2.63	64.69	-	-	-	-	-
15	7.05	30.67	4.35	3.05	67.73	7.15	33.84	4.73	2.78	67.65	-	-	-	-	-
12	7.54	30.93	4.10	3.18	69.21	7.56	34.02	4.50	2.87	69.84	-	-	-	-	-
9	7.72	30.06	3.89	3.26	68.14	7.82	33.29	4.26	2.94	69.58	8.21	36.01	4.38	69.58	2.78
6	8.32	29.33	3.52	3.16	65.67	8.22	32.42	3.95	2.83	66.71	8.26	35.07	4.25	63.30	2.62
3	8.14	29.03	3.56	2.61	49.66	7.89	30.99	3.93	2.18	46.32	8.10	32.72	4.04	45.36	2.15

The main issue with the ASHP is at lower ambient conditions when the domestic heat load increases and the HP heating capacity get reduced. The variable speed compressor-based HP system could be operated at the required heating capacity to meet varying load demands without oversizing the HP system. The number of hours for specific climatic conditions at low ambient temperature is limited and hence the oversizing could be avoided. Finding the required operating speed at different ambient conditions and keeping other variables (i.e., load, WST) constant is very crucial for the operation and installation of variable speed heat pump system for domestic heat load applications at higher efficiencies. The CoP values, & the corresponding frequency for the tested ambient conditions at WST of 35 °C is shown in Fig. 5(a, b). The frequency of operation for 12kW, 9kW,6kW, 3kW at 2 °C ambient temperature conditions, at WST of 35 °C was 92.24 Hz, 67.18Hz, 43.47 Hz, 21.1 Hz with absolute percentage difference of 53.73%, 11.96%, 27.55%, 64.85% to that of 60Hz and the corresponding impact on CoP values could be observed. The CoP values for higher WST of 45 °C & 55 °C is shown in Fig.6 (a, b).

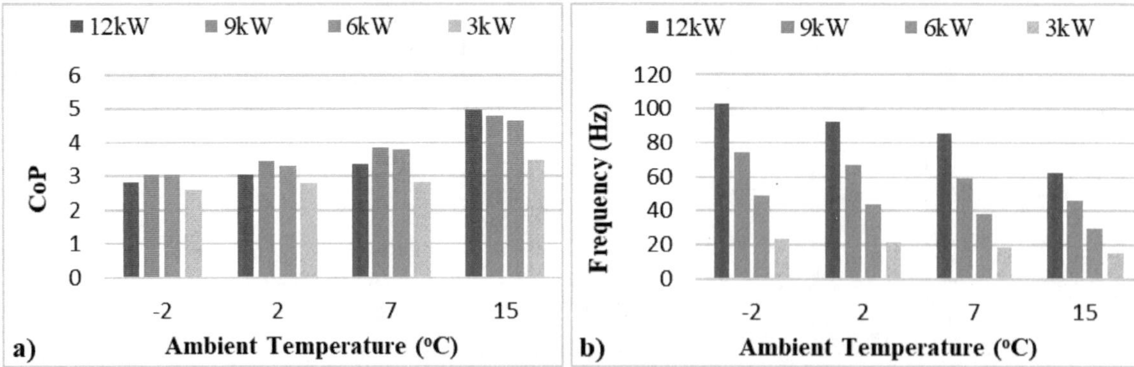

Figure 5: Ambient temperature vs. a) CoP, b) frequency at low water supply temperature (WST) of 35 °C

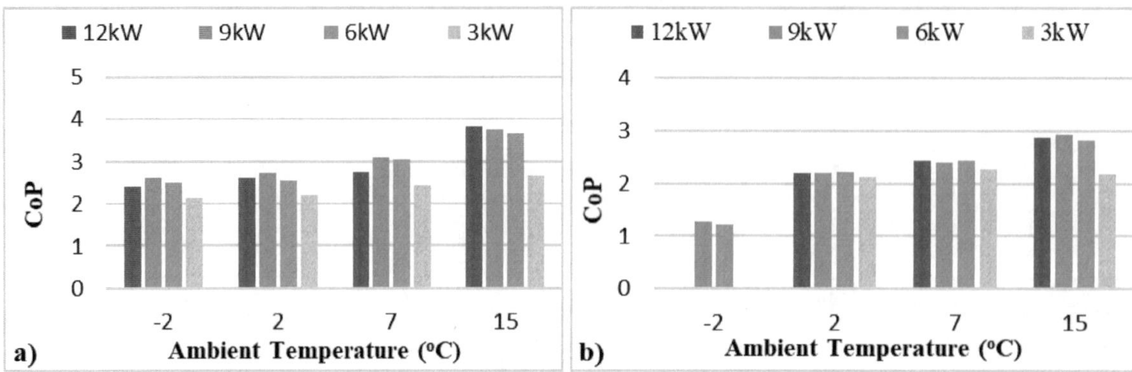

Figure 6: Ambient temperature vs. CoP at WST of a) 45 °C, b) 55 °C

4. CONCLUSION

The variable speed compressor based ASHP experimental results are presented in this paper over a range of operating conditions and heating capacities for the low to higher heat supply temperature. The aim of the experimental development and testing was to study the HP domestic retrofit and energy savings with variable speed compressor-based HP system at different heat supply temperature. The results for the developed ASHP system under controlled laboratory conditions provide a strong base for the evaluation and predictions of the system annual CoP prior to the field installations into different housing stock & climatic conditions. It was concluded that the three parameters namely ambient temperature, WST and operating frequencies strongly impact the HP performance.

NONMENCLATURE

ASHP	air-source heat pump
CoP	coefficient of performance (-)
CO_2	Carbon di-oxide (-)
delta T	difference between supplied and entering water temperature
EWT	entering water temperature (°C
WST	water supply temperature (°C)
P	power (kW)
Pr	pressure ratio (-)
PID	Proportional Integral Derivative

Q	heating load(kW)
RPM	revolution per minute (-)
HC	heating capacity (kW)
HD	Heating demand (Kw)
Φ	Frequency (Hz)

Subscripts

a	ambient
Cp	specific heat capacity
w	water
r	refrigerant

REFERENCES

Aprea, C., Mastrullo, R., & Renno, C. (2009). Determination of the compressor optimal working conditions. *Appl. Therm Eng.,* 29(10), 1991-1997.

Bundesamt, B. (2016). für Wirtschaft und Ausfuhrkontrolle. Erneuerbare Energien Wärmepumpen mit Prüfnachweis. *Eschborn: Bundesamt für Wirtschaft und Ausfuhrkontrolle.* Technical report.

Climate change act. (2008). Tech. Rep., Parliament of The United Kingdom. http://www.legislation.gov.uk/ukpga/2008/27/pdfs/ukpga_20080027_en.pdf [Accessed 21-05-2019].

Cuevas, C. & Lebrun, J. (2009). Testing and modelling of a variable speed scroll compressor. *Appl Therm Eng., 29*(2–3), 469–478.

Druckman, A. & Jackson, T. (2008). Household energy consumption in the UK: A highly geographically and socio economically disaggregated model, *Energy Pol.,* 36(8), 3177-3192.

European Standard, EN 14511–3. (2004). air conditioners, liquid chilling packages and heat pumps with electrically driven compressors for space heating and cooling–Part 3 Test Methods, AFNOR.

Energy Consumption in the UK, (2019). Tech. Rep., department for business, energy, and industrial strategy, https://www.gov.uk/government/collections/digest-of-uk-energy-statistics-dukes (Accessed 24-09-2020).

Hewitt, N. J., Huang, M. J., Anderson, M., Quinn, M. (2011). Advanced air source heat pumps for UK and European domestic buildings. *Appl. Therm Eng.,* 31(17–18), 3713–3719.

Karlsson, F. (2007). Capacity control of residential heat pump heating systems. PhD Thesis, Building Services Engineering Department of Energy and Environment Chalmers, University of Technology, Sweden.

Karlsson, F. and Fahlen, P. (2008). Impact of design and thermal inertia on the energy saving potential of capacity controlled heat pump heating systems, *Int. J. of Refrig.,* 31(6), 1094-103.

Madani, H., Claesson, J., Ahmadi, N., & Lundqvist, P. (2010). Experimental analysis of a variable capacity heat pump system focusing on the compressor and inverter loss behavior. *International Refrigeration and Air Conditioning Conference*, 1063.

Qureshi, T.Q., & Tassou, S.A. (1996). Variable-speed capacity control in refrigeration systems. *Appl. Therm Eng.,* 16(2), 103-113.

Quinn, M. (2012). The development of a capacity controlled advanced cycle air source heat pump for domestic retrofit applications, *PhD Thesis,* Ulster University, UK.

ACKNOWLEDGEMENT

The SPIRE 2 project is supported by the European Union's INTERREG VA Program, managed by the Special EU Programs Body (SEUPB). The views and opinions expressed in this paper do not necessarily reflect those of the European Commission or the Special EU Programs Body (SEUPB).

Characterizing Steady State Compressor Performance by Using Transient Test Data

Leon P. M. BRENDEL[1]*, Stephen L. CASKEY[2], James E. BRAUN[1], Eckhard A. GROLL[1]

[1]Ray W. Herrick Laboratories, School of Mechanical Engineering, Purdue University
West Lafayette, 47907-2099, IN, USA

[2]Air Squared, Inc. 510 Burbank St, Broomfield, CO 80020, USA

*brendel@purdue.edu

ABSTRACT

Compressor testing is an essential task to characterize compressor performance, but often requires significant time to be executed. This study suggests a method that could greatly reduce time needed for compressor testing by inferring steady state performance from transient data rather than waiting for true steady state conditions to be measured. The key finding is that the overall isentropic efficiency in transient operation is almost identical to its true steady state performance value after applying very simple data processing. The paper describes a simple data processing method that extracts steady-state performance from transient data. The proposed processing should not be understood as a general rule to all compressors, but as a positive result for this particular compressor and a first glimpse into the value of transient data for performance estimation.

1. INTRODUCTION

Prototype compressors are usually characterized by performing steady-state testing over an array of operating conditions. Different definitions of steady-state operation exist and most require at least 15 minutes of operation with no significant change in any measurement with time. The test matrix can prescribe the compressor speed, suction and discharge pressure, and suction superheat. The required time to collect one steady-state point of the array can take from 30 minutes up to 2 hours, highly dependent on the type of test stand, the desired operating conditions, and the experience of the testing personnel. Many studies explain the long testing time is a significant problem, both in R&D and catalogue data creation. The earliest work found (Gustafson et al., 1992) had an identical goal as set in this paper, but the approach was to set up a neural network and infer the most important operating conditions for steady-state data points from transient data. In particular for refrigeration compressors, other authors followed over the years, always seeking a testing time reduction using complex algorithms (Antonelo et al., 2018; Coral et al., 2019; Penz et al., 2012; Xu et al., 2019). Some of the aforementioned studies show good results, but the method proposed in this paper is fundamentally simpler in requiring only the removal and averaging of data by conditions. The conditions are to remove datapoints if the suction pressure falls below a certain threshold or if there is no superheat.

While targeting steady-state measurements during testing, the compressor is naturally exposed to varying suction and discharge pressures. This paper explores whether such transient data can reveal the same information as captured with time consuming, steady state measurements. Some data processing is required but limited to removing data points based on certain conditions and applying moving averages to transient data.

2. METHODOLOGY

2.1 General Approach

Steady-state measurements were collected on a positive displacement compressor by maintaining nine different operating conditions for 20 minutes. The data points over each 20-minute time window were averaged and stored as one steady-state point. The steady-state data points were compared with transient data from a single, 20-minute period operation covering a wide range of pressure ratios ($P_r = [5,35]$). The steady-state data was used as true data and the transient data was analyzed to understand which portions of the transient data align with the steady-state data.

2.2 Test Setup and Components

The compressor was an oil-free, variable-speed scroll prototype with an electric motor capacity of 800 W and a swept volume of 15.2 cm^3. For the experiment, the compressor was operated at $3900 \pm 20\ RPM$. A hot-gas bypass test stand was used for all measurements, which allowed relatively quick adjustments of suction and discharge pressures. In a hot-gas bypass test stand, the refrigerant is split after a discharge valve and a mass flow rate measurement. The hot gas is expanded to suction pressure directly via a bypass branch while the remainder of the flow is condensed and then expanded. The two branches are be mixed before the compressor suction line to reach a desired superheat. The superheat was controlled for the steady-state tests but not for the transient test data.

3. DATA

3.1 Steady-State Data

The target test matrix contained the evaporation pressures corresponding to the saturation temperatures of -35, -30, -25, -15 and -5 °C for two different discharge pressures corresponding to the saturation temperatures 30 and 40 °C at a speed of 3900 RPM and a suction superheat in the range $15 < \Delta T_{sup} < 25$. Only the operating condition (-5/30) could not be measured due to limited valve ranges on the test stand. Steady-state measurements and the overall isentropic efficiency ($\eta_{i,o}$) calculated are shown in Table 1 and the results are plotted in Figure 1 as a function of pressure ratio.

Table 1: 20-minute averaged steady-state data from the prototype compressor

T_{evap} [°C]	T_{cond} [°C]	ΔT_{sup} [K]	f [RPM]	$\dot{m}$ [g/s]	$\dot{W}$ [W]	P_r[−]	$\eta_{i,o}$ [−]
-34.4	30.6	15.1	3880	1.83	517	11.5	0.19
-30.1	29.4	20.8	3901	2.15	496	15.2	0.22
-25.1	31.9	23.5	3915	2.66	559	11.9	0.23
-14.7	30.2	24.6	3924	3.97	548	9.6	0.26
-34.9	39.7	18.2	3860	1.76	682	7.7	0.16
-29.8	39.8	16.8	3902	2.32	735	4.7	0.18
-24.8	40.5	21.7	3906	2.63	700	6.2	0.20
-14.9	40.0	23.4	3917	3.85	695	4.2	0.24
-5.3	40.1	22.2	3918	5.64	717	9.0	0.26

3.2 Transient Data

Transient data was collected to evaluate whether the steady-state performance could be reproduced. The compressor had been powered off for 1 hour and 45 minutes before starting this experiment. The compressor was restarted and used to charge the system, resulting in a runtime of approximately 10 minutes. The 20-minute period was defined to begin when the charging process was completed. The compressor frequency was between 3880 and 3920 RPM for the 20 minutes to match the steady-state compressor frequency. The sampling rate varied between 2 and 3 seconds. The pressure ratio and overall isentropic efficiency computed for each time step are displayed in Figure 2. The pressure ratio and efficiency varied greatly and quickly as the operator changed the metering valves of the test stand to cover a wide range of different pressure ratios.

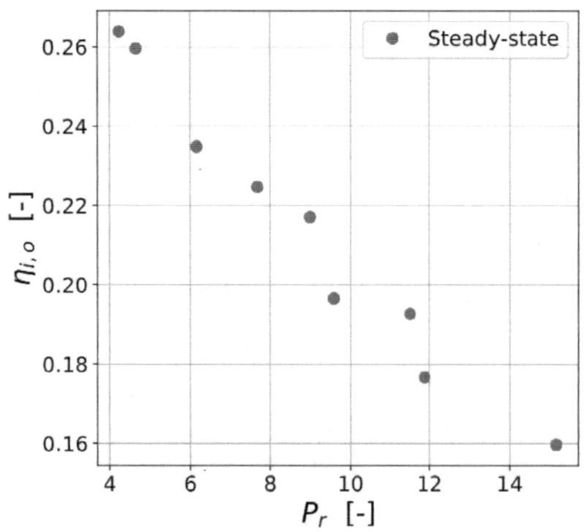

Figure 1: Steady state data

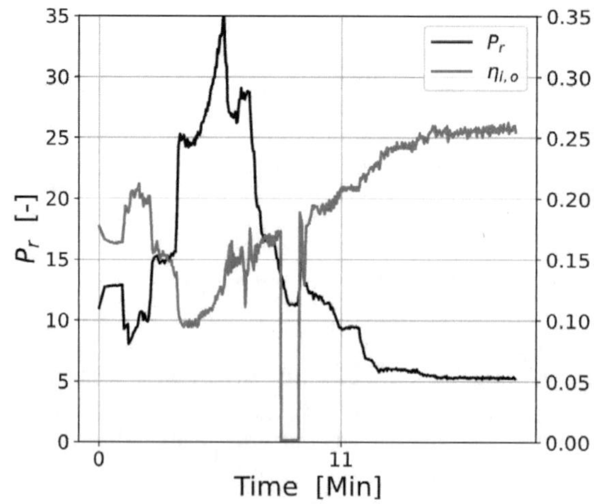

Figure 2: Transient data

4. PROCESSING

The isentropic efficiency from the transient data was plotted in Figure 3 as a function of pressure ratio (P_r) for all data samples collected by the data acquisition system (black) along with the steady-state data (red). Clearly, part of the transient data matches the steady state data, but other values are far off the linear behavior captured in the steady state data. A small group of points encircled in green is isolated from the data collected. The points did not have measured superheat making the computation of the isentropic efficiency (using property data of superheated refrigerant) not possible. The first step in processing transient data was therefore to eliminate any points that do not fulfill the criterion $\Delta T_{sup} \geq 1$ K.

A useful tool to understand the transient data is to flatten it, meaning that each data point becomes the average of a defined time window preceding it. This is sometimes also called a "rolling" or "moving" average. Figure 4 shows this result of flattening the data using 40 seconds for the moving average. As a result, the disorganized scatter plot becomes more of a line plot such that the time dependence of the collected points can be traced as indicated with the black arrows. A 54 second gap in the data acquisition is indicated with a dashed arrow and was due to stopping the code to update the output file. Points far away from one another capture quickly changing operating conditions while points close to each other have slowly changing operating conditions.

In a next step, the steady state data was overlaid over the flattened, transient data to evaluate the comparison. Varying degrees of data flattening, i.e. 10, 40 and 180 seconds were evaluated. Figure 5 shows the three different plots. With an increasing time-window for the moving average, the trend becomes clear and simple to understand. The moving average over 40 seconds was selected for the subsequent study because it gives a clear path by removing measurement fluctuations but does not remove significant information as occurred in the results shown in Figure 5 c), where the moving average window was 180 seconds. None of the three moving averages moved the transient data further away from the steady-state data. The best duration for the moving average could be a function of the compressor itself and should be revisited if this method is applied to other compressors.

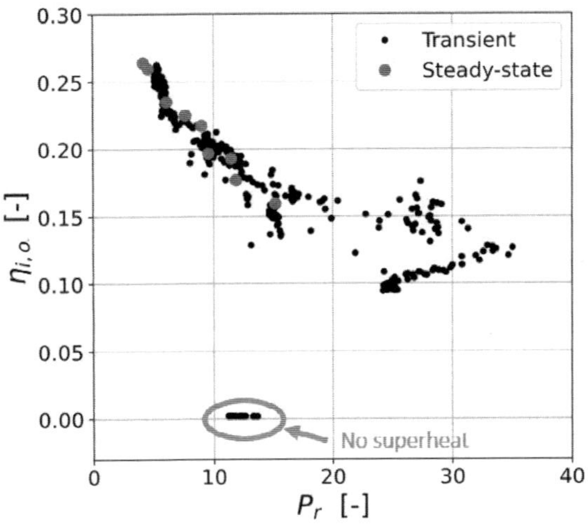

Figure 3: Overlay plot of transient and steady state data.

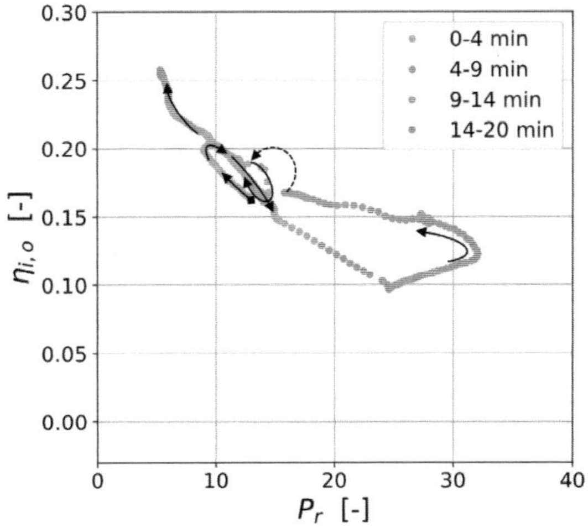

Figure 4: Flattened transient data with $\Delta T_{sup} \geq 1$ K.

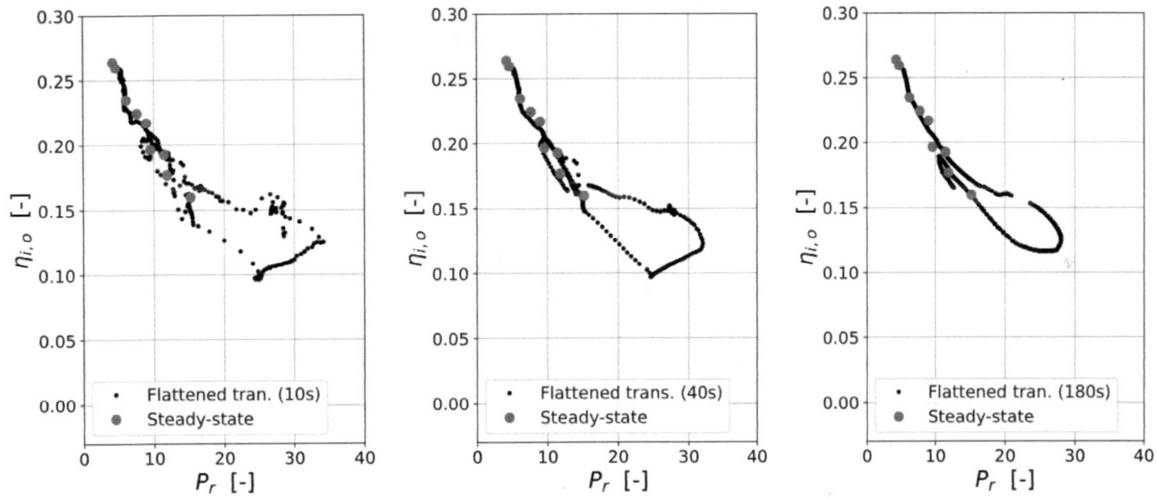

a) Flattened over 10 seconds b) Flattened over 40 seconds c) Flattened over 180 seconds

Figure 5: Transient and steady-state data with varying degrees of flattening.

After applying the moving average, one section of the transient data still had a clear disagreement with the steady-state data. For $P_r > 15$ some data is in line with the linear trend of the steady state performance while other data shows a significantly higher efficiency at the same pressure ratio. By trial and error, it was found that the data points which are off the steady state trajectory all have very low suction pressures (below 50 kPa). Figure 6 visualizes this by using the evaporation temperature (saturation temperature corresponding to the suction pressure) as a third dimension.

The colored plot identifies that all data points away from the steady-state trajectory have evaporation temperatures of $T_{evap} < -40°C$. Since the design point of the prototype compressor is a suction pressure corresponding to $T_{evap} = -35°C$ and R134a is not typically used for $T_{evap} < -40°C$, the last processing step is to disregard data points that have an evaporation temperature below a certain threshold. Figure 7 a), b) and c) apply the thresholds -50°C, -40°C and -30°C. Clearly, -50°C leaves mismatching data while a -30°C threshold eliminates more data than needed to find a good match. It can be concluded that $T_{evap} > 40°C$ is a good criterion for this compressor and the refrigerant used. At the low suction pressures, the viscosity of R134a drops quickly and the mass flow rate of the compressor decreases below the design flow rate due to the low density, both justifying a lower bound on the suction pressure. Figure 8 shows the final result after three steps of data processing:

- Disregarding data with a superheat of less than 1 K
- Flattening data using a moving average over the last 40 seconds
- Disregarding any data with an evaporation temperature lower than -40°C

With the three filter rules applied, the transient performance in terms of isentropic efficiency matches the steady state performance very well. The results are significant because collecting the steady state data resembles a time-effort of several tens of hours, while the transient data was collected in 20 minutes.

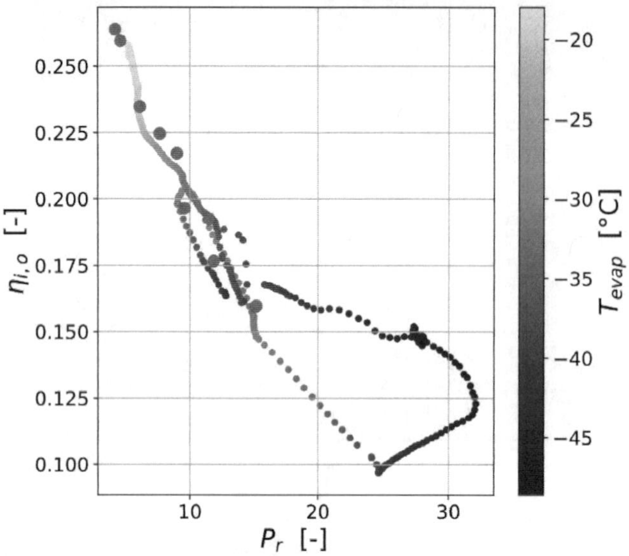

Figure 6: Overlay plot of transient and steady-state data with evaporation temperature as a third dimension for the transient data.

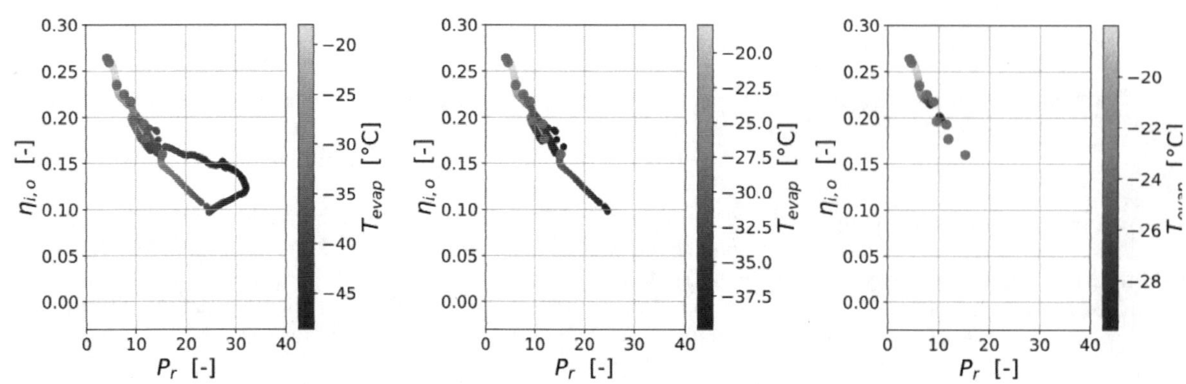

a) Data with $T_{evap} > -50°C$ b) Data with $T_{evap} > -40°C$ c) Data with $T_{evap} > -30°C$

Figure 7: Disregarding data points by varying evaporation temperature thresholds.

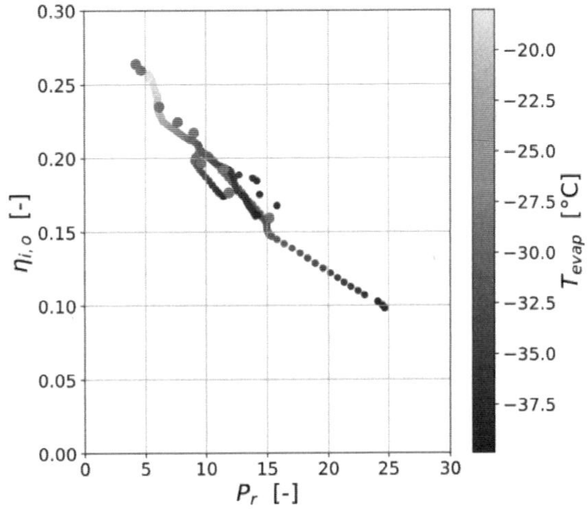

Figure 8: Steady-state and transient data after three steps of data processing.

5. DISCUSSION

The authors judge the processing steps as simple and possibly applicable to other compressor types and sizes. However, this cannot be directly inferred from this study and requires additional testing. For each compressor, a set of "true" steady-state and transient data is necessary to apply the same steps proposed in this paper and to locate additional processing steps to address disagreement.

Although the transient and steady-state data have similar ranges of pressure ratios, the operating conditions at which those pressure ratios were met are indeed different. Figure 9 shows this in a T_{cond} $vs.$ T_{evap} scatter plot. The steady state points match the array described in the "Steady-State Data" section. In comparison to that, the transient data is "all over the place" and approaches only the exact saturation suction and discharge temperatures of 2 steady state points (-30°C, 30°C) and (-30°C, -40°C).

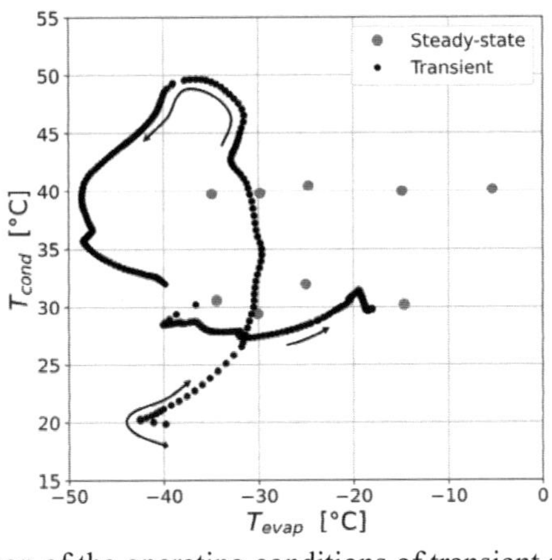

Figure 9: Comparison of the operating conditions of transient and steady state data.

The good agreement of steady state and transient measurements appears counterintuitive: During transient operation, the varying power inputs and efficiencies should result in different compressor temperatures, a process that is slow due to the high heat capacity of the metal parts. Only after the compressor is in thermal equilibrium is it expected that the correct isentropic efficiency would result from the measurements. However, the presented data in this paper shows a good match despite a very quickly changing isentropic efficiency.

Future work in the field should investigate whether the rules also apply to other performance criteria or whether additional criteria can be found. Of direct interest for compressor performance mapping are the volumetric efficiency and the heat rejection of the compressor to the ambient.

6. CONCLUSIONS

Simple data processing led to a good agreement of the isentropic efficiencies determined from 20 minutes of transient testing as compared to those determined from the testing of nine steady-state data points, each averaged over a 20-minute time period. The data processing disregarded data based on small superheat levels and extremely low evaporation temperatures and applied a moving average to smooth the transient data. The authors do not claim generalizability of the data processing steps, but believe that a larger study with different compressor sizes and types could reveal additional or other methods to find value in transient data and reduce the testing time needed to evaluate the performance of prototype compressors.

NOMENCLATURE

Symbol	Description	Unit
f	Compressor frequency	RPM
$\eta_{i,o}$	Overall isentropic efficiency $\eta_{i,o} = \dot{m}\left(h_{dis,s} - h_{suc}\right) / \dot{W}$	[-]
$\dot{m}$	Mass flow rate	[g/s]
P_r	Pressure ratio $P_r = P_{dis}/P_{suc}$	[-]
T_{evap}	Saturation temperature corresponding to suction pressure	[°C]
T_{cond}	Saturation temperature corresponding to discharge pressure	[°C]
ΔT_{sup}	Suction superheat	[K]
$\dot{W}$	Compressor power draw (measured after power supply before inverter)	[W]

REFERENCES

Antonelo, E.A., Flesch, C.A., Schmitz, F., 2018. Reservoir computing for detection of steady state in performance tests of compressors. Neurocomputing 275, 598–607. https://doi.org/10.1016/j.neucom.2017.09.005

Coral, R., Flesch, C.A., Flesch, R.C.C., 2019. Measurement of Refrigeration Capacity of Compressors with Metrological Reliability Using Artificial Neural Networks. IEEE Trans. Ind. Electron. 66, 9928–9936. https://doi.org/10.1109/TIE.2019.2898613

Gustafson, S.C., Little, G.R., Rattray, J., 1992. Correlation of transient and steady-state compressor performance using neural networks. Presented at the Conference Record AUTOTESTCON '92: The IEEE Systems Readiness Technology Conference, IEEE, Dayton, OH, USA. https://doi.org/10.1109/AUTEST.1992.270131

Penz, C.A., Flesch, C.A., Nassar, S.M., Flesch, R.C.C., de Oliveira, M.A., 2012. Fuzzy–Bayesian network for refrigeration compressor performance prediction and test time reduction. Expert Syst. Appl. 39, 4268–4273. https://doi.org/10.1016/j.eswa.2011.09.107

Xu, C., Liu, X., Wang, J., Zhang, J., Cao, J., Qin, W., 2019. An Improved GA-KRR Nested Learning Approach for Refrigeration Compressor Performance Forecasting. Presented at the 15th International Conference on Automation Science and Engineering (CASE), Vancouver, Canada.

ACKNOWLEDGEMENT

The authors are grateful for the provision of the prototype compressor by Air Squared Inc. The support of this work by NASA under SBIR contract 80NSSC18C0049 is gratefully acknowledged.

CFD Analysis of an ORC Vane Expander using OpenFOAM solver

Sham RANE[1]*, Giuseppe BIANCHI[2], Ahmed KOVAČEVIĆ[1], Roberto CIPOLLONE[3]

[1]City, University of London, Centre for Compressor Technology,
London, EC1V 0HB, UK
sham.rane@city.ac.uk

[2]Brunel University London,
Uxbridge, UB8 3PH, UK
Giuseppe.Bianchi@brunel.ac.uk

[3]University of L'Aquila, Industrial and Information Engineering and Economics Department
L'Aquila 67100, Italy
roberto.cipollone@univaq.it

* Corresponding Author

ABSTRACT

Recent studies on the use of 3D Computational Fluid Dynamics (CFD) for the analysis and design of sliding vane machines has proved beneficial for the detailed evaluation and optimisation of the vane expanders for a given working fluid and operating condition. The authors have earlier developed a customised rotor grid generator for integration with commercial CFD solvers and validated it for use in typical small-scale ORC expanders for waste heat recovery. In this paper, this customised grid generation is extended to an open source CFD solver OpenFOAM, by using a connectivity methodology originally developed for roots blower and twin-screw machines. The control of the rotor grid deformation is through a user code integrated within the flow solver. A case study of the reference ORC expander operating with R245fa was used for validation. The available experimental data for three operating conditions are compared with the results calculated with ANSYS CFX and OpenFOAM-v1912 solvers. During the filling and expansion process, the internal pressure traces are accurately captured by both the solvers and the difference is within 0.05 bar with measurements. However, between the outlet port closure and inlet port opening process the pressure and temperature prediction with OpenFOAM solver is considerably different from the ANSYS CFX solver. It was observed that the OpenFOAM solver is resulting into a non-physical low temperature zone upstream to the tangency region of the rotor and the stator that goes below 80°C. Overall, CFD solution obtained with the commercial solver ANSYS CFX is much more stable and robust than the open source OpenFOAM solver. The generic nature of the deforming grid generation used with an open source CFD solver presented in the paper allows broadening of the utilisation of CFD modelling tools for the design of vane machines.

1. INTRODUCTION

Organic Rankine Cycle (ORC) expanders for heat to power generation use sliding vane machines. Off-design conditions are well accepted in terms of inlet thermodynamic properties of the working fluid, revolution speed, working fluid flow rates. These expanders show wider margins of flexibility when compared with the other machines and such advantages make them suitable for small size ORC-based power units to recover low-medium grade heat usually wasted into mechanical energy. However, the efficiency of rotary vane expanders may be lower than other positive displacement machines (Vodicka *et al.*, 2017). This is due to high volumetric losses, particularly the ones due to leakages between adjacent vanes which take place between the blade tip and the stator inner surface. A further limiting aspect of this machine is the fixed expansion ratio. In Yan *et al.* (2019), a variable expansion ratio rotary vane expander, which adjusts its expansion ratio by acting on the outlet opening angle has been investigated. To overcome some of these drawbacks, the concept of supercharging a vane expander was proposed. Supercharging the expander requires another suction port along the closed-volume expansion phase through which an additional quantity of working fluid is supplied at the same suction conditions. In Fatigati *et al.* (2018a), using a 1-D thermo-fluid-dynamic model, validated on experiments performed using a single intake expander, an optimization analysis to define the circumferential position of the auxiliary port and its area which maximize the mechanical power produced was carried out. The results showed for the optimum configuration a 50% increase of mechanical power produced with respect the baseline case. In Fatigati *et al.* (2018b), an experimental comparison between single and dual intake expander was carried out keeping constant the intake and exhaust pressure. The results showed that the dual intake port expander can elaborate a 40% more of mass flow rate with an increase of indicated power of 50% in comparison of the original device. Nevertheless, the dual intake port expander allows to also improve the efficiency of the machine with respect to the baseline expander under certain condition and circumferential position of the auxiliary intake port. In fact, in Fatigati *et al.* (2019), an analytic relation which expresses the global efficiency difference between single and dual intake expander, keeping constant the intake and exhaust pressure, was found. Through this relation it was theoretically demonstrated if the increase of mass flow rate aspirated by the dual intake expander with respect the baseline case is lower than that of the mechanical power produced, the efficiency of the dual intake machine is higher. Fatigati *et al.* (2020), present a theoretical and experimental evaluation of a 2.0 kW vane expander and details of leakages inside the machine to quantify their effects on the expander performance. In order to achieve a deeper understanding of the supercharging process and its impact on the vane expander operation, Bianchi *et al.* (2019), performed numerical 3D CFD simulations using ANSYS CFX solver and SCORG grid generator on the machine developed and tested in earlier works. When operating at 1516 rpm and between pressures of 544 kPa at the inlet and 320 kPa at the outlet, the supercharged expander provided a power output of 325 W. The specific power output was equal to 3.25 kW/(kg/s) with a mechanical efficiency of 63.1%. For the CFD modelling strategy the essential requirement for the CFD modelling of a vane machine is the application of deforming working chamber grid and in Rane and Kovačević (2017a, 2017b), the grid generation using algebraic techniques has been extended to differential techniques in order to improve the mesh quality. The application in this study was with twin-screw compressors. This differential technique has been applied in Bianchi *et al.*, (2019) on the basic algebraic grid described in (Bianchi *et al.*, 2017a) and applied on the ORC expander analysis. The grid generation has been implemented in the customized grid generation tool SCORG (Kovačević *et al.*, 2005, 2007). The methodology developed in (Bianchi *et al.*, 2017b) was adopted to model the supercharged expander in (Bianchi *et al.*, 2019). The results show not only the successful model validation but also an extensive comparison between standard and supercharged machine. However, state-of-the-art research has been only applied to commercial CFD solvers. Initial works on application of open source CFD tools for analysis of positive displacement machines can be instead found in Casari *et al.* (2017). They presented a review of the techniques available with OpenFOAM-v1606+ solver. The application tested was a single screw expander. Casari *et al.* (2018a, 2018b, 2019) further developed an OpenFOAM interface with the customised grid generator SCORG and the application was demonstrated on Roots Blower and Twin-Screw Compressor with OpenFOAM-v1606+ solver. An oil injection study using lagrangian particle model in the Roots Blower application was presented in Casari *et al.* (2019) using with OpenFOAM-v1606+ solver and SCORG deforming grids for the rotor domains.

The objective of the presented analysis was to extend the use of customised grid generation with open source CFD solver OpenFOAM-v1912, by using a connectivity methodology originally developed for roots blower and twin-screw type of machines (Casari *et al.*, 2018a, 2018b, 2019). The rotor grid deformation was controlled through user code which works integrally with the flow solver. A case study has been presented on the reference ORC expander operating with R245fa (Bianchi *et al.*, 2019). Three operating conditions have been evaluated, for which experimental performance data are available for validation of the numerical methodology using ANSYS CFX and OpenFOAM-v1912 solvers. Through the analysis it was found that further research into the development of a robust and stable solver with positive displacement applications is required to account for the large grid deformation and volumetric changes that occur in these machines.

2. NUMERICAL ANALYSIS METHODOLOGY

Analysis of the ORC vane expander using OpenFOAM was carried out based on the available test data in literature and a corresponding, well established model in ANSYS CFX solver. Comparison with a validated ANSYS CFX model allowed for comparison of the detailed physics inside the expander for which numerical issues of the solver could be studied, together with their impact on prediction of the flow field.

2.1 ORC vane expander and test data

The supercharged expander was tested in an automotive ORC loop at the University of L'Aquila operating with R245fa. The expander is directly coupled with an asynchronous electric generator which constraints the expander to rotate at slightly higher than 1500 RPM when the machine is connected to the electric grid. Further details on the test rig can be found in (Fatigati *et al.*, 2019). From an instrumentation perspective, two Coriolis mass flow meters are installed downstream the pump and on the primary inlet of the expander. Pressure and temperature transducers are located across each component of the ORC loop. The expander performance is eventually measured through a torque meter and through a set of 3 piezo-resistive pressure transducers. The overall expander dimensions are summarized in Table 1. The dual intake phase starts immediately after the end of the main one. Three test cases were selected to enable the CFD validation. A summary is reported in Table 2. Depending on the opening of the ball valve upstream of the secondary inlet, different supercharging ratios were achieved.

Table 1: Specification of the expander geometry

Number of Chambers	7
Stator Inner Diameter	75.90 mm
Rotor Outer Diameter	65.00 mm
Eccentricity	5.45 mm
Chamber Width	60.00 mm
Blade length	17.00 mm
Blade thickness	3.96 mm
Intake port opening angle	4.4 °
Intake port closing angle	48°
Exhaust port opening angle	180°
Exhaust port closing angle	320°
Tip clearance	25 μm

Table 2: Experimental data at three operating conditions of the expander

	Test case #1			Test case #2			Test case #3		
	p	T	m	p	T	m	p	T	m
	±0.3 bar$_a$	±0.3 °C	kg/s	±0.3 bar$_a$	±0.3 °C	kg/s	±0.3 bar$_a$	±0.3 °C	kg/s
Main Inlet (m ±0.15%)	4.55	66.1	0.047	5.44	94.3	0.061	5.57	82.3	0.055
Superchar. Inlet 2 (m ±0.05%)	4.75	66.1	0.054	4.74	94.3	0.039	5.77	82.3	0.064
Outlet	3.10	58.6		3.21	85.9		3.60	73.5	
Rev. speed (±1 RPM)	1511			1516			1517		
Torque (±0.02 Nm)	1.50			2.05			2.11		

The operating point that will be selected to show the validation of the CFD results is the test case #2, even though in (Bianchi *et al.*, 2019) the validation has been successful in all the operating points. At this point, there is limited supercharging which in turn leads to a 1bar pressure difference between the suction conditions of the primary line and the ones related to the supercharging port.

2.2 CFD Model description and OpenFOAM solver implementation

Rane and Kovačević (2017a) have presented the setup requirements for SCORG grid generator with widely used commercial CFD solvers. In case of vane expanders, Bianchi *et al.* (2019) presented the CFD model with supercharger using ANSYS CFX solver and the same has been employed in the current work. The working domain is split into five main sub-domains namely rotor domain, inlet port, supercharged inlet port, outlet port and axial end leakages as highlighted in Figure 1a. A sample cross-section of rotor grid is shown in Figure 1b.

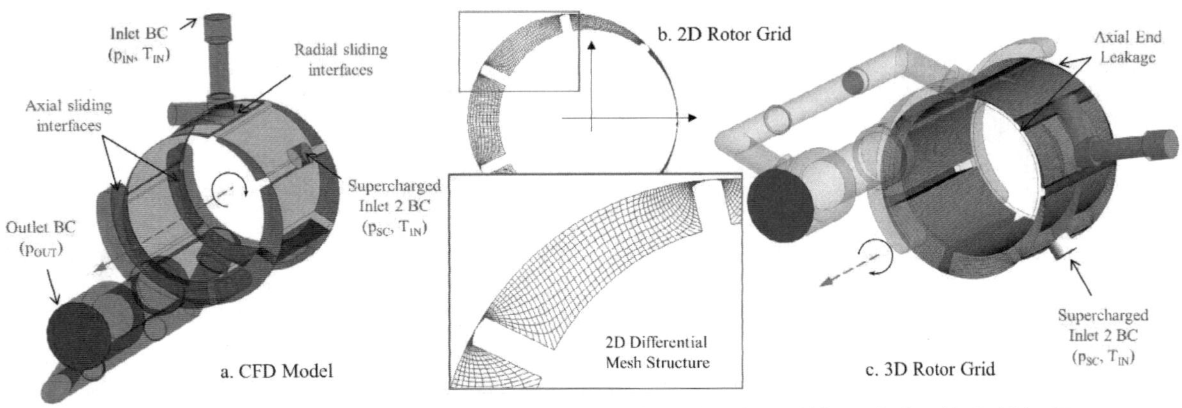

Figure 1: ORC vane expander, a) CFD Model showing various domains and boundaries, b) Grid in 2D cross-section and c) Grid of the rotor in 3D with alignment of ports and axial leakage

All sub-domains are connected within the solver using non-conformal interfaces as shown in Figure 1c. Supercharged inlet pressure was 4.74 bara with 94.3 °C gas temperature. Outlet pressure was 3.21 bara. The ideal gas model was used as the equation of state since it was found that the Peng-Robinson Real gas model was unstable in OpenFOAM-v1912 solver, although it is recommended to use a real gas model in these operating conditions (Rane *et al.*, 2021). Dynamic viscosity, thermal conductivity and specific heat properties of the fluid R245fa are set constant. Table 3 and Table 4 provide the details of the specifications of the CFD solvers ANSYS CFX and OpenFOAM-v1912 respectively.

Table 3: Specification of the numerical setup in the ANSYS CFX solver

Mesh deformation	User defined nodal displacement	Advection scheme	High Resolution
Mesh in ports	Tetrahedral with boundary layer refinements (ANSYS Mesh)	Transient scheme	Second order Backward Euler
Turbulence model	SST – k Omega (Standard Wall Functions)	Transient inner loop coefficients	Up to 20 iterations per time step
Inlet/Supercharge boundary condition	Opening (Specified total pressure and temperature)	Convergence criteria	r.m.s residual level $1e^{-03}$
Outlet boundary condition	Opening (Static pressure, backflow acts as total pressure and temperature)	Relaxation parameters	Solver relaxation fluids (0.4)

Table 4: Specification of the numerical setup in the OpenFOAM-v1912 solver

Mesh deformation	SCORG – OpenFOAM Module	Advection scheme	First order - Upwind
Mesh in ports	Polyhedral (*SnappyHexMesh*)	Transient scheme	First order Backward Euler
Turbulence model	SST – k Omega (Standard Wall Functions)	Transient inner loop coefficients	Up to 20 iterations per time step
Inlet/Supercharge boundary condition	Opening (Specified total pressure and temperature)	Convergence criteria	r.m.s residual level $1e^{-03}$
Outlet boundary condition	Opening (Static pressure backflow acts as total pressure and temperature)	Relaxation parameters	Solver relaxation pressure (0.8), velocity (0.4), energy (0.8)

One of the other differentiating factors between the two solvers is the tetrahedral grid in ANSYS CFX and polyhedral grid generated by *SnappyHexMesh* utility in OpenFOAM, as represented for the inlet port in Figure 2a and Figure 2b respectively. The number of nodes and cells in the respective solvers is also listed in Figure 2.

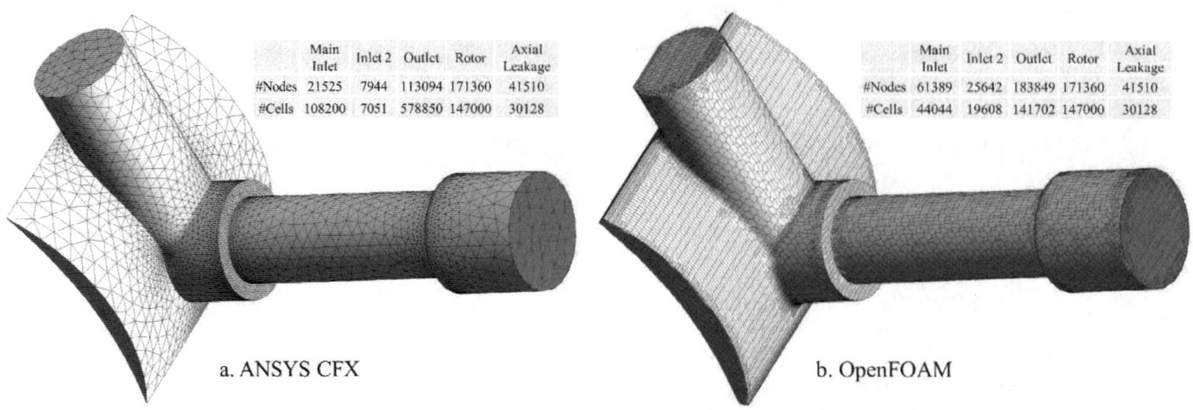

Figure 2: Grid structure in the port domain, a) ANSYS CFX – Tetrahedral, b) OpenFOAM – Polyhedral

Similarly, the supercharger port and the outlet port were both generated by *SnappyHexMesh* for OpenFOAM. The axial leakage grid was generated in ANSYS Meshing and used directly in ANSYS CFX solver. For OpenFOAM they were first converted using *fluentMeshToFoam* utility and then combined in the OpenFOAM setup. Thus, the rotor, tip and axial leakage grids are identical in both the solvers. ANSYS CFX utilized a vertex centered, pressure based coupled solution algorithm and total energy formulation. Whereas OpenFOAM solver was cell centered, set with the compressible *rhoPimpleFoam* segregated solver with *sensibleEnthalpy* energy formulation.

ANSYS CFX provides a coupling feature called Junction Box Routine that is a user defined library to specify mesh deformation from custom applications such as SCORG. The solver updates the nodes coordinates from set of pre-generated coordinate files after every crank angle step (or its submultiples). Solver time step size finally results from the selected crank angle step (at which the customized grids are generated) and revolution speed of the rotor. In this case study, the full rotation of the rotor was defined by 700 grid positions. In case of OpenFOAM, a customised SCORG module was developed (Casari *et al.*, 2018a, 2018b, 2019). Table 5 provides the implementations details of this SCORG – OpenFOAM interface in terms of the parameters of the two dictionaries that control mesh deformation in the solver based on the set of rotor grids supplied from SCORG grid generator. Solver is allowed to use time step interpolation in this interface in order to set an independent time step size, but for consistency with ANSYS CFX solver, the time interpolation of grid was not used in these case studies.

Table 5: Implementation parameters of SCORG – OpenFOAM interface

Dictionary	Parameter	Value	Usage
SCORGDict	conformalInterface	True/False	Distribution selected in SCORG is Rotor-to-Casing **Conformal** set this as True
	nonConformalInterface	True/False	Distribution selected in SCORG is Rotor-to-Casing **Non-Conformal** set this as True
	rotorToCasing	True/False	Distribution selected in SCORG is Rotor-to-Casing set this as True, otherwise set as False
	singleRotor	True/False	Calculations involving only one rotor. Vane Compressor, Set as True, Else Set as False
	lowPressurePort	True/False	Theoretical **Suction Port** from SCORG transferred to OpenFOAM, Set as True
	highPressurePort	True/False	Theoretical **Discharge Port** from SCORG transferred to OpenFOAM, Set as True
	prisms	True/False	Convert the hexahedral cells into prisms using this tag
dynamicMeshDict	frozenPointsZone	frozenPointsZone	Collector for port nodes - do not apply mesh deformation
	malePointsZone	malePointsZone	Collector for main rotor nodes - Apply mesh deformation
	femalePointsZone	femalePointsZone	Collector for gate rog nodes - Apply mesh deformation
	conformal	True/False	Set True in case of Type of Distribution selected in SCORG is Rotor-to-Casing Conformal
	numberOfGrids	700	Number of Grid files written from SCORG for one cycle
	numberOfGrooves	1	Number of Lobes on the Main Rotor. In case of Vane rotor = 1
	omega	1511	Rotor speed in revolutions per minute units

Initial conditions for the flow solver were provided as constant pressure and temperature values and the calculations were performed for 1400 time steps in every case which represent 14 expansion cycles (2 full machine revolutions). The rotor speed, inlet and outlet pressure and temperature listed in Table 2 are used as boundary condition specifications for the CFD model. The results were found to converge in the first few cycles for pressure and torque but took a little longer for outlet temperature. Data from the last few converged cycles has been used here for results presentation in the next section.

3. RESULTS AND DISCUSSION

The analysis results are presented here as comparison of operating condition of Test case #2 for each of the two solvers ANSYS CFX and OpenFOAM-v1912. Two variants of the CFD model were investigated, identified as Case A and Case B respectively for each of the three operating conditions. Case A has 25 μm tip leakage clearance and Case B additionally has two axial leakage clearances of 50 μm introduced at both the rotor ends as indicated in Figure 1c.

3.1 Internal pressure and temperature distribution

A comparison of internal pressure distribution during the filling, expansion and supercharging process followed by further expansion is presented in Figure 3 between the measured data and Case 2-A and Case 2-B with ANSYS CFX solvers.

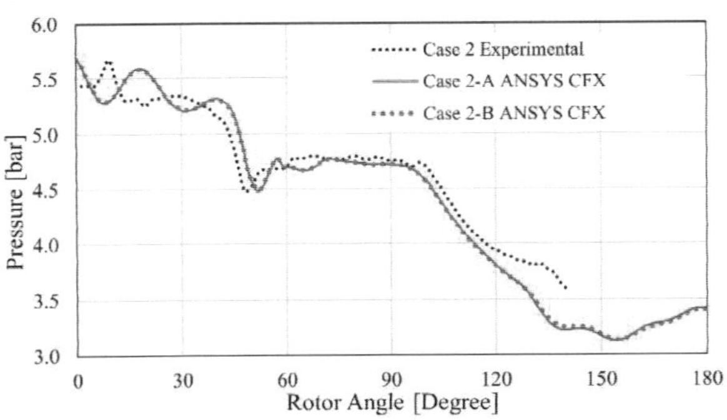

Figure 3: Comparison of pressure with vane rotation angle during filling and expansion process

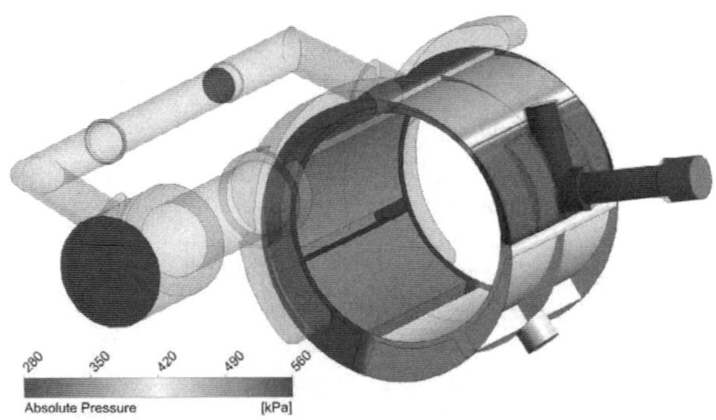

Figure 4: Instantaneous pressure distribution in the vane rotor and ports for Case 2-B ANSYS CFX

It can be seen in these stages of the expander; the internal pressure traces are accurately captured and between 30-100° rotor angle the difference is within 0.05 bar with measurements. The introduction of 50 μm end leakage in Case 2-B did not produce a significant influence on the internal pressure trace as it closely matches the Case 2-A data. The initial flow pulsations, dip in pressure at around 55° rotor angle, followed by a re-filling from the supercharged inlet is captured by the internal pressure traces. In correspondence to these data, an instantaneous distribution of pressure in the expander is shown in Figure 4 for Case 2-B with the ANSYS CFX solver. The peak pressure in the chamber at this rotor position, where one of the vane is at tangency to the stator, is 560 kPa. While the lowest pressure is seen to be close to 280 kPa which occurs at the tangency vane's tip region where flow velocity is high.

A cross-section of the expander at mid-length of the rotor is evaluated in Figure 5. Figure 5a is a contour plot for Case 2-A and ANSYS CFX solver, while Figure 5b is a corresponding plot of pressure for Case 2-A and OpenFOAM solver. Five monitor probes are located in this cross-section identified as Probe-1 to Probe-5. Data at these probes will be compared between the two solvers. On examination of Figure 5, a closeness of the pressure results between the two solvers is evident. Angular positions of significance such as closure of the main and supercharged inlet and opening of the outlet have been marked in Figure 5. As observed with ANSYS CFX solver in Figure 4, the peak pressure in the chamber at this rotor position is 560 kPa and the lowest pressure is close to 280 kPa in the OpenFOAM solver too.

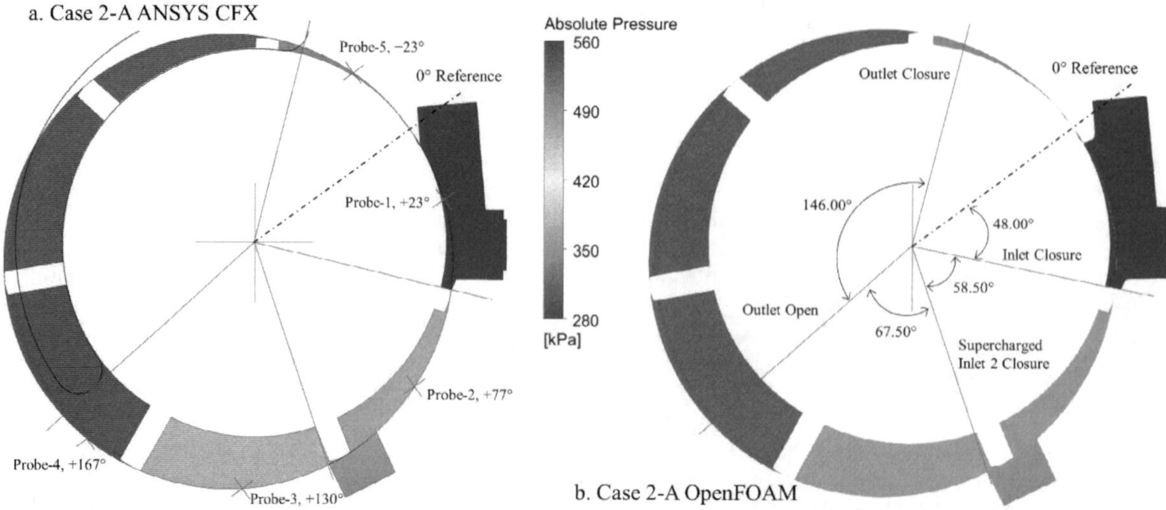

Figure 5: Monitor probe locations and pressure distribution in the expander cross-section at z=0.03 m, a) Case 2-A ANSYS CFX, b) Case 2-A OpenFOAM

Probe locations Probe-1 to Probe-4 are evaluated in the plots of Figure 6 where pressure and temperature data over one full cycle of the vanes has been compared for Case 2-A, between the two solvers. Probe-1 is overlapped on the main inlet. Hence the pressure at the location remains very close to the range of 520 – 560 kPa. Flow pulsations cause the pressure to vary from the condition of 544 kPa at the main inlet. Probe-1 is not exposed to the supercharged inlet. Probe-2 is located such that at 25° rotor angle it gets exposed to the supercharged inlet. Hence in the 0-25° range the initial expansion has caused the chamber pressure to drop to about 450 kPa and it fills back to 540 kPa supercharged inlet pressure at 25°. Probe-3 starts with a chamber pressure of 450 kPa which steadily drops to 350 kPa as expansion is progressing and at 25° rotor angle the probe gets exposed to the chamber which is open to supercharged inlet port. This results in a rise in probe pressure to 460 kPa over the remaining cycle duration. Probe-4 is located open to the outlet port from 0-10° rotor angle. Following which it gets into the expansion chamber at 390 kPa and steady drop in pressure is observed in the remaining cycle. A systematic variation of pressure in the chamber is observed from the Figure 6a plots. The results between ANSYS CFX and OpenFOAM solvers are comparable and the local differences are of the order of 10-15 kPa over the cycle.

A comparison of temperature data (Figure 6b) at these probes also follows a similar trend. But since gas temperature has significantly higher local variation, the point probes are not able to represent the chamber's average property. Nevertheless, probe data from ANSYS CFX solver and OpenFOAM solvers is close to each other over most of the cycle span. Large differences are seen at each of the Probe-1 - Probe-4 when the transition of the vane happens. At these rotor angles the points are exposed to the tip leakage gas and data generated by the two solvers is different in these regions, thus indicating a difference is accuracy of the solver in these critical regions. Typically, OpenFOAM solver is producing a dip in the gas temperature when the vane tip approaches the probe location, while the ANSYS CFX solver is producing a jump. This can be observed at Probe-2 and Probe-3 in 20-25° rotor angle and Probe-4 in 5-12° rotor angle. Probe-1 with ANSYS CFX solver is uniform as it is exposed to the inlet port, but with OpenFOAM solver there is non-physical jump at 30° rotor angle from 94°C inlet temperature to about 98°C.

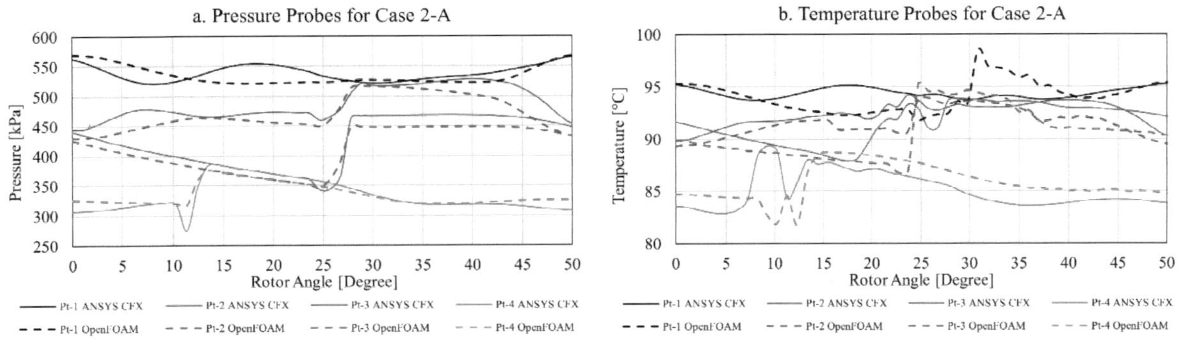

Figure 6: Comparison of pressure (a) and temperature (b) data at Probe-1, 2,3 and 4 locations for Case 2-A

The location of Probe-5 is distinct and not in the expansion cycle. It is located between the outlet port closure and inlet port opening rotor angle range as indicated in Figure 5. The pressure and temperature data over one full cycle of the vanes has been compared for Case 2-A and Case 2-B between the two solvers in Figure 7. After about 5° rotor angle the chamber containing Probe-5 disconnects from the outlet port and pressure starts to rapidly increases. In Case 2-A ANSYS CFX solver the peak pressure reaches close to about 700 kPa. In Case 2-B ANSYS CFX solver, the presence of axial leakage flow reduces this peak pressure to about 650 kPa. After 18° rotor angle the pressure starts to drop at Probe-5 and at the vane transition at 27° the least pressure of 210 kPa is observed in both Case 2-A and Case 2-B. Following this, the pressure in the chamber remains uniform at about 350 kPa till the end of the cycle. The pressure prediction with OpenFOAM solver is considerably different from the ANSYS CFX solver at Probe-5 in both Case 2-A and Case 2-B. As seen in Figure 7, the peak pressure with OpenFOAM solver is nearly 1100 kPa with Case 2-A and close to 1000 kPa with Case 2-B. Since the experimental data of internal pressure is not available at this probe location, it is difficult to validate the results in Figure 7. The temperature data plots in Figure 7 indicate a similar trend in results. Peak temperature in ANSYS CFX solver reaches close to 117°C in Case 2-A and 115°C in Case 2-B. Whereas with OpenFOAM solver the peak temperature is close to 122°C in both Case 2-A and Case 2-B. An odd effect is seen in the OpenFOAM solver that gas temperature in the 0-25° rotor angle is higher in Case-2B, which in the presence of axial leakage is expected to be lower than in Case 2-A.

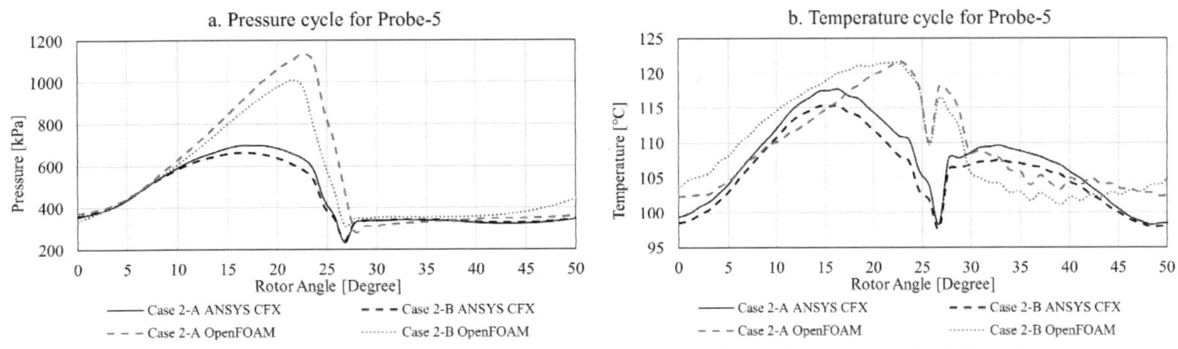

Figure 7: Comparison of pressure (a) and temperature (b) cycle data at Probe-5 location

To test these results further, the temperature distribution in the expander domain is compared between Case 2-B ANSYS CFX and Case 2-B OpenFOAM solvers in Figure 8. Figure 8a with ANSYS CFX solver indicates a uniform distribution of temperature upstream of the tangency region and a local dip at the tip of the vane to about 85°C. In Figure 8b it is seen that the OpenFOAM solver is resulting into a non-physical low temperature zone upstream to the tangency region of the rotor and the stator that goes below 80°C. Probe-5 data in Figure 7 does not capture this local low temperature zone, but it was found to be close to 65°C from data. Within the same chamber the adjacent gas temperature exceeds 105°C and this odd nature of the results can be attributed to numerical inaccuracies.

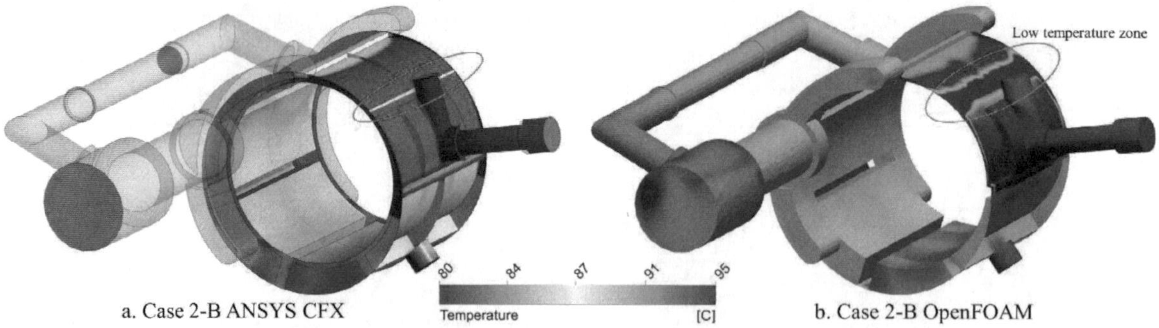

Figure 8: Comparison of instantaneous temperature distribution in the vane rotor and ports, a) Case 2-B ANSYS CFX, b) Case 2-B OpenFOAM

3.2 Comparison of mass flow rate and rotor torque cycles

A cyclic variation of the gas mass flow rate at main inlet and supercharged inlet is presented over three consecutive cycles in Figure 9. Cases 2-A with both the ANSYS CFX and OpenFOAM solver have been plotted here for comparison. The corresponding cycle averaged mass flow rates are reported in Table 6. A clear repetition of the flow is apparent from Figure 9 indicating that the solution is being well converged. Case 2-A main and supercharged inlet have different nature of variation. Main inlet being further from the rotor shows smoother variation with two pulses per cycle. Relatively, OpenFOAM Case 2-A main and supercharged inlet have a different nature of flow variation. This could be due to the difference in implementation of the total pressure boundary condition in the two flow solvers.

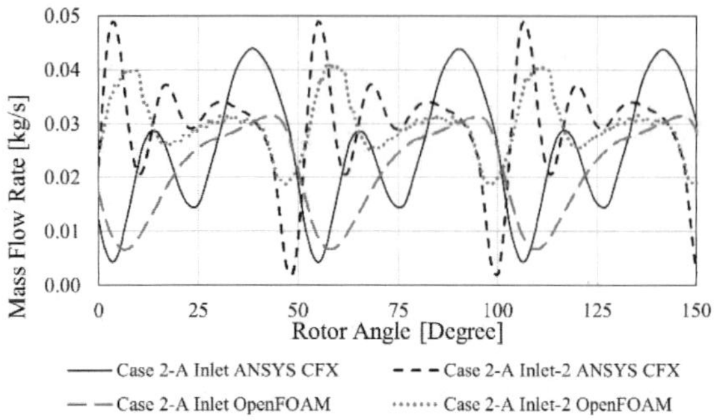

Figure 9: Comparison of cyclic mass flow rate at the two inlets (3 cycles)

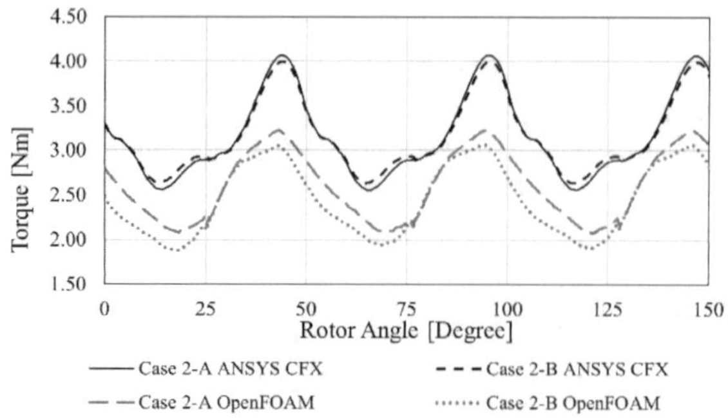

Figure 10: Comparison of cyclic torque on the vane rotor (3 cycles)

Similarly, a cyclic variation of the rotor torque over three consecutive cycles is presented in Figure 10 for Case 2-A and Case 2-B with both the solvers. Cycle averaged rotor torque is reported in Table 6. Unlike the flow variation, the nature of torque variation is similar in both the solvers. But it is clear from the plot that average torque in OpenFOAM solver is lower as compared to ANSYS CFX solver for both Case 2-A and Case 2-B. The difference between Case 2-A and Case 2-B is minor in ANSYS CFX as compared to OpenFOAM indicating a stronger influence of the axial leakage flow. The drop in peak pressure from 1100 kPa to 1000 kPa at Probe-5 indicated in Figure 7 is required to be noted here which could impact the rotor torque significantly.

3.3 Comparison of the expander performance

Performance data of the expander is obtained from the CFD model in terms of cyclically averaged mass flow rate at the main and supercharged inlets, gas temperature at the outlet, torque on the vane rotor and expander power at corresponding rotor speed. These performance data from ANSYS CFX and OpenFOAM solvers has been compared with test data available in Table 2 and the comparison has been presented here in Table 6. Expander specific power is obtained from the net mass flow rate of R245fa from both inlets and the power.

Table 6: Comparison of expander integral performance quantities

25 µm Tip, 0 µm Axial clearance	Case 1 - A			Case 2 - A			Case 3 - A		
Parameter	Measurement	ANSYS CFX	OpenFOAM	Measurement	ANSYS CFX	OpenFOAM	Measurement	ANSYS CFX	OpenFOAM
Inlet mfr [kg/s]	0.0470	0.0203 56.71%	0.0185 60.61%	0.0610	0.0251 58.92%	0.0215 64.75%	0.0550	0.0242 56.07%	0.0219 60.18%
SC inlet mfr [kg/s]	0.0540	0.0370 31.39%	0.0353 34.63%	0.0390	0.0290 25.66%	0.0294 24.73%	0.0640	0.0430 32.76%	0.0411 35.79%
Outlet T [°C]	58.60	58.93 -0.57%	60.60 -3.41%	85.90	85.50 0.47%	86.00 -0.12%	73.50	73.86 -0.49%	74.00 -0.68%
Torque [Nm]	1.50	2.76 -83.93%	2.13 -41.72%	2.05	3.17 -54.73%	2.61 -27.51%	2.11	3.78 -79.21%	3.05 -44.69%

25 µm Tip, 50 µm Axial clearances	Case 1 - B			Case 2 - B			Case 3 - B		
Parameter	Measurement	ANSYS CFX	OpenFOAM	Measurement	ANSYS CFX	OpenFOAM	Measurement	ANSYS CFX	OpenFOAM
Inlet mfr [kg/s]	0.0470	0.0204 56.67%	0.0180 61.71%	0.0610	0.0257 57.83%	0.0214 64.93%	0.0550	0.0242 55.93%	0.0214 61.12%
SC inlet mfr [kg/s]	0.0540	0.0381 29.43%	0.0360 33.39%	0.0390	0.0295 24.40%	0.0299 23.27%	0.0640	0.0443 30.79%	0.0420 34.42%
Outlet T [°C]	58.60	59.12 -0.88%	60.00 -2.39%	85.90	86.00 -0.12%	86.00 -0.12%	73.50	74.29 -1.07%	75.00 -2.04%
Torque [Nm]	1.50	2.76 -83.89%	1.96 -30.86%	2.05	3.17 -54.72%	2.45 -19.43%	2.11	3.78 -79.26%	2.88 -36.72%

In Cases A when there is no axial end leakage, the mass flow rate through the main and the supercharged inlet is consistently lower than Cases B at all the three operating conditions. This is estimated by both ANSYS CFX and OpenFOAM solvers. Within each operating condition, OpenFOAM solver is underestimating both the main inlet and supercharged inlet mass flow rate as compared to ANSYS CFX solver. Since a number of leakage paths were discarded, e.g. the clearance gaps between rotor slots and each of the seven blades, both solvers are underpredicting mass flow rate in comparison to the measurements. The difference is significantly larger for the main inlet flow. Outlet gas temperature is sufficiently accurately predicted by the CFD models as compared to the measurements. Overall, OpenFOAM is resulting into 0.5-1.5°C over estimation of the outlet temperature in all cases. This is an odd result considering that the total mass flow rate is relatively lower in the OpenFOAM cases and hence after expansion the gas temperature drop is expected to be higher. This is likely due to the simplified real gas model adopted. The rotor torque is overestimated by both ANSYS CFX and OpenFOAM models as compared to measurements. Since the peak pressure observed at Probe-5 is much higher in OpenFOAM results, this creates an opposing rotor torque. Thus relatively, OpenFOAM torque results are closer to measurements even though internal pressure variation during expansion is similar in both the solvers. Due to a lower total mass flow rate, the specific power calculated from OpenFOAM is lower than ANSYS CFX across all the cases but in both the solvers it is highly overestimated as compared to measurements. This is due to significantly higher rotor torque.

4. CONCLUSIONS

The analytical and differential grid generation methods built in the software SCORG are available for the rotating and deforming domain of vane machines and well-integrated with commercial CFD solvers. In the work presented in this paper, this customised grid generation has been extended to an open source CFD solver OpenFOAM, by using a connectivity methodology SCORG-OpenFOAM module, originally developed for roots blower and twin-screw machines. The rotor grid deformation was controlled through the user code integrated within the flow solver. A case study has been presented on a reference ORC expander operating with R245fa. Three operating conditions have been evaluated using ANSYS CFX and OpenFOAM-v1912 solvers and compared with the available experimental performance in order to validate this numerical methodology.

- The CFD solution obtained with the commercial solver ANSYS CFX is much more stable and robust than the open source OpenFOAM solver. Further, the real gas equation of state could not be used for R245fa fluid due to the instability of the OpenFOAM solver. In a separate exercise, ANSYS CFX solver obtained stable and converged results with all real gas models including REFPROP, Peng-Robinson and others.
- The indicated pressure traces obtained by the experiment and through simulation agreed well in both solvers during the filling and expansion processes. However, a large difference in the peak pressure and temperature was noticed in the process following the closure of the outlet port.

- A comparison of local temperature distribution indicated that the OpenFOAM solver is resulting into a non-physical, low temperature zone upstream to the tangency region of the rotor and the stator.
- Cycle averaged expander performance data was comparable between the two solvers. The mass flow rate through the main inlet was under predicted whiled the rotor torque was over predicted by the OpenFOAM solver.

The generic nature of the deforming grid generation together with an open source CFD solver is available to broaden the utilisation of CFD modelling tools for the design of vane machines. However, further research to improve robustness and stability of the solver for calculation of positive displacement machines is required, in order to account for the large grid deformation and large volumetric changes that occur in these machines.

REFERENCES

Bianchi, G., Rane, S., Kovačević, A. & Cipollone, R. (2017a). Deforming grid generation for numerical simulations of fluid dynamics in sliding vane rotary machines, *Advances in Engineering Software*, 112, 180-191, ISSN 0965-9978, DOI:10.1016/j.advengsoft.2017.05.010.

Bianchi, G., Rane, S., Kovačević, A., Cipollone, R., Murgia, S. & Contaldi, G. (2017b). Numerical CFD simulations on a small-scale ORC expander using a customized grid generation methodology, *Energy Procedia*, 129, 843-850, ISSN 1876-6102, DOI:10.1016/j.egypro.2017.09.199.

Bianchi, G., Rane, S., Fatigati, F., Cipollone, R. & Kovačević, A. (2019). Numerical CFD simulations and indicated pressure measurements on a sliding vane expander for heat to power conversion applications. *Designs*. 3(31). DOI: 10.3390/designs3030031.

Casari, N., Suman, A., Ziviani, D., van den Broek, M., De Paepe, M. & Pinelli, M. (2017). Computational Models for the Analysis of positive displacement machines: Real Gas and Dynamic Mesh, *Energy Procedia*, 129, 411-418, ISSN 1876-6102, DOI: 10.1016/j.egypro.2017.09.124

Casari, N., Fadiga, E., Pinelli, M., Suman, A., Kovačević, A., Rane, S. & Ziviani, D. (2018a). Full 3D numerical analysis of a roots blower with open-source software. 24th International Compressor Engineering Conference at Purdue. Paper 2620. https://docs.lib.purdue.edu/icec/2620

Casari, N., Pinelli, M., Suman, A., Kovačević, A., Rane, S. & Ziviani, D. (2018b). Full 3D numerical analysis of a twin screw compressor by employing open-source software. International Conference on Screw Machines. IOP Conf. Ser.: *Mater. Sci. Eng.*, 425, 012017. DOI: 10.1088/1757-899X/425/1/012017

Casari, N., Fadiga, E., Pinelli, M., Suman, A., Kovačević, A., Rane, S. & Ziviani, D. (2019). Numerical investigation of oil injection in a Roots blower operated as expander. International Conference on Compressors and their Systems. IOP Conf. Ser.: *Mater. Sci. Eng.*, 604, 012075. DOI: 10.1088/1757-899X/604/1/012075

Fatigati, F., Bianchi, G. & Cipollone, R. (2018a). Development and numerical modelling of a supercharging technique for positive displacement expanders, *Applied Thermal Engineering*, 140, 208-216, ISSN 1359-4311, DOI:10.1016/j.applthermaleng.2018.05.046.

Fatigati, F., Bartolomeo, M. D. & Cipollone, R. (2018b). Experimental and numerical characterization of a positive displacement vane expander with an auxiliary injection port for an ORC-based power unit, *Energy Procedia*, 148, 830-837, ISSN 1876-6102, DOI:10.1016/j.egypro.2018.08.114.

Fatigati, F., Bartolomeo, M. D. & Cipollone, R. (2019). Dual intake rotary vane expander technology: Experimental and theoretical assessment, *Energy Conversion and Management*, 186, 156-167, ISSN 0196-8904, DOI:10.1016/j.enconman.2019.02.026.

Fatigati, F., Bartolomeo, M. D. & Cipollone, R. (2020). On the effects of leakages in Sliding Rotary Vane Expanders, *Energy*, 192, 116721, ISSN 0360-5442, DOI: 10.1016/j.energy.2019.116721

Kovačević, A. (2005). Boundary Adaptation in Grid Generation for CFD Analysis of Screw Compressors, *Int. Num. Eng.*, 64(3), 401-426.

Kovačević, A., Stošić, N. & Smith, I. K. (2007). *Screw compressors - Three dimensional computational fluid dynamics and solid fluid interaction*, ISBN 3-540-36302-5, Springer-Verlag Berlin Heidelberg New York.

Qiu, G., Liu, H. & Riffat, S. (2011). Expanders for micro-CHP systems with organic Rankine cycle, *Applied Thermal Engineering*, 31(16), Pages 3301-3307, ISSN 1359-4311. DOI:10.1016/j.applthermaleng.2011.06.008

Rane, S. & Kovačević, A. (2017a). Application of numerical grid generation for improved CFD analysis of multiphase screw machines, 10th International conference on compressors and their systems, London, IOP Conf. Ser.: *Mater. Sci. Eng.*, 232, 01. DOI:10.1088/1757-899X/232/1/012017

Rane, S. & Kovačević, A. (2017b). Algebraic generation of single domain computational grid for twin screw machines. Part I. Implementation, *Advances in Engineering Software*, 107, 38-50.

Rane, S., Kovačević, A., Stošić, N. & Smith, I. (2021). Analysis of real gas equation of state for CFD modelling of twin screw expanders with R245fa, R290, R1336mzz(Z) and R1233zd(E). *International Journal of Refrigeration*. 121:313-326. DOI: 10.1016/j.ijrefrig.2020.10.022.

Vodicka, V., Novotny, V., Mascuch, J. & Kolovratnik, M. (2017). Impact of major leakages on characteristics of a rotary vane expander for ORC, *Energy Procedia*, Volume 129, 387-394, ISSN 1876-6102, DOI:10.1016/j.egypro.2017.09.249.

Yan, J., Han, Y., Tian, J., Xu, Y., Zhang, Y. & Chen, R. (2019). Performance investigation of a novel expander coupling organic Rankine cycle: Variable expansion ratio rotary vane expander for variable working conditions, *Applied Thermal Engineering*, 152, 573-581, ISSN 1359-4311, DOI:10.1016/j.applthermaleng.2019.02.103.

ACKNOWLEDGEMENT

Authors would like to thank Professor M. Pinelli, Dr. N. Casari and Mr. E. Fadiga from University of Ferrara, Italy, for development of the SCORG-OpenFOAM interface and their help in setting the CFD test case.

An Overset Mesh Methodology for CFD Modelling of Rotary Compressors

Wai Chang POH[1], Kim Tiow OOI[2], Elhadidi BASMAN[3]

[1,2,3]School of Mechanical and Aerospace Engineering,
Nanyang Technological University, Singapore
[1]WAICHANG001@e.ntu.edu.sg
[2]MKTOOI@ntu.edu.sg
[3]MBASMAN@ntu.edu.sg

ABSTRACT

Full 2D and 3D unsteady CFD simulations are gradually replacing the widely used lumped parameters simulations for rotary compressors. Lumped parameter models predict the overall thermodynamic processes; however, they cannot reveal the spatial and temporal variations in the working chamber and account for complex fluid interactions and losses. The complex and continuously deforming working chamber geometry is challenging for CFD. The overset method has been implemented here for the simulation of a 2D Coupled Vane Compressor (CVC). The method was compared to a dynamic mesh formulation using a remeshing technique and has shown; (1) it is 3-4 times faster computationally and (2) capable of modelling tight clearances not possible in the dynamic mesh approach and are difficult, if not impossible to model using the lumped parameter models. The CFD results revealed complex fluid interactions inside the chamber, accounted for leakage and can be used to optimize the CVC designs further.

1. INTRODUCTION

Historically, the lumped parameters approach was used for the performance prediction of rotary compressors. The recent advancements of computer hardware and software propelled the research of rotary compressors towards 2D or 3D CFD simulation studies. CFD modelling reveals the hidden information in the working chamber of the compressor and leads to a much better understanding of the flow physics. However, the primary challenge of CFD models is to generate the complex solution domain of the working chamber, especially for rotary compressors with its varying geometries during operation. Several meshing approaches for rolling piston compressor (Ding & Gao, 2014), screw compressor (Kovačević, 2005), and sliding vane compressor (Bianchi *et al.*, 2017) are available in the literature. However, these approaches can only cater to the relatively simple geometry deformations.

In this paper, the Coupled Vane Compressor (CVC), which was developed by Ooi & Shakya (2018) is used to showcase the complex geometry changes due to its unique dual vane configuration. Figure 1 illustrates the working principle of the CVC with the highlighted areas showing the working chamber deformations in a complete operating cycle of 540°.

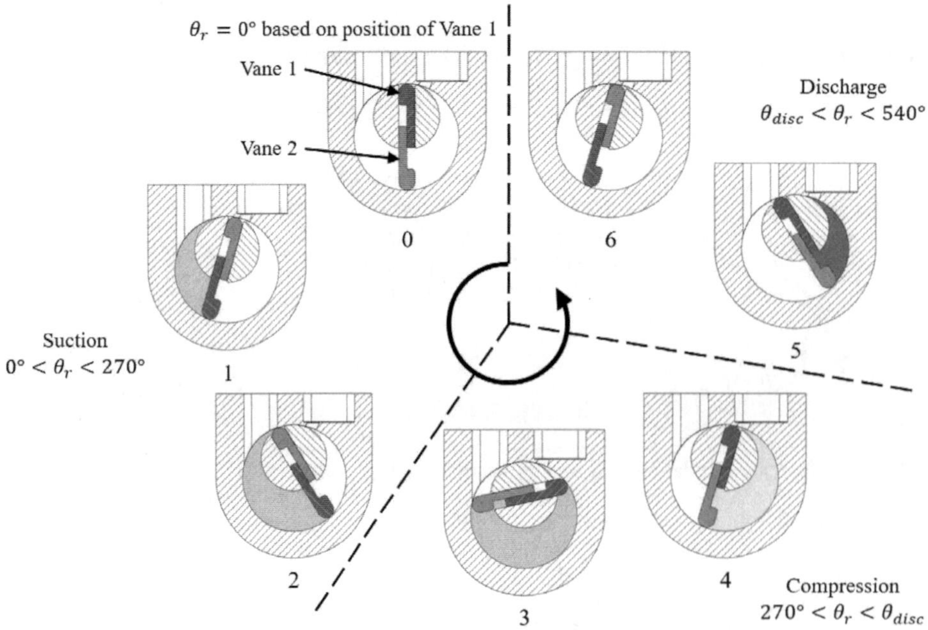

Figure 1: Operating cycle of CVC

2. GEOMETRY AND MESHING STRATEGIES

2.1 Geometry

The 2D geometry is taken at the midplane of CVC, where it encompasses the suction and discharge ports. Since the first objective is to investigate the meshing strategy, Geometry A is used as the model for the dynamic mesh and overset mesh comparison study, as shown in Figure 2. Geometry A is simplified by removing the discharge chamber and the complexity of the discharge valve from Geometry B. Furthermore, a larger gap, Gap A is introduced between the vanes to reduce the computational time by relaxing the calculations in the chamber formed by the vanes. Besides that, the feasibility of the meshing strategies used for CVC can be determined since the solution domain must remain continuous, especially at the initial clearances at the sliding regions of the two components of 100 μm, as shown in Figure 3 and Figure 5. From the initial investigations of utilizing 100 μm clearances, it was discovered that there are significant internal leakages, and the compression process could not achieve the desired discharge pressure. Thus, Geometry B is used onwards after the initial investigations comparing the meshing approaches using 50 μm clearances.

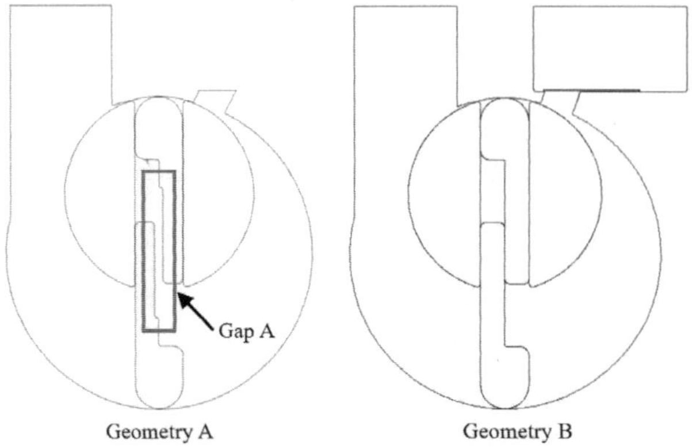

Figure 2: Geometry of the solution domains for CVC

2.2 Dynamic Mesh with Remeshing

As the remeshing feature using the dynamic mesh approach for the moving and deforming solution domain can only cater to triangular elements in the case of ANSYS computational code, the mesh is generated in most of the solution domain using quadrilateral elements. At the interface of the sliding regions, the mesh was generated using triangular elements, as shown in Figure 3. The two zones are connected using a non-conformal interface. When the elements within the deforming zone deformed due to the motion of the rotor and vanes, the solver generates new elements that conform to the specified mesh quality, which is known as the remeshing process. The limitation of this approach is the resulting large number of elements since the smallest element size at the clearance dictates the overall size of all the elements in the deforming region.

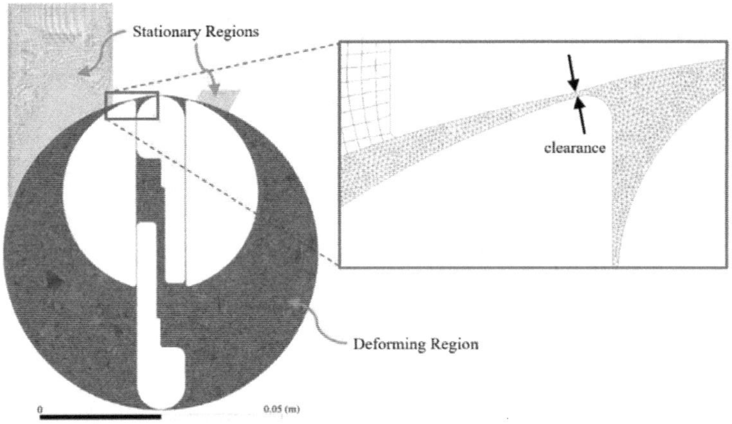

Figure 3: Mesh of Geometry A for dynamic mesh with remeshing

2.3 Overset Mesh

The overset mesh approach has been utilized successfully in other research areas with external flow systems such as ocean engineering (Hao *et al.*, 2019) and aerospace with guidelines provided by Chan *et al.* (2002). In the case of internal flow systems, Suman *et al.* (2016) attempted to model a single screw expander using such method, but further investigations are required to model the clearances and aid solution convergence for transient simulations. The overset mesh approach consists of several overlapping meshes to form the solution domain that consists of active elements to be solved. Meanwhile, the other elements (passive elements) which are not within the solution domain are turned off temporarily or permanently depending on the motion of each component. The active elements consist of donor elements that transfer information from a mesh to the receptor elements on the other mesh at the overset interface.

The individual component meshes are prepared separately with quadrilateral elements, as shown in Figure 4. Although the meshing procedure seems reasonably straightforward, special attention must be given to the boundary layer elements and the element size at the overset boundaries for each component. The boundary layer elements at the clearances of 50 μm for Geometry B is shown in Figure 5, consisting of several elements across it which can be further refined by the mesh controls. Hence, the flow details at the clearances can be captured more accurately as compared to a single cell thick element used by Ding & Gao (2014). Besides that, the overset mesh also requires a minimum number of overlapping elements at the overset boundary, especially between the near-wall regions at the clearances and to prevent orphan cells.

At the overset interfaces, the element size and mesh continuity are vital to ensure the accuracy of the solution. A 1:1 ratio of the element size at the overset interface is preferred to minimize interpolation errors from the different meshes, as shown in Figure 5. Figure 6 illustrates the case where the vane mesh is discontinuous from the vane wall and caused an anomaly in the pressure contour. This is because the priority is given to the smaller element size at the overset interface. Hence, the problem can be resolved by implementing a banded region of smaller element size in the background mesh as well as removing the outer layer of vane mesh at the vane tip, as shown in Figure 4. Similar modification is also performed at the discharge port area. Remeshing is not required since the elements do not deform during the motion of the components. Thus, the mesh quality generated initially is preserved. Figure 7 illustrates the solution domains which are formed by overlapping each meshed component onto the meshed background.

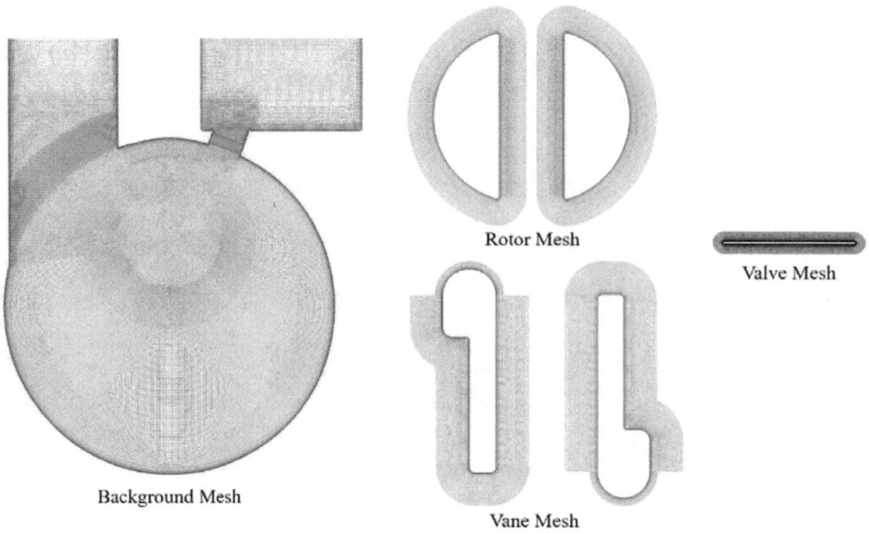

Figure 4: Component meshes for overset mesh

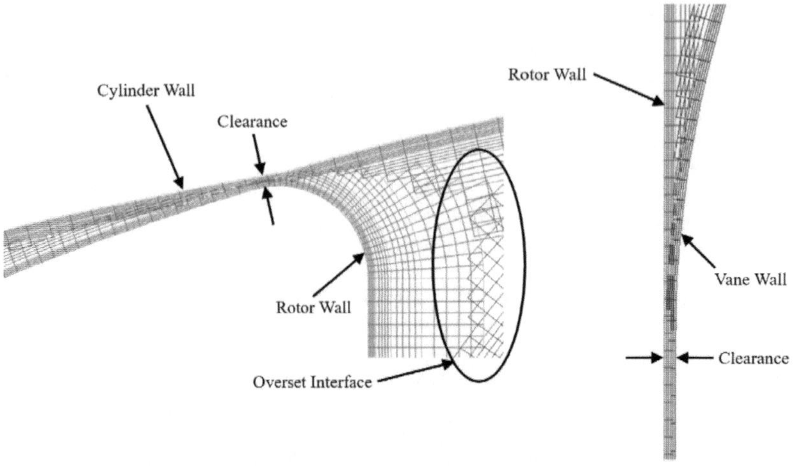

Figure 5: Elements at the clearances for overset mesh

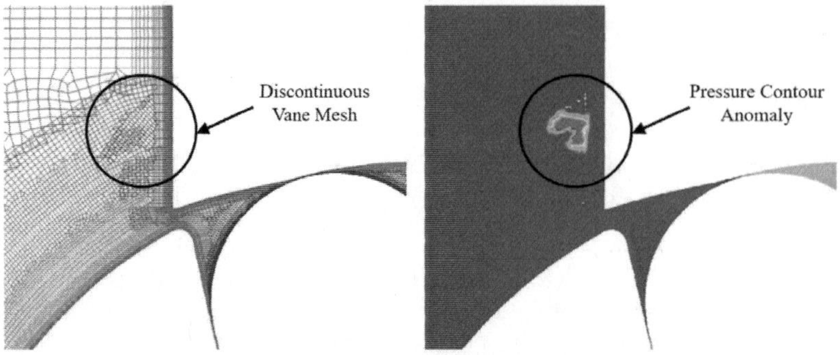

Figure 6: Pressure contour anomaly from discontinuous vane mesh

Figure 7: Generated solution domain from overset mesh

3. SIMULATION SETUP

The 2D transient adiabatic CFD simulations are carried out using the commercially available code ANSYS Fluent 2019 R3 employing the realizable $k - \epsilon$ turbulence model from the fully turbulent nature of the flow in CVC and provides a relatively accurate solution with a faster solution convergence rate as compared to the $k - \omega$ turbulence model. Air is used as the working fluid and modelled as an ideal gas. Table 1 lists the boundary conditions with the operating speed of 1500 rpm. Meanwhile, the displacements of the moving parts are controlled using user-defined equation of motions. The valve modelling is simplified as an equivalent 1D torsional spring model derived from the Euler-Bernoulli beam theory (Ding & Gao, 2014). It is defined with the six degrees of freedom motion where the valve is displaced based on the resulting fluid forces acting on it in a one-way fluid-structure interaction.

Table 1: CFD boundary conditions

	Suction Port	**Discharge Port**
Type	Pressure Inlet	Pressure Outlet
Pressure (Pa)	101,325	506,625
Temperature (K)	300	475

4. RESULTS AND DISCUSSIONS

4.1 Dynamic Mesh with Remeshing and Overset Mesh Comparison – Geometry A

Geometry A, which consists of 100 μm clearances, is used for the comparison study utilizing both the meshing approaches. The discharge port pressure indicated that the compression of the working fluid failed to achieve the specified discharge pressure at 1500 rpm. So, another set of simulation was performed at 3000 rpm, and only the overset mesh could achieve the targeted discharge pressure, as illustrated by Figure 8. The discrepancy between the pressure curves is caused by the mesh quality from the two meshing approaches. Remeshing was applied in the dynamic mesh method, which deteriorates the mesh quality as the simulation progresses, especially at the clearances. However, the overset mesh does not require any remeshing and the mesh quality is preserved throughout the simulation.

The effects of internal leakages due to the 100 μm clearances and Gap A are also perceived from the discharge port pressures that could not achieve the specified discharge pressure. Although Geometry A highlighted the issue of internal leakages, it has shown that the dynamic mesh with remeshing produced undesirable results due to the degrading mesh quality from remeshing. It is shown by the decreasing peak pressure caused by the accumulated interpolation errors from the previous mesh to the newly generated mesh. Table 2 summarizes the computational cost where the overset mesh is 3 to 4 times more cost-efficient as compared to the dynamic mesh with remeshing. Thus, Geometry B with 50 μm clearances utilizing the overset mesh is employed for the following study.

Table 2: Computational cost between dynamic mesh and overset mesh

	Dynamic Mesh	Overset Mesh
Number of nodes	218,701	195,929
Number of elements	422,826	191,126
Computational time per rotor revolution (32 cores)	36 – 48 hours	11.6 – 12.8 hours

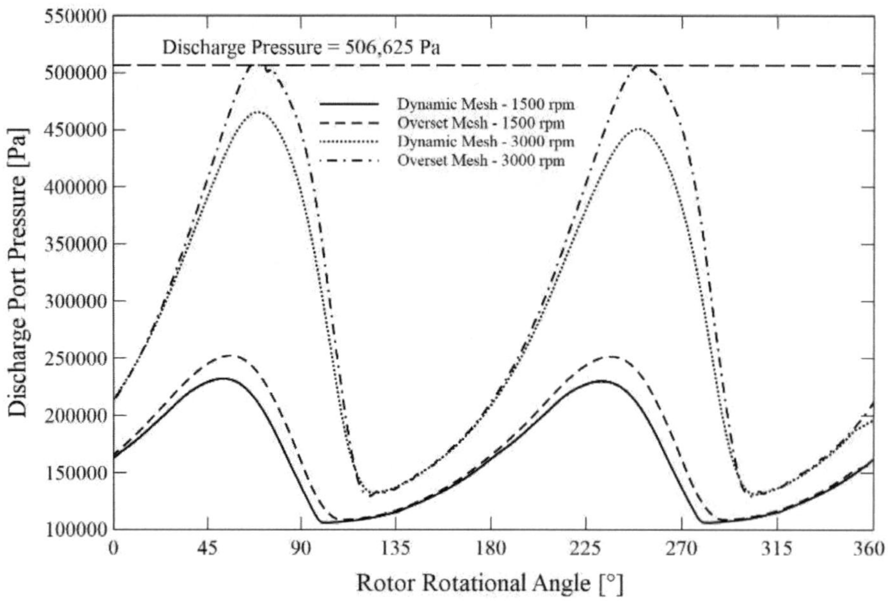

Figure 8: Variation of discharge port pressure for Geometry A

4.2 Overset Mesh Results – Geometry B

Figure 9 illustrates the variation of working chamber pressure obtained from the 2D transient CFD simulation for Geometry B utilizing the overset mesh approach. The variation of working chamber volume is used as the indicator for the processes in one operating cycle. Pressure probes were placed within the working chamber and the average working chamber pressures were obtained for each rotor rotational angle. The averaged working chamber pressures are then assembled to obtain the working chamber pressure curve. The suction process occurs as the volume of the working chamber rises to its maximum. The compression process begins at 270° rotor rotational angle until the discharge process starts at 391°. The lumped parameters model from Shakya & Ooi (2020) showed similar compression and discharge starting points.

Although there are over-compression effects due to the valve stiffness, experimental studies are required to validate the accuracy of the valve models. Conversely, the CFD result indicates the presence of internal leakage at 454° due to the vane tip passes the discharge port where the undischarged compressed working fluid in the discharge port leaks to the next cycle. However, the lumped parameters model assumed that all the working fluid would continue to be compressed and discharged after that point.

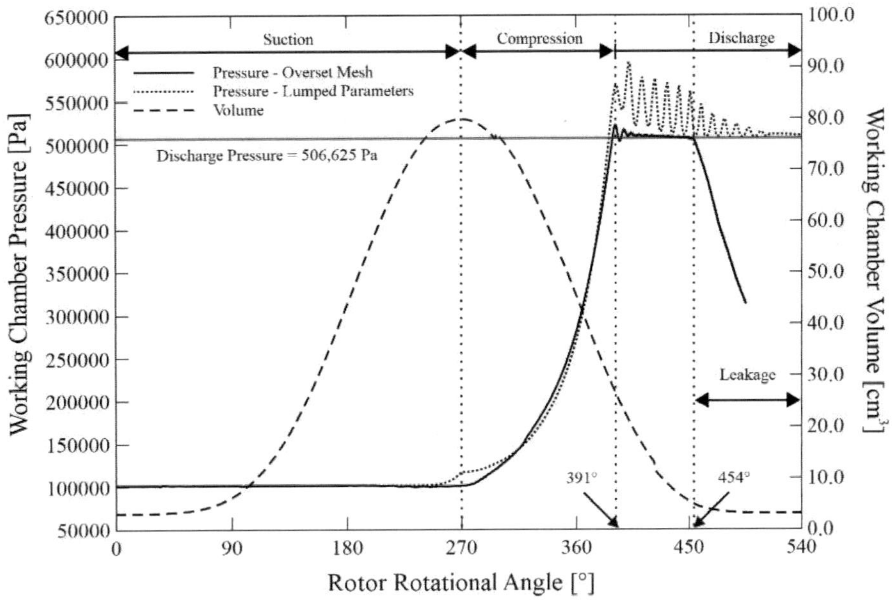

Figure 9: Variation of working chamber pressure and volume

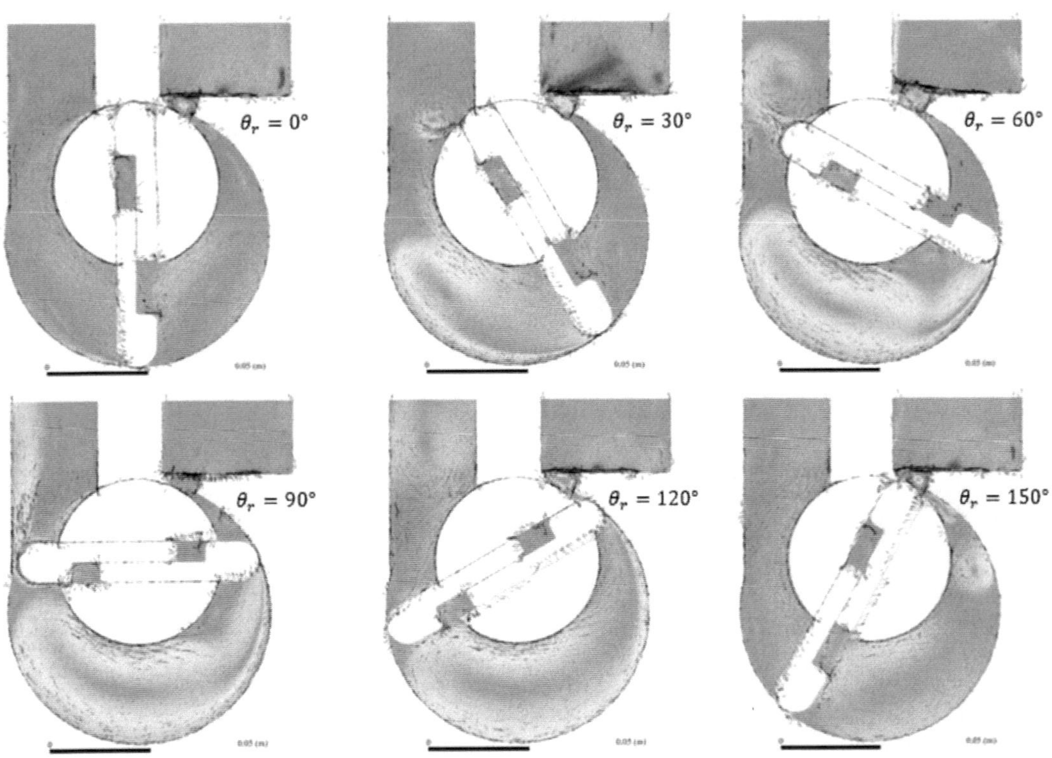

Figure 10: Velocity flow field of CVC

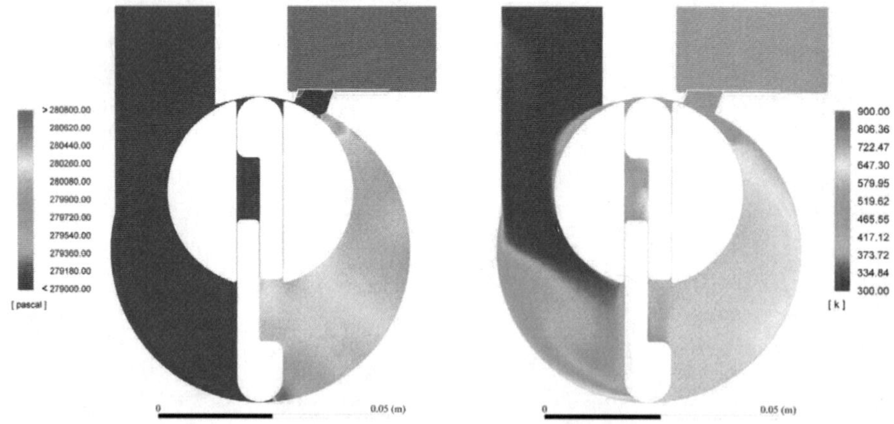

Pressure Contour Temperature Contour

Figure 11: Pressure and temperature contours spatial variations in the working chamber

Figure 10 illustrates the snapshots of the velocity flow field for every 30° rotor rotational angle. The figure reveals the presence of circulation zones due to the effects of internal leakages. Furthermore, the spatial variation of pressure and temperature in the working chamber is shown in Figure 11, where the information can be analyzed further to enhance the design and hence the performance of CVC which is currently underway.

5. CONCLUSIONS

The overset mesh method has been applied to simulate the two-dimensionally transient working processes of the Coupled Vane Compressor (CVC) and compared to that using the dynamic mesh with remeshing method. The results show that the overset meshing approach is capable of modelling tight clearances at the rubbing regions while preserving the mesh quality throughout the entire simulation process without the need to regenerate new meshes as the solution domain deforms during a complete compressor cycle. It was also noted that the computational cost was 3 to 4 times cheaper than that of the dynamic mesh approach. Furthermore, the results were validated by the lumped parameters approach with good agreement. It is concluded that the overset mesh method is superior, as compared to the dynamic meshing approach with mesh regeneration, in modelling deforming and moving boundary problems such as those found in positive displacement compressors. It is believed that this overset mesh approach is very useful and practical to be applied in improving the design of the compressor. Obviously, the 3D transient model will be able to provide an even more realistic prediction which accounts for all the flow interactions caused by the end effects.

NOMENCLATURE

θ	rotational angle	(°)

Subscript

r	rotor
disc	discharge

REFERENCES

Bianchi, G., Rane, S., Kovačević, A., & Cipollone, R. (2017). Deforming grid generation for numerical simulations of fluid dynamics in sliding vane rotary machines. *Advances in Engineering Software, 112*, 180-191.

Chan, W. M., Gomez, R. J., Rogers, S. E., & Buning, P. G. (2002). Best practices in overset grid generation. *32nd AIAA Fluid Dynamics Conference and Exhibit*, (p. 3191).

Ding, H., & Gao, H. (2014). 3-D Transient CFD Model For A Rolling Piston Compressor With A Dynamic Reed Valve. *International Compressor Engineering Conference.*

Hao, C., Ling, Q., Zhihua, M., Wei, B., Ye, L., Derek, C., & Clive, M. (2019). Application of an overset mesh based numerical wave tank for modelling realistic free-surface hydrodynamic problems. *Ocean Engineering, 176*, 97-117.

Kovačević, A. (2005). Boundary adaptation in grid generation for CFD analysis of screw compressors. *International Journal for Numerical Methods in Engineering, 64*(3), 401-426.

Ooi, K. T., & Shakya, P. (2018). A New Compact Rotary Compressor: Coupled Vane compressor. *International Compressor Engineering Conference.*

Shakya, P., & Ooi, K. T. (2020). Introduction to Coupled Vane compressor: Mathematical modelling with validation. *International Journal of Refrigeration, 117*, 23-32.

Suman, A., Ziviani, D., Gabrielloni, J., Pinelli, M., De Paepe, M., & Van Den Broek, M. (2016). Different Numerical Approaches for the Analysis of a Single Screw Expander. *Energy Procedia, 101*, 750-757.

ACKNOWLEDGEMENT

The authors would like to thank Nanyang Technological University, Singapore for the opportunity and support to carry out the study.

Development of state-of-the-art experimental technique to investigate temperature field in leakage flows of positive displacement machines

Brijeshkumar PATEL[1]*, Ahmed KOVACEVIC[2], Md Nahinul ALAM[3], Alexandros CHAROGIANNIS[4]

[1,2,3] City, University of London, London, UK, EC1V0HB

[4] LavisionUK Ltd, United Kingdom

[1]Brijeshkumar.patel@city.ac.uk
[2]A.Kovacevic@city.ac.uk
[3]Md.Alam.5@city.ac.uk
[4]ACharogiannis@lavision.com

* Corresponding Author

ABSTRACT

Investigation of leakage flows in oil-free rotary positive displacement machines (PDMs) is necessary to get a real insight into attributes of leakage flows. Experimental study of heat transfer in the leakage flows of the oil-free positive displacement machines is challenging. The primary reason is complexities associated with actual machine running condition such as extremely small (micron-size) clearance gap, high speed of rotors, optical access of flows etc. Therefore, this study presents the development of state-of-the-art experimental setup using Planar Laser-Induced Fluorescence (PLIF) technique to visualize the temperature field in the clearance flows of rotary machines. This study considers Roots blower to implement PLIF technique because it provides relatively easy optical access. Besides, it is a good representative of PDMs. Selection criteria of optical glass, suitable tracer particles, Imaging camera, optical lenses and filters are investigated in detail. Pixel Intensity was found to be conformable in the clearance gap from recorded images of temperature field, thereby confirmed the efficacy of the flow visualization and instrumentation systems developed.

1. INTRODUCTION

Rotary machines are present in the majority of energy generation and conversion systems. Today's compressed air sector utilizes almost 20% of a world generated power, which is why these machines contribute highly in carbon emission(Vittorini et al., 2015). So, this study focuses explicitly on oil-free rotary machines. As depicted in Figure 1, all rotary machines have a clearance gap to allow motion between stator and rotor, and these clearance gaps are one of the significant contributors to efficiency loss. Nowadays, it is possible to maintain optimal clearance gap in rotary machines, but it introduces question on the reliability of machines as clearance gap size varies in running condition.

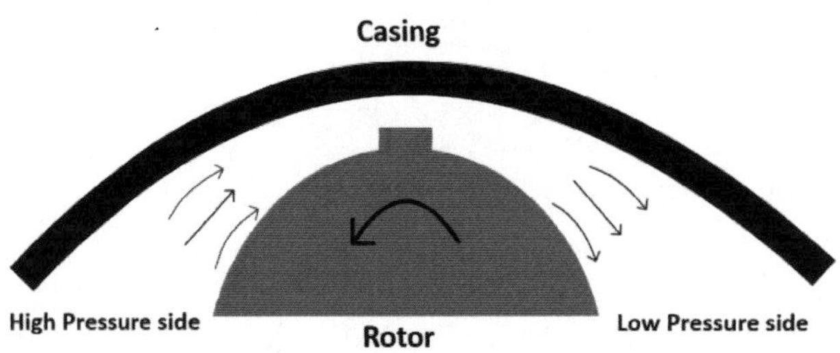

Figure 1 Leakage through the gap between casing and rotor

As per the literature on leakage flows, flow dynamics within the clearance gap is present(Coull & Atkins, 2015; Kauder & Stratmann, 2002; Sachs, 2002). These flow dynamics are greatly affected by the pressure difference across the gap and gap size, at higher pressure ratios flow separation and shocks are observed. Shock boundary layer and shearing directly affect the heat transfer rates, and these variable heat transfer can lead to uneven material growth in the machine's actual running condition (Zhang et al., 2011). Therefore, knowledge of heat transfer inside the clearance gap is vital. Experimental study of the temperature field to get an insight of the clearance gap is necessary but challenging because of the tiny micron size gap and high speed of rotors. In the present study, an oil-free Roots blower is considered as a representative of positive displacement machines because it provides comparatively easy optical access.

Recently, many researchers have used PLIF to study multiphase flows as well as organic constituents and quantitative analysis(Charogiannis et al., 2019; Richardson et al., 2017). Few of them have focused on an application of LIF for the study of temperature field in combustion engines and multiphase flows(Charogiannis & Markides, 2019; Kranz et al., 2018; Rothamer et al., 2009). The same technique can be utilized to study the temperature field in the clearance of rotary positive displacement machine. However, due to the very small size of the clearance gap and rotation of the machine, it is hard to implement a setup used by the researcher in the past. Therefore, this study focuses on developing a unique experimental setup using PLIF technique to obtain temperature field in the clearance gap of roots blower. The detailed procedure of selection of optical glass, tracer particles for LIF, optical instruments are explained.

2. EXPERIMENTAL MODELING APPROCH

2.1. Theoretical background of PLIF

The LIF technique allows for spatiotemporally-resolved, non-intrusive measurements of scalar fields such as temperature and species concentration by quantitatively interpreting the light emitted by a tracer molecule. In greater detail, upon excitation by a light/laser source, tracer molecules gain energy which is sometimes dissipated by photon emission. These photons are collected by a camera and the resulting signal is quantified according to its dependencies on any relevant flow parameters. The quantification of these dependences necessitates an in-depth understanding of the physics underlying the excitation and deexcitation of the employed molecular tracer and a calibration.

The total collected LIF (S_{LIF}) signal can be described by the following relationship,

$$S_{LIF} = \eta_{opt}\left(\frac{E_{pulse}\,\lambda}{Ahc}\right) V_{pix} v_{tracer} \sigma_{abs} \phi_f \qquad (1)$$

Where η_{opt} stands for the efficiency of the collection optics and incorporates factors such as the solid angle of the detector lens and the spectral responsivity of the detection system. E_{Pulse} is local laser fluence, h is plank constant, c is speed of light in vacuum, λ is wavelength and A is laser sheet area. Altogether the bracketed term represents the number of excitation photons per laser-sheet cross-sectional area. The product of the size of the imaged volume (V_{pix}) along the laser sheet propagation times the tracer number density (v_{tracer}) gives the number of tracer molecules available for excitation. The last two terms, the absorption cross-section (σ_{abs}) and the fluorescence quantum yield (φ_f) account for the photophysical dependencies of the fluorescence signal and represent the probability of absorption and the efficiency of fluorescence emission respectively.

The two-colour detection approach (2-colour LIF) requires the illumination of the seeded volume and the subsequent detection of the emitted fluorescence by two detectors simultaneously (each looking at a different part of the emission spectrum). The resulting signal ratio (here abbreviated as $S_{\lambda 1}/S_{\lambda 2}$) depends only on the absorption cross-section and the fluorescence quantum yield of the tracer at the respective wavelength ranges(Rothamer et al., 2009), as dependencies such as the pulse-to-pulse energy variation and tracer density cancel out, while others such as the efficiency of the detection optics are incorporated in a constant (C).

$$\frac{S_{\lambda 1}}{S_{\lambda 2}} = C\,\frac{\sigma_{abs}\varphi_{f_{\lambda 1}}}{\sigma_{abs}\varphi_{\lambda 2}} = f\left(T, v_{o_2}\right) \qquad (2)$$

In this experimental method, the fluid flow area of interest is illuminated by a laser sheet. The resulting fluorescence light from excited molecules in the light sheet is imaged through a selective filter onto a time-gated digital camera. For pulsed UV LIF applications usually, an image intensifier amplifies the LIF signal. The conversion of LIF images into a meaningful concentration or temperature fields is based on calibration measurements.

2.2. Optical roots blower

In this study, Roots blower URAI-22 from Howden is used. Optical access is required on the Roots blower to visualize the temperature field in the clearances. To get optical access of radial leakage path some portion (Area bounded by the yellow line in Figure 2) of the Roots blower is replaced by optical material. For the current application, optical material should be able to withstand high temperature and pressure up to 300 °C and 7 barg respectively. Besides, specifically for PLIF technique, the optical element should be transparent for UV light.

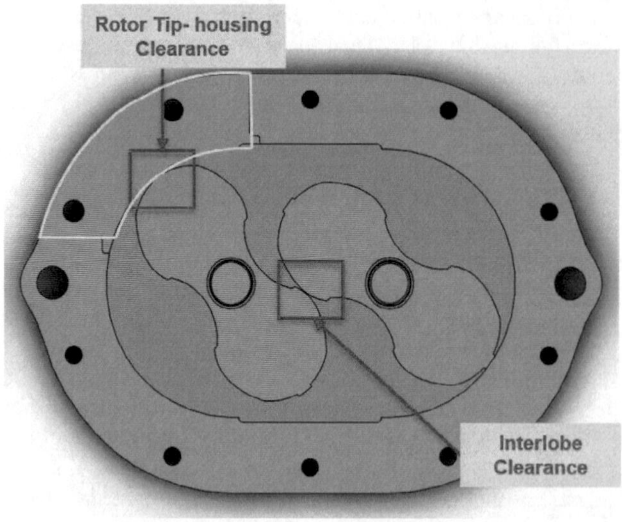

Figure 2 Clearances in Roots blower

For optical access, fused silica glass is the best suitable material for our application. Figure 3 (a) shows a complex shape of the optical Glass manufactured from the Fused silica. As shown in Figure 3 (b) & (c), Optical access from radial direction and side of the Roots blower is provided to visualize a flow in clearance gap between rotor tip and housing. A thin gasket is used between metal and Glass surface to avoid any direct contact to reduce risk of cracking in the Glass at higher working temperature. Glass is kept tight using an external metal plate to eliminate leakage through glass and metal mating surfaces.

Figure 3 (a) Optical element from Fused silica Glass (b) Radial optical access of Roots blower (c) optical access from side of Roots blower

2.3. PLIF test setup

The layout of the Roots blower test rig is depicted in Figure 4. Roots blower is connected with variable speed electric motor through pulley transmission system to run at various operating condition, and Roots blower can run up to 2700 RPM based on current pulley ratio. To monitor and control the operating parameter of the machine, pressure and temperature sensors are installed at suction (P1, T1), discharge (P2, T2) and Orifice (ΔP, T3). Shaft encoder and torque meter are used to measure the speed and power of roots blower. The flow of the machine measures using Orifice. The ball valve is installed in the discharge line of the machine require discharge pressure. By partially closing the valve, we can increase the discharge pressure of the machine. All sensors are connected with National instrument-based data acquisition system, and real-time data are recorded using LabView based programming.

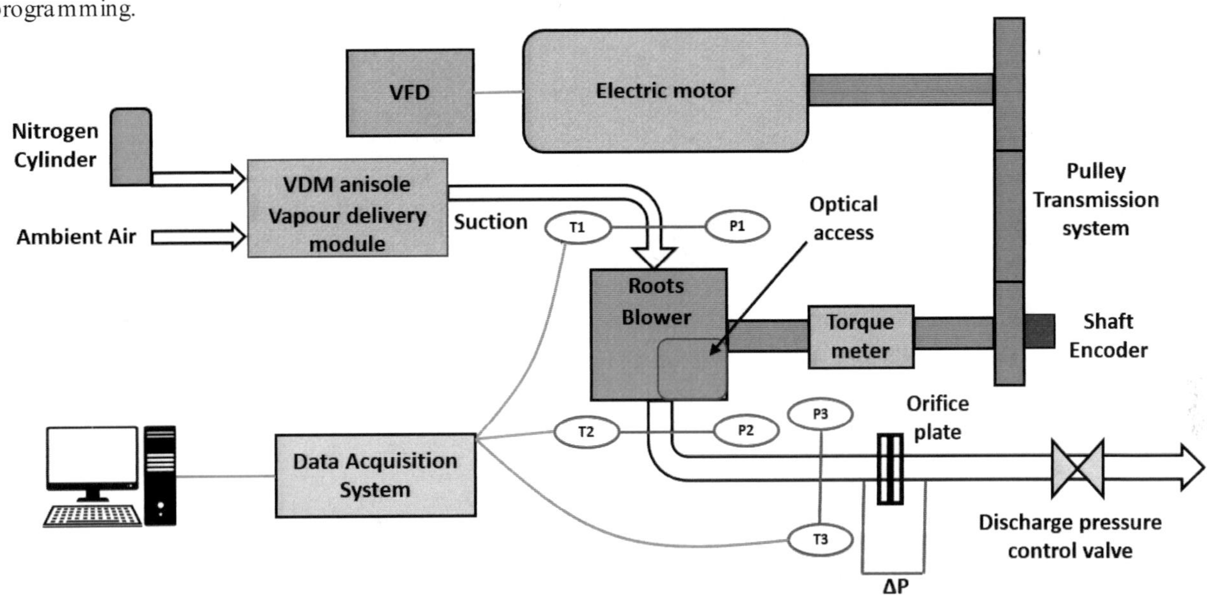

Figure 4 Layout of the Roots blower test rig

Detailed arrangement of optical access of Roots blower with laser is shown in Figure 5. The beam from a Q-smart QSM-850 Nd: YAG laser at 1064 nm is guided through 2ω and 4ω module to convert it into 266nm. A combination of optical sheet and – 50mm cylindrical lens then form a laser sheet. The sheet had a thickness of about 1mm, and it was placed at the central plane of lobes of roots blower passing through curved fused silica window. The average laser shot energy in the observation area was 100 mJ. Owing to spatial constraints, the emitted fluorescence light was rotated by 90°using a UV mirror and guided to the detection optics. Two filters, 320nm with 40nm bandwidth and 280nm with 20nm bandwidth are used. Filters with 320nm and 280nm are determined based on the fluorescence spectra obtain by Faust(Faust et al., 2014). It is necessary to change the filters alternatively at each measuring condition to capture the images with a single camera. One CMOS camera with LaVision intensified relay optics (IRO) with UV camera lenses (LaVision, f = 100 mm, f/2.8) image the field of view. Phase-locked at a particular crank angle is achieved by LaVision external triggering device with custom signal modulation. At each crack angle, 200 consecutive images are captured.

Following the two-colour strategy for measuring the temperature, the anisole fluorescence is seeded using a Bronkhorst anisole seeder. Here we choose anisole as a seeding fluorescence particle because it yields the best signal per volume in both Nitrogen and air(Faust et al., 2014). The carrier gas was Nitrogen, and the flowrate of anisole and nitrogen mixture could be adjusted using a mass flow controller of VDM vapour delivery module. For most experiments, this was kept about 5 LPM. Different flow rates are supplied at the beginning of experiment and intensity values in the captured images were checked to get the sufficient optimum flow rate value of anisole. The output of the seeder was introduced at the suction hose of the Roots blower.

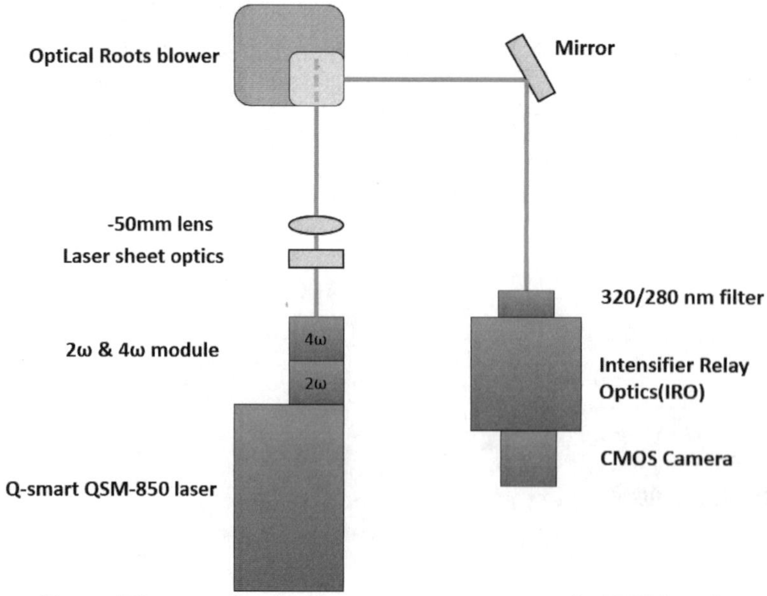

Figure 5 Layout of optical instrument arrangement for PLIF imaging

The final arrangement of PLIF optics with Roots blower test rig is shown in Figure 6, including laser, laser arm, camera, UV optics, roots blower test rig, Seeder and two sperate DAQ system. One DAQ is used to operate roots blower test rig, and another is used to operate the imaging process.

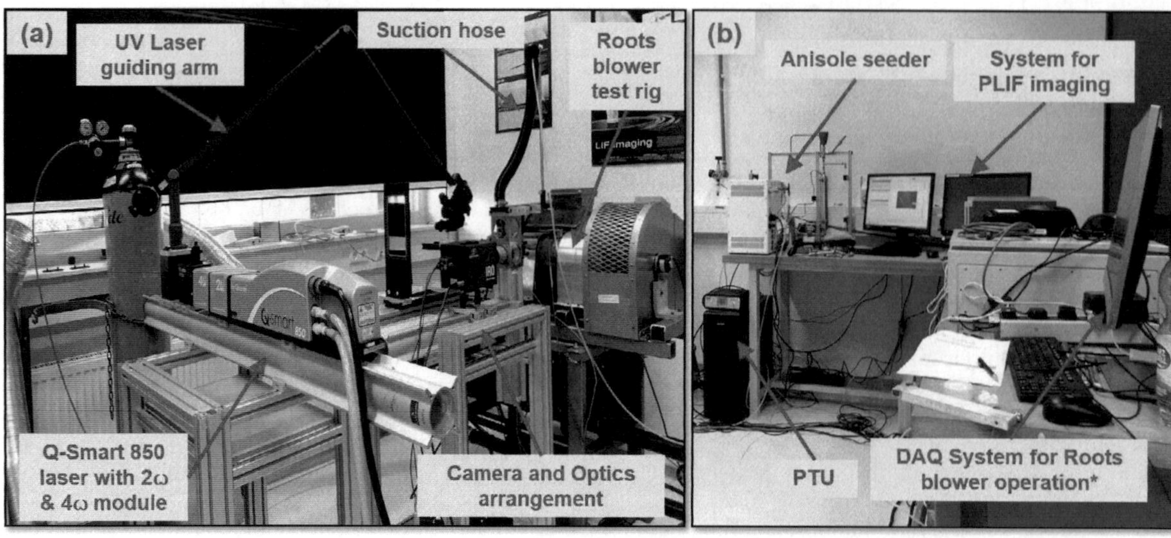

Figure 6 (a) Arrangement of PLIF optics (b) DAQ system for blower operation and PLIF imaging

3. RESULTS AND DISCUSSION

Using the experimental setup as described in the previous section, LIF images are recorded in the running condition of the machine. For particular crank angle (here 50°) of the rotor, total 200 consecutive images are recorded using 320nm and 280nm filters. The average intensity of illuminated fluorescence particles in the clearance at running speed of 300RPM and 22° C discharge temperature is shown in Figure 7. Location of clearance gap at the tip of the rotor is highlighted in Figure 7. Flow field above clearance is on the suction side while flow field below clearance is on the discharge side.

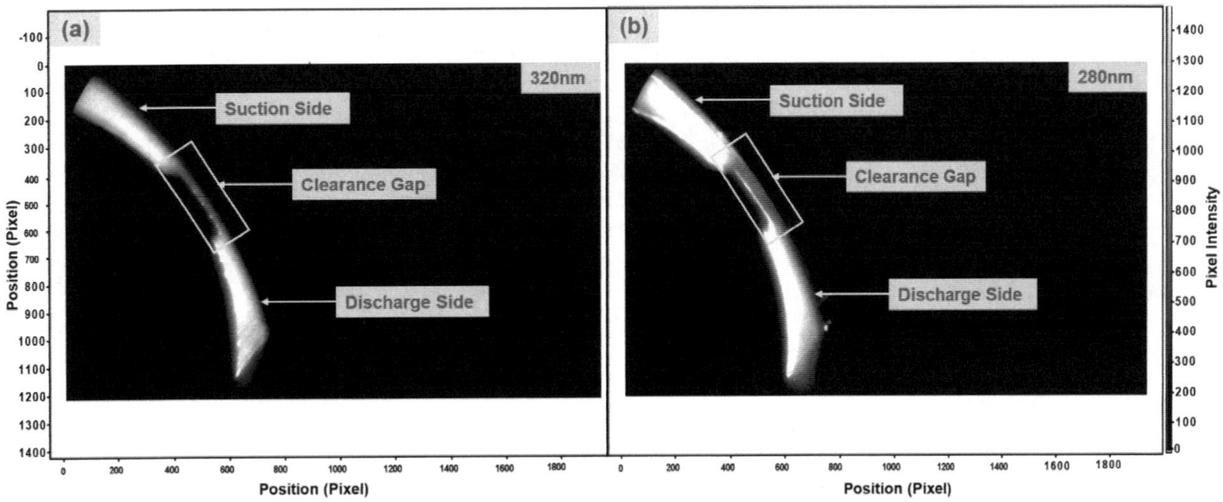

Figure 7 (a) Image captured by 320nm filter at 300 RPM (b) Image captured by 280nm filter at 300 RPM

In order to obtain the temperature from these recorded images, according to equation (2) ratio of images with 320nm and 280nm is needed, Image ratio is presented in Figure 8. Pixel values of image ratio represent the ratio of intensity which is the function of temperature. The stringent calibration process is needed to convert pixel values into temperature. Calibration is not a part of the present study, however it is assumed that measured discharge temperature by RTD is equal to temperature in discharge pocket (Discharge end in Figure 7and Figure 8) to compare the pixel intensity and determine acceptable level of intensity value. Image ratio presents reasonable intensity inside the clearance gap, which shows the capability of developed setup to get the temperature in the clearance gap of the rotary machine in running condition. However, some portion of reflection is observed on the suction side (Figure 8). Besides, few high-intensity spots are observed on the surface of the rotor; those are the noise in the captured images. Probable causes of these spots are black paint of the rotor surface and high energy of the laser. To check the quality of images, tests were also run at higher speed of 600, 900,1200 and 1500. Images found to be good except some reflection on suction side.

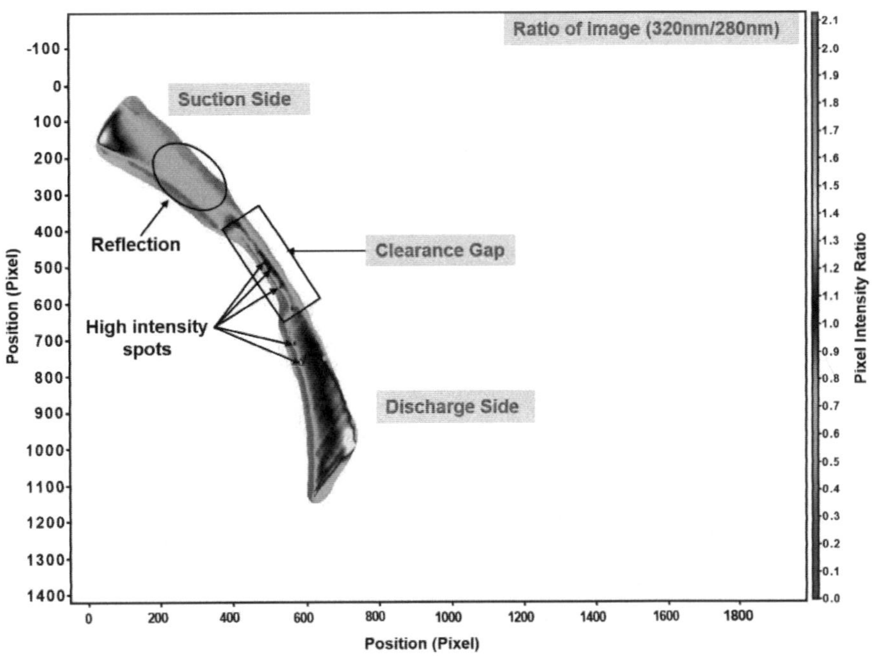

Figure 8 Image ratio (320nm/280nm) at 300 RPM

4. CONCLUSIONS

In summary, this study focuses on the application of PLIF to get temperature field in the Clearance gap of rotary machines by using a developed experimental setup. Images recorded with two different filters and image ratios obtained from those images represent an acceptable range of pixel values to produce a temperature field. Minor flaws are observed, like reflection and high-intensity spots, which can create little difficulty in getting the accurate temperature in the suction area and on the surface of the rotor. Still, these flaws can be addressed in future work. Therefore, a developed setup using the LIF technique is able to produce a reasonable temperature field in the clearance gap. This setup is helpful to reveal the physics of heat transfer in the clearance gap of the oil-free positive displacement machines.

NOMENCLATURE

A	Laser sheet area	(m^2)
C	Constant	(–)
c	Speed of light	(m/s)
E	Laser energy	(mJ)
h	Plank constant	(Js)
P	Pressure	(Bara)
S	Signal	(–)
T	Temperature	(k)
V	Imaged Volume	(m^3)
υ	Number of pixels per unit area (Tracer density)	(–)
η	Efficiency	(–)
σ	Cross section of area under investigation	(–)
φ	Quantum yield	(–)
λ	wavelength	(nm)

Subscript

LIF	Laser Induced fluorescence
opt	Optics
pulse	Laser pulse
pix	Pixel of image
abs	Absorption
f	Fluorescence

REFERENCES

Charogiannis, A., & Markides, C. N. (2019). Spatiotemporally resolved heat transfer measurements in falling liquid-films by simultaneous application of planar laser-induced fluorescence (PLIF), particle tracking velocimetry (PTV) and infrared (IR) thermography. *Experimental Thermal and Fluid Science*, *107*, 169–191. https://doi.org/10.1016/j.expthermflusci.2018.11.001

Charogiannis, A., Sik An, J., Voulgaropoulos, V., & Markides, C. N. (2019). Structured planar laser-induced fluorescence (S-PLIF) for the accurate identification of interfaces in multiphase flows. *International Journal of Multiphase Flow*, *118*, 193–204. https://doi.org/10.1016/j.ijmultiphaseflow.2019.06.002

Coull, J. D., & Atkins, N. R. (2015). *The Influence of Boundary Conditions on Tip Leakage Flow*. *137*(June), 1–10. https://doi.org/10.1115/1.4028796

Faust, S., Goschütz, M., Kaiser, S. A., Dreier, T., & Schulz, C. (2014). A comparison of selected organic tracers for quantitative scalar imaging in the gas phase via laser-induced fluorescence. *Applied Physics B: Lasers and Optics*, *117*(1), 183–194.

https://doi.org/10.1007/s00340-014-5818-x

Kauder, P. K., & Stratmann, D. D. (2002). *Theoretical Gas Flow through Gaps in Screw-type Machines*.

Kranz, P., Fuhrmann, D., Goschütz, M., Kaiser, S., Bauke, S., Golibrzuch, K., Wackerbarth, H., Kawelke, P., Luciani, J., Beckmann, L., Zachow, J., Schuette, M., Thiele, O., & Berg, T. (2018). In-Cylinder LIF Imaging, IR-Absorption Point Measurements, and a CFD Simulation to Evaluate Mixture Formation in a CNG-Fueled Engine. *SAE International Journal of Engines*, *11*(6), 1221–1238. https://doi.org/10.4271/2018-01-0633

Richardson, D. R., Roy, S., & Gord, J. R. (2017). Femtosecond, two-photon, planar laser-induced fluorescence of carbon monoxide in flames. *Optics Letters*, *42*(4), 875. https://doi.org/10.1364/ol.42.000875

Rothamer, D. A., Snyder, J. A., Hanson, R. K., & Steeper, R. R. (2009). Two-wavelength PLIF diagnostic for temperature and composition. *SAE International Journal of Fuels and Lubricants*, *1*(1), 520–533. https://doi.org/10.4271/2008-01-1067

Sachs, R. (2002). *Experimental investigation of Gas flows in screw machines*.

Vittorini, D., Bianchi, G., & Cipollone, R. (2015). Energy saving potential in existing volumetric rotary compressors. *Energy Procedia*, *81*, 1121–1130. https://doi.org/10.1016/j.egypro.2015.12.137

Zhang, Q., O'Dowd, D. O., He, L., Wheeler, A. P. S., Ligrani, P. M., & Cheong, B. C. Y. (2011). Overtip shock wave structure and its impact on turbine blade tip heat transfer. *Journal of Turbomachinery*, *133*(4), 1–8. https://doi.org/10.1115/1.4002949

A Generalized Approach for Automated Compressor Performance Mapping Using Artificial Neural Network

Jie MA[1], Xin DING[1], W. Travis HORTON[1], Davide ZIVIANI[1]

[1]Ray W. Herrick Laboratories, School of Mechanical Engineering, Purdue University

West Lafayette, 47907-2099, USA

ma319@purdue.edu; wthorton@purdue.edu; dziviani@purdue.edu

ABSTRACT

In the last decades, several research and development efforts led to new compressor technologies that have been successfully introduced into market such as hermetic compressors with variable-speed motors, compressors with economization lines (both vapor and liquid injection lines), hermetic linear compressors, novel capacity modulations, and oil-flooding among others. During the process of implementing new compressor technologies, performance mapping is essential to predict the system behavior across different operating conditions. However, current standard AHRI-540 for compressor performance rating utilizes a 10-coefficient polynomial model that has severe limitations to include compressor enhancement technologies and variable operating range. In addition, it is common practice in industry to calibrate such polynomial correlations with at least 15 to 20 compressor calorimeter test points for a single compressor to ensure a good fit, which results in extended laboratory testing time and relatively high associated costs. Therefore, an automated compressor performance mapping approach based on artificial neural network (ANN) modeling is proposed to address and overcome the limitations of the current standard including applicability to any positive displacement compressors and minimization of number of test points required to accurately predict the compressor envelope.

In this paper, the performance of a positive displacement compressor is mapped by this novel methodology, which relies on an algorithm that effectively determines the minimum set of data points required and optimizes the training/testing of ANN architecture. The accuracy and reliability of the proposed methodology are compared to the conventional 10-coefficient polynomial mapping. Lastly, the propagated uncertainties through the model and its extrapolation capabilities are also analyzed.

Keywords: compressor performance mapping, calorimeter testing, artificial neuron network

1. INTRODUCTION

Improving the performance of air-conditioning and heat pumping (ACHP) systems to address space conditioning requirements, energy savings, and environmental concerns is a particularly challenging task. Research and development efforts at both component and system levels are continuously being pursued by both industry and academia. In order to evaluate the performance of different types of compressors and predict their behaviors within systems, various mathematical models have been proposed in the literature. In industry, it is a common practice to apply current ANSI/AHRI Standard 540 (AHRI, 2015) for compressor performance rating and utilize the 10-coefficient cubic polynomial model to predict the mass flow rate and compressor power. However, to ensure a good fit, at least 15 to 20 compressor calorimeter test points are need for a single compressor and that requires extended laboratory testing time. In academic, empirical and semi-empirical models including plenty physical information have been widely used to map compressor performance and extrapolate results, which is usually suitable for compressor research and design. But this physical-based model cannot be applied for various types of compressor and is relatively cumbersome for industrial application.

In order to overcome these limitations and improve the current methodology for compressor performance map rating, a novel approach of an automated compressor performance mapping approach based on ANN modeling has been developed and proposed by Ma et al., (2020) . They applied this methodology on three different types of compressors and compared the accuracy of ANN model to 10-coefficiency polynomial model. Ziviani et al., (2019) implemented a multi-input multi-out ANN model that can achieve higher accuracy with respect to a semi-empirical model to predict the performance of a single-phase and two-phase injected scroll compressor. In these works, it illustrates that the ANN model has significant potential and benefits on compressor performance mapping and rating.

In this paper, it is demonstrated the implementation of automated compressor performance mapping methodology based on ANN modeling with various set of training data points required and ANN architecture for a hermetic dual-cylinder rolling-piston compressor. Furthermore, the accuracy and reliability of the proposed methodology are compared to the conventional 10-coefficient polynomial mapping, and the propagated uncertainties through the model and its extrapolation capabilities are also analyzed.

2. METHODOLOGY

The overall automated compressor mapping methodology is illustrated in **Error! Reference source not found.**, which is proposed by Ma et al., (2020). It is assumed that the compressor envelope has been already identified first. Hence, the compressor colorimeter (or hot gas bypass) testing data within the envelope can be used to train and test the ANN model. This simplification is necessary in order to develop and test the proposed mapping algorithm. Since the number and quality of training data impact the accuracy of prediction dramatically, the determination the training and validation dataset is critical for model regression. With respect to selecting the initial training samples, Aute et al., (2015) suggested that initial training samples can be chosen to be the vertices of a compressor envelope polygon, or the points as close to envelope boundary as possible to capture the compressor operating envelope.

In ANN model, the inputs include the ambient temperature (T_{amb}), suction pressure (P_{suc}), suction temperature (T_{suc}), and discharge pressure (P_{dis}). Whereas, the outputs from the model are the discharge temperature (T_{dis}), mass flow rate of refrigerant ($\dot{m}_r$), and compressor power consumption ($\dot{W}_{comp}$). In additions, four or five data points are selected on the compressor envelope boundary as the initial training samples, which include the lowest and highest evaporating and condensing temperatures, respectively. These initial points also present the minimum and maximum values of the output variables. After creating a training dataset with minimum number of training samples, an ANN model with single-input-layer, single-hidden-layer and single-output-layer structure is trained to predict compressor performance through iterating the weights and biases values. Additionally, the remaining data samples are used for validation. The method of determining the initial training data and ANN model structure is based on the work presented by Ma et al., (2020).

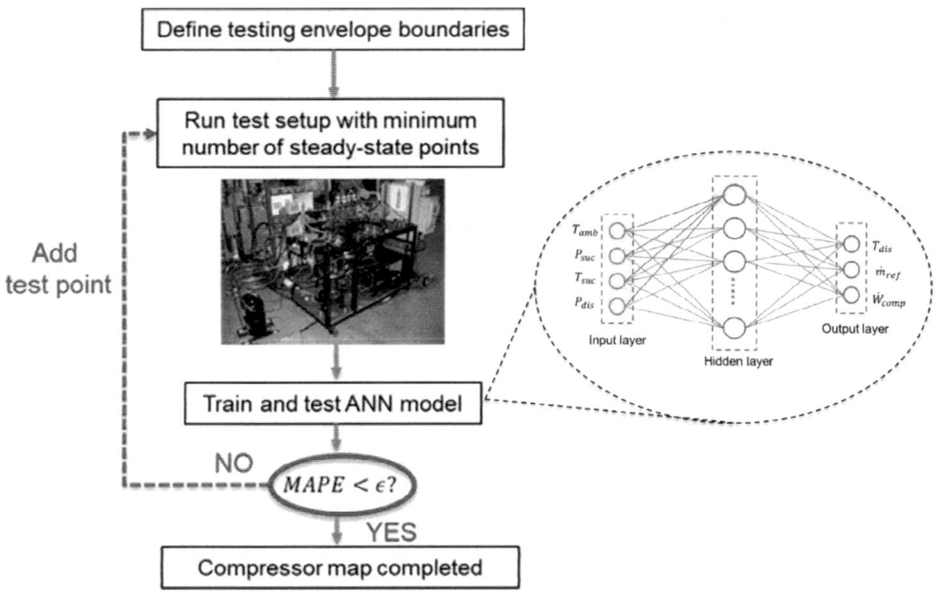

Figure 1: Conceptual schematic of the overall mapping methodology

To evaluate the accuracy of the trained ANN model, three well-known statistical quantities have been used introduced. The R^2 value is used to measure how statistically close the data are to the fitted regression line and is defined as:

$$R^2 = \frac{cov(y^{actual}, y^{predict})}{\sqrt{cov(y^{actual}, y^{actual})cov(y^{predict}, y^{predict})}} \quad (1)$$

where $cov(\cdot)$ denotes the covariance, y^{actual} is the actual experimental data, and $y^{predict}$ is the ANN model predicted data. Since the R^2 cannot determine whether the coefficient estimated and the predictions are biased, additional indicators are introduced. The mean absolute percentage error (MAPE) in Equation (2) is selected to statistically measure how accurate the ANN model.

$$MAPE = \frac{100\%}{n} \sum \left| \frac{y_i^{actual} - y_i^{predict}}{y_i^{actual}} \right| \quad (2)$$

3. COMPRESSOR MAPPING MODELS

3.1 AHRI 10-coefficient mapping and ANN models

The proposed compressor mapping approach is applied to a hermetic dual-cylinder rolling-piston compressor with R410A as the working fluid. A total of 43 steady-points were collected by colorimeter testing to train and validate the ANN model. The accuracy of the trained and validated ANN model is compared to the AHRI 10-coefficient mapping, where the 10 coefficients are trained based on Equation (3) and listed in the work done by Ma et al., (2020). Regards to the 10-coefficient polynomial regression, 11 data points, i.e. least number of regression data points for least-square, have been selected wisely based on their locations within the compressor envelope. It is proposed by Ma et al., (2020) that the selection of the points, shown in Figure 2 started by choosing points on the boundary or vertices of the sketched compressor envelope, and then the other points are selected such that all the 11 points are evenly spread across the whole data set.

$$\begin{aligned} \dot{m}_{map}[\text{kg/hr}] &= M_1 + M_2 \cdot T_e + M_3 \cdot T_c + M_4 \cdot T_e^2 + M_5 \cdot (T_e \cdot T_c) + M_6 \cdot T_c^2 + M_7 \cdot T_e^3 + M_8 \\ &\quad \cdot (T_e^2 \cdot T_c) + M_9 \cdot (T_e \cdot T_c^2) + M_{10} \cdot T_c^3 \\ \dot{W}_{map}[W] &= P_1 + P_2 \cdot T_e + P_3 \cdot T_c + P_4 \cdot T_s^2 + P_5 \cdot (T_s \cdot T_d) + P_6 \cdot T_d^2 + P_7 \cdot T_s^3 + P_8 \cdot (T_s^2 \cdot T_d) + P_9 \\ &\quad \cdot (T_s \cdot T_d^2) + P_{10} \cdot T_d^3 \end{aligned} \tag{3}$$

where T_e and T_c are evaporating and condensing temperatures in degree Celsius, respectively.

With respect to the ANN model, the architecture of the network is single-input-layer, single-hidden-layer and single-out-put layer to predict the mass flow rate and compressor power. In the first ANN model, the least number of training samples, i.e. five, are selected to train the network with 6 neurons in the single hidden layer. In the other ANN model, the training samples increase to 10 samples and another data point is chosen as testing point. All these 11 samples are the same data samples for 10-coefficiency polynomial regression. In addition, the neurons in the hidden layer are also increased to 12 in the second ANN model in order to match the expanded training data set. The connection between neurons of each layer in ANN models can be expressed as Equation (4).

$$y_k = \sum_{j=1}^{N_{neural}} (\omega_{kj}^{(2)} \varphi\left(\sum_{i=1}^{N_{input}} \omega_{ji}^{(1)} \cdot x_i + b_j^{(1)} \right) + b_k^{(2)}) \tag{4}$$

where x and y are the neural nods in input and output layers, respectively. $\omega_{ji}^{(1)}$ and $b_j^{(1)}$ are the weights and bias propagating the *i-th* input node to *j-th* neuron node in the hidden layer; and $\omega_{kj}^{(2)}$ and $b_k^{(2)}$ are the weights and bias propagating the *j-th* neuron node in hidden layer to *k-th* output node. A non-linear activation function $\varphi(\cdot)$, the hyperbolic tangent function is selected in the current work as activation function. Since the input and output variables are of different types and have different orders of magnitude, all inputs and outputs parameters are normalized by equation (Zendehboudi et al., 2017).

3.2 Uncertainty Analysis

Uncertainty of the compressor map output quantifies the potential difference between the true values of output to model predicted values of output. It starts from uncertainty of measurement values in both the model inputs and training data and propagates in the mathematical structure of compressor map or compressor ANN model. In current work, it is assumed that 10-coefficient mapping and ANN model

Based on uncertainty calculation method described in Cheung et al., (2018), the uncertainty of mass flow rate and compressor power due to inputs in 10-coefficient mapping are expressed in following equations.

$$\Delta\widehat{m}_r = \sqrt{\left(\frac{\partial \widehat{m}_r}{\partial T_e}\right)^2 \cdot \Delta T_e^2 + \left(\frac{\partial \widehat{m}_r}{\partial T_c}\right)^2 \cdot \Delta T_c^2} \tag{5}$$

$$\Delta\widehat{W}_{comp} = \sqrt{\left(\frac{\partial \widehat{W}_{comp}}{\partial T_e}\right)^2 \cdot \Delta T_e^2 + \left(\frac{\partial \widehat{W}_{comp}}{\partial T_c}\right)^2 \cdot \Delta T_c^2} \tag{6}$$

where T_e and T_c are dew-point temperatures which are calculated by the measured suction and discharge pressure.

The uncertainty of dew point temperature expressed in Equation (7) is formed as the uncertainty of pressure measurement ($\Delta P_{suc,mea}$, $\Delta P_{dis,mea}$) and uncertainty due to equation of state, since the dew-point temperatures are converted from measurements of suction and discharge pressure by the equation of state of the refrigerant.

$$\Delta T_e = \sqrt{\left(\frac{\partial T_e(P_{suc,mea})}{\partial P}\right)^2 \cdot (\Delta P_{suc,mea}^2 + \Delta P_{EOS}^2)}; \quad \Delta T_c = \sqrt{\left(\frac{\partial T_c(P_{dis,mea})}{\partial P}\right)^2 \cdot (\Delta P_{dis,mea}^2 + \Delta P_{EOS}^2)} \tag{7}$$

Uncertainty of pressure measurement (ΔP_{mea}) is usually described as a percentage uncertainty (listed in Table 1) of measurement value in the specification of a pressure transducer or thermocouple. And the uncertainty of dew-point pressure due to the equation of state is expressed by:

$$\Delta P_{EOS} = r_{EOS} P_{mea} \tag{8}$$

where r_{EOS} is the uncertainty associated with the equations of state which depends on the type of refrigerant. It is 0.5% of the dew point pressure for R410A.

Table 1: Sensor specifications

Sensor	Model	Range	Uncertainty
Thermocouple	Omega TMQSS-125T-6	0-350 C	0.75%
Pressure Transducer	Honeywell PX2AF1XX500PAAAX	0-500 psia/0-1000 psia	0.25%

According to JCGM GUM 100-2008 (JCGM, 2008), the combine uncertainty of outputs is obtained by appropriately combing the standard uncertainty of the input estimates. In ANN models, the uncertainty due to inputs can be calculated as follows:

$$\Delta \hat{m}_r = \sqrt{\begin{aligned}&\left(\frac{\partial \hat{m}_{comp}}{\partial T_{amb}}\right)^2 \cdot \Delta T_{amb,mea}^2 + \left(\frac{\partial \hat{m}_{comp}}{\partial P_{suc}}\right)^2 \cdot \Delta P_{suc,mea}^2 \\ &+ \left(\frac{\partial \hat{m}_{comp}}{\partial T_{suc}}\right)^2 \cdot \Delta T_{suc,mea}^2 + \left(\frac{\partial \hat{m}_{comp}}{\partial P_{dis}}\right)^2 \cdot \Delta P_{dis,mea}^2\end{aligned}} \tag{9}$$

$$\Delta \hat{W}_{comp} = \sqrt{\begin{aligned}&\left(\frac{\partial \hat{W}_{comp}}{\partial T_{amb}}\right)^2 \cdot \Delta T_{amb,mea}^2 + \left(\frac{\partial \hat{W}_{comp}}{\partial P_{suc}}\right)^2 \cdot \Delta P_{suc,mea}^2 \\ &+ \left(\frac{\partial \hat{W}_{comp}}{\partial T_{suc}}\right)^2 \cdot \Delta T_{suc,mea}^2 + \left(\frac{\partial \hat{W}_{comp}}{\partial P_{dis}}\right)^2 \cdot \Delta P_{dis,mea}^2\end{aligned}} \tag{10}$$

where the subscript *mea* represents a percentage uncertainty of measurement value. The partial derivative of an output with respect to an input shown in Equation (11) is calculated based on the neural network mathematical expression in Equation (4) with considering the nominalization process. In Equation (*11*), the capital X and Y represent the real measurement values of input and output, respectively; the small x and y represent the neural network inputs and outputs nominalized from measurement values.

$$\left(\frac{\partial\, Y(k)}{\partial\, X(i)}\right)^2 = \left(\sum_{j=1}^{N_{neural}} \left(\omega_{kj}^{(2)} \cdot \omega_{ji}^{(1)} \cdot \frac{\partial \varphi}{\varphi x(i)}\right) \cdot \frac{\mathrm{Y_{max}}(k) - \mathrm{Y_{min}}(k)}{\mathrm{X_{max}}(i) - \mathrm{X_{min}}(i)}\right)^2 \tag{11}$$

In ANN model, the input variables (temperature and pressure) are measured directly without converting by equation of state, the uncertainty due to state conversion are not be considered.

The other uncertainty components, such as uncertainty due to training data and uncertainty due to model random error, are not considered in this paper, since the model structure and model training methods are different between linear 10-coefficient mapping and non-linear ANN model.

2.3 Extrapolation Analysis

As aforementioned, the training samples for all three model are selected on the boundary or vertices of the sketched compressor envelope. However, the compressor envelope may not be rigorously identified in some cases that the model is necessary to predict the compressor performance beyond the range of training data set.

Thus, extrapolation analysis is conducted here to investigate the issue of compressor map performance outside the range of the training data set. Different from the previous case, the new training points in Figure 2 are selected inside the compressor envelope to avoid choosing data points located on the boundary or vertices of the sketched compressor envelope. The data points outside the training data set are validation data points for extrapolation capabilities analysis.

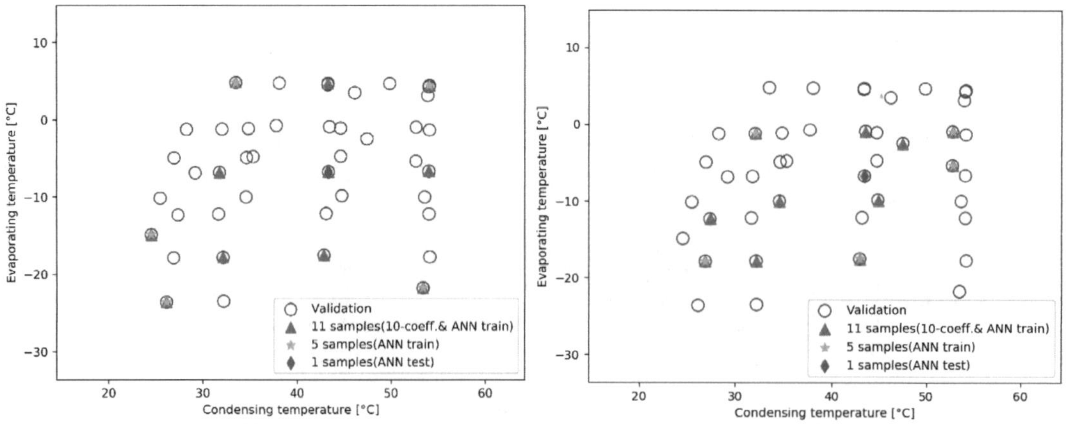

(a) (b)

Figure 2: Compressor map with training data: (a) training data selected on compressor envelope (b) training data selected within compressor envelope

4. RESULTS

To investigate the accuracy of the ANN model, a set of parity plots of are utilized to compare the 10-coefficient polynomial model and the ANN models with variable number of training samples. It is evident from Figure 3 and Figure 4 that 10-coefficient polynomial and ANN models can predict the compressor performance reasonably well. By looking at the parity plots in Figure 3, the mass flow rate is predicted by the 10-coefficient polynomial model with a $MAPE$ of 0.83% and R^2 of 99.85%, and power ouput reported a $MAPE$ of 0.26% and R^2 of 99.96%. By comparing the two ANN models, the one trained with 5 experimental samples presents a slightly larger $MAPE$ value of 2.85% and lower R^2 of 99.07%. However, the ANN model trained with 10 samples shows as high accuracy as the 10-coefficient polynomial model with a $MAPE$ of 0.75% and R^2 of 99.88%. With respect to compressor power prediction in Figure 4, both 10-coefficient polynomial and ANN model trained by 10 samples presents a good regression that nearly all data points fall on the best-fitting line with only 0.25% and 0.34% of $MAPE$ respectively. Only the ANN model trained by 5 samples report a slightly large $MAPE$ of 1.42%.

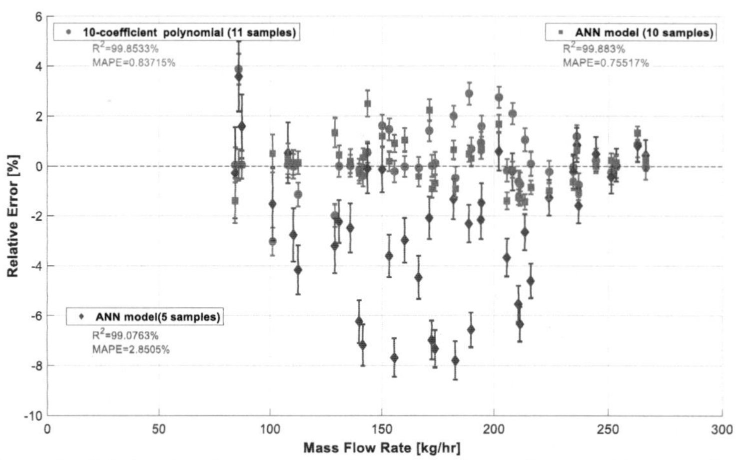

Figure 3: Comparison of validating data between models and experiment for mass flow rate (Models trained by the samples on the boundary of the compressor envelope)

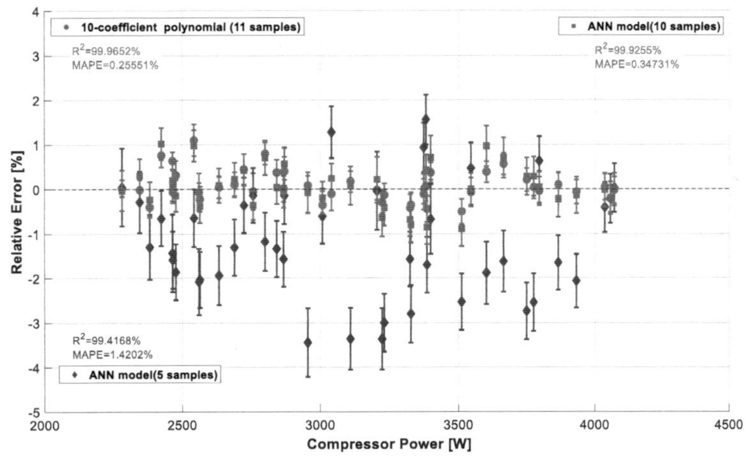

Figure 4: Comparison of validating data between models and experiment for compressor power (Models trained by the samples on the boundary of the compressor envelope)

Furthermore, the uncertainties of mass flow rate and compressor power for each data sample are represented as error bars in Figure 3 and Figure 4. In Figure 3, the ANN model trained by 5 samples has longer error bars in comparison to that in the 10-coefficient map and in ANN model trained by 10 samples, which means that the uncertainties of the ANN model trained by 5 samples are the largest among the three models in the prediction of mass flow rate. Similar results are represented in Figure 4 for compressor power prediction. It illustrates that the ANN model trained by 5 samples has larger model prediction errors and uncertainties compared with the other two models.

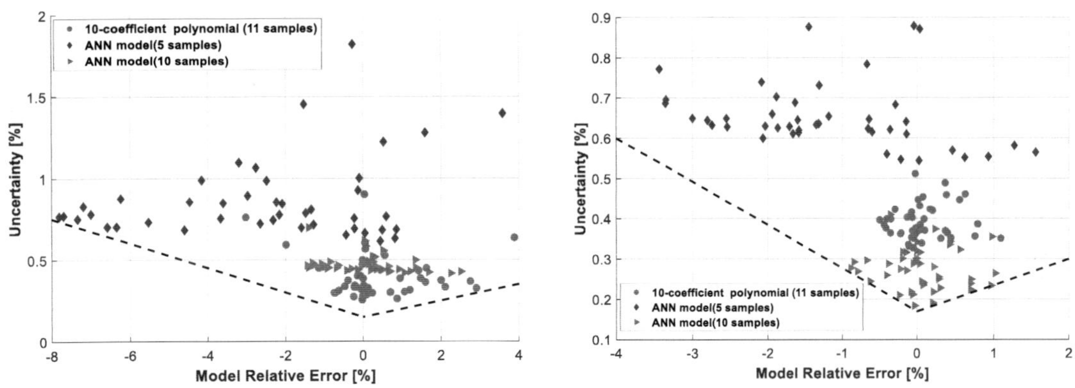

(a) Mass Flow Rate (b) Compressor Power

Figure 5: Uncertainty errors with model accuracy of all data samples (Models are trained by the samples on the boundary of the compressor envelope)

Figure 5 is plotted to investigate how uncertainty changes with model accuracy. In general, it shows that the larger the model relative error is, the higher uncertainty the model has, which means uncertainties increase with a decrease of model accuracy. In Figure 5, it is clearer to see that the ANN model trained by 5 samples represents highest uncertainty and lowest model accuracy for both mass flow rate and compressor power. In comparation ANN model with 10-coefficiency map, the ANN model with 10 training samples has higher uncertainties but lower model relative errors in the prediction of mass flow rate; but has lower uncertainties and larger model relative errors in the prediction of compressor power. In 10-coefficiency map and two ANN models, since there are several data points have lower relative error but higher uncertainties, the definite relationship between model accuracy and uncertainty cannot be identified in the current work.

Figure 6 and Figure 7 represent the model accuracy of 10-coefficiency map and two ANN models with training data samples locating inside the sketched compressor envelope. In comparation Figure 6 to Figure 3 for mass flow rate prediction, the R^2s for three models, which are 10-coefficiency polynomial, ANN model trained by 5 samples and ANN model trained by 10 samples, reduce to 98.58%, 99.56% and 98.97% respectively; the *MAPE*s increase to 2.53%, 3.73% and 3.02% respectively. In other words, all three models are not as accurate as before when the predicting data points are selected beyond the training data set. This trend also happens on compressor power prediction. In Figure 7, the R^2s slightly decrease to 99.63%, 99.26% and 97.06%, the *MAPE*s clime up to 0.63%, 1.19% and 2.44% respectively for 10-coefficient map, ANN model trained by 5 samples and ANN model trained by 10 samples.

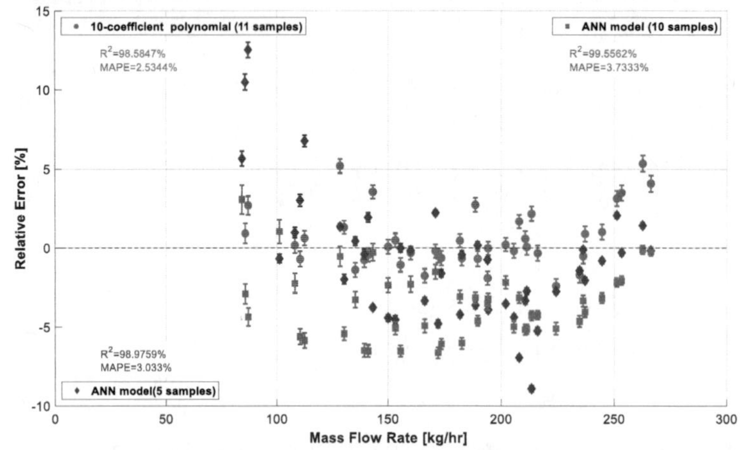

Figure 6: Comparison of validating data between models and experiment for mass flow rate (Models trained by samples inside compressor envelope)

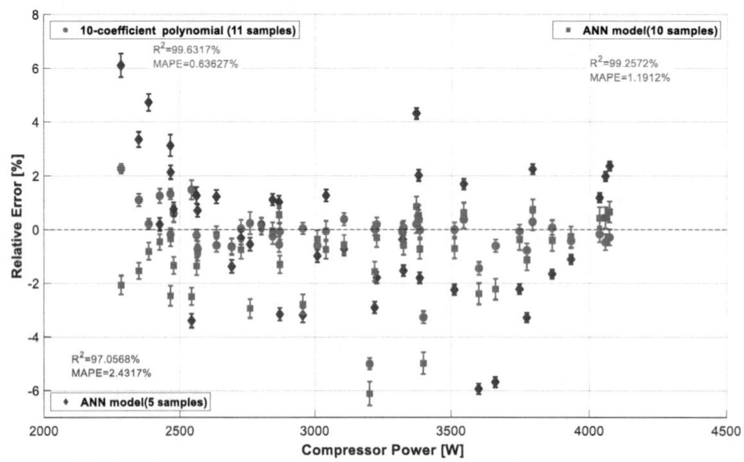

Figure 7: Comparison of validating data between models and experiment for compressor power (Models are trained by samples inside compressor envelope)

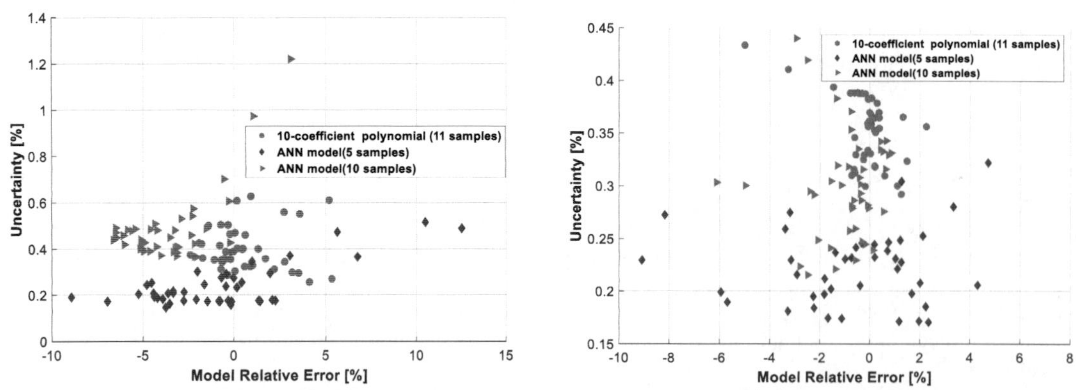

(a) Mass Flow Rate (b) Compressor Power

Figure 8: Uncertainty errors with model accuracy of all data samples (Models are trained by samples inside compressor envelope)

Figure 8 shows the relationship between the model relative error and uncertainty for 10-coefficiency map and two ANN models when the range of training samples is much smaller than compressor envelope. Comparing to Figure 5, the uncertainties of ANN model with 5 training samples are much lower than before. In contrast to ANN model with 5 training samples, the uncertainties of 10-coefficient polynomial and ANN model with 10 training samples are roughly equal to the uncertainties presented in Figure 5. Furthermore, in Figure 8, no obvious correlations is observed between the model predictive errors and uncertainty errors.

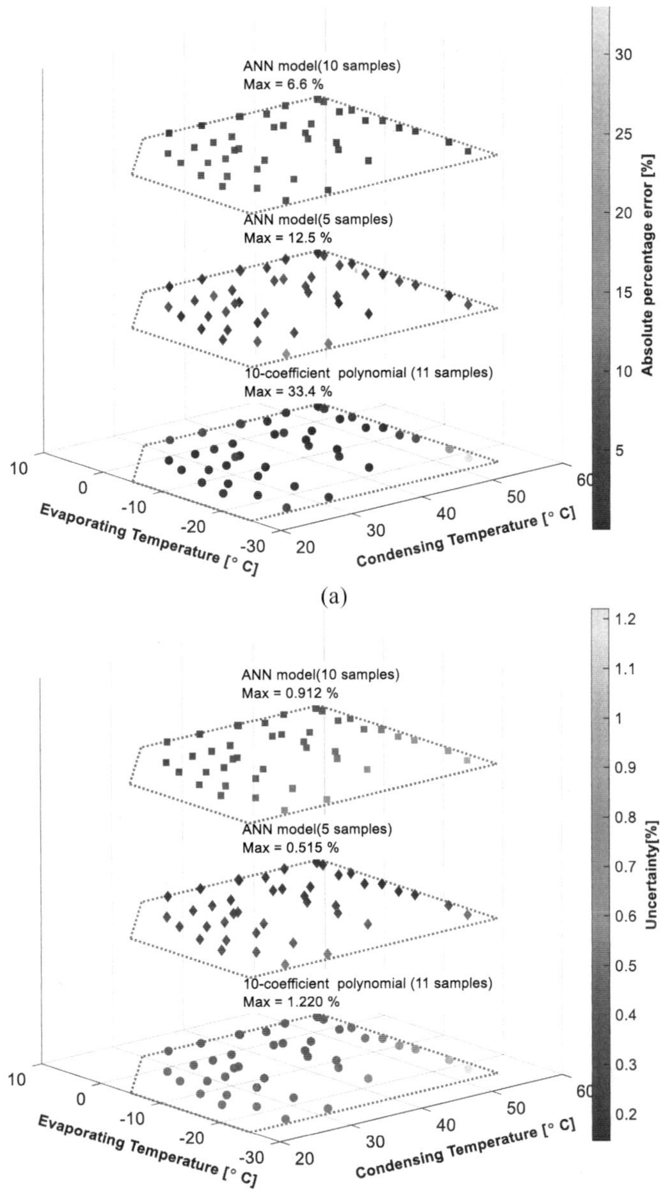

Figure 9: Absolute percentage errors and uncertainty errors of mass flow rate for all validating points of the compressor map (Models are trained by samples inside compressor envelope)

Table 2: Absolute percentage error comparison of models with different training data set

	Training samples on compressor envelope		Training samples within compressor envelope	
	Mass Flow Rate	Compressor Power	Mass Flow Rate	Compressor Power
ANN model (10 training samples)	2.5%	1.0%	6.6%	6.1%
ANN model (5 training samples)	7.8%	3.4%	12.5%	9.1%
10-Coefficient polynomial	3.9%	1.1%	33.4%	5.0%

To clearly locate the position of data samples representing larger model relative error or higher uncertainty, the plots in Figure 9 and Figure 10 represent the absolute percentage error (*APE*) and model uncertainty of mass flow rate and compressor power for each data point and their deviation distribution on compressor map when the range of model training data set is smaller than the range of validation data set. Overall, the maximum *APE*s shown in Figure 9 and Figure 10 for three models increase significantly in comparation of the corresponding results presented in Ma et al., (2020).

With respect to mass flow rate prediction, Table 2 illustrates that the maximum APE obtained by 10-coefficient map dramatically increases from 3.9% to 33.4%, where the corresponding data point shown in Figure 9(a) is on the right bottom vertex of the sketched compressor map. The data points representing maximum model prediction error is also the data point has highest uncertainty as shown in Figure 9(b). Furthermore, the maximum *APE*s for the ANN model with 10 training samples and the ANN model with 5 training samples also increase from 2.5% to 6.6%, and from 7.8% to 12.5% respectively summarized in Table 2. It also can be observed in the ANN models that the data points with higher *APE*s have higher uncertainties, where located on the vertices or boundary lines of the compressor map.

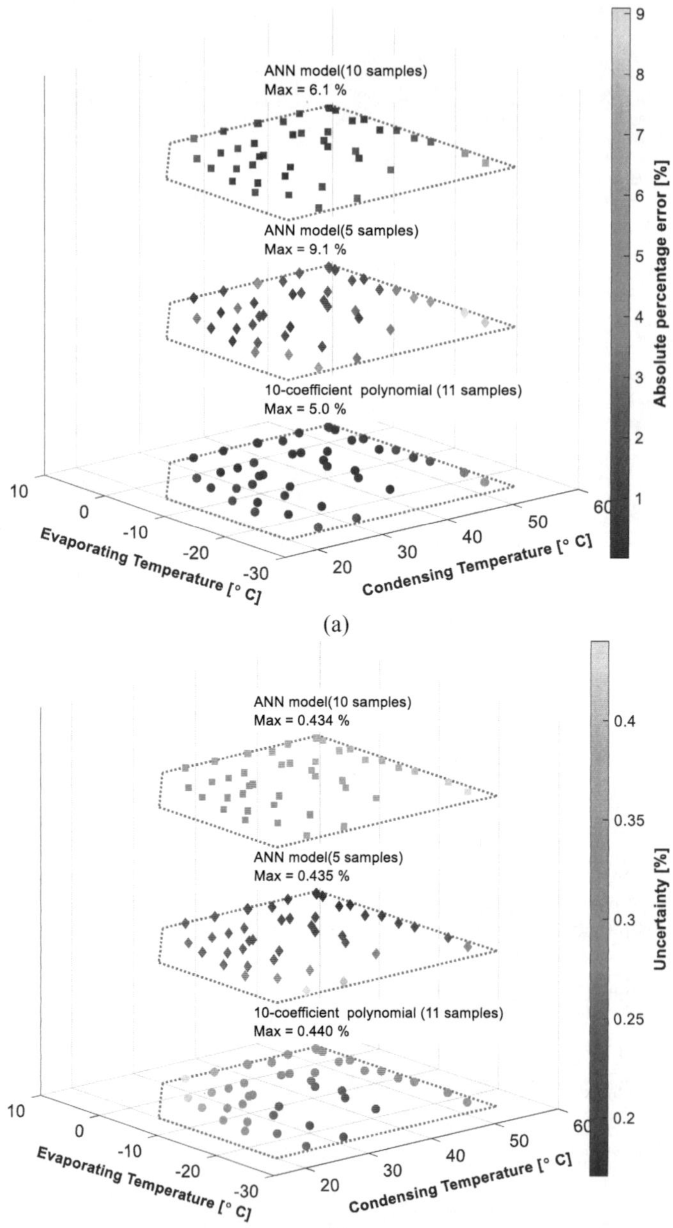

Figure 10: APEs of compressor power for all validating points of the compressor map (Models are trained by samples within compressor envelope)

To compare *APE*s in compressor power prediction summarized in Table 2, the maximum *APE*s for three models, i.e. 10-coefficiency mapping, ANN model with 5 training samples and ANN model with 10 training samples, increase from 1.1% to 5.0%, 3.4% to 9.1% and 1.0% to 6.1% respectively. It can be observed in Figure 10 that the data points showing higher *APE*s are the points having higher uncertainties, where located along the boundary lines of compressor map.

It can conclude that when the range of training data set is smaller than the compressor envelope, the model is less accurate and the uncertainty due to inputs is higher, the data points representing lower accuracy and higher uncertainty are the points beyond the range of the training data set.

5. CONCLUSIONS

In summary, the automated compressor performance mapping methodology based on ANN modeling with various training data set is implemented on a hermetic dual-cylinder rolling-piston compressor. And the accuracy and reliability of the proposed methodology are compared to the conventional 10-coefficient polynomial mapping, and the propagated uncertainties through the model and model extrapolation capabilities are also analyzed in this paper. It can conclude that the ANN model with 10 training samples has small relative error and low uncertainty when the training samples are chosen on vertex or boundary line of compressor map. With a little extrapolation, the accuracy of three models decrease and the uncertainty of ANN models also decrease.

Nomenclature

T_{amb}	Ambient temperature [°C]	P_{suc}	Suction pressure [kPa]
T_{suc}	Suction temperature [°C]	P_{dis}	Discharge pressure [kPa]
T_{dis}	Discharge temperature [°C]	$\dot{m}_r$	Mass flow rate of refrigerant [kg/hr]
$\dot{W}_{comp}$	Compressor power consumption [W]	T_e	Evaporating temperature [°C]
T_c	Condensing temperature [°C]	ω	weights
b	bias	x	Input node in ANN
y	Output node in ANN	X	Input quantity
Y	Output quantity	r	Uncertainty error
EOS	Equation of state		

References

AHRI. (2015). *NATIONAL STANDARD CAN / ANSI / AHRI 540-2015 , Performance Rating of Positive Displacement Refrigerant Compressors and Compressor Units*.

Aute, V., Martin, C., & Radermacher, R. (2015). AHRI Project 8013 : A Study of Methods to Represent Compressor Performance Data over an Operating Envelope Based on a Finite Set of Test Data. *Air-Conditioning, Heating, and Refrigeration Institute*.

Cheung, H., Sarfraz, O., & Bach, C. K. (2018). A method to calculate uncertainty of empirical compressor maps with the consideration of extrapolation effect and choice of training data. *Science and Technology for the Built Environment, 24*(7), 743–758. https://doi.org/10.1080/23744731.2017.1372805

JCGM. (2008). Evaluation of measurement data — Guide to the expression of uncertainty in measurement. *International Organization for Standardization Geneva ISBN, 50*(September), 134. http://www.bipm.org/en/publications/guides/gum.html

Ma, J., Ding, X., Horton, W. T., & Ziviani, D. (2020). Development of an automated compressor performance mapping using artificial neural network and multiple compressor technologies. *International Journal of Refrigeration, 120*, 66–80. https://doi.org/10.1016/j.ijrefrig.2020.08.001

Zendehboudi, A., Li, X., & Wang, B. (2017). Utilisation des modèles ANN et ANFIS pour prédire un compresseur à spirale à vitesse variable avec injection de vapeur. *International Journal of Refrigeration, 74*, 473–485. https://doi.org/10.1016/j.ijrefrig.2016.11.011

Ziviani, D., Bahman, A., & Groll, E. (2019). *Multi-input multi-output (MIMO) artificial neural network (ANN) models applied to economized ccroll compressors. 2016*. https://doi.org/10.18462/iir.icr.2019.1321

Acknowledgements

The authors would like to thank the Center for High Performance Buildings (CHPB) for the financial support to this project (project number CHPB-39).

Thermodynamic investigation and design concept of a bi-directional Wankel machine for compression and expansion for natural gas storage and distribution applications

Thomas W. MOESCH*, Konrad KLOTSCHE, Gotthard WILL, Christiane THOMAS, Ullrich HESSE

Technische Universität Dresden,
Institute of Power Engineering,
Bitzer-Chair of Refrigeration, Cryogenics and Compressor Technology
Dresden, Germany
thomas.moesch@tu-dresden.de

* Corresponding Author

ABSTRACT

The distribution and storage of natural gas is based on different pressure levels. The pressure is increased by process gas compressors and currently reduced by throttling valves. The isenthalpic throttling leads to exergy losses. As part of an industry project the TU Dresden and the IME Aachen GmbH developed a bi-directional Wankel machine that may be used for both compression and expansion and thus recover some of the exergy losses. The Wankel machine uses rotary valves for suction and discharge control with fixed opening and closing angles in regards to the displacer rotation. This study shows the detailed design concept of the Wankel machine and the corresponding thermodynamic model, which is based on the compressor calculation software KVA developed at the TU Dresden. The model and its results were used to conduct an in-depth analysis of the machine, including pressures and temperature curves during one displacer rotation, its expected overall efficiencies, occurring exergy losses and its economic potential. The results reveal a considerable energy savings potential for future natural gas storage and distribution application.

1. INTRODUCTION

There is a vast network of pipelines for the distribution of natural gas in Germany. Natural gas is typically stored at pressures up to 30 MPa. The gauge pressures in the transfer lines differs (e.g. long distant transfer lines vs. local distribution lines) and may reach values up to 12 MPa in high-pressure lines, up to 0.1 MPa in medium-pressure lines and up to 0.01 MPa in low-pressure lines (Adler *et al.* 2014). The pressure is reduced and/or elevated (e.g. for biogas injection) at gas transfer stations to match the previous or subsequent lines. Currently, the pressure reduction is achieved by throttling the gas, which then needs to be preheated due to the resulting Joule-Thomson effect. Möller *et al.* (2004) suggest the usage of an expansion machine for pressure reduction and a combined heat and power unit for preheating, which would lead to low CO_2 emissions for the power generation of 0.3 t_{CO_2}/MWh_{el} as compared to Germany's current CO_2 emission coefficient of 0.474 t_{CO_2}/MWh_{el} (Icha and Kuhs, 2019). Mischner *et al.* (2019) suggest the usage of a heat pump for preheating and an expansion machine as a power source for the heat pump compressor. One option for the expansion machine is a Wankel expander. Wankel machines were mainly applied as combustion engines (Yamamoto, 1981). However, applications as compressor machines and expander machines have shown to be equally appealing by recent publications of Garside (2017) and Francesconi *et al.* (2017).
This paper proposes a prototype Wankel type machine design based on a patent (Schaarschmidt and Lenth, 2016), which aims to combine both compression and expansion process in one machine by simply changing its direction of rotation. The paper includes an in-depth thermodynamic investigation of this machine for its rating conditions. The investigation is based on the in-house calculation software called KVA (Hesse and Will, 2014), which was modified for a Wankel type machine. The model is thoroughly described and its results are used to conduct an exergy loss analysis and a profitability analysis.

2. DESIGN CONCEPT

The Wankel machine is based on a 2:3 trochoid geometry and it is designed for an operation with dry natural gas (methane) and operating conditions for the compression mode as shown in Table 1.

Table 1: Operating conditions for compression mode

Parameter	Value	Unit
Low pressure	0.6	MPa (abs.)
High pressure	1.3	MPa (abs.)
Suction temperature	283.15	K
Standard volume flow (T = 283 K, p = 0.1013 MPa)	1300	m^3/h
Shaft speed	50	1/s

2.1 Geometry and components

The assembly of the machine and its main components are shown in Figure 1. Both bearings and gas seals are not shown here, however, they are included in the final machine design, which is based on Wankel type engines (Yamamoto, 1981). The geometry of the compressor housing bore is based on a trochoid curve with a base radius r_B an eccentricity e. The curve is defined in a Cartesian system by the following equation.

$$\begin{pmatrix} x \\ y \end{pmatrix}_{\text{Housing}} = \begin{pmatrix} e\cos(\varphi) + r_B \cos(\varphi/3) \\ e\sin(\varphi) + r_B \sin(\varphi/3) \end{pmatrix} \tag{1}$$

The rotor flanks are based on a circle approach with a radius r_R. Each rotor segment is defined by equation (2).

$$\begin{pmatrix} x \\ y \end{pmatrix}_{RotSeg,i} = \begin{pmatrix} r_{off} \cos(\varphi_{off,i}) + r_R \cos(\varphi_i) \\ r_{off} \sin(\varphi_{off,i}) + r_R \sin(\varphi_i) \end{pmatrix} \tag{2}$$

$$\text{with } \varphi_{off,i} = \begin{cases} \pi & \text{for } i = 1 \\ -\pi/3 & \text{for } i = 2 \\ \pi/3 & \text{for } i = 3 \end{cases} \quad \text{and} \quad \varphi_i = \begin{cases} -\alpha \ldots \alpha & \text{for } i = 1 \\ -\alpha + 2\pi/3 \ldots \alpha + 2\pi/3 & \text{for } i = 2 \\ -\alpha - 2\pi/3 \ldots \alpha - 2\pi/3 & \text{for } i = 3 \end{cases}$$

The offset of the circle centers to the rotor center r_{off} and the opening angle of the circle segments α are defined by equations (3) and (4).

$$r_{off} = r_R \cos(\alpha) - r_B \cos(\pi/3) \tag{3}$$

$$\alpha = 2\arcsin\left(\frac{r_B}{r_R}\sin(\pi/3)\right) \tag{4}$$

All geometry parameters for the Wankel machine are summarized in Table 2.

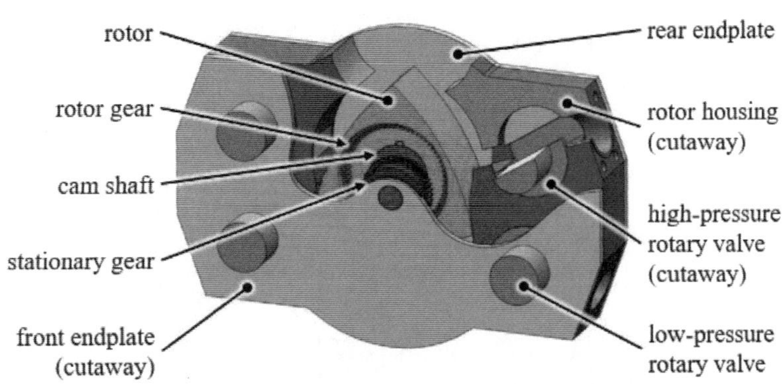

Figure 1: Wankel machine design 3D cutaway view

Table 2: Geometry parameters

Parameter	Value	Unit
Base radius r_B	120	mm
Eccentricity e	24	mm
Rotor flank radius r_R	456	mm
Offset of the circle centers to the rotor r_{off}	384	mm
Opening angle of the circle segments α	26.35	°
Outer diameter of rotary valve	104	mm
Size of flow port in low-pressure rotary valves (height x width)	62.5 x 67.5	mm
Size of flow port in high-pressure rotary valves (height x width)	62.5 x 36.0	mm
Size of flow ports between rotary valves and working chambers (height x width)	62.5 x 26.0	mm
Rotor height	72.5	mm

2.2 Mode of operation

The rotor is coupled by the cam shaft, but its rotating movement is set by its rotor gear and the stationary gear in the housing cover with a gear ratio of 2:3. This leads to a transmission of 3:1 between shaft and rotor. The rotary valves are connected to the shaft via drive belt with a transmission of 2:1. The trochoid rotor and the trochoid-shaped cylinder bore form up to six working chambers which lead to six individual suction, compression and discharge processes during one rotor rotation. Figure 2 shows the Wankel machine at different rotor angles and highlights one of its working chambers (WC11). Figure 3 shows the resulting volume curve of the two working chambers WC11 and WC12 and the opening area curves for the corresponding high-pressure and low-pressure rotary valves. The curves for volumes of WC21/WC22 and WC31/WC32 and their corresponding rotary valve opening areas are identical to the shown curves with an offset in the rotor angle of +240° and +120°, respectively. The resulting swept volume per working chamber is 1200 cm³. The compression mode and expansion mode only differ in their direction of rotation and the mirrored opening curves of the rotary valves (see Figure 3). The rotary valves control the timing and duration of the suction/discharge process and set the built-in compression and expansion ratio ($\Psi_C = \Psi_E = 1.85$). The rotor angles in Figure 2 show the points of opening and closing of the lower low-pressure rotary valve ($\varphi_{R,lrvo} = 105°$, $\varphi_{R,lrvc} = 180°$) and the high-pressure rotary valve ($\varphi_{R,hrvo} = 225°$ $\varphi_{R,hrvc} = 273°$) during compression mode.

$$\Psi_C = V_{\text{lrvc}} / V_{\text{hrvo}} \qquad \Psi_E = V_{\text{lrvo}} / V_{\text{hrvc}} \tag{5}$$

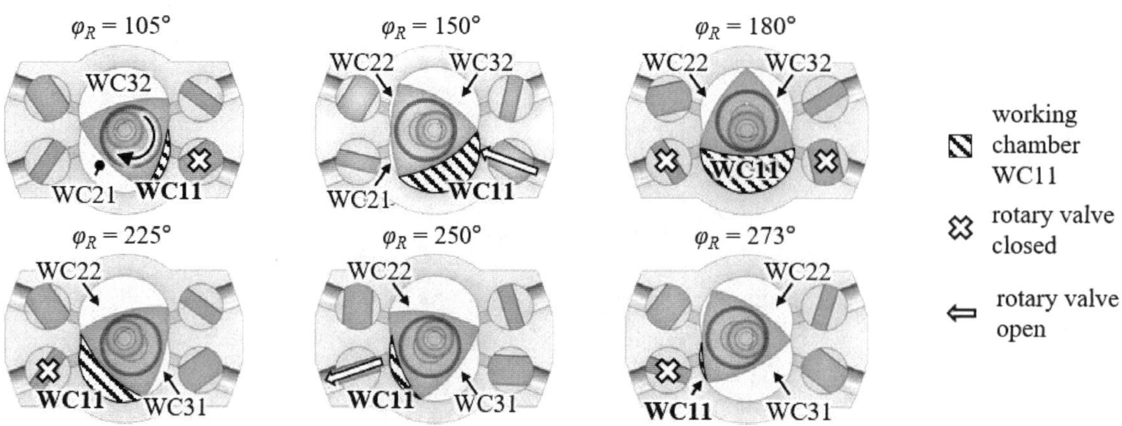

Figure 2: Wankel machine and working chambers (WC) at different rotor positions in compression mode

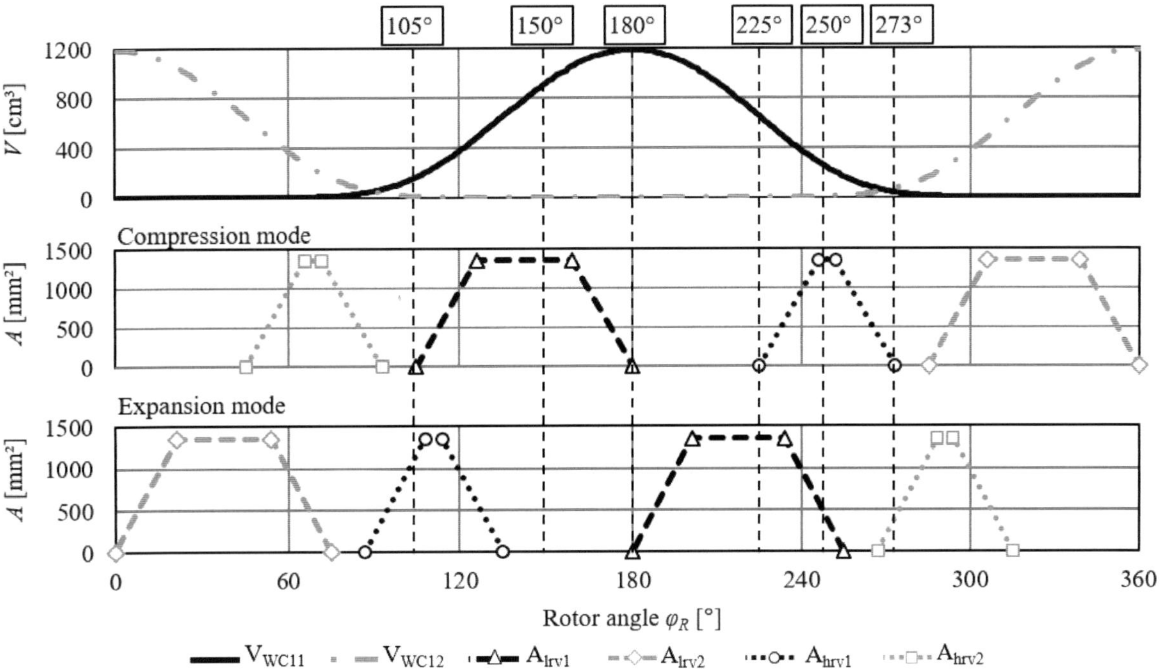

Figure 3: Volume curve of the working chambers WC11 and WC12 and the corresponding opening area curves of the high-pressure rotary valves (hrv1, hrv2) and the low-pressure rotary valves (lrv1, lrv2)

3. THERMODYNAMIC MODEL

The goal of the calculation is the determination of all relevant thermodynamic parameters of the working fluid within the Wankel machine. The calculations are carried out with an in-house program, called KVA, which computes the thermodynamic steady state of the fluid based on a 0-dimensional approach. Hesse and Will (2014) gave a detailed description of the calculation program. Klotsche *et al.* (2018) applied KVA for the first time to a Wankel machine as a household refrigerator compressor. The calculation itself requires a model of the Wankel machine, which includes all relevant volumes (working chambers, suction and pressure chambers, *etc.*) and their junctions each represented by connecting elements. This composes a network of all relevant gas-containing components. The volumes may be divided into four categories (see Table 3).

Table 3: Overview of different categories of volumes and their main properties

Category	Temperature	Pressure	Volume	Typical example
Type 1	constant	constant	constant	Suction boundary condition
Type 2	*variable*	constant	constant	Discharge boundary condition
Type 3	*variable*	*variable*	constant	Suction & discharge chamber
Type 4	*variable*	*variable*	*variable*	Working chamber

There are five different types of connection elements available, including valves, pipes, cross-sectional constrictions (which can be constant over time or time-controlled, *i.e.* controlled by edges), gaskets, and heat exchangers. The heat rejection of the Wankel machine is realized by wall elements that can be modelled to absorb, store and release heat from the gas contained in the volumes to and from the environment. Wall elements require defined contact surfaces, heat conduction properties as well as the geometric properties. The changes of state and mass in the volumes result from the properties of surrounding components, in particular the temporal behavior of the working chamber volumes, as well as the properties of the connecting elements and the given boundary conditions, *i.e.* the low- and high-pressure interface to the connected system or cycle. During the calculations, both gas pressure and gas temperatures of all volumes are determined for each time step. All other relevant thermodynamic parameters are derived from p and T.

The governing equations are based on the first law of thermodynamics and the ideal gas equation, shown by Hesse and Will (2014).

$$\frac{1}{p}\frac{dp}{dt} = \frac{K}{K-1}\cdot\left(\frac{p}{mKRT}\frac{dQ}{dt} - \frac{1}{V}\frac{dV}{dt} - \frac{1}{K}\frac{dK}{dt}\right) \tag{6}$$

$$\frac{1}{T}\frac{dT}{dt} = \frac{K}{K-1}\cdot\left(\frac{p}{mKRT}\frac{dQ}{dt} - \frac{1}{VK}\frac{dV}{dt} - \frac{1}{K}\frac{dK}{dt}\right) - \frac{1}{m}\frac{dm}{dt} - \frac{1}{R}\frac{dR}{dt} \tag{7}$$

All parameters can be found in the list of symbols, the parameters R and K are defined as followed:

$$R = \frac{p}{\rho T} \qquad K = \frac{h}{RT} \tag{8}$$

The temporal changes in pressure and temperature of all volume elements are calculated with a time-stepping method. Hesse and Will (2014) provided information about this procedure. Figure 4 shows a 2D cutaway of the Wankel design with volume and connecting elements for this time step. Figure 5 shows the corresponding calculation model with all volume and connecting elements and one wall element. Note that upper and lower-case acronyms represent volume and connecting elements, respectively. The model comprises six working chambers (WC11 – WC32) that are created between the displacer and the housing (cylinder and end plates) over one complete displacer rotation and both low- and high-pressure supply lines in the upper and lower machine part. The following list represents every volume and connecting element starting from the system interface to the working chamber (the first letter shows the pressure level: l or L = low-pressure, h or H = high-pressure):

- a low- or high-pressure line (ll1/ ll2 or hl1/ hl2) representing the pipe between the Wankel machine and the connected circuit, *i.e.* the boundary conditions of the model (L, H) (not shown in Figure 4)
- a damping volume (LDV1/ LDV2 or HDV1/ HDV2) (not shown in Figure 4)
- a rotary valve (lrv1/ lrv2 or hrv1/ hrv2)
- a valve chamber (LVC1/ LVC2 or HVC1/ HVC2),
- and six time-controlled flow ports (lfp11 – lfp16/ lfp21 – lfp26 or hfp11 – hfp16/ hfp21 – hfp26) acting as a link between the valve chamber with each of the working chambers time-dependently. Each valve chamber temporarily connects with each working chamber during one complete rotation of the displacer.

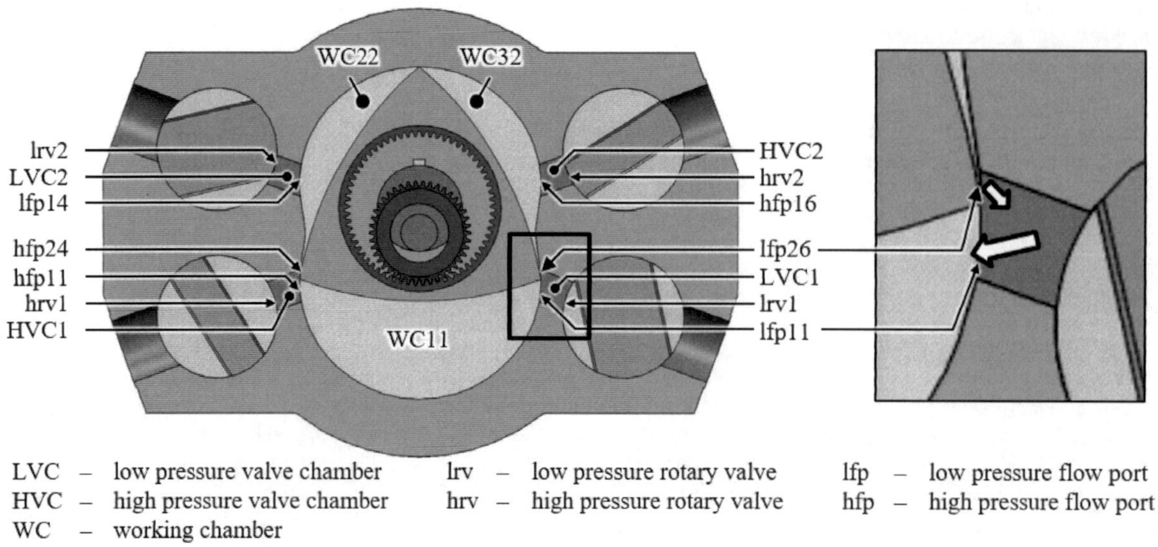

Figure 4: 2D-cutaway of Wankel design with volume elements and connecting elements

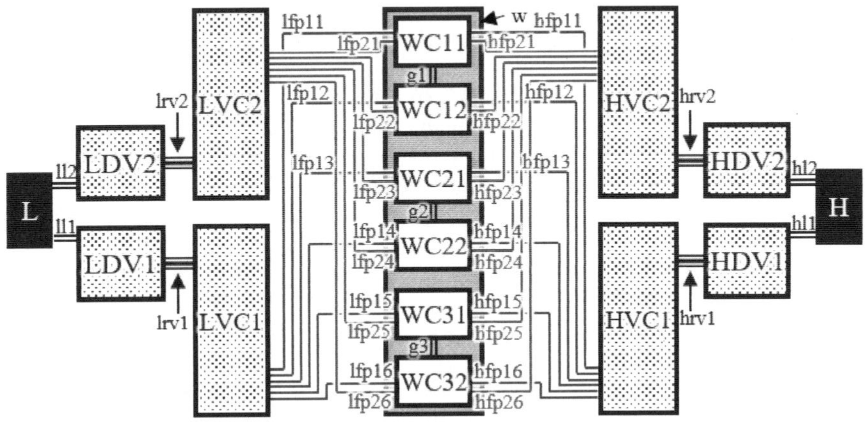

Figure 5: KVA model for the Wankel machine

The connecting elements g1 – g3 are gaps with a constant cross-section. They represent the leakage gap between the cylinder cusps and the moving displacer flank, which connects consecutive working chambers (*e.g.* WC11 and WC12). The wall element w accounts for the heat transfer between the working chambers and the environment. The flow through the connecting elements flow ports (lfp, hfp), rotary-valve (lrv, hrv), and leakage gaps (g) is based on an isentropic nozzle flow for compressible gases (Jobsen, 1955).

$$\dot{m} = A \cdot \mu \cdot \psi\left(p_1/p_2, \kappa\right) \cdot \sqrt{2\rho_1 p_1} \tag{9}$$

$$\text{with } \psi\left(p_2/p_1, \kappa\right) = \begin{cases} \left(\dfrac{2}{\kappa+1}\right)^{\frac{1}{\kappa+1}} \cdot \sqrt{\dfrac{\kappa}{\kappa+1}} & \text{for } \quad p_2/p_1 \leq \left(\dfrac{2}{\kappa+1}\right)^{\frac{\kappa}{\kappa-1}} \\ \sqrt{\dfrac{\kappa}{\kappa-1} \cdot \left[\left(p_2/p_1\right)^{\frac{2}{\kappa}} - \left(p_2/p_1\right)^{\frac{\kappa+1}{\kappa}}\right]} & \text{for } \quad \left(\dfrac{2}{\kappa+1}\right)^{\frac{\kappa}{\kappa-1}} < p_2/p_1 < 1 \end{cases}$$

Where A is the cross-sectional flow area, μ is the flow coefficient that takes friction losses into account, κ is the isentropic exponent, p_1 is the pressure at the inlet, p_2 is the pressure at the outlet, ρ_1 is the density at the inlet.

4. SIMULATION RESULTS

The calculation model is used to conduct a detailed analysis for both compression and expansion mode. The investigated operating point is similar to the operating conditions in Table 1 with an ambient temperature of 300 K. However, the expansion mode was simulated for a different inlet temperature (T = 318 K) than the compression mode (T = 283 K) to account for a preheating of the gas prior to expansion. The main results are summarized in Table 3. The results show that the machine slightly exceeds the volume flow demands of the operating conditions and achieves relatively similar exergetic efficiencies for both compression and expansion mode even though the compressor mode has a lower volumetric efficiency. The following sub-sections reveal a detailed analysis of the simulation results.

Table 3: Summarized simulation results for compression and expansion mode

Parameter	**compression mode**	**expansion mode**
Volume flow [m³/h] (T = 283 K, p = 0.1013 MPa)	1473	1885
Power input [kW]	45.1	-29.9
Volumetric efficiency [%]	57.2	75.1
Isentropic efficiency [%]	75.3	69.4
Exergetic efficiency [%]	79.5	75.1

4.1 Analysis of pressure and temperature curves

Figure 6 shows the resulting pressure and temperature curves for chamber WC11 and its connected volumes during compression and expansion mode.

The high-pressure peak during compression mode at $\varphi_R = 236°$ indicates a valve pressure loss up to 0.18 MPa due to an insufficient valve opening area. The rising temperature and pressure in the the low-pressure rotary valve chamber (LVC1) starting at $\varphi_R = 273°$ shows the compression of the remaining gas in WC11 and then its re-expansion in WC12 after the high-pressure rotary valve (hrv1) closed while the low-pressure rotary valve (lrv2) has not yet opened (see Figure 2, $\varphi_R = 273°$).

For the expansion mode, the pressure minimum in the high-pressure valve chamber (at $\varphi_R = 51°$) indicates under-expansion and recompression losses due to late control valve opening. The pressure drop during the admission starting at $\varphi_R = 90°$ indicates a restricted flow into the working chamber caused by a small high-pressure rotary valve opening area during admission. The working chamber pressure decreases below the low-pressure level at $\varphi_R = 160°$ which indicates a late opening of the low-pressure rotary valve and results in under-expansion losses. During this under-expansion the gas reaches temperatures below 283 K ($T_{min} = 274$ K at $\varphi_R = 170°$) which may lead to condensation within the machine in case of remaining moisture content in the natural gas. However, the recompression of the gas (starting at $\varphi_R = 175°$) leads to minimum gas temperatures of $T_{LVC} = 283$ K in the low-pressure valve chamber and $T_{LDV} = 285$ K in the low-pressure damping volumes. Therefore, there is no need for additional reheating of the discharged gas.

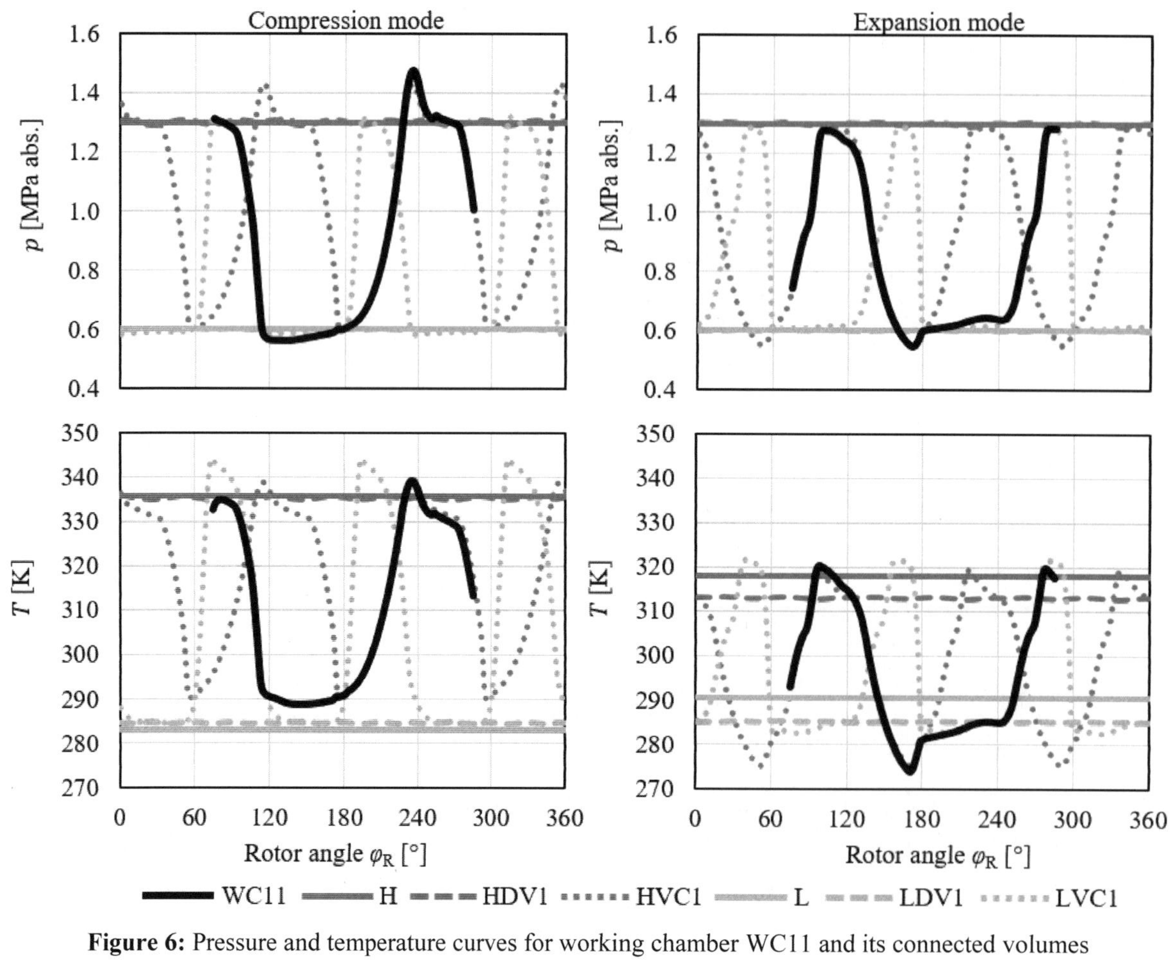

Figure 6: Pressure and temperature curves for working chamber WC11 and its connected volumes

4.2 Exergy flow loss analysis

The temperature and pressure curves only allow an evaluation of the change in state in all volumes. For a more detailed loss analysis, the exergy losses of each component are evaluated. The exergy flow loss analysis is based on the general exergy flow balance as proposed by Kretzschmar and Kraft (2016). Both potential and kinetic energy terms are neglected.

$$\dot{E}_{loss12} = \dot{E}_{\dot{Q}12} + P_{t12} + P_{diss12} - \left(\Sigma\dot{E}_2 - \Sigma\dot{E}_1\right) \tag{10}$$

with $\dot{E}_{\dot{Q}12} = \sum_i \frac{T_i - T_{amb}}{T_i}\dot{Q}_i \qquad P_{t12} = \dot{m}\int_{p_1}^{p_2} v(p)\,dp \qquad \dot{E} = \dot{m}\cdot\left[h - h_{amb} - T_{amb}\cdot\left(s - s_{amb}\right)\right]$

Where $\dot{E}_{\dot{Q}12}$ is the exergy flow loss due to heat exchange, P_{t12} is the added mechanical power to the flow, P_{Diss12} is the dissipated mechanical power through friction and $\dot{E}_2$ and $\dot{E}_1$ are the exergy flows of the entering and leaving fluid, respectively.

The results of the loss analysis are shown in Figure 7. The upper diagrams distinguish between *Working Chambers (WCs)* which account for all working chambers (WC11 to WC32), *Volumes* which account for all fixed-volume chambers (DV and VC), *Ports* which include all variable flow port openings (lfp11 to hfp26), *Valves* which include all rotary valves (lrv1 to hrv2), and *Pipes* which include both low-pressure and high-pressure lines (ll1 to hl2). The lower diagrams distinguish between *"A"-*, *"B"-*, and *"C"-ports*. *"A"-ports* include all ports that connect the working chambers with the intended low-pressure and high-pressure valve chambers (lfp11, lfp13, lfp15, lfp22, lfp24, lfp26, hfp11, hfp13, hfp15, hfp22, hfp24, hfp26). *"B"-ports* include all ports that lead to an unintentional connection between working chambers and valve chambers (lfp12, lfp14, lfp16, lfp21, lfp23, lfp25, hfp12, hfp14, hfp16, hfp21, hfp23, hfp25). *"C"-ports* include the non-sealed gap between consecutive working chambers (g1 to g3).

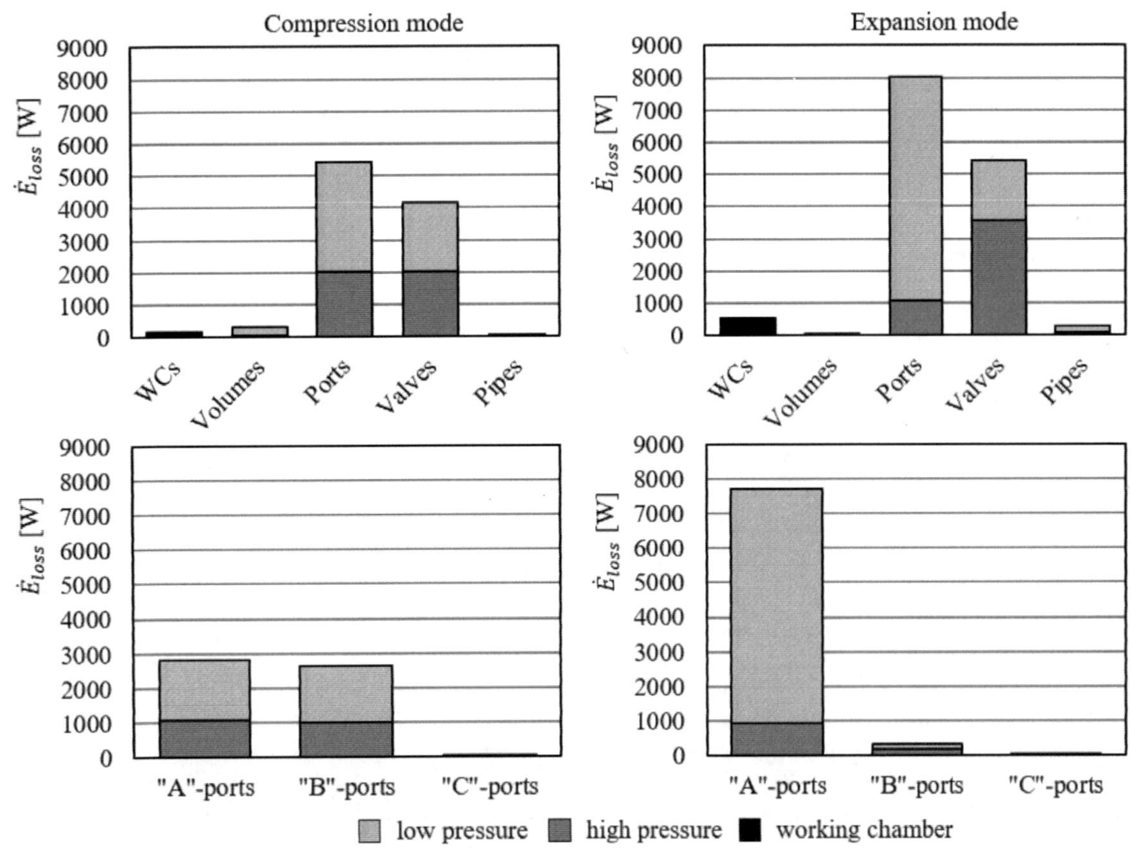

Figure 7: Exergy flow losses (by component type) during compression and expansion mode

The main exergy flow losses during compression mode occur in the low-pressure ports with $\sum \dot{E}_{loss,lfp}$= 3390 W followed by the low-pressure and high-pressure rotary valves and high-pressure ports with $\sum \dot{E}_{loss,lrv}$ = 2120 W, $\sum \dot{E}_{loss,hrv}$ = 2050 W, and $\sum \dot{E}_{loss,hfp}$ = 2040 W. These coincide with the pressure losses due to the limited size of the ports and the fixed opening phase of the rotary valves.
The main exergy flow losses during expansion mode occur in the low-pressure ports with $\sum \dot{E}_{loss,lfp}$ = 6810 W and the high-pressure valves with $\sum \dot{E}_{loss,hrv}$ = 3560 W. This coincides with both the steep pressure increase during the discharge and the pressure loss during the suction process. It ultimately shows that the opening phase of the high-pressure rotary valve and the sizing of the low-pressure ports are not ideal for the expansion mode. The leakage paths g1 through g3 only have a minimal effect on the overall exergy losses.
The exergy loss analysis shows higher exergy losses through high-pressure rotary valves during expansion mode as compared to compression mode and a significant difference in the impact of the ports. This shows the importance of the rotary valve and port design and its compromise by optimization the fixed opening and closing angles for both compression and expansion mode and reducing the overall exergy losses instead of finding the optimal design for only one mode. Future machine designs need to apply more variable rotary valve configurations

4.3 Profitability analysis

The usable electric power during the expansion mode and its economic benefit depend on the initial gas temperature, the preheating method (boiler or heat pump) and the generator efficiency. The additional preheating capacity for the calculated expansion mode volume flow can be approximated to be 27.4 kW when assuming an initial gas temperature of 283.15 K and dry natural gas with a high methane content (>93%) and a specific heat capacity of c_p = 2.23 kJ/(kg·K). For a comparison of both boiler and heat pump method, the following assumptions were applied:

- each generator efficiency and boiler efficiency equals 0.95,
- CO_2 emission factor is 0.2 t_{CO_2}/MWh for natural gas, 0.471 t_{CO_2}/MWh for electricity (Icha and Kuhs, 2019),
- the air/gas heat pump has a *SEER* of 2.5,
- yearly operating time is 6130 h (degree of utilization = 70%),
- costs of gas is 0.06 Euro/kWh and costs of electricity is 0.30 Euro/kWh.

The results are summarized in Table 4 and show that Method 1 leads to a higher electric power output and a higher economic potential. However, Method 2 does not cause any CO_2 emission and requires no natural gas consumption and is therefore the better option for a future carbon-neutral industry.

Table 4: Results of ecological and economic potential analysis for one unit

Parameter	**Method 1 (boiler preheater)**	**Method 2 (heat pump preheater)**
Usable electric power output [MWh/a]	174.2	107.0
Required natural gas [MWh/a]	176.8	0
CO_2 emission factor [kg_{CO_2}/kWh]	0.20	≈ 0
Avoided CO_2 emission [t_{CO_2}/a]	**46.7**	**50.4**
Money savings [Euro/a]	**41,648**	**32,103**

6. CONCLUSIONS

The proposed Wankel machine design allows a bi-directional operation as compressor and expander. The different operating modes (compression and expansion) were achieved by using actuated rotary valves for both inlet and outlet. The compression and expansion of natural gas with current Wankel machine design leads to isentropic efficiencies of 75.3 % / 69.4 % and volumetric efficiencies of 57.2 % / 75.1 % respectively. The largest exergy occur at the ports and the rotary valves and are particularly significant in the low-pressure ports during expansion mode. The expansion mode allows a total energy savings of 107.0 kWh/a when combined with a heat pump (with *SEER* = 2.5) as a gas preheater and 174.2 MWh/a by using a boiler preheater but with a CO_2 emission factor of 0.20 kg_{CO_2}/ kWh.

NOMENCLATURE

A	area	[m²]	R	real gas parameter (see Eq. 7)	[J·kg⁻¹·K⁻¹]
d	diameter	[m]	s	specific entropy	[J·kg⁻¹·K⁻¹]
c_p	specific heat capacity	[J·kg⁻¹·K⁻¹]	$SEER$	seasonal energy efficiency ratio	[1]
e	eccentricity	[m]	T	temperature	[K]
$\dot{E}$	exergy flow	[W]	v	specific volume	[m³/kg]
h	specific enthalpy	[J/kg]	V	volume	[m³]
K	real gas parameter (see Eq. 7)	[1]	w	width	[m]
m	mass	[kg]	x, y	Cartesian coordinates	[m]
$\dot{m}$	mass flow rate	[kg/s]	α	segment angle	[rad]
Q	heat	[J]	η	efficiency	[1]
$\dot{Q}$	heat flow	[W]	κ	isentropic exponent	[1]
p	pressure	[Pa]	μ	flow coefficient	[1]
P	power	[W]	φ	angle	[rad]
r	radius	[m]	$\Psi_{C\|E}$	compression | expansion ratio	[1]

Subscript

amb	ambient	*i*	index of element	*off*	offset
B	base	*lfp*	low-pressure flow port	*R*	rotor
diss	dissipated	*lrv*	low-pressure rotary valve	*t*	mechanical
hrv	high-pressure rotary valve	*lrvo|c*	*lrv* opens | just closed	*WC*	working chamber
hrvo|c	*hrv* opens | just closed	*min*	minimum		

REFERENCES

Adler, P., Billig, E., Brosowski, A., Daniel-Gromke, J., Falke, I., Fischer, E., Grope, J., Holzhammer, U., Postel, J., Schnutenhaus, J., Stecher, K., Szomszed, G., Trommler, M., Urban, W. (2014). *Leitfaden: Biogasaufbereitung und –einspeisung*. Fachagentur Nachwachsende Rohstoffe e.V. (FNR)

Francesconi, M., Caposciutti, G., & Antonelli, M. (2017). An experimental and numerical analysis of the performances of a Wankel steam expander. *Energy Procedia*, *129*, 395-402.

Garside, D. W. (2017). A new Wankel-type compressor and vacuum pump. In *IOP Conference Series: Materials Science and Engineering* (Vol. 232, No. 1, p. 012065). IOP Publishing.

Hesse, U., & Will, G. (2014). Thermodynamic Calculation of Reciprocating Compressors Plants. *Proceedings of the 9th Conference of the EFRC*, Vienna, Austria (158-168).

Icha, P., & Kuhs, G. (2019). Entwicklung der spezifischen Kohlendioxid-Emissionen des deutschen Strommix in den Jahren 1990-2018. Dessau-Roßlau.

Jobson, D. A. (1955). On the flow of a compressible fluid through orifices. *Proceedings of the Institution of Mechanical Engineers*, *169*(1), 767-776.

Klotsche, K., Moesch, T. W., Will, G., & Hesse, U. (2018). Thermodynamic modelling of reciprocating and Wankel type compressor for household refrigerators. Purdue Compressor Conference 2018

Kretzschmar, H. J., & Kraft, I. (2016). *Kleine Formelsammlung Technische Thermodynamik*. Carl Hanser Verlag GmbH Co. KG.

Mischner, J., Stang, R., & Gebhard, A. (2019). Energiestrombilanzen von Anlagen zur Energierückgewinnung für die Gasvorwärmung in Gasdruckminderungsanlagen. Vulkan-Verlag GmbH.

Möller, A., Niehörster, C., & Otto, F. (2004). Strom aus der Gasleitung – Entspannungsanlagen. *Gas- und Wasserfach. Gas-Erdgas*, *145*(6), 337-340.

Schaarschmidt, G. & Lenth, B. (2016). EP3091176A1. Munich: European Patent Office.

Yamamoto, K. (1981). *Rotary engine*. Tokyo, Japan: Sankaido Company.

Empirical Indicated Loss Analysis of a Semi-hermetic Light-Commercial Spool Compressor

Seth YARBOROUGH[1], Craig R. BRADSHAW[1*], Joe OROSZ[2], Greg KEMP[2]

[1]Center for Integrated Building Systems, Oklahoma State University, Stillwater, OK, USA, (405)744-5246
craig.bradshaw@okstate.edu

[2]Torad Engineering
Cumming, GA, USA, (678)366-3399
greg.kemp@toradengineering.com, joe.orosz@toradengineering.com

* Corresponding Author

ABSTRACT

An analysis of the indicated losses is presented for a semi-hermetic, light-commercial, prototype, spool compressor. The spool compressor prototype was instrumented with five high-speed pressure sensors, three in the process chamber, one in the discharge valve plenum, and one in the motor cavity. These sensors were triggered with a proximity sensor actuated by means of a custom rotary fixture attached to the compressor motor shaft. This coupling of rotational position and pressure measurements allowed for the development of an indicator (pressure v. volume) diagram for the compression process. Additionally, the added sensor in the discharge valve plenum allowed for a de-coupling of discharge valve losses and flow losses within the discharge plenum itself. The sensor in the motor cavity allowed for an analysis of the flow losses leaving the compressor shell. The compressor was tested at five motor speeds (1100, 1300, 1500, 1700 rpm and line voltage) at saturated condensing (SDT) and evaporating (SST) temperatures ranging from 37.8 – 48.9 °C (90 – 130 °F) and -3.8 – 15.6 °C(30 -60 °F), respectively at a fixed suction superheat of 11.1 K (20 °R) . Quantitative analysis shows that the suction and compression losses for this prototype compressor are relatively small compared with the discharge/valve losses. The total losses during the discharge process are generated by pressure drop and backflow through the discharge valve ports as well as when gas flows from the discharge plenum across the motor through the compressor body. It was found that a 5-6% improvement in compressor efficiency can be accomplished by redesigning the discharge plenum and motor cavity to reduce over pressurization. Further investigation into the valve dynamics need to be performed to improve the 11-12% loss in the valves. The valve losses were found to be sensitive to operating speed and SDT with maximum variations of 5% and 3%, respectively.

1. INTRODUCTION

Compressors in HVAC&R applications represent roughly 5% of total primary energy utilization[1] in the US. Therefore, to ensure maximum efficiency of new HVAC&R systems it is critical to evaluate and analyze new HVAC&R equipment, especially compressors. Novel compressors are especially critical to evaluate in great detail to ensure

[1] According to the US Energy Information Administration (EIA) , and the US Department of Energy (DOE) building data book , the consumption of electrical energy HVAC&R systems in the residential, commercial sectors corresponds to 22%, and 18%, of all primary energy annually in the US, roughly 40% in total. Compressors are estimated to account for roughly 60% of that energy consumed in cooling and refrigeration systems (Westphalen & Koszalinski, 2001). Resulting in an estimated 5% of the totally primary energy annually in the US.

adoption of new technology that creates a meaningful increase in system efficiency. Various methods over time have been used to evaluate the performance of compressors. The isentropic and volumetric efficiencies among other parameters are often used to evaluate bulk compressor performance. While those parameters can be useful for macro-analysis of compressor performance there is often a need for a more detailed breakdown of losses within a compressor.

1.1 Spool compressor background and motivation.

The rotating spool compressor is a novel rotary compressor mechanism most similar to the sliding vane compressor. Primary differences are described by Kemp *et al.* (2008) who first introduced the compressor applied to air-conditioning and refrigeration. Since its introduction, the spool compressor has undergone a large development effort including several prototype generations to represent air-conditioning applications from residential split-systems to commercial chiller products. Each prototype platform has presented additional insight into the unique attributes of this compressor and the best application of the device. This was accomplished by leveraging each prototype platform to further explore the spool compressor attributes by studying specific phenomenon/sub-systems during development.

The prototype spool compressor development progression through various application, displacements, and refrigerants has been investigated and summarized effectively in Bradshaw *et al.* (2018). The comprehensive model development and mechanism design improvements led to the 7th generation 40-ton, R134a spool compressor presented in Orosz *et al.* (2016)

Bradshaw *et al.* (2018) explored the potential for efficiency improvement by performing an indicated loss analysis on the 7th generation of prototype spool compressor, a 40-ton R134a prototype machine. The results of this study indicated the discharge plenum and valves had the largest room for improvement. A redesign of the discharge plenum was recommended for a potential 4-5% improvement in efficiency. The success of an indicated loss analysis on the 7th generation spool compressor prompted interest in exploring the potential improvement of the 8th and current generation of the prototype spool compressor, shown in Figure 1. The 8th generation of this compressor is a 30-ton R134a machine designed for air-conditioning applications and the first semi-hermetic embodiment of the mechanism. This study will present the experimental methodology, data analysis techniques, and indicated loss analysis results and suggest changes to the 8th generation prototype to maximize its efficiency.

2. Experimental Methodology

To generate an indicated loss analysis it is first necessary to measure dynamic pressure and compressor shaft position, which can be analyzed into the relevant losses. This section will focus on the collection and reduction of that data and generation of an indicator diagram. The general method to create an indicator diagram is to simultaneously measure both instantaneous chamber pressure and volume of a positive displacement compressor. The commonly used method to accomplish these measurements are often executed by mounting multiple pressure transducers into the compression chamber and measuring rotation angle of the shaft with a rotary encoder. Rotation angle is then used to calculate the volume of the compression chamber allowing for the indicator diagram to be created. Rigola *et al.* (2002) and Huang *et al. (2018)* used similar methods to produce indicated diagrams of reciprocation compressors. This is a process that follows closely with the work presented in Bradshaw *et al.* (2018) and this work continues from that effort.

The indicated loss analysis of the 8th generation spool compressor will follow similar methodology and the following section describes the installation and setup of the experiment to collect the data that creates an indicated loss analysis on the 8th generation spool compressor. This includes the sensor selection, placement, and calibration as well as an uncertainty analysis and the final test matrix of operating conditions that are collected and analyzed.

Figure 1: 8th generation, R134a, 30-ton, semi-hermetic, spool compressor prototype with high-speed pressure sensors installed showing motor cavity (MC) sensor, compression chamber (PP) sensor, discharge (DP) and discharge plenum (DC) (left) and the axial view showing the proximity sensor location. (right)

2.1 Sensor selection, location, calibration, and procedure

A 30-ton, semi-hermetic, prototype spool compressor is fitted with three Meggitt 8530B-500 high-speed pressure sensors at locations which will allow for easy installation of the sensors and for most of the process pressures to be measured at all rotor angles, called SP, PP, and DP, respectively. Additionally, two additional sensors were placed downstream of the discharge valve assemblies but upstream of the discharge manifold plumbing connection of the test stand, called DC and MC, respectively. Figure 1 shows the physical locations of these sensors relative to the suction and discharge of the compressor and the axial view of the physical location of the proximity sensor relative to the suction of the compressor. The angular locations of the three sensors in the process section are shown in Figure 2, which is an axial view schematic of the compressor cylinder with the sensor angles relative to Top-Dead-Center (TDC). The only regions which cannot be measured are near the TDC area of the compression process on both the suction and discharge side of the compressor. The volume in this region is insignificant when compared to the other volumes it is assumed that any losses or phenomenon that is unable to be measured is also small, therefore that loss in considered negligible.

The piezoresistive pressure transducers used in this analysis are susceptible to a change in the zero measured output as a result of electrical installation and mechanical installation (*i.e.* applied torque and wire strain). To mitigate the influence of error due to mechanical installation each of the five sensors was carefully installed with the recommended 1.69 N-m of torque and the strain on the sensors were minimized. To reduce influence as a result of electrical installation each sensor was electrically insulated at each of the connection points and were wired to ground at the signal conditioner. Each sensor was also calibrated in-place using dry nitrogen and a pressure reference measured using a Druck [DPI 612] with an accuracy of .0086 (0.125) bar (psia). The calibration was additionally repeated at the end of each set of tests to ensure no significant changes to the calibration occurred.

To measure shaft speed and position an inductive proximity sensor (Sensor Solutions S50FW-18ADSO-ODSB5) is used. The proximity sensor is arranged such that the sensor triggered when the vane of the compressor was at compressor Top Dead Center (TDC). This was used as a datum, or index, and all angles were measured in reference to this location. From the trigger, time is measured using the Data Acquisition Equipment (DAQ) and the shaft speed is assumed to be constant within one rotation. Using the measured time and the aforementioned assumption regarding speed allowed for an inference of the shaft position at any instant in the rotation.

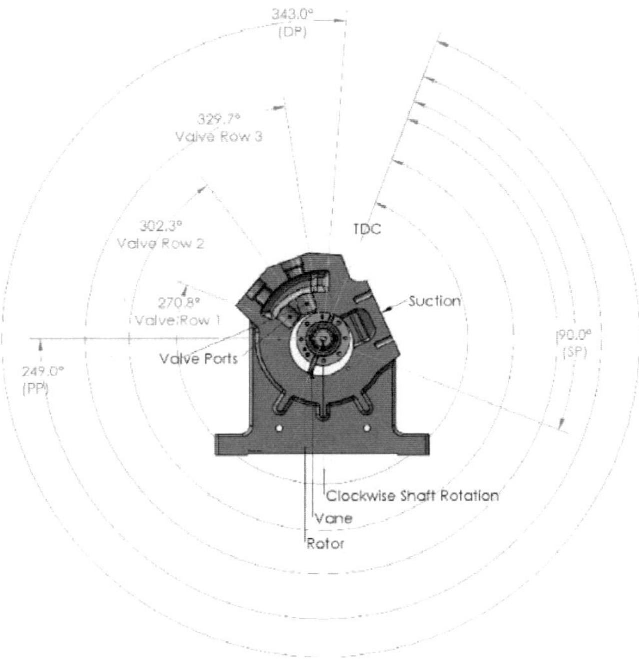

Figure 2: Axial view of compressor cylinder block highlighting the angular location of in-pocket sensors and valve location relative to vertical and the compressor top-dead-center (TDC).

2.2. Experimental procedure and test conditions

The compressor is first operated until it reaches steady-state conditions at a prescribed operating condition using the hot-gas bypass load stand environment described in Orosz *et al.* (2016). This load stand environment includes bulk pressure, temperature, and refrigerant mass flow rates as well as a means to control the conditions at steady-state. The pressure sensors are sampled at 70,000 samples per second (Hz), triggered using the inductive proximity sensor to start sampling at TDC, and sampled for a length of time that ensures at minimum of one complete rotation of the shaft for each test condition. This process was repeated 20 times per operating condition and these samples were averaged into a single sample which reflects the behavior at the current condition.

The prototype compressor was operated using refrigerant R134a at five shaft speeds, various Saturated Discharge Temperatures (SDT) and Saturated Suction Temperatures (SST) at a fixed compressor inlet superheat of 20 °R for a total of 36 data points. The final test matrix collected for this study is presented in Table 1. The limited SST range for certain speeds and SDT are constrained by the limits of the test environment. Test condition 8, bolded and underlined in Table 1, is used as the illustrative example, throughout this work as it represents the typical behavior of the compressor. The final pressures as a function of crank angle are presented in Figure 3 for Test condition 8. This figure also highlights the angular location of the different rows of valves in this compressor prototype, for reference.

3. DATA REDUCTION AND ANALYSIS

This section will present the procedures used to reduce the collected data, estimate the uncertainty associated with the calculated losses and calculate the suction, compression, and discharge indicated losses within the compressor. Additionally, two external indicated losses, the plenum and motor cavity, are also estimated as shown in this section.

3.1 Data reduction and Uncertainty

The uncertainty of the loss measurements is represented as a total relative uncertainty of the boundary work, as shown in the section below, this includes contributions from the high-speed pressure measurements and the volume calculation. The total uncertainty of the pressure measurement is calculated as a quadrature addition of the random and systematic uncertainty. The systematic uncertainty of the pressure sensors is 0.172 (2.5) bar (psi) as verified by calibration and is assumed to be fixed. The random uncertainty in the pressure measurement is calculated using the 20 sample points taken per operating condition and calculated throughout the rotation of the shaft. The calculated total uncertainty of the pressure measurement ranged from 0.172 bar to 0.405 bar depending on the sensor and operating condition. The largest uncertainty consistently came from the discharge pressure sensor (DP). Test condition 8, from Table 1, has a total uncertainty of the pressure ranging 0.172 bar to 0.177 bar having a relatively small random uncertainty.

Table 1: Final test matrix of 36 operating conditions (presented with various saturated suction and discharge temperatures, SST and SDT, respectively) with a fixed 11.1 K superheat and shaft speeds high-speed measurements.

Speed	SST	Test #	SDT	Speed	SST	Test #	SDT
rpm	°C	-	°C	rpm	°C	-	°C
1100	4.4	5	37.8	1300	4.4	6	37.8
		4	43.3			3	43.3
		11	48.9			10	48.9
		12	54.4			13	54.4
1500	4.4	7	37.8	**1700**	1.67	34	48.9
		2	43.3		**4.4**	**8**	**37.8**
		9	48.9			1	43.3
		14	54.4			16	48.9
LV	-1.1	17	37.8			15	54.4
		18	43.3			32	32.2
		19	48.9			31	35.0
		20	54.4			33	48.9
	4.4	30	37.8		7.22	35	48.9
		27	43.3		12.78	36	48.9
		26	48.9				
		21	54.4				
	10	29	37.8				
		28	43.3				
		25	48.9				
		22	54.4				
	16	24	48.9				
		23	54.4				

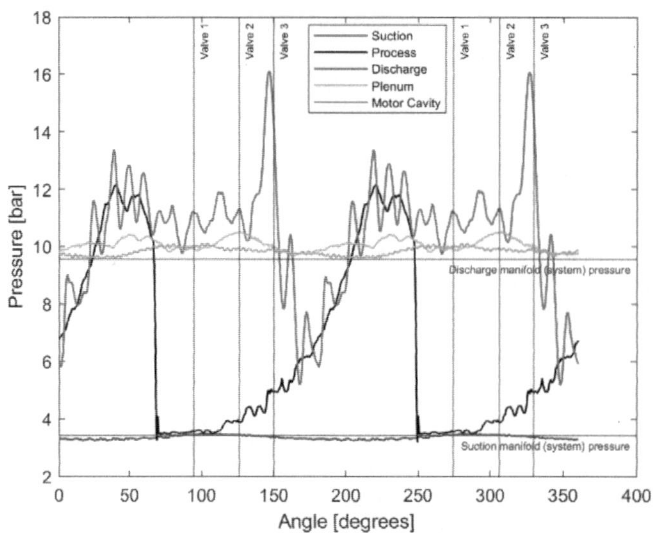

Figure 3: The five resulting pressure signals for test condition 8 from Table 1 as a function of compressor crank angle, highlighting the suction and discharge pressures measured at the load stand and overlaying the starting location of the discharge valve rows.

3.1.1 Uncertainty associated with fixed compressor speed assumption on the volume calculation

The inductive proximity sensor installed on the compressor will trigger once every rotation. To calculate the volume of the compressor at each instant requires an assumption that the rotational speed of the compressor remains constant throughout a single rotation. Large changes in shaft torque throughout the compression process necessitate that the shaft speed is not strictly uniform throughout a single rotation. To quantify the uncertainty of this assumption an analysis was performed using high-fidelity shaft position information collected from a similar spool compressor prototype, with data presented in Bradshaw *et al.* (2018). Bradshaw *et al.* (2018) collected data on an open-drive spool compressor prototype, with similar geometric characteristics and operating parameters as the compressor in this work, equipped with a 4096 steps/revolution rotary encoder that was also tested with R134a. Using the position data from the rotary encoder it is possible to estimate the variation in shaft speed over a rotation. For an operating condition similar to test condition 14, it was found that the shaft speed in the open-drive prototype varied by no more than 20 rpm.

The resulting differences in instantaneous volume are calculated using a fixed speed and the encoder speed was then estimated using the geometric model presented in Bradshaw and Groll (2013). It was found that the assumption of fixed speed would result in errors in volume averaging 0.67%. An additional assumption is then made that the open-drive and semi-hermetic mechanism operation is similar enough that the uncertainty in the both compressors volume calculations will be roughly the same.

The resulting total uncertainty from each of the two sources, volume and pressure, is used as an inputs to an uncertainty propagation analysis for the calculation of the boundary work as calculated following the procedure in equation 1. This analysis results in an error propagation of 0.85% for test condition 8 boundary work. This method follows ASME PTC 19.1 to combine the total uncertainty based on the respective loss expression described in the following sections.

3.2. Analysis of discharge losses

The discharge loss is a result of flow losses associated with the discharge port, valves, plenum, and flow into the motor cavity. Pressure drop associated with these areas result in a chamber pressure that is higher than the discharge manifold (system) pressure (p_{dis}) to overcome these losses. This loss is illustrated in Figure 4 and can be calculated using Equation 1 where the chamber pressure is equal to the discharge chamber (DP) during the discharge process, therefore the total discharge loss is calculated as,

$$L_{discharge} = \underbrace{-\int_{V_{min}}^{V_{dis,start}} p_{DP} dV}_{w_{BW,dis}} - \underbrace{\left(p_{dis}\left(V_{dis,start} - V_{min}\right)\right)}_{w_{ideal}}. \quad (1)$$

where the boundary work is integrated between the minimum compression volume and the volume corresponding with the cylinder pressure exceeding the discharge manifold (system) pressure (p_{dis}). The location of the discharge sensor allowed for the entire discharge process to be captured by the discharge sensor.

The discharge loss can be further separated due to the additional pressure transducers in the discharge plenum and motor cavity. These two sensors allow for the discharge loss to be expanded into three separate losses, motor cavity loss, plenum loss, and valve loss. To separately capture the losses associated with the fluid leaving the discharge plenum the same technique is used using pressure data collected from the DC sensor. This assumes that the boundary work required to push fluid from the discharge plenum to the discharge manifold (system) pressure requires the same change in volume as the discharge process itself. Therefore, the total losses from discharge plenum can be evaluated as,

$$L_{dis,plenum} = \underbrace{-\int_{V_{min}}^{V_{dis,start}} p_{DC} dV}_{w_{BW,dis}} - \underbrace{\left(p_{dis}\left(V_{dis,start} - V_{min}\right)\right)}_{w_{ideal}}. \quad (2)$$

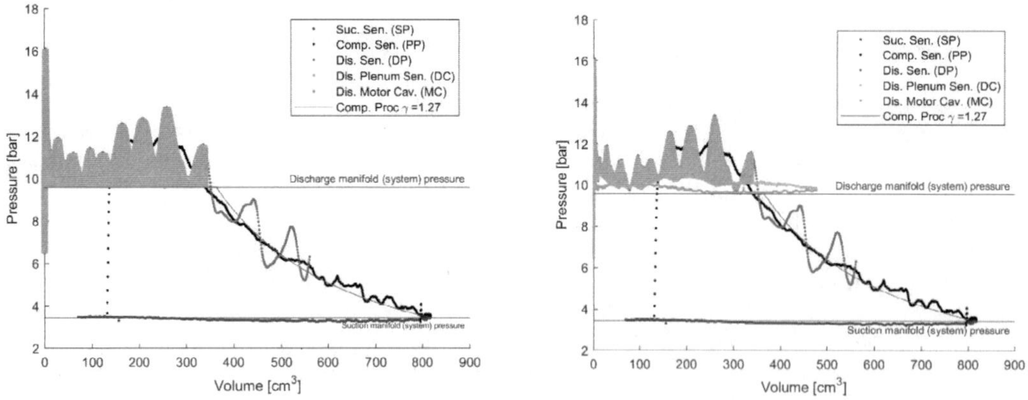

Figure 4: Indicator diagram of Test condition 8 highlighting the areas representing the total discharge loss (left) and the valve losses (right) as calculated by equations 1 and 3.

Following the same procedure at the discharge plenum the motor cavity losses are calculated using the MC sensor. Finally, taking the difference between the plenum losses and the total discharge losses the remainder is assumed to be dominated by the discharge valves/ports as shown by the shaded portion of Figure 4. Therefore, a derived loss is defined to capture this as,

$$L_{\text{valves}} = L_{\text{discharge}} - L_{\text{dis,plenum}}. \tag{3}$$

3.3 Analysis of compression losses

The compression process losses are associated with pressure during the closed compression process and calculated relative to an isentropic compression process. An isentropic compression process is modeled using a polytropic compression process where the polytropic exponent is assumed to be the ratio of specific heats of the refrigerant calculated at the various suction conditions of the compressor. The general expression to find the loss for the compression process is defined as,

$$L_{comp} = -\int_{V_{max}}^{V_{dis,start}} (p_{PP} - p_{ideal})dV, \tag{4}$$

where p_{ideal} is the isentropic compression process, $V_{dis,start}$ is the volume where the compression stops and discharge process begins, and $V_{\max}$ is the maximum volume of a single compression pocket.

3.4 Analysis of suction losses

Suction losses are associated with flow losses within the suction port as well as leakage that occurs during the suction process. These values are calculated using a similar procedure as the discharge process. The loss is therefore given as,

$$L_{comp} = -\int_{V_{max}}^{V_{dis,start}} (p_{SP} - p_{suction})dV. \tag{5}$$

4. RESULTS

This section presents results of the experimental campaign including a detailed breakdown of losses for the 30-ton spool compressor at test condition 8 and an analysis of the trends in the losses. Additionally, the losses are globally analyzed across the breath of conditions presented in the test matrix. Finally, a more detailed analysis of the compression losses is explored. The losses in this section are presented as percentage of total work measured from the compressor.

4.1 Loss breakdown at 40 °F SST, 100 °F SDT, and 1700 rpm (Test condition 8)

Figure 5 (left) shows the indicator diagram of operating condition Test condition 8 (left). Additionally, the suction manifold (system) pressure, discharge manifold (system) pressure and calculated isentropic work pressure is also overlaid on top of the data as a basis of comparison. The indicator diagram is also presented in this figure (right) with the three main loss areas shaded to reflect the areas used in the analysis presented in the previous section. The results of the analysis for Test condition 8 are collected in Table 2 and broken down by percentage loss associated with each loss mechanism as a percentage of total compressor work.

Table 2: Collection of losses for Test condition 8 presented as a percent of total work.

Test condition 8 Results					
Discharge	**Valves**	**Plenum**	**Motor Cavity**	**Suction**	**Compression**
%	%	%	%	%	%
18.04	12.16	3.21	2.67	1.78	.68

This table shows that the total discharge loss is by far the largest loss at 18.04% with a portion of that loss being associated with the plenum losses (3.21%) and the motor cavity losses (2.67%), leaving 112.16% of valve losses. Both the plenum and motor cavity losses are significant but neither one is as dominant as the discharge valve losses.

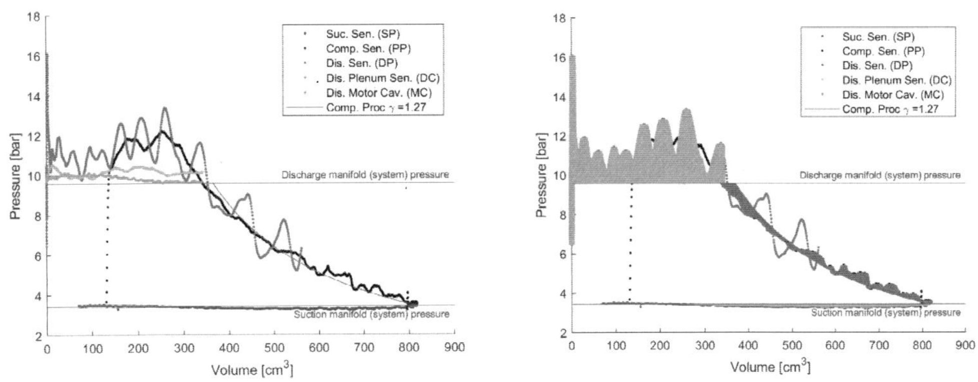

Figure 5: Indicator diagram of Test condition 8 with system suction and discharge pressures as well as the estimated isentropic compression process overlaid (left) and the loss areas shaded (right).

The suction loss (1.78 %) reflects the suction pressure measured blow the system pressure. Suction loss occurring in this manner is the typical expected loss that occurs at the suction port. The spool compressor tested doesn't have suction valves therefore this loss likely occurs as a result of pressure drop in the suction port.

The compression loss is relatively small despite not qualitatively matching the ideal compression process as shown in Figure 5. The measured process is closer to isothermal than the isentropic. Therefore, the specific work required is reduced and this results in a 'negative' loss for the portion of the process that approaches the start of the discharge process. This overcomes some of the additional work required to overcome the portion of the process just at the beginning of the compression process. Overall, the analysis from Test condition 8 suggested that the discharge plenum, motor cavity, and valve losses were the largest in the compressor and should be further explored.

4.2 Discharge plenum and valve losses

This section will expand on the single operating condition analysis by presenting the suction, compression, discharge, discharge plenum, motor cavity, and valve losses with various operating conditions with an emphasis of the three discharge-related losses.

Figure 6 presents the losses for the suction, compression, and discharge (left) and the plenum, motor cavity, and discharge (right) for speeds ranging from 1100 rpm to line voltage (approx. 1750 rpm) compressor shaft speeds at all the SST and SDT reported in Table 1. The suction, compression, motor cavity and plenum losses for all the speeds and operating conditions appear to show unremarkable trends. Figure 6 highlights that these four losses present with minimal trends with shaft speed. A separately analysis was also explored with SST and SDT and a trend wasn't discovered. The analysis does show a persistent loss of roughly 2-4% for suction, 0-1% for compression, 3-5% for plenum and 1-3% for the motor cavity. The addition of the motor cavity sensor has allowed for this distinction between the flow losses in the plenum and the motor cavity.

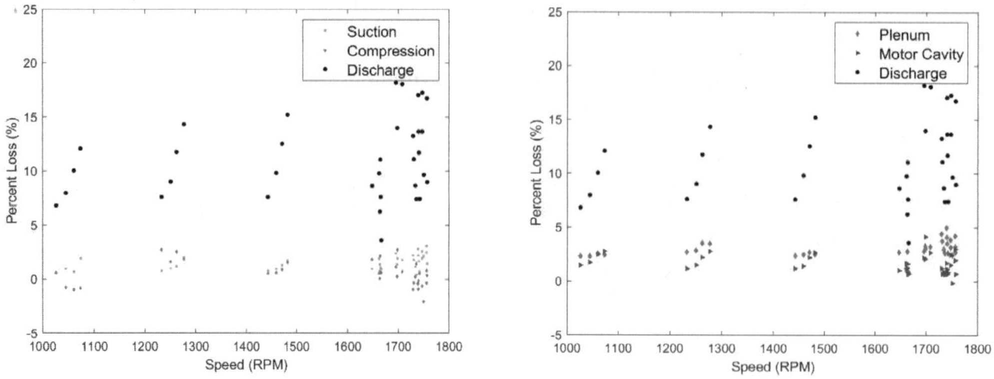

Figure 6: Indicated discharge plenum, motor cavity and valve losses shown for various speeds (right) and Indicated suction, discharge, and compression losses for various speeds (left). The marker size includes the experimental loss uncertainty.

In contrast, the valve losses show some significant trends as presented in Figure 7. The shaft speed and SDT correlate strongly with the valve loss. The valve loss increases by as much as 5% with speed while the loss decreases by as much as 3% at higher SDT conditions.

It is hypothesized that these losses are associated with either the discharge valves or the valve port/port placement. The valves are generally still not well understood. The results suggest that a significant performance improvement is possible with further optimization of the discharge process, most notably the valve performance..

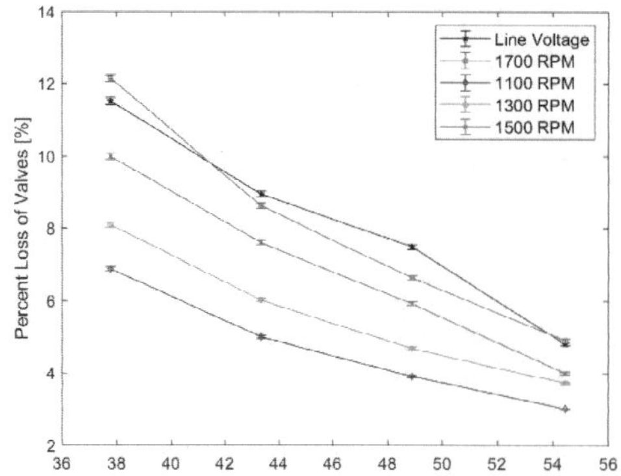

Figure 7: Percent Loss of discharge valves v. SDT at a constant 4.4 C (40F) SST for various speeds with error bars.

5. CONCLUSIONS

This paper presents an indicated loss analysis for the 8th generation 30-ton prototype spool compressor. The losses were collected using high-speed pressure measurements from five locations within the compressor and synchronized with a proximity sensor as the trigger mechanism. The results suggest that the largest simple opportunity for improvement in this prototype is a re-design of the discharge plenum. This re-design has the potential to result in 3-4% improvement in the overall compressor efficiency. Additionally, a redesign of the motor cavity could add an additional potential improvement of 2-3%.

The valve losses presented an increasing trend with increasing speed and decreasing SDT. The valve improvements presented are the largest potential improvement with losses ranging from 9-18% depending on speed and operating condition. The valve losses presented as somewhat independent of suction conditions but relatively sensitive to discharge condition. The indicated losses of the valves indicate that more study of the dynamics of the valves is necessary to ensure the losses are mitigated across the entire operating range.

NOMENCLATURE

L	indicated loss	(Btu, kJ)
p	pressure	(psia, kPa)
V	volume	(in^3, m^3)
w	specific work	(Btu, kJ)

Subscript

BW	boundary work
comp	compression process
dis, plenum	discharge plenum portion of discharge
dis, start	start of discharge process
discharge	total discharge process
DC	discharge cover/discharge plenum
DP	discharge pocket
MC	motor cavity
ideal	ideal process
PP	process/compression pocket
suc, suction	suction process
SP	suction pocket
valves	valve portion of discharge

REFERENCES

ASME PTC 19.1-2018, Test uncertainty, ASME. 2018.

Bradshaw, C. Spool compressor tip seal design considerations. In 8th International Conference on Compressors and their Systems (2013), p. 341.

Bradshaw, C. R., and Groll, E. A. A comprehensive model of a novel rotating spool compressor. International Journal of Refrigeration 36, 7 (2013), 1974 – 1981. New Developments in Compressor Technology.

Bradshaw, C. R., Kemp, G., Orosz, J., and Groll, E. A. Development of a loss pareto for a rotating spool compressor using high-speed pressure measurements and friction analysis. Applied Thermal Engineering 99 (2016), 392–401.

Kemp, G. T., Garrett, N., and Groll, E. A. Novel rotary spool compressor design and preliminary prototype performance. In International Compressor Engineering Conference at Purdue University (2008), no. 1328.

Orosz, J., Bradshaw, C., Kemp, G., and Groll, E. Updated performance and operating characteristics of a novel rotating spool compressor. In International Compressor Engineering Conference (2016), no. 1377.

Rigola J., Pérez-Segarra C. D., Raush G., Oliva A., Escribà M., Jover J., and Escanes F. (2002). Experimental Studies of Hermetic Reciprocating Compressors with Special Emphasis On PV Diagram. International Compressor Engineering Conference, Paper 1506. https://docs.lib.purdue.edu/icec/1966

US Energy Information Administration (EIA) Electric Power Monthly, Table 5.1 (2011)

US Department of Energy (DOE) 2010 Buildings Energy Data Book, Sections 2.1.5 and 3.1.4 (2011)

Westphalen, D. and S. Koszalinski 2001. Energy Consumption Characteristics of Commercial Building HVAC Systems Volume I : Chillers, Refrigerant Compressors and Heating Systems.

Electrochemical Compression of Ammonia/Hydrogen Blends for Heat Pump Application

Ashish CHOUHAN[a], Utsav R. ARYAL[a], Premanand O. SIVAKUMAR[b], Bamdad BAHAR[c*],

Ajay K. PRASAD[a]

[a]Department of Mechanical Engineering, University of Delaware, Newark DE 19716, USA
[b]Department of Chemical Engineering, University of Delaware, Newark DE 19716, USA
[c*]Xergy Inc., 299A Cluckey Dr., Harrington DE 19952, USA (bamdad.bahar@xergyinc.com,

Tel. 302 218 4279)

ABSTRACT

Ammonia is widely used as a refrigerant in air conditioning, food preservation and for fuel storage purposes. Ammonia is an attractive candidate for electrochemical compression (ECC). Electrochemical compressors can be an ideal replacement for mechanical compressors because of noiseless operation, higher efficiencies at lower voltages and the ability to operate without the use of Global Warming Potential (GWP) working fluids. This paper explores the electrochemical compression of ammonia with hydrogen as a carrier gas, using fuel cell type hardware. A perfluorosulfonic acid (PFSA) membrane allows the transfer of ammonia, using hydrogen as a carrier gas, by the conduction of NH_4^+ ions from anode to cathode. The ECC performance was evaluated using electrochemical methods using a single cell at STP. The effects of back-diffusion in the compression process are highlighted and suggestions to improve the ECC of ammonia-hydrogen blends are proposed.

Keywords: Ammonia; Hydrogen; Electrochemical compression; Nernst voltage; Diffusion; Compression Ratio

1. INTRODUCTION

Global energy demand is rising rapidly as we face the depletion of conventional sources of energy. There is a need to look for sustainable alternatives to traditional mechanical energy conversion devices such as heat pumps that account for a large and growing portion of the global demand for energy. The developments in fuel cell technology to generate electricity in a clean and environment-friendly manner has placed hydrogen as an important energy vector to meet both stationery and automobile power generation applications. Ammonia is a rich source of hydrogen and could also play a significant role as an energy vector in a hydrogen economy. Moreover, ammonia itself is widely used as a refrigerant in large scale industrial heat pump applications particularly in applications involving the food chain. Electrochemical compression of ammonia therefore can be used for conveying ammonia when used as a bulk energy vector but could also be an alternative to mechanical compressors for heat pump applications.

Electrochemical compression is superior to mechanical compression as it offers the potential to compress a working fluid at higher efficiency, noiseless operation, usage of green refrigerants, and greater reliability due to a lack of moving parts.

The first validation of the concept of an electrochemical compressor based on hydrogen was done by Sedlak et al. [1], who applied a potential difference to transfer protons across a proton exchange membrane and achieved higher pressures. Rohland et al. [2] and Strobel et al. [3] carried the idea forward to electrochemically transfer and purify hydrogen. All these studies were only focused on hydrogen compression until 2009, when the general idea of using electrochemical compression for heat pump applications was pioneered by Bahar et al. [4]. Bahar et al. [5] then proposed the use of ammonia in electrochemical compressors (ECC) using hydrogen as the carrier gas by engaging perfluorosulfonic acid (PFSA) membranes to electrochemically compress ammonia under a potential difference. An overall system analysis was favorably reported by Tao et al. [6] in 2017.

Ammonia is a well-established contaminant for PFSA membranes in fuel cell applications and is actively scrubbed out of reformed hydrogen gas streams used for fuel cells. Our challenge in this project was to understand and then develop transportation mechanisms for ammonium ions in ion-exchange membranes.

The hardware of an ECC is similar to that of a fuel cell and consists of a membrane-electrode-assembly (MEA) sandwiched between two graphite plates with flow fields for gas distribution, held together by end plates on either side. The entire assembly is held together by eight bolts, all maintained at a same torque (Figure 1).

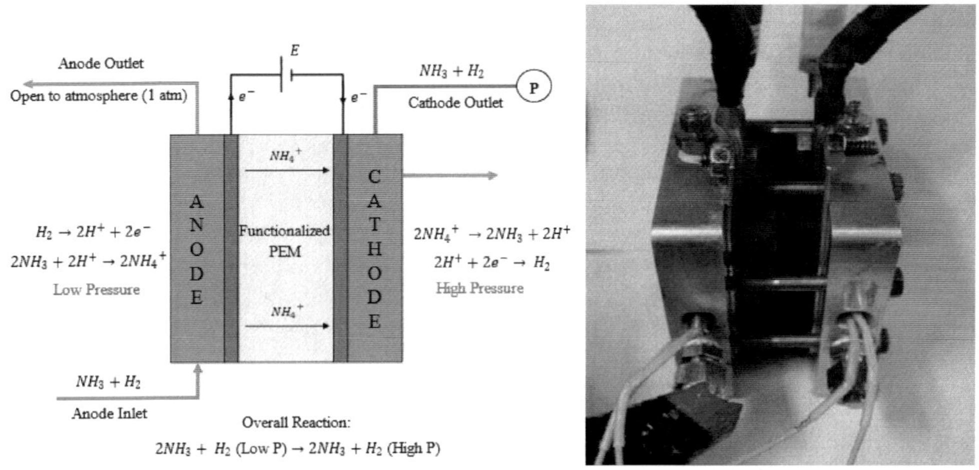

Fig. 1. Schematic of an electrochemical ammonia compressor (left); ECC hardware (right)

A mixture of ammonia and hydrogen gases is fed to the anode, where, by applying a DC voltage, hydrogen gets oxidized to protons (H^+), which combine with ammonia to form ammonium ions (NH_4^+), as in Equations 1 and 2.

$$H_2 \rightarrow 2H^+ + 2e^- \tag{1}$$

$$2NH_3 + 2H^+ \rightarrow 2NH_4^+ \tag{2}$$

The ammonium ions get transferred across the membrane to the cathode. Two electrons are required for this reaction to take place, as per Equation 1. The ammonium ions are reduced at the cathode to regenerate ammonia and hydrogen gases at a higher pressure, as shown in Equations 3 and 4.

$$2NH_4^+ \rightarrow 2NH_3 + 2H^+ \tag{3}$$

$$2H^+ + 2e^- \rightarrow H_2 \tag{4}$$

The overall electrochemical reaction can be written as:

$$2NH_3 + H_2\,(low\ pressure) \rightarrow 2NH_3 + H_2\,(high\ pressure) \tag{5}$$

The production of protons facilitates the formation of ammonium ions which is the species conducted by the PFSA membrane. The membrane needs to be functionalized by soaking in aqueous ammonia solution overnight. In this paper, the use of ammonia as a working fluid in an electrochemical compressor (ECC) is reviewed. The performance of the ECC is presented in the form of output pressures reached within the phase conversion pressure limits of gaseous ammonia at room temperature; cathode pressures and current densities are also plotted for all voltages tested.

The membrane used for experiments was a sulfonated tetrafluoroethylene-based fluoropolymer-copolymer membrane, 50 µm thick. The platinum (Pt) catalyst loading on the electrodes was 0.3 mg/cm^2. The membranes were functionalized for ion exchange from H^+ to NH_4^+ by soaking in ammonia solution overnight, after which, they were thoroughly rinsed with deionized water and dried gently using paper towels. The membrane surface has residual ammonia and does not readily stick to the electrodes, so the membrane and electrodes were stacked together inside the ECC and assembled by tightening the bolts. The ECC was assembled with 25 cm^2 hardware.

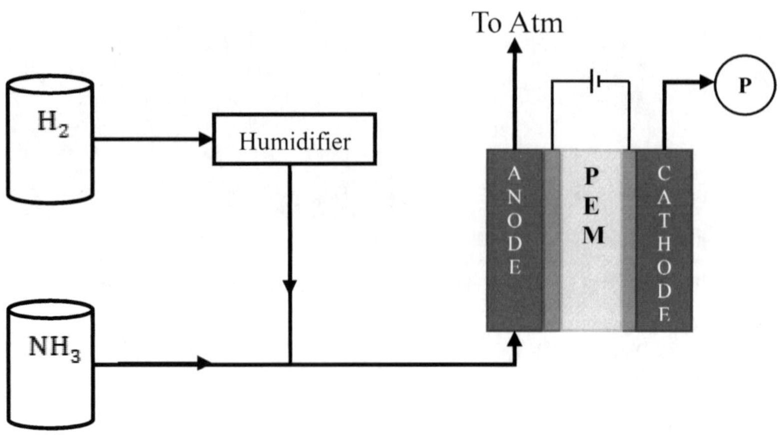

Fig. 2. Experimental setup of the electrochemical ammonia compressor

The ECC was tested at room temperature (22°C) and pressure (1 atm). The PFSA membranes used for experiments only function when properly hydrated [7, 8]. Hence, it is important to maintain adequate humidity in the system. The inlet gas stream at the anode consisted of a 50-50 mixture of fully hydrated hydrogen (100% RH at a flow rate of 100 cm^3/min) and dry ammonia (0% RH at a flow rate of 100 cm^3/min), resulting in an overall humidity around 50%, whereas, the cathode inlet was closed and cathode outlet port terminated with a pressure transducer (OMEGA PX309-10KG5V). A schematic of the experimental setup is shown in Figure 2. The ammonia inlet stream was kept dry as ammonia reacts with water to form ammonium hydroxide, which can precipitate inside the humidifier chamber. The anode outlet was open to the environment, hence the anode pressure was fixed at 1 atm. The ECC cell was tested using the fuel cell test station (FCTS) from Arbin Instruments. The experiments were performed at four different operating voltages: 0.025, 0.05, 0.075 and 0.1 V.

2. RESULTS & DISCUSSION

Before preparing the MEA, the membranes were soaked in ammonia solution, which makes them alkaline in nature implying that the contribution of H^+ to the generated current can be neglected, which is also evident by the almost zero current when the ECC was tested with only hydrogen at the anode. Also, the total amount of ammonia and hydrogen gases generated at the cathode can be expressed in terms of total current I:

$$\dot{n}_{H_2} = \frac{I}{2F} \tag{6}$$

$$\dot{n}_{NH_3} = \frac{I}{F} \tag{7}$$

where $\dot{n}_{H2}$ is the mol/s of hydrogen transferred across the membrane according to Faraday's law [9], $\dot{n}_{NH3}$ is the mol/s of ammonia transferred, and F is Faraday's constant.

The applied cell voltage E_{cell} is the sum of the Nernst potential, the activation overpotential, and ohmic (IR), mass transport and back diffusion losses [10]:

$$E_{cell} = E_{Nernst} + \Delta E_{Activation} + \Delta E_{Ohmic} + \Delta E_{Transport} + \Delta E_{Back\ Diffusion} \tag{8}$$

The activation overpotential $\Delta E_{Activation}$ is a sum of the overpotentials at the anode and the cathode. The ohmic losses ΔE_{Ohmic} arise mainly due to the membrane resistance and contact resistance between different components of the ECC. At high current densities, mass transport losses denoted by $\Delta E_{Transport}$ could become significant due to the inability to maintain adequate reactant gas supply to the anode catalytic sites; however this loss term is not expected to be significant in the case of ammonia compression due to the small operating currents. $\Delta E_{Back\ Diffusion}$ represents losses due to back diffusion of ammonia and hydrogen from the pressurized cathode to the anode through the membrane. All of these losses cause the required voltage to exceed the Nernst voltage for a given compression ratio. The Nernst potential E_{Nernst} defined by the Nernst equation, Equation 9, depends logarithmically on the pressure ratio between the cathode (high) and anode (low). The total pressure at the anode and cathode can be expressed in terms of the partial pressures of hydrogen and ammonia.

$$E_{Nernst} = \frac{RT}{F} \ ln\left\{\left(\frac{p_{H_{2,out}}}{p_{H_{2,in}}}\right)^{\frac{1}{2}}\left(\frac{p_{NH_{3,out}}}{p_{NH_{3,in}}}\right)\right\} \tag{9}$$

where R is the universal gas constant, and T is the cell temperature in K.

The ECC performance at different operating voltages is presented in Figure 3. For each applied voltage, the current density (red curve) is the highest at the start of the operation and declines rapidly at first before gradually reaching an equilibrium value after a few hours. Concurrently, the cathode pressure (blue curve) rises rapidly at the start and eventually plateaus to an equilibrium value. Both the initial and equilibrium current densities increase with the applied voltage as shown in Figure 3. Similarly, the equilibrium pressures also increase with the applied voltage. The fact that the current is non-zero although the cathode pressure has equilibrated indicates that back-diffusion is occurring.

The backward diffusive flux $(\dot{n}_b)$ as given by Fick's law is directly proportional to the difference in the concentration of individual species (ΔC) across the membrane and inversely proportional to the membrane thickness (d):

$$\dot{n}_{b,NH_3} = \mathcal{D}_{NH_3} A \frac{\Delta C_{NH_3}}{d}, \quad \dot{n}_{b,H_2} = \mathcal{D}_{NH_3} A \frac{\Delta C_{H_2}}{d} \tag{10}$$

where $\dot{n}_{b,NH_3}$ and $\dot{n}_{b,H_2}$ are the backward diffusive fluxes for ammonia and hydrogen, respectively. $\mathcal{D}_{NH_3}$ and $\mathcal{D}_{NH_3}$ are the diffusivities of ammonia and hydrogen, respectively, in the membrane. The concentration of ammonia and hydrogen at either face of the membrane is directly related to their partial pressures through Henry's law. Henry's constant (H) depends on the type of membrane and the gas in contact with the membrane [10].

$$C = HP \tag{11}$$

Combining Equations 10 and 11, we can express the diffusive fluxes as a function of cathode and anode pressures as:

$$\dot{n}_{b,NH_3} = (\mathcal{D}H)_{NH_3} A \frac{(P_c - P_a)_{NH_3}}{d} \tag{12}$$

$$\dot{n}_{b,H_2} = (\mathcal{D}H)_{H_2} A \frac{(P_c - P_a)_{H_2}}{d} \quad (13)$$

where $\mathcal{D}H$ is the permeability constant and depends on the nature of the gas and membrane. Denoting $\dot{n}_n$ as the net molar flow rate to the cathode (= forward flux – back diffusive flux), and assuming that both ammonia and hydrogen follow the ideal gas law, we can write:

$$(\dot{n}_n)_{NH_3} = \dot{n}_{f,NH_3} - \dot{n}_{b,NH_3} = \frac{\dot{p}_{c,NH_3} V_c}{RT} \quad (14)$$

$$(\dot{n}_n)_{H_2} = \dot{n}_{f,H_2} - \dot{n}_{b,H_2} = \frac{\dot{p}_{c,H_2} V_c}{RT} \quad (15)$$

where $\dot{P}_c$ is the rate of cathode pressure rise ($\mathrm{Pa/s}$), and V_c is the pressurized volume on the cathode side ($\mathrm{m^3}$). Substituting the expressions for forward and backward fluxes in Equations (14) and (15), the compression equations can be written as:

$$\text{Ammonia: } \frac{I}{F} - (\mathcal{D}H)_{NH_3} A \frac{(p_{NH_3,c} - p_{NH_3,a})}{d} = \frac{\dot{p}_{NH_3,c} V_c}{RT} \quad (16)$$

$$\text{Hydrogen: } \frac{I}{2F} - (\mathcal{D}H)_{H_2} A \frac{(p_{H_2,c}(t) - p_{H_2,a})}{d} = \frac{\dot{p}_{H_2,c} V_c}{RT} \quad (17)$$

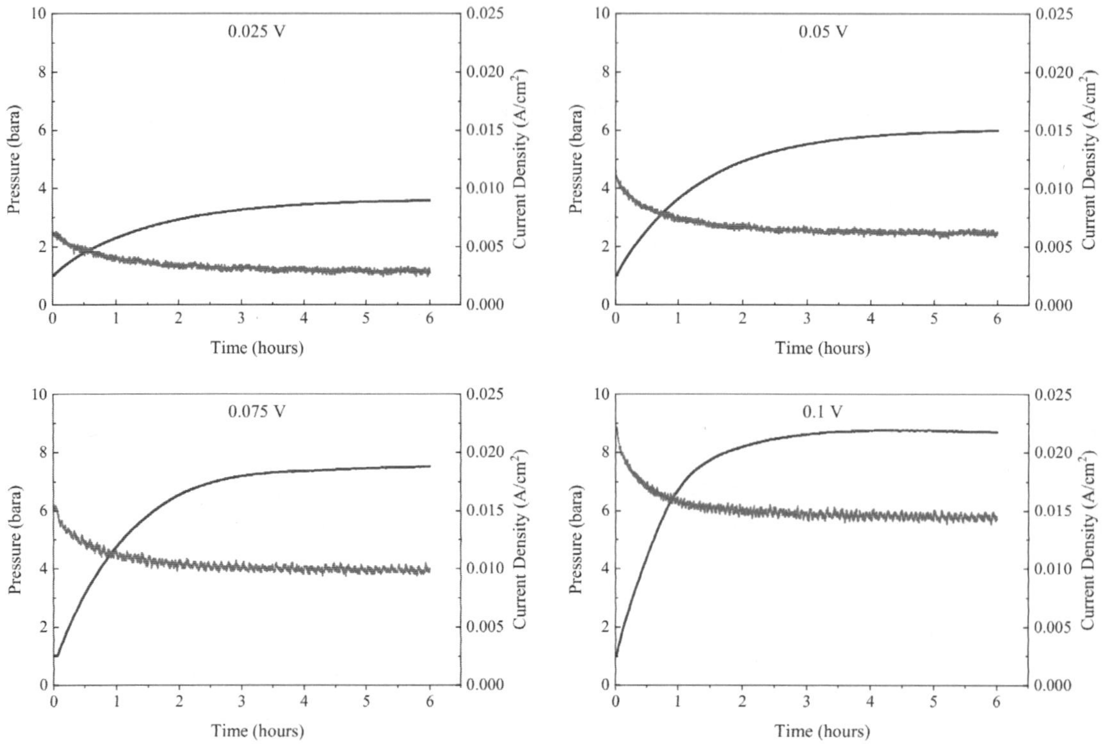

Fig. 3. Pressure (blue) and current density (red) vs. time for Nafion 212 at different applied voltages

Since ammonium is a much bulkier ion than a proton, the current density observed in this case is much less than in the case of a pure hydrogen compressor [7].

When the pressure reaches its equilibrium value, $\dot{P}_c = 0$, in an ideal scenario without back-diffusion, the equilibrium current would also reduce to zero. However, in reality, the forward flux equals the backward flux at steady state, and hence, there is an equilibrium current which maintains the cathode at equilibrium pressure $(P_{c,eq})$ given by Equations (18) and (19).

$$\frac{I_{eq}}{F} = \frac{(DH)_{NH_3} A \left(P_{c,eq} - P_a\right)_{NH_3}}{d} \tag{18}$$

$$\frac{I_{eq}}{2F} = \frac{(DH)_{H_2} A \left(P_{c,eq} - P_a\right)_{H_2}}{d} \tag{19}$$

As the pressure builds at the cathode, with the anode still at atmospheric pressure, it creates a pressure difference across the membrane, which in turn forces ammonia to permeate back through the membrane to the anode by back-diffusion. This poses a barrier in achieving higher pressures. The back-diffusion losses increase as the pressure difference across the membrane increases, hence lowering the ECC efficiency at higher pressures.

The composition of the ammonia-hydrogen gas mixture at the cathode side was measured using the Nesslerization method at two different stages of compression – (i) cathode open and (ii) cathode closed. The cathode outlet stream containing the ammonia-hydrogen mixture was passed into a graduated cylinder (250 mL) which was filled with and held inverted in a 1250 mL beaker containing Nessler's reagent. Ammonia in the outlet stream immediately and completely reacts with Nessler's reagent to change its color from pale yellow to yellowish-brown [11,12], which was analyzed using a spectrophotometer (Vernier SpectroVis Plus). The hydrogen in the mixture is inert with respect to Nessler's reagent and was collected in the graduated cylinder using the downward displacement method.

First, the quantification of the cathode outlet stream was carried out when the cathode was open to the atmosphere. At this stage, both anode and cathode were maintained at the same pressure $(P_a = P_c)$ and hence, back-diffusion losses were zero. In this case, the ratio of ammonia to hydrogen measured was 2, which is also the stoichiometric ratio of NH_3 and H_2 at the cathode as per Equations 1-4.

Next, the cathode outlet was closed and ECC was allowed to pressurize. After the pressure attained equilibrium, the quantification of the cathode outlet stream was carried out by slowly opening the outlet valve and passing the constituents to the Nessler's reagent solution. The ammonia-hydrogen ratio obtained in this case is plotted along with the results from the cathode open case for all four operating voltages in Figure 4.

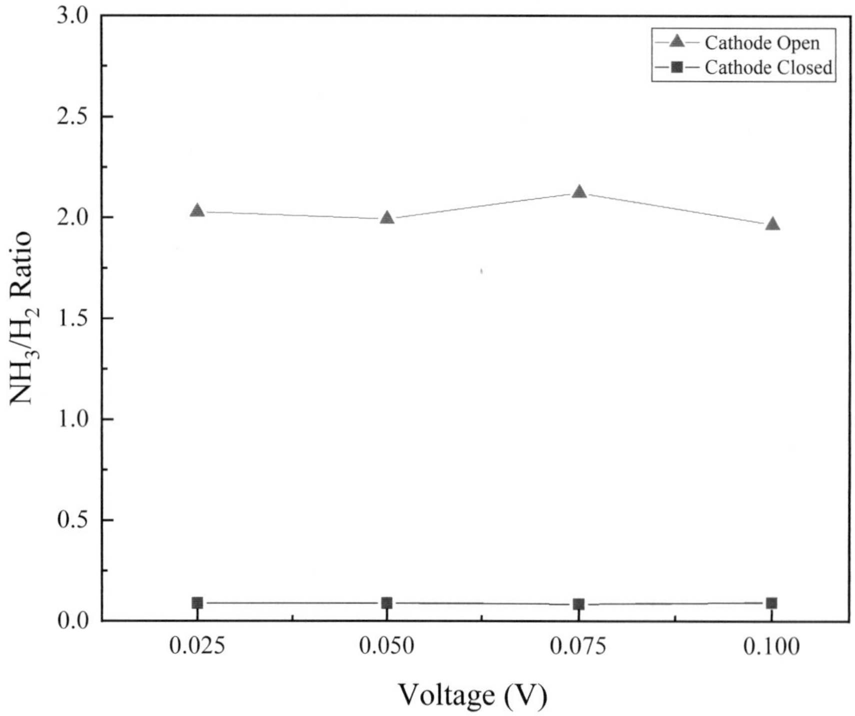

Figure 4. $NH_3{:}H_2$ volumetric ratios for 'cathode open' and 'cathode closed' cases for the four voltages tested

The $NH_3{:}H_2$ for cathode open case was measured to be 2. However, the ratio obtained in the 'cathode closed' case is only about 0.085. This means that the cathode chamber is predominantly composed of hydrogen in the 'cathode closed' case. Thus, one can conclude ammonia diffuses back to the anode at a significantly faster rate than hydrogen.

Furthermore, separate depressurization experiments were carried out to evaluate the permeability coefficients of both hydrogen and ammonia through the Nafion membrane. The cathode was initially pressurized by directly connecting it to the respective gas cylinders. The applied voltage was set to zero and the cathode was allowed to depressurize. The gas diffusing back from cathode to anode was detected by the formation of gas bubbles when the anode outlet tube was immersed in oil.

The rates of depressurization can be determined by setting $I = 0$, $P_{NH_3,a} = 0$, and $P_{H_2,a} = 0$ in Equations (16) and (17), which gives exponential expressions for the rates of depressurization:

$$P_{NH_3,c}(t) = P_{NH_3,c,i}\exp\left(-\frac{(DH)_{NH_3}ART}{dV_c}t\right) \tag{20}$$

$$P_{H_2,c}(t) = P_{H_2,c,i}\exp\left(-\frac{(DH)_{H_2}ART}{dV_c}t\right) \tag{21}$$

where $P_{NH_3,c,i}$ and $P_{H_2,c,i}$ are the initial cathode pressures of ammonia and hydrogen, respectively, at the start of depressurization.

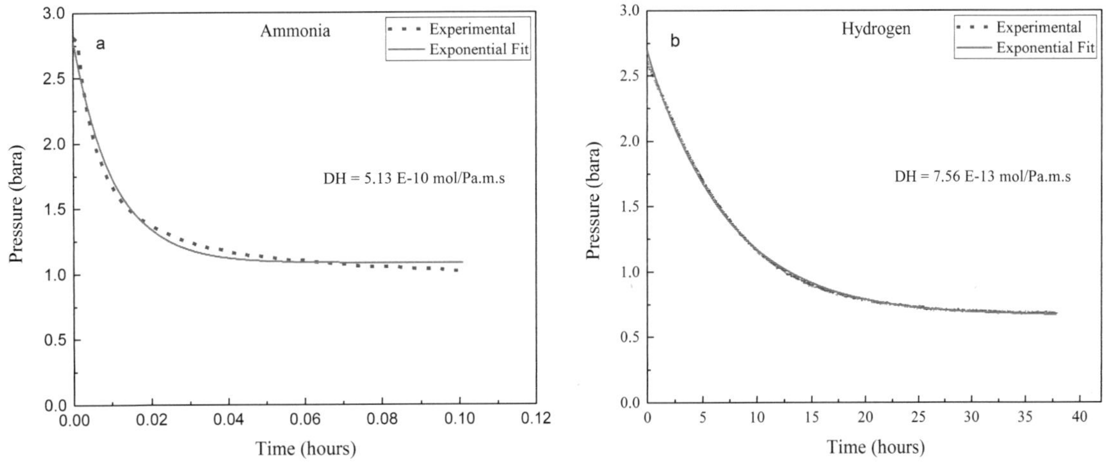

Figure 5. Cathode depressurization profiles for (a) ammonia, and (b) hydrogen

Figure 5 shows the measured depressurization curves (blue) obtained for both ammonia and hydrogen. The exponential curve fits (red) closely match the experimental curves for both gases as shown in Figure 5. Using the curve fits and substituting the known values of $\boldsymbol{A, R, T, d}$, and $\boldsymbol{V_c}$, into Equations (20) and (21), the values of the two permeability coefficients are obtained as:

$$(\mathcal{D}H)_{NH_3} = 5.13\,E-10\ \mathbf{molPa^{-1}m^{-1}s^{-1}}$$

$$(\mathcal{D}H)_{H_2} = 7.56\,E-13\ \mathbf{molPa^{-1}m^{-1}s^{-1}}$$

The permeability coefficients obtained here are similar to the values found for Nafion membranes for ammonia ($\sim 9E-10\ \mathrm{molPa^{-1}m^{-1}s^{-1}}$) in Refs. [12-14] and hydrogen ($\sim 3.5\ E-14\ \mathrm{molPa^{-1}m^{-1}s^{-1}}$) in Refs. [15–17]. Figure 5 shows that when $(\mathcal{D}H)_{NH_3} \gg (\mathcal{D}H)_{H_2}$, then the cathode chamber is mostly composed of hydrogen and ammonia is only present in trace quantities. The Nafion membrane functionalized in ammonia solution is highly permeable to molecular ammonia and hence, it is very difficult to pressurize ammonia using an ECC with the given set of materials and conditions. Nafion being a PFSA membrane, requires moisture to function properly. However, higher hydration levels also exacerbate back-diffusion. Therefore, a better candidate for ammonia compression could be a membrane that can operate at low humidity levels. Also, membrane chemistry could be optimized to enhance the conduction of ammonium ions while mitigating the diffusion of molecular ammonia. A thicker membrane could also reduce the rate of back-diffusion and improve the working process of compression.

3. CONCLUSION

This work discusses the advantages of an electrochemical ammonia compressor over a conventional mechanical compressor. Electrochemical compression hardware was used for the ECC of an ammonia-hydrogen blend; hydrogen was used as the carrier gas, which facilitates the generation of ammonium ions. The ECC was operated at various voltages using a single cell setup. When the cathode outlet is open, the amount of ammonia produced at the cathode is twice that of hydrogen. When the cathode is closed, the pressure rises rapidly and eventually plateaus to an equilibrium value. Similarly, the current is maximum at the start of the compression and drops to attain an equilibrium value. The quantification of cathode constituents during equilibrium resulted in an ammonia:hydrogen ratio to be around 0.085, which is much smaller than the ideal value of 2. Separately conducted depressurization experiments for ammonia and hydrogen showed that the permeability coefficient of ammonia is about three orders of magnitude larger than that of hydrogen. This implies that ammonia back-diffuses much faster than hydrogen and hence, hydrogen contributes almost the entire cathode pressure. Successful ECC of ammonia would require an alternate membrane formulation that allows the forward transfer of ammonium ions while simultaneously reducing the back diffusion of molecular ammonia.

ACKNOWLEDGEMENTS

This work was supported by a research award DE-EE-0008218 funded by the U.S. Department of Energy, Building Technologies Office.

REFERENCES

[1] J. M. Sedlak, J. F. Austin, and A. B. Lacont, "Hydrogen recovery and purification using the solid polymer electrolyte electrolysis cell," International Journal of Hydrogen Energy, Vol. 6, pp. 45-51, 1981.
[2] B. Rohland, K. Eberle, R. Ströbel, J. Scholta, and J. Garche, "Electrochemical Hydrogen Compressor," Electrochimica Acta, vol. 43, pp. 3841-3846, 1998.
[3] R. Ströbel, M. Oszcipok, M. Fasil, B. Rohland, L. Jörissen, and J. Garche, "The Compression of Hydrogen in an Electrochemical Cell Based on a PE Fuel Cell Design," Journal of Power Sources, vol. 105, pp. 208-215, 2002.
[4] B. Bahar, US Pat., 8,769,972 2009; 8,627,671 2010; 9,464,822 2011.
[5] B. Bahar, US Pat., 9,909,224 2016.
[6] Y. Tao, W. Gibbons, Y. Hwang, R. Radermacher, and C. Wang, "Electrochemical ammonia compression," Chem. Commun., 2017,53, 5637-5640.
[7] A.V. Anantaraman, CL, Gardner, "Studies on ion-exchange membranes. Part 1. Effect of humidity on the conductivity of Nafion®", Journal of Electroanalytical Chemistry, vol. 414, pp 115-120, 1996.
[8] J. Zhang, Y. Tang, C. Song, Z. Xia, H. Li, H. Wang, and J. Zhang, "PEM fuel cell relative humidity (RH) and its effect on performance at high temperatures," Electrochimica Acta, vol 53, pp 5315-5321, 2008.
[9] J. Larminie, and A. Dicks, "Fuel cell systems explained", John Wiley & Sons LTD, 2001.
[10] Chouhan A, Bahar B, Prasad AK. Effect of back-diffusion on the performance of an electrochemical hydrogen compressor. International Journal of Hydrogen Energy, 45(2020), pp. 10991-99. https://doi.org/10.1016/j.ijhydene.2020.02.048.
[11] Jeong H, Park J, Kim H. Determination of NH 4 + in Environmental Water with Interfering Substances Using the Modified Nessler Method. Journal of Chemistry 2013;2013. https://doi.org/10.1155/2013/359217
[12] Koch FC, McMeekin TL. A new direct nesslerization micro-kjeldahl method and a modification of the nessler-folin reagent for ammonia. Journal of the American Chemical Society 1924;46:2066–9. https://doi.org/10.1021/ja01674a013.
[13] He Y, Cussler EL. Ammonia permeabilities of perfluorosulfonic membranes in various ionic forms. Journal of Membrane Science 1992;68:43–52. https://doi.org/10.1016/0376-7388(92)80148-D.
[14] Lee YT, Iwamoto K, Seno M. Gas permeabilities of perfluorocarboxylate membranes in various ionic forms. Journal of Membrane Science 1990;49:85–93. https://doi.org/10.1016/S0376-7388(00)80779-6.
[15] Sakai T, Takenaka H, Torikai E. Gas Diffusion in the Dried and Hydrated Nafions. Journal of The Electrochemical Society 2019;133:88–92. https://doi.org/10.1149/1.2108551.
[16] Broka K, Ekdunge P. Oxygen and hydrogen permeation properties and water uptake of Nafion® 117 membrane and recast film for PEM fuel cell. Journal of Applied Electrochemistry 1997;27:117–23. https://doi.org/10.1023/A:1018469520562.
[17] Bessarabov D, Millet P. PEM Water Electrolysis, Volume 2. Academic Press; 2018. https://doi.org/10.1016/c2018-0-02127-1.

Investigation of Back-Diffusion Losses in an Electrochemical Hydrogen Compressor

Ashish CHOUHAN[1], Bamdad BAHAR[2*], and Ajay K. PRASAD[1]

[1]Department of Mechanical Engineering, University of Delaware, Newark, DE 19716

[2]Xergy Inc., Harrington, DE 19952 (bamdad.bahar@xergyinc.com, Tel: 302 218 4279)

ABSTRACT

Hydrogen offers the potential to decarbonize the automotive and stationary power sectors and is therefore expected to play an increasingly significant role in meeting global energy demand. However, due to its low volumetric and gravimetric energy densities, it is important to find methods to efficiently store hydrogen in order to grow the hydrogen economy. Storing hydrogen as a compressed gas could be achieved by electrochemical compression (ECC), which is a membrane-based alternative to conventional mechanical compressors. ECC can be superior to mechanical compressors because of its higher efficiency, lack of moving parts and noiseless operation. Here, we report on the ECC of hydrogen using a Nafion 115 membrane at room temperature. Pressure vs. time curves have been collected at various operating voltages, and a compression ratio of 150 has been achieved with a single cell at an operating voltage of 0.1 V. This work focuses on analyzing the contribution of back-diffusion to the performance of an electrochemical hydrogen compressor.

1. INTRODUCTION

Greenhouse gas emissions from the combustion of fossil fuels in responsible for climate change. With a rapid increasing in the world energy demand, it is imperative to develop energy generation and conversion technologies that are not based on fossil fuels. One viable clean-energy alternative is hydrogen which can be produced renewably in a carbon-free manner and used to generate electricity with environmentally friendly and energy-efficient devices like fuel cells. Hydrogen is therefore an attractive fuel for both stationary and automotive applications. However, being the lightest gas, hydrogen is difficult to store either as a compressed gas or as a liquid. According to the DOE [1], the ultimate targets for gravimetric and volumetric efficiency of hydrogen storage are 2.2 kWh/kg and 1.7 kWh/L, respectively. In comparison, the state-of-the-art efficiencies are 1.4 kWh/kg and 0.8 kWh/kg. Creating and expanding the hydrogen economy of the future requires advances in the field of hydrogen storage.

Storing hydrogen as a compressed gas is often preferred because of the simplicity of the process. Hydrogen is typically compressed using a mechanical compressor which has moving parts and requires frequent maintenance, adding to cost and complexity. One alternative to mechanical compression is electrochemical compression (ECC). Electrochemical compression is superior to mechanical compression because it ensures noiseless operation due to a lack of moving parts. Moreover, for lower power applications, ECC is more efficient than mechanical compression and is less prone to failure. ECC is also attractive for refrigeration and air-conditioning systems because it can replace conventional mechanical compressors while permitting the use of environmentally friendly refrigerant gases.

These multi-fold advantages over mechanical compression make ECC an attractive candidate for pressurization applications. In this paper, we will focus on ECC of hydrogen.

ECC of hydrogen (i.e. hydrogen pumping) based on polymer electrolyte membranes (PEM) was originally developed several decades ago [2, 3]. ECC is accomplished with an electrochemical cell, whose assembly is shown schematically in Figure 1. The ECC consists of a membrane electrode assembly (MEA) and gaskets, two bipolar plates with gas flow channels, and two end plates for cell compression. When an external voltage is applied across the electrodes, hydrogen entering the ECC at the anode undergoes oxidation to generate protons and electrons. The protons migrate through the PEM to the cathode, while the electrons pass through an external circuit and recombine with the protons at the cathode to produce hydrogen at an elevated pressure [4, 5, 6]. The anode and cathode reactions are shown below:

$$H_2 \text{ (low P)} \rightarrow 2H^+ + 2e^- \qquad \text{(Anode)} \qquad (1)$$

$$2H^+ + 2e^- \rightarrow H_2 \text{ (high P)} \qquad \text{(Cathode)} \qquad (2)$$

The overall electrochemical reaction can be written as:

$$H_2 \text{ (anode, low pressure)} \rightarrow H_2 \text{ (cathode, high pressure)} \qquad (3)$$

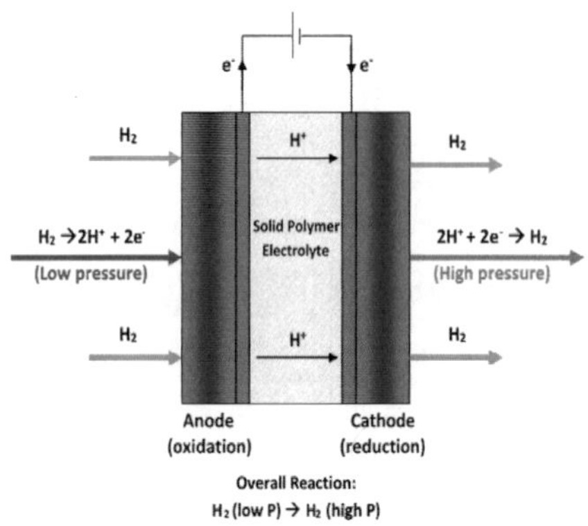

Figure 1. Schematic of an electrochemical cell for the compression of hydrogen.

Figure 1 reveals that the hardware for ECC is identical to PEM fuel cells. While PEM fuel cells have been investigated extensively over the past decades, the literature for ECC is rather sparse. Rohland et al. [4] used Nafion membranes to conduct ECC experiments at STP and showed a somewhat linear behavior between voltage and current density. Strobel et al. [5] demonstrated the ability to reach high cathode pressures and evaluated the back-diffusion rate and diffusion coefficients for different membranes tested. Sedlak et al. [7] demonstrated the use of PEM for electrochemically pumping and purifying hydrogen. Wong et al. [8] used a stack assembly of multiple cells connected in series to achieve high pressures up to 85 MPa. Similarly, Bouwman et al. [9, 10] used a series configuration to compress hydrogen to pressures as high as 100 MPa. More recently, Nordio et al. [11] undertook experimental and modeling studies to investigate the electrochemical separation and compression of hydrogen from nitrogen and methane. Suermann et al. [12] studied electrochemical losses related to gas crossover due to hydrogen pressurization up to 5 MPa. Grigoriev et al. [13] employed PEM to pressurize hydrogen to 5 MPa in a single step using a small voltage of 0.140 V. Nguyen et al. [14] investigated the performance of an electrochemical hydrogen compressor at different operating conditions and pointed out its limits at low inlet hydrogen concentrations. Perry et al. [15] used a high temperature polybenzimidazole membrane for electrochemical hydrogen pumping and tested its durability under typical operating conditions for 4000 hrs. Kee et al. [16] used a thermodynamics-based approach to compare the performance of electrochemical hydrogen compressors at low temperature using a polymer-based PEM (e.g. Nafion) and at high temperature using a ceramic-based PEM (e.g. yttrium-doped barium zirconates).

Perfluorosulfonic acid (PFSA) membranes like Nafion have been extensively used for both fuel cell and ECC applications because of their superior mechanical and chemical durability and high proton conductivity. However, Nafion also permits the back-diffusion of molecular hydrogen which can limit the maximum achievable pressure. Most of the previous studies focused on achieving higher pressures. In this paper, we focus on the effect of back diffusion and develop a theoretical basis for the effect of back diffusion on the cathode pressure and the overall performance of the ECC.

2. EXPERIMENTAL PROCEDURE

The experiments were carried out using a single cell apparatus with an active area of 25 cm^2 provided by Xergy, Inc. The end plates (2.5 cm thick) and bipolar plates (0.6 cm thick) with parallel flow field channels were machined out of aluminum. Each flow field channel was 50 mm long, 2 mm wide and 2 mm deep, and adjoining channels were separated by a 2.5 mm wide rib. Nafion 115 (thickness 125 μm) was used as the membrane, and the electrodes used to fabricate the membrane electrode assemblies (MEA) had a platinum (Pt) catalyst loading of 0.3 mg/cm^2 on Vulcan-carbon cloth, purchased from Fuel Cell Store. Mylar gaskets were used between the MEA and the flow fields to ensure a good seal. The MEAs were prepared by hot-pressing under a load of 0.5 metric tons at 80°C for 2 minutes. All of the electrochemical compression tests reported here were carried out using an Arbin Fuel Cell Test Stand.

The ECC experiments reported here were conducted at room temperature (22°C). Nafion only conducts protons when properly hydrated. Therefore, the inlet hydrogen stream to the anode was fully humidified. The anode flow rate was set to 100 cm^3/min, which was sufficiently high to preclude mass transport losses even at the highest currents. The anode outlet was open to the atmosphere and so the anode pressure was maintained at 1 atm. The cathode port was closed off with a pressure transducer (OMEGA PX309-10KG5V). A stainless-steel mesh was placed between the MEA and the bipolar plates to provide structural support to the MEA and ensure that it was not deformed under the very high differential pressures developed during the experiments, thereby mitigating the possibility of MEA failure. To ensure proper sealing at high pressures, O-rings were installed in the bipolar plates. The end plates were held together with eight bolts, each tightened to a torque of 90 N-m. Experiments were performed at four operating voltages: 0.025, 0.05, 0.075, and 0.1 V. Each experiment was continued until the current and pressure reached equilibrium.

3. RESULTS AND DISCUSSION

The rate of hydrogen production at the cathode or forward flux $(\dot{n}_f)$ is directly related to the current drawn by the compressor according to Faraday's law [17]:

$$\dot{n}_f = \frac{I}{zF} \tag{4}$$

where $\dot{n}_f$ is the mol/s of hydrogen generated, I is the current in amperes, $F = 96485\ C/mol$ is Faraday's constant, and z is the number of electrons per molecule of hydrogen. As shown by Equations 1 and 2, $z = 2$ for the ECC of hydrogen. The Nernst equation specifies the ideal voltage required to attain the desired compression ratio:

$$E_{Nernst} = E_o + \frac{RT}{2F} \ln\left(\frac{P_c}{P_a}\right) \tag{5}$$

where R is the universal gas constant (8.314 $Jmol^{-1}K^{-1}$), T is the operating temperature in K, P_c is the cathode pressure, and P_a is the anode pressure. E_o is the standard electrode potential and is equal to zero as the half-cell reactions at the anode and cathode are the same.

The variation of pressure and current density with time during the compression process is presented in Figure 2 for all four voltages. For each applied voltage, the cathode pressure (blue curve) rises rapidly at the start of the operation and eventually plateaus to an equilibrium value. Concurrently, the current density (red curve) is the highest at the start and declines rapidly at first before gradually reaching an equilibrium value after a few hours. As shown in Figure 3, the equilibrium compression ratio $(P_{c,eq}/P_a)$ increases with the applied voltage as predicted by Equation 5. Figure 3 also shows that the initial current density increases linearly with voltage. Likewise, the equilibrium current density also increases roughly linearly with voltage. The equilibrium current density is about an order of magnitude smaller than the initial value.

The fact that the equilibrium current density is non-zero although the cathode pressure has equilibrated indicates that back-diffusion is occurring. In fact, the equilibrium current when converted to a molar flux by Faraday's Law is an exact measure of the backward flux due to Fickian diffusion. The rate of back diffusion is proportional to the pressure differential across the membrane and therefore, at elevated pressures, a higher current is required to overcome the back-diffusive flux.

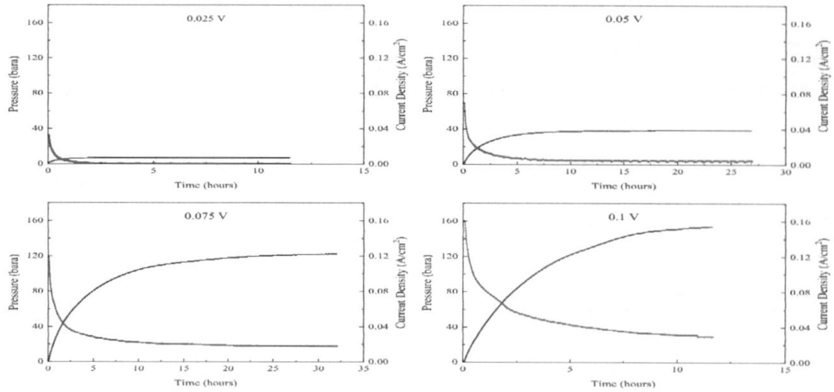

Figure 2. Pressure (blue) and current density (red) vs. time for Nafion 115 – 4 Voltages

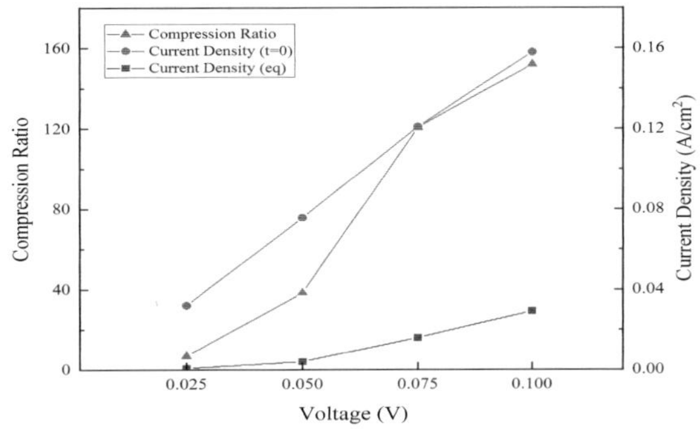

Figure 3. Initial current density, equilibrium current density, and equilibrium compression ratio - 4 applied voltages.

According to Equation 5, the ideal cathode pressures corresponding to 0.025, 0.05, 0.075, and 0.1 V should be 7.24, 51.8, 370, and 2647 bara, respectively. In contrast, the corresponding steady-state pressures are measured as 6.84, 38.8, 122, and 154 bara. Clearly, the steady-state pressures shown in Figures 2 and 3 cannot match the predicted values. For a given compression ratio, the supplied voltage E_{cell} has to be larger than the Nernst voltage in order to overcome all the irreversibilities associated with ECC operation including activation overpotentials at the anode and cathode, ohmic resistance, mass transport losses, and back diffusion:

$$E_{cell} = E_{Nernst} + \Delta E_{Activation} + \Delta E_{Ohmic} + \Delta E_{Transport} + \Delta E_{Back\ Diffusion} \quad (6)$$

where $\Delta E_{Activation}$ is the sum of activation overpotentials η_a and η_c at the anode and cathode, respectively. ΔE_{Ohmic} is the sum of ohmic losses associated with the membrane resistance and contact resistances between the cell components, with membrane impedance being the dominant factor. At high current densities, mass transport losses denoted by $\Delta E_{Transport}$ could become significant due to the inability of the reactant gas to approach the anode catalytic sites fast enough. Mass transport losses could also arise because of water flooding the flow field channels and/or saturating the gas diffusion layers, low porosity of the electrodes or any other factor that restricts the reactant flow. $\Delta E_{Back\ Diffusion}$ represents losses due to back diffusion of hydrogen from the pressurized cathode to the anode through the membrane. All of these losses cause the required voltage to exceed the Nernst voltage for a given compression ratio.

In the ECC of hydrogen, $\Delta E_{Activation}$ is expected to be small because of the fast reaction kinetics of hydrogen oxidation and reduction. Since the compressor operates at fairly low voltages, the current generated is relatively small so reactant starvation is not expected to be significant, and thus $\Delta E_{Transport}$ is also small. ΔE_{Ohmic} could be reduced by employing thinner membranes, however, thinner membranes may not be able to withstand the high pressures being generated and could be more susceptible to failure. Thus, it is essential to find an optimal membrane thickness for the hydrogen compressor. The need to optimize membrane thickness becomes even more important when we consider back-diffusion. If we convert the measured steady-state pressures listed above at the applied voltages of 0.025, 0.05, 0.075 and 0.10 V to their corresponding Nernstian voltages using Equation 5, we would obtain values of 0.024, 0.046, 0.061, and 0.064 V, respectively. The difference between the applied and Nernstian voltage increases rapidly with the compression ratio and clearly indicates that back-diffusion is playing a dominant role. As the pressure builds at the cathode with the anode still being at atmospheric pressure, the pressure differential forces hydrogen to permeate through the membrane back to the anode. This poses a barrier in achieving higher pressures. The backward diffusive flux is $(\dot{n}_b)$ as given by Fick's law is directly proportional to the difference in hydrogen concentration (ΔC) across the membrane and inversely proportional to the membrane thickness (d):

$$\dot{n}_b = -\mathcal{D}A\,\frac{\Delta C}{d} \tag{7}$$

where $\mathcal{D}$ is the diffusivity of hydrogen in the membrane, and A is the active area of the membrane. The concentration of hydrogen at either face of the membrane is directly related to its pressure through Henry's law. Henry's constant (H) depends on the type of membrane and the gas in contact with the membrane.

$$C = HP \tag{8}$$

Combining Equations 7 and 8, we can express the diffusive flux as a function of cathode and anode pressures as:

$$\dot{n}_b = -\mathcal{D}HA\,\frac{(P_c - P_a)}{d} \tag{9}$$

where $\mathcal{D}H$ is the permeability constant and depends on the nature of the gas and membrane. The procedure to obtain the value of $\mathcal{D}H$ will be presented shortly. Hydrogen can be assumed to follow the ideal gas law, giving:

$$\dot{n}_n = \dot{n}_f - \dot{n}_b = \frac{\dot{P}_c V_c}{RT} \tag{10}$$

where $\dot{n}_n$ is the net molar flow rate of hydrogen to the cathode $(\mathrm{mol/s})$, $\dot{P}_c$ is the rate of cathode pressure rise $(\mathrm{Pa/s})$, and V_c is the pressurized volume on cathode side $(\mathrm{m^3})$.

As shown in Figure 4a, hydrogen flux across the membrane changes with time during the compression process. The forward flux (red curve, obtained from Equation 4) decreases rapidly with time at the start of operation and eventually reaches a plateau, the backward flux (blue curve, obtained from Equation 9) increases with time until it attains an equilibrium value, and the net flux (green curve) also decreases with time as shown. When equilibrium is attained, the net flux of hydrogen through the membrane drops to zero and that is also when the cathode pressure reaches its equilibrium value $(\dot{P}_c = 0)$ as predicted by Equation 10. The flux curves in Figure 4a are integrated in time to obtain the respective moles transferred in Figure 4b. Figure 4 pertains to the supplied voltage of 0.075 V; similar curves are observed for all the other voltages tested here.

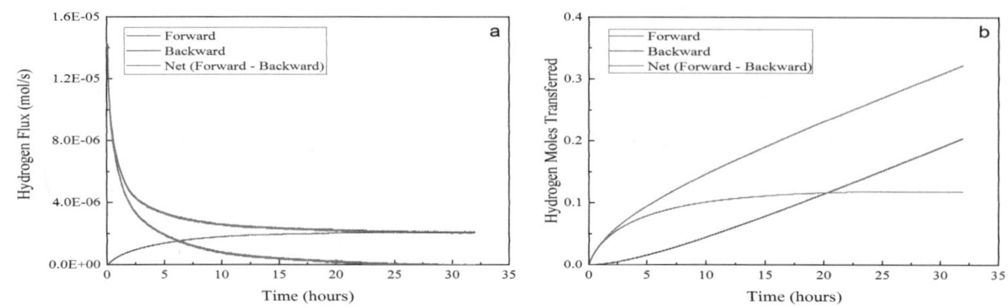

Figure 4. (a) Hydrogen flux through the membrane, and (b) moles transferred at 0.075 V.

Using Equations 4 and 9, Equation 10 can be rewritten to express both current and pressure as functions of time:

$$\frac{I(t)}{2F} - \frac{DHA\,(P_c(t)-P_a)}{d} = \frac{\dot{P}_c(t)V_c}{RT} \tag{11}$$

Equation 11 accounts for the role of back diffusion in the compression process by incorporating the backward flux term $(\dot{n}_b)$. It also provides insight about the electrochemical compression process in three distinct stages:

Initial Condition: At the start of the compression process $(t = 0)$, both anode and cathode are at atmospheric pressure, hence there is no back diffusion $(\dot{n}_b = 0)$. We can then rewrite Equation 11 as:

$$V_c = \frac{IRT}{2F\dot{P}_c} \tag{12}$$

Using the known values of I and $\dot{P}_c$ at $t = 0$ from Figure 2, we obtain the value of V_c as 23.87, 24.98, 24.28 and 23.49 cm^3, respectively, for 0.025, 0.05, 0.075 and 0.10 V. The strong agreement between the V_c values across the different voltages is reassuring. The mean value of V_c is 24.15 cm^3 which is a reasonable estimate for the cathode chamber volume.

Equilibrium: When the pressure reaches its equilibrium value, $\dot{P}_c = 0$. In an ideal scenario without back-diffusion, the equilibrium current would also reduce to zero. However, in reality, the forward flux equals the backward flux at steady state, and hence, there is an equilibrium current which maintains the cathode at equilibrium pressure $(P_{c,eq})$ given by:

$$I_{eq} = \frac{2FDHA\,(P_{c,eq}-P_a)}{d} \tag{13}$$

Using the measured equilibrium currents at the four applied voltages, Equation (13) can be rewritten to calculate the permeability constant:

$$DH = \frac{I_{eq}d}{2FA\,(P_{c,eq}-P_a)} \tag{14}$$

The average value of DH for the four voltages tested was $8.61E-15\ \mathrm{molPa^{-1}m^{-1}s^{-1}}$ which closely matches the values for Nafion 115 found in the literature [18-21].

Depressurization: When the equilibrium pressure is attained and the applied voltage is reset to zero ensuring there is no forward flux $(I = 0)$, the system is then allowed to depressurize. The hydrogen diffusing from the cathode to the anode can be detected by the formation of gas bubbles when the anode outlet tube is immersed in water. In this situation, the rate of cathode depressurization is determined solely by the backward flux and can be obtained by setting $I = 0$ in Equation 11:

$$\dot{P}_c = -\frac{DHART\,(P_c-P_a)}{dV_c} \tag{15}$$

Equation 15 can be integrated in time to show that the cathode pressure follows an exponential behavior while depressurizing:

$$P_c(t) = P_a + (P_{c,eq} - P_a)\exp\left(-\frac{DHART}{dV_c}t\right) \tag{16}$$

where $P_c(t)$ is the time-varying cathode pressure during depressurization, P_a is the (constant) anode pressure, and $P_{c,eq}$ is the equilibrium cathode pressure at the start of depressurization $(t = 0)$.

As shown in Figure 5, the cathode was depressurized after attaining an equilibrium pressure of 60 bara by setting the applied voltage to zero. The anode was continuously supplied with fully humidified hydrogen during depressurization to mirror the operating conditions during pressurization and ensure that the membrane does not dry out during depressurization thereby altering the value of $\mathcal{D}H$. Figure 5 shows that the measured rate of depressurization exactly matches the exponential curve fit from equation (16). From the curve fit, the value of the constant $\frac{\mathcal{D}HART}{dV_c}$ is found to be $1.76E-5\ \mathrm{s}^{-1}$. Substituting the known values of A, R, T and d, we obtain the value of $\frac{\mathcal{D}H}{V_c}$ as $3.59E-10\ \mathrm{molPa^{-1}m^{-4}s^{-1}}$. Substituting $\mathcal{D}H = 8.61E-15\ \mathrm{molPa^{-1}m^{-1}s^{-1}}$ as found earlier, we obtain $V_c = 23.98\ \mathrm{cm}^3$, which is consistent with the values obtained from Equation 12.

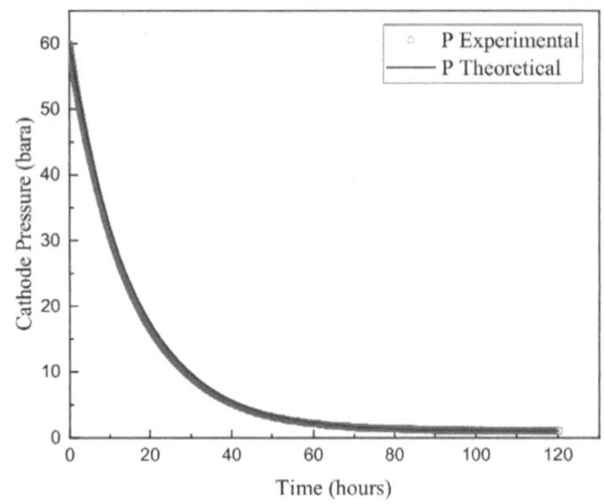

Figure 5. Pressure profile during cathode depressurization (experimental curve and exponential curve fit).

As stated earlier, a thicker membrane is better able to mechanically withstand the large pressure differentials obtained during ECC. At the same time, a thicker membrane would proportionally reduce back-diffusion. Therefore, there is a double benefit to increasing membrane thickness. However, a thicker membrane would increase ohmic losses, but this may be tolerable under the low voltage conditions employed during ECC. Future studies should be aimed at controlling back diffusion by either reducing the membrane's hydrogen permeability or increasing the membrane's thickness. The results obtained in this work also offer valuable insights into the effect of back-diffusion in the ECC efficiency [22] and stack design [23].

4. CONCLUSION

Electrochemical compression of hydrogen was demonstrated using a single-cell apparatus. It was possible to achieve pressures exceeding 150 bar with a small operating voltage of 0.1 V. This work focused on the role of back-diffusion in limiting the performance of an electrochemical hydrogen compressor. At higher pressures, the rate of back-diffusion increases, which in-turn creates a barrier in achieving even higher pressures. A theoretical formulation for the electrochemical compression process including back-diffusion was proposed and experimentally validated. The ability to achieve high pressures at modest voltages using an electrochemical hydrogen compressor makes it an attractive alternative to conventional mechanical compressors. However, back diffusion limits performance, therefore novel membranes with lower hydrogen diffusivities are highly desirable.

REFERENCES

1. DOE Technical Targets for Onboard Hydrogen Storage for Light-Duty Vehicles | Department of Energy [Internet]. Available from: https://www.energy.gov/eere/fuelcells/physical-hydrogen-storage.
2. S.H. Langer, R.G. Haldeman, "Electrolytic hydrogen purification and recovery of same," US Patent 3,475,302A (1969).
3. H.J.R. Maget, "Process for gas purification," US Patent 3,489,670A (1970).
4. B. Rohland, K. Eberle, R. Ströbel, J. Scholta, J. Garche, "Electrochemical Hydrogen Compressor," *Electrochimica Acta*, 1998, 43, 3841-3846.
5. R. Ströbel, M. Oszcipok, M. Fasil, B. Rohland, L. Jörissen, J. Garche, "The Compression of Hydrogen in an Electrochemical Cell Based on a PE Fuel Cell Design," *Journal of Power Sources*, 2002, 105, 208-215.
6. G. Sdanghi, G. Maranzana, A. Celzard, V. Fierro, "Review of the current technologies and performances of hydrogen compression for stationary and automotive applications," *Renewable and Sustainable Energy Reviews*, 2019, 102, 150–170.
7. J.M. Sedlak, J.F. Austin, A.B. LaConti, "Hydrogen recovery and purification using the solid polymer electrolyte electrolysis cell," *International Journal of Hydrogen Energy*, 1981, 6(1), 45–51.
8. T. Wong, F. Girard, T. Vanderhoek, "Electrochemical hydrogen compressor," US Patent US20040211679 (2004).
9. P.J. Bouwman, J. Konink, D. Semerel, L. Raymakers, M. Koeman, W. Dalhuijsen, E. Milacic, M. Mulder, "Electrochemical hydrogen compression," *Journal of the Electrochemical Soc*iety, 2014, 64(3), 1009–18.
10. P.J. Bouwman, "Advances In Electrochemical Hydrogen Compression And Purification," *Journal of the Electrochemical Soc*iety, 2016, 75 (14), 503-510.
11. M. Nordio , F. Rizzia , G. Manzolini , M. Mulder , L. Raymakers , M. V. S. Annaland , F. Gallucci, "Experimental and modelling study of an electrochemical hydrogen compressor," *Chemical Engineering Journal,* 2019, 369, 432-442.
12. M. Suermann, T. Kiupel, T. J. Schmidt, F. N. Buchi, "Electrochemical Hydrogen Compression: Efficient Pressurization Concept Derived from an Energetic Evaluation" *Journal of The Electrochemical Society*, 2017, 164 (12), F1187-F1195.
13. S.A. Grigoriev, I.G. Shtatniy, P. Millet, V.I. Porembsky, V.N. Fateev, "Description and characterization of an electrochemical hydrogen compressor/concentrator based on solid polymer electrolyte technology," *International Journal of Hydrogen Energy,* 2011, 36(6), 4148-4155.
14. M.T. Nguyen, S.A. Grigoriev, A.A. Kalinnikov, A.A. Filippov, P. Millet, V.N. Fateev, "Characterisation of a electrochemical hydrogen pump using electrochemical impedance spectroscopy," *Journal of Applied Electrochemistry*, 2011, 41, 1033–1042.
15. K.A. Perry, G.A. Eisman, B.C. Benicewicz, "Electrochemical hydrogen pumping using a high-temperature polybenzimidazole (PBI) membrane," *Journal of Power Sources*, 2008, 177(2), 478-484.

16. B. L. Kee, D. Curran, H. Zhu, R. J. Braun, S. C. DeCaluwe, R. J. Kee, S. Ricote, "Thermodynamic Insights for Electrochemical Hydrogen Compression with Proton-Conducting Membranes," *Membranes* 2019, 9(7), 77
17. Larminie, James (1 May 2003). Fuel Cell Systems Explained, Second Edition. SAE International. ISBN 978-0-7680-1259-0.
18. M. Schalenbach, T. Hoefner, P. Paciok, M. Carmo, W. Lueke, D. Stolten, "Gas Permeation through Nafion. Part 1: Measurements", *The Journal of Physical Chemistry C*, 2015, 119(45), 25145–25155.
19. T. Sakai, H. Takenaka, E. Torikai, "Gas diffusion in the dried and hydrated Nafions." *Journal of the Electrochemical Soc*iety, 1986, 133(1), 88–92.
20. J. Jiang, A. Kucernak, "Investigations of fuel cell reactions at the composite microelectrode| solid polymerelectrolyte interface. I. Hydrogen oxidation at the nanostructured Pt|Nafion® membrane interface." *Journal of Electroanalytical Chemistry*, 2004, 567(1), 123–137.
21. K. Broka, P. Ekdunge, "Oxygen and hydrogen permeation properties and water uptake of Nafion® 117 membrane and recast film for PEM fuel cell," *Journal of Applied Electrochem*istry, 1997, 27(2), 117–123.
22. A. Chouhan, B. Bahar, A. K. Prasad, "Effect of back-diffusion on the performance of an electrochemical hydrogen compressor," *International Journal of Hydrogen Energy,* 2020, 45(19), 10991-10999
23. A. Chouhan, U. R. Aryal, B. Bahar, A. K. Prasad, "Analysis of an electrochemical compressor stack," *International Journal of Hydrogen Energy,* 2020, 45(56), 31452-31465

ACKNOWLEDGEMENT

This work was supported by the US Department of Energy through the Buildings Energy Efficiency Frontiers and Innovation Technologies (BENEFIT) grant. Portions of this study were originally published in the International Journal of Hydrogen Energy [22].

Development of a Mechanistic Chamber Model of a Novel Peristaltic Compressor for Air-conditioning and Refrigeration Applications

Mazharul ISLAM[(a)], Craig R. BRADSHAW[(b)]

Oklahoma State University, Mechanical & Aerospace Engineering,
Stillwater, OK 74075, USA
[(a)]mazharul.islam10@okstate.edu, [(b)]craig.bradshaw@okstate.edu.

ABSTRACT

The air-conditioning and refrigeration industry are constantly searching for efficiency improvements to vapor compression refrigeration systems. A valve-less compressor with variable volume ratios can significantly increase efficiency and add flexibility to these systems. The novel peristaltic compressor is introduced, which has the ability to operate at variable volume ratios and without valves. This compressor operates by means of a progressively actuated diaphragm that compresses vapor and stimulates flow through a cylindrical chamber. In this study, we will present a discretized thermodynamic model of a peristaltic compressor by splitting the mechanism into a series of conjoined segments. Mass and energy balance equations for each compression segment analyzed to create a mechanistic chamber model of the compressor with the segments interacting through flow of mass between them. This model includes a geometric model of the compression chamber created by the interface between the flexible diaphragm and the cylinder. The geometric model is coupled with the mass and energy balance to predict the compressor performance metrics. The preliminary model results show that the peristaltic compressor presents unique attributes that will be explored in future work.

Keywords: Peristaltic movement, Valve-less compressor, Novel Compressor, Geometric Model, Thermodynamic model.

1. INTRODUCTION

In the United States, 24% of the nation's total energy consumption is used by refrigeration and air-conditioning systems (U.S. EIA, 2018). These two applications use the vapor compression cycle under different operating conditions. Consequently, efficient and innovative technology in the vapor compression equipment has the potential to reduce energy consumption and reduce operating cost. The energy consumption of a vapor compression cycle is highly dependent on the compressor. Energy consumption of vapor compression cycles increases with inefficiencies (losses) in cycles and components. Significantly more losses of vapor compression cycles occur in the compressor compared against the condenser, expansion valve, and evaporator (Ahamed *et al.*, 2011; Pearson, 2015). Improvements to leakage, porting and over/under-compression losses can reduce these losses. A scroll compressor was introduced to eliminate the losses associated with valves and leakage but the compressor has internal volume ratios, which results in over/under-compression losses (Diniz and Deschamps, 2016); similar is true for screw compressors. Reciprocating and rolling piston compressors do not have over/under compression losses but require valves, which results in leakage and/or porting losses associated with those components. A compressor that is able to operate without valves and over/under-compression losses will provide a significant advantage over legacy technologies.

A peristaltic compressor is a promising technology that provides an effective mechanism to eliminate the aforementioned drawbacks. A peristaltic compressor includes a series of segmented compression pockets, similar to a scroll compressor. However, in contrast to the radial compression of a scroll compressor, the peristaltic compressor performs its compression along a linear axis. The segments are formed via a sandwiched diaphragm material (or tube) that is actuated via electromechanical actuation to isolate adjacent portions of the diaphragm. Compression then occurs by reducing the number of actuated electromechanical devices along the linear direction. The plurality of segments allows for isolation of the suction and discharge ports from one another, which removes the need for valves and allows for adjustment of the volume ratio on the fly. These features, combined, have the potential for large improvements in compressor efficiency.

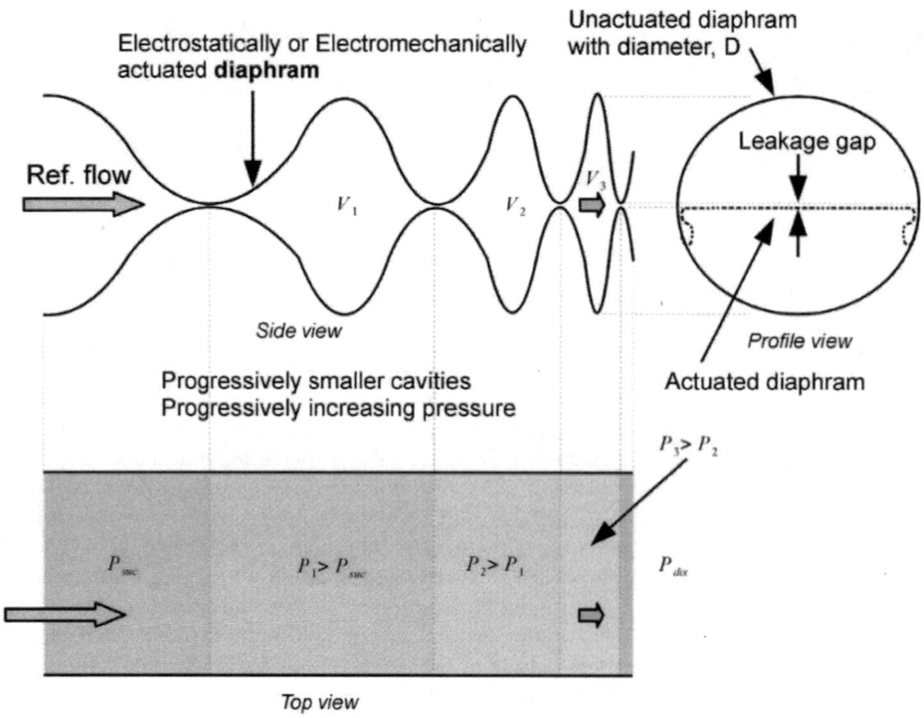

Figure 1: Side, top, profile, and schematic views of peristaltic compressor highlighting mode of compression and leakage gap between pockets (Bradshaw, 2017).

Figure 1 schematically presents the compression mechanism of the peristaltic compressor. While previous studies have explored the peristaltic compressor (Lawless *et al*., 1987; Sulfridge and Miller, 2001) to date, a feasibility study using a validated thermodynamic model has not been presented applied to Heating, Ventilation, Air-Conditioning, and Refrigeration (HVAC&R) industry.

The global objective of this project is to determine thermodynamic advantages of the novel peristaltic compressor, applied to HVAC&R technologies. The approach is to focus on the volumetric performance of this novel technology as it relates to legacy compressor technologies. To that end, a reconfigurable prototype compressor is designed to study this and validate a separately developed thermodynamic model for further study. The work presented here is a component of this global effort and will present a thermodynamic model of the novel peristaltic compressor, which will later be validated against data from the prototype and utilized to explore the advantages compared against legacy compressors.

2. LITERATURE REVIEW

2.1. Peristaltic and Diaphragm-based Compression

The peristaltic compressor as a concept is very unique (Bradshaw et al., 2020) but has been explored, in limited capacity, by a few studies in the past. Lawless *et al.* (1987) designed a miniature compressor which introduced the peristaltic actuation mechanism in the compressor. The system used a hollow deformable elastic tube to flow the gas and electrostrictive ceramic blocks for gas compression. The ceramic blocks were motion amplified that helps them to move back and forth to the tube channel by controlling the voltage to the blocks. The synchronized compression of the blocks on the tube operates the gas to flow and the process reduced the requirement of an inlet and discharge valve. However, 71% of energy is absorbed by the circuit to run the process and the size of the capacitors is very large which reduces efficiency.

Sulfridge and Miller (2001) presented a new model for a mesoscopic peristaltic compressor. A diaphragm placed in a center of a chamber and connected with a programmable logic device. The membrane was driven by an electric field which helped to make a wave pattern and flow the fluid throughout the channel. The compression ratio can be controlled by changing the depth of the cavity which would be helpful to operate the compressor for variable volume ratios. The computational analysis of the paper showed that the peristaltic compressor can perform the required work for a compressor. The analysis of physical testing is missing to evaluate the result and the author did not examine the heat transfer of the system which may influence the behavior and results of the compressor. In the peristaltic process, the pressure rise and volume of the discharge gas depend on the forward stroke of the diaphragm. A maximum pressure rise can be obtained if the stroke operates both compress and discharge of the gas. On the other hand, the flow rate and performance of the system can be maximized when the forward stroke of the diaphragm cover all the compression chamber. (Mathew and Hegab, 2013)

The peristaltic mechanism relies on deformation of a flexible membrane, most similar to a diaphragm in pump and compressor applications. In recent years, diaphragms have been examined as a mechanism to be used as a compressor in vapor compression cycles. This synergy to the peristaltic compressor has shown some useful outcomes to leverage for this work. Previous work showed that the diaphragm offered a less leaky mechanism and the ability to generate higher pressures. Hartley (1999) offered a pump model where the peristaltic actuation mechanism used to move the fluid. The pump had an electrically conductive membrane that was actuated using a voltage applied between a conductive diaphragm layer and a base layer, separated by a non-conductive material. The process is reversed when the voltage is reversed. The process creates pressure and when coupled with an electrostatically actuated valve, creates a mechanism to move fluid. This work was restricted to pumping applications and only considered the electrostatic actuation mechanism which has limited displacement potential.

The existing studies of the diaphragm compressors and pumps highlight the potential for a compressor with low leakage and others suggest it is possible to operate similar mechanisms without valves. However, the existing studies have some limitations and they are all either only a theoretical model of pumps/compressors or are not specifically applied to HVAC&R applications.

2.2. Mechanistic Chamber Models

Mechanistic Chamber models have been used for modeling positive displacement compressors by various individuals (rotary -Mathison (2011), scroll -Bell (2011), linear -Bradshaw (2012), spool -Bradshaw and Groll (2013)). It was first generalized by Bradshaw (2012) and further formalized in recent work and accompanying software by Bell *et al.* (2019). The model is a discretization of a compressor into a series of control volumes. These control volumes represent the various compression chambers at a given instant in a compressor.

Navarro *et al.* (2007) proposed a phenomenological model for a reciprocating compressor to predict the overall isentropic and volumetric efficiency. The approach to the model uses an unsteady mass and energy balance on the compression chamber. The volumetric and isentropic compressor efficiencies were presented as a system of implicit equations and several correction factors are needed in this model to evaluate the results. Xie *et al.* (2004) applied a comprehensive model for the purposes of studying the effect of structural parameters and peristaltic pumping operation on the performance of a hermetic reciprocating compressor. Mainly, bearing frictional power loss, valves, leakage and heat transfer models are presented as sub-models. The predicted results are compared with the published compressor map values. The model was restricted to measure the effects of structural parameters on the compressor performance. Mathison *et al.* (2008) developed a comprehensive model that investigates the pressure and temperature at different crankshaft angle and overall efficiency of the system by considering the effects of compressor chambers leakage and heat transfer between the wall and fluid. Later, Mathison (2011) used the same method to characterize the performance of a novel spool compressor with multiple injection points. The prototype and comprehensive model measured the mass flow rate and power consumption accurately where error of the results varied within 5%. However, error of the predicted discharge temperature and intermediate pressure was very large. Bell (2011) offered a geometric model and overall compressor model to investigate the liquid flooded compression in a scroll compressor. The numerical method helped to improve the compressor performance for use in oil-flooding. Recently, Peng *et al.* (2017) presented a comprehensive thermodynamic model with differential geometry to study an oil-free scroll compressor. The changes of the different parameters calculated by using the improved Euler method. This work presented that under different discharge pressure, the temperature and power change range is large. Bradshaw (2012) presented a miniature-scale linear compressor, which is a free-piston mechanism where the piston is attached to a linear motor and mechanical springs and operated at a resonant frequency. The free-piston embodiment of this technology coupled the dynamics of the piston movement with the thermodynamic model in a unique way. These mechanistic chamber models require geometric and a series of sub-models to solve the compression process equations and evaluate the target parameters to design a novel compressor. The present work develops a mechanistic chamber model for a peristaltic compressor to get better understanding on the actuation technology and improve overall efficiency.

3. PERISTALTIC COMPRESSOR PROTOTYPE

The segmentation of the peristaltic compressor allows to reduce pressure differences between each compression pocket and improve leakage characteristics due to valve-less operation. The main purpose of developing a peristaltic compressor is to get better volumetric efficiency and an optimized actuation technology to improve overall efficiency. The mechanical actuation technology has never been studied to operate a peristaltic compressor. A series of linear actuators are considered in the design to put load on the diaphragm for actuating the flow in the compressor. Figure 2 presents a schematic of the proposed re-configurable peristaltic compressor prototype. The diaphragm is constrained and sealed against a female die which would create a cylindrical compression chamber. A male die is attached with the linear actuator to press the diaphragm against the female die to provide compression by sealing the chamber synchronously. In the model, 10 linear actuators are used to control the mass flow of the compression chamber. Preliminary testing will be done using air as a working fluid in the compression chamber.

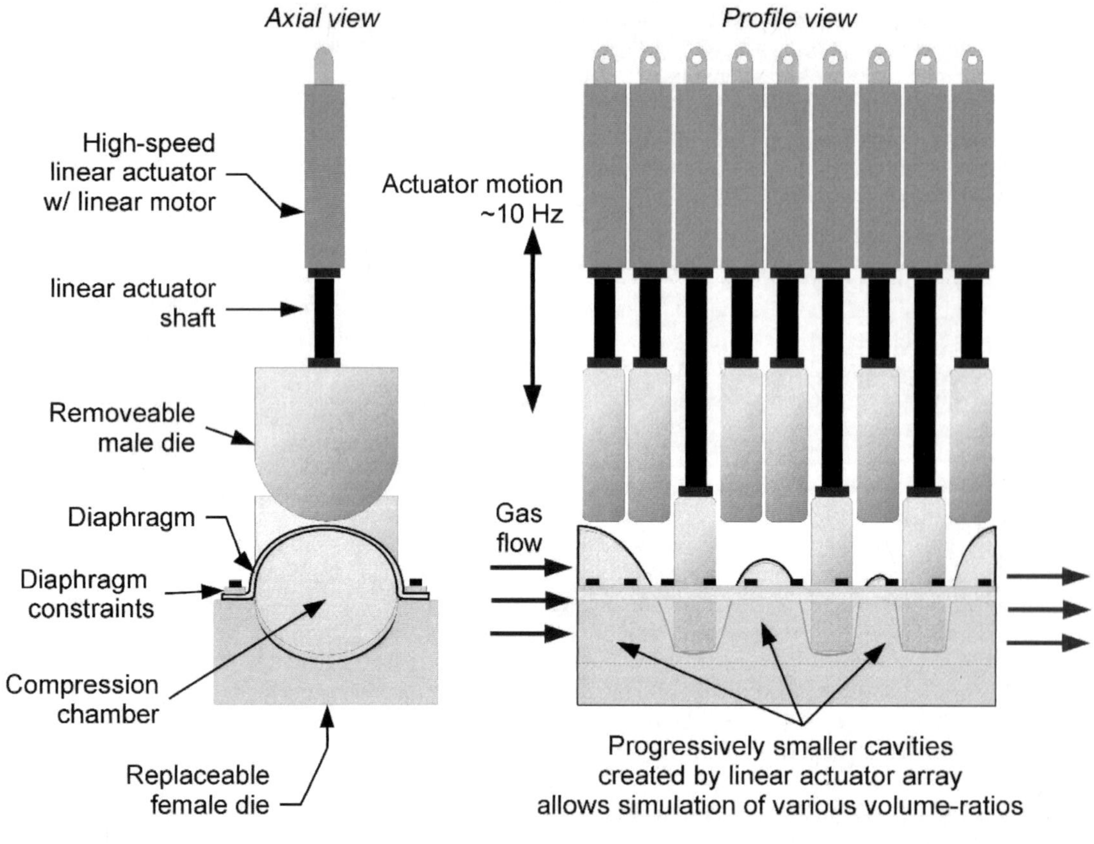

Figure 2: Reconfigurable prototype platform of a peristaltic compressor that has been designed for this study. (Bradshaw, 2017)

The 1st and 10th piston will operate as a suction and discharge valve of the compressor. Gas is compressed, and different volume ratios are achieved, by using a pivotal piston which indicates when the fluid from the compression chamber will begin to discharge. For example, consider a selection of the 7th piston as the pivotal piston, this is illustrated in Figure 3. Initially, the 10th piston block the compression chamber at the discharge side and the working fluid fills the diaphragm. Then, at step 2, the 1st piston isolate the fluid within the compression chamber. At the start of compression in step 3, the 2nd to 6th piston will press the diaphragm sequentially and compress the fluid to a significantly smaller volume. Then in steps 4-5, when the system hits the 7th piston, simultaneously 7th to 9th piston will press the diaphragm in a sequence and the 10th piston will rise to allow fluid to discharge into the system.

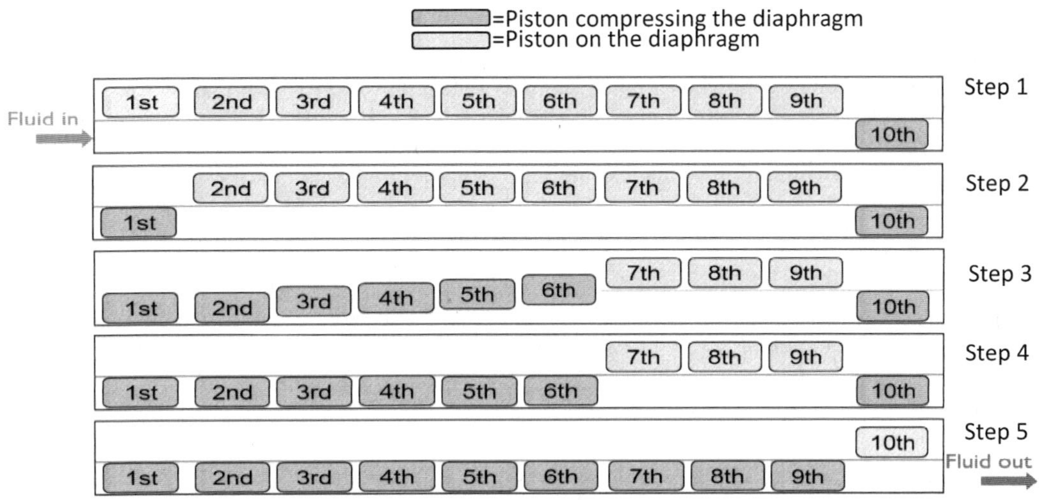

Figure 3: Schematic diagram of the prototype compressor operation with pivotal piston 7 selected.

This process can be controlled to achieve a large variety of different volume ratios of operation by selecting different pivotal pistons. By adjusting the pivotal piston from the 3rd to the 9th the prototype can achieve volume ratios from 1.143 to 8. The results from this large variety of volume ratios will allow for an efficient study of the influence on the volume ratio on relevant compressor metrics such as the volumetric efficiency.

4. MECHANISTIC CHAMBER MODEL OF A PERISTALTIC COMPRESSOR

The thermodynamic model of the peristaltic compressor utilizes the unsteady mass and energy balance of a control volume assuming all work is done by boundary work. Density (ρ) and temperature (T) are the two unknown parameters in the thermodynamic model, which are related using the mass and energy conservation relation. To account for the plurality of embodiments and actuation techniques the peristaltic compressor may encounter, the thermodynamic relationships that govern it was developed as a function of vertical displacement, Z. The actuation process gradually increases the pressure (P) of the fluid by reducing the volume (V) of the compression chamber. Figure 4 shows the computational algorithm for the mechanistic chamber model of the peristaltic compressor highlighting the compression process solver and its required sub-models, which are discussed in the following sections.

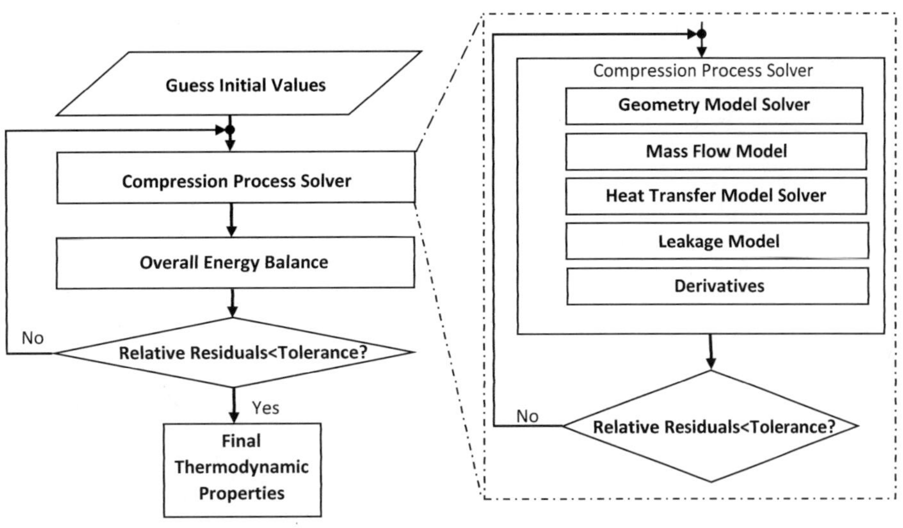

Figure 4: Computational algorithm of the mechanistic chamber model for the peristaltic compressor.

4.1. Compression Process Equations

The compression process equations are derived using an unsteady mass and energy balance equations. The equations provide the state of the working fluid in the compression chamber at any point during the compression process and are solved for each control volume at a particular displacement of the actuator (Z). The density of the system can be determined by differentiating the equation with displacement, which is computed as,

$$\frac{d\rho}{dZ} = \frac{1}{V}\left(\frac{1}{S}\left(\sum m_{in} - \sum m_{out}\right) - \rho\frac{dV}{dZ}\right), \tag{4.1}$$

For the model we have considered actuation speed, S is 0.254 m/s and compression chamber volume, V which will be changed for each compression step. The parameters from the mass balance equations would solve the energy balance equations in order to find the other derivatives. The conservation of energy for a general control volume can be written as follows,

$$\frac{dE_{cv}}{dt} = \dot{Q} - \dot{W} + \sum m_{in}h_{in} - \sum m_{out}\ h_{out}, \tag{4.2}$$

The equation can be expended in terms of changes in internal energy and mass. The following relation is the energy balance equation, which can be obtained by differentiating the expanded equation with displacement,

$$\frac{dT}{dZ} = \frac{\frac{1}{S}(\dot{Q}+\sum m_{in}\ h_{in}-\sum m_{out}h_{out})-uV\frac{d\rho}{dZ}-\left(u\rho+T\left(\frac{\partial P}{\partial T}\right)_v\right)\frac{dV}{dZ}}{\rho V C_v}, \tag{4.3}$$

Here, the only temperature is the independent variable. Several numerical integrators are used to calculate the temperature and density for each step during the compression. Pressure and enthalpy of the system calculated with the known temperature and density using CoolProp (Bell *et al.*, 2014),which is an open-source database of fluid and humid air properties.

Equations 4.1 and 4.3 require a series of inputs that are also functions of the instant in time, called sub-models as highlighted on the right-hand side of Figure 4. The following sections outline the sub-models required to provide complete solutions to the compression process equations including a sub-model for geometry, mass flow, heat transfer and leakage.

4.2. Geometric Model

The mechanistic chamber model requires an instantaneous volume of the peristaltic compressor. In the designed reconfigurable prototype peristaltic compressor, linear actuators operate pistons sequentially to generate volume change, as described in Section 3. Various diaphragm shapes can have impact on the overall efficiency of the compressor. To explore this, two types of diaphragm arrangement are considered, circular and semi-circular compression chamber. During the protype test run, it is shown that the circular compression chamber doesn't able to rebound the diaphragm properly after the linear actuation. For that reason, we have chosen semi-circular compression chamber to analyze the model for the peristaltic compressor.

If we seal a flat diaphragm on the female die. The diaphragm creates a semi-circular compression chamber which is illustrated in Figure 5. The following equation is used to determine the deflected volume (V_d) of the chamber during the actuation:

$$V_d = (R^2\cos^{-1}\left(1-\frac{Z}{R}\right) + (R-Z)\sqrt{2ZR-Z^2})Y, \tag{4.4}$$

Here, Y is depth of the piston and Z is changing from 0 to 0.0127m, meaning at the initial the displacement is 0 and sequentially when the piston touches the female die, at the end the maximum deflection would be 0.0127m.

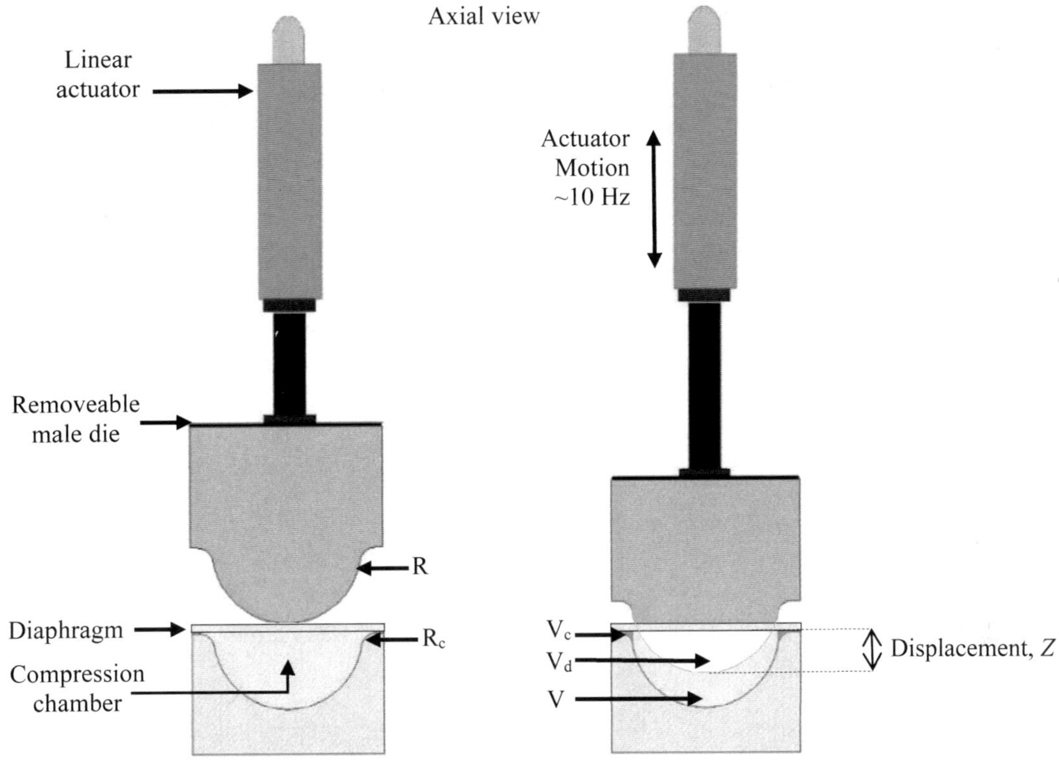

Figure 5: Geometric schematic depicting the geometry model approach of a single-piston is deflecting the diaphragm which is creating a semi-circular compression chamber.

However, there is a curve in both sides of the compression chamber with radius 0.00254m (R_c) which we have to consider during the volume change. The following equation is used to derive the volume of the curve segment (V_c),

$$V_c = \left(R_c{}^2 \cos^{-1}\left(1 - \frac{h}{2R_c}\right) - \frac{1}{4}(2R_c - h)\sqrt{(4R_c - h)h}\right) Y, \tag{4.5}$$

We considered the displacement of the curve section, h is 0 to 0.00127m. To calculate the semi-circular compression chamber's final volume (V) we can subtract the V_d and V_c from the total volume and the following equation is

$$V = \left(\frac{\pi}{2}R^2 + \left(2R_c^2 - \frac{1}{2}\pi R_c^2\right)\right)Y - V_d - V_c, \tag{4.6}$$

Figure 6 (a), shows how the volume of the compression chamber change during one compression cycle for semi-circular diaphragm arrangements. Here, 1st to 9th pistons are pressing the diaphragm in sequence to reduce the compression chamber volume and release the fluid. After that, all pistons rise together to pump the fluid into the channel and reach the initial compression chamber volume. However, the chamber volume for the discharge flow area would be different for the different pivotal piston. For example, figure 6 (b), demonstrates how the discharge volume will change for the pivotal piston 7. Here, 1st to 6th piston will press the diaphragm sequentially to compress the fluid. For the pivotal piston 7, when the system hits the 7th piston the discharge will start. Sequentially 7th to 9th piston will press the diaphragm to move out all fluid from the chamber, throughout this segment, we will get the discharge volume. During compression and suction, there would be no discharge, so the discharge volume would be zero.

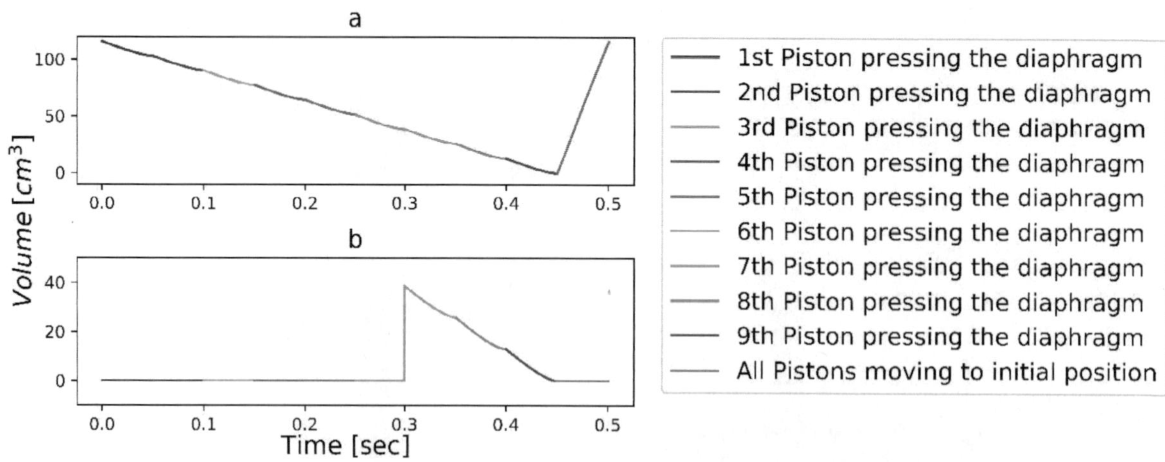

Figure 6: (a) Volume of the compression chamber as a function of time, (b) Discharge volume during the compression for pivotal piston 7.

4.3. Mass Flow Model

When the actuators load the diaphragm, the mass flow changes with respect to the area of the compression chamber. The basic equation of mass flow rate for a fluid with a specific heat (γ) is:

$$\dot{m} = \rho\vartheta A, \quad (4.7)$$

Here the density of the fluid is ρ, working fluid velocity ϑ and area A. To calculate the fluid velocity the Mach number of the flow is calculated using the following relationship,

$$Ma = \sqrt{[(\frac{P_{low}}{P_{high}})^{(-\frac{\gamma-1}{\gamma})} - 1].\frac{2}{\gamma-1}}, \quad (4.8)$$

We consider the compression chamber as a nozzle where pressure ratio is the difference in pressures between compression chamber and exit control volume. The next unknown function is fluid velocity, which we can get from the following equation:

$$\vartheta = \sqrt{\gamma . R_g . T} . Ma, \quad (4.9)$$

Here, R_g is the gas constant and T is the temperature of the fluid at each step. The temperature of the fluid can be evaluated from the energy balance equation.

4.4. Heat Transfer Model

Heat transfer is modeled between the working fluid and the walls of the compression chamber at each instant. The quantity is calculated using the approach of Adair et al., 1972,

$$\dot{Q} = h_t A(T_w - T), \quad (4.10)$$

where Heat transfer coefficient is defined as,

$$h_t = 0.0278\,(1 + 0.38\vartheta)\sqrt{(P^2T^2)}. \quad (4.11)$$

Here, P is the fluid pressure at the compression and T_w is the wall temperature. The compression process solver and overall energy balance which are shown in the Figure 4, converge and specify the guess wall temperature during the iteration by using the secant method.

4.5. Leakage Model

The leakage model only focused on the leakage between the diaphragm and the female die when the piston is fully engaged. 1% area of the total compression chamber for single piston is considered as leakage area ($A_{leakage}$). Following equation is used to determine the leakage mass low rate:

$$\dot{m}_{leakage} = \rho \vartheta A_{leakage} \qquad (4.12)$$

$$\text{Where, } A_{leakage} = \frac{Total\ compression\ chamber\ volume\ (V)}{Y} \times 0.01$$

5. PRELIMINARY MODEL RESULTS AND ANALYSIS

The thermodynamic and the geometric model of the peristaltic compressor from Section 4 implemented in a computational model. Air is used as the working fluid to run the model. The suction pressure 104 kPa, and discharge pressure 364 kPa are considered. Figure 7 shows, for the pivotal piston 7 how the thermodynamic properties are behaving during the compression given a corresponding volume change associated with the semi-circular diaphragm as shown in Figure 6.

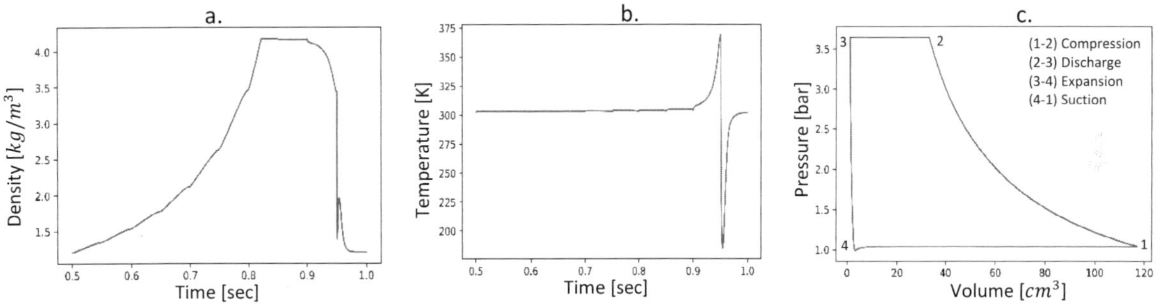

Figure 7: (a) Density of the compressor for one cycle, (b) Temperature of the compressor for one cycle, (c) Pressure-Volume diagram for the peristaltic compressor.

As Figure 7(a) shows, the density of the compression chamber increases until the pressure of the fluid reaches the discharge pressure. As discharging of gas occurs the mass velocity and heat-transfer of the chamber start decreasing. Therefore, the density begins to decrease when discharge start and returns to its initial position when the fluid starts entering the compression chamber again. Figure 7(b) highlights that the temperature of the compressor rises modestly until the final piston compresses when the temperature increases exponentially. Similar to the density, when the pistons start open the inlet segment for the fluid flow, the pressure drops significantly and volume of the compression chamber start increasing from nil. For that reason, the temperature starts dropping during the expansion and reaches to its initial condition when the suction completed. Figure 7(c) shows the fluid pressure as a function of compression chamber volume for the peristaltic compressor. This pressure-volume diagram is also referred to as an indicator diagram.

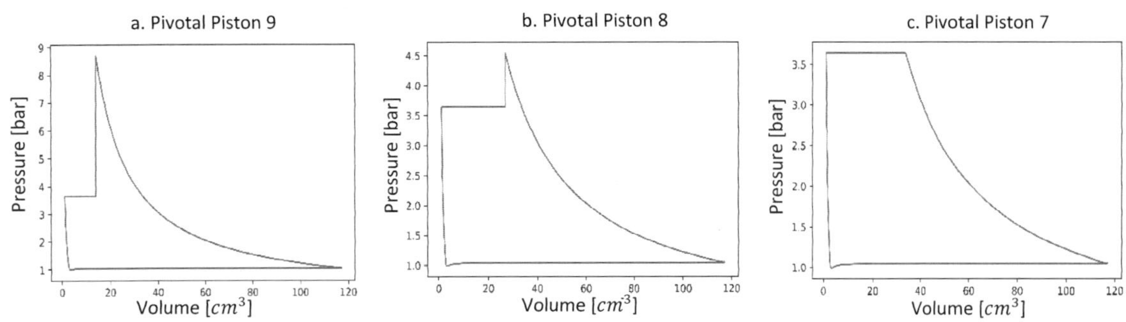

Figure 8: PV diagram for different actuation mechanism for semi-circular compression chamber.

Figure 8(a) shows that the chamber pressure for pivotal piston 9 greatly exceeds the system discharge pressure. This over-compression is a result of the volume change when compressing pistons 1 to 8 exceeding the volume change required to achieve the system discharge pressure. This requires more work (power) to execute and is therefore a loss. The re-configurable nature of this prototype allows study of these changes, as Figures 8(b) and 8(c) show with the results for the same system conditions with two different pivotal pistons with sequentially lower volume ratios. Pivotal piston 8 (Figure 8(b)) is also over-compressed but to a lesser degree than with pivotal piston at 9. When the pivotal piston is at 7 (Figure 8(c)) the system presents with neither over or under compression. This will result in minimal power and maximize compressor efficiency. A scroll compressor, for example, does not have the ability to modify its volume ratio, so if an operating condition results in significant over-compression the scroll compressor will be less efficient. In contrast, this result suggests that a peristaltic compressor could overcome that inefficiency by adjusting its volume ratio, on the fly.

6. CONCLUSIONS

The total system energy of a vapor compression system can be reduced by improving the efficiency of the compressor. The peristaltic compressor presents an adaptable technology potentially increase the efficiency of these systems. This study presents the mechanistic chamber model of a peristaltic compressor including the relevant sub-models including geometry to match a re-configurable prototype compressor, heat transfer, and mass flow. Preliminary results of the model are presented and highlight the influence of adjusting the volume ratio by adjusting the pivotal piston. These preliminary results will allow to predict most appropriate mechanical activation technology and various design parameters such as different aspect ratios of the diaphragm, different operating speed and conditions for the re-configurable prototype peristaltic compressor. The mechanistic chamber model of the peristaltic compressor will next be validated against data from the reconfigurable prototype compressor and then leveraged to explore the most useful geometric arrangements of this device applied to HVAC&R applications.

NOMENCLATURE

A	Area of the compression chamber	R_c	Radius of the curve segment in the compression chamber
C_V	Specific volume	R_g	Gas Constant
h	Displacement of the piston when acting on the curve segment	S	Actuation speed
h_{in}	Enthalpy in	T	Temperature of the fluid
h_{out}	Enthalpy out	T_w	Wall temperature
h_t	Heat transfer constant	V	Volume of the semi-circular compression chamber
Ma	Mach number	V_c	Volume of the curve segment in the compression chamber
$\dot{m}_{in}$	Mass flow in	V_d	Volume of the deflected segment
$\dot{m}_{out}$	Mass flow out	Y	Depth of the piston
$\dot{m}_{leakage}$	Mass flow leakage	Z	Displacement of the piston
P	Pressure of the fluid	ρ	Density of the fluid
$\dot{Q}$	Heat transfer rate	ϑ	Fluid velocity
R	Radius of the piston/compression chamber	γ	Specific heat

REFERENCES

Adair, R.P., Qvale, E.B., J.T., P., 1972. Instantaneous Heat Transfer to the Cylinder Wall in Reciprocating Compressors. Int. Compress. Eng. Conf. 86.

Ahamed, J.U., Saidur, R., Masjuki, H.H., 2011. A review on exergy analysis of vapor compression refrigeration system. Renew. Sustain. Energy Rev. 15, 1593–1600. https://doi.org/10.1016/j.rser.2010.11.039

Bell, I.H., 2011. Theoretical and Experimental Analysis of Liquid Flooded Compression in Scroll Compressors. Policy 9, 2003–2006. https://doi.org/UMI 3402307

Bell, I.H., Wronski, J., Quoilin, S., Lemort, V., 2014. Pure and Pseudo-pure Fluid Thermophysical Property Evaluation and the Open-Source Thermophysical Property Library CoolProp. Ind. Eng. Chem. Res. 53, 2498–2508.

Bradshaw, C.R., 2017. Development of a Novel Peristalitc Compressor for Air Conditioning Applications. ASHRAE 019-1819. Funding Source: ASHRAE Innovative Research Grant. Project Dates: S. 2019-present.

Bradshaw, C.R., 2012. A Comprehensive model of a Miniature-Scale Linear Compressor For Electronics Cooling. Purdue Univ. https://doi.org/10.1016/j.ijrefrig.2010.09.016

Bradshaw, C.R., Groll, E.A., 2013. A comprehensive model of a novel rotating spool compressor. Int. J. Refrig. 36, 1974–1981. https://doi.org/10.1016/j.ijrefrig.2013.07.004

Bradshaw, C.R., Islam, M., Tubbs, C., 2020. Peristaltic Mechanism for Compression of Compressible Fluids. US Provisional Patent Application 63/019,707.

Diniz, M.C., Deschamps, C.J., 2016. Comparative Analysis of Two Types of Positive Displacement Compressors for Air Conditioning Applications. Int. Compress. Eng. Conf.

Hartley, F.T., 1999. <title>Micropump technology and applications</title>. Electron. Struct. MEMS 3891, 403–409. https://doi.org/10.1117/12.364470

Lawless, W.N., Arenz, R.W., Cross, L.E., Lawless, W.N., Arenz, R.W., 1987. Miniature , solid-state gas compressor Miniature , solid-state gas compressor 1487.

Mathew, B., Hegab, H., 2013. Analytical modeling of microscale diaphragm compressors. Appl. Therm. Eng. 51, 130–136. https://doi.org/10.1016/j.applthermaleng.2012.08.052

Mathison, M.M., 2011. Modeling and Evaluation of Advanced Compression Techniques for Vapor Compression Equipment Vol:9.

Mathison, M.M., Braun, J.E., Groll, E.A., 2008. Modeling of a Two-Stage rotary compressor. HVAC R Res. 14, 719–748. https://doi.org/10.1080/10789669.2008.10391036

Navarro, E., Granryd, E., Urchueguía, J.F., Corberán, J.M., 2007. A phenomenological model for analyzing reciprocating compressors. Int. J. Refrig. 30, 1254–1265. https://doi.org/10.1016/j.ijrefrig.2007.02.006

Pearson, A., 2015. What does the future hold for compressor manufacturers? Proc. Inst. Mech. Eng. Part E J. Process Mech. Eng. 229, 88–95. https://doi.org/10.1177/0954408914564814

Peng, B., Zhao, S., Li, Y., 2017. Thermodynamic Model and Experimental Study of Oil-free Scroll Compressor. J. Phys. Conf. Ser. 916. https://doi.org/10.1088/1742-6596/916/1/012048

Sulfridge, M.A., Miller, N.R., 2001. The Mesoscopic Peristaltic Compressor 61801.

U.S. EIA, 2018. Residential Energy Consumption Survey (RECS) [WWW Document]. URL https://www.eia.gov/consumption/residential/

Xie, J., Shih, J., Lin, Q., Yang, B., Tai, Y.C., 2004. Surface micromachined electrostatically actuated micro peristaltic pump. Lab Chip 4, 495–501. https://doi.org/10.1039/b403906h

ACKNOWLEDGEMENTS

The authors gratefully acknowledge support for this work from the ASHRAE Innovative Research Grant (019-1819) and Mechanical and Aerospace Engineering department of Oklahoma State University.

Analysis of Dynamic Heat Transfer Coefficient in a Reciprocating Compressor using the Immersed Solid Method in CFD

Rakesh Eettisseri[1]*, Marcelo Real[2], Sidnei Olivera[2]

[1]Tecumseh Products India Pvt Ltd.
Hyderabad, Telangana, India
rakesh.pe@tecumseh.com

[2]Tecumseh Products Company LLC
5683 Hines Drive, Ann Arbor, MI, USA

*Corresponding autor

ABSTRACT

Accurate mapping of heat transfer inside the Hermetic compressor has been a major challenge for the designing and evaluation of performance. Various methodologies have been adopted for the thermal mapping such as experimental analysis, numerical methods, CFD etc. Mapping is further complex when the motions of multiple components are considered. In this work heat transfer is mapped inside a reciprocating hermetic compressor by considering the rotation of crankshaft and associated motion of connecting rod and piston. The refrigerant gas and oil in the sump are used as the working fluids for this conjugate heat transfer model. The motion is imposed on the moving parts through trigonometric relations using immersed body technique in Ansys CFX. The transient analysis is carried out on a simplified model with few thermally sensitive components to save computation cost. The heat transfer from the omitted components are imposed as boundary conditions in the simplified model. A steady simulation has been done with the simplified model providing 'zero' motion as a reference for comparison and as an initial condition for the transient run. Finally, transient run was carried out and results were compared with the steady state. It has been found that heat transfer between the fluids and components increased up to 600% with the imposed motion. The transient heat transfer coefficients were used in a 1D simulation model and compared with experimental results.

1. INTRODUCTION

Heat transfer inside a hermetic reciprocating compressor is a complex phenomenon. Compressor has two levels of temperatures at inlet and outlet, and heat generation at the motor, heat generation during gas compression process and frictional heat generation at the bearings. Various components are encapsulated inside the housing leading to a complex flow among heat transfer areas, making the accurate thermal mapping tedious. The trade-off between motor cooling by suction gas and resulting increase in suction superheat make the thermal management in compressor design very critical.

.

Various methodologies have been used for thermal mapping by various researchers around the world in past years. Which involves actual measuring using thermocouples inside the compressor and numerical approaches- like computational fluid dynamics and hybrid methods by combining CFD and 1D heat transfer. Shiva Prasad (1998) and Ribas et al. (2008) have consolidated some of the progressive developments.

An earlier attempt for the actual measurement was done by Meyer and Thompson (1988). Dutra and Deschamps (2010) measured heal flux and validated by energy balancing. Silva et al. (2016) has measured heat flux and temperature using thin film heat flux sensors and thermocouples during ON/OFF cycle operation.

Various researchers considered flow and solid domains of compressors as lumped masses and solved for better understanding of heat and temperature distribution. Meyer and Thompson (1988) developed an analytical heat transfer model in which the energy and mass flow balance equations were solved on the flow area split into various control volumes. Todescat et al. (1992) also adopted similar methods in energy balancing and numerical calculations. Both considered the Heat transfer coefficient through various correlations and experimental data.

Trying to save computational cost from full 3D model, researchers developed hybrid methods. A combination of lumped and 3D methods helped to provide better results by better capturing the heat and temperature distributions at locations where the thermal gradients are higher. Almbauer et al. (2006) developed a thermal network model solving heat transfer equations. He solved the fluid flow in 1D formulation and the conductive heat transfer in 3D formulation. Sanvezzo and Deschamps (2012) also considered a hybrid model in which the conduction heat transfer was solved in 3D and the convection in lumped modelling approach. Posch et al. (2016) worked on CFD model for suction and discharge gases and 3D numerical on solid part but lumped analysis for gas in shell and lubrication oil.

3D numerical analysis became the favorite among researchers because of its accuracy, despite its high computational cost. Sivaprasad (2004) discussed about the needs, benefits and risks of using CFD approach for positive displacement compressors. He concluded by reminding the users that CFD is not a magic wand and it has to be trusted only after verification and validation. Numerical 3D simulations focused on certain areas of interests such as suction superheat done by Lacerda and Takemori (2014). Chikurude et al. (2002) and Birari et al. (2006) modelled the entire compressor with conductive and convective heat transfer through CFD approach. Raja (2003) had also modelled the full compressor considering oil at the bottom. Lacerda and Takemori (2016) also performed a detailed simulation by discretizing solid and fluid components as 3D with motor and frictional heat generation.

In this work a full transient CFD model of a compressor is developed regarding crankshaft, connecting rod and piston as moving parts, which induced motion to the oil/gas inside the shell. The simulation was performed in Ansys CFX, where the crankshaft and connecting rod were assumed as immersed bodies and piston motion has been captured through moving mesh as it moves linearly whereas the motion of other two moving components are non-linear and unsymmetrical. For this evaluation it was considered R404A refrigerant gas and POE 32 oil. In order to reduce computational cost in transient simulation, we only considered the domains, which are highly influenced by the oil dynamics and heat transfer from other components were imposed as boundary conditions which are obtained through a complete steady model. A few iterations with 'zero' motion run was carried out and used to initiate the transient run before imposing motion. The compressor had to run for a few cycles to understand the dynamics of heat transfer characteristics. Due to the thermal inertia of components, achieving steady temperatures of components must bear high computational cost. So, a 1D model of compressor with all the components was simulated in GT-Suite platform, using the heat transfer coefficients from CFD. The results were compared with experimental results and the variations are lying under 10%.

2. METHODOLOGY

2.1 Motion Equations

In the model, motion was imposed to 3 components: crankshaft, piston and connecting rod. Crankshaft in rotational motion, piston in linear motion and connecting rod in semi-circular motion at an angular velocity about the smaller end, which also moves with piston. The model is developed based on the work of Nigus (2016).

From the Figure 1, the angular velocity of the crankshaft is,

$$\omega = \frac{d\theta}{dt} \tag{1}$$

The displacement of piston from TDC,

$$x = l(1 - \cos\phi) + r(1 - \cos\theta) \tag{2}$$

From the figure, the value of $\cos\phi$ can be obtained after factorising and neglecting higher powers of λ,

$$\cos\phi = 1 - \frac{(\lambda \sin\theta)^2}{2} \tag{3}$$

The angular velocity of connecting rod about small end,

$$v = \frac{d\phi}{dt} = \lambda\omega\cos\theta \tag{4}$$

Equations (1), (2) and (4) defines the motion equations of crankshaft, piston and connecting rod respectively with known values.

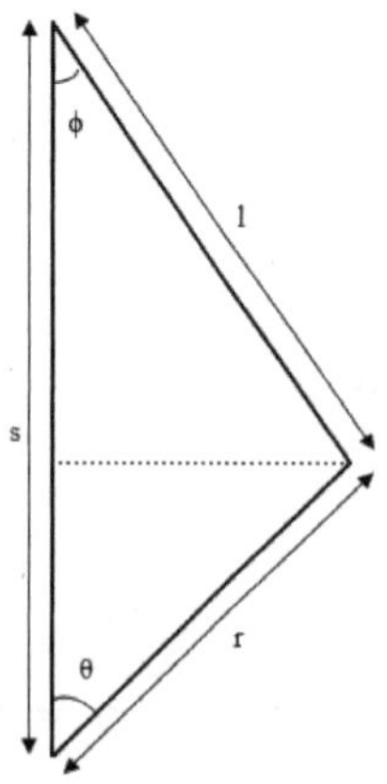

Figure 1: schematic of crank mechanism

2.2 Simulation Methodology

The motion is imposed over crankshaft and connecting rod which are assumed as solids immersed in a fluid medium representing its shape, orientation, position and velocity (Ansys CFX theory guide). The solver imposes a source of momentum to the fluid overlapped by the solid domain at its boundary. In effect, the fluid surrounding the body will behave as if it is flowing around it, although the immersed bodies cannot take the heat transfer. Therefore, in this method, the heat transfer from/to these moving bodies are neglected. As the piston is moving linearly, hexagonal moving meshes can be used to represent the piston motion without much complexity unlike other unsymmetrical, nonlinear moving bodies.

The oil-refrigerant flow field is solved as a Homogeneous multiphase flow. In this Eulerian- Eulerian model, both the fluids share the same flow fields. With free surface flow interface, the flow domain will have a clear and distinguishable interface. Buoyancy is activated to capture the density induced flow and heat transfer. The simulation is solved with forward upwind method with SST turbulent model. The total number of elements including the boundary layers is 5.8 million.

Crankshaft and connecting rod are used as the immersed bodies and will impose momentum into the flow field. Even though the motion of piston is imposed through moving mesh, piston was also considered as an immersed body for better visualisation in post processing. Initially it is solved the entire flow field with oil sump, cavity gas, discharge gas and suction gas with all the solid components in CFD for steady Heat Transfer Coefficient. Oil lubrication circuit is not considered. Heat transfer inside the cylinder is also not considered. Instead, a flow outlet from suction side and flow inlet to discharge side was given with appropriate conditions. The domains which are not highly influenced by oil dynamics are disabled in the transient run and heat transfer from these domains would serve as boundary conditions. The simplified model was run for few iterations with no motion imposed, to initiate the transient run. Then the transient run was started with the motion equations above defined on respective components.

Some minor components like brackets, mounting springs etc. are not considered for the simulation. The full geometry is simplified at the blends and small gaps to reduce the number of meshes. Airgaps, gaskets, insulators etc. are replaced with corresponding thermal contact resistance values before the simulation. Heat generated by the motor is also added as volumetric heat generation in the motor parts. Cross sectional views of the model considered for full steady and transient simulations are shown in Figure 2.

The compressor simulation was run at 60Hz and ASHRAE LBP condition: -23.3°C evaporator temperature, 54.4°C condenser temperature and 32.2°C return gas temperature. Mass flow, discharge temperature for R404A were imposed accordingly. The ambient temperature was at 32.2C and ambient-housing heat transfer coefficient was kept at 40

W/m^{2}K as the compressor was exposed to an air flow in the calorimeter setup. The matrix of Interfaces and boundary conditions for the transient is marked in Table 1 like Lacerda and Takemori (2016) with domain description in the first column and row. In the matrix, the heat transfer from/to the disabled domains are marked as boundary conditions and no connection assigned from/to the immersed body as it cannot take any heat transfer. A cross sectional view of flow domains in full model and transient model are shown in Figure 2. The steady model was meshed with 8.1 million elements which is taking 45 hours for 1500 iteration for a 32Gb 4 core simulation. For transient run with 5.8 million elements is taking 17 hours to complete one cycle with 2degree steps for same configuration with 12 cores. The average courant number for transient simulation is under 10.

Table 1: Matrix of heat transfer interface and boundaries between parts for transient run

	CG	SO	SG	DG	ST	HS	SM	CH	VP	CC	DM	SL	DT	SR	RO	WD	CS	CR	PN
CG																			
SO																			
SG																			
DG																			
ST																			
HS																			
SM																			
CH																			
VP																			
CC																			
DM																			
SL																			
DT																			
SR																			
RO																			
WD																			
CS																			
CR																			
PN																			

Enabled domain
interface
Disabled domain
Boundary condition
immersed body
no connection assigned

Because of the thermal inertia of bulky compressor components, it will be computationally very costly to achieve a repeatable transient heat transfer at each cycle. So, a few cycles only simulated to understand the variation of heat transfer and the maximum percentage of variation of heat transfer from previous cycle of any heat transfer coefficient was reduced to 5% at 15th cycle. A 1D model also created in GT-Suite with the same full geometry and boundary conditions and average of transient Heat transfer coefficients from steady and transient CFD cases are imposed on the interfaces. The final temperatures from 1D models and experimental results are compared and found within the range of 10%.

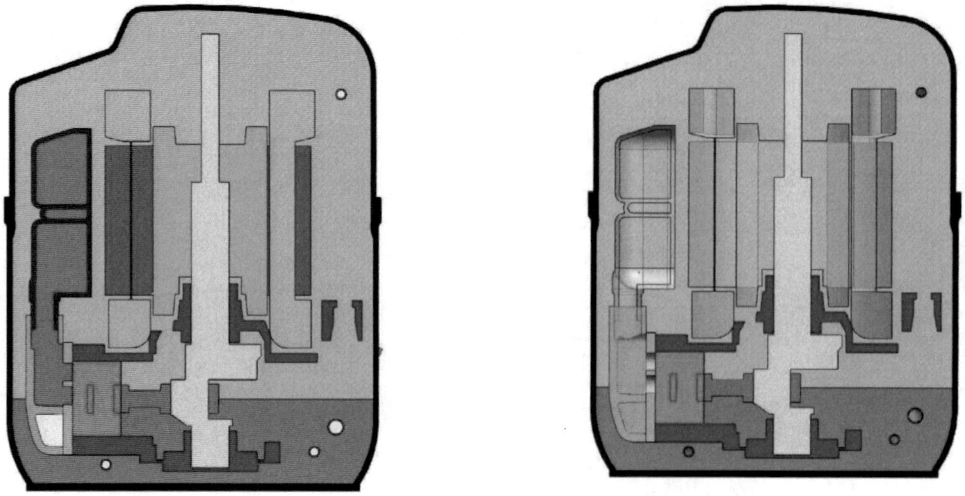

(A) (B)

Figure 2: Cross sectional front view of domains on (A) steady full model and (B) Transient model

2.3 Experimental Setup

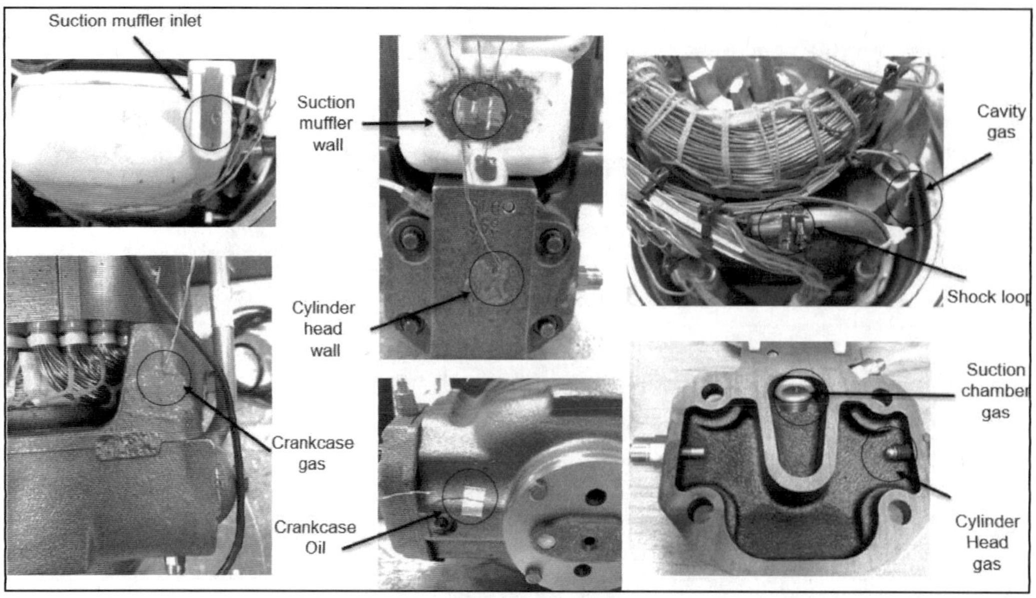

Figure 3: Thermocouples position - Compressor kit

T-type Thermocouples were installed to measure the compressor temperatures at different positions: suction muffler inlet, suction muffler wall, suction chamber, cylinder head gas, cylinder head wall, shock loop, crankcase (oil side and gas side), cavity gas, oil sump, suction and discharge tube and also upper and lower housing. The T-type thermocouples range are from -200°C to 350°C and have the accuracy of ±1C or ±0.75%. National Instruments CDAQ NI9178 chassis, NI9212 temperature module and software LabView were used to acquire the data. The location of thermocouples installed are in the Figure 3.

3. RESULTS AND DISCUSSION

A Cross sectional view of the compressor with gas and oil domains at the beginning of the transient or steady case and at the end of the 15th cycle is shown in Figure 4. Some oil waves can be observed due to the moving components into the fluids. These movements increase the heat transfer inside the gas/oil domains. Figure 5 shows the volume fraction of oil at different positions of crank rotations during different cycles. The waves will spread inside the compressor after few cycles and the wave fronts may bounce back from the surfaces, which also may generate secondary, tertiary wave fronts. The simulation will be computationally very costly to track all the waves generated. So, the simulation is stopped at 15 cycles (0.3 sec) when the magnitude of average heat flux difference through various interfaces are near or less than 5% of the previous cycle.

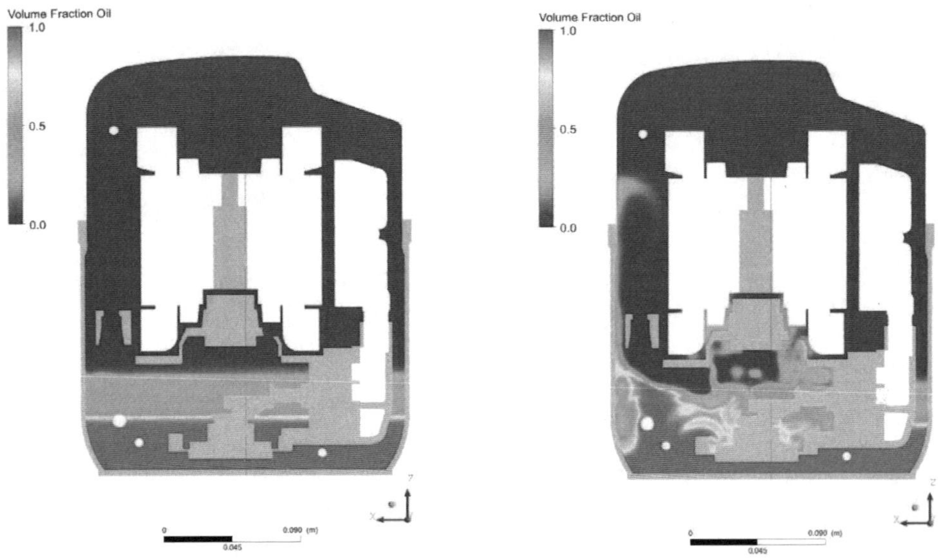

(A) (B)

Figure 4: Cross sectional front view of fluid domains at (A) steady and (B) at the end of 15 cycles

The average heat transfer over a cycle from the beginning to the 15th cycle is plotted in Figure 6. The development of heat transfer over cycles are in good agreement with the shape of flow domains. The crankcase has a shape which bounds the fluid from all direction except one which is opposite to the piston side. As a result, the sudden increase in the heat transfer of crankcase gas/oil interfaces are very evident. Shock loops and discharge mufflers are lying opposite to the piston where the velocity is high, consequently the heat transfer is also very high. The cylinder head and valve plate are the farthest from the sources of motion. As the waves had to travel around the crankcase, the heat transfer is increased very slowly and lightly.

The variation of the heat transfer coefficients over a cycle at the 15th cycle for different interfaces are in Figure 7. After a few cycles a periodic pattern of heat flux started to appear showing some variations in the magnitude, however the temperature changes happen at a very slow rate due to the thermal inertia. Till the end of last cycle maximum increase in any component was less than 0.3% except cavity which is 1.75% higher. It is evident due to the mixing of oil in the gas domain.

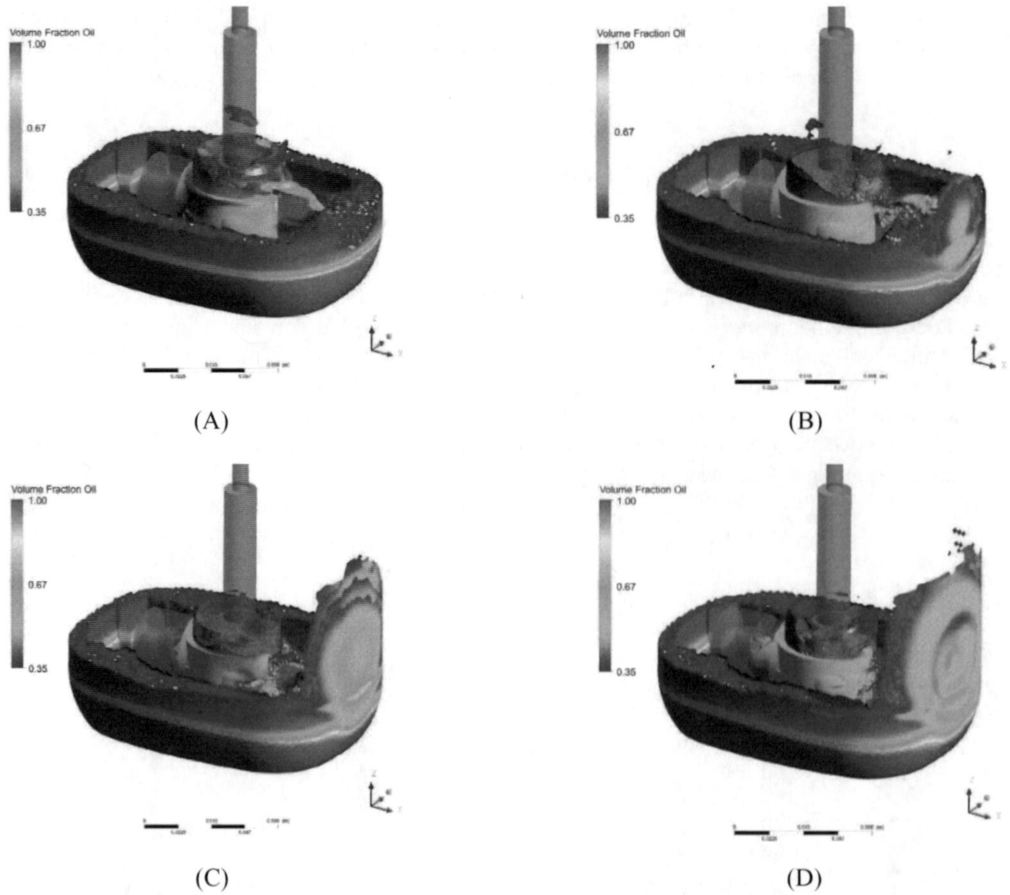

Figure 5: Oil volume fraction at (A) 270° of 3rd cycle (B) 180° of 7th cycle (C) 90° of 11th cycle and (D) end of 15th cycle

of oil in the gas domain. Change in sump domain is only 0.28%. The contact area between oil/gas interfaces with other components also will change, but as the calculation was based on dimensions fixed in CFX domains, contact area was also considered as constant. The average Heat Transfer Coefficient was calculated from the equation 5. The average heat transfer coefficient over the last cycle is calculated and compared with steady heat transfer coefficient in Table 2.

$$Q = hA\Delta T \tag{5}$$

Two 1D simulations were done, one with steady and other updated with transient heat transfer coefficients. In the transient model, whichever coefficients are not calculated by transient CFD were kept as it is in steady. GT-Suite can calculate the average temperature as well as the temperature at the interface called as port temperature. As per the thermocouple locations in actual test, for large components like crankcase and housing port locations were considered for comparison and for other components average temperatures were compared. The results are plotted in Figure 8.

Immersed bodies can be assumed as virtual bodies which can only transfer momentum to the fluid it surrounds through its boundaries but it itself replaced by the fluid in which it is immersed. So, this body cannot take part in the heat transfer process which can contribute deviations in the actual prediction of heat transfer. Also, the presence of additional fluid which replace the immersed bodies can also contribute. Introducing moving mesh can eliminate these chances of errors. But the asymmetric shape and non-linear motion of moving components will make the meshing a hectic task. But the flow dynamics and heat transfer behaviour were captured through this method for initial few cycles. Simulation with a greater number of cycles can be performed with high computational cost.

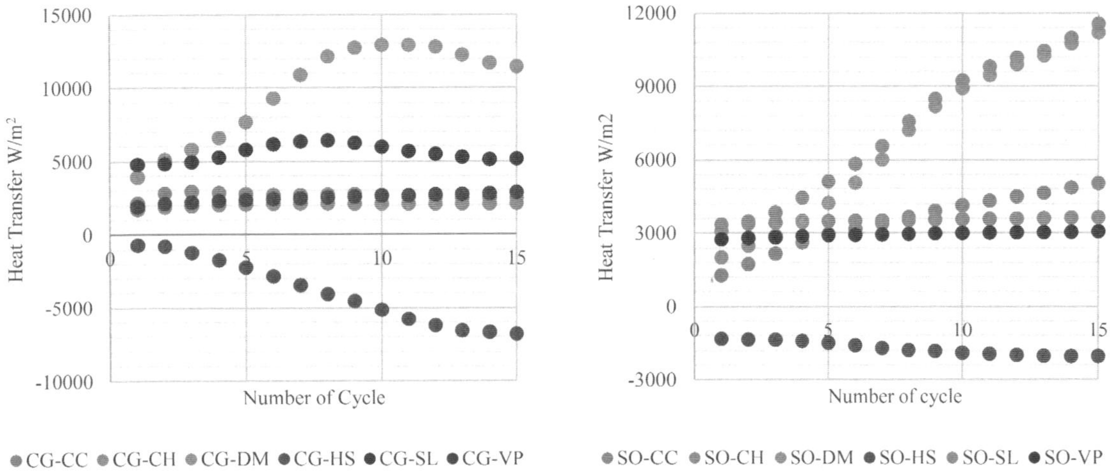

(A) (B)

Figure 6: Cycle Averaged Heat transfer of various interfaces with (A) Cavity domain and (B) sump domain

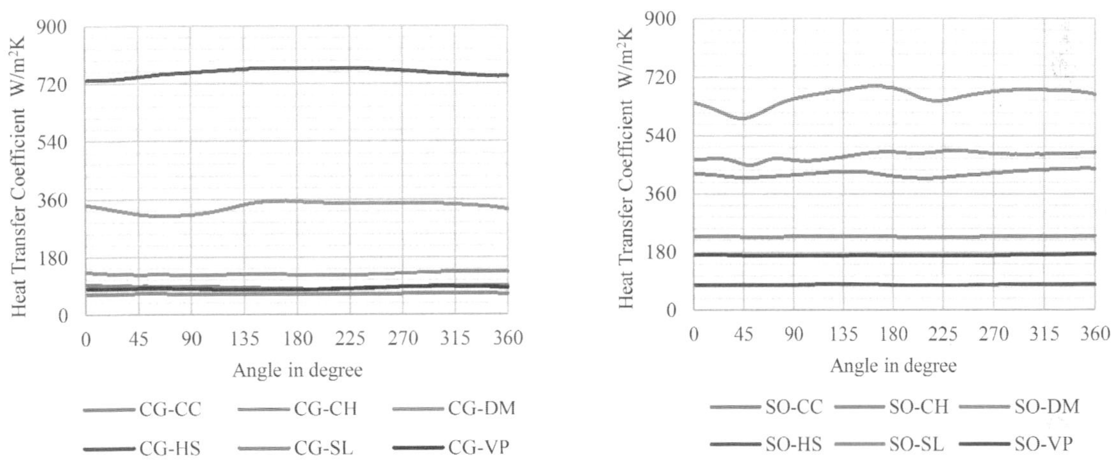

(A) (B)

Figure 7: Variation of Heat Transfer Coefficients in a cycle of various interfaces with (A) Cavity domain and (B) Sump domain

4. CONCLUSION

In this work, a CFD approach was used to estimate the transient heat transfer coefficients inside the compressor and used them on a 1D simulation model, having the model validated by experimental results. Moving components were considered as immersed bodies with induced motion into the fluid domain of refrigerant gas and oil in the sump. A comparison of steady and transient heat transfer coefficients was done and applied on a 1D model to obtain the simulated results. The computational time of transient solution is high compared to steady analysis. To save computational time, only few key components were considered in transient CFD simulation. At some interfaces, heat transfer coefficient increased up to 6 times with transient analysis. The final output temperatures from the 1D were compared with experimental results and it was found a difference under 10%. The Immersed body in Ansys CFX can only transfer momentum through its boundaries without replacing any fluid or taking up heat. To save computational time simulation was continued only for 15 cycles and transient simulation was limited to just a few components. These might be some reason for the variation. By considering the immersed bodies as regular solid bodies, the flow domain as moving meshes and by considering all the components into picture, a better prediction can be performed but at the cost of huge computational time.

Table 2: Steady heat transfer coefficient

	CG						SO					
W/m²K	CC	CH	DM	HS	SL	VP	CC	CH	DM	HS	SL	VP
Steady	34.9	40.6	75.8	162.8	100.9	42.3	85.5	218.6	94.9	46.4	122.6	170.0
Transient	86.2	63.9	336.7	754.4	125.4	82.0	420.8	228.0	659.4	78.7	476.6	170.1
change	147.3%	57.1%	344.4%	363.6%	24.3%	93.6%	392.3%	4.3%	594.7%	69.3%	288.8%	0.1%

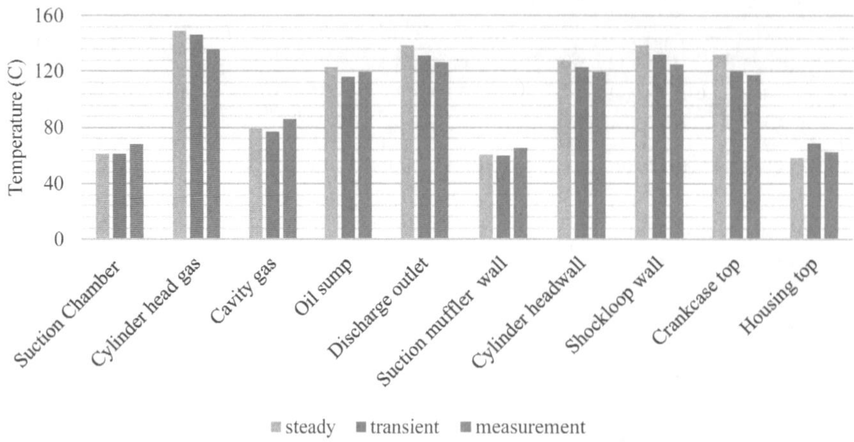

Figure 8: Comparison chart of temperature from 1D simulation

NOMENCLATURE

CG	Cavity gas	DT	Discharge tube
SO	Sump oil	SR	Stator
SG	Suction gas	RO	Rotor
DG	Discharge gas	WD	Winding
ST	Suction tube	CS	Crank shaft
HS	Housing	CR	Connecting rod
SM	Suction muffler	PN	Piston
CH	Cylinder head	l	Length of connecting rod
VP	Valve plate	r	Radius of throw
CC	Crankcase	θ	Crank angle
DM	Discharge mufflers	ϕ	Angle of rotation of conrod
SL	Shock loops	λ	r/l

REFERENCES

Ansys CFX theory guide, 2018

Almbauer, R., Burgstaller, A., Abidin Z. & Nagy D. (2006), *3-Dimensional Simulation for Obtaining the Heat Transfer Correlations of a Thermal Network Calculation for a Hermetic Reciprocating Compressor*, Proc. Int. Compressor Eng. Conf. Purdue, paper 1794

Birari, Y. V., Gosavi, S.S & Jorwekar, P. P. (2006), *Use of CFD in Design and Development of R404A Reciprocating Compressor*, Proc. Int. Compressor Eng. Conf. Purdue, paper 1727

Chikurude, R.C., Longanathan, E., Dandekar, D. P. & Manivasagam, S. (2002), *Thermal Mapping of Hermetically Sealed Compressors Using Computational Fluid Dynamics Technique*, Proc. Int. Compressor Eng. Conf. Purdue, paper 1520

Dutra, T. & Deschamps, C. J. (2010), *Experimental Investigation of Heat Transfer in Components of a Hermetic Reciprocating Compressor*, Proc. Int. Compressor Eng. Conf. Purdue, paper 2002

Lacerda, J. F. & Takemori. C. K. (2014), *Predicting the Suction Gas Superheating in Reciprocating Compressors*, Proc. Int. Compressor Eng. Conf. Purdue, paper 2332

Lacerda, J. F. & Takemori, C. K. (2016), *CFD Approach to Evaluate Heat Transfer in Reciprocating Compressors*, Proc. Int. Compressor Eng. Conf. Purdue, paper 2487

Meyer, W. A. & Thompson, H. D. (1988), *An Experimental Investigation into Heat Transfer to the Suction Gas in a Low-Side Hermetic Refrigeration Compressor*, Proc. Int. Compressor Eng. Conf. Purdue, paper 661

Meyer, W. A. & Thompson, H. D. (1988), *An Analytical Model of Heat Transfer to the Suction Gas in a Low-Side Hermetic Refrigeration Compressor*, Proc. Int. Compressor Eng. Conf. Purdue, paper 662

Nigus H.,2015, *Kinematics and Load Formulation of Engine Crank Mechanism, Engineering Journal*, Magnolithe, 10.13140/RG.2.1.3257.1928

Posch, S., Hopfgartner, J., Heimel, M., Berger E. & Almbauer R. (2016), *Thermal Analysis of a Hermetic Reciprocating Compressor Using Numerical Methods*, Proc. Int. Compressor Eng. Conf. Purdue, paper 2429

Raja, B., Sekhar, S. J., Lal D. M. & Kalanidhi, A. (2003), *A numerical Model for Thermal Mapping in a Hermetically Sealed Reciprocating Refrigerant Compressor*, International Journal of Refrigeration 26 (2003) 652–658

Ribas Jr., F. A., Deschamps, C. J., Fagotti, F., Morriesen, A., & Dutra, T. (2008), *Thermal Analysis in Reciprocating Compressors – A Critical Review*, Proc. Int. Compressor Eng. Conf. Purdue, paper 1907

Sanvezzo Jr., J. & Deschamps, C.J. (2012), *A Heat Transfer Model Combining Differential and Integral Formulations for Thermal Analysis of Reciprocating Compressors*, Proc. Int. Compressor Eng. Conf. Purdue, paper 2210

Shiva Prasad, B.G. (1998), *Heat Transfer in Reciprocating Compressors - A Review*, Proc. Int. Compressor Eng. Conf. Purdue., paper 1349

Shiva Prasad, B.G. (2004), *CFD for Positive Displacement Compressors*, Proc. Int. Compressor Eng. Conf. Purdue, paper 1689

Silva, L.M., Dutra, T. & Deschamps, C. J. (2016), *Experimental Investigation of Heat Transfer in Components of a Hermetic Reciprocating Compressor under Thermal Transient*, Proc. Int. Compressor Eng. Conf. Purdue, paper 2464

Todescat, M. L., Fagotti, F., Prata., A. T. & Ferreira, R. T. S. (1992), *Thermal Energy Analysis in Reciprocating Hermetic Compressor*, Proc. Int. Compressor Eng. Conf. Purdue, paper 936

ACKNOWLEDGEMENT

To Tecumseh Products Company LLC and Tecumseh Products India Pvt Ltd. for the financial support.

FORCE FAILURE DIELECTRIC TEST DESIGN FOR STATORS OF INVERTER COMPRESSORS

Tuğba Çetintürk

Arçelik A.Ş., Compressor Plant R&D Department,
Eskişehir, TURKEY
+902222134657, +902222134344, tugba.cetinturk@arcelik.com

ABSTRACT

(PMAC) Electric motor is used at Variable Speed Refrigeration Compressors. Stator is the stationary part of a PMAC motor and stator wires directly carry the current, which energizes the motor and runs the refrigerator compressors. These wires are wound on stator stack which is made of electrical steel. The windings can be also close to other metal parts of the compressor. Stator stack is connected to compressor metal parts by metal screws. So, if a leakage current occurs through the stator stack, it is directly conveyed to compressor housing and then to refrigerator which can cause electric shock on the customer. That's why, not only the quality but also the reliability of insulation between stator wires and metal parts is so critical.

Standard dielectric strength tests are applied to both stators and compressors during production in order to check the insulation quality for product safety in the field. But it is not possible to detect the weakness and aging in this way. Most insulaton faults arise in time by repetitive voltage increase, thermal, mechanical, environmental and electrical stresses.

This study explains the test which is carried out to detect potential electrical safety, leakage current and compressor failure problems which can occur on inverter stators in process of time. The development of Forced Failure Dielectric Test for the stators aims to uncover the insulation weakness which cannot be find out by Quality Tests and which can cause functional and safety problems in the field.

1. INTRODUCTION

In order to prevent electrical safety problems of a compressor, both internal and external insulation of the stator should be provided. Internal insulation means that the stator wires should be insulated from the stator stack. It is done by using paper insulation or plastic holders and/or by providing a minimum 2,5 mm clearance between wires and stack laminations. The enamel coating quality of stator wires is also critical and it shouldn't get damaged during the stator winding or compressor assembly process.

External insulation means that the stator wires and lead terminals should be insulated from other metal parts of the compressor again by using an insulator and/or by providing a minimum 2,5 mm clearance. This should paid attention to during the structural design of the compressor. Otherwise, unworking compressors and/or electrical safety problems can occur at the site.

Stone and Culbert (2018a) mentioned the diagnostic tests such as insulation resistance, polarization index, dissipation factor, capacitance and partial discharge that can be useful to diagnose the condition of the stator winding insulation, and identify many problems that could lead to insulation failure. From all these tests; Hi-pot, Insulation and Leakage Current tests are applied to each produced compressor for 1 second each, at the end of the compressor assembly lines. But these tests cannot guarantee to reveal the ground faults that can occur in time while working on the site. Stone and Culbert (2018b) also examined the literature and practical experience using the remaining life estimation tools for stator winding. The purpose of this study is finding a force failure test method in order to prevent compressor failures, leakage current and electrical safety problems by determining the potential electrical faults on stators of inverter compressors in time. The development of the forced failure dielectric test for PMAC stators aims to find insulation defects, which cannot be detected by quality tests, and can cause functional and safety problems in the field. So, if any potential and sleeping faults or weakness exist at winding insulation, the new test can reveal it and make it possible to prevent usage of the weak insulated stators on compressors.

2. PHYSICS OF INSULATION FAILURES AT HOUSEHOLD COMPRESSORS

Factors effecting insulation resistance include temperature, humidity, insulation thickness, mechanical stress and applied voltage. Since household refrigerating compressors are hermetic and include no moisture, we skip the effect of moisture on the insulation at this paper. Instead; temperature, insulation dimensions and voltage can be used as a stress factor in order to accelerate the test for rapid aging and failing of test stators.

Ambient Temperature: Tommasini (2009a) stated Montsinger's rule, that a temperature rise of 10K halves the expected lifetime of an insulation system. ABB Industrial Systems (1997) showed the relation between temperature and insulation life at Figure 1. That is, increase of temperature decreases the insulation life.

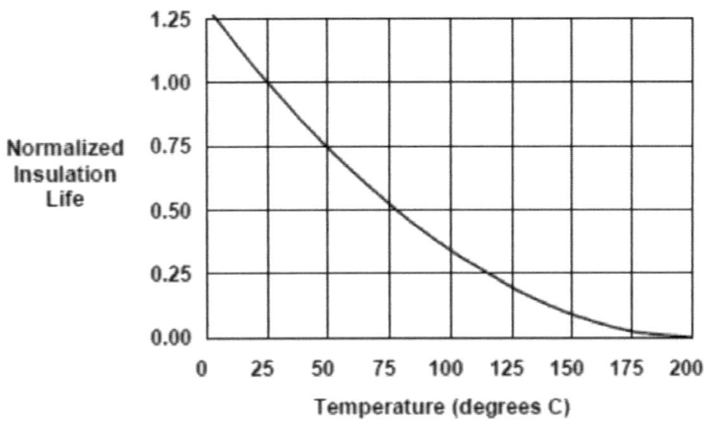

Figure 1: Insulation Life vs. Temperature

Ritamäki *et al.* (2017) said that the exposure to elevated temperature may have caused material degradation leading to more low field breakdowns. It is also pointed out that the measurements conducted in the temperature range from room temperature to 100 °C revealed a linearly decreasing trend in the characteristic 63.2 % breakdown probability after 60 °C and the presence of a smaller weak spot subpopulation.

High Dielectric Voltage: Electrical appliances are designed to operate within a defined range of voltage. If the operating temperature is increased gradually at some value of voltage, the breakdown of the insulating materials will occur. Applying a gradually increasing voltage to the stator winding or any conductor will definitely destroy the insulation. That's why breakdown voltage is always a working stress factor at stator force failure tests. ZVEI (2016) pointed out that the penetration through the insulation material, a voltage increase of 500V/sec. is commonly used. However, it is difficult to perform a Reliasoft analysis because the time data of insulation failure cannot be obtained by applying breakdown voltage.

Gupta *et al.* (2009) reported that Hipot tests fail marginal, not good windings; and AC Hipot test is better in detecting winding defects than DC hipot test. Grubic (2008) suggests the development of an online-monitoring method applicable to low-voltage machines, which is capable of diagnosing the deterioration of the turn-to-turn insulation prior to a fault and is also reasonable from a cost standpoint. In this study, both the fixed high-potential voltage and the gradually increasing high-potential voltage have undergone different test designs.

Insulation Dimensions: Tommasini (2009b) reported that failure of dielectric insulation is often determined by a direct or indirect mechanical failure and a good designer ensures the design correctly considers operation & fault conditions. These expressions can be paraphrased that in order to overcome mechanical stress that the windings are exposed during production or working conditions, choosing the correct insulation height, width and thickness is critical. For this reason, the effects of insulation thickness and height along the stator slots can be also used to prepare test samples with different insulation qualities and to check if the test properly distinguishes the good and bad stators.

Metal Burrs: Insulation thickness should be considered as not only the thickness of insulating material, but also the amount of clearance between enameled wires and metal parts. It is easy to control the design and production tolerances. But if a metal burr occurs and falls on the stator, it can provide a connection between wires and metal parts within the compressor. Preventing metal burrs inside the compressor is very difficult. That's why it should be used as a force failure effect on this test design.

3. TEST DESIGN STUDIES

The connection between Stator Life and Stress is used to design this force failure test. Two types of stress is used in order to accelerate the test. One is the internal stress, that is caused by weak insulation of the stator itself, which simulates design faults. The other type is the external stress, that occurs by stator working conditions inside the compressor.

3.1 Preparing Test Samples

In order to generate the internal stress, two different stator models A and B are choosen as test sample, one of which has lowest and the other has highest power shaft at current product range. Arc failures occured in the past especially on these two models which couldn't have catched by Electrical Safety tests at the production line. After some detailed examinings, it is found out that, arc failure problems are caused by low slot insulation height, thickness and high slot fillness of these stators. That's why; insulation thickness, insulation height and slot fillness parameters are used to prepare stator samples for validation of this test design. 3 stress factors with two levels and half- fractional factorial experiment design is done for Stator A as shown at Table 1. In order to prepare samples of Stator B, only 2 levels of slot fillness is used as shown at Table 2.

Table 1: Stress Factors Applied to Test Samples of Stator A

Stress Factors of Stator A	Stress Level (-)	Stress Level (+)
Thickness of Slot Insulation (mm)	0,25	0,35
Height of Slot Insulation (mm)	24,5	25,5
Slot Fillness %	70	85

Table 2: Stress Factors Applied to Test Samples of Stator B

Stress Factors of Stator B	Stress Level (-)	Stress Level (+)
Slot Fillness %	81	94

Stator A samples, with (-) level of thickness and height and (+) level of slot fillness are expecting to fail during the designed test. B stator samples with (+) level of slot fillness are expecting to fail during the test. The remaining samples are expecting to pass the test. Test conditions and test method, which will provide the expected test results will be assigned and fixed accordingly. Table 3 explains the coding of prepared test samples according to their properties described at Table 1 and Table 2. **A_2, A_3** and **B_6** Stators should certainly fail, **A_1** and **A_4** Stators should certainly pass the test.

Table 3: Coding of Test Samples According to Stress Factors

Sample Code	Insulation Thickness (mm)	Insulation Height (mm)	Slot Fillness %
A_1	0,25	25,4	70
A_2	0,25	24,5	70
A_3	0,25	24,5	85
A_4	0,35	25,4	70
B_5	0,25	34,5	81
B_6	0,25	34,5	94

In order to see the repeatability of the test results, 2 different samples are tested from each sample code. For example; "**A_1-1**" and "**A_1-2**" are the first and second sample that both have 0,25 mm insulation thickness; 25,4 mm insulation height and 70% slot fillness.

3.2 Test Equipments

Figure 1 shows the oven which is used to heat the stators at a fixed temperature for aging.

Figure 1: Regulated oven in order to heat the stators up to 200°C.

Adjustments of electrical safety tester are done by setting AC Hipot Voltage, AC Frequency (50 Hz), leakage current limit and testing time as shown in Figure 2.

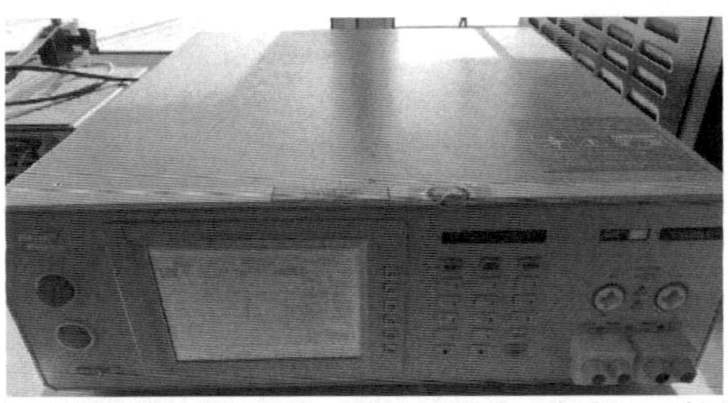

Figure 2: Electrical Safety Tester for Hi-pot (Dielectric Strength) Measurement

In order to do necessary cabling between electrical safety tester and stator sample inside the oven; additional test cables are used, which hang down from top to inside of the oven. Phase terminal of electrical safety tester is connected to stator cluster plug of which 3 phase windings are short circuited. Notr terminal of electrical safety tester is connected to stator stack from one of the screw holes as shown in Figure 4.

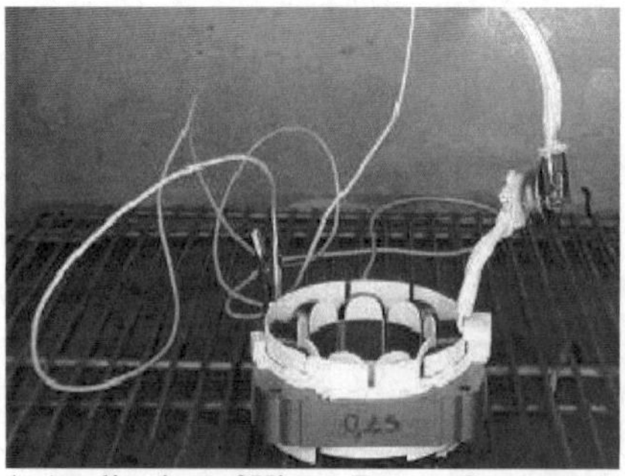

Figure 4: Application of Hi-pot Test to Compressor Stators

3.3 Determiming Test Method and Test Conditions

Temperature exposure and applying Hi-pot Voltage is chosen first in order to accelerate the failures of the stator samples. Table 4 shows the stres levels applied to stator samples for rapid aging. Design of Experiment (DOE) with 2x2 Factorial Design is generated according to Six Sigma Methodology and four different test conditions are applied to stator samples in order to accelerate the aging and failure.

Table 4: Stress Factors and Levels used at Test Design 1

Stress Factors	Stress Level (-)	Stress Level (+)
Temperature (°C)	150	200
Hi-pot Voltage (V_{ac})	2000	2500

Hi-pot Voltage is applied between live ends of stators samples and stator stack for 1 minute, before heating. 1 hour period is fixed to make the stator samples wait at the defined high temperatures. After aging the stators by heating and applying Hipot voltage, 2800 Vac constant dielectric voltage continuously applied between stator conductors and chasis (stator stack) in order to record the duration until the stator fails. Failing criteria is chosen maximum 1,8 mA leakage current. But the samples never failed. That's why, time dependent datas couldn't be generated for Reliasoft Weibull Analysis by Test Design 1.

Table 5: Stress Factors and Levels used at Test Design 2

Stress Factors	Stress Level (-)	Stress Level (+)
Iron Sand splashed on stators (grams)	5	10
Temperature (°C)	150	200
Hi-pot Voltage (V_{ac})	2000	2500

Table 5 indicates the stress levels applied to the stator samples in order to fail the stators in the second test design. This time, in addition to the same stress factors and levels of Test Design 1, some specific amount of iron sand is randomly splashed on stator samples in order to simulate metal burrs that can reach stator wires inside the compressor. After aging the samples by the stress levels shown in Table 4, 2000V ac voltage is applied continuously between stator conductors and chassis (stator stack) in order to record the duration until the stator fails. Failing criteria is again chosen maximum 1,8 mA leakage current. This time, all the samples immediately failed by the effect of metal burrs.

Since time dependent datas couldn't be generated by the first two Test Designs, a different test method is studied. In Test design 3, a fixed amount of iron sand was sprayed on all stator samples to accelerate the test, and a high voltage of 2000V was applied to the stator again to age it, and then heated. After heating the stators at 3 temperature levels for 1 hour; a high dielectric voltage is gradually applied to the stator and the leakage current is recorded at each voltage level until it reaches 1.8 mA. If a leakage current of any sample is measured higher than 1,8 mA; it is defined as failure and test stopped at that Hi-pot voltage level. "Hi-pot" is written to show the failure of each tested stator. Table 6 shows the leakage current values obtained at each Hi-pot voltage and heating temperature level.

Table 6: Results Obtained By Test Design 3.

Test Samples	Temperature (°C)	Measured Ground Leakage Currents (mA) during application of Hi-pot Voltages for 1 Minute (Max limit: 1,8 mA)						
		500V	750V	1000V	1250V	1500V	1750V	2000V
A_1-1	25	0,068	0,103	0,139	0,179	0,225	Hi-pot	
	150	0,073	0,11	0,164	Hi-pot			
	200	0,077	0,12	Hi-pot				
A_1-2	25	0,067	0,103	0,0137	0,177	0,237	0,294	Hi-pot
	150	0,088	0,132	0,182	0,247	Hi-pot		
	200	0,09	0,137	0,188	0,251	Hi-pot		
A_2-1	25	0,081	0,124	0,167	Hi-pot			
	150	0,086	0,13	0,179	Hi-pot			
	200	0,088	0,133	0,184	0,245	Hi-pot		
A_2-2	25	0,084	0,127	0,172	0,219	0,274	0,342	Hi-pot
	150	0,083	0,127	0,174	0,237	Hi-pot		
	200	0,087	0,131	0,179	Hi-pot			
A_3-1	25	0,065	0,1	0,135	0,172	0,23	Hi-pot	
	150	0,081	0,124	0,169	Hi-pot			
	200	0,083	0,125	0,17	Hi-pot			
A_3-2	25	0,067	0,103	0,138	0,177	0,236	Hi-pot	
	150	0,075	0,114	0,156	0,212	Hi-pot		
	200	0,08	0,122	0,167	0,226	Hi-pot		
A_4-1	25	0,212	0,425	0,746	Hi-pot			
	150	0,164	0,332	Hi-pot				
	200	Hi-pot						
A_4-2	25	Hi-pot						
	150	Hi-pot						
	200	Hi-pot						
B_5-1	25	0,307	0,596	Hi-pot				
	150	Hi-pot						
	200	Hi-pot						
B_5-2	25	Hi-pot						
	150	Hi-pot						
	200	Hi-pot						
B_6-1	25	0,1	0,151	0,204	0,26	0,339	0,505	Hi-pot
	150	0,107	0,161	0,22	0,306	0,433	Hi-pot	
	200	0,112	0,168	0,229	0,31	0,419	Hi-pot	
B_6-2	25	Hi-pot						
	150	Hi-pot						
	200	Hi-pot						

The test results in Table 6 are checked to determine the temperature level and Hi-pot voltage limit that can distinguish a good insulated stator from a weakly insulated stator. **A_1** and **A_4** Stators are supposed to fail at high voltages and, **A_2, A_3** and **B_6** Stators are supposed to fail at low voltages. But on the contrary; A_4 and B_6 fail quickly, A_1, A_2 and A_3 fail at higher voltages. This complicated results are attributed to using iron sand, it is removed from stress factors of the test design.

Table 7: Results Obtained By Test Design 4.

Test Samples	Temperature (°C)	Measured Ground Leakage Currents (mA) during application of Hi-pot Voltages for 1 Minute (Max limit: 1,8 mA)										
		500V	750V	1000V	1250V	1500V	1750V	2000V	2250V	2500V	2750V	3000V
A_1-1	25	0,078	0,118	0,159	0,206	0,268	0,34	0,43	0,522	0,605	Hi-pot	
	150	0,082	0,124	0,175	0,254	0,358	0,46	0,579	0,756	Hi-pot		
	200	0,084	0,127	0,175	0,258	0,347	0,456	0,597	Hi-pot			
A_1-2	25	0,071	0,108	0,144	0,187	0,256	0,342	0,434	0,523	0,597	Hi-pot	
	150	0,08	0,121	0,177	0,253	0,35	0,454	0,558	0,668	Hi-pot		
	200	0,084	0,127	0,175	0,26	0,352	0,456	0,583	Hi-pot			
A_2-1	25	0,078	0,118	0,159	0,21	0,285	0,375	0,468	0,549	0,605	Hi-pot	
	150	0,084	0,127	0,182	0,26	0,363	Hi-pot					
	200	0,084	0,127	0,174	0,26	Hi-pot						
A_2-2	25	0,076	0,115	0,154	0,207	0,276	0,362	0,456	0,542	0,598	Hi-pot	
	150	0,081	0,124	0,176	0,253	0,35	0,453	0,555	0,699	Hi-pot		
	200	0,083	0,126	0,171	0,255	0,346	0,448	0,585	Hi-pot			
A_3-1	25	0,073	0,11	0,147	0,187	0,26	0,344	0,435	0,524	0,606	Hi-pot	
	150	0,078	0,118	0,163	0,229	0,317	0,417	0,511	Hi-pot			
	200	0,079	0,119	0,164	0,224	0,307	0,402	Hi-pot				
A_3-2	25	0,071	0,108	0,145	0,188	0,257	0,342	0,433	0,521	Hi-pot		
	150	0,077	0,116	0,159	0,225	0,315	0,414	0,501	Hi-pot			
	200	0,077	0,116	0,161	0,22	0,302	0,325	Hi-pot				
A_4-1	25	0,069	0,105	0,142	0,177	0,222	0,288	0,365	0,443	0,51	0,572	Hi-pot
	150	0,073	0,112	0,151	0,206	0,277	0,36	0,439	0,519	Hi-pot		
	200	0,075	0,114	0,152	0,211	0,278	0,354	0,44	0,558	Hi-pot		
A_4-2	25	0,066	0,101	0,136	0,171	0,214	0,276	0,35	0,43	0,495	0,554	Hi-pot
	150	0,071	0,107	0,145	0,203	0,274	0,354	0,436	0,522	Hi-pot		
	200	0,073	0,11	0,15	0,204	0,271	0,347	0,438	0,617	Hi-pot		
B_5-1	25	0,091	0,137	0,184	0,231	0,322	0,429	0,552	0,677	0,739	Hi-pot	
	150	0,098	0,149	0,209	0,304	0,425	0,563	0,705	Hi-pot			
	200	0,103	0,155	0,215	0,315	0,433	0,572	0,719	Hi-pot			
B_5-2	25	0,086	0,129	0,174	0,219	0,299	0,4	0,516	0,645	0,759	0,879	Hi-pot
	150	0,093	0,14	0,194	0,284	0,4	0,534	0,665	0,812	Hi-pot		
	200	0,097	0,147	0,203	0,293	0,409	0,544	0,68	Hi-pot			
B_6-1	25	0,095	0,143	0,192	0,241	0,331	0,442	0,578	0,712	Hi-pot		
	150	0,104	0,156	0,217	0,324	0,448	0,595	0,746	Hi-pot			
	200	0,109	0,165	0,225	0,331	0,449	0,601	Hi-pot				
B_6-2	25	0,093	0,141	0,189	0,237	0,325	0,434	0,564	0,702	0,826	Hi-pot	
	150	0,101	0,152	0,214	0,314	0,436	0,581	Hi-pot				
	200	0,105	0,159	0,218	0,315	0,436	Hi-pot					

When we check the test results in Table 7; after heating at 200°C, most of the weakly insulated stators will fail, and the good insulated stators will pass 2000 V high voltage application. Leakage current values of all samples at 2250, 2500 V and 2750V are also checked. Decreasing leakage current limit from 1.8 mA to 0.75 mA, seems to distinguish the good and weak stators more clearly at 2000 V dielectric voltage.

4. RELIASOFT ANALYSIS

"Temperature-Nonthermal" method is used for analysis at Reliasoft ALTA module, because voltage and insulation quality is used as a stress factor together with temperature. The graph of "Stator Reliability change vs Time" is given in Figure 5, according to metal burr and temperature exposing. The figure shows that the effect of high temperature is small, but no matter whether the insulation quality of the stator is good or weak, metal burrs will have a significant impact on the life of the stator.

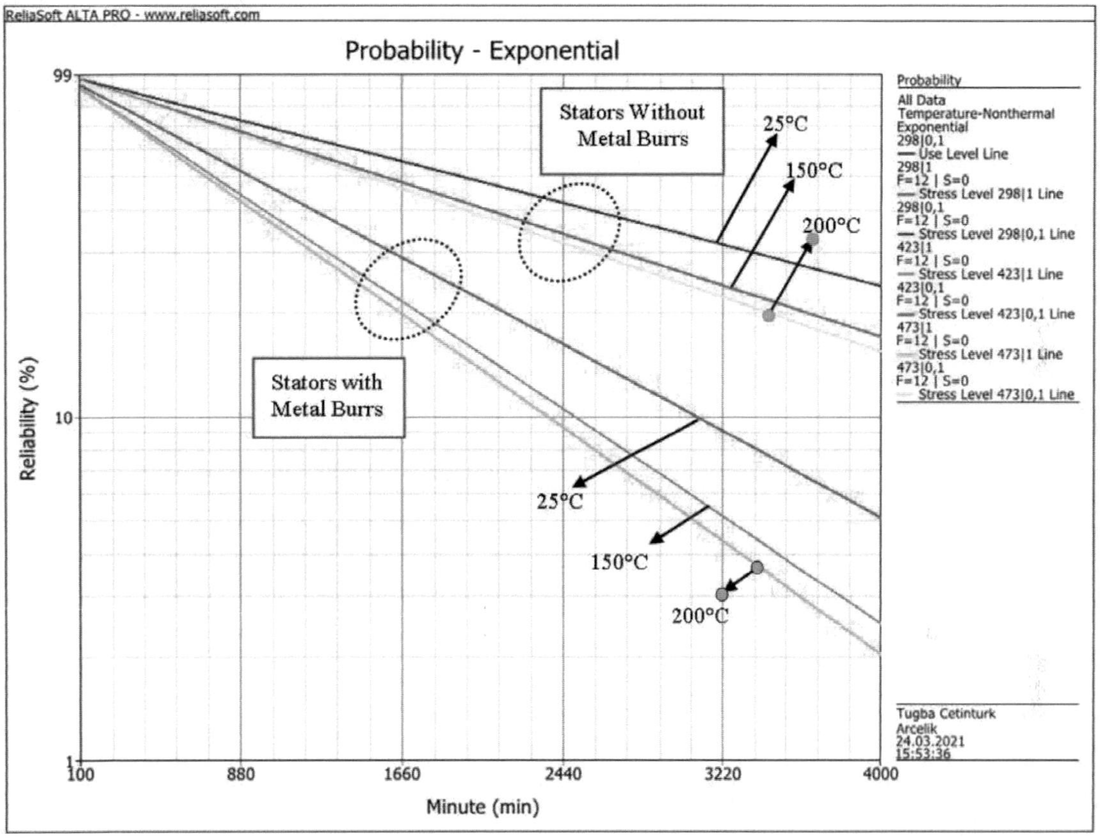

Figure 5: Effect Of Metal Burrs and Temperature on Stator Reliability

The graph in Figure 5 also verifies that a temperature level of 200°C should be used as the stress factor to accelerate the aging of the stator in the forced failure dielectric test. On the other hand, metal burrs shouldn't be used for this test, because it fails both good and weak insulated stators.

Figure 6 compares the reliability of the stators with good and weak insulation under abnormal conditions used for the Forced Failure test design. The purpose of this test is failing all the stators with a distinct aging speed at good and weak insulated stators. The graph at Figure 6 shows that it is achieved. So that the reliability slope of the weak insulated stators is sharper than the reliability slope of the good insulated stators.

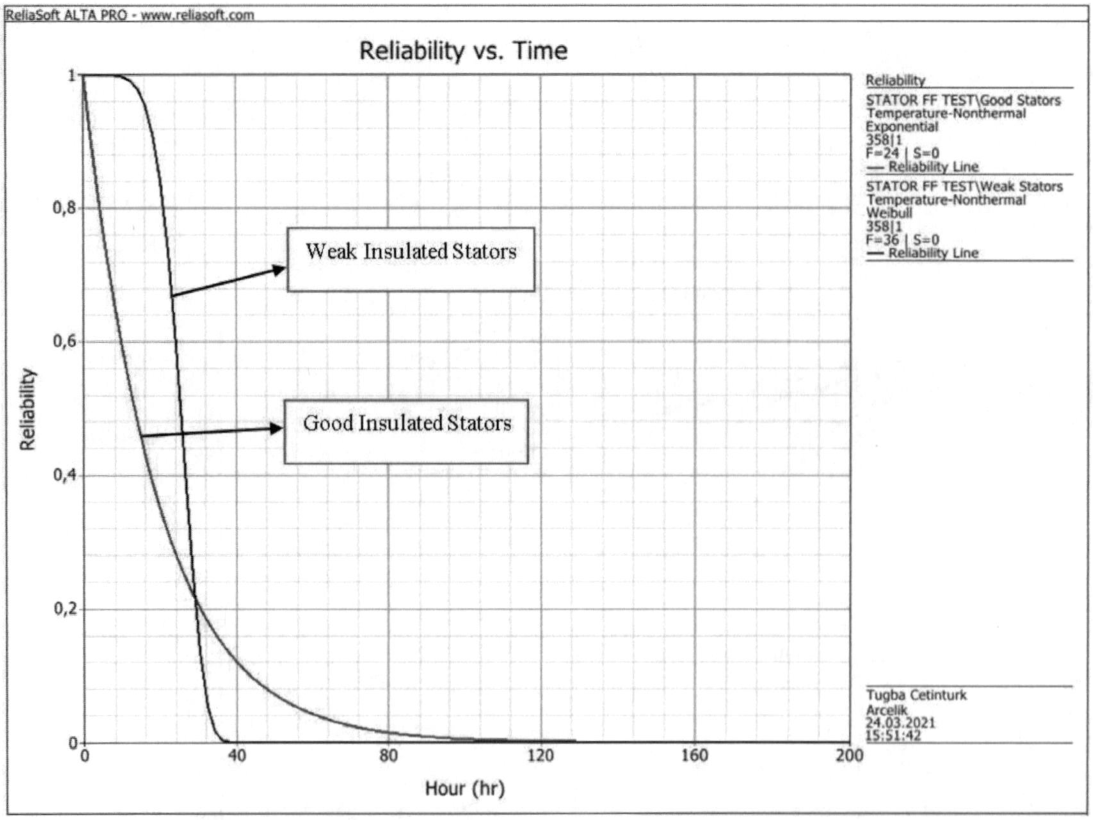

Figure 6: Effect of Insulation Dimensions & Slot Fillness On The Stator Reliability

5. CONCLUSION

The most important parameter on stators is electrical insulation in order to provide both compressor quality and safety. If the stators have not apparent insulation faults but weaknesses that can emerge after working for a while at site, the standard tests can not reveal them. There is no time for long life tests for ongoing productions either. That's why, determining the potential or sleeping insulation problems of stators in a short time is very important.

By evaluating the results of 4 different testing methods and reliasoft analysis, the forced failure test of the inverter compressor stator is shaped and designed. The test instruction is finalized as follows:

At ambient temperature, a fixed 2500 Vac dielectric voltage will be applied to the stator sample for 1 minute. If the stator fails during this test, test result is recorded as "negative" and the test is stopped. If the stator passes the 1 minute 2500 Vac Dielectric voltage test, it is kept inside a 200°C oven for 1 hour. After heating for 1 hour; without taking it out of the oven, apply 500 V/min gradually increasing high voltage voltage to the stator, up to 2000V. Maximum 0,75 mA leakage current limit is defined for each 500 V, 1000 V, 1500 V and 2000 Vac voltage application to the tested stator. If the stator fails at any of these voltages, test result is negative.

This Force Failure Dielectric Test is put into use systematically on control plans of both inhouse stator production and outsourced stators for inverter compressors.

This test study showed that heating stators at high temperature values, which are higher than thermal resistance of the stator wires and paper insulations do not provide rapid aging. Continuous application of high dielectric voltage will not accelerate stator failures in isolation and reveal their insulation defects. But applying gradually increasing dielectric voltage has a big effect to find out the stator sleeping weaknesses.

If the size of the insulating layer is not large enough to cover the steel wall in the slot, the dielectric strength of the stator will decrease in a short time. Also, if the slot fillnesses of stators are so high, mechanical stress occurs on stator insulation, wires and paper insulations can be damaged or torn; plastic insulations can be deformed.
It should also be noted that if any number of compressor metal burrs fall or reach the stator, both good and weak stators will fail quickly rather than in time. Since permanent magnets are used aţ inverter compressors, it is difficult to keep the metal burrs away from compressor components, but it is essential to provide compressor safety.

6. REFERENCES

ABB Industrial Systems, Inc. (1997). *Technical Guide No. 102 - ABB.* https://library.e.abb.com/public/fec1a7b62d273351c12571b60056a0fd/voltstress.pdf.

Failure Reasons for Insulations - ZVEI. (2016). https://www.zvei.org/fileadmin/user_upload/Presse_und_Medien/Publikationen/2015/mai/Failure_Reasons_for_Insulations/Failure_Reasons_for_Insulations-Rev-May2016.pdf.

Grubic, S. (2008). *A Survey on Testing and Monitoring Methods for Stator Insulation Systems of Low-Voltage Induction Machines Focusing on Turn Insulation Problems.* IEEE Xplore. https://ieeexplore.ieee.org/document/4624549/.

Gupta, B. K., Stone, G. C., & Stein, J. (2009). *Stator winding hipot (high potential) testing.* IEEE Xplore. https://ieeexplore.ieee.org/document/5166381.

Ritamäki, M., Rytöluoto, I., & Lahti, K. (2017). Temperature Effect on Breakdown Performance of Insulating Polymer Thin Films. *Proceedings of the Nordic Insulation Symposium*, (24). https://doi.org/10.5324/nordis.v0i24.2288

Stone, G. C., & Culbert, I. (2018, June 5). *Prediction of Stator Winding Remaining Life From Diagnostic Measurements.* Iris Power. https://irispower.com/learning-centre/prediction-stator-winding-remaining-life-diagnostic-measurements/.

Tommasini, D. (2009). *Dielectric Insulation & High Voltage Issues.* https://cas.web.cern.ch/sites/cas.web.cern.ch/files/lectures/bruges-2009/tommasini.pdf.

RESTART TEST DESIGN FOR INVERTER COMPRESSORS

Tuğba Çetintürk[1], Bilgin Hacıoğlu[2]

[1]Arçelik A.Ş., Compressor Plant R&D Department,
Eskişehir, TURKEY
+902222134657, +902222134344, tugba.cetinturk@arcelik.com

[2]Beko LLC, Production Engineering,
Kirzach, Russia
+74923726091, +79636249803, bilgin.hacioglu@beko.ru

ABSTRACT

When a short power cut or electricity failure occurs on an electric supply system, which the household refrigerator is connected to, the suction and discharge pressures of the system become unbalanced, and the compressor exposes to higher loads than rated conditions. If the power recovers very quickly before the pressures get balanced, the compressor may not overcome the load and fails to start up. If the inverter control of the compressor has Class B protection software instead of mechanical overload protector; the inverter software forms a fault signal, cuts the energy through the compressor and waits for a while that provides pressure balancing before the next compressor start up trial.

Provided that the compressor starts up, or the inverter control cuts energy by a fault signal, the compressor periodically tries running in an infinite loop. This situation prevents compressor running and cooling of the refrigerator. In order to prevent this, the inverter board and the control software should be controlled at every load condition if it works properly or not.

Generally, a standard restart test is done to refrigerators by cutting the electricity off and energizing after a few seconds in order to check if the compressor starts up or a fault signal occurs. This test only checks one load condition and every different suction-discharge pressure value after power cut cannot be simulated that can occur in the field. Simulation of all loads on the refrigerator is a hard situation because providing variable suction-discharge pressure values on the refrigerator system is impossible and waiting for pressure balancing takes a long time for every load condition.

For this paper; a compressor restart test is studied without a refrigerator and an array of suction-discharge pressures that can occur against the compressor is determined in case of a short power off-on situation in order to check all load conditions of the refrigerator. The test is carried out by compressor- electronic control pairs on a cooling system simulator. The compressor test results are verified by testing the same compressors on the refrigerator.

1. INTRODUCTION

Refrigerator cooling cycle occurs by the heat transfer performed by the refrigerant. The refrigerant absorbs the heat from the cooling area and transmits to the environment. In this period, the compressor is in charge of compressing the refrigerant via its piston and increasing its pressure. Figure 1 shows the refrigerator cooling cycle and flow direction of the refrigerant with numbers '1' to '4'. Summation of the cooling process is given below in order, with the numbers in Figure 1:

- 1-2 Compression of the refrigerant from low to high pressure .
- 2-3 Concentration of the refrigerating fluid by heat transfer at condenser .
- 3-4 Expansion and pressure decreasing at capillary tube.
- 4-1 Vaporization of the refrigerant at evaporator, by heat transition from the cooled area to the refrigerant.

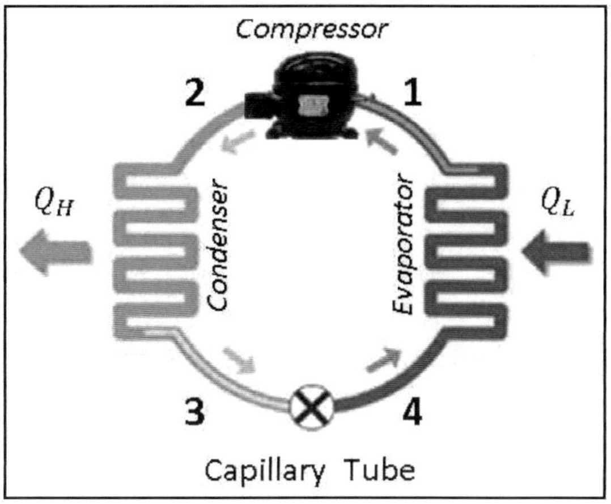

Figure 1: Refrigerator Cooling Cycle

'1' indicates suction port, '2' indicates dicharge port of the compressor. As Kim *et al.* (2009a) briefly described the start and restart operations of the reciprocating compressor, the start-up usually occurs when a pressure difference between suction and discharge port is nearly zero. In order to restart the compressor after it stops running, a temporary pause of starting is required until the pressure of the suction and the discharge port goes to an equilibrium state since the presence of the pressure difference between the suction and the discharge port acts as an excessive load to the motor for driving the compressor.

2. THE RELATIONSHIP BETWEEN RESTART OF INVERTER COMPRESSOR AND CLASS B SOFTWARE

An overload protector opens its contacts by sensing high current and/or temperature. The contacts close again in definite time at the protector's datasheet. While choosing an overload protector for an Inverter Compressor, closing time of the contacts should be long enough to balance suction and discharge pressures.

If the inverter compressor has Class B Protection, a time delay should occur on the motor control software in order to provide enough time for pressure balancing before energizing compressor. Kim *et al.* (2009b) mentioned that, the temporary waiting time for restarting is generally fixed to 5 minutes in a commercial refrigerator so that the compressor may start its operation before the pressure between the suction and the discharge port reaches a balanced state.
At Class B protection, motor control software has "Safety Related" functions that can detect abnormal and/or unsafety conditions to cut power and protect both the compressor and the control board. Some of these Class B functions are:

- Locked rotor fault
- Running Overload
- Loss of Phase
- Short circuit or open circuit failures of related electronic components.

By sensing any abnormal or unsafe condition, the software turns off the electronic control board in order to stop the compressor running. After waiting for the defined waiting time (in minutes), the electronic control board turns on and try to run the compressor again. This waiting and trial cycle repeats for a determined number of times until the compressor properly starts up. If the compressor still cannot start-up, the inverter board completely turns off without any other trials. If the power outage and recovery occur from the mains, the suction and discharge pressures may not have enough time to get balanced and the compressor may be exposed to an excessive load. In this case, either the compressor should start-up and run or Locked Rotor Alarm should occur by Class B software in order to start the waiting and trial cycle for restart. If start-up or locked Rotor Alarm do not occur, an infinite loop of trial of motor control software is in question which prevents cooling until unplugging the refrigerator and plugging again.

Class B Software is not only more reliable but also more complicated. That's why, it should be tested carefully because while it provides protection, it can cause functional problems of the compressor or inverter board in any missing working condition.

Kim *et al.* (2008) verified the effectiveness of a starting method using the reciprocating compressor load change system. A simplified refrigeration system is used to provide 10 pressure differences between suction and discharge ports of the compressor and starting current waveforms in order to check the starting method.
Kim *et al.* (2009c) also tested another improved starting method by using a similar test bed in order to adjust some specific suction and discharge pressure differences.

3. CURRENT START-UP AND RESTART TESTS APPLIED TO COMPRESSORS AND REFRIGERATORS

3.1 Restart Test of Refrigerators

Generally, a refrigerator is tested by cutting the electricity off and energizing after some seconds in order to check if the compressor starts up or a fault alarm occurs on the inverter board. But only one certain time gap between power cut and re-energizing refrigerator is applied at Restart Test in order to determine if the compressor properly starts up or how long or how many trials does it take for the compressor to start up at one condition.

But in real life, because the time interval between power outage and recovery is countless, the compressor load during restart is not clear. Since power recovery can be either in 1 second or 10 seconds or in minutes after outage; the suction-discharge pressures can be at any level during restart. But this test checks only one load condition that occurs after 10 seconds and cannot simulate other possibilities of the pressure difference between suction and discharge outlets during restart, which may occur on site.

3.2 Start-up and Stall Tests of Single Compressors

The test system shown in Figure 2, is used for starting or stall testing of a single compressor. In order to simulate the cooling cycle of a refrigerator during start-up or stall conditions. After fixing the compressor, R600a refrigerant is pumped into the system. The valves and manometers at both suction and discharge port provides adjusting the pressures.

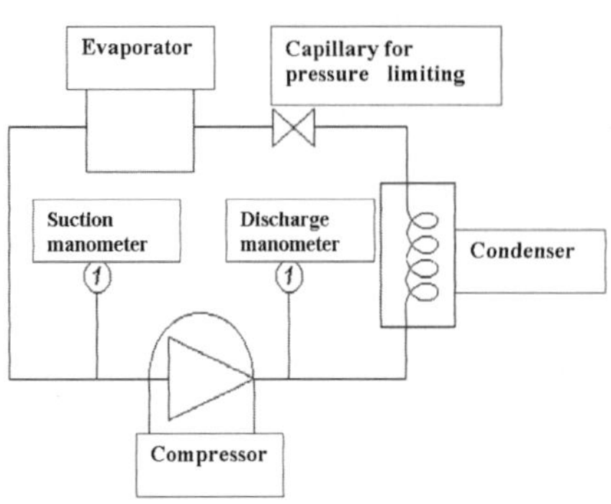

Figure 2: Compressor Start Up ans Stall Test System

Unlike start-up test, the compressor stall test is done by setting unequal pressures at suction and discharge ports. Discharge pressure levels that the compressor can reach and cannot reach can be seen at stall curves. If the stall curve of the compressor is below the stall curve of a refrigerator, the compressor is not compatible for that refrigeration system. The compressor should have enough shaft power to overcome the refrigerator load particularly at Pull Down point of the refrigerator at minimum working voltage and maximum working temperature. Stall curves of 2 different compressors and a refrigerator is seen at Figure 3, which shows that the 2 compressors can work properly on that refrigerator at the suction-discharge pressures indicated in the graph.

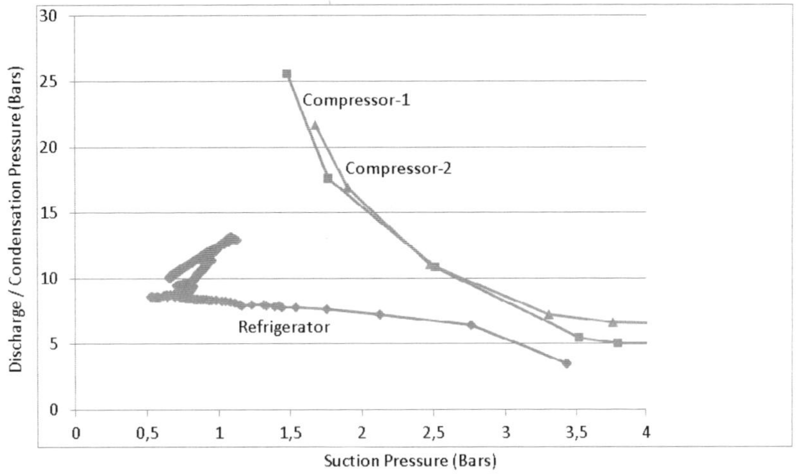

Figure 3: Comparison of Compressor and Refrigerator Stall Curves

3.3 The Reason Why a Restart Test is Necessary For Inverter Compressors

It is not possible to apply Stall Test to inverter compressors used by Class B protection, because the safety functions prevent the compressor to reach high suction pressures. The compressor start-up test only checks the compressor load at rated equilibrium pressure. Refrigerator Restart Test is the best way to simulate power off-on condition. But, simulation of all possible restart loads on a refrigerator is a hard situation because adjusting different suction-discharge pressure values on the same refrigerator system is not possible and waiting for pressure balancing takes a long time for every restart checking by power off – on. As Andrade and Negrao (2013a) explained, the experimental tests in order to check the performance of household refrigeration systems are expensive and time-demanding, e.g., a single start-up experiment can take more than 12 h to be performed. Guzella and Gomez (2017), Hermes and Melo (2018), Andrade and Negrao (2013b) studied on numerical simulations of household refrigerators during start up and cycling operations. Even if their results were close to actual measurements, discrepancies exist at some conditions. These models are useful for preliminary examinations but still needs to be verified by tests.

However, it is easy to test the start and restart compatibility of a single compressor and inverter board pairs without refrigerator. Therefore, a new test need arised to check if inverter compressors with Class B protection will start-up properly or not at any load when the refrigerator restarts.

4. STUDIES FOR TEST DESIGN AND DETERMINATION OF TEST METHOD

Every time a refrigerator runs and stops, rebalancing values of pressures at suction and discharge ports change. The balanced pressures of a newly produced and packaged refrigerator is different from that of a refrigerator that has been operated for a long time. Or balanced pressures of refrigerators with different capacities or designs are not the same even if they worked and stopped at the same condition. That's why, there are too many suction-discharge pressure pairs at refrigerator restart after a power outage and back on case. However, the number of possible pressure pairs are limited by the refrigerator design.

The first task of the test design is to determine the test conditions by covering all possible loads, and to determine the boundary of suction and discharge pressure. Thereby; cooling, start – up, pull down and restart test results of current household refrigerator series are examined and all possible suction – discharge pressure pairs are listed within the limits of the cooling system design. Since the compressor loads during start-up are critical; not only the unbalanced pressure pairs that occur after a sudden power off-on case is determined, but also the balanced pressure pairs are listed. While suction pressure levels are obtained within 0.5 – 5.0 Bars, discharge pressure levels are obtained within 2.0 to 8.0 Bars. 38 pressure pairs are determined as testing condition.

It is also possible to design and establish a separate test system specifically to specifically check the restart compatibility of compressor- electronic control pair by simulating the working characteristics of the refrigeration system. In order to use the current source, the semi-anechoic Ccmpressor noise test chamber is used to adjust all 38 test conditions. The picture of the test chamber is shown in Figure 4.

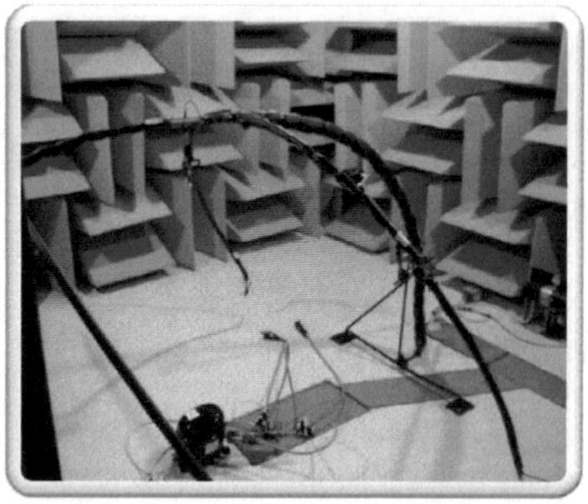

Figure 4: Compressor Noise Testing Room

5. TESTS CONDUCTED AND THEIR RESULTS

At least 15 pcs of compressors were tested under the defined pressure pairs, and the test results were recorded on the Test format shown at Table 1.

Table 1: Compressor Restart Test Format

Explanation	Pressure Pair (Bars)		Compressor/inverter responses after inverter is energized at defined suction-discharge pressures		
	Suction Pressure	Discharge Pressure	No Start-Up, No Locked Rotor Alarm Occured (Infinite Loop)	Locked Rotor Alarm occured	Compressor Start-Up and run
Balanced Pressures	a	a			X
	b	b			X
	c	c			X
Un-Balanced Pressures	Min	Min		X	
	.	.			X
	.	.		X	
	.	.	X		
	Max	Max		X	

By evaluating and analyzing the relation between test results and pressure differences between suction and discharge ports, some test conditions that provide the same compressor load and have the same test results are removed. So that, number of test conditions are minimised in order to accelerate the test.

At first, temperature is also evaluated in order to simulate the compressor load for the compressor restart test. When the compressor is heated, the phase currents of the compressor motor change under each load. But since all possible pressure pairs are listed as a test condition, heating the compressors removed from test conditions.

The test is applied to compressors with different designs, cooling capacities, motors and inverters. In order to check the repeatability of the test, at least 3 compressors with the same properties are tested and 15 test results are compared.

Figure 5 shows two different examples of negative test results. At any suction-discharge pressure pairs of the test; if the compressor fails to start up, or if its inverter software does not issue a locked rotor alarm to cut off the energy passing through the compressor, it means that the test has failed. The compressor of motor, inverter hardware and/or software should be checked and revised until there will be no marking at "No Start No Locked Rotor" column shown at Figure 5.

Compressor Model: G, Inverter HW: H, Inverter SW: K

Compressor Suction Pressure (Bar)	Compressor Discharge Pressure (Bar)	No Start No Locked Rotor Alarm	Locked Rotor Alarm	Start Up
2,0	2,0			X
...	...			X
3,5	3,5			X
...	4,0			X
....	4,8	X		
...	3,5			X
0,5	4,5			X
...	....		X	
...	...		X	
...	5,0	X		
1,2	....			
1,2	8			
....	....	X		
...	5,2	X		
1,7	6,5		X	
....	5,5	X		
...	...	X		
2,2	7,2	X		
2,8	....	X		
....	...	X		
...	...	X		
3,3	...			X
3,6	4,5			X
3,7	...	X		
4,0	4,5			X
....	....		X	
4,0	6,3		X	

Compressor Model: J, Inverter HW: K, Inverter SW: L

Compressor Suction Pressure (Bar)	Compressor Discharge Pressure (Bar)	No Start No Locked Rotor Alarm	Locked Rotor Alarm	Start Up
2,0	2,0			X
...	...			X
3,5	3,5			X
...	4,0			X
....	4,8			X
...	3,5			X
0,5	4,5			X
...	....			X
...	...			X
...	5,0			X
1,2	....		X	
1,2	8	X		
....	....		X	
...	5,2		X	
1,7	6,5	X		
....	5,5		X	
...	...	X		
2,2	7,2		X	
2,8	....		X	
....	...	X		
...	...		X	
3,3	...		X	
3,6	4,5			X
3,7	...		X	
4,0	4,5			X
....	....			X
4,0	6,3		X	

Figure 5: Examples of Negative Test Results done to different Compressor, Inverter HW and SW Combinations

"No Start No Locked Rotor" columns are completely empty at Figure 6, which means the test results are positive. So the combination of compressor motor, inverter hardware and software can be reliably tested on refrigerator cabinets.

Compressor Model: A, Inverter HW: B, Inverter SW: C

Compressor Suction Pressure (Bar)	Compressor Discharge Pressure (Bar)	No Start No Locked Rotor Alarm	Locked Rotor Alarm	Start Up
2,0	2,0			X
...	...			X
3,5	3,5			X
...	4,0			X
....	4,8		X	
...	3,5			X
0,5	4,5			X
...	....			X
...	...			X
...	5,0			X
1,2	....			X
1,2	8		X	
....	....			X
...	5,2			X
1,7	6,5		X	
....	5,5			X
...	...		X	
2,2	7,2		X	
2,8	....			X
....	...		X	
...	...		X	
3,3	...		X	
3,6	4,5			X
3,7	...		X	
4,0	4,5			X
....	....			X
4,0	6,3		X	

Compressor Model: D, Inverter HW: E, Inverter SW: F

Compressor Suction Pressure (Bar)	Compressor Discharge Pressure (Bar)	No Start No Locked Rotor Alarm	Locked Rotor Alarm	Start Up
2,0	2,0			X
...	...			X
3,5	3,5			X
...	4,0			X
....	4,8			X
...	3,5			X
0,5	4,5			X
...	....			X
...	...			X
...	5,0			X
1,2	....			X
1,2	8			X
....	....			X
...	5,2			X
1,7	6,5		X	
....	5,5		X	
...	...		X	
2,2	7,2		X	
2,8	....		X	
....	...		X	
...	...		X	
3,3	...		X	
3,6	4,5			X
3,7	...		X	
4,0	4,5			X
....	....			X
4,0	6,3		X	

Figure 6: Examples of Positive Test Results done to different Compressor, Inverter HW and SW Combinations

Lee *et al.* (2008) monitored the starting current waveforms of the BLDC Motor used in the reciprocating compressor of household appliances by providing a pressure differences between the suction and the discharge port of the compressor, thereby optimized the motor control software to improve the starting. At this study, motor phase currents are also monitored by osciloscope during start-up, locked rotor and neither start-up nor locked rotor (infinite loop of start-up trial and fail). The differences between phase current waves of three different cases are shown in Figure 7, 8 and 9.

At start-up condition, the phase current fluctuates for a while until start up is done. After the start-up is done, the current flows around a stable level to make the compressor run, as shown in Figure 7.

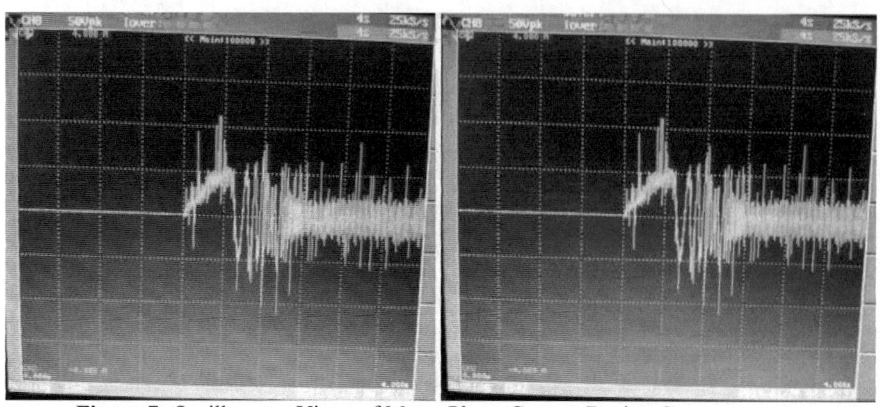

Figure 7: Oscilloscope Views of Motor Phase Current During Compressor Restart Test
Start up and Run Condition

At locked rotor condition, the software creates a false alarm, cuts off the energy through the compressor and calculates the time to wait for the next trial. That's why, after locked rotor detection, the phase current directly goes to zero value as shown in Figure 8.

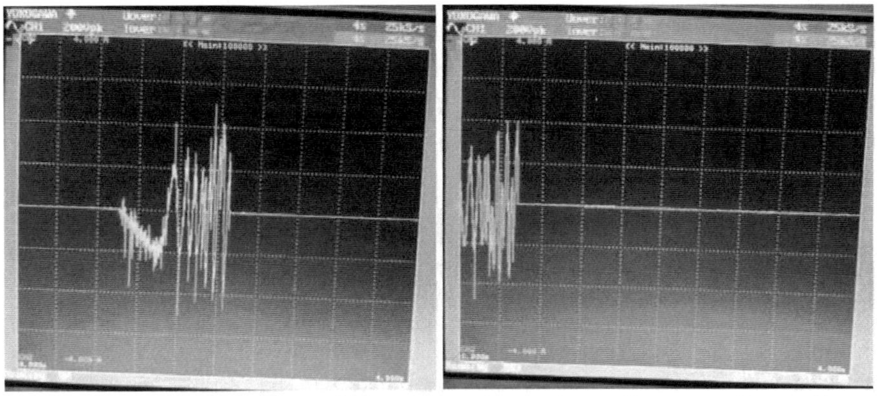

Figure 8: Oscilloscope Views of Motor Phase Current During Compressor Restart Test
Locked Rotor Condition

In the case of neither start-up nor locked rotor condition (That is, the test fails), the phase current distortion continues because the compressor cannot overcome the load for start-up as shown in Figure 9. In addition, the amplitude and duration of the current have not reached the locked-rotor current limit, so the motor control software cannot provide locked-rotor alarm. In this case, the motor control software continuously energizes the compressor without waiting of pressure balancing, and an infinite loop occurs which prevents cooling process.

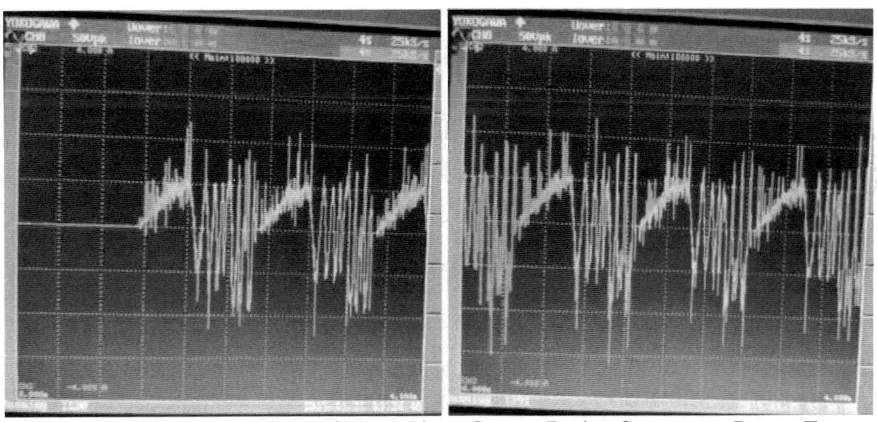

Figure 9: Oscilloscope Views of Motor Phase Current During Compressor Restart Test
No Start-up, No Locked Rotor Alarm (Infinite Trial Loop) Condition

6. CONCLUSIONS

In the beginning, some faults found in the refrigerator restart test triggered this research, and there was a need for compressor restart tests to reduce the workload and time allocated to refrigerator tests. Due to a software error, an infinite loop of continuous running trials were detected in the power off/on simulation. In particular, these malfunctioning compressor and inverter pairs are used to check whether the compressor restart test works properly, and detects the same malfunction seen in the refrigerator restart test. Therefore, the compressor restart test matches the refrigerator restart test, and the new test design is successful. With this new test, the motor software can be corrected and improved under test conditions that the compressor cannot pass. Tests are repeated by the improved software and good results of mutual verification will be obtained in the restart test of the compressor and the refrigerator.

The listed conditions determined for the new compressor restart test cover all possible compressor loads that may occur on the refrigeration system during any restart case of a household refrigerator.

In order to perform systematically, the new compressor restart test has become a standard procedure before engaging new platform compressor designs, the derivative of the existing designs, motor design change and motor control software change. Therefore, it is now possible to simulate all possible restart loads on the compressor in a short period of time without waiting for the pressure of the refrigerator to equalize. Through the compressor restart test, the detected faults can be seen and resolved in the early stages of the works, and then tested on the refrigerator to avoid wasting time.

7. REFERENCES

Andrade, D. E., & Negrão, C. O. (2013). Analysis of start-up tests of household refrigeration systems. A potential for on-line predictions of test results. *Applied Thermal Engineering, 54*(1), 255-263. doi:10.1016/j.applthermaleng.2013.02.008

Guzella, M. D., & Gómez, L. C. (2017). *Numerical Simulation Of The Performance Of a Household Refrigerator During Start-up And Cycling Operations.* Manuscript submitted for publication, Engineering School of São Carlos, University of São Paulo, São Paulo.

Hermes, C. J., & Melo, C. (2008). A first-principles simulation model for the start-up and cycling transients of household refrigerators. *International Journal of Refrigeration, 31*(8), 1341-1357. doi:10.1016/j.ijrefrig.2008.04.003

Kim, D., Rhyu, S., Lee, K., Kim, B., Chung, D., & Kwon, B. (2008). Comparison of Starting Method for Position Sensorless BLDC Motor Driven Reciprocating Compressor. *2008 IEEE Industry Applications Society Annual Meeting*. doi:10.1109/08ias.2008.17

Kim, D., Shin, D., Lee, S., Kim, H., Kwon, B., Kim, B., & Lee, K. (2009). Novel position sensorless starting method of BLDC motor for reciprocating compressor. *INTELEC 2009 - 31st International Telecommunications Energy Conference*. doi:10.1109/intlec.2009.5351997

Lee, K., Kim, D., Kim, B., & Kwon, B. (2008). A Novel Starting Method of the Surface Permanent-Magnet BLDC Motors Without Position Sensor for Reciprocating Compressor. *IEEE Transactions on Industry Applications, 44*(1), 85-92. doi:10.1109/tia.2007.912734

AN ANALYTICAL TOOL TO DETERMINE THE OPTIMUM COUNTERWEIGHTS FOR MULTI-CYLINDER RECIPROCATING COMPRESSORS

Salih Güvenç USLU [1], Kaan ŞENGÜL [2]

[1]Kırpart Automative, Research & Development, Bursa, Turkey
guvenc.uslu@kirpart.com.tr

[2]Dalgakıran Makina, Research & Development, Kocaeli, Turkey
kaan.sengul@dalgakiran.com

ABSTRACT

The industrial air compressor market has been growing fast, thereby compelling the manufacturers to produce competitive products with less vibration and noise, which also better meet the customers' expectations and related regulations. Both mechanical and hydrodynamical factors induce vibrations which have substantial negative effects in the maintenance periods and the lifetime of certain components. This study focuses on the balancing of the crankshaft by modifying the geometry of the counterweights. It aims at reducing the overall vibration of a compressor with the outputs of an analytical model which investigates the dynamics of the crankshaft of a W-type reciprocating compressor. The model predicts the entire motion of the compressor. An interface is created on MATLAB to ease the use of the model. The theoretical results are validated by both a series of tests and rigid body dynamics, RBD, simulations on Ansys Workbench. The tests encompass the progressive change of the outer diameter of the counterweights on the crankshaft. The RMS (root mean square) velocities on several locations on the compressor head are obtained with a piezoelectric triaxial accelerometer for each outer diameter. The analytical model, RBD model and experimental results match each other with a maximum deviation of 5%. The conclusion of the study is that not only the optimization of the resultant forces acting on the crankshaft take a role in reducing the overall vibration on the compressor head but also the moments on the crankcase bearings alter the vibrational amplitude. The reciprocating and the centrifugal motions of the crank mechanism, the geometry of the crankshaft in axial direction and gyroscopic effect due to the crankshaft inertia tensor are considered in the calculation of the resultant moments on the bearings.

1. INTRODUCTION

The vibration levels are of primary cause that determine the compressor life and accordingly its noise-vibration-harshness (NVH) characteristics (Hanlon, 2001). This paper focuses on the balanced inertial forces and the inertial moments created by the vibration in an air compressor which can be caused by natural frequencies, unbalanced inertial forces and hydrodynamic impacts. The unbalanced inertial forces in a crank mechanism are caused by reciprocating behavior of components, such as piston and connecting rod, and by centrifugal forces of rotating components, such as crank webs and counterweights (Olgun, 2010).
There are several methods to reduce the vibration levels in an air compressor. Modifying the counterweights, adding balance shafts addressing to first and/or first and second order reciprocating forces are of them. This study only adjusts the counterweights due to the feasibility of this method.
This paper is a complementary study to that of Pisirici *et al.* (2018). Pisirici *et al.* studied on the dynamics of a W-type, three stage reciprocating compressors and presented a mathematical method to calculate the unbalanced inertial forces on the steady and transient conditions. The recent study contributes to it by evaluating the inertial moments acting on the crank bearings, creating a tool on MATLAB software, performing RBD analyses on Ansys Workbench and further experiments to determine the optimum counterweights.
Both this study and Pisirici *et al.*'s have been shedding light on the on-going design updates of single and multi-stage reciprocating compressors in Dalgakiran Compressor.

2. MATHEMATICAL MODEL

Pisirici *et al.* built a methodology to calculate the resultant inertial forces acting on the crankshaft of the air compressor. In addition to this finding, in the recent study, using the inertial forces in different directions, the inertial moments acting on each crank bearing are calculated. The moment along which the crankshaft rotates are ignored as it is assumed that the moment in this direction was neglectable and it does not contribute to the vibrant effect to the compressor body.

There are two impacts to calculate the moments on the bearings. One is a component of resultant inertial velocity field and its axial level arm to the crankcase bearings, the other is due to the gyroscopic effect that arises from the shape of the crankshaft and accordingly the inertial reference coordinate frame. As seen in Figure 1, the deviation of z axis from the rotational axis of the crankshaft causes the additional inertial moment as gyroscopic torque on the crankcase bearings. The gyroscopic torque is calculated using equation (1). The resultant inertial moment on the crankcase bearings is calculated as in Equations (2-4).

$$M_{gy} = J_{yz} * \omega * \omega_p \tag{1}$$

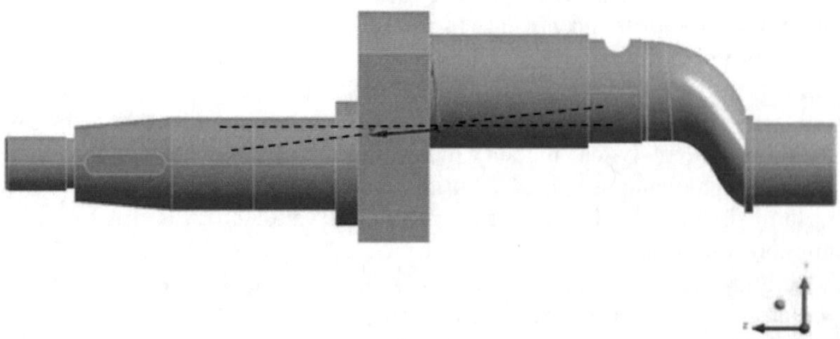

Figure 1: The deviation of the inertial reference coordinate system from the rotational axis of the crankshaft in z direction

$$M_{ax} = -F_c * cos(\theta) * x_1 - F_a * x_2 * cos(\theta \pm \varphi) + M_{gy} * cos(\theta) \tag{2}$$

$$M_{ay} = F_c * sin(\theta) * x_1 + F_a * x_2 * sin(\theta \pm \varphi) - M_{gy} * sin(\theta) \tag{3}$$

$$M_r = \sqrt{(M_{ax})^2 + \left(M_{ay}\right)^2} \tag{4}$$

The centrifugal force produced by the crankshaft rotation is calculated using equation (5). The variable parameter *mh,* which is aimed to be found in the software, is calculated using equation (6).

$$F_c = -[mh - m_{bk} * r] * \omega^2 \tag{5}$$

$$mh = m_k * h_k \tag{6}$$

The mathematical model computes the resultant terms over one cycle for one *mh* parametric variable and it repeats the computation for other *mh* variables within a given range and an increment. After the model scans the interval of *mh,* it highlights the *mh* variable/s with which the minimum resultant inertial force and moments are acquired.

3. THE SOFTWARE

A part of the interface of the software is depicted in Figure 2. The tool was created on MATLAB.

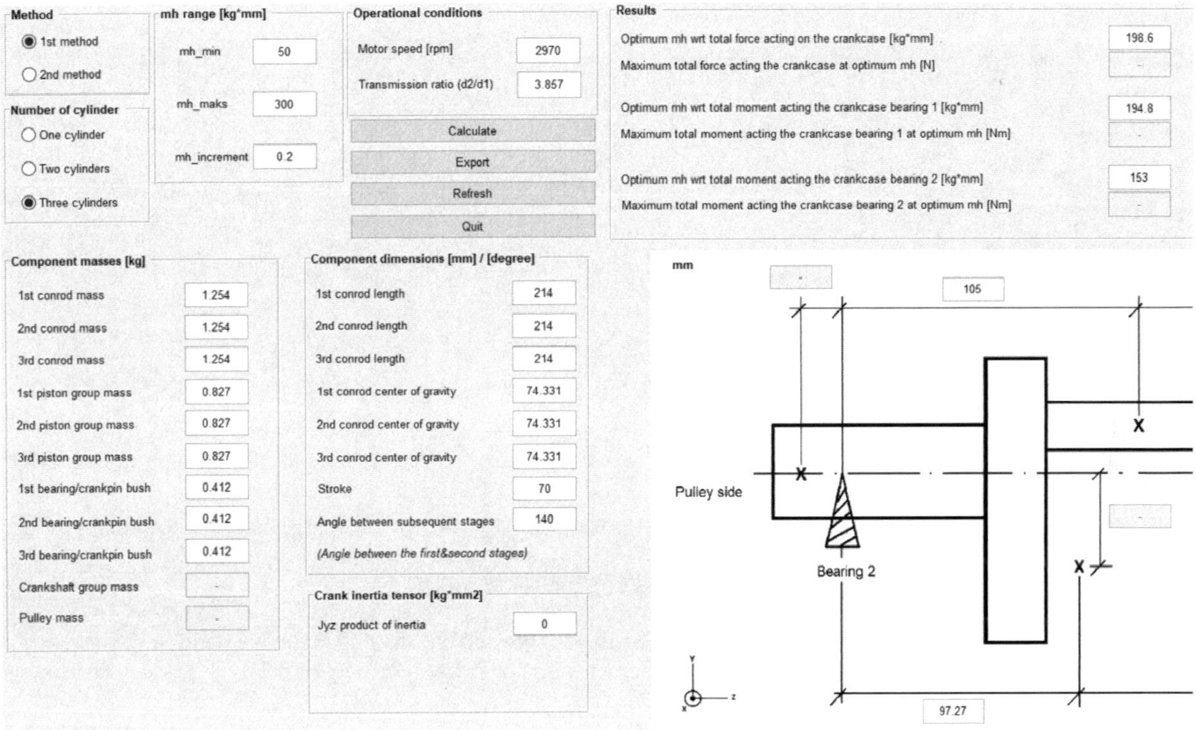

Figure 2: The interface of the software which helps to determine the optimum *mh* parametric variables

The software consists of six sections. There are two types of solutions under the method section. The first method determines the optimum *mh* parametric variable/s. The second one calculates the absolute inertial force and moments acting on the crankcase bearings. The solution with the second method becomes a guideline in selecting the proper bearings. The solution can be performed for single-cylinder, V-type and W-type compressor heads. To set the range of *mh* parameter, the extremes and the increment are defined. The transmission ratio and the motor speed are defined in the operational conditions section. It should be noted that the model assumes that the compressor runs steadily. The other inputs are component masses, component dimensions, the related product of the crankshaft inertia tensor and the lever arms. An example of a result for the first method can be seen in Figure 2 in the results section.

4. MODEL VALIDATION

The model mentioned in Section 2 was evaluated with two approaches which were the validation with a computer-based analysis and an experimental method. In the validation process Dalgakiran DBK 30 reciprocating compressor model was utilized, which is a booster compressor generally coupled with a screw compressor that pressurizes air from 7 bar to 35 bar and runs at a fixed-speed electric motor. The torque generated by the motor is transmitted to the compressor head with two V-belts.

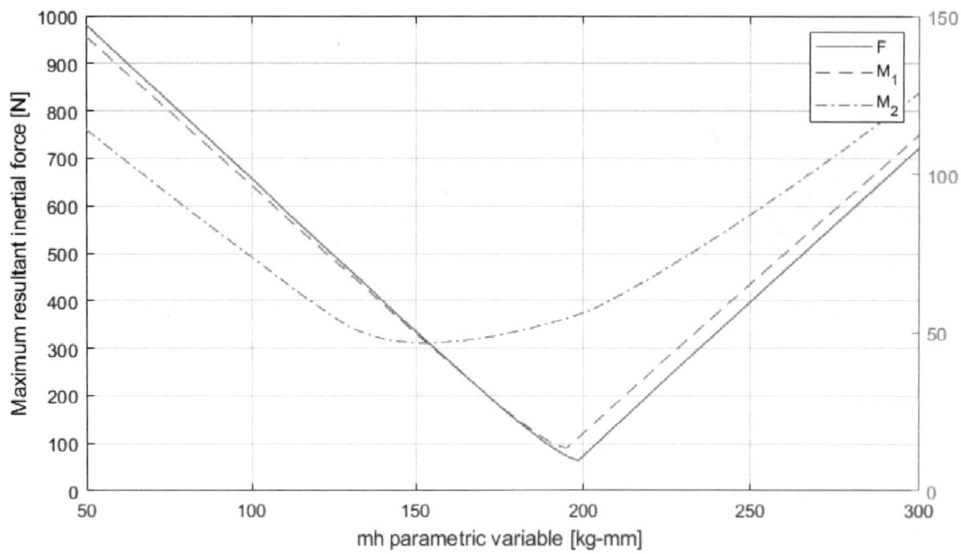

Figure 3: The optimum *mh* values calculated with the software created on MATLAB (F, M_1 and M_2 stand for the maximum inertial force, the maximum inertial moments on crankcase bearings, respectively, as *mh* varies).

Figure 3 exhibits the ideal *mh* values with respect to the total inertial forces on the crankcase and the moments by different bearings in DBK30 booster compressor. The ideal *mh* based on the total inertial forces was 198.6 kg-mm. The ideal *mh* depending the crankcase bearings were 194.8 kg-mm and 153 kg-mm.

4.1 Rigid Body Dynamics Analysis

DBK 30 has a W-type compressor head. The cylinder heads are identical, but the heads pressurize the sucked air with a phase shift of 70°. The crank mechanism of the compressor can be seen in Figure 4.
RBD analysis was run at steady state condition at the fixed speed of 770 rpm. The rotational velocity was defined on the revolute joint between the crankshaft and the ground.

Figure 4: DBK 30 booster compressor on ANSYS Workbench RBD module

In order to find the optimum radius of the counterweights, a parametric optimization process was used with a definition of parametric variable. Figure 5 demonstrates the parts of the counterweight which were split and suppressed step by step over a given range of radius. At each radius, the analysis was repeated automatically, and the total inertial forces were obtained on the revolute joint of the crankshaft.

Figure 5: The gradually split and suppressed parts of the counterweights

The total inertial moment on the revolute joint of the crankshaft depended on the origin of the reference axis. The crankshaft was mounted on the crankcase at two locations. These locations were where the bearings were. As depicted in Figure 5, one bearing was in the side of the pulley, the other was in the side of radiator. The bearings were suppressed in Figure 5. Eventually, two total inertial moments were obtained depending on the reference location.

Table 1: RBD results with respect to different design points

Design point	Counterweight radius	Total maximum crankshaft force	Total maximum crankshaft moment
Units	m	N	Nm
DP 0 (Current)	0,104	189,17	64,524
DP 1	0,1035	152,84	59,419
DP 2	0,103	118,39	54,563
DP 3	0,1025	84,9	49,725
DP 4	0,102	57,249	44,919
DP 5	0,1015	81,782	40,151
DP 6	0,101	115,79	35,438
DP 7	0,1005	149,56	30,769
DP 8	0,1	183,07	26,145
DP 9	0,0995	216,36	21,572
DP 10	0,099	249,38	17,193
DP 11	0,0985	282,06	13,832
DP 12	0,098	314,43	15,7
DP 13	0,0975	346,48	20,172
DP 14	0,097	378,22	24,603
DP 15	0,0965	409,64	28,992
DP 16	0,096	440,75	33,34

Table 1 displays the results of the parametric solution obtained through the rigid body dynamics analysis. The column on the left indicates the design points, the second column lists the radius of the counterweights, the others are the total inertial forces and moments, respectively. Starting from 104 mm radius, the radius of the counterweights was reduced by 0.5 mm down to 96 mm. The minimum resultant inertial force was obtained at the third design point where the radius of the counterweights was 102 mm. Similarly, the minimum resultant moment was obtained at 98.5 mm radius. 98.5 mm radius and 102 mm radius correspond to 195.7 kg-mm and 198.9 kg-mm, respectively. The difference between the mathematical model results and the RBD results stem from the increment of the parametric solution and the assumption that in the mathematical solution the inertial coordinate system is fixed at each radius of the counterweight within the given *mh* range on MATLAB.

4.2 Experimental Setup

Modelling on MATLAB and Ansys Workbench ran simultaneously, and they validated each other. The computer-based studies were supported by a series of vibration test using the equipment seen in Figure 6-7.
The tests were performed with a progressive method. The diameter of the counterweights seen in Figure 5 was reduced by 1 mm in the sequential vibration tests. The ideal *mh* interval had been determined with the mathematical and the RBD models prior to testing phase. The number of tests was determined and reduced considering these models. Figure 8 demonstrates the process of metal removing on the counterweights.

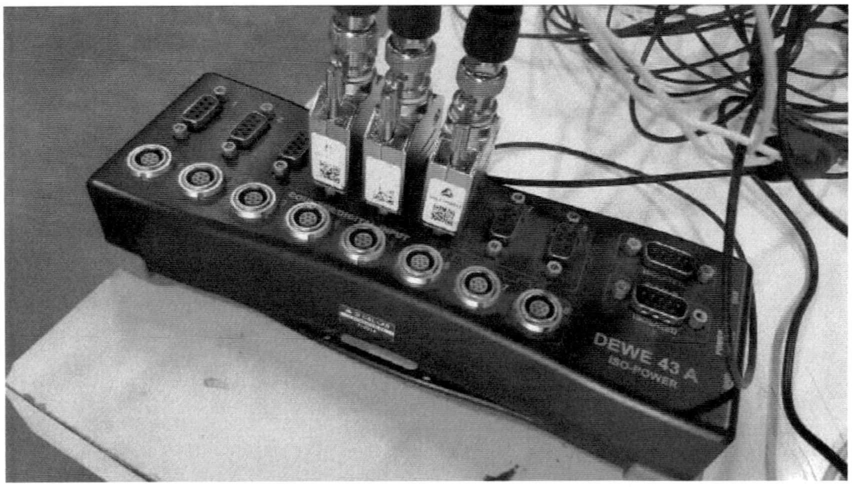

Figure 6: Dewe 43A data logger

Figure 7: Dytran 3263A2 accelerometer and Dytran 6272 magnetic base used in the tests to measure vibration levels

Figure 8: Metal removing process within the successive vibration tests

Figure 9 demonstrates the velocity amplitudes of vibration with respect to the counterweight outer radius on DBK 30 at 12.83 Hz which is equivalent to 770 rpm. 770 rpm is the first order of rotational speed of the compressor head. A modification on the counterweights directly impacts on the first order, thus the velocity amplitudes at this speed was studied primarily. One curve in Figure 9 belongs to the tests performed on the crankcase. This location is seen in Figure 6. The other curve and the mean velocity amplitudes on it belong to the measurements taken in several locations on the compressor head. As seen in Figure 9, the amplitudes were minimized at 99 mm radius. This learning complies with the one obtained in the mathematical and the RBD models.

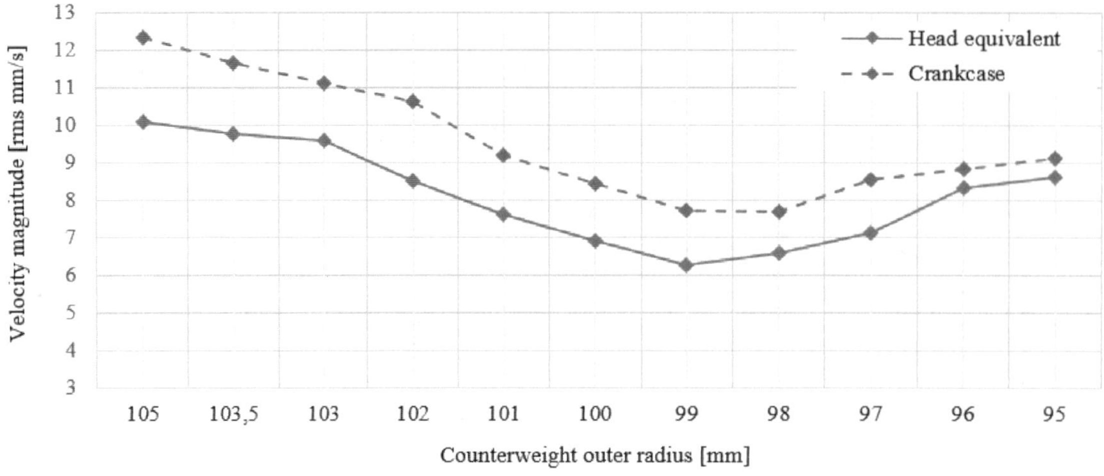

Figure 9: The velocity amplitude of vibration with respect to the counterweight outer radius on DBK 30 at 12.83 Hz (770 rpm)

It should be highlighted that the sequential tests were carried out by 1 mm radius increment. Therefore, the optimum counterweight outer radius might have been any value within 98 - 100 mm range. It was certain that the optimum one was not by 102 mm radius which the resultant inertial force points at.

Table 2 summarizes the results. The mathematical model obtained three different *mh* values. One of them was 153 kg-mm. As seen in Figure 3, this *mh* did not lead to a sharp absolute minimum in the curve. Moreover, at 153 kg-mm *mh,* the resultant inertial parameters were greater in comparison with those obtained at around 190 kg-mm. Therefore, 153 kg-mm *mh* was not involved in this study. Table 2 emphasizes that the experimental result best matches with those which considered the resultant inertial moment with respect to the bearing in the radiator side. The pulley seen in Figure 4-5 does not only transmit torque from the motor but also plays the role of flywheel. It stores inertia, thus preventing velocity fluctuations during an operation. In DBK30, the pulley approximately 50% heavier than the whole crank mechanism, that is, that its mass is 34.7 kg. In the other side of the crankshaft an oil filter is mounted as seen in Figure 7. It is named as the radiator side in Table 2. The bearing in this side is exposed to a lighter mass.

Table 2: *mh* values and counterweight outer radius with different validation methods

Method	Reference parameter	*mh* [kg-mm]	Counterweight outer radius [mm]
Mathematical model	Resultant inertial force	198.6	101.8
	Resultant inertial moment with respect to the bearing position in the radiator side	194.8	97.7
	Resultant inertial moment with respect to the bearing position in the pulley side	153	62
RBD analysis	Resultant inertial force	198.9	102
	Resultant inertial moment with respect to the bearing position in the radiator side	195.7	98.5
Experiment	-	196.1	99

5. CONCLUSION

The study was aimed at determining the optimum counterweight for a W-type reciprocating compressor using a mathematical model and an application created on MATLAB. The model was validated with RBD analysis on Ansys Workbench and consecutive vibration tests. In this way, an approach which had been created in Dalgakiran was extended with the conclusion that not only the inertial forces but also the inertial moments on the crankcase have a significant role in determining the counterweights.
The experimental results supported the findings of the computer-based models with a maximum deviation of 5%. It can be concluded that the method, which regards the resultant inertial moment calculated with respect to the bearing position in the lighter side of the crankshaft, best matches with the test results.
Dalgakiran Compressor has over thirty different reciprocating compressor heads and this study will be a guideline in re-evaluating their vibrational performance. Moreover, this learning will be a tool to be used in the design phase of new reciprocating compressor crankshafts within the company in the future.

NOMENCLATURE

M_{gy}	Gyroscopic torque (N-m)
J_{yz}	Product of crankshaft inertia tensor with respect to global axis (kg-m)
ω	Rotational speed of crankshaft (rad/sec)
ω_p	Velocity of precession (rad/sec)
F_c	Centrifugal force produced by the rotation of the crankshaft (N)
θ	Crank angle (deg)

x_1	Axial lever arm from the center of gravity of the crankshaft to one of the crankcase bearings (m)
F_a	Reciprocating force (N)
x_2	Axial lever arm from the center of gravity of the crankpin to one of the crankcase bearings (m)
φ	Angle between subsequent cylinder heads (deg)
M_{ay}	Moment by the horizonal axis in Figure 1 at one of the crankcase bearings (N-m)
M_{ax}	Moment by the vertical axis in Figure 1 at one of the crankcase bearings (N-m)
M_r	Resultant inertial moment (N-m)
mh	Parametric variable to vary the centrifugal force (kg-m)
m_{bk}	Substitute mass of the connecting rod at the crank end (kg)
r	Crank radius (m)
m_k	Mass of the crankshaft including the counterweight/s (kg)
h_k	Distance between the center of gravity of the crankshaft and the rotational axis of it (m)

REFERENCES

Hanlon, P.C. (2001). Compressor handbook. New York, Mc-Graw-Hill Book Co.

Olgun, M., Kutlar, O.A. (2010). "Tek silindirli bir dizel motorun atalet kuvvetlerinin analizi ve dengeleme hesaplamaları (Unpublished master's thesis)". Istanbul Technical University Institute of Science and Technology.

Pisirici, S. Et Al. (2018). "On the dynamics of a three-stage single acting reciprocating compressor". International Compressor Engineering Conference. 1700:1-9.

A Hybrid Simulation Model to Predict the Steady-State Thermal Profile of Hermetic Reciprocating Compressors

Fernando A. AARDOOM, Victor H. P. ROSA, Cesar J. DESCHAMPS*

[1]POLO Research Labs, Federal University of Santa Catarina
Florianopolis, SC, Brazil
fernando.aardoom@polo.ufsc.br, victor.rosa@polo.ufsc.br, deschamps@polo.ufsc.br

* Corresponding Author

ABSTRACT

Numerical simulation models are paramount to design compressors that follow reliability and efficiency requirements. This paper presents a simulation model to predict the steady-state thermal profile of hermetic reciprocating compressors. A finite element method is used to compute the temperature distribution of the solid components and the fluid in the suction and discharge lines, whereas a lumped-parameter formulation is used to evaluate the internal environment temperature and the gas temperature at the end of the compression cycle. The heat transfer between the gas and the solid components is predicted using imposed convective heat transfer coefficients; some of which are estimated using heat transfer correlations, and others calibrated using experimental data and a genetic optimization algorithm. The numerical results were validated by comparisons with experimental data for different operating conditions and rotation speeds, showing that the model can be used to predict the compressor thermal profile in the entire application envelope. The low computational cost of the model enables its application to carry out sensitivity analysis and to assess thermal management alternatives to improve the compressor reliability or thermodynamic performance.

1. INTRODUCTION

Suction gas superheating can significantly reduce the thermodynamic performance of hermetic reciprocating compressors used in refrigeration systems, as it reduces the refrigerant gas density at the inlet of the compression chamber and therefore the mass flow rate supplied by the compressor. The three main heat sources of superheating in reciprocating compressors are: the electric motor, which generates heat due to electrical losses; the mechanism bearing, which generates heat due to mechanical friction; and the refrigerant gas at the discharge line of the compressor, which leaves the compression chamber at high pressure and temperature. In addition to the effect on gas superheating, high operating temperatures can affect the integrity and reliability of key components, such as the electric motor windings. Therefore, the prediction of the temperature distribution in hermetic reciprocating compressors and the assessment of thermal management solutions are important aspects of compressor design.

The temperature profile of hermetic reciprocating compressors can be measured directly by thermocouple wires, for example. Measurements are very accurate, but they are costly and time consuming, which can make it impractical to assess thermal management alternatives to improve the compressor performance. Numerical simulation tools, therefore, can be applied as an alternative to experimental activities. As they are cheaper and less time consuming, and if they are accurate enough, they can be very useful to perform such analyses.

Several numerical methodologies have been developed to predict the temperature distribution of reciprocating compressors. Sim et al. (2000) and Dutra and Deschamps (2015), for example, applied lumped-parameter formulations to predict the steady-state thermal profile of reciprocating compressors. In their methodology, the equations of conservation of mass and of energy are applied to several control volumes that represent the gas chambers in the compressor, which are linked through equivalent thermal conductances obtained from experimental calibration or correlations available in literature. These models can be applied for a low-cost estimation of the thermal behavior of reciprocating compressors, but are not accurate enough to capture the changes that may arise from subtle geometric changes.

More complex computational tools to predict the thermal behavior of hermetic reciprocating compressors have been presented by Chikurde et al. (2002) and Birari et al. (2006). In these distributed-parameter models based on the finite volume method, velocity, pressure and temperature profiles are predicted for both solid and fluid domains. The main difference between them is that the model by Chikurde et al. (2002) couples the fluid at the suction and discharge lines considering a polytropic process, whereas the model by Birari et al. (2006) solves the fluid flow in the compression chamber using a dynamic mesh. Oliveira et al. (2017) also applied a finite volume method to predict the temperature distribution and the fluid flow for a linear compressor, whose operation is based on a linear motor and a spring. Despite of their accuracy, distributed-parameter models present high computational cost and difficulty to model the lubricating oil flow inside the compressor, which directly impacts its thermal behavior.

Hybrid simulation models such as those presented by Ribas (2007), Schreiner et al. (2009) and Sanvezzo Jr. and Deschamps (2012) combine lumped-parameter and distributed-parameter formulations, coupling them iteratively. In the hybrid approach, the thermal interactions between the different control volumes are predicted using equivalent thermal conductances, while the heat conduction through the solid components is calculated using a differential model. The model described by Ribas (2007) applies global heat transfer coefficients obtained from experimental calibration and calculates temperatures for the fluid and solid domains using energy balances. This simulation provides the boundary conditions (temperatures and heat transfer coefficients) required by the distributed-parameter model. Once solved, the heat conduction model provides the heat fluxes through the solid components, which return to the lumped-parameter formulation to adjust the thermal profile. In order to increase the flexibility of this kind of approach, Sanvezzo Jr. and Deschamps (2012) proposed the calculation of the global heat transfer coefficients using heat transfer correlations from the literature and considered the effect of the lubricant oil flow inside the compressor shell. In order to quantify this effect, Sanvezzo Jr. and Deschamps (2012) considered that 80% of the lubricant oil mass flow rate is directed to the shell surface, while the rest of it flows to the crankcase region.

Comparing the studies available in literature concerned with numerical simulation models for the thermal management of hermetic reciprocating compressors, it can be noted that the lumped-parameter formulations are accurate to predict the thermal interactions between the refrigerant gas and solid components through the application of energy balances and thermal conductances, making them particularly useful when low computational costs are required at initial design steps. However, lumped models present relevant limitations as they are not capable of predicting the heat conduction and the temperature stratification at the solid components and gas chambers. On the other hand, distributed-parameter models can accurately simulate the fluid flow through the fluid chambers and the heat conduction through the solid components, but they present a high computational cost, reducing their applicability when a large design space must be considered to improve compressor efficiency.

One of the advantages of a hybrid approach is that the thermal interactions between fluid and solid domains can be modelled without solving the fluid flow, so that a low simulation time is achieved when compared to distributed-parameter models that solve the temperature, velocity and pressure fields for the fluid domains. Additionally, the heat conduction through the solid components, which is not solved in lumped-parameter formulations, can be efficiently predicted using a finite element method, since the mesh requirements are less strict compared to models based on a finite volume method. The hybrid models found in literature, however, do not simulate the temperature stratification at the refrigerant gas, applying lumped models for the gas.

This paper presents a hybrid simulation model to predict the steady-state thermal profile of the solid components and the refrigerant gas of hermetic reciprocating compressors. The model couples distributed-parameter models to predict the temperature stratification on the solid components and the refrigerant gas with lumped-parameter formulations to calculate the internal environment temperature and the temperature rise in the compression cycle. Due to its low computational cost, the model can be applied to assess thermal management solutions and to perform sensitivity analysis, therefore aiding the design of hermetic reciprocating compressors.

2. METHODOLOGY

In the model presented in this paper, the compressor CAD model is used to describe the solid domain. The fluid domains are represented by tubes that are drawn across the chambers in the suction and discharge lines. The following hypotheses were made regarding the refrigerant gas flow: unidimensional, incompressible, inviscid and in steady-state regime. Therefore, for the two fluid domains (the suction and discharge lines), as the cross-section of the tubes is constant, the velocity distribution is trivial and the numerical solution used to obtain the temperature distribution along the domain. For that, fixed mass flow rate and temperature-dependent specific heat coefficients must be assigned.

The heat conduction in the solid components is modelled by the thermal energy equation and the Fourier's Law, and is solved numerically by a finite element method. Considering a steady-state operation and homogeneous and isotropic materials, the heat conduction equation is

$$k\nabla^2 T + \dot{q} = 0, \quad (1)$$

where k is the thermal conductivity, ∇^2 is the Laplace operator, T is the temperature, and $\dot{q}$ is the volumetric heat generation.

The thermal interactions between the refrigerant gas and the solid components is predicted by the application of a coupled solution of the heat transfer problem on the fluid-solid interfaces. Figure 1 shows a representation of the different methods that are coupled to predict the thermal interactions in the compressor. The pipes shown in black represent the fluid domain, which is discretized in elements that interact with the surface elements of the solid domain based on the proximity between them. As a result of the thermal interactions between the fluid and solid elements, the temperature along the fluid line vary, and this stratification affects the local heat transfer with the solid components.

In order to calculate the heat transfer between the external surfaces of the solid components and the internal environment, and between the internal environment and the compressor shell, convective heat transfer coefficients and an internal environment temperature are defined. Since the internal environment temperature is not known at the beginning of the simulation, it is calculated iteratively based on a lumped formulation and a thermal circuit that represents the heat transfer between the internal components and the shell, as shown in Figure 2.

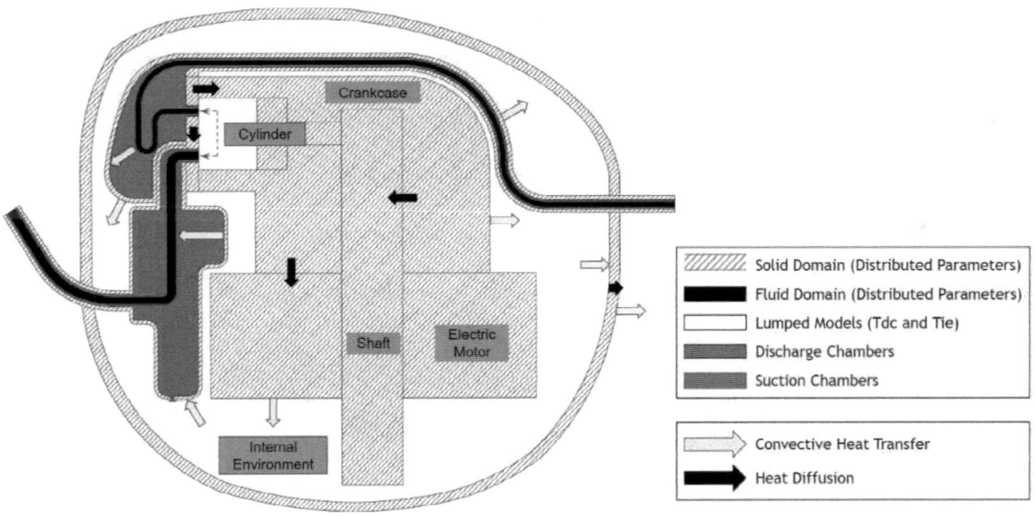

Figure 1: Methods that are coupled to predict the thermal interactions in the compressor.

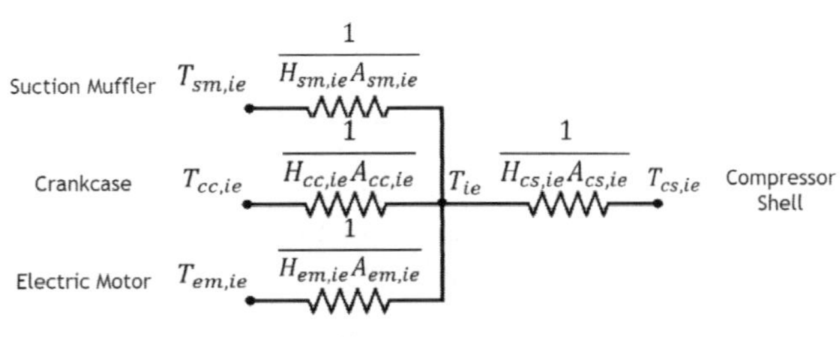

Figure 2: Thermal circuit between the internal components and the compressor shell.

Considering a steady-state operation, the sum of the heat exchanged between the internal components and the internal environment must be transferred to the internal surface of the compressor shell before leaving the compressor to the external ambient. Therefore, the average temperatures at the surface of each solid component ($T_{i,ie}$) can be related to the internal environment temperature (T_{ie}) and to the average temperature at the internal surface of the compressor shell ($T_{cs,ie}$) by performing an energy balance. After algebraic manipulation, an explicit equation for T_{ie} is written:

$$T_{ie} = \frac{\sum H_{i,ie} A_{i,ie} T_{i,ie} + H_{cs,ie} A_{cs,ie} T_{cs,ie}}{\sum H_{i,ie} A_{i,ie} + H_{cs,ie} A_{cs,ie}} \tag{3}$$

where the convective heat transfer coefficients between each internal component and the internal environment ($H_{i,ie}$) and the convective heat transfer coefficient between the internal environment and the shell internal surface ($H_{cs,ie}$) are adjusted using experimental data. The surface areas of the components ($A_{i,ie}$) and the shell internal surface area ($A_{cs,ie}$) are obtained from the CAD model. The surface-average temperatures $T_{i,ie}$ and $T_{cs,ie}$ are obtained from the distributed-parameter model.

The heat transfer between the compressor external surface and the external ambient is computed using values of temperature and heat transfer coefficients. The external environment temperature must be prescribed, while an equivalent heat transfer coefficient (H_{eq}) that considers convection and radiation is calculated as

$$H_{eq} = H_{ee} + \left[\sigma\epsilon\left(T_{cs,ee}^4 - T_{ee}^4\right)\right]\left(T_{cs,ee} - T_{ee}\right)^{-1} \tag{4}$$

where σ is the Stefan-Boltzmann constant, ϵ is the emissivity of the compressor external surfaces, $T_{cs,ee}$ is the average temperature on the external surface of the shell and T_{ee} is the external ambient temperature. The convective heat transfer coefficient between the compressor shell and the external environment is calculated using the correlation proposed by Yovanovich (1987) for arbitrarily-shaped immersed bodies, in which the Nusselt number is given by

$$Nu_L = 3.47 + 0.51 Ra_L^{1/4} \tag{5}$$

where Ra_L is the Rayleigh number with the square root of the exposed surface area as characteristic length. For the internal flow in the discharge tube and in the inlet and outlet pipes, the heat transfer coefficients are calculated using the Dittus-Boelter correlation for turbulent flow.

The two fluid lines that represent the refrigerant gas in the suction and discharge lines of the compressor require an inlet temperature as a boundary condition. The inlet temperature of the suction line must be prescribed as a boundary condition for the problem, whereas the inlet temperature of the discharge line is calculated based on the gas temperature at the end of the suction line. The refrigerant gas enthalpy at the end of the compression process, h_{dc}, is calculated using the first law of thermodynamics

$$h_{dc} = h_{sc} + \frac{1}{\dot{m}}\left(\dot{W}_{th} - \dot{Q}_{cc}\right), \tag{6}$$

where h_{sc} is the refrigerant gas specific enthalpy at the inlet of the compression chamber, $\dot{m}$ is the mass flow rate, and $\dot{W}_{th}$ is the thermodynamic power. For the results presented in this paper, the compression process is considered adiabatic, so that $\dot{Q}_{cc}$ is set to zero.

The refrigerant gas mass flow rate attributed to the fluid domain must be imposed as a boundary condition. Additionally, the electric power must be known in order to compute the heat generated due to electrical and mechanical losses and the thermodynamic power delivered to the refrigerant gas in the compression process. In order to define these values, the following three basic options are available: i) imposing the mass flow rate and the electric power based on the compressor datasheet or experimental data; ii) predicting the mass flow rate and the electric power using the semi-empirical model described by Li (2012); iii) imposing a volumetric flow rate and the electric power and predicting the mass flow rate based on the refrigerant gas density at the compression chamber inlet. For the second option, the semi-empirical model coefficients must be adjusted using the compressor datasheet or experimental data.

For the third option, a reference volumetric flow rate ($\dot{V}_{ref}$) can be obtained by simulating the model with reference input parameters and based on the following equation:

$$\dot{m} = \rho_{sc}\dot{V}_{ref} \tag{7}$$

The numerical solution is obtained by solving the distributed-parameter model using a commercial finite element solver. The calculation of the internal environment temperature and the temperature of the refrigerant gas at the end of the compression process is based on lumped models implemented in programming scripts that are coupled internally to the finite element solver. The first step in the solution procedure consists on the definition of input parameters and initial estimates for the two variables that depend on the lumped formulations: the discharge temperature (T_{dc}) and the internal environment temperature (T_{ie}). Once $\dot{m}$ is attributed for the fluid domain, T_{dc} is set as a boundary condition at the first node of the discharge line and T_{ie} is set as condition to calculate the heat transfer between the internal components and the internal environment. Then, the distributed-parameter model is solved, hence updating the suction temperature (T_{sc}), the average temperatures at the components surfaces ($T_{i,ie}$) and the average temperature on the compressor shell's internal surface ($T_{cs,ie}$). These updated variables are used to calculate new values for T_{dc} and T_{ie} using the lumped-parameter formulations. Additionally, if the volumetric flow rate is defined, the mass flow rate is updated using the new gas density at the inlet of the compression chamber. A flowchart of the numerical procedure is shown in Figure 3.

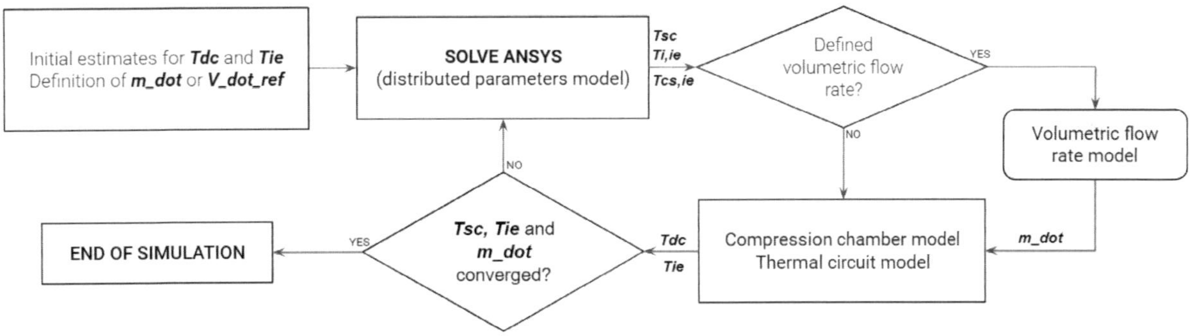

Figure 3: Flowchart of the solution procedure.

3. RESULTS

The experimental results presented hereafter were obtained for a low-capacity reciprocating compressor with a displacement of 9.04 cm³ operating with R600a in a hot-cycle calorimeter bench (Dutra and Deschamps, 2009). The numerical results were obtained using the compressor CAD model assuming constant electrical and mechanical efficiencies of 86% (except for the sensitivity analysis of the electrical efficiency) and 90%, respectively. Additionally, the compressor's external surface emissivity was defined at 0.9. Values for the temperature of the gas at the compressor inlet and for the external environment temperature were defined based on the experimental tests.

3.1. Model calibration and validation

The model calibration is performed by simulating the compressor with prescribed electric power and mass flow rate taken from experiments and comparing the predicted temperature with measured data at 8 locations. Then, 7 heat transfer coefficients and a multiplication factor for external heat transfer coefficient are adjusted by using the 8 selected temperatures as targets of an optimization routine using a genetic algorithm. The operating condition used for the calibration consists of an evaporating temperature of -20°C and a condensing temperature of 45°C, with the compressor operating at 4000 rpm. The parameters that were calibrated in the numerical model, their respective constraints and final results are shown in Table 1.

When applying the model for other conditions, the adjusted heat transfer coefficients in the suction and discharge chambers (H_{sl} and H_{dl}) are updated based on the functional dependence of the Nusselt number with the mass flow rate given by the Dittus-Boelter correlation, i.e.

$$H_{sl} = \left(\frac{\dot{m}}{\dot{m}_{calib}}\right)^{\frac{4}{5}} H_{sl,calib}; \qquad H_{dl} = \left(\frac{\dot{m}}{\dot{m}_{calib}}\right)^{\frac{4}{5}} H_{dl,calib} \tag{8}$$

The model validation was performed by analyzing numerical predictions for 9 operating conditions that consist in combinations of the evaporating temperatures of -30°C, -20°C and -10°C and condensing temperatures of 35°C, 45°C and 55°C. In Table 2, the 8 temperatures that were considered as targets for the calibration procedure are shown for the calibration condition and for two other extreme conditions (with a high mass flow rate, *Validation 1*; and with a low mass flow rate, *Validation 2*). The numerical results were obtained with the model using experimental results for the mass flow rate and electric power. The calibration is considered adequate, since all the predicted temperatures are within ±1°C of the measured data for the calibration condition. For both validation conditions, the cylinder temperature presented the highest deviations in relation to experimental results, reaching a maximum absolute value of 4.2°C in *Validation 2*, but most of the predicted temperatures are within ±2°C of the experimental data.

Figure 4 shows a comparison between experimental and numerical results obtained with the numerical model by imposing measured values for the mass flow rate and the electric power for the 9 operating conditions, all for a rotational speed of 4000 rpm. As can be seen, the numerical results are in good agreement with the experimental data in magnitude and trend. The largest deviations in relation to experimental results are observed for the cylinder temperature (maximum deviation around 4°C), followed by the electric motor temperature. The smallest differences between experimental and numerical results are observed for the internal environment temperature (T_{ie}) and for the temperatures of the refrigerant gas at the outlet of the suction muffler and at the discharge chamber.

Figure 5 shows a comparison between numerical results obtained with imposed experimental values for $\dot{m}$ and $\dot{W}$ and numerical results obtained with $\dot{m}$ and $\dot{W}$ predicted using the semi-empirical model described by Li (2012). By noting that the temperature predictions using both methods are similar, it is shown that the semi-empirical model can be used to predict the compressor thermal profile for conditions without measured mass flow rate and electric power. For these predictions, the model by Li (2012) has been implemented and calibrated with measurements of $\dot{m}$ and $\dot{W}$.

Figure 6 presents a comparison between experimental and numerical results obtained by imposing measured $\dot{m}$ and $\dot{W}$ at a rotational speed of 2000 rpm. Larger deviations compared to experimental data were obtained for the evaporating temperature of -30°C, especially for the discharge temperature (maximum around 5°C) and the cylinder temperature (maximum around 7°C). An interesting fact is that these deviations get smaller when the evaporating temperature increases, showing that the model becomes more accurate when the mass flow rate gets closer to the calibration condition (the model was calibrated at 4000 rpm). Additionally, Figure 6 presents values of T_{dc} obtained with the model without correcting the heat transfer coefficients in the suction and discharge chambers (Eqs. (8) and (9)), showing that the correction improves the accuracy of the model.

Table 1: Calibrated parameters and results of the calibration procedure.

Parameter	Bounds	Result
Hee_{factor} (-)	[1, 20]	7.4
$H_{sm,ie}$ (W m^{-2} K^{-1})	[50, 150]	115
$H_{dt,ie}$ (W m^{-2} K^{-1})	[50, 150]	40
$H_{cc,ie}$ (W m^{-2} K^{-1})	[50, 150]	140
$H_{em,ie}$ (W m^{-2} K^{-1})	[50, 150]	65
$H_{cs,ie}$ (W m^{-2} K^{-1})	[50, 300]	250
$H_{sl,calib}$ (W m^{-2} K^{-1})	[100, 400]	250
$H_{dl,calib}$ (W m^{-2} K^{-1})	[100, 800]	720

Table 2: Experimental and numerical results for the targets of the optimization procedure.

	Calibration (-20°C/45°C)			Validation 1 (-10°C/35°C)			Validation 2 (-30°C/55°C)		
Results	**Exp.**	**Num.**	**Dif.**	**Exp.**	**Num.**	**Dif.**	**Exp.**	**Num.**	**Dif.**
T_{mo} (°C)	45.9	45.1	-0.8	35.3	34.9	-0.4	44.9	44.4	-0.5
T_{dc} (°C)	93.1	93.0	-0.1	75.7	77.0	1.3	94.8	96.2	1.4
T_{em} (°C)	60.0	59.7	-0.3	50.9	52.3	1.4	57.2	54.1	-3.1
T_{cw} (°C)	66.8	66.8	0.0	55.6	57.9	2.3	65.9	61.7	-4.2
T_{sm} (°C)	50.9	51.2	0.3	40.8	42.2	1.4	50.5	48.0	-2.5
T_{ie} (°C)	54.6	55.4	0.8	47.0	47.6	0.6	50.7	50.6	-0.1
T_{dl} (°C)	66.3	65.4	-0.9	58.3	58.2	-0.1	58.6	58.1	-0.5
T_{cs} (°C)	52.9	52.5	-0.4	45.8	44.9	-0.9	49.1	47.8	-1.3

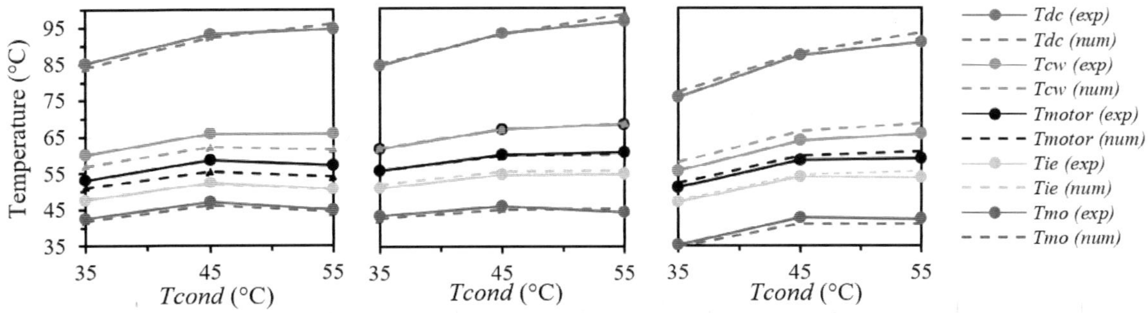

(a) *Tevap* = -30°C (b) *Tevap* = -20°C (c) *Tevap* = -10°C

Figure 4: Comparison between experimental data (*exp*) and numerical predictions (*num*) using experimental data for $\dot{m}$ and $\dot{W}$ for the compressor operating at 4000 rpm.

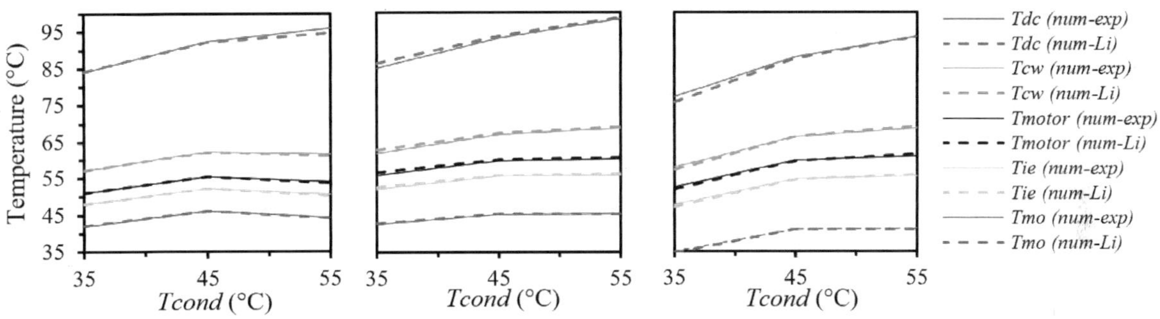

(a) *Tevap* = -30°C (b) *Tevap* = -20°C (c) *Tevap* = -10°C

Figure 5: Comparison between numerical predictions using measured $\dot{m}$ and $\dot{W}$ (*num-exp*) with numerical predictions with $\dot{m}$ and $\dot{W}$ predicted with the model by Li (2012) (*num-Li*) for the compressor operating at 4000 rpm.

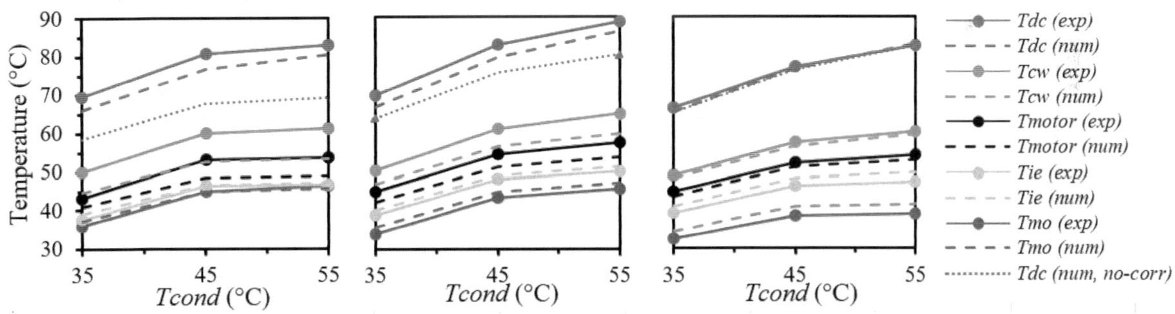

(a) *Tevap* = -30°C (b) *Tevap* = -20°C (c) *Tevap* = -10°C

Figure 6: Comparison between measured data and numerical predictions for the compressor operating at 2000 rpm. The effect of updating H_{sl} and H_{dl} according to the mass flow rate for the prediction of T_{dc} is highlighted by comparing *"Tdc (num)"* with *"Tdc (num, no-corr)"*.

3.2. Temperature contours

The refrigerant gas exchanges heat with the inlet pipe, the suction muffler, and the orifice in the valve plate before entering the compression chamber. As these components receive heat from the compressor internal environment and other components, superheating takes place at the suction line. Figure 7a presents the temperature distribution of the refrigerant gas at the suction line of the compressor for the calibration condition, showing an increase of approximately 13°C due to a heat transfer of approximately 16 W to the gas in the suction line. As a result of the compression process, the refrigerant gas leaves the compression chamber at 116°C and loses approximately 71 W of heat to the valve plate, the discharge chamber, the discharge muffler, the discharge pipe, and the outlet pipe, leaving the compressor at 65.4°C.

The temperature distribution of the refrigerant gas until it reaches the discharge tube is shown in Figure 7b. A large part of the gas cooling occurs right after it leaves the compression chamber, when it flows through the valve plate and the discharge chamber.

Figure 8a shows the internal thermal profile of the compressor for the calibration condition. The coldest regions are those in direct contact with the refrigerant gas at the suction line, whereas the hottest region comprises the discharge chamber. The maximum temperature observed for the electric motor (which is a critical component in terms of reliability) is approximately 60°C, which is considered acceptable. Figure 8b shows the thermal profile of the compressor shell for the calibration condition. This thermal profile is influenced by the contacts between the shell and the inlet and outlet pipes, which generate the regions with the extreme temperatures in blue and red.

3.3. Sensitivity analysis

A sensitivity analysis was performed to estimate the influence of the electric motor efficiency and the external ambient temperature on the thermodynamic performance and the reliability of the compressor. For this analysis, the model that considers a reference volumetric flow rate to predict the compressor mass flow rate was used, with the reference volumetric flow rate being computed by simulating the model for the calibration condition with an electrical efficiency of 86% and an external environment temperature of 33.2°C.

Figure 9a shows the change in temperatures of interest as a function of the motor electrical efficiency. The temperature of the electric motor varies with the greatest slope, indicating that the electrical efficiency is an important parameter in terms of reliability related to temperature. The influence of the electrical efficiency on the mass flow rate is also shown as a deviation from the reference mass flow rate for an electric efficiency of 86%. The effect of the ambient temperature on selected temperatures of interest and on the mass flow rate is shown in Figure 9b. The model predicts a reduction of approximately 0.7% in mass flow rate when the external environment temperature rises from 33.2°C to 38.2°C.

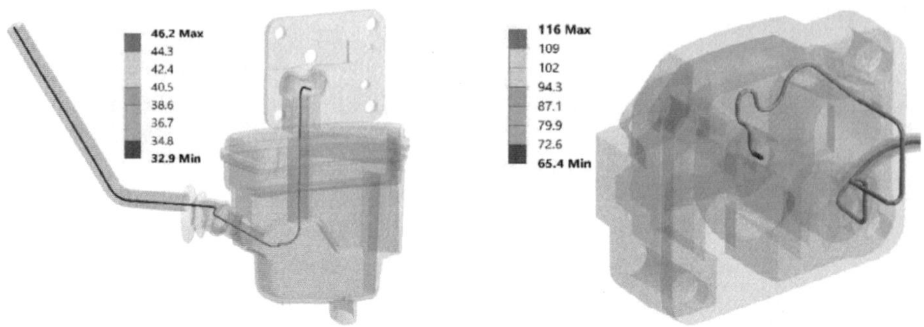

(a) (b)

Figure 7: Refrigerant gas temperature (a) throughout the suction line, and (b) at the beginning of the discharge line.

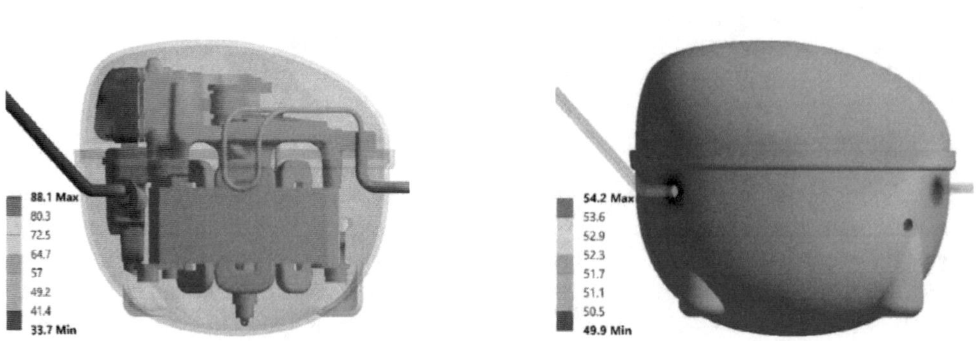

(a) (b)

Figure 8: Thermal profile of (a) the internal components and (b) the compressor shell.

In order to determine which is the main source of superheating and therefore reduction of volumetric efficiency, four simulations for the calibration condition were performed with different modifications in the compressor: i) with a perfectly insulated discharge line; ii) without mechanical losses; iii) without electrical losses; iv) with the valve plate's thermal conductivity divided by 100. Table 3 presents the results of these simulations, showing that the main source of superheating in the suction line is the gas in the discharge line. Insulating the discharge line reduced the superheating by approximately 9°C, whereas the other modifications yielded smaller superheating reductions.

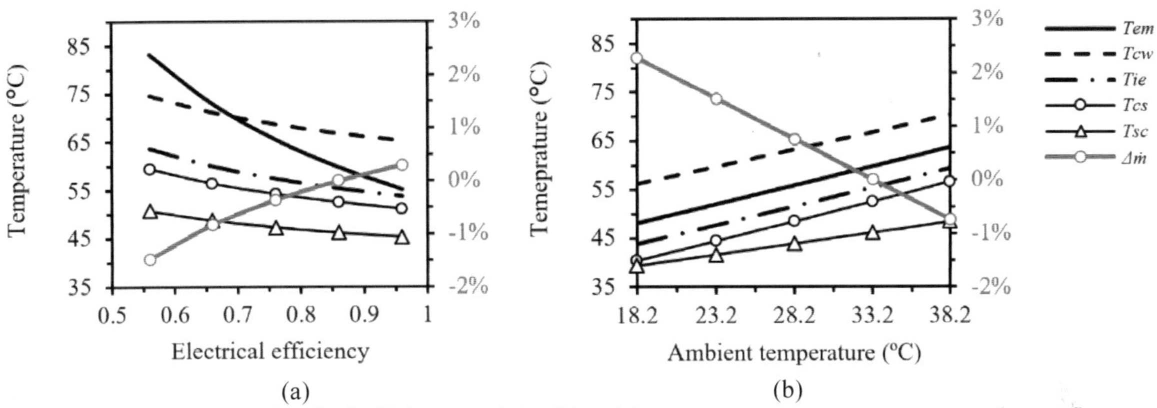

Figure 9: Impact of the (a) electrical efficiency and the (b) ambient temperature on temperatures and mass flow rate.

Table 3: Suction temperature and heat flow rates from/to the gas in the suction and discharge lines.

Condition	ΔT_{sc} (°C)	Heat flow rate from the gas in discharge line (W)	Heat flow rate to the gas in suction line (W)
Baseline	0.0	70.7	15.6
Insulated discharge line	-9.1	0.0	5.0
Without mechanical losses	-1.2	72.3	14.2
Without electrical losses	-1.3	71.8	14.1
Valve plate with lower thermal conductivity	-0.9	69.4	14.5

4. CONCLUSIONS

A simulation model to compute the thermal profile of hermetic reciprocating compressors was presented. The model applies a finite element method to solve the tridimensional heat conduction in the solid components of the compressor and the fluid flow in its tubes and chambers. The solid is considered homogeneous, but interfaces between different materials are allowed, and the fluid flow is considered inviscid, incompressible and unidimensional. To compute the convective heat transfer between the fluid and the solid, convective heat transfer coefficients were adjusted with an optimization algorithm and experimental data taken in a hot-cycle test rig. Additionally, the temperature distribution given by the finite element method is coupled to a lumped-parameter formulation for the fluid temperature in the internal environment and at the compression chamber outlet. To assess the model accuracy, its predictions for both solid surfaces and fluid regions were compared to thermocouple measurements at 9 operating conditions and 2 different rotational speeds, showing that the predictions are within 2°C for most of the measured data. Besides its accuracy, the distributed-parameter nature of the finite element method allows the visualization of the temperature profile inside the compressor, which can be used to understand the heat path in the compressor and improve its design.

NOMENCLATURE

Symbols

A	Area	(m^2)	$\dot{Q}$	Heat flux	(W)
ϵ	Emissivity		$\dot{q}$	Volumetric heat generation	(W m^{-3})
h	Specific enthalpy	(J kg^{-1})	Ra	Rayleigh number	
H	Convective heat transfer coeff.	(W m^{-2} K^{-1})	ρ	Density	(kg m^{-3})
k	Thermal conductivity	(W m^{-1} K^{-1})	σ	Stefan-Boltzmann constant	(W m^{-2} K^{-4})
$\dot{m}$	Mass flow rate	(kg s^{-1})	T	Temperature	(°C)
Nu	Nusselt number		$\dot{V}$	Volumetric flow rate	(m^3 s^{-1})
			$\dot{W}$	Power	(W)

Subscripts

		eq	Equivalent
calib	Related to the calibration condition	*exp*	Experimental
cc	Crankcase	*ie*	Internal environment
cs,ee	External surface of the shell	*mo*	Suction muffler outlet
cs,ie	Internal surface of the shell	*motor*	Electric motor
cw	Cylinder wall	*num*	Numerical
dc	Discharge chamber	*ref*	Reference
dl	Discharge line	*sc*	Suction chamber
dt	Discharge tube	*sl*	Suction line
ee	External environment	*sm*	Suction muffler
em	Electric motor	*th*	Thermodynamic

REFERENCES

Birari, Y. V., Gosavi, S. S., & Jorwekar, P. P. (2006). Use of CFD in design and development of R404a reciprocating compressor. *Proc. Int. Compressor Eng. Conf. at Purdue,* West Lafayette, USA.

Chikurde, R. C., Longanathan, E., Dandekar, D. P., & Manivasagam, S. (2002). Thermal mapping of hermetically sealed compressors using computational fluid dynamics technique. *Proc. Int. Compressor Eng. Conf. at Purdue,* West Lafayette, USA.

Dutra, T., & Deschamps, C. J. (2015). A simulation approach for hermetic reciprocating compressors including electrical motor modeling. *Int. J. Refrig., 59.*

Li, W. (2012). Simplified steady-state modeling for hermetic compressors with focus on extrapolation. *Int. J. Refrigeration, 35, 1722-1733.*

Meyer, W. A., & Thompson, H. D. (1988). An analytical model of heat transfer to the suction gas in a low-side hermetic refrigeration compressor. *Proc. Int. Compressor Eng. Conf. at Purdue,* West Lafayette, USA.

Oliveira, M. J., Diniz, M. C., & Deschamps, C. J. (2017). Predicting the temperature distribution and suction gas superheating of an oil-free linear compressor. *Journal of Process Mechanical Engineering, 231(1), 47-56.*

Ribas Jr, F. A. (2007). Thermal analysis of reciprocating compressors. *Proc. Int. Conf. on Compressors and Their Systems,* London, UK.

Sanvezzo Jr, J., & Deschamps, C. J. (2012). A heat transfer model combining differential and integral formulations for thermal analysis of reciprocating compressors. *Proc. Int. Compressor Eng. Conf. at Purdue,* West Lafayette, USA.

Schreiner, J. E., Deschamps, C. J., Ribas Jr, F. A., Barbosa Jr, J. R., & Rosa, V. H. P. (2009). Thermal management of a commercial reciprocating compressor through numerical simulation. *Int. Conf. on Compressors and Coolants,* Castá Papiernicka, Slovakia.

Sim, Y. H., Youn, Y., & Min, M. K. (2000). A study on heat transfer and performance analysis of hermetic reciprocating compressors for refrigerators. *Proc. Int. Compressor Eng. Conf. at Purdue,* West Lafayette, USA.

ACKNOWLEDGEMENTS

The authors appreciate the support from Nidec-GA (Embraco) and EMBRAPII through Grant No. 202001416 ("Development of new technologies for refrigeration compressors" Project). Additional funding was provided by CAPES (Talents for Innovation Program - Grant No. 88887.194773/2018-00) and by the National Institutes of Science and Technology (INCT) Program (CNPq Grant No. 404023/2019-3; FAPESC Grant No. 2019TR0846).

Miniature Oil-less Reciprocating Compressor for High-Lift, High-Temperature Heat Pump Applications

Daniel L. MURPHY[1*], Michael S. KATANCIK[1]

[1]Mainstream Engineering Corporation, Rockledge, FL, USA
(321) 631-3550, dmurphy@mainstream-engr.com

*Corresponding Author

ABSTRACT

In the extreme temperature environment of the moon, the regolith surface temperature can reach 100°C during the 354-hour lunar day. This often exceeds the maximum allowable temperature of critical electronics onboard small lunar rovers designed for long-term science missions. In environments with such high heat sink temperatures, active vapor compression cooling solutions allow greater mission flexibility than traditional passive spacecraft thermal control. The objective of this research was to design and experimentally evaluate a critical enabling technology for effective lunar rover class heat pumps: the high-lift, high-temperature, microgravity compressor. The main technical challenges in the design of a compressor for these conditions include operation in a microgravity environment, relatively high compression ratio requirements, extreme temperatures during transport and operation in space and on the moon, and the low equipment vibration requirements for small lunar rovers. A twin-piston oil-less reciprocating compressor was designed to meet these requirements. The oil-less design mitigates the challenges of oil distribution in microgravity environments. A twin-piston configuration was designed to balance the dynamic forces and lower the vibration of the compressor. Compressor components were sized to operate with new high-temperature, environmentally friendly refrigerants at compression ratios between 3 and 7. The compressor was designed for a cooling capacity on the order of 100 W. A prototype compressor was fabricated and experimentally evaluated. The total mass of the compressor was 2.86 kg. Maximum exported vibration was estimated to be 0.04 N at the base of the compressor based on kinematic calculations and measured motor torque. Experimental capacity and efficiency measurements were lower than predicted. This was likely due to manufacturing and tolerance deficiencies in the first prototype compressor resulting in blowback around the piston seal, higher friction than predicted, suboptimal bearing performance, suboptimal reed valve design, and tolerance stack up in the crankshaft and cylinders. These manufacturing issues can be overcome in next generation prototypes.

1. INTRODUCTION

Unmanned lunar landers and rovers used for science missions at the Moon's equator house critical electronics, instrumentation, and batteries in a warm electronics box (WEB) (Bugby et al., 2010). To enable long-term lunar missions, these electronics must survive the 354-hour (approximately 15-earth day) lunar days and nights in which equatorial surface temperatures swing from 100°C to –183°C (373 K to 90 K). Spacecraft and rover thermal management is typically achieved through passive two-phase systems such as variable conductance heat pipes, capillary pumped loops, and loop heat pipes as well as active single-phase or two-phase mechanically pumped fluid loops (Bugby et al., 2010). Passive thermal management strategies rely on heat sink temperatures below the electronics temperatures such as those in deep space. However, lunar science mission profiles may require small rovers to traverse terrain where a favorable view factor between the radiator and deep space is impossible to maintain. During such mission profiles, an active lifting cycle such as a vapor compression (VC) heat pump is needed to reject heat to high temperature lunar regolith.

Lunar landers have traditionally applied various approaches to surviving the temperature extremes of lunar operation. Some probes were designed for unimodal operation (only hot or only cold environments) either by continuous circumnavigation following the sun or by careful landing site selection. Other landers use natural or artificial thermal mass such as boulders or thermal wadis to survive the lunar night. True bimodal operation is achieved by some landers by using electric resistance heaters to survive the cold nighttime temperatures at the cost of a high-energy and mass penalty from additional batteries (Thornton et al., 2010). The rover-class heat pump designed in this research enables efficient thermal management with a lower mass and energy penalty than state-of-the-art landers by shutting down the VC system at night and relying on variable conductance thermal switches to regulate the heat loss to the environment without requiring additional heaters.

Of the primary VC heat pump components, the greatest technological risk is the compressor. In microgravity environments, oil entrained in the refrigerant flow tends to pool in locations other than the compressor sump because conventional oil separation is accomplished through gravity-dependent or inertial methods. Thus, the compressor can be oil starved using terrestrial oil management techniques. Oil-lubricated compressors have been proposed for space-based systems; however, proper refrigerant/oil management in two-phase heat exchangers can be challenging and the high ambient temperature exceeds the maximum operating temperature of common refrigerant lubricants. Additionally, current refrigeration lubricants have poor performance at extremely low temperatures due to high viscosity and/or insolubility with refrigerants. Designing an oiled VC system for the extremes of space-based mission profiles would require developing higher temperature lubricants compatible with atypical working fluids that could also survive exposure to temperatures much colder than what a commercial refrigeration unit would experience. Given the difficulties that oil within the refrigerant system presents in microgravity environments, previous research and niche production efforts have developed space-based oil-less compressors (Grzyll & Cole, 2000).

The current effort focused on the design and demonstration of the key enabling technology for a gravity-insensitive lunar rover heat pump: a miniature oil-less compressor. The design requirements included: compressor mass less than 3 kg, exported vibration less than 0.1 N in all axes, and coefficient of performance (COP) of 2.5 at 50°C lift with 100 W cooling capacity.

2. THERMODYNAMIC DESIGN CONDITIONS

2.1 Refrigerant Trade Study

It was first necessary to select the working fluid to determine the geometrical and operational parameters of the compressor. A lunar rover heat pump needs a refrigerant with a high critical temperature and high heat of vaporization because of the extreme temperatures encountered in the equatorial lunar environment. A trade study of high temperature refrigerants was performed to select the working fluid. Fluorocarbon (FC), hydrofluoroether (HFE), hydrofluoroolefin (HFO), hydrochlorofluorocarbon (HCFC), hydrofluorocarbon (HFC), and hydrocarbon refrigerants were evaluated. Hydrocarbons have a high heat of vaporization but high flammability (ASHRAE 34 class A3). Even though there is no oxygen on the moon, safety during handling and transportation could be an issue with these refrigerants. Some refrigerants such as R514a and R245fa could also have safety and handling concerns due to toxicity (ASHRAE 34 class B1).

Figure 1 shows a schematic of the lunar rover heat pump. A VC cycle model was developed to consider the effects of different refrigerants, rover configuration, heat exchanger size, and compressor efficiency on WEB cooling efficiency in typical lunar equatorial operating environments. The performance of the different refrigerants was evaluated by comparing the predicted cooling coefficient of performance (COP_c) of the system under two representative lift conditions (50°C and 70°C lift) defined as the difference between the condenser and evaporator saturation temperatures. For the initial refrigerant evaluation, the compressor isentropic efficiency was assumed to be 50% at all conditions, and the evaporator temperature set at 38°C. This analysis assumed 5°C superheat and 5°C subcooling. The

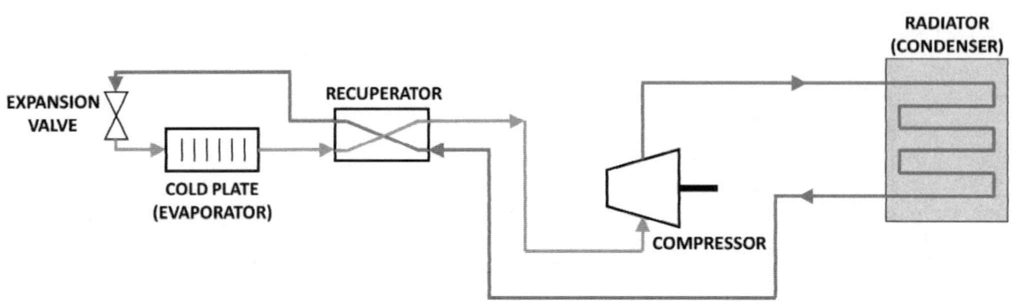

Figure 1: Schematic of the lunar rover heat pump

Table 1: Fluid properties and predicted performance of selected refrigerants

Refrigerant	ASHRAE 34 Safety Class	Critical Temperature (°C)	Heat of Vaporization at 38°C (kJ/kg)	Predicted COP_c	
				50°C Lift	70°C Lift
FC-72	A1	176	90	2.84	1.95
Novec 7000	A1	165	132	2.80	1.88
R1336mzz(E)	A1	130	135	2.60	1.68
R1336mzz(Z)	A1	171	164	2.80	1.89
R514a	B1	178	185	2.83	1.91
R601 (n-Pentane)	A3	197	356	2.85	1.94

model also included a recuperator between the subcooled liquid on the high-pressure side and the superheated vapor on the low-pressure side.

Table 1 shows a selection of the evaluated refrigerants and the predicted COP at 50°C and 70°C lift conditions. New HFO isomers R1336mzz(Z) and R1336mzz(E) were recently introduced to the market as a green alternative for high-temperature heat pumps and foam blowing. These refrigerants have potential for non-NASA applications because they are non-toxic and non-flammable (unlike pentane and R514a). They also have good material compatibility with POE oil, copper, steel, aluminum, and common plastics and elastomers (Juhasz, 2017). R1336mzz(E) has a lower critical temperature and higher normal boiling point than the (Z) isomer. Although the thermodynamic performance of R1336mzz(E) is lower than R1336mzz(Z), the fluid properties could be more favorable for efficient compressor operation. The higher suction density of R1336mzz(E) will also improve the size, mass, and power (SMaP) of the overall system relative to low pressure refrigerants such as FC-72, Novec 7000, or R1336mzz(Z) because the refrigerant lines can be smaller. Although fluids such as Novec 7000 and FC-72 have more established history as heat transfer fluids in space-based missions, R1336mzz(Z) and R1336mzz(E) were ultimately selected for evaluation to determine if they would also be suitable for future NASA applications. Figure 2 shows a map of predicted COP_c for R1336mzz(Z) and R1336mzz(E) for different lift conditions set by varying the saturated suction temperature (SST) and saturated discharge temperature (SDT) as well as expected operating points (OPs).

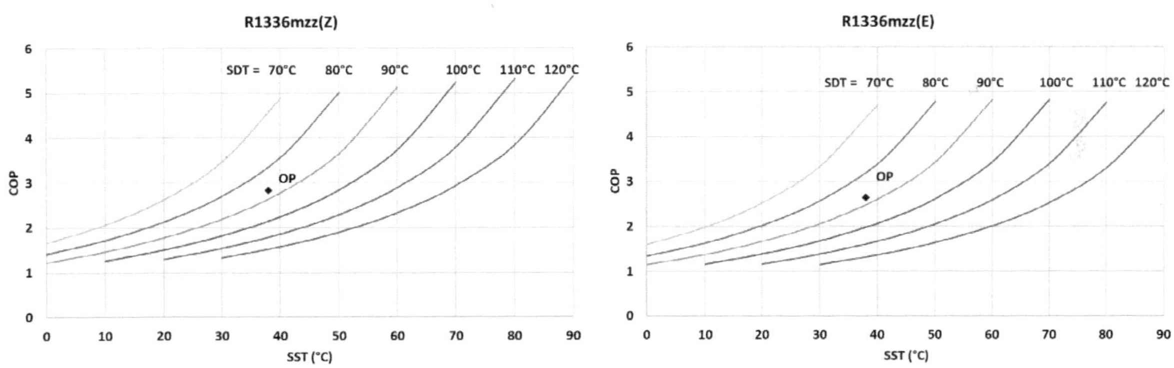

Figure 2: Predicted heat pump COP for different lift conditions with R1336mzz(Z) and R1336mzz(E)

2.2 Lunar Rover Thermal Model

A thermal model of a small lunar rover was developed and coupled to the VC cycle model described above to determine the required operating conditions for the compressor. The lunar rover model was based on a 65 kg rover developed at Carnegie Mellon University for lunar equatorial exploration (Jones et al., 2011; Thornton et al., 2010). Figure 3 shows a picture of this lunar rover and a heat flow schematic. The major components of the rover are the WEB, the radiator, the rover chassis, and the solar panels. The radiator angle was set to 90 degrees to simulate an unfavorable view factor (0.5) between the radiator and the lunar regolith. The thermal model considers conduction heat transfer between the rover structural components and radiation heat transfer with the environment. Performance was simulated at worst case cooling conditions (1300 W/m^2 solar flux, 100°C regolith temperature).

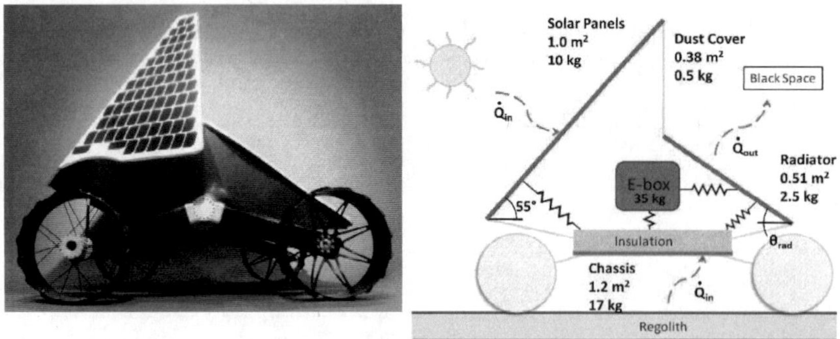

Figure 3: Lunar equatorial rover designed by Carnegie Mellon (left) and schematic showing thermal paths, component mass and surface area (right) (Jones et al., 2011; Thornton et al., 2010)

The lunar rover model was used to determine expected operating points and to set compressor design targets. The SST was set to 38°C based on the expected closest approach temperature between the WEB and the refrigerant in the evaporator cold plate. With a 0.51 m^2 radiator angled at 90 degrees, the SDT at the OP was 86.9°C (48.9°C lift) with R1336mzz(Z) and 87.4°C (49.4°C lift) with R1336mzz(E). The COP_c of the cycle at this OP is 2.82 and 2.64 for R1336mzz(Z) and R1336mzz(E), respectively. Thus, the heat pump can operate with good efficiency with a radiator surface area of 0.51 m^2 angled at 90 degrees. A smaller radiator may be desirable to reduce size and mass but would require a higher rejection temperature (higher lift), and thus lower COP_c. Table 2 shows a comparison of predicted thermodynamic operating conditions required to cool 100 W electronics load in the WEB with different refrigerants and with different radiator sizes. This analysis assumed 50% compressor isentropic efficiency. These conditions were used as a guide to design the compressor geometry and nominal operating conditions.

Table 2. Predicted compressor operating conditions

Refrigerant	Radiator Area (m^2)	Lift (°C)	Suction/ Discharge Pressure (kPa)	Compression Ratio	Suction Density (kg/m^3)	Refrigerant Mass Flow Rate (kg/h)	Suction Volume Flow Rate (cm^3/s)
R1336mzz(Z)	0.51	48.9	119.3 / 513.8	4.3	6.8	2.8	114.6
R1336mzz(E)	0.51	49.4	303.7 / 1138	3.7	17.8	3.5	54.3
R1336mzz(Z)	0.30	72.7	119.3 / 900.3	7.5	6.4	3.0	132.5
R1336mzz(E)	0.30	75.3	303.7 / 1978	6.5	16.4	4.0	67.3

3. COMPRESSOR MECHANICAL DESIGN AND FABRICATION

The design of the rover-class microgravity oil-less compressor was inspired by a previous reciprocating oil-less compressor developed by Mainstream Engineering that is currently onboard the International Space Station (Grzyll & Cole, 2000). While the previous design used a single-piston, the current compressor used an opposed twin-piston approach to meet the low-vibration requirements and to reduce the operating speed while using similarly sized mechanical components. The opposed cylinders balance the forces leaving only a small yaw moment due to the axial offset of the cylinders along the crankshaft. This configuration resulted in a significant reduction in transmitted vibration. A commercial reed valve and cylinder head were used in the prototype design. The calculated operating conditions as well as the geometry of the commercial components influenced the piston design. The compressor is driven by a brushless DC (BLDC) motor selected to provide variable speed control for improved cycle efficiency over a wide range of operating conditions. For the purposes of design, a nominal compressor speed of 1500 rpm was selected. Based on this nominal speed, the displacement of each cylinder was set to be 5.2 cm^3.

Figure 4 shows a section view of the compressor highlighting the key dimensions and components. A pinless piston configuration was selected to reduce the number of components requiring lubrication and failure points. The rod/stroke (R/S) ratio is important in a pinless piston design because of piston rock inside the cylinder. Greater R/S ratio leads to smaller angles of the connection rod, and therefore the smaller the angle of the piston head. A lower angle leads to better sealing between the cylinder and the bore. An R/S ratio of 4 results in a maximum connection rod angle of 7.6°. The piston rock requires a small amount of additional clearance added at the top of the piston bore to ensure there is no interference between the reed valve assembly and the piston head near top dead center (TDC). The nominal clearance between the reed valve and the piston at TDC was 0.8 mm, and the total clearance volume was 7.4% of the piston displacement volume. The bore/stroke (B/S) ratio directly influences the piston speed. An oversquare piston (B/S > 1) reduces piston speeds and results in less friction, but it also requires a larger piston head with more potential leak area (i.e. lower volumetric efficiency) if not sealed properly. Reducing friction in the bore is paramount in an oil-less design; therefore, a B/S ratio of 1.4 was selected. A PTFE piston ring seal was clamped between the top piston plate and the piston-connecting rod body. The seal wraps up around the piston head to prevent blowback at all rocking angles. PTFE was selected for the piston seal due to its low friction, flexibility, and easy manufacturability. A 0.5 µm thick WS_2 coating was also applied to the cylinder bore using a dry air application method at 35°C. The low coefficient of friction (0.03 to 0.07 μk) reduces sliding friction between the seal and the cylinder wall.

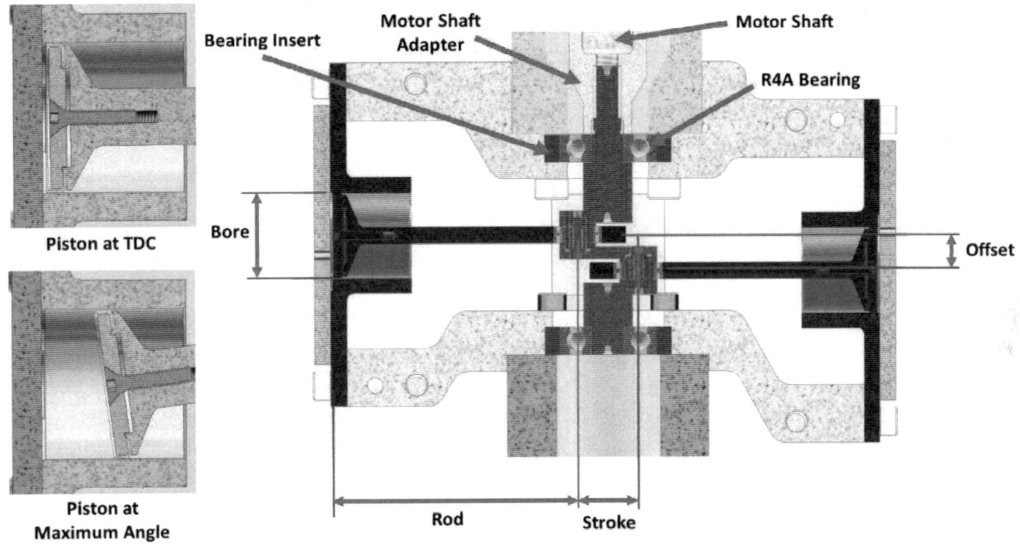

Figure 4: Opposed twin-piston compressor section view

A three-piece crankshaft assembly was necessary for assembly because of two connection rod ball bearings (R168) that were press fit onto the crank shaft center leaf (Figure 5). To ensure alignment and balance, a symmetrical arrangement of two bolts and four locating pins was used. The two larger ball bearings (R148) that the crankshaft runs on were press fit down to the shoulders sized to the abutment inner diameter specified by the manufacturer. The motor connection portion of the crankshaft is threaded. This threaded portion interfaces with a custom adapter that was press fit onto the motor shaft. During operation the motor was spun so the threads on the adapter and the threaded portion of the crankshaft are tightening during operation. Both the small, connection rod ball bearings (R148) and the large, crankshaft ball bearings (R4A) were coated with WS_2 to reduce friction in the dry rolling elements. A 4340 bearing insert was designed as an intermediary between the crankshaft bearing outer race and the housing bore to account for tight tolerance requirements and coefficient of thermal expansion (CTE) mismatches.

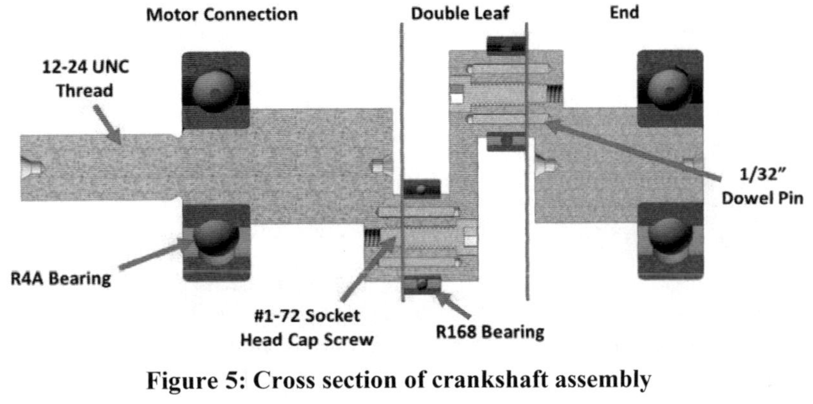

Figure 5: Cross section of crankshaft assembly

Materials were selected based on mechanical properties, thermal properties, and machinability. Because of the large temperature differential that the compressor will experience during transportation, cold storage, and operation (-183°C during lunar night hibernation to 190°C compressor discharge temperature at lunar noon), we selected materials with similar CTE. A summary of the materials used in the compressor along with key properties is presented in Table 3.

Table 3: Material properties of materials used in the oil-less piston compressor

Material	Components	Density (g/cm^3)	Yield Strength (MPa)	CTE (μm/m-K)
4032 Aluminum	Housing, Pistons	2.68	300	19.4
4340 Steel	Crankshaft, Bearing inserts	7.85	470	12.7
52100 Steel	Bearing races	7.75	2034	11.6
304 Stainless Steel	Motor-to-crankshaft adapter	8.00	215	17.3

A kinematic analysis was performed on the compressor pistons and crankshaft to determine the expected forces and torques on the components. The main loads acting on the crankshaft included pressure forces, inertia of the piston-connecting rod assembly, and friction loads. Friction loads on the piston were not included in this analysis. In the case of using R1336mzz(Z) as the working fluid at 38°C SST and 50°C lift conditions, a refrigerant mass flow rate of 2.82 kg/h and compressor speed of 1670 rpm is needed for a cooling capacity of 113.8 W (100 W electronics load + 13.8 W conduction from other rover components). At this operating point, the ideal volumetric efficiency was predicted to be 79% based on standard thermodynamic relationships:

$$\eta_v = 1 - \left(r_p^{1/n} - 1\right)c_{\text{clearance}} \tag{1}$$

where r_p is the compression ratio, n is the polytropic index, and $c_{\text{clearance}}$ is the percent clearance defined as the ratio of the clearance volume to the swept volume. The connecting rod angle, angular velocity, and angular acceleration as well as the piston position, velocity, and acceleration were calculated over the entire crank cycle. These values were used to calculate the forces and torques on the crankshaft in each axis (as defined in Figure 6). For this operating point, the maximum piston acceleration was 205.4 m/s^2 at TDC, and the maximum piston velocity was 1.1 m/s at 85° and 275° crank angle. The mean piston speed was 0.66 m/s.

In an opposed twin-piston compressor, the forces from each piston are balanced, but there are unbalanced moments in the x- and - and y-directions due to the 7.14 mm offset between the two pistons. The moment about the z-axis is the torque that must be supplied by the motor. Figure 7 shows the crankshaft torque around each axis at this operating point. The maximum motor torque (z-axis) is 0.86 N·m, and the root mean square (RMS) average motor torque is 0.28 N·m. In the x-axis, the maximum and RMS moments are 0.06 N·m and 0.02 N·m, respectively. In the y-axis, the maximum and RMS moments are 0.65 N·m and 0.31 N·m, respectively.

The moment arm between the crank shaft center and the mounting point at the base of the compressor is 61.5 mm. Based on these measurements and predicted moments, the maximum exported vibration was predicted to be 0.04 N in the y-direction. This is less than the maximum allowed vibration for lunar science rovers.

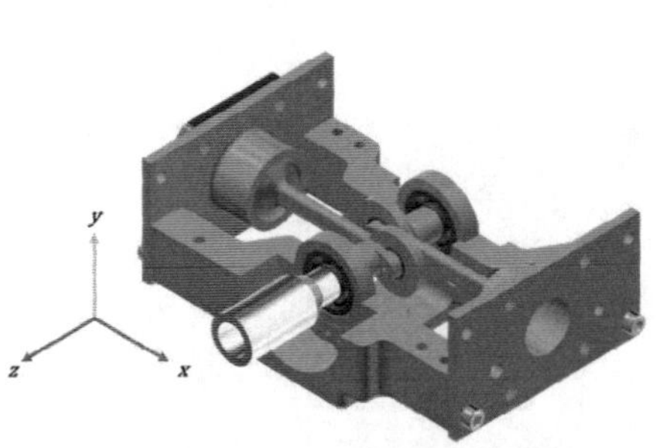

Figure 7. Cut view of the compressor crankshaft assembly indicating coordinate directions

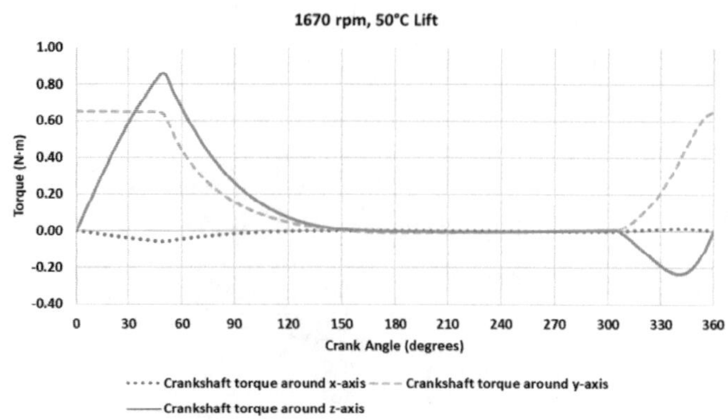

Figure 6. Compressor crankshaft torque around each axis over the entire crank cycle

Figure 8 shows a computer aided design (CAD) model and Figure 9 shows photographs of the compressor prototype. The total compressor mass was 2860 g including a custom hermetic housing that was designed but not yet manufactured. This is 5% below the target mass of the compressor. The commercial BLDC motor makes up 35% of the total compressor mass. The motor controller was not included in this estimate. Suction vapor is drawn from the hermetic shell through a suction port at each piston, and compressed refrigerant discharges through a copper tube brazed to the cylinder plate. The two discharge tubes are joined and expanded to a high-pressure reservoir to limit pulsation in the discharge line before being routed outside the hermetic shell.

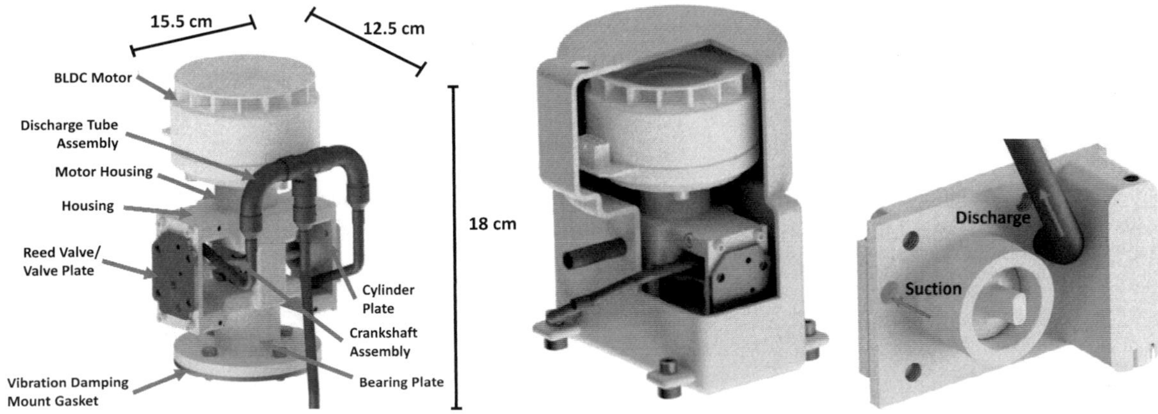

Figure 8: (left) Twin piston oil-less compressor internal components (cylinder heads not shown); (center) Concept rendering of the compressor inside a hermetic shell; (right) cylinder suction and discharge

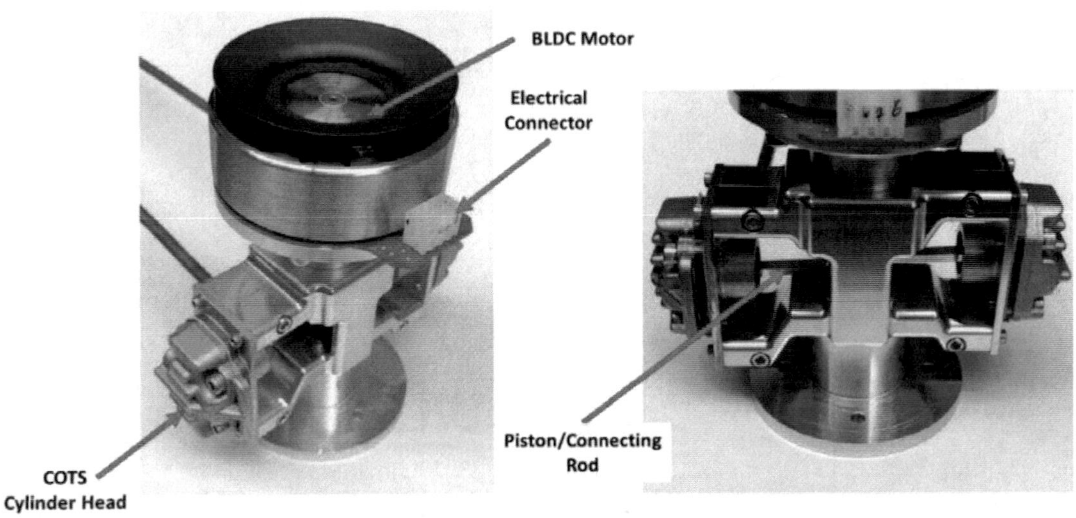

Figure 9. Two views of the fully assembled oil-less compressor prototype

4. EXPERIMENTAL SETUP, RESULTS, AND DISCUSSION

A test stand was fabricated to evaluate the performance of the prototype compressor. The test loop schematic is shown in Figure 10, and a photo of the test stand is shown in Figure 11. We measured the refrigerant temperature at six locations using resistance temperature detectors (RTDs) and refrigerant absolute pressure at four locations to fully characterize the thermodynamic states in the loop. We measured the refrigerant mass flow rate using a Coriolis mass flow meter (Siemens FC300 DN4). All sensors with their ranges, positions, and accuracy are listed in Table 4. All sensors produced an analog 4-20 mA signal read by a National Instruments (NI) Data Acquisition (DAQ) system.

The evaporator was a brazed plate heat exchanger connected to a secondary deionized water loop. A 500 W cartridge heater in the loop connected to a variable autotransformer was used to control the heat load. The water temperature at

Table 5: Sensor information for Mainstream's oil-less compressor test loop

Sensor Type	Manufacturer	Position	Range	Accuracy
RTD	Evolution Sensors	All Temp Sensors	-50 – 260 °C	Class A
Pressure Transducer	Omega	High Side	0 – 300 psia	± 0.25% FS
Pressure Transducer	SSI Technologies	Low Side	0 – 100 psia	± 1.00% FS
Mass Flowmeter	Siemens	Upstream of TXV	0 – 350 kg/h	± 0.1% reading
Volumetric Flowmeter	Omega	Evaporator Loop	0.2 – 2 LPM	± 1.0% FS
DC Voltage	CR Magnetics	Compressor Input	0 – 50 VDC	± 1.0% FS
DC Current	CR Magnetics	Compressor Input	0 – 20 ADC	± 1.0% FS

the inlet and outlet of the evaporator and the volume flow rate were used as a secondary capacity measurement. The condenser was a microchannel aluminum heat exchanger with a variable speed DC fan to adjust the condenser heat load. A manually controlled needle valve was used for the expansion valve because commercial thermostatic expansion valves (TXVs) are not designed for the small capacity of this system or the new refrigerants that we tested.

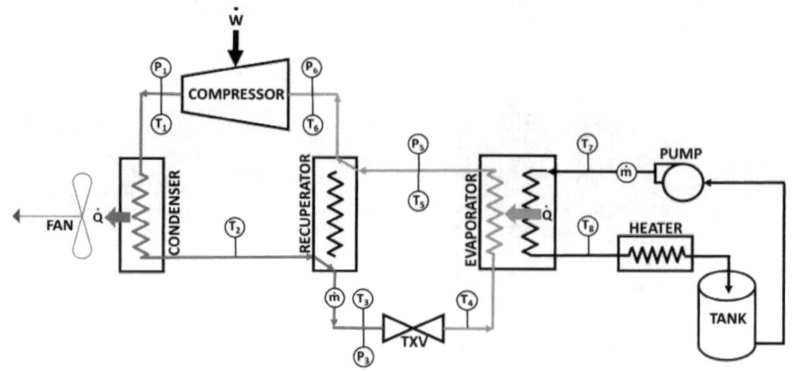

Figure 10: Test loop schematic for testing Mainstream's oil-less compressor

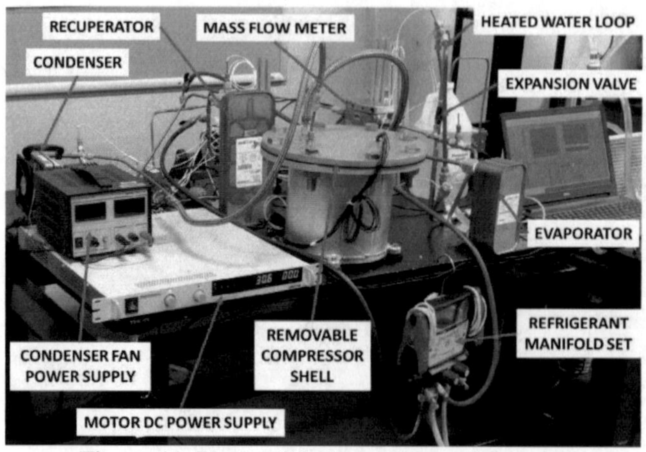

Figure 11. Photo of the compressor test stand

The compressor was operated with both R1336mzz(Z) and R1336mzz(E) working fluids. Compressor speed was varied up to 2000 rpm limited by the motor input voltage (30 VDC). The system was unable to produce the desired state points with either of these refrigerants; however, Table 5 shows a lower lift data point collected using R1336mzz(E). The thermal model was exercised at these thermodynamic conditions (i.e. suction temperature and pressure, pressure ratio) to calculate anticipated performance at this condition and to compare with the experimental results. At these conditions, the compression ratio was 2.61 (26.73°C lift). The cooling capacity was 49.6 W ± 0.1 W and the COP_c was 2.5 ± 0.4. This capacity and COP_c were significantly lower than the predicted values. The low volumetric efficiency (i.e. capacity) is likely due to poor sealing, blowback around the piston during compression, and inefficiencies in the reed valves and discharge geometry through the cylinder heads. The low isentropic efficiency (24.3%) was caused

Table 4. Compressor Performance, R1336mzz(E)

	Measured	Predicted
Compressor Speed (rpm)	1000	
SST (°C)	9.65	
SDT (°C)	36.38	
Lift (°C)	26.73	
Pressure Ratio	2.61	
Volumetric Efficiency	51.0% ± 0.8%	91.6%
Refrigerant Mass Flow (kg/h)	1.25 ± 0.001	2.22
Cooling Capacity (W)	49.6 ± 0.1	87.2
Compressor Power (W)	19.7 ± 3.0	19.5
COP_c	2.5 ± 0.4	4.5
Isentropic Efficiency	24.3% ± 3.8%	50.0%
Crankshaft Torque (N·m)	0.16 ± 0.03	0.17

by factors including high friction loads and blowback around the piston.

The friction torque was quantified by measuring motor power and crankshaft torque with the cylinder heads removed (i.e. no compression). Figure 12 shows the friction power and crankshaft torque of the compressor at speeds between 150 rpm and 1500 rpm. At the operating point shown in Table 5 (1000 rpm), the friction torque was 0.06 N·m and accounts for a substantial portion (35%) of the total crankshaft torque.

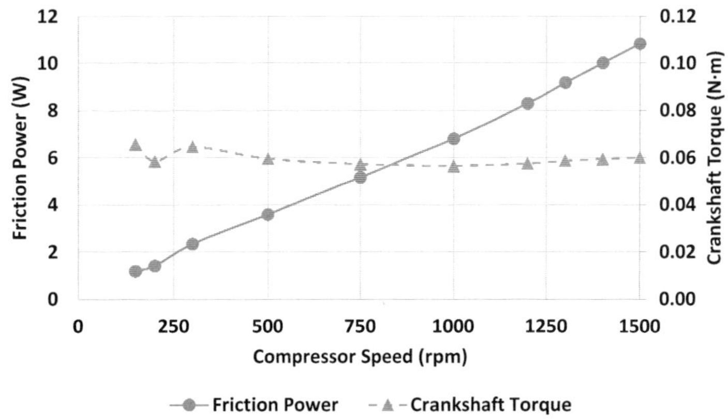

Figure 12. Compressor friction power and crankshaft torque

To achieve the predicted performance of the heat pump, the compressor isentropic efficiency and the volumetric efficiency must be increased. A breakdown analysis of the compressor revealed several areas that may have contributed to the poor performance. Several different piston seal diameters and thicknesses were evaluated to minimize friction and blowback. A 0.79 mm thick PTFE seal was not flexible enough to operate with low friction and resulted in visible wear around the cylinder plate during dry run conditions (Figure 13). While switching the piston seal to a 0.25 mm thick ring improved friction, some PTFE residue was still visible after operation. The ring outside diameter was 18.51 mm. Other materials such as graphite impregnated PTFE could also provide low friction sealing with better dry lubrication. A more thorough trade study on piston seal materials and geometry should be performed. Additionally, reed valve design can play an important role in improving compressor efficiency. This compressor used commercial reed valves and cylinder heads. A more targeted valve design effort could improve the performance and reduce losses.

When the compressor was disassembled, crankshaft misalignment was also discovered. This was likely due to tolerance stack-up in between the crankshaft, the cylinder plate, and the housing. The crankshaft position was set by a shoulder machined into the housing block, and the cylinder plate attached to the housing with four bolt holes. The tolerance stack-up caused angulation of the crankshaft resulting in additional friction and wear (Figure 14). Side loading forces on the piston also caused the connecting rod to walk down the bearing on the crankshaft and rub against the leaf of the crankshaft. The housing block shoulder dimension was modified, but there was still some remaining crankshaft misalignment. It was also observed that the bearings did not spin as smoothly as before operation. This could also have been due to the misalignment and part wear. In future designs, the cylinder position should be fixed more tightly in the horizontal direction while allowing room in the vertical direction to match the crankshaft position.

Figure 13. PTFE residue on the cylinder

Figure 14. Evidence of crankshaft rubbing due to piston-cylinder misalignment

The connecting rod bearings should also be installed with a press fit on the bearing-to-connecting rod interface and a slip fit on the crankshaft-to-bearing interface (the opposite of the current construction) to ensure no slippage of the piston on the bearing. The position of the bearing on the crankshaft will be set by machined shoulder features.

5. CONCLUSION

In this research effort, a miniature oil-less opposed twin-piston reciprocating compressor for lunar equatorial rovers was designed, fabricated, and tested. Two new HFO refrigerants were evaluated that show promise for both NASA and terrestrial heat pump applications. The twin-piston configuration is promising for this application because it enables reduced size, lower piston speeds, and low vibration. The total mass of the compressor was 2.86 kg. Maximum exported vibration was estimated to be 0.04 N at the base of the compressor based on kinematic calculations and agreement between measured and predicted motor torque.

During testing, the prototype compressor was not able to operate at the desired state points and did not meet the performance predictions. Sample data using R1336mzz(E) showed a cooling capacity of 49.6 W and COP of 2.5. The poor performance was most likely due to a number of factors including high friction, manufacturing and tolerance issues, and suboptimal design of reed valves and piston seals. A breakdown analysis of the compressor showed several areas of wear on the piston seals and the crankshaft-connecting rod assembly. It is likely that many of the mechanical issues with the compressor can be addressed in future iterations of the design. Although the performance of this prototype was lower than desired, valuable progress has been made towards a working compressor for extreme environment application such as lunar equatorial science missions.

NOMENCLATURE

$c_{\text{clearance}}$	percent clearance	η_v	volumetric efficiency
n	polytropic index	θ_{rad}	radiator angle from horizontal
r_p	pressure ratio		

REFERENCES

Bugby, D. C., Farmer, J. T., O'Connor, B. F., Wirzburger, M. J., Abel, E. D., & Stouffer, C. J. (2010). *Two-Phase Thermal Switching System for a Small, Extended Duration Lunar Surface Science Platform.* Paper presented at the AIP Conference Proceedings.

Grzyll, L., & Cole, G. (2000). *A prototype oil-less compressor for the international space station refrigerated centrifuge.* Paper presented at the Proceedings of the 2000 International Compressor Engineering Conference at Purdue.

Jones, H., Thornton, J., Balasubramaniam, R., Gokoglu, S., Sacksteder, K., & Whittaker, W. (2011). *Enabling Long-Duration Lunar Equatorial Operations With Thermal Wadi Infrastructure.* Paper presented at the 49th AIAA Aerospace Sciences Meeting including the New Horizons Forum and Aerospace Exposition.

Juhasz, J. R. (2017). *Novel Working Fluid, HFO-1336mzz (E), for Use in Waste Heat Recovery Application.* Paper presented at the Proceedings of the twelfth IEA heat pump conference. 2017. Rotterdam.

Thornton, J., Whittaker, W., Jones, H., Mackin, M., Barsa, R., & Gump, D. (2010). *Thermal strategies for long duration mobile lunar surface missions.* Paper presented at the 48th AIAA Aerospace Sciences Meeting Including the New Horizons Forum and Aerospace Exposition.

ACKNOWLEDGEMENTS

This research was funded by a Small Business Innovation and Research (SBIR) grant through NASA Jet Propulsion Laboratory. The authors would like to thank Eric Sunada for his input and advice during this research project.

An Approach to Detach the Thermodynamic Losses of a Reciprocating Compressor

Ernane SILVA[1]*, Thiago DUTRA[2]

[1]TEG Thermofluids Engineering Group, Federal University of Santa Catarina, 89219600, Joinville, SC, Brazil
Phone:+ 55 48 37217307
ernane.silva@teg.ufsc.br

[2]TEG Thermofluids Engineering Group, Federal University of Santa Catarina, 88905120, Ararangua, SC, Brazil
Phone:+ 55 48 999177691
thiago.dutra@teg.ufsc.br

* Corresponding author

ABSTRACT

Due to several incentives for reducing energy consumption worldwide, there is an increasingly demand for high-efficiency compressors for refrigeration systems. During the design stage of a refrigeration compressor, the designer needs to identify the main energy losses so as to propose alternatives to increase the compressor efficiency. The energy losses associated with the compression cycle are typically evaluated from the indicated diagram, which allows an accurate estimate of viscous flow losses through the suction and discharge systems, but no clear identification of piston-cylinder leakage and heat transfer losses. This paper presents a theoretical method to assess the energy losses associated with the compression chamber based on the analysis of the first law of thermodynamics, in which losses are detached into heat transfer, leakage, valve backflow and others. For this purpose, a compressor simulation model based on a transient and lumped formulation of mass and energy conservation equations is applied to the compression chamber and is validated with experimental data. A study case is performed in which the energy losses of a baseline compressor is compared with those of a 4% higher isentropic efficiency compressor. It is concluded that the typical indicated diagram analysis does not allow a clear identification of the benefits that cause the increase of 4% in isentropic efficiency, whereas the approach proposed herein and based on the first law of thermodynamics indicates that the increase of isentropic efficiency is associated with reduction in heat transfer and leakage losses during expansion and discharge processes.

1. INTRODUCTION

Refrigeration and air conditioning systems consume about 17% of the overall electricity used worldwide (Coulomb *et al.*, 2015). Many of these systems are based on the vapor compression cycle and the compressor is responsible for most part of the energy consumption. According to Pérez-Segarra *et al.* (2005), the compressor energy losses can be classified as: (i) electrical losses caused mainly by ohmic resistance in the electric motor; (ii) mechanical losses generated by friction in the bearings and; (iii) thermodynamic losses associated with irreversibilities during the compression of the gas. Currently, the thermodynamic and electrical efficiencies of a small reciprocating compressor are around 82-83%, whereas the mechanical efficiency is nearly 91-92% (Diniz *et al.*, 2018). Further improvements of compressor performance require the detailed detachment of these energy losses.

The thermodynamic losses are generally assessed via the pressure-volume diagram. Pérez-Segarra *et al.* (2005), for instance, split the compression work in one compression cycle according to its physical sub-processes (suction, compression, discharge, and expansion). They define efficiencies associated with each sub-process taking the ideal (isentropic) compressor as reference. However, due to heating effects and valves closing/opening delays the efficiency associated with the expansion sub-process is generally higher than one and the authors suggest that the compression and expansion processes would be more conveniently expressed by a combined efficiency. Therefore, it is possible to account for three sources of energy losses: (i) underpressure during the suction process, (ii) overpressure during the discharge process, and (iii) remaining losses of the compression cycle. Diniz et al. (2018) adopted the same approach, except that they dealt with the energy rates and separated the remaining losses into suction superheating and “others”. In the view of the authors, the latter term would account for leakage and heat transfer during the expansion and compression processes and cylinder heat transfer during the suction process. Posch et al. (2018) proposed an alternative approach based on the formulation of the second law of thermodynamics, describing the compressor losses in terms of entropy generation and exergy losses.

In the present study, we propose to detachment of the thermodynamic losses following the first law of thermodynamics, replacing the ordinary “process-oriented” approach by an “energy transfer-oriented” one. The applicability of this technique is demonstrated with a theoretical comparative analysis of two reciprocating compressors applied for household refrigeration. Both compressors have the same geometric characteristics, but differ in how the compression chamber volume changes with time. In the first one, a sinusoidal piston displacement is produced by a crank-rod mechanism, while in the second one an optimized piecewise linear piston trajectory is adopted. Prior to the aforementioned analysis, we describe and validate the comprehensive compressor simulation model employed in this study.

2. COMPRESSOR SIMULATION MODEL

The compressor simulation model is based on an unsteady-state lumped-parameter formulation of the mass and energy conservation equations applied to the compression chamber (Bradshaw *et al.*, 2011; Engel and Deschamps, 2019), leading to:

$$\frac{dm}{dt} = \dot{m}_s - \dot{m}_{sb} - \dot{m}_d + \dot{m}_{db} - \dot{m}_l + \dot{m}_{lb} \, , \tag{1}$$

$$\frac{dU}{dt} = \dot{Q} - \dot{W} + \dot{m}_s h_{sc} - \dot{m}_{sb} h - \dot{m}_d h + \dot{m}_{db} h_{dc} - \dot{m}_l h + \dot{m}_{lb} h_{ie} \, . \tag{2}$$

In these equations, m is the mass and U is the internal energy of the gas in the compression chamber, t is the time and $\dot{Q}$ is the instantaneous heat transfer rate. The instantaneous compression power is $\dot{W} = p\,dV/dt$, p being the pressure and V the volume of the gas. The variations of kinetic and potential energy wee neglected. The mass flow rates entering and leaving the compression chamber are represented by $\dot{m}$ in which the subscripts s, d and l stand for flow through the suction and discharge ports and leakage through the piston-cylinder gap, whereas sb, db and lb denote backflow through the same passages, respectively. The variable h stands for the specific enthalpy of the gas in the compression chamber whereas h_{sc}, h_{dc} and h_{ie} refer to the specific enthalpies at the suction chamber, discharge chamber and internal environment, which is occupied by gas at low pressure. These equations are integrated using the Euler explicit method and the required thermodynamic properties are evaluated using the thermodynamic properties library CoolProp (Bell *et al.*, 2014). Figure 1 presents a schematic of the compression chamber.

The flow of gas through the suction and discharge orifices are modeled assuming a one-dimensional isentropic compressible flow through a nozzle, that is,

$$\dot{m}_i = C_{ee,i} A_i p_{up} \sqrt{\frac{2\gamma}{(\gamma - 1) R T_{up}} \left(\Pi^{\frac{2}{\gamma}} - \Pi^{\frac{\gamma+1}{\gamma}} \right)} \, , \tag{3}$$

where $\dot{m}_i$ stands for the direct or reverse flow through the orifice, A_i is the orifice area, $C_{ee,i}$ is the effective flow area coefficient, γ is the specific heat ratio, R is the specific gas constant and $\Pi = p_{down}/p_{up}$ is the pressure ratio, which is limited to

$$\Pi = \left(\frac{2}{\gamma + 1} \right)^{\frac{\gamma}{\gamma - 1}} \tag{4}$$

for choked flow condition. The subscripts up and $down$ indicate variables evaluated at the upstream and downstream thermodynamic states, respectively. The effective flow area coefficient is a correction parameter to account for the real flow restriction imposed by the orifice and valve geometry, which is normally obtained from experiments or CFD simulations. Generally, this parameter is described as a function of the valve displacement.

The heat transfer rate between the gas and the walls of the compression chamber is calculated using the Newton's law of cooling:

$$\dot{Q} = HA(T_w - T) \quad , \tag{5}$$

where H is the convective heat transfer coefficient between the gas and the walls, which is obtained from the correlation proposed by Annand (1963). The heat transfer area A includes all surfaces that enclose the gas in the compression chamber. The temperature of the cylinder walls, T_w, is obtained from experimental data and T is the temperature of the gas in the compression chamber.

Additional equations are needed for the valve dynamics, the leakage rate through the piston cylinder gap and the volume of the compression chamber. The valve dynamics is treated as a one-degree-of-freedom mass-spring-damper system, using the concept of effective force areas (Pérez-Segarra *et al.*, 2003; Pereira and Deschamps, 2011). The leakage rate through the piston-cylinder gap is calculated assuming an incompressible fully developed laminar flow of pure gas (Ferreira and Lillie, 1984). Finally, the volume of the compression chamber is either imposed by the optimization algorithm or computed from the driving mechanism kinematics equation for a crank-rod compressor (An *et al.*, 2002).

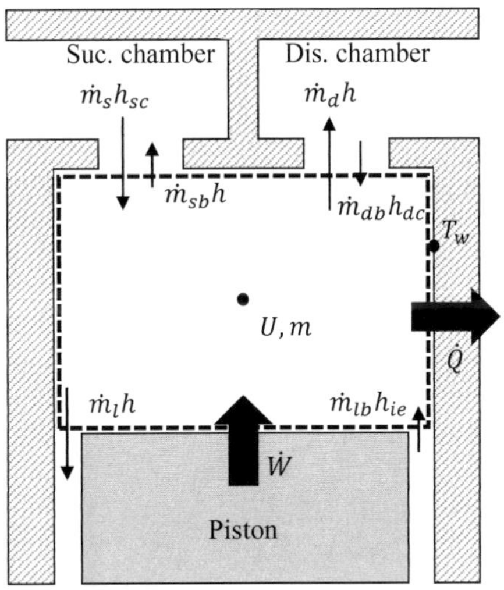

Figure 1. Schematic of the compression chamber.

3. DETACHMENT OF THE THERMODYNAMIC LOSSES

The thermodynamic efficiency is defined as the ratio between the theoretical power $(\dot{W}_{th})$ estimated considering an isentropic process, and the indicated power $(\dot{W}_i)$, obtained from the pressure-volume diagram. Such a definition is also well known in the literature as isentropic efficiency and is expressed as:

$$\eta_t = \frac{\dot{W}_{th}}{\dot{W}_i} = \frac{\dot{m}(h_{d,s} - h_{sc})}{\dot{W}_i} \quad , \tag{6}$$

in which $h_{d,s}$ is the specific enthalpy evaluated with the discharge pressure and the specific entropy of the gas in the suction chamber. The actual mass flow rate $\dot{m}$ is calculated as a function of the instantaneous mass flow rates through the discharge port:

$$\dot{m} = f \int_0^{1/f} (\dot{m}_d - \dot{m}_{db}) dt \ , \tag{7}$$

where f is the operation frequency of the compressor. The indicated power $\dot{W}_i$ is computed as

$$\dot{W}_i = -f \oint p dV \ . \tag{8}$$

The coefficient of performance is a parameter commonly adopted to assess the energetic performance of refrigeration systems and is directly proportional to the thermodynamic efficiency. For this reason, the determination of the main energy losses associated with the compression cycle is paramount. Two approaches are adopted herein to perform this task: (i) one based on the analysis of the indicated diagram and (ii) another derived from the first law of thermodynamics.

3.1 Analysis via the Indicated Diagram

In this traditional approach (Pérez-Segarra et al., 2005), the energy losses due to the suction and discharge processes are computed with reference to ideal isobaric processes at suction line and discharge line pressures p_s and p_d, respectively. In this study, the evaporating pressure p_e is assigned as p_s and the condensing pressure p_c as p_d. Therefore, $\Delta\dot{W}_s$ is estimated by integrating the area of the indicated diagram below the evaporating pressure

$$\Delta\dot{W}_s = -f \oint \min(p, p_s) dV \ , \tag{9}$$

whereas $\Delta\dot{W}_d$ is determined from the area of the indicated diagram above the condensing pressure

$$\Delta\dot{W}_d = -f \oint \max(p, p_d) dV \ . \tag{10}$$

Finally, the energy loss associated with the compression and expansion processes is computed by:

$$\Delta\dot{W}_{ce} = \dot{W}_i - \Delta\dot{W}_s - \Delta\dot{W}_d - \dot{W}_{th} . \tag{11}$$

3.2 Analysis via the First Law of Thermodynamics

By combining Eqs. (1) and (2), the following expression is obtained for the instantaneous compression power:

$$\begin{aligned} \dot{W} = (\dot{m}_d - \dot{m}_{db})(h_{sc} - h_{d,s}) + (\dot{m}_d - \dot{m}_{db})(h_{d,s} - h) + \dot{Q} + \dot{m}_l(h_{sc} - h) + \dot{m}_{sb}(h_{sc} - h) \\ + \dot{m}_{lb}(h_{ie} - h_{sc}) + \dot{m}_{db}(h_{dc} - h) + \frac{dm}{dt} h_{sc} - \frac{dU}{dt} \end{aligned} \ , \tag{12}$$

which is integrated along time for one complete cycle and multiplied by $-f$, yielding

$$\dot{W}_i = \dot{W}_{th} + \Delta\dot{W}_{\dot{Q}} + \Delta\dot{W}_l + \Delta\dot{W}_{lb} + \Delta\dot{W}_{sb} + \Delta\dot{W}_{db} + \Delta\dot{W}_o \ . \tag{13}$$

The terms $\Delta\dot{W}$ are the energy losses due to heat transfer ($\Delta\dot{W}_{\dot{Q}}$), leakage ($\Delta\dot{W}_l$), backflow through the piston-cylinder gap ($\Delta\dot{W}_{lb}$), backflow through the suction port ($\Delta\dot{W}_{sb}$), backflow through the discharge port ($\Delta\dot{W}_{db}$) and other losses ($\Delta\dot{W}_o$). All relationships used to calculate the energy losses are presented in Table 1. The term $\Delta\dot{W}_o$ corresponds to the energy loss brought about by the deviation of the gas enthalpy in the compression chamber h with reference to $h_{d,s}$

during the discharge process. Thus, $\Delta\dot{W}_o$ is affected by all losses aforementioned, since they contribute to change the gas enthalpy.

Table 1 – Definition of the energy losses (W).

$\Delta\dot{W}_{\dot{Q}}$	$-f\int_0^{1/f}\dot{Q}dt$	$\Delta\dot{W}_l$	$f\int_0^{1/f}\dot{m}_l(h-h_{sc})dt$	$\Delta\dot{W}_{sb}$	$f\int_0^{1/f}\dot{m}_{sb}(h-h_{sc})dt$
$\Delta\dot{W}_{db}$	$f\int_0^{1/f}\dot{m}_{db}(h-h_{dc})dt$	$\Delta\dot{W}_{lb}$	$f\int_0^{1/f}\dot{m}_{lb}(h_{sc}-h_{ie})dt$	$\Delta\dot{W}_o$	$f\int_0^{1/f}(\dot{m}_d-\dot{m}_{db})(h-h_{d,s})dt$

4. RESULTS

4.1 Compressor Model Validation

Initially, the numerical model was validated against experimental data provided by Diniz et al. (2018) for a compressor with volumetric displacement V_{sw} = 5.96 cm^3, operating with R600a at 50 Hz in a household refrigeration system, under on-off cycling conditions. The compression cycle was simulated considering the particular pairs of evaporating and condensing temperatures just before the compressor shut-down for two operating conditions, namely 16B and 32B. The input parameters are shown in Table 2, where T_e, T_c, T_{sc} and T_{dc} represent the evaporating, condensing, suction chamber and discharge chamber temperatures. Due to the lack of information about the thermodynamic state of the gas in the internal environment of the compressor, it was assumed the same as that of the suction chamber. The numerical predictions and experimental data of mass flow rate and indicated power are presented in Table 3. Larger deviations are found for condition 32B, where the numerical predictions of mass flow rate and indicated power are 5.4% and 3.5% lower than the measured ones, respectively. These deviations are acceptable given the simulation model simplicity.

Table 2. Input data for compressor simulation.

Parameter	Condition 16B	Condition 32B
T_e [°C]	-27.1	-24.5
T_c [°C]	31.7	45.6
T_{sc} [°C]	27.0	48.5
T_{dc} [°C]	63.5	92.2
T_w [°C]	42.7	69.9

Table 3. Comparison between measurements and predictions.

Parameter	$\dot{m}$ [kg/h]	$\dot{W}_i$ [W]
	Condition 16B	
Experimental	1.00	29.24
Numerical	0.97	29.37
Deviation	-3.0%	0.4%
	Condition 32B	
Experimental	0.93	34.60
Numerical	0.88	33.38
Deviation	-5.4%	-3.5%

4.2 Thermodynamic Losses Detachment

After the model validation, an optimization procedure was adopted to determine the piston trajectory that maximizes the compressor thermodynamic efficiency under condition 32B, but retaining the baseline volumetric displacement V_{sw}. Valve dynamics was disregarded for the computation of the optimum piston trajectory, which means valves open immediately with the minimum flow resistance when subjected to any positive pressure difference and do not experience backflow. Such an assumption intended to avoid an optimized piston trajectory constrained by any particular set of suction and discharge valves. Moreover, valve design is typically carried out after the driving mechanism is specified, i.e. after the piston trajectory is defined. For comparison purposes, ideal valves were also adopted for the compressor with the crank-rod mechanism. More details about the piston trajectory optimization are described in Silva and Dutra (2021).

Table 4 presents results for the thermodynamic losses detachment using the indicated diagram analysis. All losses are normalized with respect to the theoretical power, $\Delta W^* = \Delta W/W_{th}$. Baseline crank-rod compressor and optimized piston trajectory compressor are denoted by CR and OPT, respectively. The optimization of the piston trajectory brings about a slight increase in the discharge losses, which are overcome by a decrease of losses associated with compression and expansion processes. As a result, the indicated power is reduced and the thermodynamic efficiency increases 3.8% with the OPT configuration (from 88.3% to 92.1%). However, it is not possible to identify the mechanisms behind the compression and expansion processes that are responsible for such an efficiency improvement.

The approach based on the application of the first law of thermodynamics gives a different insight of losses detachment. Table 5 presents the thermodynamic losses for CR and OPT cases obtained from a stratification using the First Law Approach. Note that only heat transfer, piston-cylinder leakage and other losses are included, since leakage backflow is zero ($\Delta\dot{W}_{lb} = 0$) and ideal valves were adopted in this section for both configurations, resulting in no backflow losses ($\Delta\dot{W}_{sb} = \Delta\dot{W}_{db} = 0$). The optimization of the piston trajectory provides a reduction of the energy loss due to the piston-cylinder leakage of 43.0% (from 2.6 W to 1.5 W) and a decrease on the net heat transfer rate by 40.6% (from 2.6 W to 1.6 W) with reference to the CR configuration. Moreover, the term $\Delta\dot{W}_o$ acts as a gain for both OPT and CR, which means that h is lower than $h_{d,s}$ during the discharge process. Such a deviation is larger for the CR (-1.1 W) because heat transfer and leakage losses are also larger for this compressor and they contribute to reduce the gas enthalpy in the compression chamber.

Table 4. Thermodynamic losses detachment Indicated Diagram Analysis / Condition 32B.

Parameter	CR	OPT
$\dot{W}_{th}$ [W]	31.5	32.0
$\Delta\dot{W}_s$ [W] / $\Delta\dot{W}_s^*$[%]	0.1 / 0.3	0.1 / 0.3
$\Delta\dot{W}_d$ [W] / $\Delta\dot{W}_d^*$[%]	0.2 / 0.6	0.6 / 1.9
$\Delta\dot{W}_{ce}$ [W] / $\Delta\dot{W}_{ce}^*$[%]	3.8 / 12.1	2.0 / 6.3
$\dot{W}_i$ [W]	35.6	34.7
η_t [%]	88.5	92.2

Table 5. Thermodynamic losses detachment First Law Approach / Condition 32B.

Parameter	CR	OPT
$\dot{W}_{th}$ [W]	31.5	32.0
$\Delta\dot{W}_{\dot{Q}}$ [W]	2.6	1.6
$\Delta\dot{W}_l$ [W]	2.6	1.5
$\Delta\dot{W}_o$ [W]	-1.1	-0.3
$\dot{W}_i$ [W]	35.6	34.7
η_t [%]	88.5	92.2

It is necessary to analyze the piston trajectory to understand the reasons behind the reduction of leakage and heat transfer losses associated with the OPT. Figure 2 shows the volume of the compression chamber along the compression cycle for both the (a) conventional compressor with a crank-rod mechanism and (b) the optimized compressor. It is interesting to note that the bottom dead center is anticipated in the optimized configuration, which means that the time required to execute the expansion and suction processes is shorter than that comprising the compression and discharge processes. In addition, the expansion and discharge strokes in the OPT case are nearly 50% shorter than those in the CR compressor.

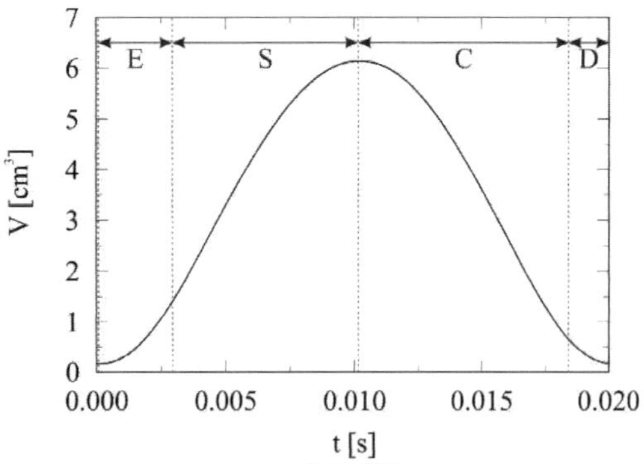

(a) CR

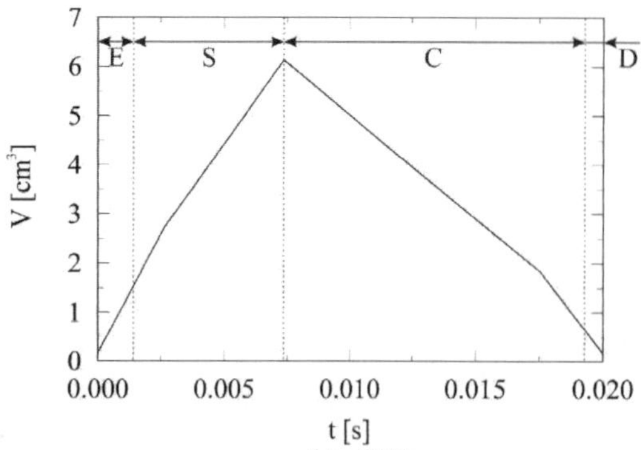

(b) OPT

Figure 2. Volume of the compression chamber (V) as a function of time (t). E: expansion, S: suction, C: compression, D: discharge.

The thermodynamic losses assessed via the First Law Approach are normalized with respect to the theoretical power and detached into the four strokes of the compression cycle, as shown in Fig. 3. There is an increase in the heat transfer losses related to the compression stroke in the OPT case, since the time required to carry out this process is 43% longer than that observed in the CR mechanism (Fig. 2). However, the decrease in the thermodynamic losses associated with the expansion and the discharge processes of the optimized configuration overcomes the increase in the compression process losses, leading to a gain in terms of thermodynamic efficiency.

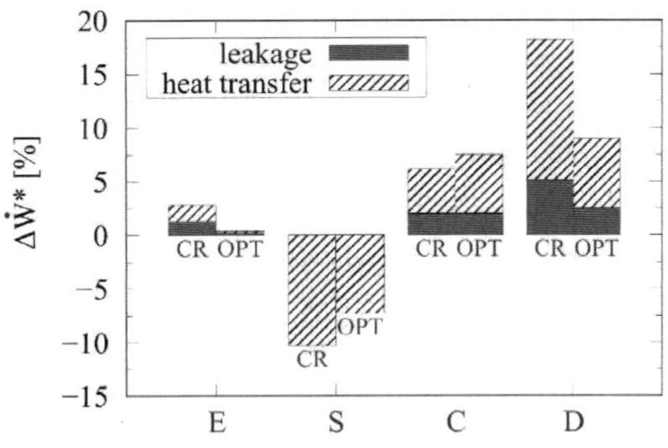

Figure 3. Normalized energy losses along different processes.

5. CONCLUSION

This paper presented an approach to assess the thermodynamic losses of a reciprocating compressor based on the application of the first law of thermodynamics. For this purpose, a compressor simulation model based on a transient and lumped formulation of mass and energy conservation equations applied to the compression chamber was developed and validated with experimental data. A study case was performed in which the thermodynamic losses of a baseline compressor was compared with those of an optimized piston trajectory compressor. Loss assessment based on the indicated diagram analysis indicates the efficiency of the optimized piston trajectory compressor increases nearly 4% due to the reduction of losses associated with compression and expansion processes, but the mechanisms that lead to such a loss reduction are not identified. By applying the First Law Approach, we found out that the optimized piston trajectory led to reductions of 43.0% and 40.6% in the piston-cylinder leakage and heat transfer losses, respectively, which were due to 50% shorter expansion and discharge processes with respect to the baseline crank-rod compressor. Although the First Law Approach have been presented for a reciprocating compressor, it can be extended to other compressor types, such as rotary and scroll.

NOMENCLATURE

			Subscript	
A	area	(m^2)		
f	frequency	(Hz)	c	condensing
h	specific enthalpy	(J/kg)	d	discharge
H	convective heat transfer coefficient	$(W/m^2.K)$	db	discharge backflow
m	mass	(m)	dc	discharge chamber
$\dot{m}$	mass flow rate	(kg/s)	e	evaporation
p	pressure	(Pa)	i	Indicated
$\dot{Q}$	heat transfer rate	(W)	l	Leakage
t	time	(s)	lb	leakage backflow
T	temperature	(°C)	o	Other
V	volume	(m^3)	$\dot{Q}$	heat transfer
V_{sw}	volumetric displacement	(m^3)	s	isentropic, suction
$\dot{W}$	power	(W)	sb	suction backflow
U	internal energy	(J)	sc	suction chamber
$\Delta\dot{W}$	thermodynamic loss	(W)	t	Thermodynamic
$\Delta\dot{W}^*$	normalized thermodynamic loss	(-)	th	theoretical
η	efficiency	(-)	W	cylinder wall
ρ	density	(kg/m^3)		

REFERENCES

An, K.H., Lee, J.H., Lee, I.W., Park, S.C. (2002). Performance prediction of reciprocating compressor. *In: Proc. Int. Compressor Eng. Conf. at Purdue*, West Lafayette, Paper 1525.

Annand, W.J.D. (1963). Heat transfer in the cylinders of reciprocating internal combustion engines. *Proc. Instn. Mech. Engnrs*., 117, pp. 973-996.

Bell, I.H., Wronski, J., Quoilin, S., Lemort, V. (2014). Pure and pseudo-pure fluid thermophysical property evaluation and the open-source thermophysical property library coolprop. *Ind. Eng. Chem. Res.,* 53 (6), pp. 2498-2508.

Bradshaw, C.R., Groll, E.A., Garimella, S.V. (2011). A comprehensive model of a miniature-scale linear compressor for electronics cooling. *Int. J. Refrigeration*, 34, 63-73.

Coulomb, D., Dupont, J.L., Pichard, A. (2015). The role of refrigeration in the global economy. 29th Informatory Note on Refrigeration Technologies, *Technical Report, International Institute of Refrigeration*: Paris, France.

Diniz, M.C., Melo, C., Deschamps, C.J. (2018). Experimental performance assessment of a hermetic reciprocating compressor operating in a household refrigerator under on–off cycling conditions. *Int. J. Refrigeration*, 88, 587-598.

Engel, R.C., Deschamps, C.J. (2019). Comparative analysis between the performances of reciprocating and rolling piston compressors applied to a domestic heat pump water heater. *Int. J. Refrigeration*, 102, 130-141.

Ferreira, R.T.S., Lillie, D.E.B. (1984). Evaluation of the leakage through the clearance between piston and cylinder in hermetic compressors. *In: Proc. Int. Compressor Eng. Conf. at Purdue, West Lafayette*, Paper 424.

Pereira, E.L.L, Deschamps, C.J. (2011). Influence of piston on effective areas of reed-type valves of small reciprocating compressors. *HVAC&R Res*., 17:2, 218-230.

Pérez-Segarra, C.D., Rigola, J., Sòria, M., Oliva, A. (2005). Detailed thermodynamic characterization of hermetic reciprocating compressors. *Int. J. Refrigeration*, 28, 579-593.

Pérez-Segarra, C.D., Rigola, J., Oliva, A. (2003). Modeling and numerical simulation of the thermal and fluid dynamic behavior of hermetic reciprocating compressors-part 1. *HVAC&R Res*., 9:2, 215-235.

Posch, S., Hopfgartner, J., Dürr, L., Eichinger, M., Stangl, S. & Almbauer, R. (2018). Thermal loss analysis of hermetic compressors using numerical simulation. *Appl. Therm. Eng.,* 130, 1580–1589.

Silva, E., Dutra, T. (2021). Piston trajectory optimization of a reciprocating compressor. *Int. J. Refrigeration*, 121, 159-167.

Improving Lifetime of Domestic Compressor Subjected To Repetitive Internal Stresses

Seong-woo Woo[1*], Dennis L. O'Neal[2], Dereje Engida Woldemichael[1], Samson Mekbib Atnaw[1], Muluneh Mekonnen Tulu[1]

[1]Mech. Eng., College of Electrical & Mechanical Engineering, Addis Ababa Science & Technology University, Addis Ababa, Ethiopia
Contact Information (Phone: +251-90-047-6711, Email: twinwoo@yahoo.com)

Baylor University, Engineering and Computer Science, Waco, TX 76798-7356
Contact Information (Phone: +1-254-710-3871, E-mail: Dennis_ONeal@baylor.edu)

* Corresponding Author

ABSTRACT

To enhance the lifetime of a mechanical system such as compressor, parametric accelerated life testing (ALT) as a systematic reliability method is proposed. It consists of (1) a parametric ALT plan formed on BX lifetime, (2) load examination for ALT, (3) a customized parametric ALTs with design alternatives, and (4) assessment of the compressor design to secure the objective BX lifetime is fulfilled. As an experiment instance, newly designed reciprocating compressors for French-door refrigerators returned from the marketplace were investigated. The refrigerators had been making disagreeable noise and vibrations, creating the consumer to ask for replacing their refrigerators. As the vibration level of the problematic refrigerators was recorded in an anechoic chamber, the result was 0.35g. Upon closer inspection of the refrigerator, the noise originated from the reciprocating compressor where there was interference between the compressor upper shell and the stator frame. To reproduce the compressor problem from the marketplace, a parametric ALT was performed. The failure mode during the ALTs for the compressor was alike to those of the failed refrigerators from the field. As a corrective action, the stator frame in the compressor system was redesigned to increase the minimum gap between the compressor upper shell and the stator frame. During the second ALT, there were no issues. After parameter ALTs were used to develop corrective action plans, the lifetime of the compressor was reassured to have a B1 life 10 years.

1. INTRODUCTION

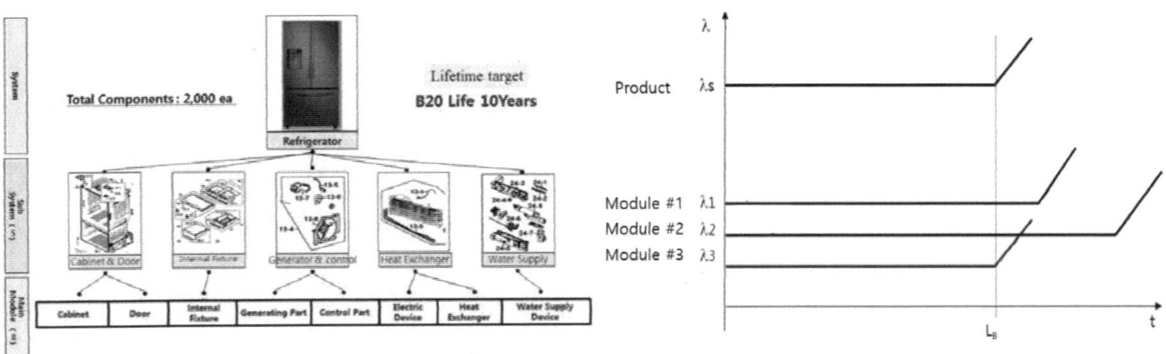

(a). Breakdown of refrigerator with multi-modules (b). Product lifetime L_B and failure rate λ_s

Figure 1: Product lifetime with multi-modules decided by critical module such as compressor

A refrigerator supplies chilled air from the evaporator to both the freezer and refrigerator sections where frozen or fresh food is stocked. It utilizes a vapor-compression refrigeration cycle (Sonntag, 2007). A refrigerator is composed of approximately 3,000 components, including cabinet, doors, sensors and controls, internal fixtures (shelves and drawers), generating parts (motor or compressor), heat exchanger (condenser and evaporator), a water supply device, and other components. A refrigerator's lifetime is anticipated to be no less than B20 life 10 years. The product lifetime is determined by module #3 such as compressor, which has design faults (Figure 1).

To improve the efficiency of a French-door refrigerator, it was fitted with a newly designed, high efficiency compressor. A compressor expands the refrigerant pressure in a vapor-compression refrigeration cycle. In the process, compressor systems are subjected to repeated stress due to internal pressure loads. If there is a design fault in the structure that causes unexpected vibrations when dynamic loads are applied, the compressor start making noises before it reaches its expected lifetime. In such a case, a consumer may then request to replace their refrigerators.

Robust design methods, including statistical design of experiments (DOE) (Montgomery, 2013) and the Taguchi approach (Taguchi, 1978), were evolved to conduct optimal design for mechanical systems. Because numerous design parameters for a mechanical structure have to be examined, the Taguchi product array demands an enormous experimental design over a range of conditions to ensure a reliable of mechanical system. To resolve rare product failures in the field, another engineering approach includes the finite element method (FEM) (Hrennikoff, 1941). Numerous engineers believe uncommon product failures can be evaluated by: (1) mathematical modeling utilizing Newtonian methods, (2) estimating the time response of the system for dynamic loads and discovering the product stress, (3) utilizing the rain-flow counting method for von Mises stress, and (4) approximating system damage which uses the Palmgren-Miner's rule. However, utilizing an analytical methodology that can create a closed-form, precise consequence often requires inducing numerous presumptions that are not capable of recognizing multi-module product failures due to design defects.

Most global manufacturers have evolved a reliability testing method such as parametric ALT. This methodology can help clarify diverse design faults within shortening test cycles. However, there are some things to be considered in using ALT. Any failures after ALT may not designate those occurring under field states. This problem usually occurs because of the inconsistency of the direction and magnitude of the load. Furthermore, the number of test samples and the test times are inadequate to expose rare failure modes under accelerated conditions. ALT equipment also must be properly designed to match product loads.

This study develops a new methodology as parametric ALT for improving the lifetime of domestic compressor subjected to repeated internal stresses. As product lifetime is targeted, parametric ALT for mission time can identify the design faults of a product such as vibration at a specific frequency from the induced pressure loads and modify them. The lifetime of a newly designed compressor might be improved. This systematic reliability method includes: (1) a parametric ALT plan formed on product BX lifetime, (2) a load examination for ALT, (3) a customized parametric ALTs with the design alternatives, and (4) an assessment of whether the design(s) of the product fulfill the objective BX lifetime. As an experiment instance, newly designed reciprocating compressor in French-door refrigerators returned from the marketplace will be investigated.

2. PARAMETRIC ALT FOR A COMPRESSOR

2.1 Index of product lifetime such as BX life

A suitable index of product lifetime, such as BX life, needs to be chosen for running parametric ALT. BX life is expressed as the time at which X percent of the items in a population will have been unsuccessful. For instance, if a product has B20 life 10 years, then 20% of the population will have failed during 10 years of working time. 'BX life Y years' helps to correctly decide the cumulative failure rate of a product and its lifetime replying to market circumstances. The meantime to failure (MTTF) as the reverse of the failure rate cannot represent the lifetime, because it means around B60 life and its failure rate until lifetime is too large. BX life reflects the designer's feelings more adequately than MTTF.

2.2 Putting an Whole Parametric ALT Plan

Reliability can be explained as the ability of a system to work under stated conditions for a specified period of time. It can be clarified by a diagram named a "bathtub curve" that consists of three regions (Klutke, 2015). First, there is a decreasing failure rate in the premature life of the product (□ □ 1). Then, there is a constant failure rate (□ = 1) in its middle. Eventually, there is an increasing failure rate to the end of the product's life (□ □ 1). If a product pursuits the bathtub curve, it may have difficulties succeeding from the marketplace because of the lofty failure rates and shortened lifetimes due to design defects. Consequently, manufacturers incur financial losses for the whole product life cycle. They can definitely enhance the design of a product by putting reliability targets for products to (1) lessen early failures, (2) decrease random failures during the product working time, and (3) increase product lifetime. As the design of a mechanical product is improved, the failure rate of the product from the marketplace decreases and its lifetime increases. For such circumstances, the conventional bathtub curve can be changed to a linear line with the shape parameter β (Figure 2).

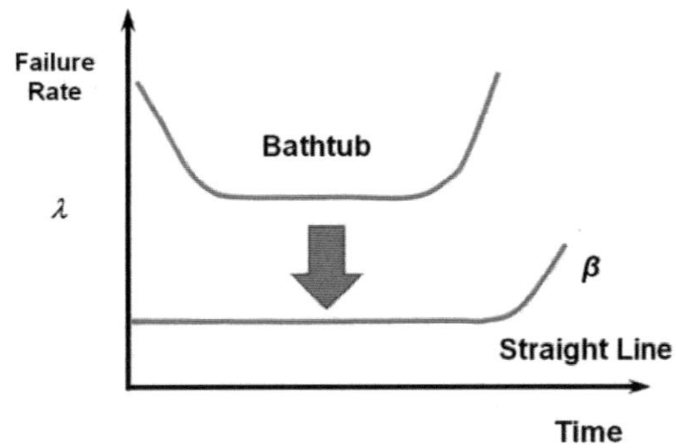

Figure 2: Bathtub curve and straight line with slope β

Because the linear line with low failure rate pursuits the exponential distribution, the reliability of the mechanical product might be quantified from the multiplication of product lifetime L_B and failure rate λ as follows:

$$R(L_{BX}) = 1 - F(L_{BX}) = e^{-\lambda L_{BX}} \cong 1 - \lambda L_B \tag{1}$$

where $R(\cdot)$ is reliability function, $F(\cdot)$ is cumulative distribution function, λ is failure rate, and L_B is BX lifetime

A refrigerator compressor is a mechanical device that increases the refrigerant pressure by reducing the cylinder volume through crank mechanisms. To keep the intended function, the product for BX lifetime should be designed to properly operate under the range of conditions subjected to it by consumers. After targeting the product lifetime L_B, the engineer should recognize design faults and alter them through parametric ALT.

In targeting the lifetime of a compressor in refrigerator, there is 1) a new module, 2) a modified module, and 3) a similar module where there is no alternation to the earlier design on the basis of market data. The compressor discussed in section 2.4 can be considered as an altered module because consumers want to enhance the energy efficiency and compressor parts is redesigned. To reply to customer request, a compressor lifetime might be put to B1 life 10 years.

2.3 Failure Mechanics and Accelerated Testing

A compressor increases the refrigerant pressure in a vapor-compression refrigeration system. Thus, compressor systems are subjected to repeated stress due to internal pressure loads. If there is design fault in the structure that causes vibration where the loads are exerted, the compressor may suddenly fail over its anticipated lifetime. After recognizing the product failure by reliability test such as parametric ALT, an engineer can optimally design the compressor.

The most important issue in the reliability testing of compressor is how fast the potential failure mode can be attained. To do this, it is required to express a failure model and determine the associated coefficients. First of all, we can configure the life-stress (LS) model, which incorporates stresses and reaction parameters. This equation can explain various failures in the mechanical structure. Because product failure originates from a micro-depletion (void) regardless of whether it is a mechanical or electronic system, the life-stress model might be expressed as the following processes. They are utilized for solid-state diffusion of impurities in silicon: 1) electro-migration-induced voiding; 2) build-up of chloride ions; 3) trapping of electrons or holes.

So the junction function for life-stress model might be expressed as (Grove, 1967):

$$J \cong \Phi(x,t,T)\sinh(a\xi)\exp\left(-\frac{Q}{kT}\right) \tag{2}$$

Contrastingly, a reaction process that is depending on speed might be defined as:

$$K = K^{+} - K^{-} = a\frac{kT}{h}e^{-\frac{\Delta E - \alpha S}{kT}} - a\frac{kT}{h}e^{-\frac{\Delta E + \alpha S}{kT}} = 2\frac{kT}{h}e^{-\frac{\Delta E}{kT}}\cdot\sinh\left(\frac{\alpha S}{kT}\right) == B\sinh(aS)\exp\left(-\frac{\Delta E}{kT}\right) \tag{3}$$

So the reaction rate, *K,* from Equations (2) and (3) can be outlined as:

$$K = B\sinh(aS)\exp\left(-\frac{E_a}{kT}\right) \tag{4}$$

where *B* is constant

If the reaction rate in Equation (4) takes a reverse function, the generalized stress model can be obtained as

$$TF = A[\sinh(aS)]^{-1}\exp\left(\frac{E_a}{kT}\right) \tag{5}$$

The hyperbolic sine stress term in Equation (5) increases the stress as follows: (1) S^{-1} initially in low stress effect, (2) S^{-n} in medium stress effect, and (3) $(e^{as})^{-1}$ in high stress effect. Because accelerated testing will be performed in the medium range, Equation (5) is expressed as follows:

$$TF = A(S)^{-n}\exp\left(\frac{E_a}{kT}\right) \tag{6}$$

As the compressor stress is hard to be quantified in accelerated testing, we need to alter Equation (6). The stress of mechanical system comes from the acting force. Especially, we know that the compressor stress may obtain from the pressure difference between suction pressure, P_{suc}, and discharge pressure, P_{dis} in the refrigerating cycle (Woo and O'Neal, 2006). For a compressor system, the time-to-failure, TF, can be expressed as

$$TF = A(S)^{-n}\exp\left(\frac{E_a}{kT}\right) = A(F)^{-\lambda}\exp\left(\frac{E_a}{kT}\right) = A(\Delta P)^{-\lambda}\exp\left(\frac{E_a}{kT}\right) \tag{7}$$

The accelerated factor (AF) from Equation (7) can be redefined as

$$AF = \left(\frac{S_1}{S_0}\right)^{n}\left[\frac{E_a}{k}\left(\frac{1}{T_0}-\frac{1}{T_1}\right)\right] = \left(\frac{F_1}{F_0}\right)^{\lambda}\left[\frac{E_a}{k}\left(\frac{1}{T_0}-\frac{1}{T_1}\right)\right] = \left(\frac{\Delta P_1}{\Delta P_0}\right)^{\lambda}\left[\frac{E_a}{k}\left(\frac{1}{T_0}-\frac{1}{T_1}\right)\right] \tag{8}$$

To get the mission cycle of a parametric ALT from the objective BX lifetime on the test plan, the sample size equation combined with acceleration factors might be derived as (Woo *et al.*, 2020):

$$n \geq (r+1)\cdot\frac{1}{x}\cdot\left(\frac{L_{BX}^{*}}{AF\cdot h_a}\right)^{\beta} + r \tag{9}$$

The lifetime of a refrigerator compressor is targeted to have B1 life 10 years. If the objective number of life cycles L_{BX} and AF are given, the mission time can be acquired from Equation (9). The ALT equipment can then be used in accordance with the work of the refrigeration cycle (Figure 3).

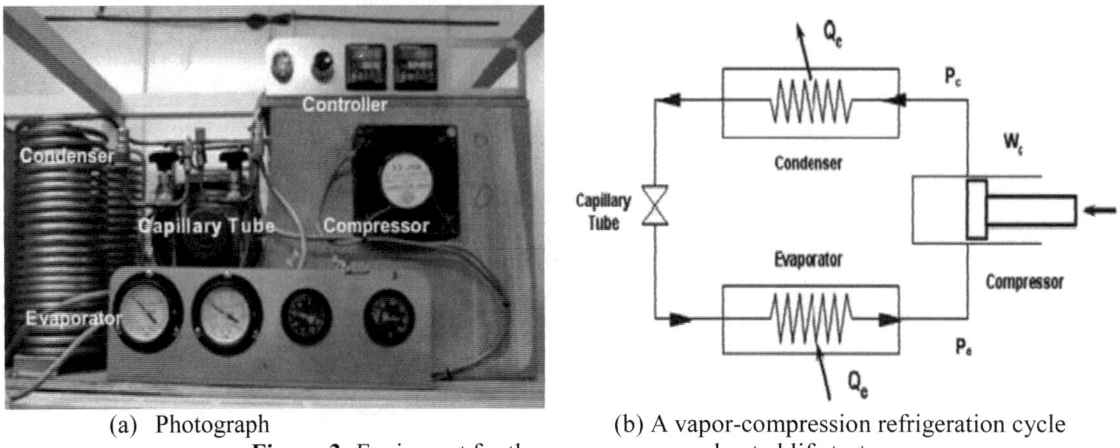

(a) Photograph (b) A vapor-compression refrigeration cycle

Figure 3: Equipment for the compressor accelerated life tests

In parametric ALTs, the design flaws of the mechanical system can be recognized to fulfill the lifetime target (Woo *et al.*, 2008&2009&2010).

2.4 Case Study: Improving the noise of mechanical compressor subjected to repetitive pressure loading

To improve the energy efficiency of a refrigerator, a reciprocating compressor with newly designed rotor and stator was redesigned (Figure 4a). In the field, the refrigerators had been making noise, causing the consumer to request replacement of them (Figure 4b). When the vibration in the problematic refrigerator was measured, the peak vibration level shown in the chart was 0.35g. Because the vibration specifications requirements were for less than 0.2g, the refrigerators with the newly designed compressor did not meet the specifications. Upon closer inspection, we found interference marks on the compressor shell for the failed compressor. We suspected that these marks came from a possible design flaw in the compressor, i.e., the newly designed stator frame. When the compressor operates, the piston moved back and forth within the cylinder assembly that was directly connected to the stator frame. It also supported the piston-cylinder assembly. The cylinder assembly and the stator frame could interfere with the compressor shell because it was not properly designed with enough clearance. To decide the root cause(s) of the failure of this refrigerator experimentally, it was necessary to be reproduced and corrected by a proper ALT. So the compressor might be robustly designed to work under a wide span of consumer usage conditions (Figure 5).

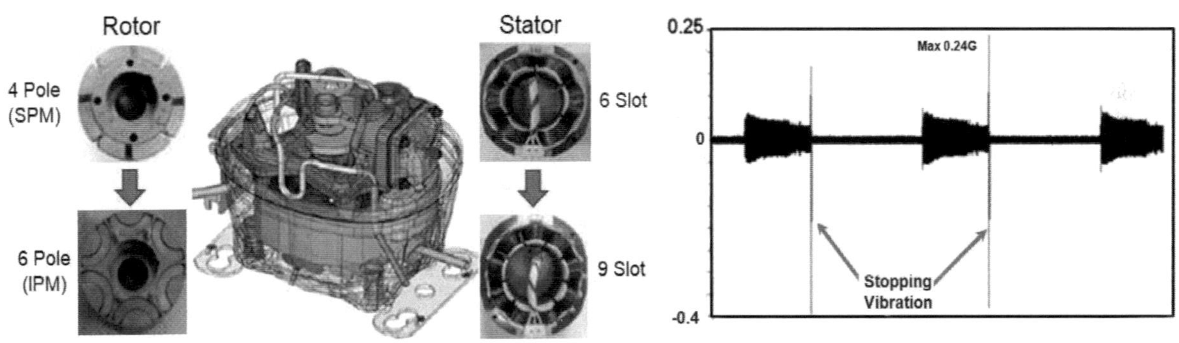

(a) Reciprocating compressor with redesigned rotor and stator (b) A noised (or vibrated) refrigerator

Figure 4: Reciprocating compressor and its noise

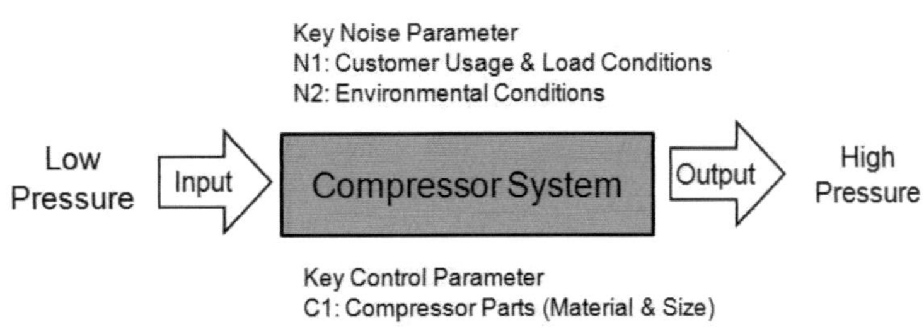

Figure 5: Parameter diagram of compressor system

To assess the ride quality of the new designed stator/rotor frame and piston-cylinder assembly, the most practical mathematical model for a vehicle suspension system installed on compressor shell was a quarter car model (Jazar, 2008). Although quarter car model was two degree of freedom (DOF) and four state variables, it served the purpose of determining the vehicle motion mounted on compressor. For all values of the forcing frequency, the presumed model of the vehicle consisted of the sprung mass and the un-sprung mass, separately. The sprung mass, m_s, represents 1/4 of the body of the vehicle, i.e., the piston-cylinder assembly and stator frame. The un-sprung mass, m_{us}, represents the wheel of the vehicle, i.e., the rotor-stator assembly. The cucial suspension was modeled as a spring k_s and a damper c_s in parallel, which attaches the un-sprung to the sprung mass. The suspension spring on compressor shell was modeled as a spring k_{us} and represented the transfer of the road force to the un-sprung mass (Figure 6)

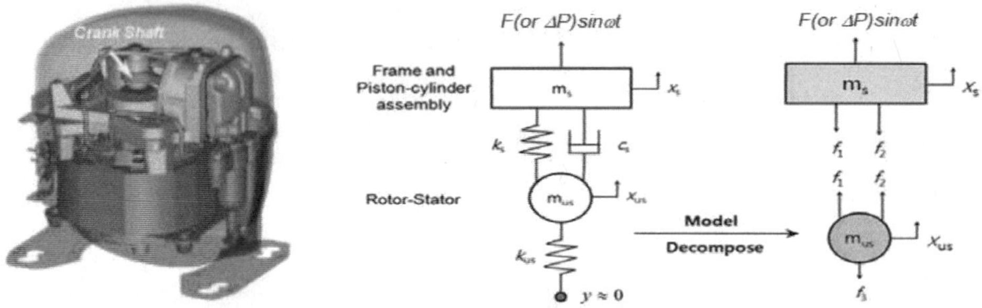

(a) Reciprocating compressors (b) A quarter car model of compressor assembly and its decomposition

Figure 6: Compressor assembly and its quarter car model

The governing differential equations of motion for the quarter car model can be expressed as:

$$m_s\ddot{x}_s + c_s(\dot{x}_s - \dot{x}_{us}) + k_s(x_s - x_{us}) = F(or\ \Delta P)\sin\omega t \tag{10}$$

$$m_{us}\ddot{x}_{us} + c_s(\dot{x}_{us} - \dot{x}_s) + (k_{us} + k_s)x_{us} - k_s x_s = 0 \tag{11}$$

So the above equations of motion can be briefly expressed as:

$$\begin{bmatrix} m_s & 0 \\ 0 & m_{us} \end{bmatrix}\begin{bmatrix} \ddot{x}_s \\ \ddot{x}_{us} \end{bmatrix} + \begin{bmatrix} c_s & -c_s \\ -c_s & c_s \end{bmatrix}\begin{bmatrix} \dot{x}_s \\ \dot{x}_{us} \end{bmatrix} + \begin{bmatrix} k_s & -k_s \\ -k_s & k_{us}+k_s \end{bmatrix}\begin{bmatrix} x_s \\ x_{us} \end{bmatrix} = \begin{bmatrix} F(or\ \Delta P)\sin\omega t \\ 0 \end{bmatrix} \tag{12}$$

As a result, Equation (12) can be represented in a matrix form

$$[M]\ddot{X} + [C]\dot{X} + [K]X = F(or\ \Delta P)e^{i\omega t} \tag{13}$$

When Equation (13) was numerically integrated, we obtained the time response of the state variables (displacement, velocity, and acceleration) due to the compressor operation, i.e., excitation forces (or pressure difference) *F(or ΔP)*$e^{i\omega t}$. And the force transmissibility (Q) for all values of the forcing frequency was defined as (Rao, 2017):

$$Q = \frac{F_T}{kY} = r^2\left[\frac{1+(2\zeta r)^2}{(1-r^2)^2+(2\zeta r)^2}\right]^{1/2} \approx \frac{r^2}{(r^2-1)} \quad \text{for small values of damping ratio} \tag{14}$$

The acceleration factor (AF) from Equation (8) is expressed as the ratio between the normal condition and the accelerated condition. When most accelerated testing is carried out in room temperature, AF can be defined as the multiplication of the excitation forces (or pressure difference) and force transmissibility Q. In other words,

$$AF = \left(\frac{S_1}{S_0}\right)^n = \left(\frac{F_1}{F_0}\right)^\lambda = \left(\frac{\Delta P_1}{\Delta P_0}\frac{F_T}{kY}\right)^\lambda = (R\times Q)^\lambda \approx \left(\frac{\Delta P_1}{\Delta P_0}\right)^\lambda \quad \text{for } r \geq 2 \tag{15}$$

For the worst case, the differential pressure anticipated in the compressor would be an operating pressure of 0.38MPa. For the accelerated testing, the differential pressure was increased to 1.35MPa. With a cumulative damage exponent, λ, of 2, the accelerated factor (AF) was roughly 12.6 utilizing Equation (13) (Table 1).

Table 1: ALT conditions in refrigeration cycle for refrigerant R600a

System Conditions		**Worst Case**	**ALT**	**AF**
Pressure (MPa)	High-side	0.40	1.39	12.6
	Low-side	0.02	0.04	
	ΔP	0.38	1.35	

The lifetime target is put to have more than the B1 life 10 years. The reciprocating compressor works cycles on and off between 22 and 74 times per day. With product life cycles for 10 years, L_B, the compressor is expected to have up to 270,100 usage cycles. The assumed shape parameter β was 2.0. The actual test cycles computed in Equation (9) were 21,400 cycles for a sample size of 100 units. If the parametric ALT for the compressor system fails less than once during 21,400 cycles, it would be reassured to have a B1 life 10 years with about a 60% level of confidence.

3. RESUTS AND DISCUSSION

We found two samples with noise and vibrations at 100 cycles in the 1st ALT (Figure 7). The failure sites from the marketplace and the first ALT happened in the upper shell of compressor. Figure 8 is a graphical analysis if the ALT results and market data show on a Weibull distribution. For the shape parameter, the approximated value on the chart was 2.0. For the last design, the shape parameter was confirmed to be 1.9.

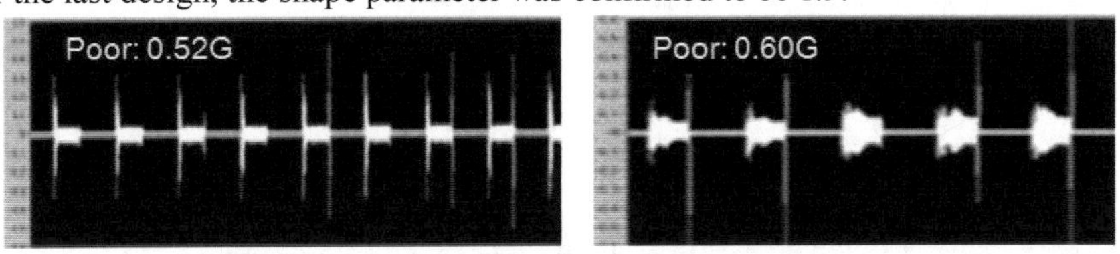

(a) Vibration of failed samples

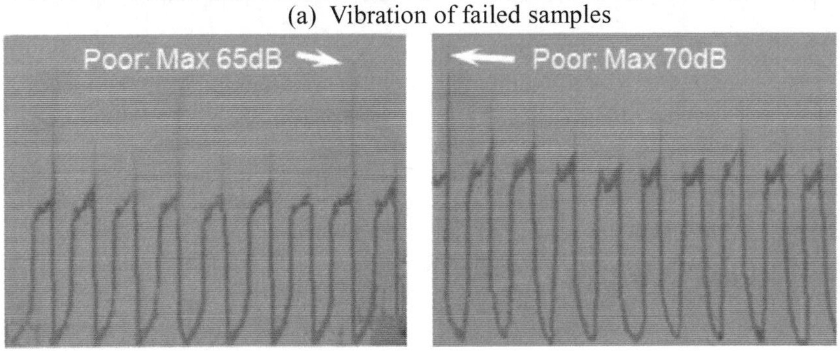

(b) Noise of failed samples

Figure 7: The vibration and noise of failed compressors in 1st ALT

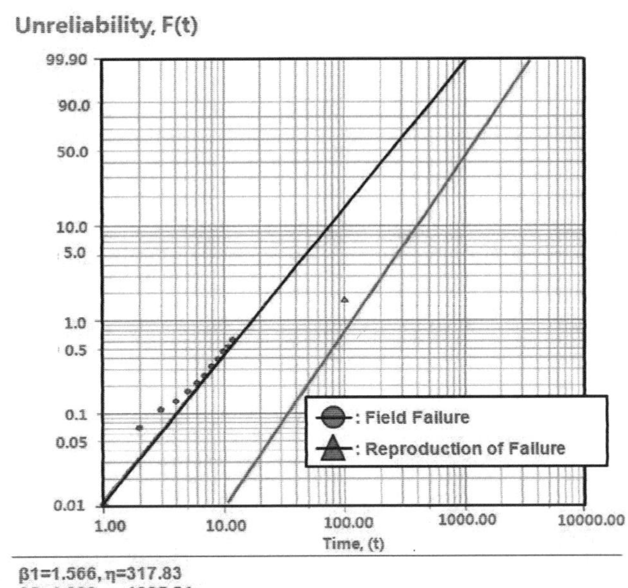

Figure 8: Market data and 1st ALT on Weibull chart

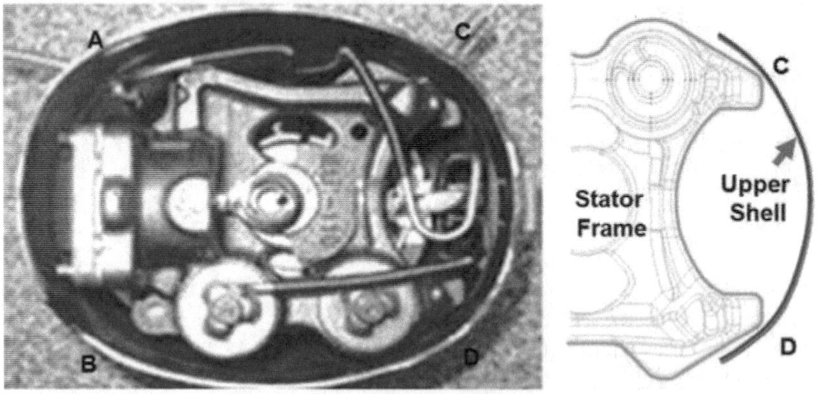

(a) Stator frame mounted in compressor

A	B	C	D
8.5	9.1	3.2	2.9

(b) Gap between frame and upper shell (Spec.: 6mm↑)

Figure 9: Structure of noised compressor in field

As seen in Figure 7 and Figure 9, the noise and vibration in the first ALT originated from the interference between the stator frame and the compressor upper shell. The gap between the frame and the shell was calculated to be approximately 2.9 mm even though the design specification was 6mm.

To prevent the noise and vibration of the compressor, the stator frame in compressor was redesigned and the gap increased from 2.9 mm to 7.5 mm (Figure 10). As the design flaw was rectified, the parametric design basis of the newly designed samples was shown to have more than the lifetime target - B1 life 10 years. The affirmed value, β, on the Weibull chart was 1.9. For the second ALT with sample size 100 units, the real test cycles in Equation (9) were 21,400. In the second ALT, there were no design issues in the compressor system until the test was performed to 25,000 cycles. We thus deduced that the altered design parameters obtained from 1st ALT were successful.

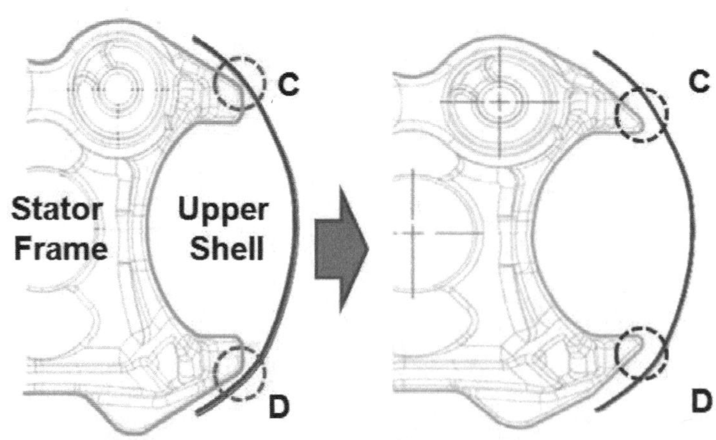

(a) Modified stator frame mounted in compressor

	A	B	C	D
Old Part	8.5	9.1	3.2	2.9
New Part	8.3	8.8	7.7	7.5

(b) Final gap between frame and upper shell (Spec.: 6mm↑)

Figure 10: Improved structure of noised compressor in field

Table 2 abridges the ALT results. With the altered design parameters, the last samples of the compressor system were reassured to have the lifetime target – B1 life 10 years.

Table 2: Results of ALTs

Parametric ALT	1st ALT	2nd ALT
	Initial Design	Final Design
In 21,400 cycles, there is no noise and vibration	100 cycles: 2/100 noise	21,400 cycles: 100/100 OK 25,000 cycles: 100/100 OK
Photo	<Compressor> 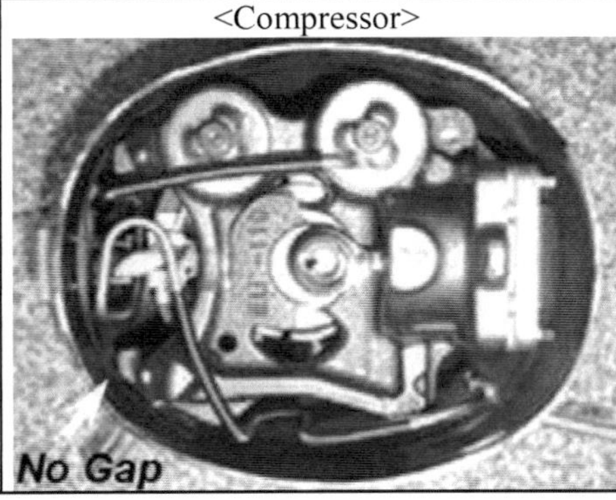	-

4. CONCLUSION

We suggested the reliability design methodologies that can find and modify the problems of mechanical system such as the noise of a newly designed reciprocating compressor in French-door refrigerator. It consists of the inspection of the failed products from the marketplace, a parametric ALT plan formed on BX lifetime, load examination, parametric ALTs with design alternatives, and assessment of the compressor design to assure the objective BX lifetime is fulfilled. As an experiment instance, newly designed compressor returned from field was studied. At that time, the lifetime of compressor was targeted to have a B1 life 10 years.

In the field and the 1st ALT, the compressors were noisy and vibrated because of a design flaw – there was interference between the stator frame and the compressor upper shell. By reshaping the stator frame in the compressor, the problematic compressor system was corrected. As a result of this altered design, there were no issues in the 2nd ALT. Consequently, the newly designed compressors were expected to have the lifetime requirement of B1 life 10 years. Through the inspection of returned products in field, load examination, and parametric ALTs with action plans, we knew that the study of the missing design parameters of compressor system was successful in redesigning more reliable parts with notably longer life.

This systematic reliability method might be relevant to other mechanical systems such as airplane, automobiles, refrigerators, washing machines, construction equipment, etc. To utilize this systematic reliability method, engineers should comprehend why mechanical products such as compressor fail. That is, if there is a design flaw in the structure that causes insufficient stiffness (or strength) when the loads are exerted, the product will fail during its lifetime. Engineers also need to recognize the load attributes of a product so that the parametric ALT can be conducted to meet the required number of mission cycles under accelerated conditions. Eventually, engineers can use parametric ALT to recognize and modify the design problems of a mechanical system such as compressor.

NOMENCLATURE

a	distance between (silicon) atoms
BX	time which will be X percent of the accumulated failure rate, durability index
E_a	activation energy, 0.56 eV
$F(t)$	unreliability
h	testing cycles (or cycles)
k	Boltzmann's constant, 8.62×10^{-5} eVdeg-1
L_B	target BX life and $x = 0.01X$, on the condition that $x \leq 0.2$
n	number of test samples
ΔP	pressure difference, MPa

P	pressure, MPa
r	failed numbers
r	frequency ratio (=ω/ω_n)
S	stress
t_i	test time for each sample
T	absolute temperature, K
X	accumulated failure rate, %
x	$x = 0.01 \cdot X$, on condition that $x \leq 0.2$

Greek symbols

η	characteristic life
λ	cumulative damage exponent in Palmgren-Miner's rule
ζ	damping ratio (=c/c_c)
ξ	applied field
α	confidence level

Superscripts

β	shape parameter in a Weibull distribution
n	stress dependence, $n = -\left[\dfrac{\partial \ln(T_f)}{\partial \ln(S)}\right]_T$

Subscripts

0	normal stress conditions
1	accelerated stress conditions
a	actual

REFERENCES

Grove, A. (1967). *Physics and Technology of Semiconductor Device*. 1st ed., New York, NY: Wiley International Edition.

Hrennikoff, A. (1941). Solution of problems of elasticity by the framework method, *Journal of Applied Mechanics*, 8, 169–175.

Jazar, R. (2008). *Vehicle Dynamics: Theory and Application*. New York: Springer, pp. 931-975.

Klutke, G., Kiessler, P.C., Wortman, M.A. (2015). A critical look at the bathtub curve, *IEEE Transactions on Reliability*, *52 (1)*, 125–129.

Montgomery, D. (2013). *Design and Analysis of Experiments*, 8th Ed., Hoboken, NJ: John Wiley.

Rao, S. S. (2017). *Mechanical Vibration*. 6th ed., New York: Pearson Prentice Hall.

Sonntag, R.E., Borgnakke, C. (2007). *Introduction to Engineering Thermodynamics*. York, PA: John Wiley & Sons, Inc,.

Taguchi, G. (1978). Off-line and on-line quality control systems. *Proceedings of the International Conference on Quality Control*, Tokyo, Japan: JUSE.

Woo, S., Pecht, M., O'Neal, D. (2020). Reliability design and case study of the domestic compressor subjected to repetitive internal stresses, *Reliability Engineering & System Safety*, *193(C)*, 106604.

Woo, S., Pecht, M. (2008). Failure analysis and redesign of a helix upper dispenser, *Engineering Failure Analysis*, 15(4), 642-53.

Woo, S., O'Neal, D., Pecht, M. (2009). Design of a hinge kit system in a Kimchi refrigerator receiving repetitive stresses, *Engineering Failure Analysis*, 16 (5), 1655–1665.

Woo, S., O'Neal, D., Pecht, M. (2010). Failure Analysis and Redesign of the Evaporator Tubing in a Kimchi Refrigerator, *Engineering Failure Analysis*, 17 (2), 369–379.

Woo, S., O'Neal, D. (2006). Reliability Design of the Newly Designed Reciprocating Compressor for a Refrigerator, *Proceedings of International Compressor Engineering Conference*. Purdue, USA: Purdue University.

Experimental Investigation of an Optimized Mechanically Assisted Suction Reed Valve Design Concept in a Small Hermetic Reciprocating Compressor

Andreas EGGER[1]*, Raimund ALMBAUER[1], Michael LANG[1], Christoph ZAINER[1]

[1]Graz University of Technology, Institute of Internal Combustion Engines and Thermodynamics,
Inffeldgasse 19, 8010 Graz, Austria
egger@ivt.tugraz.at
almbauer@ivt.tugraz.at
michael.lang@ivt.tugraz.at
zainer@ivt.tugraz.at

* Corresponding Author

ABSTRACT

Reed valves are widely used in hermetic reciprocating compressors for domestic refrigeration. They are crucial components in terms of efficiency, cooling performance, noise and reliability of the compressor. While reed valves already cause a significant proportion of the thermodynamic losses in fixed speed compressors, they induce even more challenges in variable speed compressors. Especially in variable speed compressors, a further improvement of the reed valve dynamics requires the consideration of a new valve concept.
In this work, a new design concept of a mechanically assisted suction valve is experimentally investigated. A mechanically actuated spiral spring generates a variable supporting force on the surface of a conventional reed valve. Experiments are conducted over a wide compressor speed range to determine how the new valve design contributes to the thermodynamic requirements of a variable speed compressor, which are high efficiency from low to medium compressor speed and high cooling capacity at high compressor speed. In addition, acoustic measurements will show the influence of the new design concept on noise, vibration and suction gas pulsation of the compressor.

1 INTRODUCTION

The refrigeration sector (including air conditioning) is estimated to require approximately 20 % of the global electrical energy consumption (IIR, 2019). Up to now, most of this electrical energy is produced by burning fossil fuels, which is associated with the emission of the greenhouse gas carbon dioxide. Making refrigeration systems more energy efficient would thus have a considerable influence on the reduction of the anthropogenic greenhouse effect and thereby help to meet the Paris Agreement from 2016.

In domestic refrigeration, mainly fixed speed hermetic reciprocating compressors are used to drive the refrigeration cycle. Variable speed operation is one effective approach to increase the energy efficiency of the entire refrigeration system. However, variable speed operation is associated with challenges concerning the reed valve dynamics. Reed valves are characterized by a simple and inexpensive design and by their automatic adaption to different operating conditions, as they are pressure actuated only. Reed valves are crucial compressor components in terms of efficiency, cooling capacity, noise and reliability. The main sources of thermodynamic losses are opening delays due to oil sticking and inertia effects, intermediate valve closings (valve flutter) and gas backflow due to valve closing delays.

Great efforts have been made in the past to analyze the relationship between reed valve dynamics and the compressor efficiency. The suction reed valve in particular shows a high flutter intensity, since the suction phase requires almost half a revolution of the crank shaft. Bhakta et al. (2012) conducted a sensitivity analysis of different suction valve parameters using a 1d compressor model. They emphasized the importance of reduced valve flutter for the reduction of valve losses. Similar findings were obtained by (Burgstaller et al., 2008). They investigated the potential of a theoretical valve without any flutter. The suction power consumption decreased by 36 % leading to a COP increase of 2 %. In variable speed operation, the suction time and discharge time varies with the compressor speed while the valve flutter frequency remains almost constant (Tan et al., 2014; Wang et al., 2016; Wang et al., 2012). This may lead to gas backflow due to delayed valve closure and thereby reduce the volumetric efficiency of the compressor (Nagata et al., 2010; Tao et al., 2018). Unlike the discharge valve, the suction valve lift is not limited by a stopper. Thus, increasing compressor speed leads to increased valve lifts and higher valve impact velocities, potentially provoking serious reliability issues.

Further improvements of the valve dynamics require the consideration of new valve concepts. An ideal suction reed valve opens fast, remains open during the entire suction phase and finally closes timely with a low valve impact velocity. Nevertheless, the automatic adaption to different operating conditions must be maintained. A first experimental attempt in this direction was made by Hopfgartner et al. (2017). The authors attempted to reduce the suction losses of a hermetic reciprocating compressor by using a negative preloaded suction reed valve in combination with an electromagnetic coil to support valve closure. Calorimeter measurements showed a COP improvement potential of up to 1.7 %. However, the COP improvement was strongly influenced by the operating condition and the energy consumption of the electromagnetic coil was not considered in the calculation. A different valve concept was proposed by (Egger et al., 2019). They investigated a mechanically assisted suction reed valve (MASV). Preliminary experimental investigations indicated a favorable behavior of the valve dynamics and calorimeter measurements at different operating conditions showed an average COP improvement of 1.9 %. Although the study has proven the functionality of the new valve concept, the proposed design is too costly and therefore not suitable for serial application. Therefore, Egger et al. (2020) proposed a new and cost-effective design of a MASV. The authors optimized the design using a simulation based multi-response optimization approach. In addition, some predicted improvements have already been confirmed in preliminary measurements of COP, cooling capacity and valve dynamics.

The present study provides an experimental investigation of the new and cost-effective MASV design. Not only calorimeter measurements, but also noise, vibration and suction gas pulsation measurements under various operating conditions provide a comprehensive insight into how the proposed MASV design affects the performance of a hermetic reciprocating compressor.

2 MECHANICALLY ASSISTED SUCTION REED VALVE

Figure 1 shows the components of the MASV. Although the design of the mechanism has been optimized for a small hermetic reciprocating compressor for domestic refrigeration, the principle can be applied to any other reciprocating compressor. The principle is based on an additional supporting force which acts on the conventional suction reed

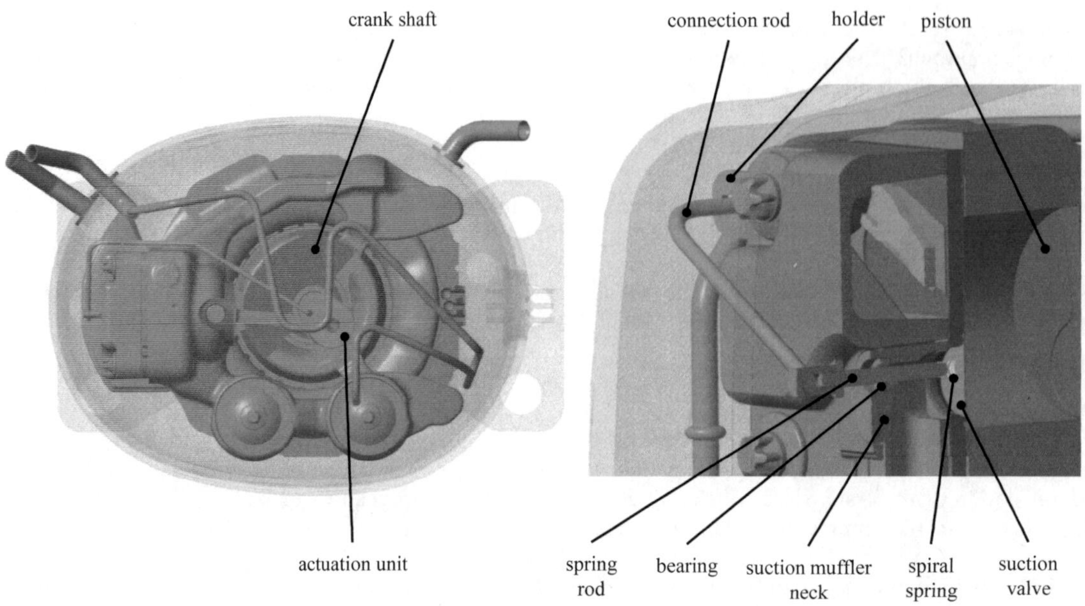

Figure 1: Components of mechanically assisted suction reed valve

valve, while the automatic adaption of the valve to different operating conditions is maintained. The magnitude of the supporting force must vary to ensure that the valve opens quickly, remains open during the entire suction phase and finally closes timely and with a low impact velocity. The valve support mechanism basically consists of a mechanical actuation unit, a connection rod and a spiral spring. The actuation unit drives the connection rod through an eccentric bore and thus leads to a sinusoidal motion of the spiral spring in valve opening direction. The compression of the spiral spring generates a variable force on the surface of the suction valve. The illustrated mechanism was designed in a way that enables fast changes of individual design parameters for experimental purposes. For example, the insert of the actuation unit can be changed in order to modify the amplitude and phase angle shift of the actuation. Furthermore, the distance offset between spiral spring and suction valve can be adjusted. No offset leads to a valve support duration of 180 °ca, while positive or negative offsets decrease or increase the support duration. Once all parameters are set, the mechanism design can be further simplified, e.g. the connection rod and the spring rod can be combined in one component.

3 EXPERIMENTAL APPROACH

The goal of this experimental investigation is to determine the influence of a new MASV design on the performance values i.e. COP, cooling capacity and acoustics over a wide compressor speed range and different operating conditions. The MASV design was selected on the basis of the investigations in (Egger et al., 2020), where several design variants are proposed. The underlying investigation is based on a MASV design which is composed of a valve support mechanism and the standard suction reed valve. Although this design variant does not lead to the best results over the entire compressor speed range, results similar to the optimal design variant are expected from low to medium compressor speeds (Egger et al., 2020) while, at the same time, requiring fewer modifications to the experimental unit.

3.1 Experimental Unit

The suction valve support mechanism was assembled on a commercially available hermetic reciprocating compressor for R600a with 9.6 cm^3 displacement and a cooling capacity of 171 W (ASHRAE test conditions, 220 V, 50 Hz). Figure 2 shows the experimental unit (without the upper shell cover) which was used for the experimental investigations. The most important design parameters of the tested valve support mechanism are shown in Table 1. The suction reed valve design remained the same in each measurement.

Figure 2: Hermetic reciprocating compressor equipped with a MASV (without upper shell cover)

Table 1: Design parameter of valve support mechanism

rod diameter (mm)	actuation amplitude (mm)	actuation phase angle shift [1)] (°ca)	spiral spring offset [2)] (mm)	spiral spring stiffness (N/mm)
2.0	1.5	0	0	2.0

[1)] crank angle offset to tdc at first contact between spiral spring and suction valve (having zero spiral spring offset)
[2)] distance offset between spiral spring and suction valve at tdc (having zero actuation phase angle shift)

3.2 Calorimeter measurements

A R600a secondary fluid calorimeter which complies with (ASHRAE, 2005) was used to determine the performance data i.e. cooling capacity, electrical power consumption and coefficient of performance (COP) of the compressor. The calorimeter automatically ensures stabilized operating conditions via an internal control unit. However, small deviations from the set point – still within the limits specified in (ASHRAE, 2005) – cannot be avoided. To account for this source of uncertainty, the measurement results were time averaged over 30 minutes measuring time. Furthermore, each operating condition was replicated three times. In addition to the standard compressor, two different variants based on the valve support mechanism in Table 1 were tested, see Table 2. Table 3 shows the different test conditions. Two different operating conditions in terms of evaporating and condensing temperature were tested at nominal compressor speeds starting from 2000 rpm up to 4500 rpm. The ambient temperature was kept constant at 32 °C. Based on the measurement results of the cooling capacity and electrical power consumption of the compressor, the COP is calculated according to equation 1. The results obtained with the MASV are given as relative change compared to the standard compressor, e.g. the COP change ΔCOP is calculated according to equation 2.

Table 2: Tested variants colorimeter measurements

variant name	MASV	suction bore diameter
standard	no	standard
var. C1	yes (see Table 1)	standard
var. C2	yes (see Table 1)	+ 7.5 %

Table 3: Test conditions calorimeter measurements

condition name (-)	evaporating temperature (°C)	condensing temperature (°C)	compressor speed (rpm)
-23.3°C/45°C	-23.3	45	2000 – 4500 (500rpm steps)
-23.3°C/55°C	-23.3	55	2000 – 4500 (500rpm steps)

$$COP = \frac{\dot{Q}_0}{P_{el,comp}} \quad (1)$$

$$\Delta COP_{variant} = 100 * \frac{COP_{variant} - COP_{standard}}{COP_{standard}} \quad (2)$$

3.3 Acoustic measurements

Since the compressor is designed for domestic refrigerators and freezers, the acoustic behavior is another important quality criterion. In order to quantify the influence of the MASV on the acoustic behavior of the compressor, a total of three different variants as listed in Table 4 were tested with the compressor being operated in a reverberation room. In addition to the measurements of standard valve and MASV, a combination of both (var. A2) was tested to determine the acoustic influence of the small gap between the spring rod and the bearing in the suction muffler neck. Therefore, the spring rod was fixed in a position where no contact between the spiral spring and the suction valve could occur. A total of three different acoustic quantities were measured i.e. sound power level, shell vibration and suction gas pulsation. The sound power was determined according to DIN EN ISO 3741. Sound power and suction gas pulsation were measured simultaneously, while the shell vibration had to be measured separately, since the vibration sensors would influence the sound power level. All three measurements were conducted from 2000 rpm to 4500 rpm at two typical operating conditions for acoustic measurements, see Table 5.

Table 4: Tested variants acoustic measurements

variant name	MASV	suction bore diameter
standard	no	standard
var. A1	yes (see Table 1)	standard
var. A2	disabled [1)]	standard

[1)] to determine the acoustic influence of the small gap between the spring rod and the bearing in the suction muffler neck, the spring rod was fixed in a position where no contact between the spiral spring and the suction valve could occur.

Table 5: Test conditions acoustic measurements

condition name (-)	evaporating temperature (°C)	condensing temperature (°C)	compressor speed (rpm)
-25°C/55°C (LBP)	-25	55	2000 – 4500 (500rpm steps)
-10°C/40°C (MBP)	-10	40	2000 – 4500 (500rpm steps)

4 RESULTS

4.1 Calorimeter measurements

This chapter outlines the calorimeter measurement results i.e. COP, cooling capacity and electrical power consumption between 2000 rpm and 4500 rpm at two different operating conditions. The results are given for var. C1 and var. C2 as relative change compared to the standard compressor according to equation 2.

Figure 3 and Figure 4 show the relative COP change related to the compressor speed. Except for 4500 rpm at the second operating conditions, significant COP improvements are visible at both operating conditions over the entire compressor speed range. The first operating condition which corresponds to a condensing temperature of 45 °C shows slightly higher COP improvements than the second operating condition with 55 °C condensing temperature. This could be attributed to the fact that the optimization was carried out for the first operating condition in (Egger et al., 2020). Var. A2 shows higher COP improvements than var. A1 which is attributed to the enlarged suction bore. Furthermore, the difference between var. A1 and var. A2 increase with increasing compressor speed, which is plausible because the flow losses through the suction bore become more dominant at higher flow speeds. The largest COP improvement is about 3.6 % and can be observed at the first operating condition at 4000 rpm.

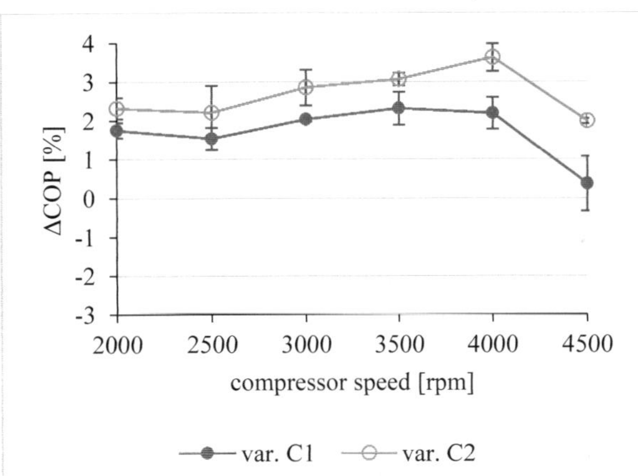

Figure 3: Change of COP over compressor speed at first operating condition (T_{evap} = - 23.3 °C, T_{cond} = 45 °C, T_{amb} = 32 °C, R600a)

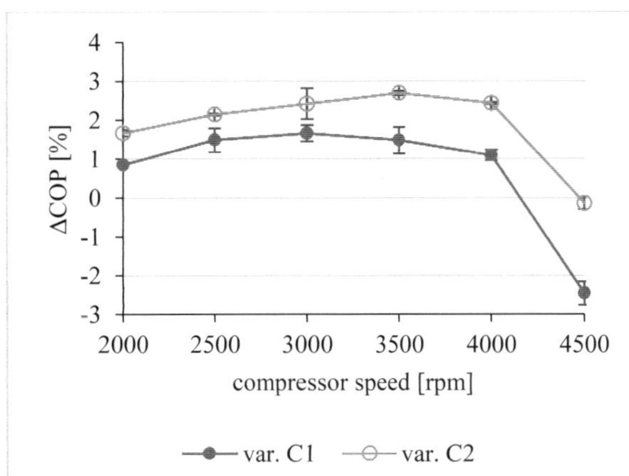

Figure 4: Change of COP over compressor speed at second operating condition (T_{evap} = - 23.3 °C, T_{cond} = 55 °C, T_{amb} = 32 °C, R600a)

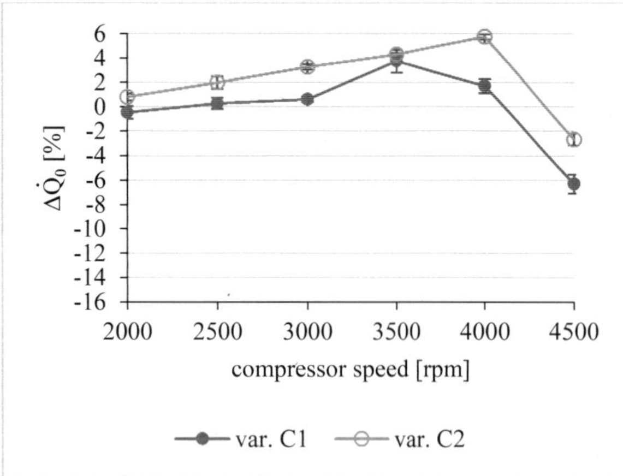

Figure 5: Change of cooling capacity at over compressor speed at first operating condition (T_{evap} = - 23.3 °C, T_{cond} = 45 °C, T_{amb} = 32 °C, R600a)

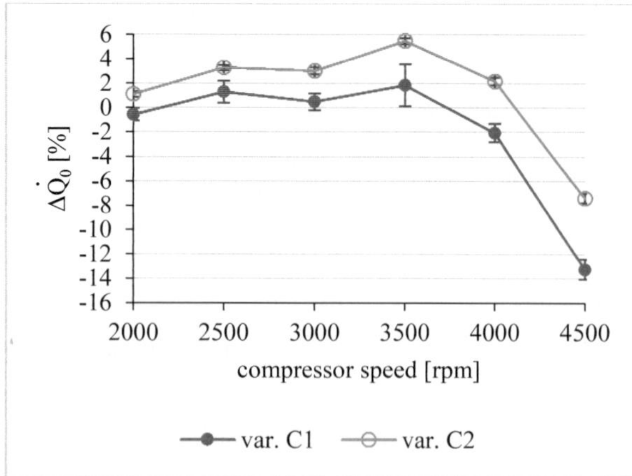

Figure 6: Change of cooling capacity over compressor speed at second operating condition (T_{evap} = - 23.3 °C, T_{cond} = 55 °C, T_{amb} = 32 °C, R600a)

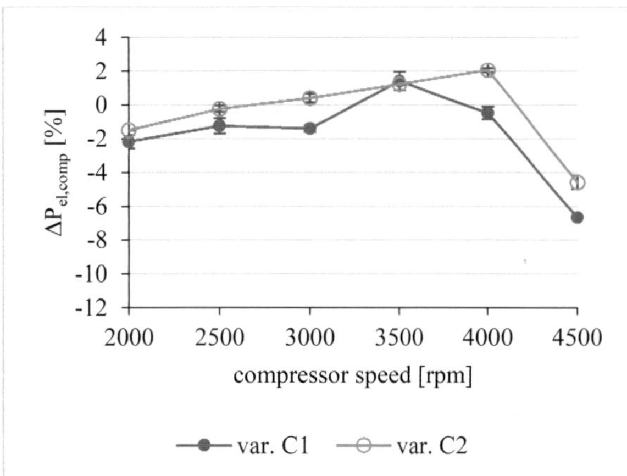

Figure 7: Change of electrical power consumption over compressor speed at first operating condition (T_{evap} = - 23.3 °C, T_{cond} = 45 °C, T_{amb} = 32 °C, R600a)

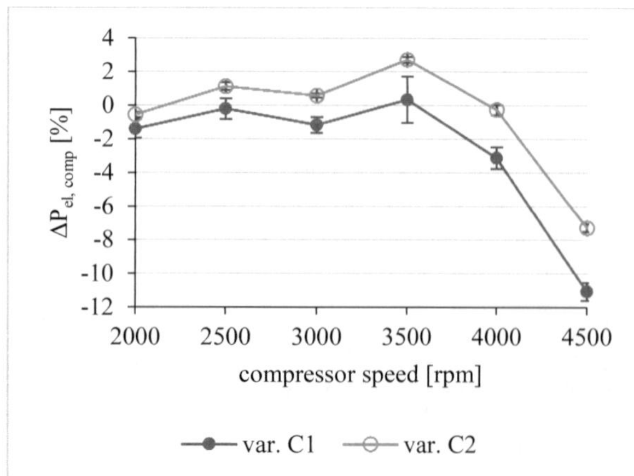

Figure 8: Change of electrical power consumption over compressor speed at second operating condition (T_{evap} = - 23.3 °C, T_{cond} = 55 °C, T_{amb} = 32 °C, R600a)

Figure 5 and Figure 6 illustrate the relative change of the cooling capacity related to the compressor speed. In general, the cooling capacity curves of both variants show a similar trend as the corresponding COP curves in Figure 3 and Figure 4. This is not surprising as the cooling capacity is directly included in the calculation of the COP according to equation 1. At the first operating condition, var. C2 shows a continuous increase in cooling capacity improvement from 2000 rpm up to 4000 rpm, with a maximum improvement of 5.8 % at 4000 rpm. At maximum compressor speed however, var. C1 and var. C2 show lower cooling capacities than the standard compressor. This is in accordance with the findings in (Egger et al., 2020), where the simulation indicated weak valve dynamics at high compressor speeds if the valve support mechanism is combined with the standard suction reed valve.

Figure 7 and Figure 8 show the relative change of the electrical power consumption of the compressor related to the compressor speed. The electrical power consumption depends strongly on the mass flow and thus also on the cooling capacity of the compressor. An increase in cooling capacity is associated with an increase in electrical power consumption. The effect of the MASV is clearly visible when comparing the electrical power consumption in Figure 7 and Figure 8 with the cooling capacity in Figure 5 and Figure 6. Some results with cooling capacity changes close to zero nevertheless show significant reductions in electrical power consumption, e.g. var. C1 at first operating condition and 2500 rpm.

4.2 Acoustic measurements

a) Sound power level

Figure 9 and Figure 10 show the compressor sound power level in the frequency domain (one-third octave spectrum) at LBP and MBP at low (2000 rpm) and high (4500 rpm) compressor speed. In general, higher compressor speeds lead to higher sound power levels and smaller differences between the individual variants. At 2000 rpm, var. A1 shows significant higher sound power levels between 2500 Hz and 6300 Hz than var. A2 and the standard compressor. Since var. A2 has similar levels as the standard compressor, it is likely that the increased noise of var. A1 is caused by the mechanism movement and not by the small gap in the suction muffler neck. Figure 11 and Figure 12 show the total sound power level related to the compressor speed at LBP and MBP. It turns out, that at LBP, from low to medium compressor speeds, var. A1 leads to considerable higher total sound power levels than the standard compressor, which is in accordance with Figure 9. At MBP, however, the differences between the individual variants are small. Furthermore, at a given compressor speed, var. A1 shows smaller differences in sound power level between LBP and MBP than the standard compressor. In general, it can be said that the influence of the compressor speed on the sound power level is much greater than the influence of the MASV.

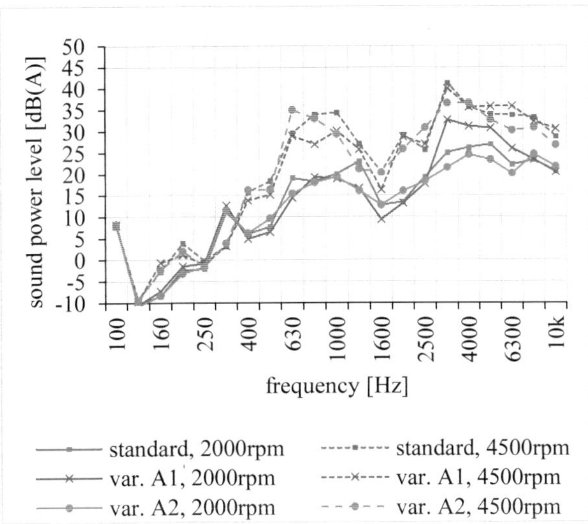

Figure 9: Compressor sound power level at LBP, 2000 rpm and 4500 rpm

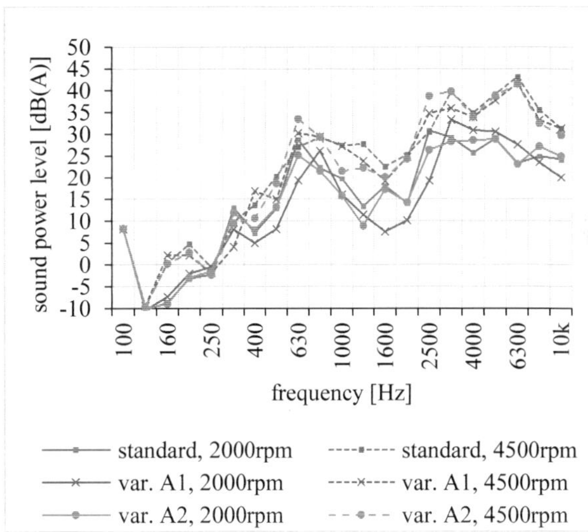

Figure 10: Compressor sound power level at MBP, 2000 rpm and 4500 rpm

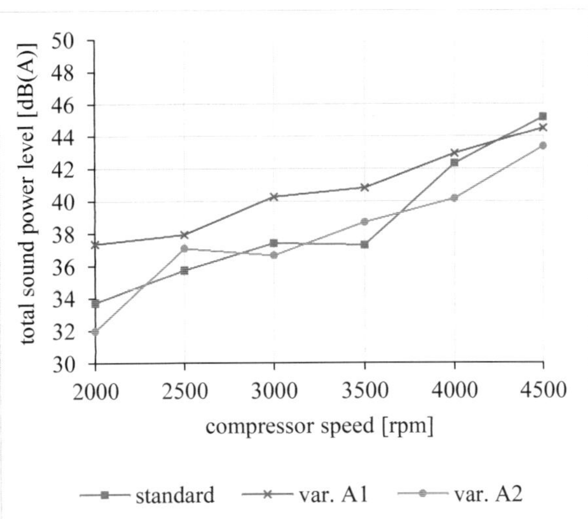

Figure 11: Total sound power level of compressor from 2000 rpm to 4500 rpm at LBP

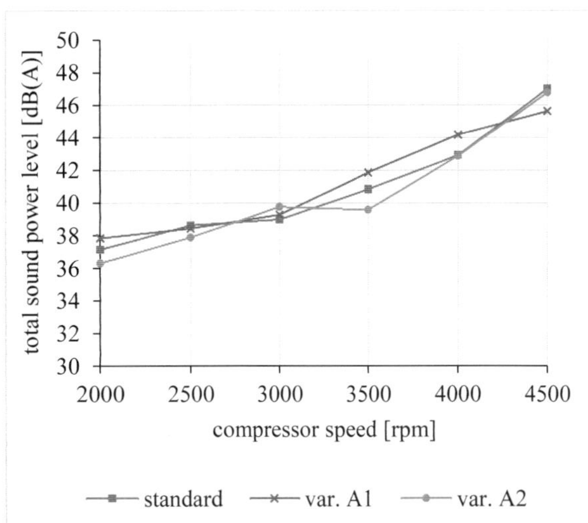

Figure 12: Total sound power level of compressor from 2000 rpm to 4500 rpm at MBP

b) Vibration

Figure 13 and Figure 14 show the compressor shell vibration in the frequency domain at LBP and MBP at low and high compressor speed. Both plots have a logarithmic ordinate due to the large differences in the size of the vibration values. The highest peaks occur near the operating frequency of the compressor, i.e. 33.3 Hz and 75 Hz for 2000 rpm and 4500 rpm respectively. No difference between the tested variants can be observed at these operation frequency related vibration peaks. The remaining frequency range of var. A1 is characterized by slightly higher and slightly lower vibration values compared to the standard compressor. The largest difference occurs at MBP and 2000 rpm near 630 Hz, where var. A1 has a significantly lower vibration value. However, considering the whole frequency spectrum, no general improvement or worsening of the MASV (var. A1) can be observed. The small gap in the suction muffler neck caused by the bearing of the spring rod seems to have a neglectable effect on the vibration, because var. A2 has similar vibration values as the standard compressor.

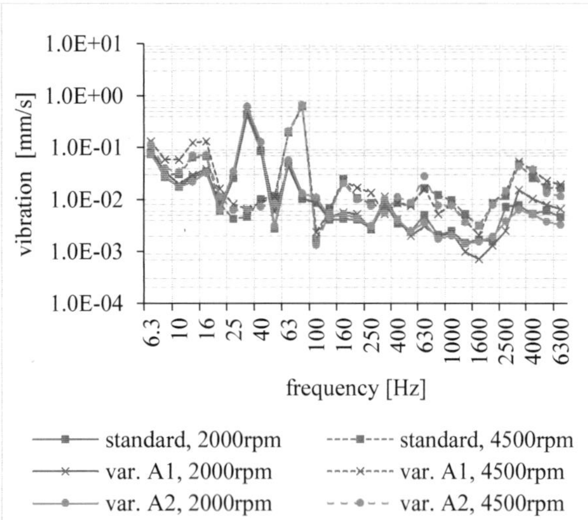

Figure 13: Shell vibration at LBP, 2000 rpm and 4500 rpm

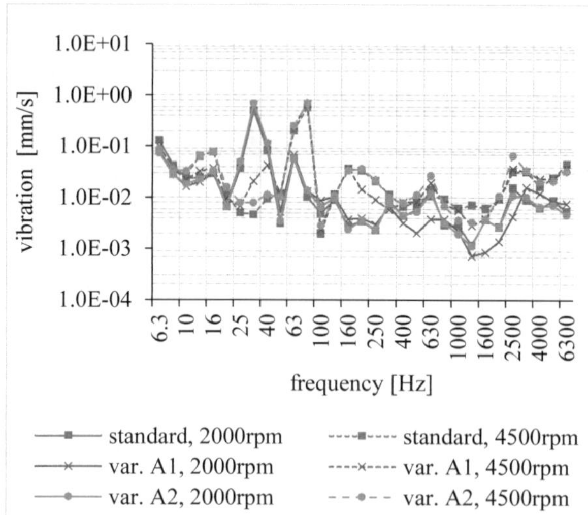

Figure 14: Shell vibration at MBP, 2000 rpm and 4500 rpm

c) Suction gas pulsation

Figure 15 and Figure 16 show the suction gas pulsation in the frequency domain at LBP and MBP at low and high compressor speed.

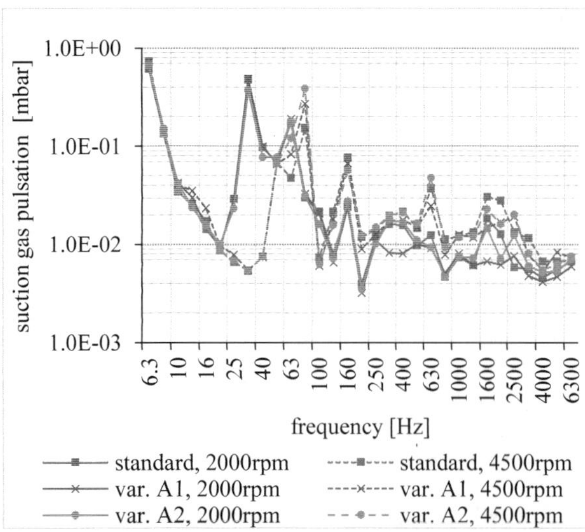

Figure 15: Suction gas pulsation at LBP, 2000 rpm and 4500 rpm

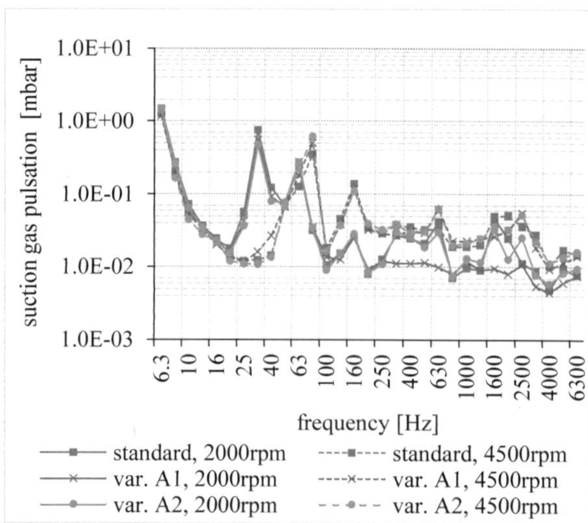

Figure 16: Suction gas pulsation at MBP, 2000 rpm and 4500 rpm

As with the shell vibration measurement, the largest peaks in suction gas pulsation occur at the operating frequency of the compressor. At 2000 rpm a significant reduction of the suction gas pulsation between 250 Hz and 4000 Hz can be observed at both operating conditions. At 4500 rpm the reduction is slightly less pronounced. This could be due to the fact that higher compressor speeds lead to fewer intermediate valve closings, as the suction time decreases while the valve flutter frequency remains the same. Also, with regard to suction gas pulsation, the small gap in the suction muffler neck seems to be neglectable.

5 CONCLUTION AND OUTLOOK

The influence of a new and cost-effective MASV design on the performance values of a hermetic reciprocating compressor has been experimentally investigated. Not only calorimeter measurements, but also acoustic measurements were conducted from 2000 rpm up to 4500 rpm at various operating conditions to determine the effect on COP, cooling capacity and electrical power consumption, noise, vibration and suction gas pulsation of the compressor.

Calorimeter measurements indicate a significant increase of the COP at almost every compressor speed and operating condition when using the MASV. The highest COP improvements, up to 3.6 %, are obtained if the MASV is combined with an enlargement of the suction bore. In the standard compressor, a suction bore enlargement is hardly feasible, as it usually increases the valve impact velocity, which is already close to the limit from reliability perspective. However, as proved in (Egger et al., 2020; Egger et al., 2019) a MASV significantly reduces the valve impact velocity, making a suction bore enlargement feasible. Not only the COP, but also the cooling capacity can be increased with the MASV. As with the COP, the highest cooling capacity improvements are obtained if the MASV is combined with a suction bore enlargement. From low to medium compressor speeds, the cooling capacity improvements tend to increase with increasing compressor speed. At 4500 rpm, however, a strong decline in cooling capacity can be observed at both tested operating conditions. In principle, the MASV concept would be able to improve COP and cooling capacity even at high compressor speeds, but this would require a suction reed valve with different valve parameters (Egger et al., 2020).

The acoustic measurements indicate that, when using a MASV, the sound power level of the compressor increases considerably at LBP from low to medium compressor speeds. This can be attributed to the motion of the valve support mechanism and most likely has its origin in the bearing points. At MBP, which is usually the noisier operating condition, the total sound power level of the compressor equipped with the MASV remains almost the same as at LBP. This is why at MBP no significant differences in total sound power level between standard and MASV can be observed. The shell vibration measurement indicated no fundamental changes when using the MASV, while the suction gas pulsation between 250 Hz and 4000 Hz at 2000 rpm was significantly reduced.

Future work will be focused on a further investigation of the noise coming from the valve support mechanism as well as a measurement-based Design of Experiment approach to determine the effect of different design factors on COP and cooling capacity of the compressor.

NOMENCLATURE

COP	coefficient of performance
ΔCOP	relative change of coefficient of performance
$P_{el,comp}$	electrical power consumption of compressor
$\Delta P_{el,comp}$	relative change of electrical power consumption of compressor
$\dot{Q}_0$	cooling capacity
$\Delta\dot{Q}_0$	relative change of cooling capacity

Abbreviations

amb	ambient
ca	crank angle
cond	condensing
evap	evaporating

LBP low back pressure
MASV mechanically assisted suction reed valve
MBP medium back pressure
rpm revolutions per minute
var variant
1d one-dimensional

ACKNOWLEDGEMENTS

This research was supported by the Austrian Research Promotion Agency (FFG) under the reference number 861517 and by Secop GmbH.

6 References

American Society of Heating, Refrigerating and Air-Conditioning Engineers, 2005. ANSI/ASHRAE Standard 23-2005: Methods of Testing for Rating Positive Displacement Refrigerant Compressors and Condensing Units. Atlanta, Ga.

Bhakta, A., Dhar, S., Bahadur, V., Angadi, S., Dey, S., 2012. A Valve Design Methodology For Improved Reciprocating Compressor Performance. International Compressor Engineering Conference.

Burgstaller, A., Nagy, D., Almbauer, R., Lang, W., 2008. Influence of the Main Parameters of the Suction Valve on the Overall Performance of a Small Hermetic Reciprocating Compressor. International Compressor Engineering Conference.

Egger, A., Almbauer, R., Dür, L., Hopfgartner, J., Lang, M., 2020. Multi-Response optimization applied to a mechanically assisted reed valve of a hermetic reciprocating compressor. International Journal of Refrigeration 119, 119–130. https://doi.org/10.1016/j.ijrefrig.2020.07.008.

Egger, A., Almbauer, R., Dür, L., Hopfgartner, J., Zuber, B., Stangl, S., 2019. Experimental Investigation of a Mechanically Assisted Suction Reed Valve in a Small Hermetic Reciprocating Compressor. IOP Conference Series: Materials Science and Engineering 604, 12018. https://doi.org/10.1088/1757-899x/604/1/012018.

Hopfgartner, J., Posch, S., Zuber, B., Almbauer, R., Krischan, K., Stangl, S., 2017. Reduction of the suction losses through reed valves in hermetic reciprocating compressors using a magnet coil. IOP Conference Series: Materials Science and Engineering 232, 12034. https://doi.org/10.1088/1757-899x/232/1/012034.

IIR, 2019. The Role of Refrigeration in the Global Economy. 38th Note on Refrigeration Technologies. https://iifiir.org/en/fridoc/142028.

Nagata, S., Nozaki, T., Akizawa, T., 2010. Analysis of Dynamic Behavior of Suction Valve Using Strain Gauge in Reciprocating Compressor. International Compressor Engineering Conference.

Tan, Q., Pan, S.-l., Feng, Q.-k., Yu, X.-l., Wang, Z.-l., 2014. Fluid–structure interaction model of dynamic behavior of the discharge valve in a rotary compressor. Proceedings of the Institution of Mechanical Engineers, Part E: Journal of Process Mechanical Engineering 229 (4), 280–289. https://doi.org/10.1177/0954408914530122.

Tao, W., Guo, Y., He, Z., Peng, X., 2018. Investigation on the delayed closure of the suction valve in the refrigerator compressor by FSI modeling. International Journal of Refrigeration 91, 111–121. https://doi.org/10.1016/j.ijrefrig.2018.05.004.

Wang, F., Mu, G., Guo, Q., 2016. Design optimization of compressor reed valve based on axiomatic design. International Journal of Refrigeration 72, 132–139. https://doi.org/10.1016/j.ijrefrig.2016.07.019.

Wang, Y., Feng, J., Zhang, B., Peng, X., 2012. Modeling the valve dynamics in a reciprocating compressor based on two-dimensional computational fluid dynamic numerical simulation. Proceedings of the Institution of Mechanical Engineers, Part E: Journal of Process Mechanical Engineering 227 (4), 295–308. https://doi.org/10.1177/0954408912465366.

Experimental and numerical analysis of reed valve movement in an impact fatigue test system and reciprocating compressors

Muhammad Waqas TOFIQUE[1]*, Alexander LÖF[1], Eugenio SCHILLACI[2], Pablo CASTRILLO[3], Joaquim RIGOLA[3]

[1]voestalpine Precision Strip AB, Research & Development,
Munkfors, Sweden

[2]Termofluids,
Av. Jacquard 97-E, 08227 Terrassa, Barcelona, Spain

[3]Universitat Politècnica de Catalunya - Barcelona Tech (UPC),
ESEIAAT, C/ Colom, 11 - 08222 Terrassa (Barcelona), Spain.

* Muhammad Waqas Tofique
E-mail: waqas.tofique@voestalpine.com

ABSTRACT

During operation of a reciprocating compressor, its flapper valve opens to allow the passage of gases and closes by striking against the valve impact plate. This reed valve movement and impact is repeated billions of times. This cyclic movement has a significant influence on the impact fatigue life of the reed valve and, hence, the lifetime of a compressor.

Inside a reciprocating compressor, a number of parameters including: the valve design, valve material, compressor operating frequency and suction/exhaust pressure, influence the reed valve movement. The valve movement can be defined in terms of valve frequency, valve lift, valve velocity and impact velocity. These response parameters heavily influence the compressor efficiency and impact fatigue life of reed valves.

In this paper, we first studied the valve movement parameters for a reed valve design using an experimental test setup. In experimental testing, the valves were excited into movement using air pressure pulses at 100 Hz frequency (air pulse width of 5 milliseconds). The valve movement was recorded by a laser sensor at 10 000 frames per second. The operating conditions such as the operating frequency, air pulse width, applied pressure and airflow rate were measured. The valves were not tested to failure but only to collect the dynamic data of valve movement such as the valve movement curve (valve displacement vs time), average valve lift and reed velocity data.

The experimental results were employed to validate a complex in-house 3D CFD finite-volume model aimed at studying in detail the valve and gas dynamics, whose interaction is solved by means of a Fluid-Structure Interaction (FSI) algorithm. The complete valve movement curve obtained from simulations – reed valve displacement vs time – correlated with that obtained from the experimental tests with small error. Similarly, the difference in experimental and numerically obtained average valve lift and impact velocity values were quite small for practical purposes.

Finally, fluid-dynamic results for pressure and impact forces were employed to feed a finite-element based code aimed at studying the structural behavior of the reed vale. Bending fatigue and impact fatigue stresses induced in the reed valve during its movement cycle were calculated. The magnitude of stresses and their positioning was determined, which correlated with the commonly observed areas of fracture initiation in these valves.

The numerical models, as well as the information obtained from this study will help the compressor manufacturers to design their compressors to enhance efficiency and reed valve's reliability. This type of information is not readily available from a working compressor.

1. INTRODUCTION

Significance of the reed valve movement can be judged by its influence on the impact fatigue and bending fatigue life of the reed valve. Long impact fatigue and bending fatigue lives are essential for the reliable operation of a compressor and, hence, the operation of any appliance that the compressor is installed in. In addition, the valve movement parameters, such as valve lift, affect the coefficient of performance (COP) of a compressor. On the other hand, higher valve lift would mean higher reed velocities due to the elastic restoring force, hence, higher impact intensities when the valve shuts down and strikes against the valve seat. The higher impact intensities would mean higher impact fatigue stresses are generated inside the reed valve body after the impact. Similarly, higher bending fatigue stresses would be generated in the reed valve body when the valve lift is higher. There have been numerous theoretical studies in the past such as Pandeya (1978) and Böswirth (1980) that discussed the dynamic behavior of a generic geometry of reed valves and the impact stress state generated due to the valve striking against the impact seat. However, a deeper analysis of the reed valve movement and the resultant stresses generated in the reed valve body is required to design safe and reliable compressors.

The interaction between a reed valve and the gas flowing through it can be characterized entirely by computational fluid dynamics (CFD), and, more specifically, by means of fluid structure interaction (FSI) algorithms. In addition, valve internal stresses due to external forces can be evaluated in detail by means of Computational Solid Dynamics (CSD).

A numerical set-up capable of fully simulating the behavior of the gas and its mutual interaction with a valve includes a certain number of numerical issues whose complexity makes the methodology not yet fully reliable. Some of the problems that must be solved concern: (1) the difficulty in mutually coupling a structural solver for studying the stresses operating in the valve with the external forces exerted by the fluid; (2) the representation of the impact between the valve and the base, with the corresponding 'disappearance' of control volumes which requires the use of numerical techniques such as re-meshing or immersed boundary for the representation of the solid; (3) detailed representation of all the physical phenomena that characterize the process, as flow topology, turbulent effects, backflow, fluttering and collisions, among others. The previously mentioned problems also introduce strict requirements from the point of view of computational cost, making such simulations expensive as well as complex. Some examples of resolution of valve dynamics by means of CFD and FSI techniques already exist in the literature. For instance, Mayer et al. (2014) and Tao (2018) carried out a validation of FSI computational methods by reproducing experimental results of reed valves. Furthermore, the FSI numerical prediction has allowed investigating the effect of gas velocity, valve thickness and other geometrical parameters on the process efficiency as shown by Coskun et al.(2016), Gasche et al. (2016) and Wu et al. (2016). The numerical method used in this work was developed on the Termofluids' in-house platform and verified for the framework of reed valves in Gonzalez et al. (2016) and González et al. (2019). The numerical method consists of a combination of a finite volumes solver for the resolution of the flow on an unstructured mesh with a solver for the valve movement. The interface between gas and valve is represented by a moving mesh while the forces are integrated on the valve through a normal mode superposition method with penalty forces to represent the impact between valve and base.

Regarding CSD, it is possible to find works in the literature that simulate and analyze valves using the Finite Element Method (FEM), in some cases commercial software are used such as in Yu et al. (2017), Lajús Junior et al. (2013) and Yu et al. (2018); while in others in-house codes are developed such as in Lee & Son (2008) and Nilsson et al. (1980). Generally, the displacement of the valve, the impact velocity of the valve when it strikes the seat and the impact stress are points of interest in those works. Regarding the impact stresses, in some cases the Von Mises stresses are studied, see Nilsson et al. (1980); and in others the principal stresses, see Yu et al. (2017) and Lajús Junior et al. (2013). In this work a code is developed using the FreeFEM ++ tool (Hecht, 2012) to analyze the stresses within the valve, how these stresses are modified by the impact with the seat and the characterization of the impact force when it appears. In this simulation, only the solid is studied using the pressure generated by the fluid as a boundary condition. Likewise, in this code the impact force is implemented using the penalty method (Armero & Petőcz, 1998).

In the current work, the parallel development of an experimental campaign together with the numerical one, with experiments designed ad-hoc and easily replicable, allow a sure advance towards maturity in the use of CFD-FSI methods for the study of valve dynamics, as well as interesting insights about valve dynamics and stresses distribution, carried out by means of CSD.

2. EXPERIMENTAL

Reed valve movement experiments were conducted on a custom-built impact fatigue test rig that, in normal operation, uses air pulses to produce movement of the valves at a range of frequencies (Hz) and pulse widths (milliseconds). The frequency, and pulse-width, is provided as input to the control software on the connected computer.

A schematic sketch of this custom-built impact fatigue test rig is shown in Figure 1. A dedicated compressor provides compressed air at up to 13 bars pressure. This compressed air gets stored in a storage tank. The compressed air is transferred through tubes of 4 mm inner diameter and passes through a flow meter that measures the airflow rate in liters per minute. The air pressure regulator regulates the magnitude of the compressed air pressure before it passes through a high frequency solenoid valve. There is a provision to increase or decrease the magnitude of pressure supplied to the solenoid valve, and hence the airflow rate, through the pressure regulator. The compressed air pressure reported in this paper is measured at the solenoid valve opening.

The reed valve was made to strike the valve plate repeatedly inducing impact fatigue stresses in them, hence, mimicking the movement of a reed valve in an actual compressor. A laser sensor measures the reed valve's displacement, and thus velocity, and its operating frequency at a sampling rate of 10000 per second. This data is displayed and stored on the control software.

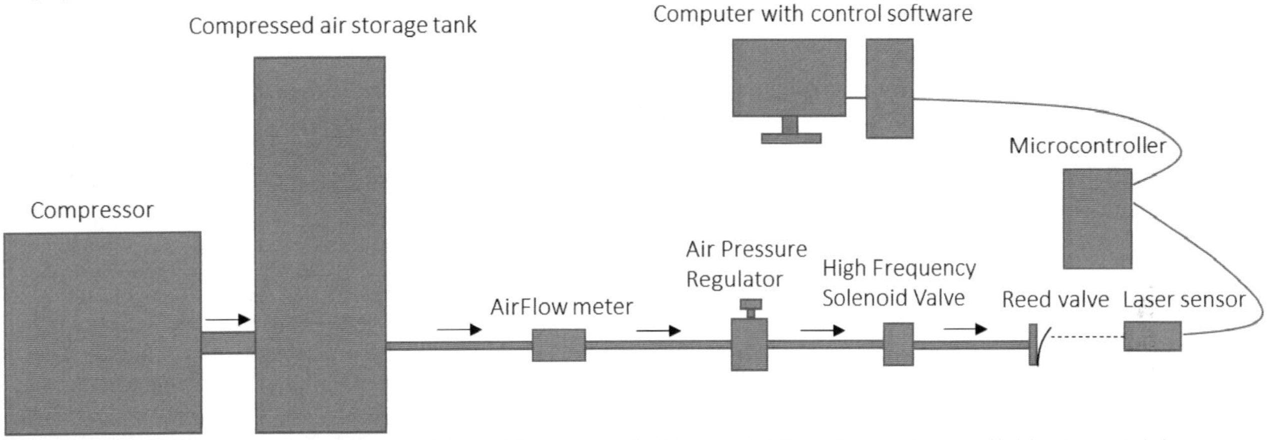

Figure 1: Schematic drawing of the working of the custom-built impact fatigue test system available at voestalpine Precision Strip AB, Sweden that was used to test the reed valve (test specimen) movement in this study.

In the custom-built impact fatigue test system, one crucial component that influences the reed valve movement is the high frequency solenoid valve. It is a two-way, in-line valve that is electrically actuated with moulded-in cable. Installed in the impact fatigue test system, it receives compressed air after the pressure regulator and through its fast switching action coverts it into pressurized air pulses of fixed frequency and pulse-width. These pressurized air pulses are then supplied to a reed valve/test specimen that moves in response. A schematic representation of the switching behavior of the solenoid valve for a trigger pulse-width of 5 milliseconds is shown in Figure 2. It shows that the solenoid valve takes about 1 millisecond to switch on as the electric current flows through the moulded-in wire and into the solenoid coil. As the solenoid valve opens the compressed air pushes through a small opening as the pressure starts to rise to reach its maximum value of 6 bars; the air pressure then stabilizes for approx. 3 milliseconds before the valve switches off; the valve switching off time delay is approx. 0.6 milliseconds; as the solenoid valve starts to close the pressure starts to reduce to its minimum value. This pulsing cycle was repeated a few thousand times as the reed valve movement data is recorded on the control software.

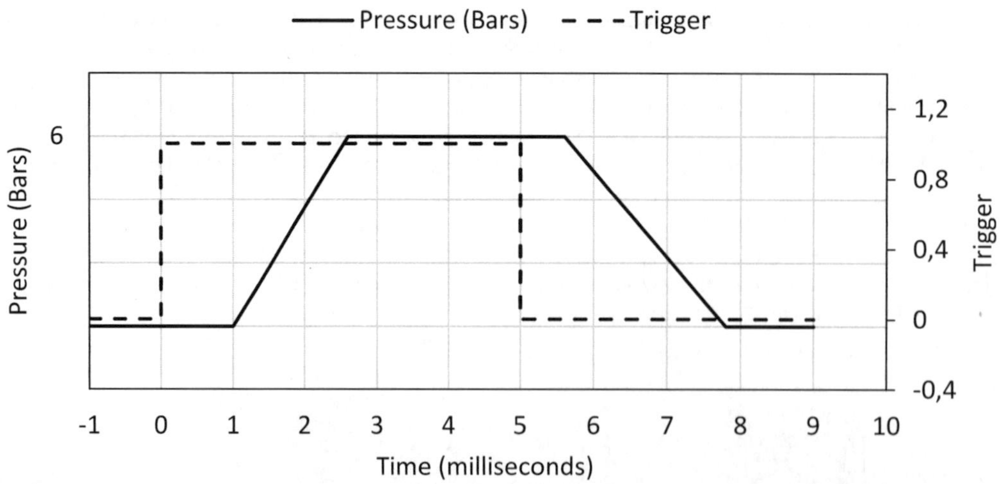

Figure 2: Schematic representation of the switching behavior of the high frequency solenoid valve provided by its manufacturer for compressed air pressure of 6 bars.

2.1 Test specimen design

The geometry of the reed valves tested in this work is presented in Figure 3. This reed valve design was chosen as it is a simple and efficient design that is quite common in the compressor industry. The reed valves were produced by photo-chemical etching process along the rolling direction. The reed valves were not subsequently tumbled. In the impact fatigue test system; the reed valve was mounted using a steel die that covered a part of its length. The measuring point of the laser detecting valve movement data was 2.5 mm from the top edge of the valve head along the centerline.

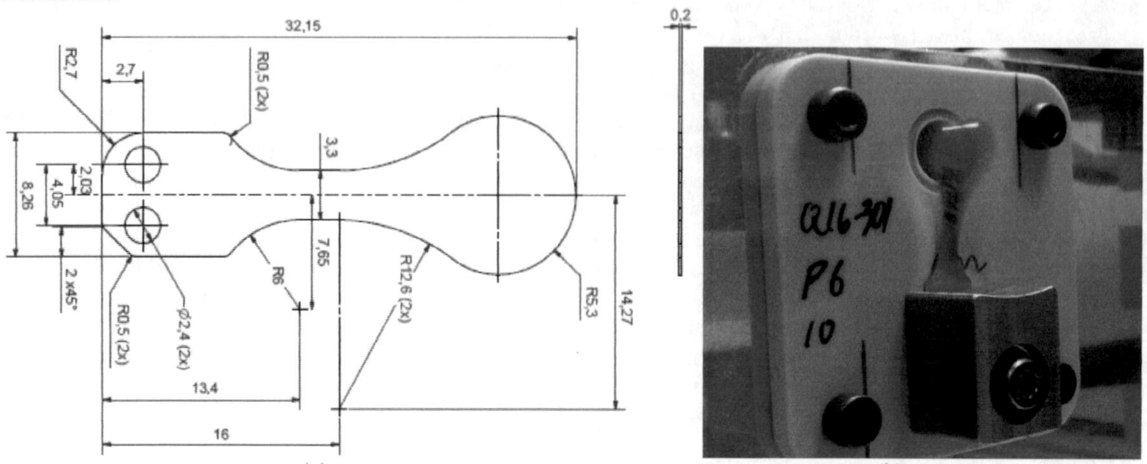

(a) (b)

Figure 3: a) The geometry and dimensions of the reed valve specimens used in this study, (b) the reed valve as tested in the impact fatigue test system.

The reed valves were manufactured from 0.203 mm thick strip of Flap-X steel grade. Flap-X is a martensitic stainless steel that is hardened and tempered to achieve a combination of high tensile strength, ductility and high cleanness that are vital for good impact fatigue and bending fatigue properties. The nominal chemical composition of Flap-X is shown in Table 1.

Table 1. Nominal chemical composition (wt. %) of the Flap-X steel grade used for the reed valve specimens

Steel grade	C	Si	Mn	Cr	Mo	P	S
Flap-X	0.38	0.45	0.55	13.5	1.00	≤0.025	≤0.015

3. NUMERICAL APPROACH

3.1 CFD-FSI methodology

The CFD problem consists in solving the Navier-Stokes equations in the incompressibility limit. In the model, a Newtonian behavior of the fluid and constant thermophysical properties was assumed. With these hypotheses, the governing equations can be written as follows.

$$\rho\frac{\partial\vec{v}}{\partial t}+\rho(\vec{v}\cdot\nabla)\vec{v}=\mu\nabla^2\vec{v}-\nabla p \tag{1}$$

The various magnitudes are listed here: $\vec{v}$ air velocity [m/s], t time [s], p air pressure [Pa], ρ air density [kg/m^3], μ air viscosity [Pa·s].] The compressed air is characterized by ρ= 8.5 kg/m^3, μ = 18.3e-6Pa·s. The numerical simulations described in this section are carried out with the unstructured parallel CFD code TermoFluids (see Termofluids webpage). Large-Eddy Simulations (LES) are carried out: a Sub-Grid Scale (SGS) model is applied in order to represent the smallest scales, while only the larger eddies are explicitly solved. The chosen SGS scheme is the Wall-adapting local eddy-viscosity (WALE) model. The Navier-Stokes equations are solved by means of a Fractional Step Method and advanced in time with an explicit scheme, yielding the air velocity field at each time step. The domain is represented in a simplified way in Figure 4(a), accounting for an inlet port where a variable inlet flow rate is imposed, and lateral outlets with pressure-based conditions are set to mimic the flow discharge. The other walls are solid walls boundaries. The base mesh is an unstructured tetrahedral mesh composed of around 300k cells and represented in Figure 4(c). At each time step, the mesh undergoes deformations to adjust the position of the internal points, thus, accommodating changes in the domain boundaries (upper and lower ones) without involving topological alterations. In order to avoid a remeshing procedure when valve is closed, some elements of the mesh perform as solid by using an immersed boundary method (IBM), as represented in Figure 4(b). The motion of the valve (i.e. the boundaries limiting its surface) is modeled by means of the Kirchoff-Love plate theory, valid for arbitrarily shaped plates with uniform thickness h. The displacement of this vibrating system can be obtained from a combination of M free vibration modes, as $u_z(x,y,t)=\sum_{m=1}^{M}q_m(t)\phi(x,y)$, where $\phi(x,y)$ and q_m are the normal deformation pattern and the generalized coordinates corresponding to the vibration mode m, respectively. Normal deformation patterns are obtained by means of a commercial software by solving the eigenvalue problem of the free vibration equation of the system while q_m are obtained from the Kirchoff-Love equation by applying the external stresses to the plate (fluid pressure and shear stress, gravity, impact force). More details about the whole CFD-FSI method are given in González et al. (2019). A semi-implicit approach for strongly coupled problems developed by Naseri et al. (2018) is employed to solve the fluid-solid interaction.

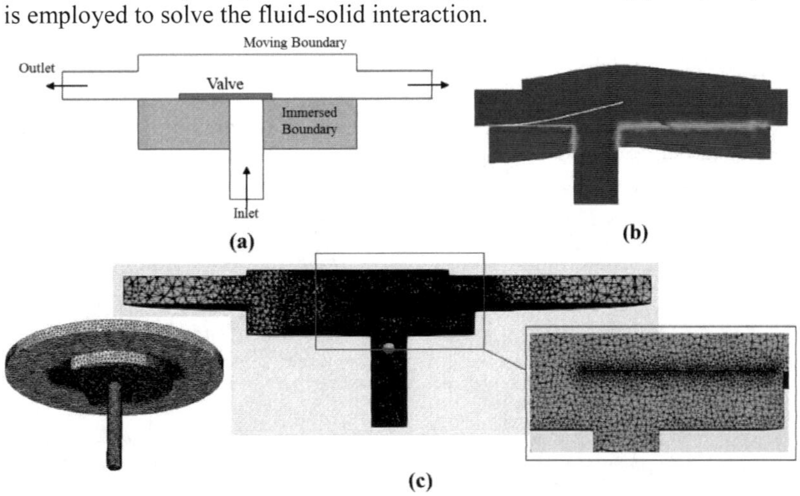

Figure 4: (a) Schematic of the CFD-FSI set-up; (b) Deformed domain with valve fully open; (c) unstructured mesh and detail around the valve boundaries.

3.2 Solid analysis methodology

The Cauchy moment equations are solved in order to study the solid dynamics and its internal stresses. In this case it is assumed that the material is homogeneous, has a linear elastic behavior and geometric non-linearity effects are not considered. Likewise, volume forces are not considered. With these hypotheses the equation reduces to:

$$\rho \frac{\partial^2 \vec{u}}{\partial t^2} = \nabla \cdot \boldsymbol{\sigma} \tag{2}$$

where $\rho = 7700\ \mathrm{kg/m^3}$ is the density, $\vec{u}$ is the displacement and $\boldsymbol{\sigma}$ is the Cauchy tensor. With the aforementioned hypotheses Cauchy tensor is related to the displacement by means of the constitutive equation $\boldsymbol{\sigma} = 2\mu\, \boldsymbol{\varepsilon} + \lambda\, \boldsymbol{\varepsilon}$ where $\boldsymbol{\varepsilon} = \frac{1}{2}(\nabla\vec{u} + \nabla\vec{u}^{\,T})$ is the infinitesimal strain tensor and μ and λ are the Lamé's material parameters related to the most common Young's modulus $E = 220264\ \mathrm{MPa}$ and the Poisson's ratio $\nu = 0.28$; $\left(\lambda = \frac{E\nu}{(1+\nu)(1-2\nu)} \text{ and } \mu = \frac{E}{2(1+\nu)}\right)$.

In order to solve this differential equation, boundary conditions are necessary. In that sense, it is known that the valve is held at one end (displacement conditions), while Neuman conditions are imposed on the other surfaces, in terms of fluid pressures and impact forces. For the fluid pressure $\vec{p}$, obtained from the CFD-FSI simulation, the boundary condition is $\boldsymbol{\sigma} \cdot \vec{n} = \vec{p}$ where $\vec{n}$ is the outward normal of the solid. The condition for the impact force is similar to the fluid pressure but it depends on the valve displacement that makes it a non-linear condition. To solve this problem it is necessary to solve the unilateral contact constraints and the persistency condition (Armero & Petőcz, 1998). For this formulation it is necessary to choose two penalty parameters k_p and m_p, in this work $k_p = 5000\ \mathrm{kPa/mm}$ and $m_p = 1000\ \mathrm{kg/mm^2}$. For the simulation, a code was developed in the FreeFEM++ tool using an HHT scheme (α-method) for the temporal integration with the parameters $\alpha = 0.51$, $\beta = 0.555025$ and $\gamma = 0.99$ (Armero & Petőcz, 1998).

4. RESULTS

4.1 Experimental results from the impact fatigue test system

In the impact fatigue test system, a frequency of 100 Hz and pulse-width of 5 ms was specified in the control software. This meant that compressed air pulses of 100 Hz frequency were created by the solenoid valve whose characteristic switching behavior is shown by the schematic representation in Figure 2, albeit for slightly higher maximum pressure of 7.2 bars.

The reed valve displacement that occurred in response to the compressed air pulses of 100 Hz is shown in Figure 5. As the solenoid valve was about to open the compressed air pressure started to mount that reached its maximum value when the solenoid valve was fully open and then reduced to lower pressure magnitude as the solenoid valve closed. In response to the applied compressed air pulses, the reed valve started to move away from the valve seat – see the valve seat position at approx. -3,90 mm indicated by a dashed line in Figure 5 to reach its maximum opening amplitude. The reed valve then closed due to elastic spring force as it acted against the compressed air, since the solenoid valve was still open. As the reed valve struck against the valve seat, it bounced up twice partly under the action of the compressed air. As another compressed air pulse arrived at the reed valve, this movement cycle was repeated again. However, it can be seen in Figure 5 that there are slight differences in the magnitude of the reed valve amplitude that occur due to local differences in air pressure, airflow rate and obliqueness of the reed valve during its movement – especially after it bounces back from the impact seat.

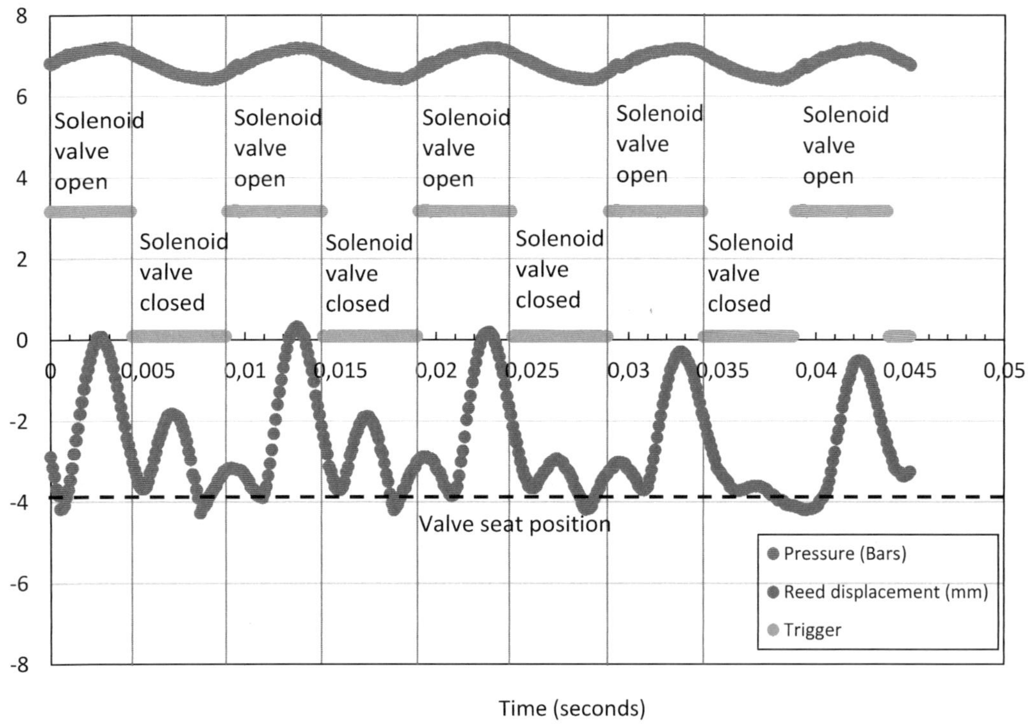

Figure 5: Reed valve displacement plotted as a function of time for input frequency of 100 Hz and pulse-width of 5 milliseconds specified to the solenoid high frequency valve.

4.2 Numerical simulation results

4.2.1 CFD-FSI results:

The study simulated the evolution of the flow and the movement of the valve during an entire valve stress cycle, which lasts 10 ms (Airflow pulse 100 Hz). The flow imposed at the entry of the domain mimics the air pulse of the experimental set-up. It grows rapidly and remains constant for a certain period, and then gradually decreases to 0 at t=4 ms from the beginning. The inlet flow rate evolution along a cycle is reported in Figure 6 (a). During the rest of the cycle, the valve opens and closes twice, rebounding against the base. The results in Figure 6 show a good agreement between numerical (CFD-FSI) and experimental data, both in terms of valve lift and displacement. This is especially true when considering that the experimental signal is a 'typical' one, as certain variability is visible between successive cycles, as visible in the displacement signal reported in Figure 5. Differences between the two results can be seen in the descent phase after the first opening. In fact, the numerical curve shows a deflection, probably due to the fact that the incompressible gas in the inlet duct acts as an obstacle during the valve closure. Another difference can be noticed (in the valve velocity curve in Figure 6b) at the point of the first impact: in the experimental signal a certain plateau can be noticed, probably due to the flexion of the valve when it comes in contact with the base, while in the simulation the contact is practically 'instantaneous' (sudden velocity increase).

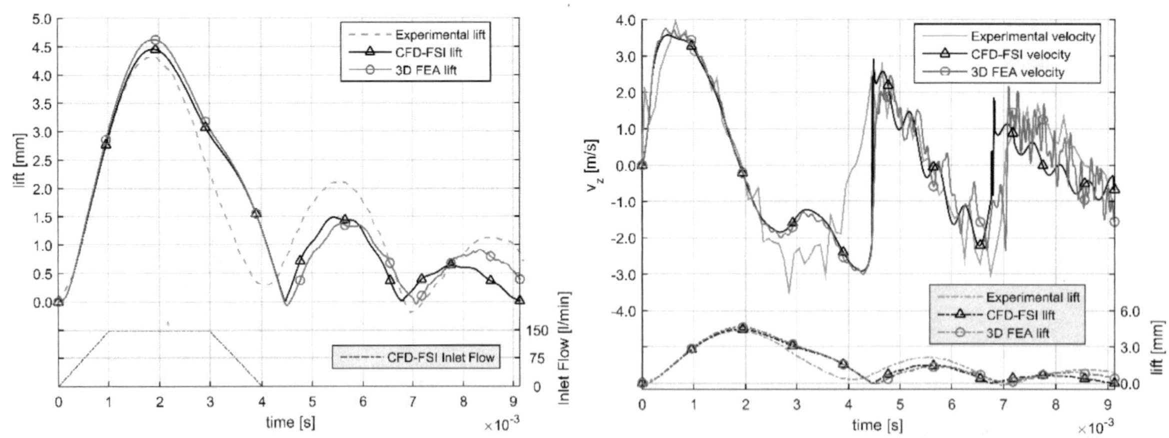

Figure 6: Numerical (CFD-FSI) vs experimental results for: (Left) valve lift, (Right) valve velocity.

4.2.2 Results of Finite Element Analysis:
The solid simulation in FreeFEM++ uses the pressure obtained from the FSI simulation carried out in TermoFluids. A verification of the outputs of the CSD method i.e. valve lift and velocity, also shown in Figure 6, are calculated. The first impact occurs in the time interval $t = (4.453, 4.579)\mathrm{e}^{-3}\mathrm{s}$ and the second impact at $t = (6.923, 7.122)\mathrm{e}^{-3}\mathrm{s}$. These time intervals correspond to the instants $(460 - 480)$ and $(768 - 790)$ of the simulation respectively. Figure 7 show the impact pressure Cp and the impact force FCp for each of these instants.

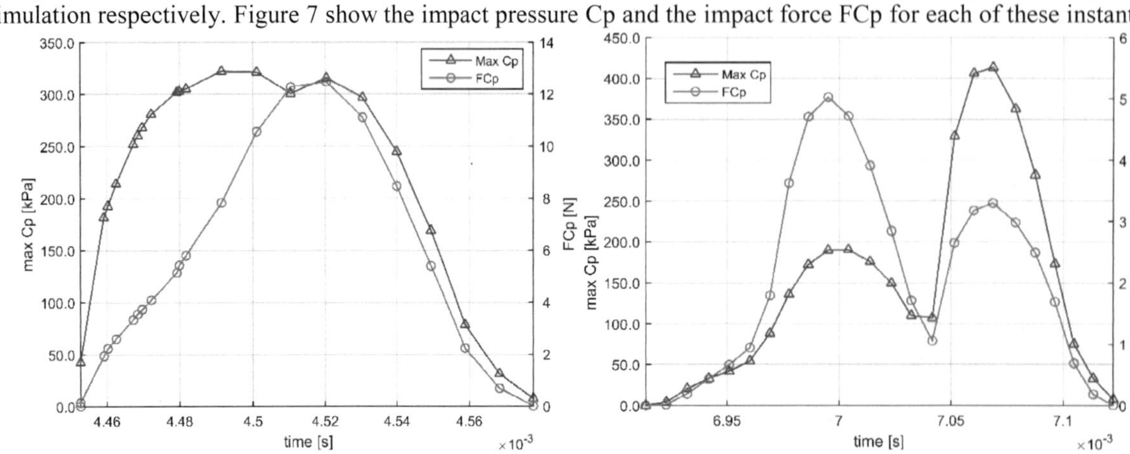

Figure 7: Maximum value of Cp and impact force FCp for each instant: (Left) first impact; (Right) second impact.

Before impact, the maximum stresses in the valve are due to its bending provoked by the fluid pressure and they can be found in the neck of the valve. Subsequently, due to the impact, it is observed that the maximum and minimum stresses are found at the tip of the valve increasing the Von Mises stress. Figures 8-9 show the Von Mises equivalent stress (σ_{vm}) in the lower fibres for both impacts.

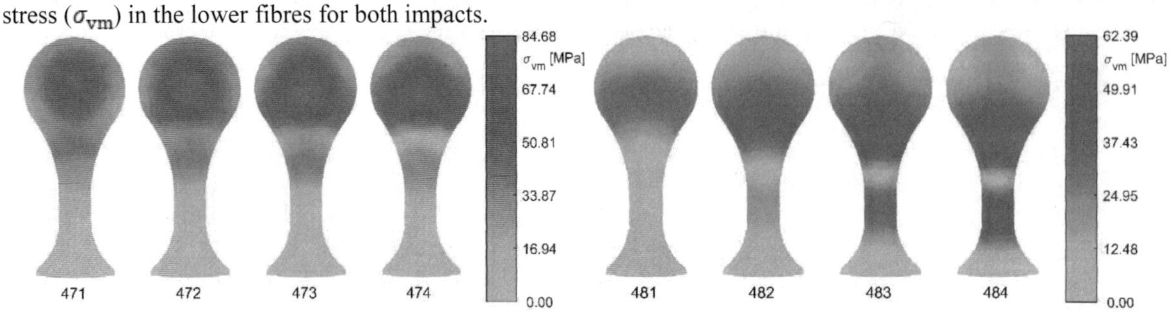

Figure 8: Von Mises stress in the first impact.

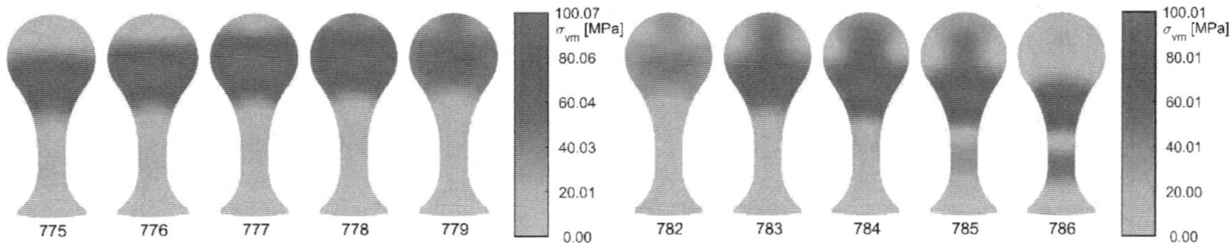

Figure 9: Von Mises stress in the second impact.

Four points on the valve tip were analysed to study the temporal evolution of the aforementioned stresses in the lower fibres of the valve. In Figure 10 it is possible to see that in all the points studied, the impact clearly generates a change in the stresses. Likewise, at the points 2 & 3 in Figure 10a where the valves usually fail – see Figure 10c – relatively higher stresses are noted, see Figure 10b.

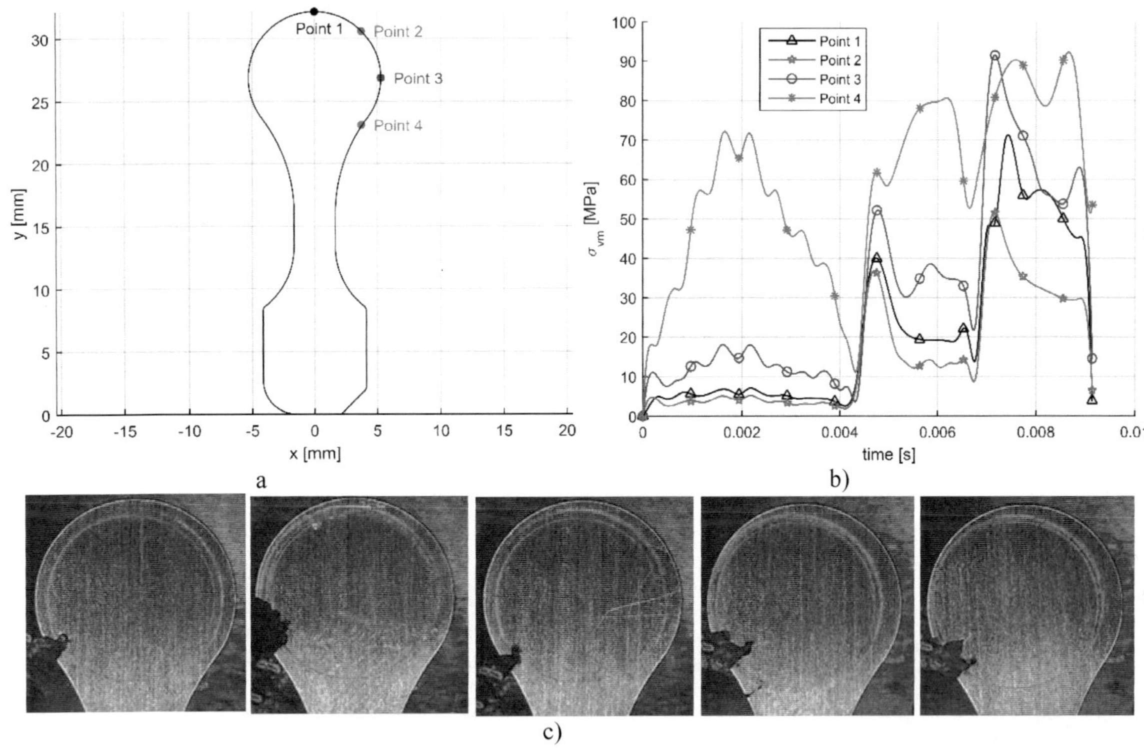

Figure 10: (a) Points studied at the valve tip; (b) Envelope of Von Mises stress σ_{vm}; c) failed reed valves in impact fatigue experiments.

6. CONCLUSIONS

Based on the work done in this study it can be concluded that the model delivered reasonably accurate results for the reed valve operating at 100 Hz frequency, taking into account the high variation in the experimentally measured results. The valve displacement and velocities obtained with the simulations correlate accurately enough for practical purposes with the experimental results. The obtained bending stress has a reasonable magnitude and is located in the expected area in the valve. For the impact stresses, it is seen qualitatively how the stresses increase significantly at the impact as the stresses travel in the valve body in the form of waves starting from its tip. The position of failures of the valves in the experimental results correlates with concentration of the higher magnitude of impact fatigue stresses in the valve after the point of time of the valve impact against the valve seat. The CFD and CSD models, in their current form, can be employed by compressor designers to study valve dynamics for other parameters such as valve thickness, valve design and physical properties.

ACKNOWLEDGEMENT

This research was made possible through funding and support by voestalpine Precision Strip AB, Sweden. Special thanks is extended to our research partners in this project at Termofluids and Universitat Politècnica de Catalunya - Barcelona Tech (UPC).

REFERENCES

Armero, F., & Petőcz, E. (1998). Formulation and analysis of conserving algorithms for frictionless dynamic contact/impact problems. *Computer Methods in Applied Mechanics and Engineering, 158*(3-4), 269-300.

Böswirth, L. (1980). Hypothesis on the failure of spring loaded compressor valve plates. *Proceedings of the International compressor engineering conference at Purdue, Purdue University (198-206), Paper 332.* Indiana, USA.

Coskun, U. C., Gunes, H., & Sarioglu, K. (2016). A numerical model of fluid structure interaction of a fluttering valve. *In Proceedings of the 5th International Conference on Jets, Wakes and Separated Flows (ICJWSF2015)* , 421-429.

Gasche, J. L., Dias, A. D., Bueno, D. D., & Lacerda, J. F. (2016). Numerical simulation of a suction valve: comparison between a 3D complete model and a 1D model. *International Compressor Engineering Conference (Indiana).*

Gonzalez, I., Lehmkuhl, O., Naseri, A., Rigola, J., & Oliva, A. (2016). Fluid-structure interaction of a reed type valve. *International Compressor Engineering Conference (Purdue).*

González, I., Naseri, A., Rigola, J., Pérez-Segarra, C. D., & Oliva, A. (2019). Detailed prediction of fluid-solid coupled phenomena of turbulent flow through reed valves. *In IOP Conference Series: Materials Science and Engineering*, Vol. 604, No. 1, p. 012064.

Hecht, F. (2012). New development in FreeFem++. *Journal of Numerical Mathematics , 20*(3-4), 251--265.

Lajús Junior, F. C., Deschamps, C. J., & Alves, M. (2013). Numerical analysis of seat impact of reed type valves. *8th International Conference on Compressors and their Systems*, 555-563.

Lee, Y., & Son, S. (2008). Study on the Fatigue Strength of a Suction Flapper Valve Used in a High Efficient Reciprocating Compressor. *International Compressor Engineering Conference*(1913).

Mayer, J., Bjerre, P., & Brune, F. (2014). A comparative study of different numerical models for flapper valve motion. *International Compressor Engineering Conference (Indiana).*

Naseri, A., Lehmkuhl, O., Gonzalez, I., Bartrons, E., Pérez-Segarra, C. D., & Oliva, A. (2018). A semi-implicit coupling technique for fluid–structure interaction problems with strong added-mass effect. *Journal of Fluids and Structures, 80, 94-112.*

Nilsson, J. O., Nilsson, L., & Oldenburg, M. (1980). Impact Stresses in Flapper Valves - A Finite Element Analysis. *International Compressor Engineering Conference*(364).

Pandeya, P. N. (1978). Analysis of the influence of seat-plating or cushioning on valve impact stresses in high speed compressors. *Proceedings of the International compressor engineering conference at Purdue, Purdue University.* Indiana, USA.

Tao, W., Guo, Y., He, Z., & Peng, X. (2018). Investigation on the delayed closure of the suction valve in the refrigerator compressor by FSI modeling. *International Journal of Refrigeration, 91, 111-121.*

Termofluids s.l., webpage: www.termofluids.com . (u.d.).

Wu, W., Jian, Z., Song, M., & Zhang, Z. (2016). Study of Valve Motion in Reciprocating Refrigerator Compressors based on the 3-D Fluid Structure Interaction Model. *International Compressor Engineering Conference (Indiana)*.

Yu, X., Ren, Y., Tan, Q., Lu, Z., Jia, X., & Wang, X. (2018). Study on the torsional movement of a reed valve in a rotary compressor. *Advances in Mechanical Engineering, 10*(6), 1-10.

Yu, X., Tan, Q., Ren, Y., Jia, X., & Jin, L. (2017). Numerical study of the reed valve impact in the rotary compressor by FSI model. *Energy Procedia, 105*, 4890 – 4897.

A New Alloy for High Performance Valve Steel for High Efficiency Compressors

Anders HOEL[1]*, Guocai CHAI[2,3], Jonas NILSSON[1], Stefan JONSSON[1], Guofan ZHANG[1]

[1]Strip, Sandvik Materials Technology, Sandviken, Sweden
(+46 26 9190, anders.hoel@sandvik.com)
(+4626264916, Jonas.nilsson@sandvik.com)
(+4626263046, stefan.jonsson@sandvik.com)
(+86 13813916176, guofan.zhang@sandvik.com)

[2]Strategic Research, Sandvik Materials Technology, Sandviken, Sweden
[3]Engineering Materials, Linköping University, Linköping, Sweden
(+46 26 263534, guocai.chai@sandvik.com)

* Corresponding Author

ABSTRACT

In 2018 the International Energy Association IEA reported that air conditioning will be a major contributor for future electricity consumption (Birol, F. 2018). As recently seen in China, tougher legislation on energy efficiency for air conditioning and refrigeration is being introduced all the time. It is therefore crucial that new compressor models perform at the highest levels in terms of efficiency to reduce global energy consumption. The prerequisites to fulfill this legislation vary for the different applications such as rotary, reciprocating, scroll and linear compressors. However, to succeed in developing more energy efficient compressors, the ability to improve the performance of the flapper valves will play an important role. Furthermore, the current commercial alloys available for flapper valves may not be sufficient to meet these increasing demands. In the past, to improve the material fatigue properties, the development of high-performance flapper valve material has been focused around optimizing the metallurgical cleanliness, microstructure, surface finish and mechanical properties. Now research has been completed to develop a new alloy composition aiming to develop a material with further improved fatigue properties for flapper valve applications. The results from internal and external testing of the new alloy show significantly improved fatigue properties. Within this paper, some of the critical properties of the new alloy are presented and the test results are compared with other existing flapper valve steels on the market.

1. INTRODUCTION

Energy and environmental aspects have been the incentives for regulations to reduce energy consumption, which have led to a development of more energy efficient cooling devices globally. The past few years of increased investments by the manufacturers to design newer and better performing cooling equipment have made us witness some rapid technological breakthroughs within the compressor industry. New platforms show lower weight and noise as well as increased number of the highly efficient inverter technology. In general, todays compressor system needs a lower amount of energy to run with more environmentally friendly cooling medias which result in reduced green-house emissions and increased energy savings.

A technical roadmap for future compressor development has been formulated in China (CHEAA, 2019, CHEAA, 2019) For example, the road map states requirements for the future compressor development concerning compressor efficiency in terms of coefficient of performance COP, weight and fraction of inverter type of compressors where a variable speed is employed. Until year 2025 and improvement of the COP should be achieved with 25% and 10% for refrigeration and Room AC application, respectively. In addition, as inverter type of compressors generally has a higher COP in comparison to traditional compressors, an increased fraction of above 50% is demanded. Furthermore, the weight of compressors in Room AC shall be reduced by 10% which is a driver for miniaturization and sustainability due to a reduced material need. These demands will only be met with innovative solutions from compressor manufacturers and all parts of the supply chain.

In this study Sandvik, a raw material researcher and manufacturer of flapper valve steel, has attempted to set a new direction for developing higher performing flapper valves which is in line with the future challenges for compressor development.

2. ALLOY DEVELOPMENT

2.1 Literature Review

Suction and discharge flapper valves are critical components for the function of a compressor. However, they will suffer from both cyclic bending stresses and impact stresses during service, which can cause the valves to fail (Soedel, 1984, Chai *et al.*, 2014). High bending fatigue strength and high impact fatigue strength are therefore the basic requirements for valve materials (Olsson, S., 1992). AISI 1095 and AISI 420 types of two martensitic steels are traditional flapper valve materials. These materials show both high tensile strength and high fatigue strengths under bending and impact stress conditions (Sandvik, 2003). Recent years, Sandvik Hiflex™ had been developed and has been widely used as flapper valve material for advanced compressors. This material exhibits up to 10% improvement in bending fatigue strength and as much as 25% improvement in impact fatigue strength compared with the traditional modified AISI 420 material. It was initially designed particularly for use in carbon dioxide compressors for automotive air conditioning, where higher pressures and temperatures place increasing demands on the flapper valves (Chai *et al.*, 2004). Today, it is widely used within refrigeration and residential AC applications. To develop a compressor with even higher efficiency, however, the valve material is now becoming a limiting factor since even higher fatigue strengths are required due to the increase of the flapper valve lift or a shorter flapper valve with the same lift (Chai *et al.*, 2014).

It is known that the fatigue strength of a metallic material initially increases, more or less linearly, with increasing tensile strength. However, no further increase or even a decrease in fatigue strength is followed by further increasing the tensile strength above a critical value. This is generally attributed to the increase in the sensitivity of the stress-raising effect at material defects by increasing the tensile strength (Forrest P. G, 1962). This observation indicates that it may be possible to have a further improvement of the fatigue strength of a material with a high tensile strength by retaining a low sensitivity to the stress-raising effect. Actually, this can be done by increasing the material´s strength and ductility simultaneously (Chai *et al.*, 2014). These properties are also the principles for the new alloy development presented in this paper. Strengthening was done by solid solution hardening with alloying elements and precipitation hardening with new stable nano-carbides. The ductility of the alloy was improved by refinement of the microstructure and introducing multiphases in the alloy.

2.2 Composition of the new alloy

In search for a higher strength material without significant need of changing the process parameters a material with higher amount of C is developed. In addition, the Cr – content is increased in order to follow the C – content and to allow additional carbides to be formed. Thereby the amount of free carbon can be limited which could lead to a brittle material. The composition for the new alloy, named Freeflex™, is given in Table 1. The typical chemical composition of a number of different flapper valve steels are also shown. A solid solution strengthening by use of Cu provides new possibilities to obtain superior properties.

Table 1 The typical chemical composition of the new alloy. For comparison other existing flapper valve steels are also given.

Sandvik Grades	Material	C	Si	Mn	S	P	Cr	Mo	Cu
20C	A	1.0	0.30	0.40	<0.010	<0.025	-	-	-
7C27Mo2	B	0.38	0.40	0.55	<0.010	<0.025	13.5	1.0	-
Hiflex™	C	0.38	0.40	0.55	<0.010	<0.025	13.5	1.0	-
Freeflex™	D	0.53	0.40	0.68	<0.010	<0.025	14.0	1.0	0.70

3. EXPERIMENTAL SETUP

3.1 Microstructure and wear

The materials are put into a mount and ground and polished carefully. The materials are furthermore etched in Kallings for microstructural depicting. Prior the AxioVision investigation, the materials are color etched, depicted by SEM and image processed. In addition, in search for Cu particles, the samples of material D are analyzed in SEM. The atomic number contrast is pronounced, and the images are further processed to be analyzed by AxioVision. As a complement to the image processing, the Cu – particle sizes are also determined manually. The smallest particles are likely to be too small for detection which implies the number of particles is reasonably higher and thereby the average size is likely to be overestimated.

The wear testing is conducted by use of a CSM Instruments TRIBOMETER operating on a linear reciprocating path of 5 mm. The total length of wear is 150 m. The counter material is a cemented (tungsten) carbide ball of 5 mm diameter. The wear mark is being evaluated using an interferometer. The depth of the wear mark is determined at 5 positions along the path.

3.2 Residual stress

The residual stress analysis of the samples was performed using XRD- Bruker axs D8 DISCOVER diffractometer (35kV,50mA) according to the $sin^2\psi$-method with Cr - Kα (λ = 2.2879 Å) radiation.

3.3 Tensile properties and hardness

The tensile testing is conducted according to standard ISO6892. The hardness measurements are conducted onto the surface by means of the standards ISO6507 and 6508.

The thermal stabilities of hardened and tempered materials have been investigated by additional tempering at a temperature from room temperature and then 200 up to 450 C in step of 50C for 2h. When the additional tempering results in a change in hardness the thermal stability level is passed.

3.4 Fatigue testing

Samples are blanked and ground to the proper dimensions. Fatigue testing is conducted onto samples with different designs for different type of fatigue testing. The samples are tumbled in order to improve the edge condition and to build in the proper amount of residual stresses to ensure the material properties are tested. The different samples are shown in Figure 1.

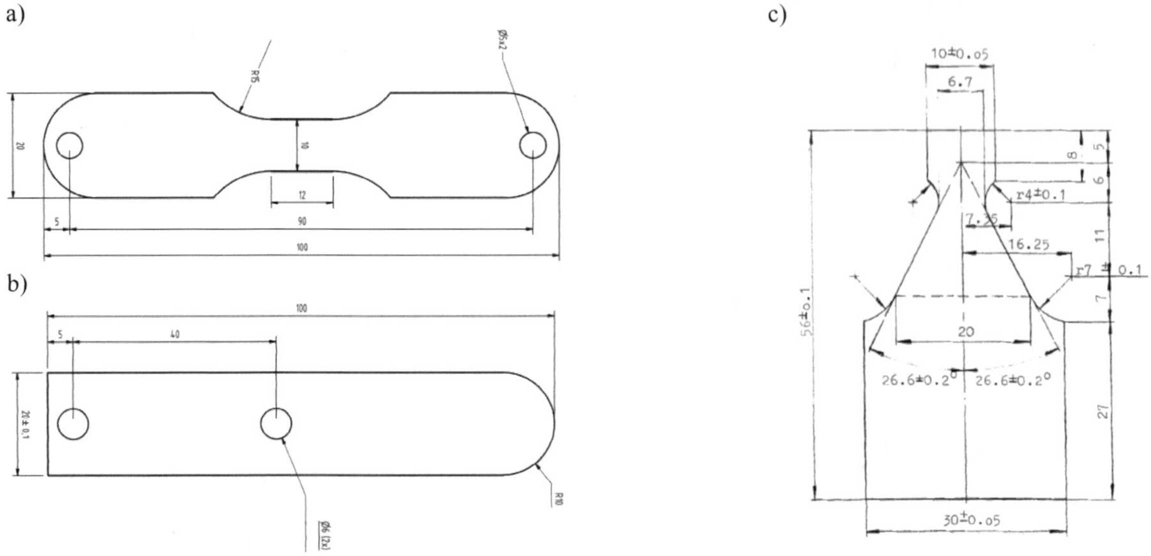

Figure 1 The design of the samples used for a) Fluctuating tensile testing b) Bending fatigue testing and c) Impact Fatigue testing, respectively.

4. RESULTS

4.1 Microstructure and wear

The microstructure of material C and D is given in Figure 2. For both alloys the microstructure consists of a martensitic structure with a fine dispersed distribution of carbides. The carbide distribution is investigated by use of image analysis using AxioVision. The distribution for material D is to be compared with the results from material B and C. The data is presented in Figure 3. A summary of the data is given in Table 2. The new alloy has a carbide density approximately twice of what is observed for material B and C. The maximum observed carbide is smaller, but this is however not a statistical measure. Instead, the volume fraction of carbides with a diameter above 0,70 µm is used as a measure. The volume fraction for carbides larger than 0,70 µm is 6-15 % for material D whilst 21-48 % for material C and B.

In addition to the effect of solid solution strengthening, an oversaturation of Cu, which is ensured in the interval around and above the composition of Cu for alloy D, enables a cluster strengthening and also a precipitation hardening. In the graph in Figure 5, an effect is observed where the A1 – transition temperature is reduced with addition of Cu for up to about 0.55 wt% after which the reduction is flattening out. The composition is chosen to be above the flattening threshold and furthermore to be kept low to avoid unnecessary addition of Cu for scrap handling reasons, possible limitations in hot processing as well as for economic reasons.

The A1 – temperature is the eutectoid temperature and defined as the minimum temperature for the austenite in equilibrium. This is illustrated in the Phase diagram in Figure 4. The presence of Cu – precipitates are shown in Figure 5b and the corresponding data is shown in Table 3, respectively. The Thermo Calc data is purely calculations and some deviations from reality is expected, however the results on A1 reduction from ThermoCalc – calculations are confirmed with dilatometry investigations even though the effect is not as pronounced. Whether there exist Cu – clusters as a pre-condition to precipitates or particles has not been confirmed but if there are Cu – particles it is reasonable that there are not yet formed particles, i.e. clusters and often referred to as Guinier – Preston or GP zones (Reed – Hill and Abbaschian, 1994). The Cu -particle density is found to be around 7 particles / 100 µm^2. The fine dispersed carbides together with the reasonably presence of Cu clusters, maximal solid solution strengthening, precipitates of Cu particles are indicators of an efficient pinning of dislocations resulting in elevated mechanical properties in comparison to the other alloys.

Figure 2 The microstructure of the materials a) C and b) D, respectively

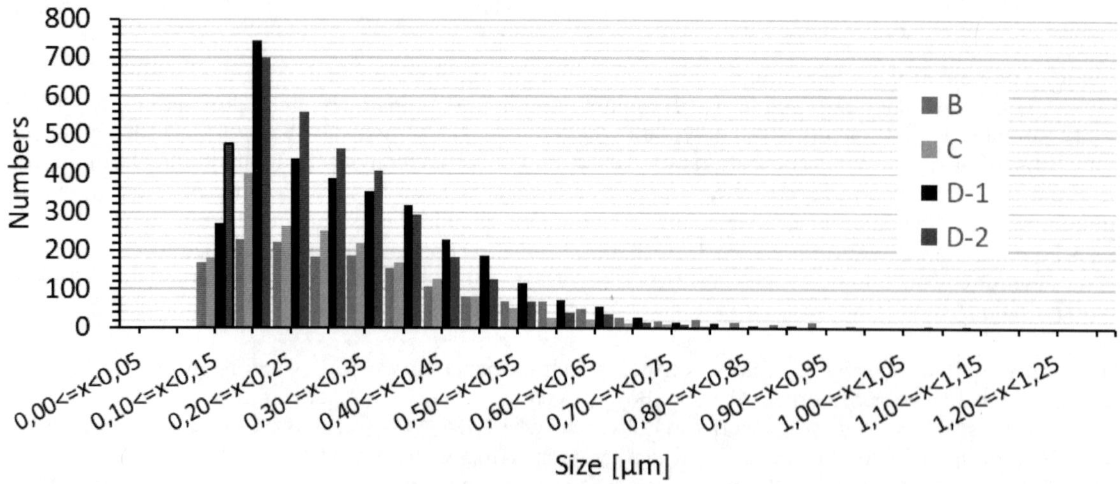

Figure 3 The carbide distribution for the different materials.

Table 2 A summary of the carbide distribution

ID	No of particles	Particle density No/100 µm²	Particle area fraction %	Average particle diameter µm	Max particle diameter µm	Average particle volume µm³	Volume fraction above 0,70 µm %
Sample D-1	3249	56	5,0	0,30	1,07	0,03	15
Sample D-2	3393	58	4,1	0,27	0,97	0,02	6
Sample C	1839	32	2,6	0,29	1,36	0,02	21
Sample B	1663	30	3,7	0,35	1,30	0,05	48

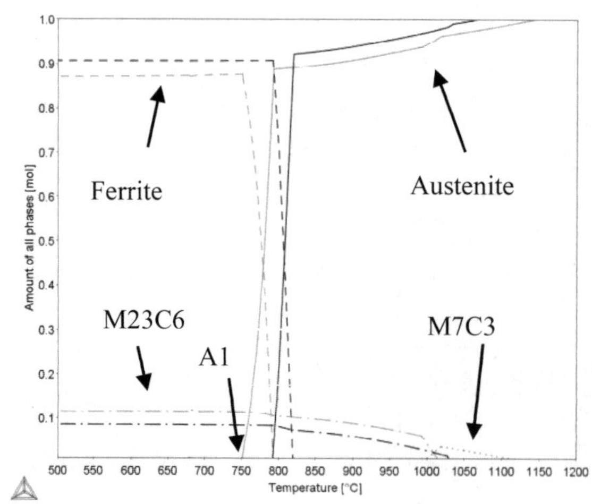

Figure 4 A ThermoCalc diagram showing the differences of the alloys based on material B/C and material D.

Table 3 A summary of investigation of Cu- particles.

Method	No of particles	Average Area [µm²]	Average Diameter [µm]	Inspected area [µm²]	Particles / 100 µm²
Image processing	92	0,012	0,12	1370	7
Manual	96	0,011	0,09	1370	7

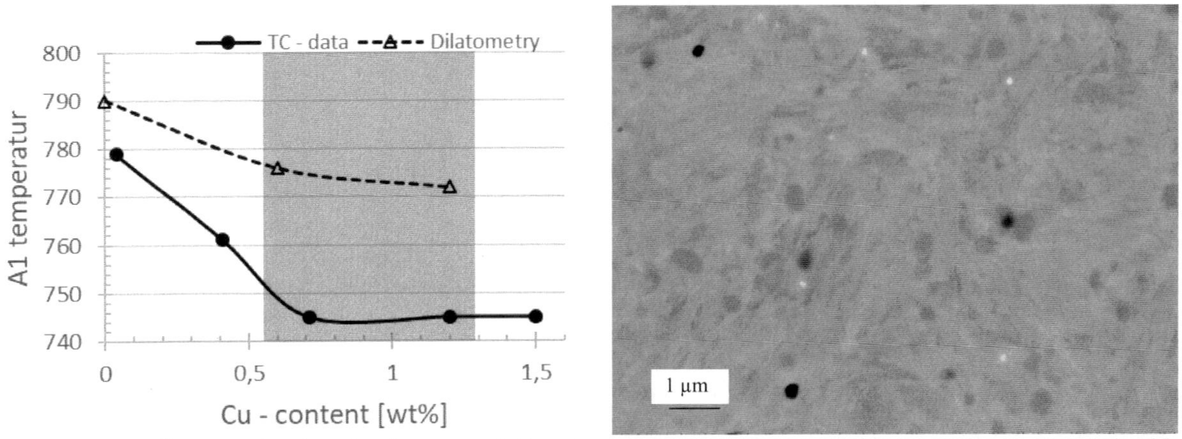

Figure 5 a) A diagram showing the decrease of A1 – temperature as a function of Cu – content as determined by Thermo – Calc as well as dilatometry, respectively. b) In the image, the presence of Cu precipitates (white) is observed besides Cr -carbides (grey).

The results from the wear mark analysis is shown in Figure 6. In a) the interferometry images are shown. In b) the average depth and the corresponding standard deviations are given. In c) a SEM image of material A is shown as an example. The ranking of the wear resistance according to this specific test is clear where material D exhibits the best wear resistance closely followed by C and then with a significant gap to material B and A.

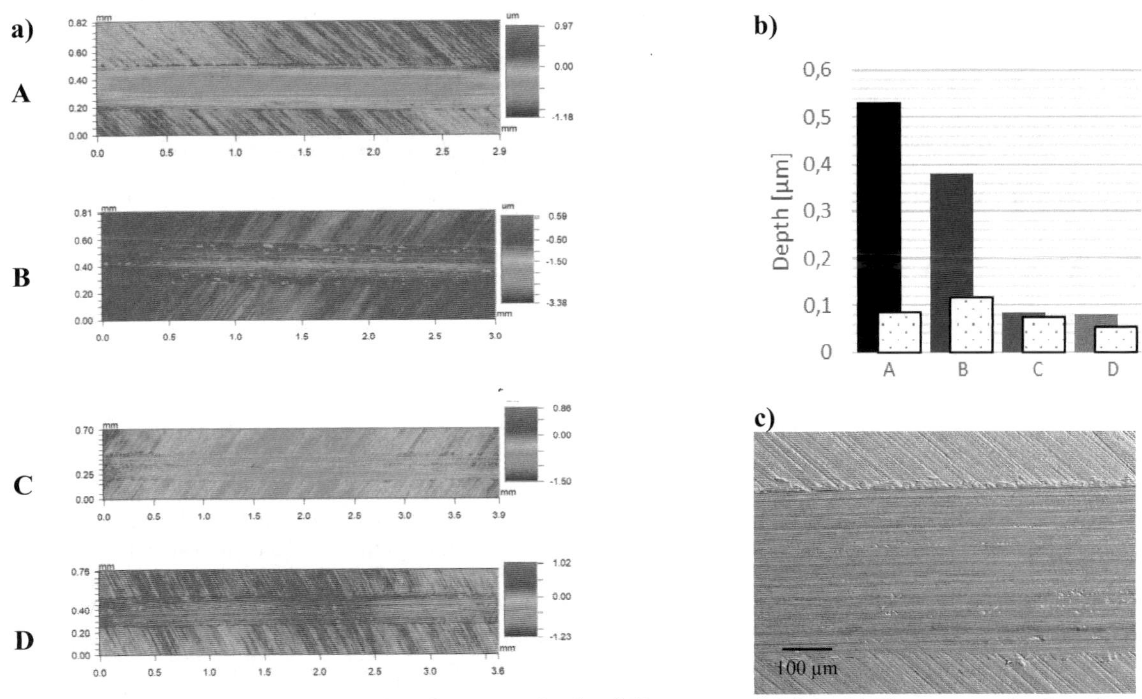

Figure 6 a) The wear mark from the interferometer for the different material A-D. b) The average depth and standard deviation of the wear mark. c) As an example, the wear mark of material A as seen by means of secondary electron microscopy in SEM.

4.2 Residual stress

The tumbling is conducted in a centrifugal tumbling unit in a batch of 30 samples. The residual stresses for as produced and tumbled material are shown in Figure 7. The values shown represents the average value of the components in the rolling direction and transversal direction, respectively.

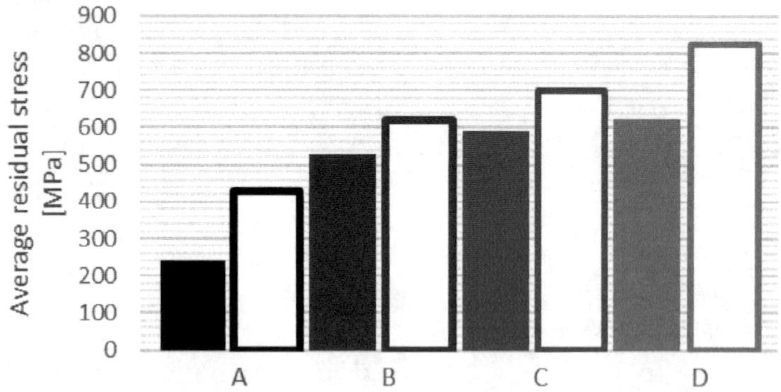

Figure 7 Residual stress for the different materials. The filled column corresponds to samples in as produced condition whilst the open column corresponds to tumbled condition, respectively.

4.3 Tensile properties and hardness

The mechanical properties are shown in Figure 8. For the materials C – D the values are of general measure for thicknesses in a wide range up to 1.0 mm. For material A, the tensile strength varies from 1650 – 2200 MPa depending on thickness. The shown values are valid for a material of thickness 0.305 mm. In a) the tensile strength is given together with the corresponding value for yield strength. In b) the hardness interval for the different materials are given. In c) the average elongation is defined by A25%.

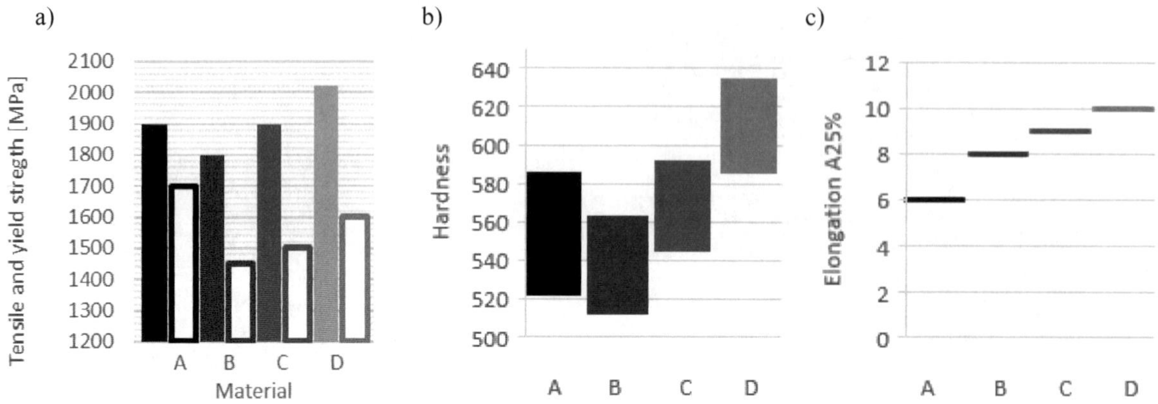

Figure 8 The mechanical properties in terms of a) Tensile strength R_m and yield strength $R_{p0.2}$ b) Hardness HV and c) Average Elongation A25%, respectively.

The thermal stabilities of the materials are evaluated. The results are shown in Figure 9. The chromium steels are almost unaffected by the additional tempering at these different elevated temperatures. The carbon steel on the other hand becomes softer as the temperature exceeds 300 °C. The maximum utility temperature recommended for the carbon steel is 150 °C. For the chromium steels discussed in this paper the corresponding temperature is up to 300 °C. It is reported that the temperature in commercial compressors can exceed up to 240 °C which indicate the need for a high thermal stability.

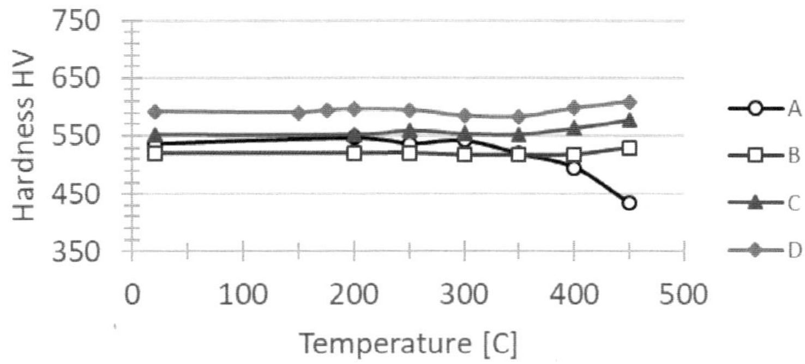

Figure 9 Thermal stability in terms of tempering resistance upon 2 h treatment.

4.4 Fatigue testing results

The fatigue properties of the above-mentioned materials are shown in Figure 10. The presented values are based on staircase method results with a probability to failure of 50 %. For fluctuating tensile strength, the load condition of R=0.1 is applied with a run out level of 5 M cycles. The maximum value is given. For bending fatigue testing the load condition R=-1 is applied with a run out level of 2 M cycles. For the fluctuating tensile strength and for the bending fatigue the capability to build surface residual stress is essential for the results. For impact fatigue a run out level of 10 M Cycles are used and the fatigue limit is given in terms of index normalized to 10 for material C.

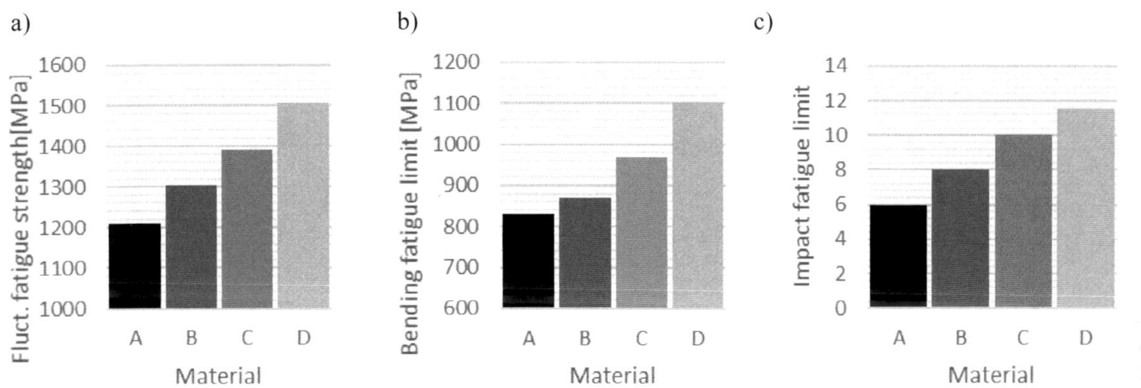

Figure 10 The fatigue properties of the various materials. a) Fluctuating fatigue strength b) Bending fatigue strength and c) Impact fatigue strength by index.

5. CONCLUSIONS

In this paper a new material is presented. The material is a further development from the well-known AISI 420 -based chromium steels. The modifications of the chemical composition enable an enhanced capability to pin the dislocations by use of solid solution strengthening, possible cluster strengthening and hardening effects by use of various precipitates. The new composition enables an increase of carbide particle density with slightly smaller sizes leading to a high wear resistance which is believed to be a crucial property for a superior performance in lifetime perspective in the compressor application. The increase of the surface residual stress acts as a further protection against surface related fractures. The development of the new material enables elevated material properties in terms of tensile strength, yield strength and elongation. These properties lead to a further increase of reliability through the improved fatigue properties, which in general terms can be stated to be about 10 % higher in comparison to the second-best material presented in this paper. The new material is clearly believed to be an important tool to allow further development of the compressor technology through enabled miniaturization and by further design developments increasing energy efficiency (COP) and thereby meeting the needed improvements to fulfill the continuously stricter legislations.

REFERENCES

Birol, F., (2018), "The future of cooling – Opportunities for energy – efficient air conditioning", IEA - https://webstore.iea.org/the-future-of-cooling

Chai, G., Zetterholm, G., Walden, B., (2004), "Flapper valve steels with high performance, *Proceedings of International Compressor Engineering Conference(C132)*, Purdue University

CHEAA (2019). Handbook of China domestic refrigerator industry technology roadmap. China household electrical appliance technology conference. Foshan, China, October, 2019.

CHEAA (2019). Handbook of China room air conditioner industry technology roadmap. China household electrical appliance technology conference. Foshan, China, October, 2019.

Hareland, M., Hoel, A., Jonsson, S., Liang, D., Chai, G. (2014). "Selection of flapper valve steel for high efficient compressor", *Proceedings of International Compressor Engineering conference (Paper 1408)*, Purdue University

Forrest, P. G. (1962). Fatigue of metals. Oxford, Pergamon Press Ltd.

Olsson, S. (1992), "Improved characteristics of stainless compressor valve steel", *Proceedings of International Compressor Engineering conference (pp. 909-918),* Purdue University

Soedel, W., (1984), Design and mechanics of compressor valves, OFFICE PUBLICATION, Purdue University

Sandvik Data sheet, 2003, *Strip steel for flapper valves,* S-343-Eng. AB Sandvikens Tryckeri, Sweden.

Reed – Hill, R. E., Abbaschian, R. (1994). Precipitation hardening. In *Physical Metallurgy Principles* (515-537). Boston, MA: PWS Publishers

Study of High-Performance Engineering Polymers Applied in Reciprocating Hermetic Refrigeration Compressors

Roger Nelson[1]*, Patrick Kunc[1], Laurent Hazard[1], Celso Kenzo Takemori[2], Diego Willian da Silva[2], Edmar Baars[2], Gustavo Rafael Bernardes[2], Luis Miguel Valdes Lopez[2]

[1]Solvay Specialty Polymers
Alpharetta, GA, United States of America
roger.nelson@solvay.com

[2]Vibroacustica Research and Development,
Joinville, SC, Brazil
celso@vibroacustica.com.br

*Corresponding Author

ABSTRACT

Manufacturers of hermetic refrigeration compressors for domestic and commercial applications have been increasing the energy efficiency of their products systematically over the last decades through the continuous efforts of in-house research groups, universities and suppliers. This joint effort encompasses the technical fields of design, materials, processes, manufacturing, and simulation techniques, to achieve a multidisciplinary approach to compressor optimization as applied to refrigeration systems. Gains such as increased efficiency of the electric motor, reduction of bearing friction power loss, minimizing back flow on valves have already been addressed and the continuous search for new approaches have been proposed. This work presents a study of energy efficiency gain by comparing low conduction heat transfer plastic materials to metallic materials used in current compressor designs. The components compared include the cylinder head, discharge muffler, shock loop, valve plate and crankcase. The energy efficiency gain was evaluated using the Coefficient of Performance (COP) through the use of the Conjugate Heat Transfer (CHT) model using superheating temperature. The CHT model temperature results were compared with the values measured by an instrumented compressor. To complement the study, the parts and assemblies were structurally evaluated using the finite element method under performance and durability test operating conditions. The feasibility of the mold filling process was verified, because the final design of the components must be a compromise between functional requirements and process solution. In conclusion, the results show that significant efficiency gains can be achieved through the use of engineering thermoplastic. This is achieved through careful material selection and proper design that considers not only material properties, but material and process costs using the flexibility of injection molding to group parts and subsystems.

1. INTRODUCTION

After several years of progress of increasing electric motor efficiency and reducing valve losses, the superheating inside the compressor remains an area where energy efficiency improvement can be made. Refrigerant superheating from the suction tube to the cylinder decreases the cooling capacity and the efficiency of a compressor. An excessive temperature can also degrade oil lubrication properties and may result in catastrophic failure of the compressor. There are three main heat sources inside the compressor: 1). The conversion of the electric energy into mechanical energy is not 100% efficient, so a portion is converted into heat inside the compressor, mainly in the stator and the coils. 2) The friction between moving/rotating parts and stationary parts, for example the crankshaft and crankcase and the piston and cylinder walls. 3) The compression process itself. It is not exactly a heat source, but the compression process heats up the refrigerant that enters inside the cylinder, and after being discharged the hot gas flows inside the compressor exiting the compressor and releasing heat energy to the compressor internal parts, i.e., cylinder head, valve plate, discharge muffler, shock loop, and crankcase.

One method to improve the thermal performance of a compressor, or decrease the superheating, is increase the thermal insulation between the hot and cold parts. This can be achieved by decreasing the thermal conductivity of the solid parts like the cylinder head and the discharge tube. There are several methods to estimate the effect of the thermal conductivity on the temperature distribution on the compressor. Lacerda and Takemori (2016) summarized the main works about the heat transfer models applied to compressors in twelve methods, that are a combination of (i) steady state or transient, (ii) partial compressor or full compressor and (iii) lumped model (0D), full three-dimensional model (3D) and 0D & 3D model.

In this work a full three-dimensional heat transfer model is employed, where the fluid domain is modeled by the tridimensional transient Navier-Stokes equations, while the solid domain is governed by the Fourier's Law. Based on that mathematical model, this work studies the efficiency variation if the thermal conductivity is decreased by replacing the current metallic components with polymeric parts. As mentioned before, some data is needed to perform the heat transfer simulation including the mass flow rate, the friction losses and the electric motor losses. For the mass flow rate, the method described in Ussyk (1984) is implemented in the commercial code GT-SUITEmp (GT-SUITE 2018). The friction losses are calculated using the established model of Reynolds equation for journal bearings as described in Raimondi and Boyd (1958) and Hamrock (1994). The electric motor losses are based on values found on the specialized literature, Diniz (2018).

2. METHODOLOGY

2.1 Compressor Selection and Reverse Engineering

For this work an after-market compressor was purchased. The selection criterion was a high efficiency household freezer refrigerator system based on the Brazilian energy efficiency rating program Inmetro (2018). The selected model is a variable speed R600a compressor whose speed ranges from 1620 to 4800 RPM. To build the 3D geometries needed for the numerical simulations, the compressor was disassembled, and parts were 3D scanned using optical and tomography methods. The part dimensions were measured using calipers and micrometers. Part weights were measured using a precision scale. Figure 1 shows the 3-D scan for the crankcase. The measured and calculated weights for all parts are shown in Table 1.

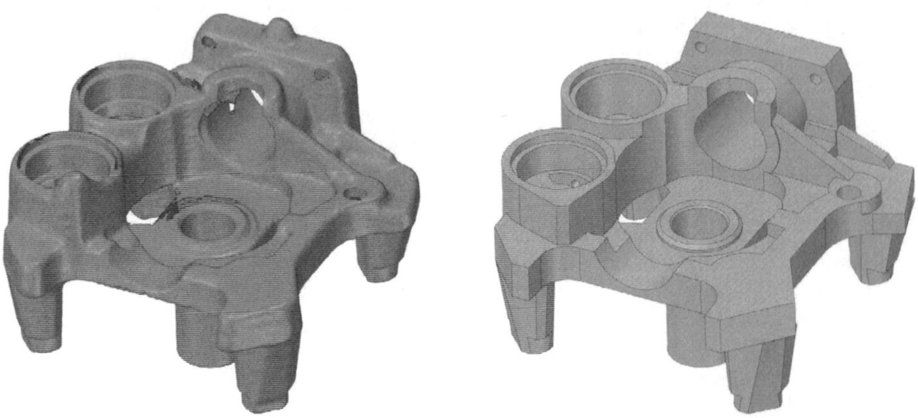

Figure 1: Crankcase. Left: 3D scan STL. Right: CAD geometry.

Table 1: Compressor Parts Mass

Part	Measured [g]	CAD [g]
Crankcase	1438	1421
Connecting Rod	17.5	20
Cylinder Head	45	46
Piston	44	45.6
Shell	2502	2536

2.2 Polymer Selection

The process of selecting suitable polymers for compressor components requires evaluating all the operating requirements, environmental conditions, ease of processing, and cost. As most polymers are thermal insulators, with thermal conductivities several orders of magnitude less than aluminum, the other factors become paramount. From a structural standpoint, the polymer must be able to withstand the high pressures and loads at elevated temperature over long lifetimes. This requires a polymer with good retention of mechanical properties at high temperature and good dimensional stability for creep and fatigue resistance. These requirements dictate that fillers be used. In addition, these properties must be maintained when exposed to the various refrigerants encountered in a compressor, so chemical resistance is required. Pai-Paranjape (2014) has demonstrated that certain filled engineering thermoplastics have good chemical compatibility with R-134A and R-410A compressor environments. Another material characteristic is the ability to be easily injection molded into complex components and maintain tight dimensional tolerances. The selected polymer must not be overpriced as to exclude its use.

When all the above are considered, the candidate materials are limited to a few engineering thermoplastics. Two semi-crystalline aromatic polymers which are excellent candidates are polyphenylene sulfide (PPS) and polyphthalamide (PPA). The selected compounds studied in detail are a 45% glass filled PPA and a 65% glass/mineral filled PPS. Not only do they meet the requirements needed for compressor components, but they are also cost competitive. Relevant mechanical and physical properties are listed in Table 2. [Solvay Specialty Polymers 2020]

Table 2: Properties Relevant to Compressor Components

Material	45% GF PPA	65% G/M PPS	Cast Aluminum
Thermal Conductivity (W/m.K)	0.328	0.49	160
Mold shrinkage (Flow / Cross flow) (%)	0.2 / 0.6	0.2 / 0.4	
Density (g/cc)	1.56	1.94	2.7
Tensile Modulus (GPa) @ 23°C	16	19	69
Tensile Modulus (GPa) @ 75°C	13	16.5	
Tensile Modulus (GPa) @ 120°C	10.5	10.3	
Tensile strength break (MPa) @ 23°C	263	165	317 ult, 160 yield
Tensile strength break (MPa) @ 75°C	185	141	
Tensile strength break (MPa) @ 120°C	125	100	
Poisson's ratio	0.4	0.34	0.33
Fatigue strength at 23°C 10E6 cycles (MPa)	82	83	105

2.3 Three-dimensional heat transfer model

As described in Lacerda and Takemori (2016), this model considers a full three-dimensional compressor geometry, Figure 7, with a steady state mass flow rate. This model also considers both fluid flow and solid parts. The solid parts are the suction tube, electric motor windings, connecting rod, counterweight, crankcase, crankshaft, cylinder gasket, cylinder head, cylinder head gasket, discharge covers, discharge valve stopper, discharge tube, piston, piston pin, rotor magnets, rotor, shell, shock loop, stator, suction muffler, valve plate, and oil (it is considered as a solid part, considering only the heat conduction). The suction fluid domain starts at the beginning of the suction tube and ends at the suction port through the suction muffler, Figure 2. The discharge fluid domain starts at the discharge port and ends at the end of the discharge tube, Figure 3. All interfaces are considered in such a way that the heat can flow between both fluid and solid domains. A total of 66 interfaces were considered.

A steady mass flow rate is applied to the suction port and to the beginning of the suction tube based on a pressure and temperature of 0.63 bar and 32 °C, respectively. The same mass flow rate is applied to the discharge port with a temperature T_D that follows Equation 1. The compression process is not solved but modeled evaluating the temperature increase from the beginning of the compression process when the refrigerant enters inside the cylinder until the time the gas is discharged out of the cylinder through the discharge port following an isentropic process. The temperature T_S is the mass flow averaged temperature on the suction port. The discharge pressure at the end of the discharge tube is 7.62 bar.

$$T_D = 0.98 \cdot T_S + 73.717 \tag{1}$$

To consider the induced flow by the rotating parts and the oil circulation inside the compressor, a momentum source term was used inside the compressor cavity, Figure 6. Using the momentum source term, it was possible to analyze the system without the need of solving a transient and multiphase model and increase the computational cost by orders of magnitude. An equivalent heat transfer coefficient, h_{eq}, is used on the shell external surface to consider both the convection and radiation effects, Equation 2. The value h_0 is calculated by setting the variable h_{eq} based on Dutra (2010).

Frictional losses are added to the pressure-volume power diagram to obtain the shaft power with the electric motor and inverter (controller) efficiency, making it possible to estimate the power consumption of the compressor. The values for the electric motor efficiency were extracted from Diniz (2018). The electric motor losses, Figure 4, are equally divided into the stator and windings losses. The friction losses, Figure 5, are applied on the following interfaces: crankshaft and crankcase, crankshaft and connecting rod, connecting rod and piston pin, piston pin and piston, and piston and cylinder, Lacerda and Takemori (2016). Figure 10 and 11 shows the results for the current research project.

$$h_{eq} = h_0 + \varepsilon\sigma\left({T_{shell}}^4 - {T_\infty}^4\right) \quad (2)$$

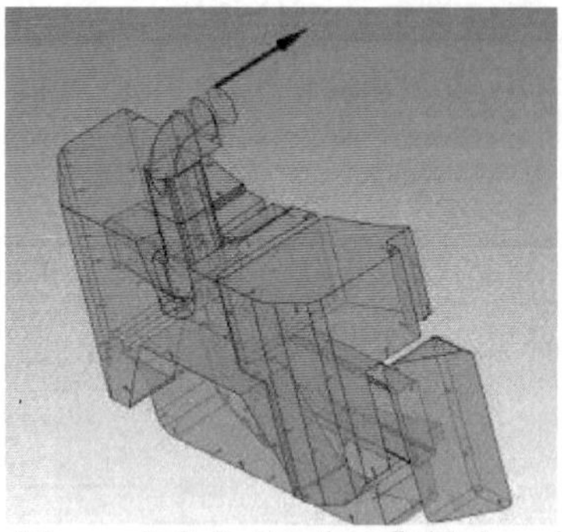
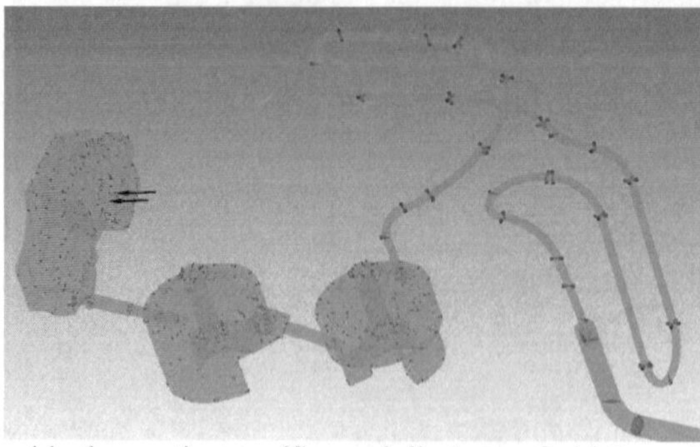

Figure 2 & 3: Suction fluid domain inside the suction muffler and discharge fluid Domain

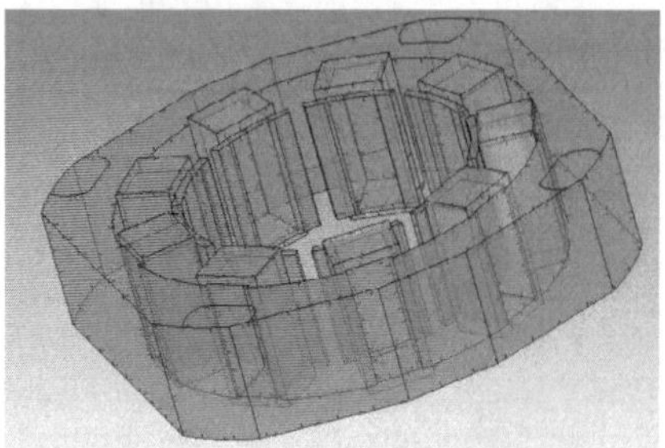

Figure 4 & 5: Electric motor solid domain, showing the windings and stator heat sources and the heat from friction losses are applied to the shaft, connecting rod and piston.

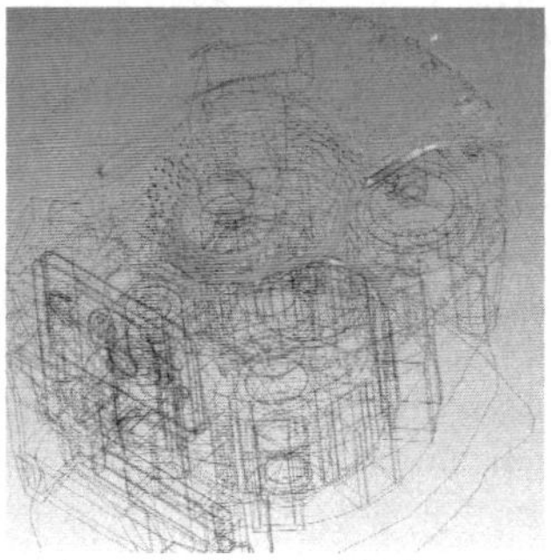
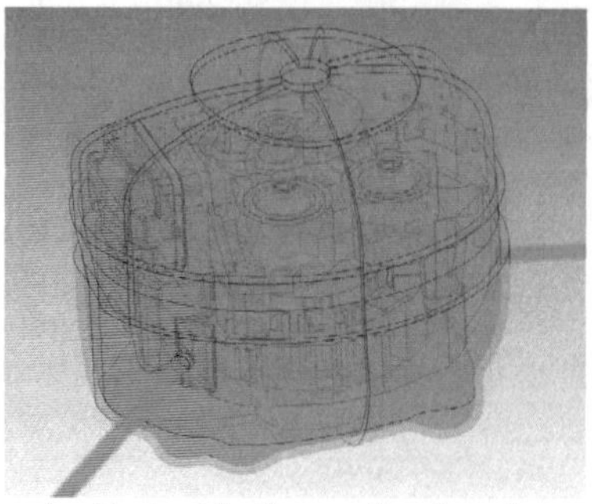

Figure 6 & 7: Momentum source term induces flow velocity inside the cavity showing velocity vectors on a horizontal plane. Full three-dimensional compressor geometry.

2.4 Experimental Setup for Temperature Measurement

Temperatures were measured at 14 locations on the compressor using K type thermocouples. Figure 8 shows an overview of the instrumented compressor. An accelerometer, the black wire connection in Figure 8, was attached to the shell to measure the rotational speed using the acceleration signal. Figure 9 shows how the thermocouples were attached to the compress.

Figure 8: Overview of the compressor with the thermocouple wires.

Figure 9: Thermocouples attached to the crankcase, suction muffler and cylinder head.

3. RESULTS AND DISCUSSION

Simulations showing the various effects of the momentum source term can be visualized in Figure 10 for a given rotational speed. In both charts the blue line are the measured values, and the orange ones are the simulation results. The chart on the left shows the simulation results without the momentum source term, while the chart on the right shows the results with it. Differences are expected as the experiment has uncertainties and a large metal flange that is absent from the simulation. The momentum source term improved the correlation between the simulation and the experiment. Figure 11 shows the comparison for three different rotational speeds with the momentum source term.

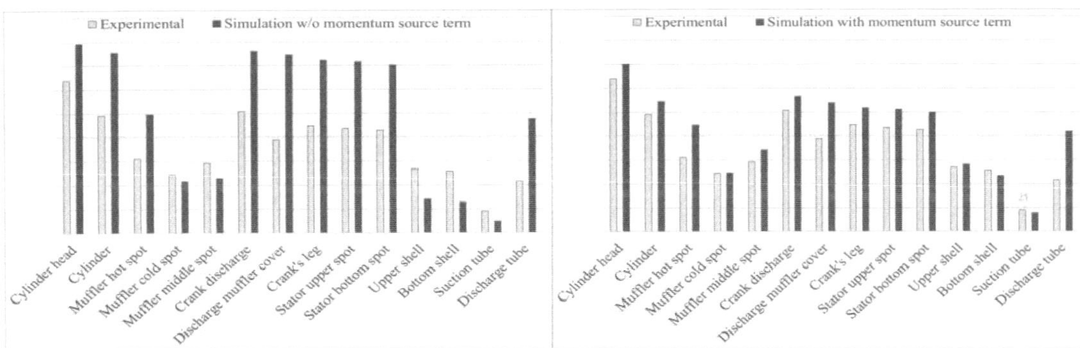

Figure 10: Temperature comparison between the measured and simulated values: Left chart is simulation **without** momentum source term. Right chart is simulation **with** momentum source term.

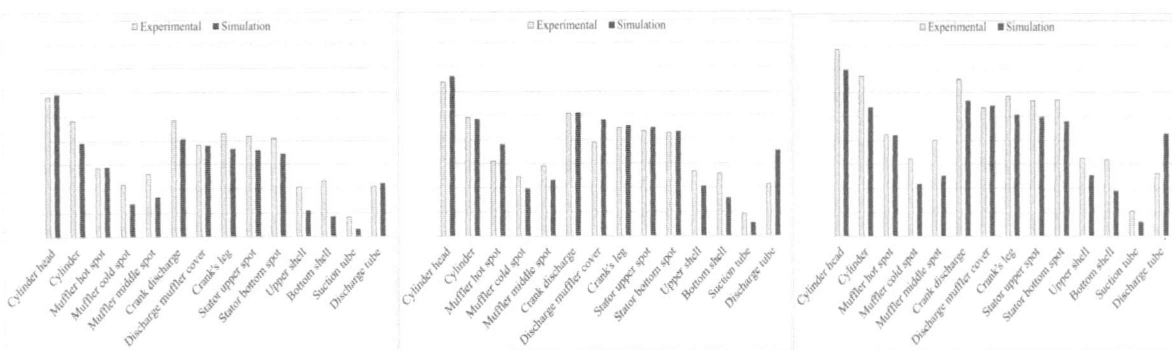

Figure 11: Temperature comparison between instrumented experimental compressor and simulation with momentum source term at three different rotational speeds: Low speed-left chart, Mid speed-middle chart, High speed-right chart

The 3D Heat Transfer Model temperature predictions closely mimic the temperatures measured in the instrumented compressor. This provides confidence, within limits, that this model's simulations can accurately predict temperatures that can be used to estimate the performance parameters in a compressor. Based on these thermal predictions, compressor efficiencies can be calculated and compared to estimate efficiency gains in a compressor by replacing metal components with polymeric materials, PPS and PPA, as suggested in the Polymer Selection section. Table 3 shows the efficiency gain when the thermal conductivity of metallic parts is replaced by the thermal conductivity of polymers. With polymers, the density of the fluid refrigerant that enters in the cylinder increases, resulting in an improvement of the cooling capacity and efficiency. The efficiency gain presented in the table below is estimated based upon the increase in the fluid refrigerant density.

Table 3: Polymer studies and results

Compressor part	Baseline	Case 01	Case 02	Case 03	Case 04	Case 05	Case 06
Cylinder Head	Metal	Polymer	Polymer	Polymer	Polymer	Polymer	Polymer
Shock loop	Metal	Metal	Polymer	Polymer	Polymer	Polymer	Polymer
Crankcase	Metal	Metal	Metal	Polymer	Metal	Polymer	Polymer
Valve Plate	Metal	Metal	Metal	Polymer	Polymer	Polymer	Polymer
Discharge Tube	Metal	Metal	Metal	Metal	Metal	Polymer	Polymer
Discharge Muffler Covers	Metal	Metal	Metal	Metal	Metal	Metal	Polymer
Efficiency Gain	0.00%	0.47%	0.64%	1.65%	0.83%	1.95%	2.10%

4. STRUCTURAL ANALYSIS

The cylinder manifold consists of the cylinder head, cylinder head gasket, valve plate, suction reed valve, valve plate gasket and crankcase. Those components are assembled to ensure tightness and define the cylinder's clearance volume, which is closely related to the efficiency of the compressor. The temperatures of the compressor components were determined by the CHT analysis for the baseline and were compared with experimental data. The plastic material properties were reduced to account for the elevated temperatures as shown in Table 2. The effect of temperature is shown for selected properties such as tensile modulus, strength, and the fatigue limit. The characterization of the NI-2085, an NBR binder gasket material, which is largely used by compressor manufacturers, is made by a special compression rig test where the deformation versus load is determined and used in finite element programs through the deformation versus pressure relationship.

In the analysis, three load steps were applied (see Figure 12):

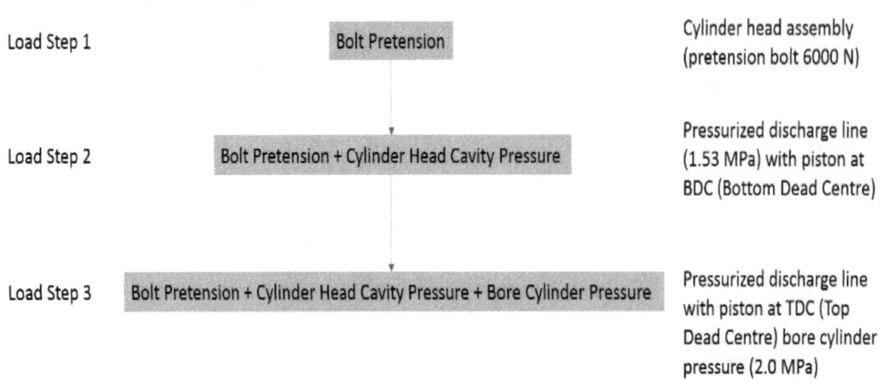

Figure 12: Load steps for structural analysis

The structural simulation of the baseline compressor, Figure 13, verified stress concentrations in the region of the fillet radius of the bolt head seat, Figure 14. A sensitivity analysis was performed and it was found that even by reducing this stress concentration by increasing the radius the resulting stresses were still too high for the plastic material. This indicated a geometrical change would be necessary.

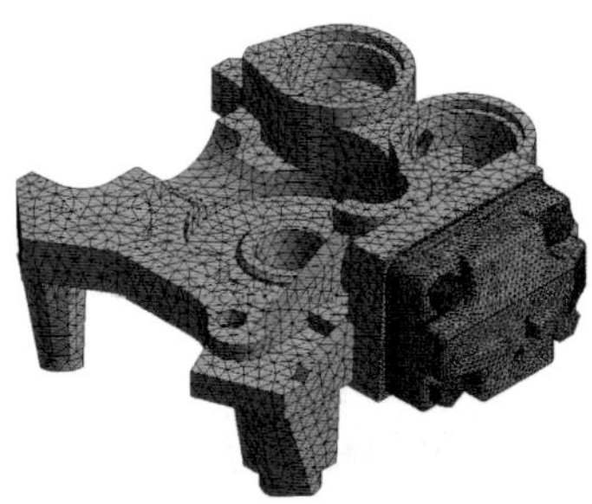

Figure 13: Baseline version - Mesh of cylinder manifold components.

Figure 14: Stress concentration at fillet radius on recess of bolt head seat.

The strategy for obtaining a plastic cylinder head was to reduce the stress concentration in the filleted region since the stress values in the rest of the part were compatible with the plastic material properties considered for the application. In Figure 15 we show the evolution of the alternatives, starting with the simplest model (Version 1), in which the cylinder head was a rectangular solid object with a flat top surface where the stress concentrator is eliminated. The following designs were optimized for stiffness, improved processability, and weight reduction.

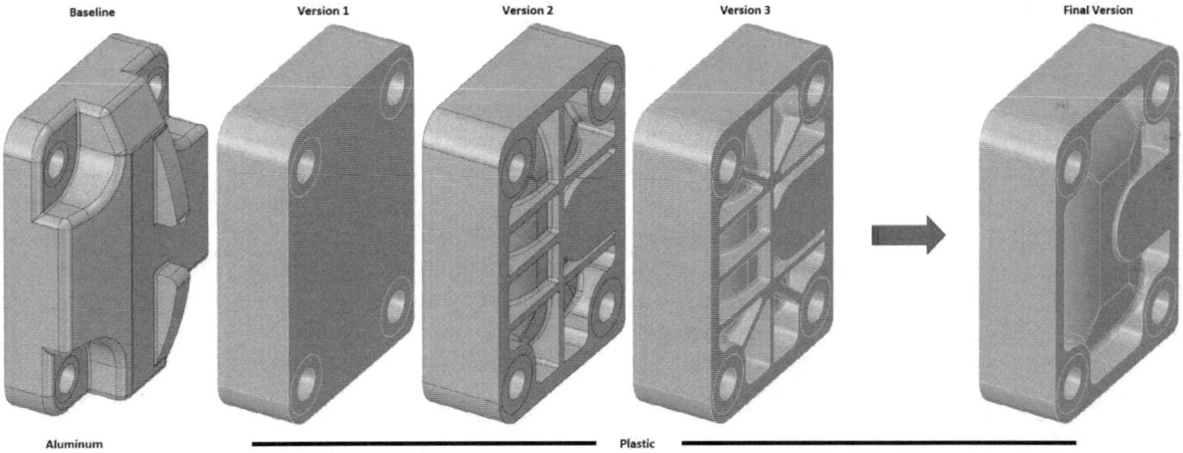

Figure 15: Geometry of cylinder head from baseline to plastic final version

The same loads were applied to the optimized model of the plastic cylinder head. The region of maximum tensile stress has changed, and the value is much lower than that found in the aluminum cylinder head, see Figure 16. The results in the Goodman-Haigh diagram, Figure 17, show that the cylinder head with the proposed geometry (Final Version) can be made of plastic and the stress level is compatible with the material. The average tension is high, but three times less than the aluminum cylinder head, and the alternating stress is still low. The adoption of a plastic cylinder head is a compromise between the geometry necessary to obtain the stiffness, contact pressure and the low mechanical stresses compatible with the material, and a robust process that can obtain parts without defects.

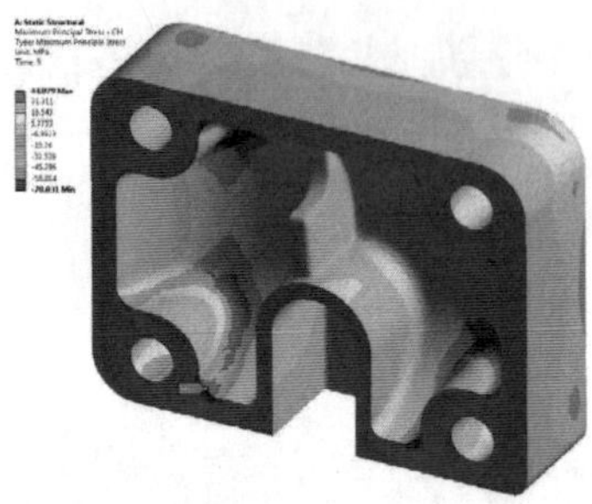

Figure 16: Stress results for load step 3 of plastic cylinder head

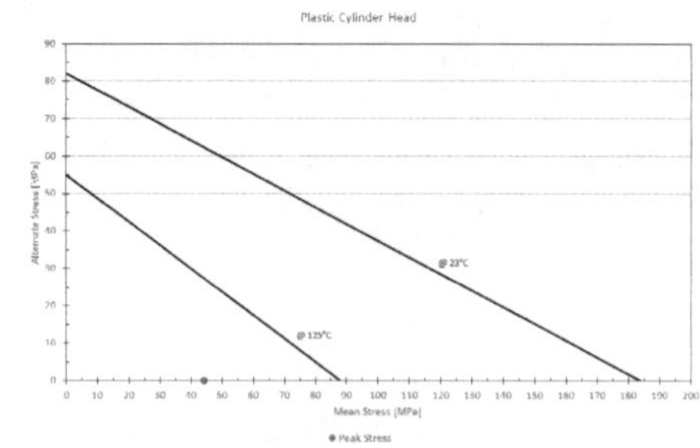

Figure 17: Goodman-Haigh diagram for cylinder head made of plastic

5. PROCESS MOLDING INJECTION SIMULATION

Since the most cost-efficient plastic production method for structural plastic parts is injection molding, this process was chosen for this study. Currently, cylinder heads are cast aluminum and in general the geometries used for casting are not suitable for a robust injection molded part. Injection molding simulations were performed on different geometries in order to find an optimized configuration that could both sustain the structural requirements and be easily injection molded. The original baseline geometry is shown in Figure 15. The wide variation in part thickness was not suitable for injection molding, so a cored-out version with ribs was initially analyzed, Figure 18. The injection point is shown. The partially filled cavity is shown in Figure 19. An uneven fill pattern is observed in Figure 19 caused by the thickness variation resulting in converging flow fronts at high stress areas. The converging flow fronts are called weld lines and the resulting strength in these locations are less than the nominal material. Several design iterations were made to reduce the severity of the weld lines and relocate them in lower stress areas, see Figure 15. The resulting flow pattern is shown in Figure 21.

The fiber orientation is shown in Figure 22. Areas of higher orientation are shown in red. Since the strength and stiffness of an injection molded part are influenced by the fiber orientation, it is important to know how the fibers are distributed throughout the part. This information can be readily transferred from the flow simulation to the structural analysis to fine tune the design. The predicted deformation or warpage of the part after molding is shown in Figure 23. Of special interest for the cylinder head is the flatness of the seal area. The variation should not be so great that it inhibits the gasket from sealing. As shown in this example, the overall variation in the out-of-plane displacement of the seal or flatness is .0025 mm. Note that this variation can be reduced by changing processing parameters, geometry, materials, and even accounted for with tooling modifications. The results show that it is possible to develop a geometry which can be readily injection molded and still withstand the service loads imposed upon it.

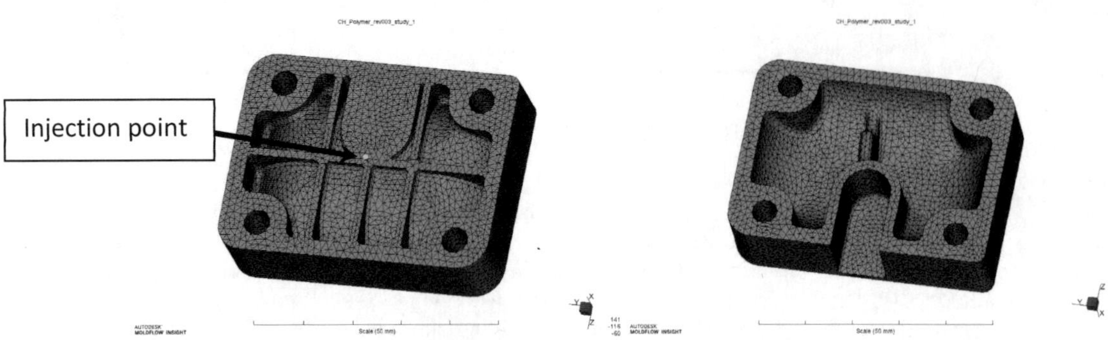

Figure 18: Finite element model

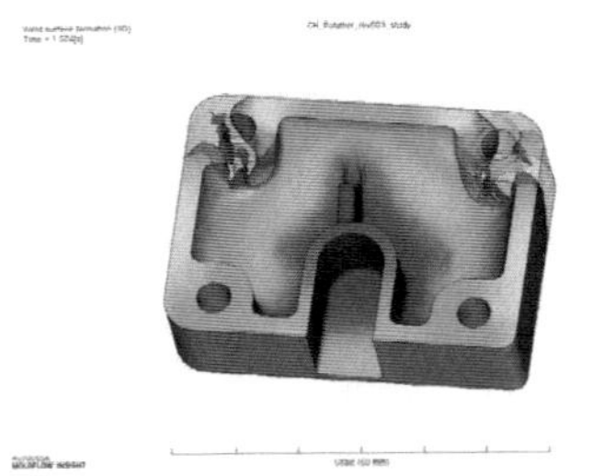

Figure 19: Fill pattern at 1.5 seconds with weld lines shown

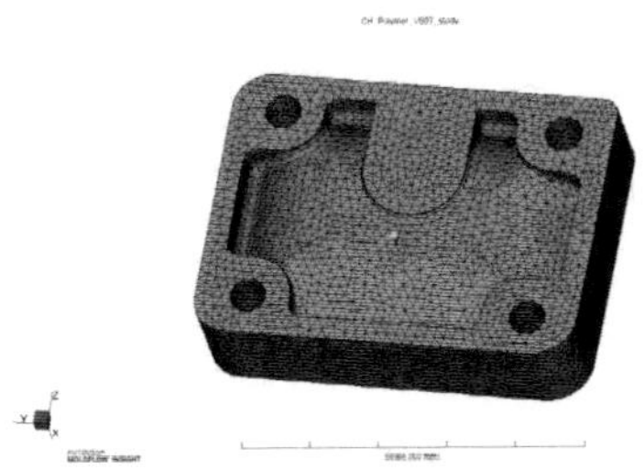

Figure 20: Optimized design eliminates critical weld line

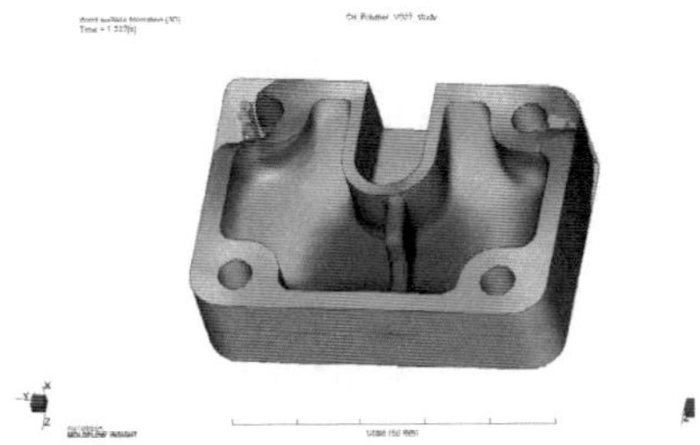

Figure 21: Fill pattern at 1.5 seconds with weld lines shown

Figure 22: Fiber orientation

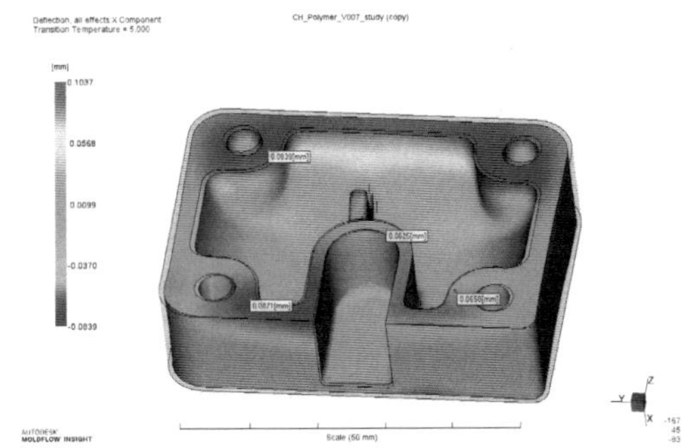

Figure 23: Deformation of seal surface (flatness)

4. CONCLUSIONS

Properly selected polymers, in particular those that maintain their high stiffness, strength and chemical resistance when exposed to compressor environments over long times, combined with the proper design to meet the physical properties requirements of the application, can replace metal components in reciprocating compressors, and increase a compressor efficiency. In this article, plastic alternatives for the cylinder head, valve plate, shock loop, crankcase, discharge tube, and discharge muffler cover were all evaluated for efficiency improvement. The use of structural analysis and mold flow simulation software can substantiate design modifications and increase the success of metal to plastic replacement. This was demonstrated by applying those techniques to a cylinder head made of die cast aluminum and replacing it with a 45% glass reinforced polyphthalamide PPA polymer.

The 3D Heat Transfer model accurately predicted the temperature profile of our instrumented reciprocating compressor and the addition of the momentum source term improved the temperature correlation between the instrumented compressor and the 3D Heat Transfer Model, thus providing a method to study the effects of semi-crystalline aromatic polymer replacement of metal components in reciprocating compressors and estimating the associated efficiency gains. This heat transfer model predicted compressor efficiency gains when the metal components were replaced with PPS and PPA materials. These polymers, primarily due to their lower thermal conductivity, increased the density of the refrigerant entering the cylinder resulting in an improvement of the cooling capacity and efficiency.

NOMENCLATURE

STL	Stereolithography file format
A_p	Pressure area
A_{eff}	Flow area
$\dot{m}$	Mass flow rate

ρ_{is}	Density at the throat
ρ_0	Upstream stagnation density
U_{is}	Isentropic velocity at the throat
C_D	Discharge coefficient
A_R	Reference flow area
P_R	Absolute pressure ratio (static outlet pressure/total inlet pressure)
R	Gas constant
T_0	Upstream stagnation temperature
γ	Specific heat ratio
h_{eq}	Equivalent heat transfer coefficient
h_0	Heat transfer coefficient
T_{shell}	Local shell external surface temperature
T_∞	External ambient temperature

REFERENCES

Raimondi, A. A., Boyd, J. (1958). A solution for the finite journal bearing and its application to analysis and design. ASLE Transactions.

Hamrock, B. J. (1994). Fundamentals of fluid film lubrication.

Ussyk, M. S. (1984). Simulação numérica do desempenho de compressores herméticos alternativos (in portuguese). Federal University of Santa Catarina, Department of Mechanical Engineering.

Lacerda, J. F., Takemori, C. K. (2016). CFD Approach to Evaluate Heat Transfer in Reciprocating Compressors. Proceedings of the International Compressor Engineering Conference.

INMETRO (2018), http://www.inmetro.gov.br/consumidor/pbe/refrigeradores.pdf, visited in March 2020.

Pai-Paranjape, V. (2014). Chemical Compatibility of High-Performance Engineering Thermoplastics in Compressor Environments. Proceedings of the International Refrigeration and Air Conditioning Conference

Solvay Specialty Polymers (2020)-Amodel® Design Guide https://content.solvay.com/amodel-ppa-design-guide

GT-SUITE Flow Theory Manual Version 2018 (2018), Gamma Technologies.

Diniz, M. C. (2018). Análise numeric-experimental de compressores alternativos em transientes periódicos típicos de refrigeradores domésticos (in portuguese), Federal University of Santa Catarina, Department of Mechanical Engineering.

ACKNOWLEDGEMENT

Design Optimization for Discharge Valve in a Rotary Compressor

Joon-hyung Kim[1]*, Mun-seong Kwon[1], Hyo-lim Heo[1]

[1] Advanced C&M R&D Lab., SAMSUNG ELECTRONICS Co., Ltd.,
Suwon-si, Gyeonggi-do, Republic of Korea
E-mail: joonh82.kim@samsung.com

* Corresponding Author

ABSTRACT

Rotary compressor used for refrigerant compression is the core equipment for room air conditioners(RACs). For an energy efficient RACs, energy efficiency metrics is moving to seasonal performance evaluation. And the compressor efficiency at low load conditions is getting more important than high load conditions. Therefore, compression efficiency under low load conditions is a major indicator of overall operational efficiency. This paper presents a design optimization procedure to improve the efficiency of a rotary compressor under low load condition using response surface method(RSM) and fluid-structure interaction(FSI) technique. Three shape dimensions associated with the discharge valve system were selected as design parameters, fifteen design points were planned using Box-Behnken scheme. The performance of each design point was evaluated by numerical analysis using FSI for the composition of response surface. In order to ensure the reliability, FSI techniques used in the performance evaluation were established through verification with PV test data. The optimum model for the valve system was derived by analyzing the constructed response surface. The responsibility of the optimum model was improved because it had relatively low stiffness compared to the base model. As a result, compression efficiency increased as the energy loss caused by the unstable behavior of the base model was reduced.

1. INTRODUCTION

A discharge valve in the rotary compressor is a unit that controls the discharge of compressed refrigerant, which has a significant effect on the efficiency and reliability of the compressor. The excessive stiffness in the discharge valve causes over-compression and reduces compressor efficiency. On the other hand, too little stiffness in the discharge valve leads to reduced efficiency and compressor reliability problems due to unstable valve behavior. Therefore, valve design with suitable stiffness for operating conditions is a key component of compressor design.

Both refrigerant (fluid) flow and valve (structural) behavior in the compressor should be considered when carrying out the numerical evaluation of rotary compressors. To accomplish this, a three-dimensional numerical method using the Fluid–structure Interaction (FSI) technique may be applied. However, because of the reliability problems of FSI schemes and the need for high computing resources, performance predictions were performed primarily in the past using theoretically accumulated formulas (Matsuzaka and Nagatomo, 1982). As the reliability of FSI schemes has been verified and the computing power has evolved, numerical studies using three-dimensional FSI techniques for rotary compressors have been gradually increasing (Ding and Gao, 2014, Yu et al., 2017, Jang et al., 2019).

In this study, design optimization was carried out for discharge valve in a rotary compressor using 3D FSI analysis technique and DOE (design of experiment). Numerical reliability was verified through PV test of the base model and comparative analysis with the results of the FSI analysis. The design variables and range of variation for discharge valves were selected to perform design optimization, and the optimum model was derived using the response surface method (RSM). The performance of derived optimum model was evaluated using FSI analysis to validate the optimization.

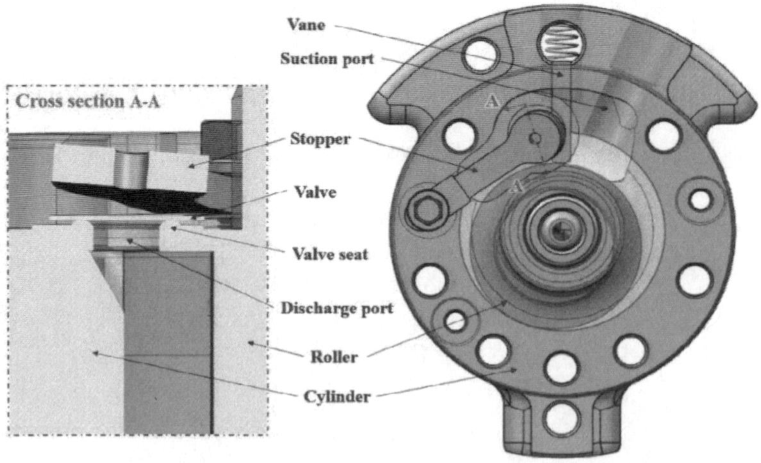

Figure 1: Schematic diagram of the discharge valve in a rotary compressor

2. COMPUTATIONAL METHODS

The rotary compressor in this study is a twin type, and the computational domain applied to the interpretation is the same as the one in Fig. 2. Because the design optimization need many test sets, the computational domain was limited to the pump section of the compressor (excluding the accumulator and the motor section) in consideration of the computing resources.

A grid system for performing the numerical analysis was created using ANSYS Mesh ver. 19.1. The grid system consists of structured and unstructured grid with the total number of nodes generated being approximately 200,000. The volume inside the cylinder is composed of a structured grid, and the following user function allows the position of each node to simulate the rotation of the roller as it moves in the x and y directions.

$$x_n = r_{dis} \cdot cos\theta,\ y_n = r_{dis} \cdot sin\theta \tag{1}$$

$$r_{dis} = r_r{}' + (r_c - r_r) \cdot \alpha_n \tag{2}$$

$$\alpha_n = \frac{r_n - r_r}{r_c - r_r} \tag{3}$$

where r_r is radius of the roller outside, r_c is radius of the cylinder inside, r_n is radius of each node point.

ANSYS CFX was used to calculate the internal flow field of the compressor and ANSYS Mechanical was used for the analysis of the valve behavior. Fluid and structure analysis data according to time step were made to be Fully Coupled in real time through ANSYS system coupling module. The governing equations used in the numerical analysis are defined as follows.

Computational Fluid dynamics:

$$\frac{\partial \rho}{\partial t} + \nabla \cdot (\rho u) = 0 \tag{4}$$

$$\frac{\partial (\rho u)}{\partial t} + \nabla \cdot (\rho u u - \tau) = f_b \tag{5}$$

where t is the time, ρ the fluid density, τ the stress tensor of the fluid region, u the velocity vector and f_b is the body force vector of the fluid medium.

Structural dynamics:

$$\mathrm{M}\ddot{U}^{t+\Delta t} + \mathrm{C}\dot{U}^{t+\Delta t} + K^t U = R^{t+\Delta t} - F^t \tag{6}$$

where M is the mass matrix, $\ddot{U}$ is the nodal point acceleration vector, C is the damping matrix, $\dot{U}$ is the nodal point velocity vector, K is the stiffness matrix, U is the nodal point displacement, R is the external load, F is the nodal point force vector.

Fluid–Structural interaction:

(Displacement compatibility) $$\dot{d}_f = \dot{d}_s \quad (7)$$

(Traction equilibrium) $$n \cdot \dot{\tau}_f = n \cdot \dot{\tau}_s \quad (8)$$

where $\dot{d}$ is the displacement and n is the unit boundary normal vector outwards from computational domain.

The pressure conditions were assigned to the inlet and outlet as a boundary condition. The low load conditions that are used in the operation of the air conditioner are targeted, and the following detailed condition is as shown in table 1. The working fluid is R410A refrigerants and the real gas properties are set to simulate compressibility. The valve material is Sandvik 20C steel.

Table 1: Operating condition

Evaporating Temp. [°C]	**Condensing Temp.** [°C]	**Superheat Temp.** [°C]	**Sub-cool Temp.** [°C]	**Rotating speed** [rev./s]
7.2	37.8	11.1	8.3	30

Table 2: Specifications of the sensors

Measuring range [mm]	**Output voltage** [V]	**Resolution** [μm]	**Linearity**	**Temperature range** [°C]
0 ~ 1	0 ~ 5	0.3	less than ±1 %	-20 ~ 180

Table 3: Specifications of the dynamic pressure sensor

Measuring range [bar]	**Natural frequency** [kHz]	**Sensitivity** [pC/bar]	**linearity**	**Temperature range** [°C]
0 ~ 250	160	-20	less than ±0.3 %	-20 ~ 350

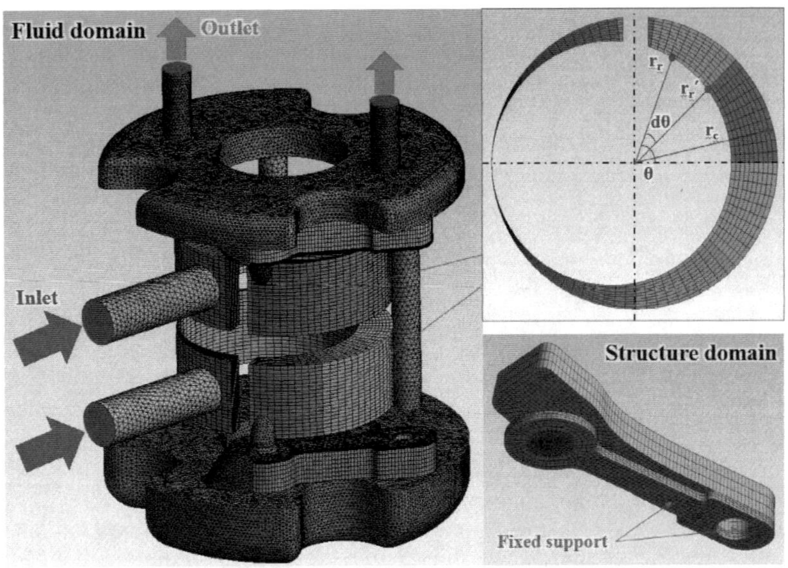

Figure 2: Computational methods to perform the FSI analysis

3. EXPERIMENTAL METHODS

To verify reliability of the computational methods, a PV test was performed along with a performance evaluation using a calorie meter. The PV diagrams derived by PV tests represent the pressure changes resulting from volumetric variation within the rotary compressor cylinder. This is an indicator of volumetric and indicated efficiency, and it is possible to identify the loss components that occur inside the cylinders in the compression process.

The volume inside the cylinder varies periodically with the rotation of the roller, so it can be calculated from the location data of the roller. Since a vane reciprocate according to the roller rotation, the location of the roller can be checked by measuring the movement of the vanes. To measure the vane position data, a gap sensor was installed on the vane as shown in Fig. 3. In addition, dynamic pressure sensors were installed by making holes on the cylinder suction side and discharge side for measuring pressure inside the cylinder. The detailed specifications of the sensors used during the PV test are described in Table 2 and 3.

Finally, the assembled compressor pump unit was installed in the bolted case and the sensor data cables were connected to external amplifiers and DAQ equipment through feed-through as shown in Fig. 4. The base model set up in a calorie meter to conduct the compressor performance evaluation and PV testing at the same time.

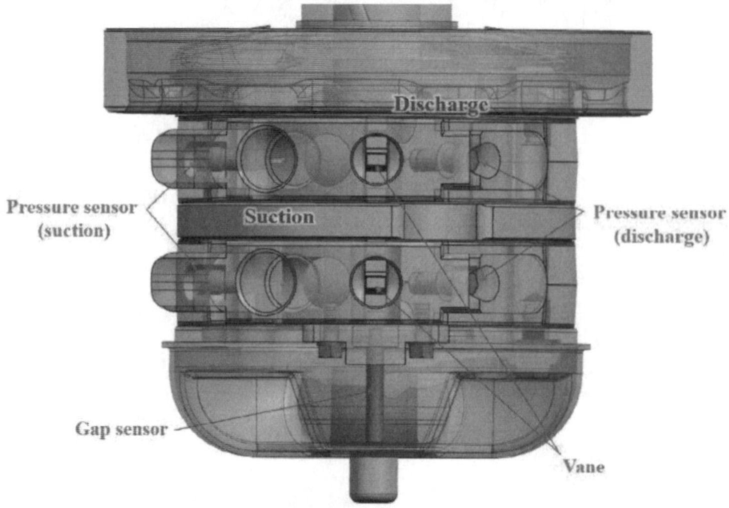

Figure 3: Installation of the PV test sensors

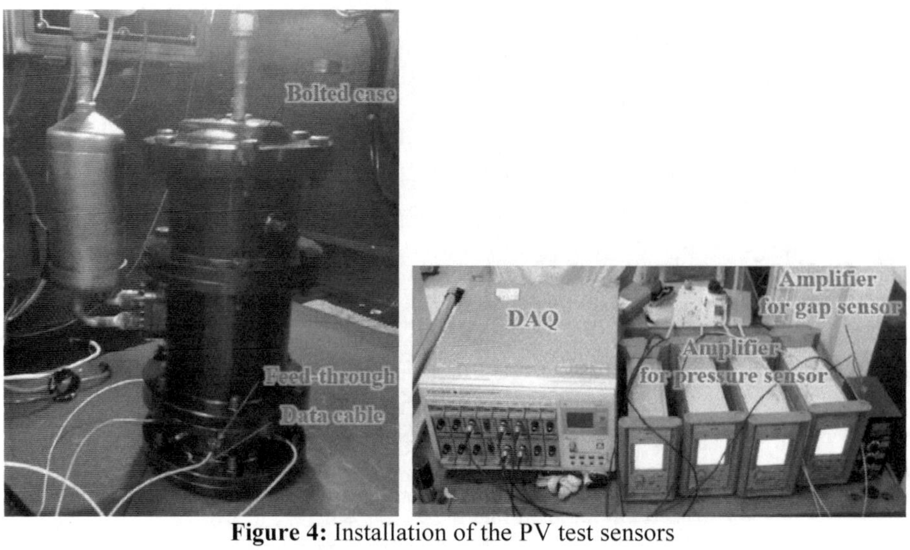

Figure 4: Installation of the PV test sensors

4. OPTIMIZATION TECHNIQUES

4.1 Design Variables

As shown in Fig. 5, discharge valves and stoppers have various design variables. As the chosen design variables increases in performing design optimization using the DOE, the number of test sets for deriving regression approaches increases exponentially. As a result, it is very important to select design variables that are highly influential because it is practically impossible to review all design variables.

Three design variables of the discharge valve and stopper were selected for independent change as they had a significant impact on the valve stiffness and were not correlated with other design variables. The selected design variables and the range are shown in Table 4.

4.2 Response Surface Method (RSM)

RSM is one of the experimental planning methods that produce the optimal solution by constructing the response surface for the design variables and the objective function. There are central position method and box-behnken method as representative methods for constructing the response surface. The review scope of the central position method is defined in the following equation.

$$R_{CCM} = R_V \times \sqrt{n_V} \tag{9}$$

where R_{CCM} refers to the generation range of the response surface, R_V refers to the range of each variable, and n_V refers to the number of variables. This means that it is possible to review areas wider than the range of variable changes selected. The box-behnken method, on the other hand, performs an evaluation of the center points of the range of design variable changes. Accordingly, the scope of review is narrower than the central composition method but has the advantage of relatively small test sets, so it is suitable to apply to the ones with limited range of changes in design variables.

Among the design variables chosen to perform this study, the valve thickness (T_v) was related to reliability and had to be reviewed within the range of the selected changes. Therefore, the box-behnken method was applied as a method to construct the response surface.

Table 4: Design variables and ranges for the design optimization

Design variable	Range (relative dimension to the base model)	
	Min.	Max.
Stopper bending radius (R_s)	0.82	1.18
Valve neck width (W_v)	0.6	1.4
Valve thickness (T_v)	0.525	1.050

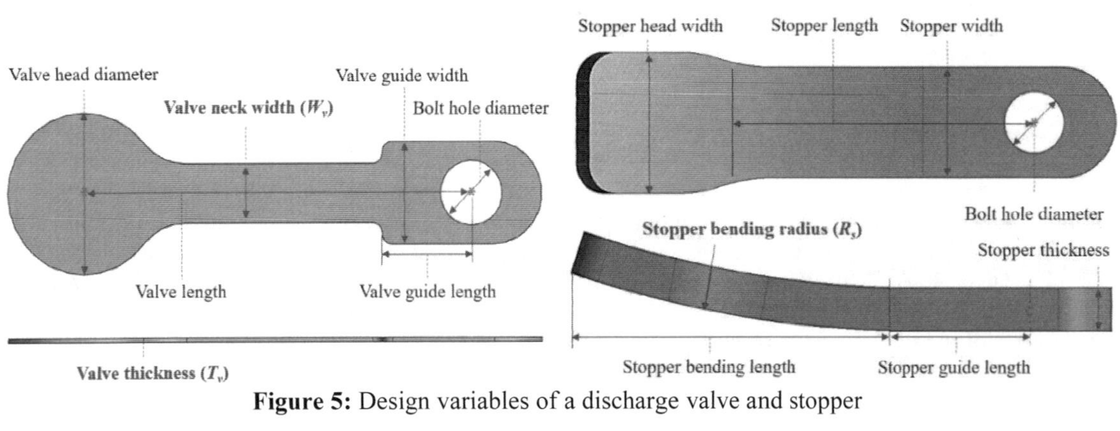

Figure 5: Design variables of a discharge valve and stopper

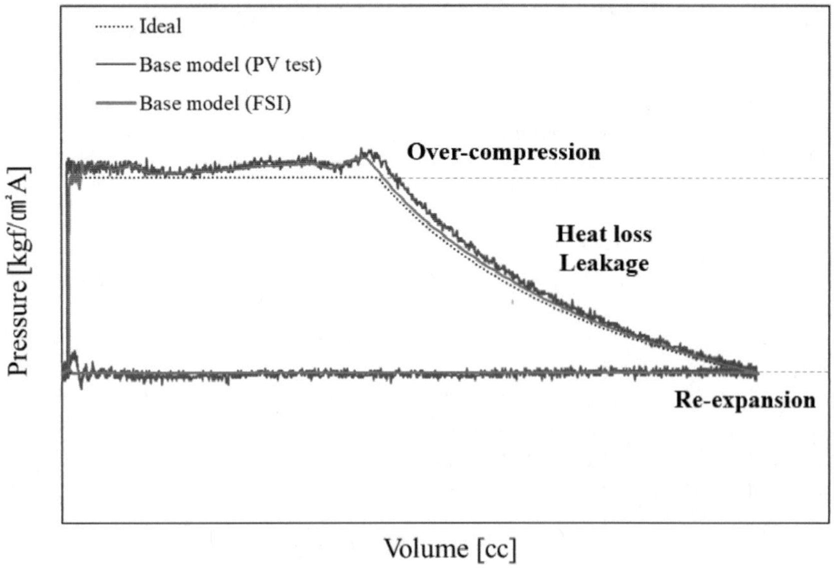

Figure 6: PV diagram for the base model

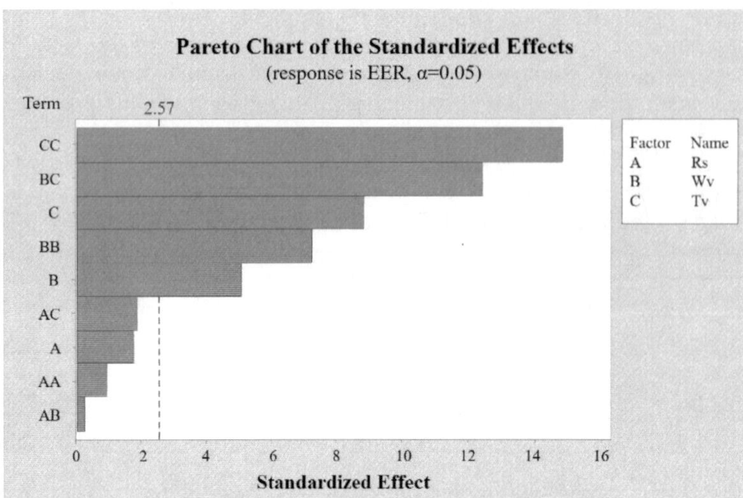

Figure 7: Pareto chart for the design effect

5. RESULTS AND DISCUSSION

5.1 Validation Results for FSI Analysis

To verify reliability of the FSI analysis, the two data were compared and verified after the numerical analysis and the PV test of the base model were carried out. The verification result is shown in Fig. 6. The calculated data from the FSI analysis showed a tendency to be very similar to the PV test data.

In this study, the FSI analysis limited the computational domain to the compressor pump section, taking into account the computing load. Accordingly, the condition for heat exchange was assumed as adiabatic condition, and only the amount of sweep clearance leakage was considered. As a result, there was some error caused by heat loss and other leakage components. However, it was determined that the error caused by heat, other leakage would not affect the optimization results because the optimization targets are ones for valve and stopper. Previous studies evaluated for the discharge system using FSI analysis techniques were also performed excluding the effect on heat loss and other leakage components (Tan et al., 2014, Lemke et al., 2016).

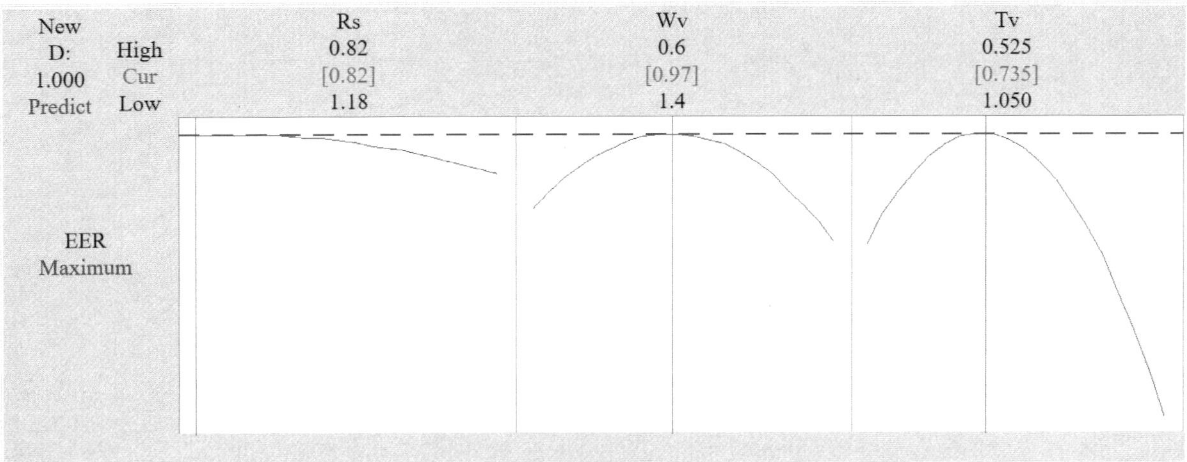

Figure 8: Optimization result for the design variables

5.2 Results of the Design Optimization

As mentioned in Section 4.1, three design variables and range of variation for the discharge valve and stopper were selected for performing optimization. A total of 15 test sets were created for constructing the response surface using the box-behnken method. Two sets of them are the ones for calculating the error range that occurs during repeated tests. The set for repeated performance was excluded from the assessment because the numerical analysis was used to evaluate the performance of each test set.

After performing a performance evaluation of 13 test sets, the optimization analysis was performed using the commercial program MiniTab 18. The objective function for the optimization was targeted at the compressor efficiency (energy efficiency ratio, EER). Figure 7 presents the impact assessment result for each variable and interaction. The interaction for valve thickness were shown to have the greatest influence on compressor efficiency, and the influence of interaction between valve neck width and valve thickness was also shown to be high. On the other hand, the impact of the stopper bending radius was minimal. This is judged to be because the amount of valve displacement is not large due to optimization for low load conditions.

Finally, the optimal values were derived to maximize compressor efficiency through a regression analysis of the calculated response surface as shown in Fig. 8. The selected optimal design variable values (relative dimension to the base model) were 0.82 in stopper bending radius (R_s), 0.97 in valve neck width (W_v) and 0.735 in valve thickness (T_v), when the EER increase rate (relative performance to the base model) predicted 0.99%. To verify the generated optimum model, this model was evaluated using the same FSI analysis method as the base model. The relative dimensions and EER increase rate for the optimum model were calculated. The calculated performance for the optimum model (EER increase rate: 1.01%) was consistent with predicted value.

PV diagram and valve behavior characteristics calculated by the FSI analysis were created to confirm the performance improvement as shown in Fig. 9 and Fig. 10. Data on valve behavior were dimensionless based on maximum valve opening level for the base model. The optimum model designed by optimization has reduced the neck width and thickness of the discharge valve compared to the base model. It means restoring force reduction due to reduced stiffness of the valve. The initial valve opening level increased during the discharge of compressed refrigerant due to reduced stiffness of the valve. In addition, the valve displacement after the initial opening was stable in the optimum model. The valve of the base model was closed before the refrigerant discharge was completed, and secondary behaviors were observed for the residual refrigerant discharge. The stabilization of the valve behavior resulted in the smooth refrigerant discharge, reducing the loss of over-compression along with the rise of the cooling capacity, thus increasing the efficiency of the compressor. Based on these results, it was verified that the design optimization of the discharge valve and stopper for low load condition was properly achieved.

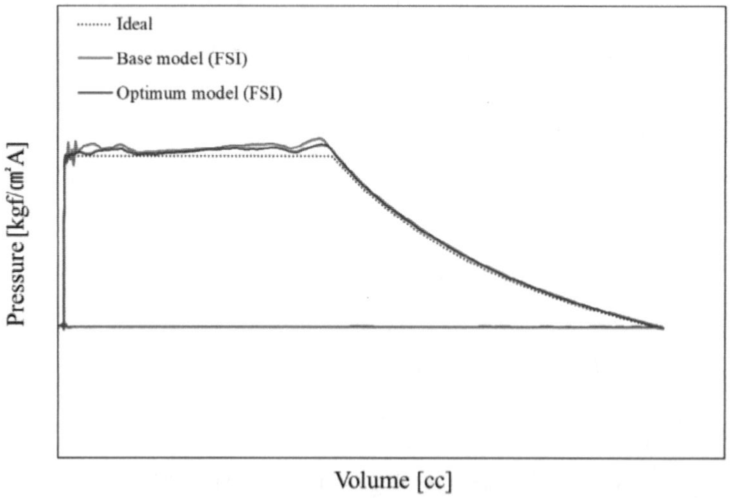

Figure 9: PV diagram for each model

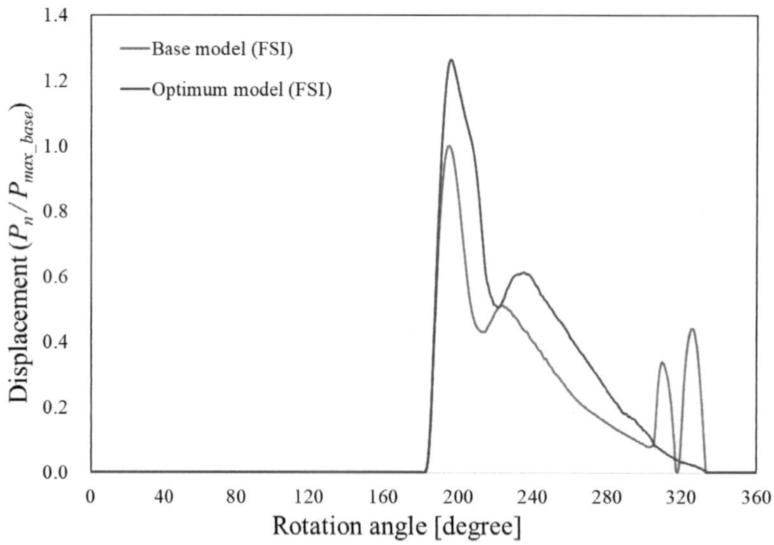

Figure 10: Valve displacement characteristic for each model

6. CONCLUSIONS

In this study, the design optimization of the discharge valve and stopper was carried out to improve compressor efficiency under low load conditions using FSI numerical analysis and DOE. The results of the study are as follows.

- Full FSI analysis methods were established for compressor performance evaluation and reliability was verified by PV test. Although heat loss is excluded considering computing time, it is appropriate to use them in evaluating design variables that affect volumetric and indicated efficiency.

- Design optimization has resulted in an optimum model for the discharge valve with lower stiffness compared to the base model. The efficiency of the optimization model was increased by 1% over the base model. This means that if the compressor operating load change, a review for the valve stiffness to improve compressor efficiency will be effective.
- In future works, a reliability review of the optimum valve derived from this study will be carried out and the review design variables will be expanded as well.

NOMENCLATURE

C	damping matrix	
f_b	body force vector of the fluid medium	
K	stiffness matrix	
M	mass matrix	
P_n	valve opening level at each point	(mm)
P_{max_base}	maximum valve opening level for base model	(mm)
r_c	radius of cylinder inside	(mm)
r_n	radius of each node point	(mm)
r_r	radius of roller outside	(mm)
R_s	stopper bending radius	(mm)
T_v	vale thickness	(mm)
u	fluid velocity vector	
U	nodal point displacement	
$\dot{U}$	nodal point velocity vector	
$\ddot{U}$	nodal point acceleration vector	
W_v	valve neck width	(mm)
x_n	displacement(x-direction) of each node point	(mm)
y_n	displacement(y-direction) of each node point	(mm)
ρ	fluid density	(kg/m^3)
τ	stress tensor	

REFERENCES

MatSuzaka, T., & Nagatomo, S. (1982). Rolling piston type rotary compressor performance analysis. *In: International Compressor Engineering Conference at Purdue*. Paper 386.

Tan, Q., Liu, Z., Cheng, J., & Feng, Q. (2014). Effective flow and force areas of discharge valve in a rotary compressor. *In: International Compressor Engineering Conference at Purdue*. Paper 2266.

Ding, H., & Gao, H. (2014). 3-D Transient CFD Model For A Rolling Piston Compressor With A Dynamic Reed Valve. *In: International Compressor Engineering Conference at Purdue*. Paper 2362.

Lemke, N. C., König, M., Hennig, J., Försterling, S., & Köhler, J. (2016). Transient experimental and 3D-FSI investigation of flapper valve dynamics for refrigerant compressors. *In: International Compressor Engineering Conference at Purdue*. Paper 2415.

Yu, X., Tan, Q., Ren, Y., Jia, X., & Jin, L. (2017). Numerical study of the reed valve impact in the rotary compressor by FSI model. *Energy Procedia*, *105*, 4890-4897.

Jang, S., Choung, H., Park, S., & Lee, S. (2019). Investigation on noise of rotary compressors using fluid-structure interaction. *Journal of Mechanical Science and Technology*, *33*(11), 5129-5135.

Research on Noise Reduction of Variable Speed Rotary Compressor with Large Capacity

Haijun Wang, Zhibin Xie, Yinxiao Lu, Weiyan Chu, Zheng Li

Shanghai Highly Electric Appliance Co., Ltd.
Shanghai, China
xiezb@shec.com.cn

ABSTRACT

With the increasing speed and capacity of variable-speed rotary compressors, the problem of noise especially low and medium frequency noise in the air conditioning system which can't be solved by wrapping soundproof cotton has became more serious. In this paper, based on the noise problem of the rotor compressor with a working capacity of more than 80CC, the main frequency and the position of the noise source within 1000Hz are confirmed by simulation and experiment. Then on the base of this, the muffler and accumulator are respectively optimized and improved combining with Computer Aided Engineering (CAE) means. The final application results show that the optimized scheme can reduce noise by 6.1dB in 160Hz and 8.9dB in the frequency range of 500Hz to 800Hz, achieving good results.

1. INTRODUCTION

The rotor compressor has good refrigeration performance and variable condition ability, which makes it have great application advantages in small refrigeration device. It constantly breaks through the upper limit of refrigeration capacity to replace small scroll compressor, and is widely used in household air conditioner and small commercial refrigeration device. Nowadays, rapid increasing of refrigeration and air conditioning system in modern industries brings attention to the urgency of core technology development in this area. (Qian and Zhang(2018)) A variable speed rotor compressor with a working capacity of 80CC has been designed and developed, but its noise also increases significantly with the increase of refrigeration capacity, especially the medium and low frequency noise which is difficult to be suppressed by external sound-absorbing materials. At present, great progress has been made in the research on vibration and noise reduction of rotary compressors at home and abroad. Shang Zhiwu(2003) researched the identification method of the noise source of the rotary compressor and achieved the goal of vibration and noise reduction by optimizing the structure of the muffler and accumulator. Kwang ha Suh and Jin Dong Kim(2000) studied the muffler of the rotary compressor, and explored the influence of the exhaust hole and cavity capacity of the muffler in the concerned frequency band on the noise reduction. In order to reduce the noise of the newly developed rotor compressor, noise analysis and low-noise design were carried out in this study.

2. MECHANISM of NOISE GENERATION

In order to define the target of noise reduction and the mechanism of noise generation, the noise of the compressor in the single machine state and the state of wrapping sound-absorbing material in the simulation air-conditioning system are tested respectively in the semi anechoic room, and the sound power spectrum is shown in Figure 1. It can be seen that there are peaks in the frequency bands of 160Hz, 500-800Hz and 1600-2500Hz for single compressor under test, while the noise attenuation above 1000Hz is obvious in the state of analog air conditioning system, so the main noise reduction target is the peaks in the frequency bands of 160Hz and 500-800Hz.

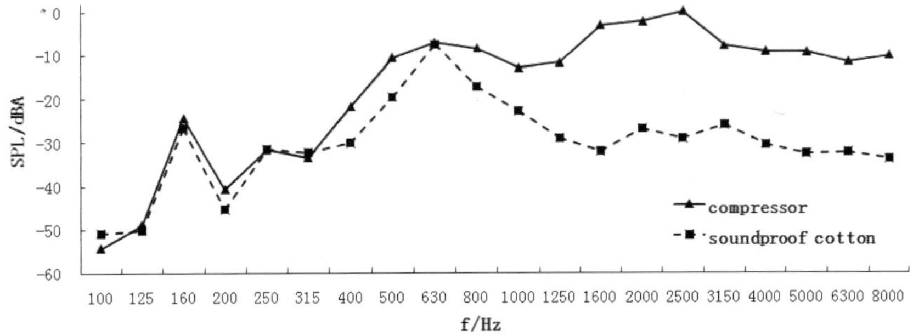

Figure 1: Noise level of rotary compressor

2.1 The mechanism of Medium-frequency Noise Generation

The research method of the noise source location of the rotary compressor is relatively mature that shown in Yang Weibiao (2000). In order to confirm the location of the noise source in the frequency range of 500 to 800Hz, the test of near-field noise source is carried out for the compressor. The location of the measurement point and the test results in 630Hz are shown in Figure 2. It can be confirmed that the medium-frequency noise is mainly located in the upper part of the compressor cavity, which includes cavity resonance and motor noise, etc.

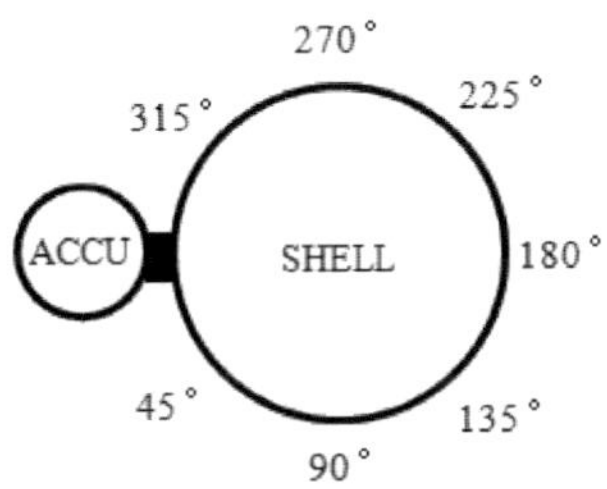

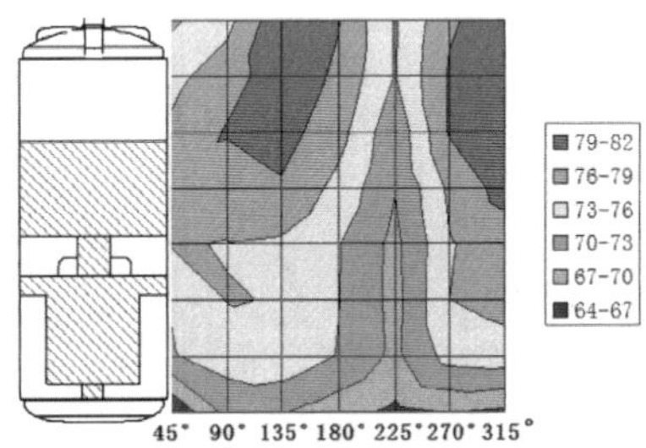

Figure 2: (a)Schematic view for measurement point in detail, (b) result of experiment

In order to further confirm the mechanism of noise generation, the acoustic Finite Element Method (FEM) is used to calculate the acoustic mode of the compressor cavity. It is found that there are multi-stage cavity acoustic modes in the frequency range of 500 to 800Hz. The calculation results are shown in Figure 3. In order to verify the accuracy of the numerical calculation results, the acoustic mode of the compressor was tested. The frequency response curve obtained from the test is shown in Figure 4. It can be seen that there are multiple peaks of sound pressure response in the concerned frequency band, which indicates that there are multiple acoustic modes. The excitation frequency of fluid pulsation in compressor is broadBand, which is easy to excite acoustic mode and cause noise problems.

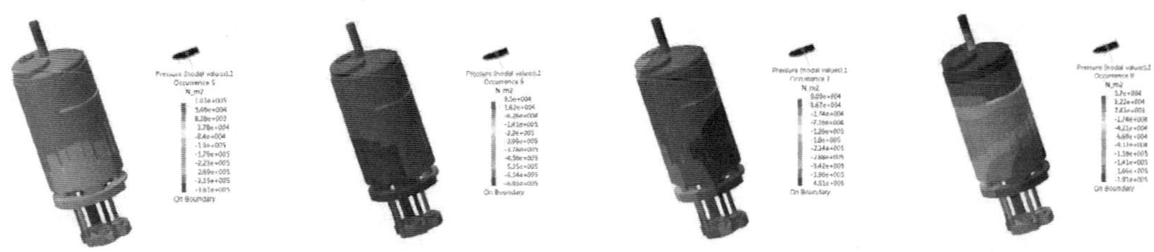

(a) 521 Hz (b) 551 Hz (c) 564 Hz (d) 577 Hz

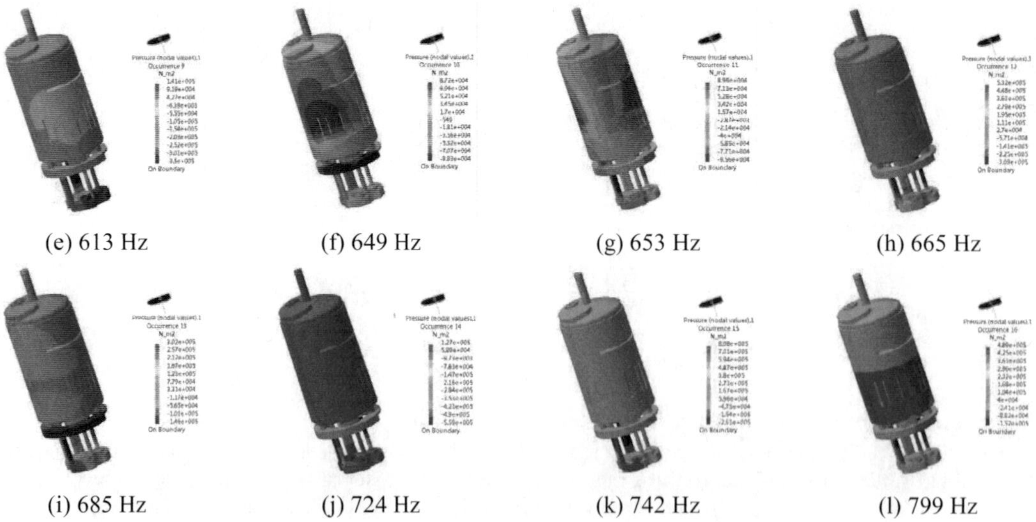

Figure 3: The schematic view mode shape in a rotary compressor

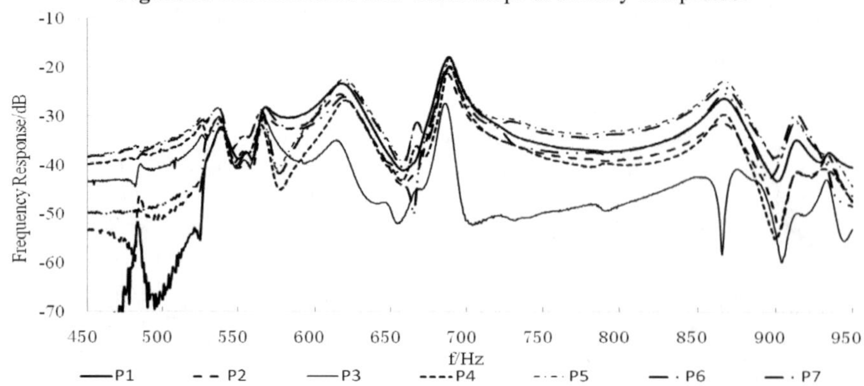

Figure 4: Frequency response function of cavity

2.2 The Mechanism of Low-frequency Noise Generation

Due to the poor effect of microphone on the sound source positioning test in 160Hz, the sound source analysis is carried out by testing the acceleration of different positions on the surface of the whole machine, and the position of the test point is shown in Figure 5. As shown in Figure 6, it can be found that the vibration points on both sides of the accumulator in the tangential direction of the compressor are significantly higher than the other measurement points, and the values of the other measurement points are similar, which indicates that the noise source in the 160Hz is located in the accumulator.

As we all know, some natural modes of the structure itself will be excited due to the working load, which will generate corresponding responses at the excited modal locations. These superimposed responses are the mechanical noise we are concerned about. In order to clarify the mechanism of mechanical noise in 160Hz, the structural mode of the accumulator was tested by the impact testing. The results are shown in Figure 7. At the same time, because the compressor uses a two-cylinder structure, the pulsating frequency of its load torque is 2 times the rotational speed frequency. The double-speed frequency in 160Hz has a strong excitation force when running at 4800 rpm. The operating load of the compressor caused the first-order tangential swing mode of the accumulator to be excited, and the structural resonance caused the generation of noise, which explained the mechanism of noise at 160Hz.

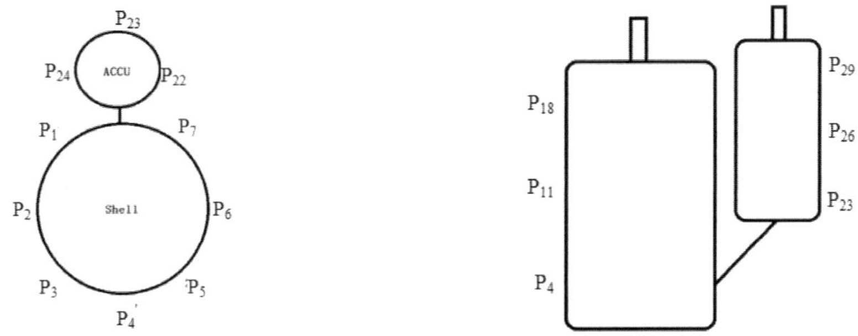

Figure 5: Schematic view location of vibration sensor

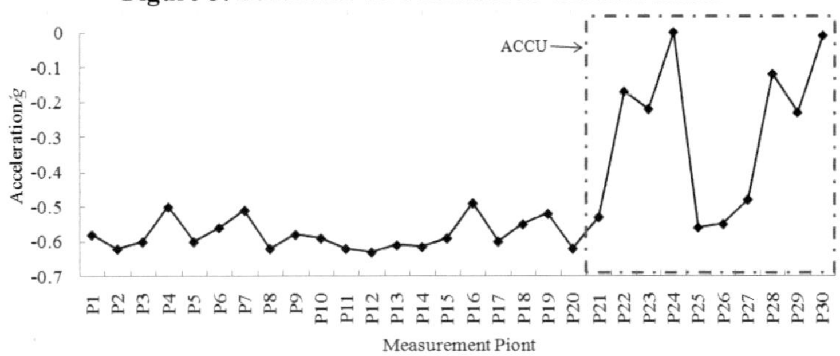

Figure 6: Schematic view experimental result of vibration

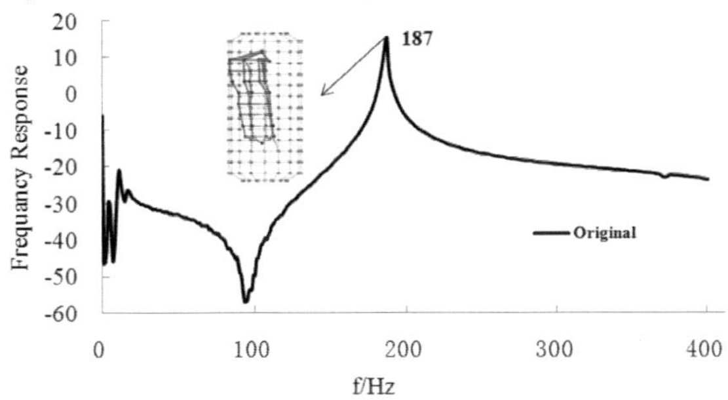

Figure7: Schematic view mode shape and frequency response function (FRF) of the accumulator

3. OPTIMITZATION OF MUFFLER

The previous section confirmed that the noise in the frequency range of 500 to 800Hz is caused by the resonance which caused by the acoustic mode of the compressor's cavity, which can be reduced by optimizing the structure of the compressor's muffler to reduce the exhaust pulsation and change the excitation frequency.(Huang and Liu(2007), Li and Jia(2009), Liu(2004)) The evaluation method of muffler is established to evaluate the accuracy of optimization design and confirm the optimization effect. Based on the muffler model, Finite Element Method(FEM) is used to calculate the transmission loss, which is also used as the basis of muffler performance evaluation.

The muffler of the rotor compressor belongs to the resistant muffler, which can suppress the noise in a specific frequency band through the reflection and superposition of sound waves. The noise reduction performance of this muffler is closely related to its volume, cavity distribution, location and area of the inlet and outlet. In this study, these parameters are optimized. Figure 8 shows the three-dimensional model of the original structure and the improved structure. The original muffler is a single-layer structure, which can be optimized by adding a layer of partition in the cavity of the muffler. At the same time, the outlet structure of the muffler is also improved.

Figure 9 shows calculation results of the transmission loss of the muffler before and after improvement. It is found that by changing the structure of the muffler, the acoustic mode of the cavity inside the muffler is adjusted, and the transmission loss performance of the muffler is improved. In the target frequency band of 500 to 800Hz, the transmission loss of the optimized muffler is significantly improved.

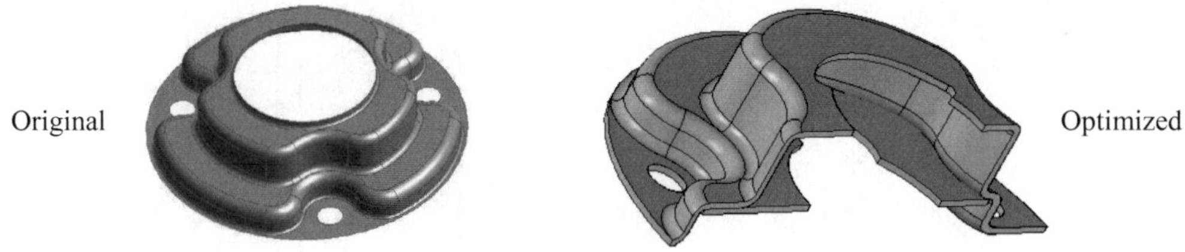

Figure8: Modification of muffler shape

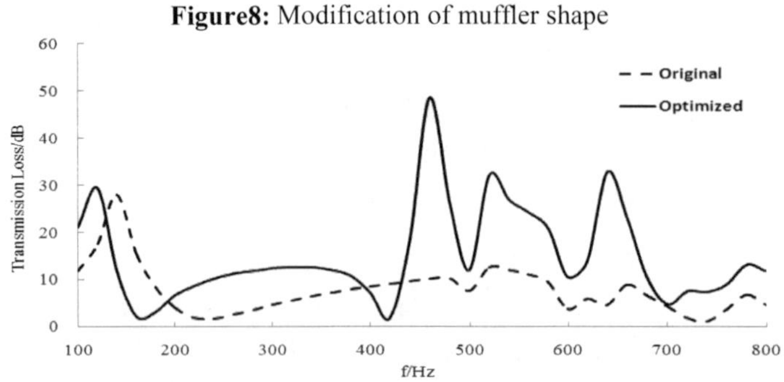

Figure 9: The schematic view transmission loss of the muffler

4. OPTIMITZATION OF ACCUMULATOR

The low-frequency noise in 160Hz is high, which has a great correlation with the tangential mode of the accumulator. Therefore, the natural frequency of the main structural mode of the accumulator is far from the range of second harmonic frequency of rotational speed by optimizing the structure of the accumulator, so as to achieve the goal of noise reduction.(Zhu and Gao(2008), Xiang and Ye(2004)). Considering that the operating speed range of the variable frequency compressor is broadband, which can reach 3000 to 6000 RPM (50 to 100 rps), the natural frequency can be adjusted by optimizing the tangent stiffness of the accumulator, and the first-order tangential swing mode can be increased to the area far away from the rotational speed frequency.

The natural frequency of the structure is positively related to its own stiffness. The increase of the stiffness of the accumulator system results in the increase of its modal frequency. The tangential stiffness of the accumulator is improved through the optimization design of the tank-stay of the accumulator. The CAE method is used to analyze and evaluate the optimization schemes. The results of the optimization scheme and its modal natural frequencies are compared with that of FIG 10. After the tank-stay is optimized, the first order tangential swing mode frequency of the optimal scheme is increased from 193Hz to 267Hz.

Impact testing is used to verify the structural mode of the accumulator before and after improvement. Figure 11 shows the frequency response characteristic curve of the accumulator structure. The test results are similar to the simulation results, and the modal frequency of the accumulator has been significantly improved.

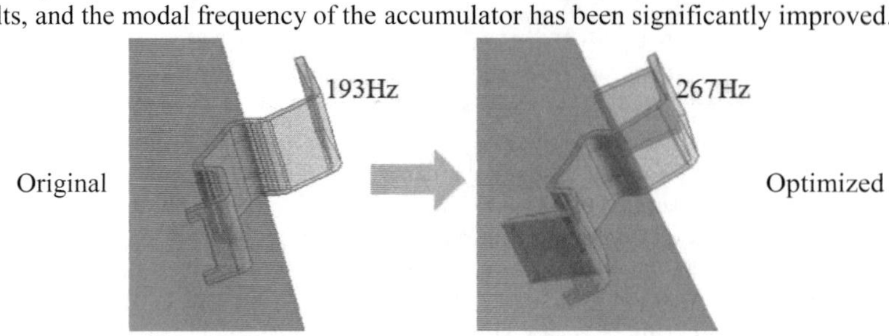

Figure10: Modification of tank-stay shape

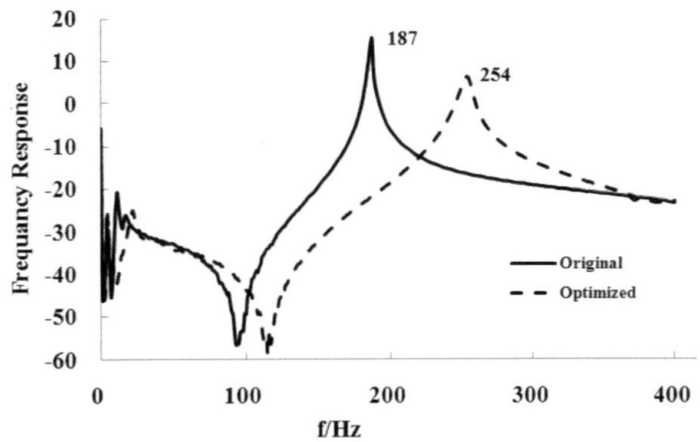

Figure 11: The frequency response function (FRF) of the accumulator

5. EFFECT of NOISE REDUCTION

Based on the optimization of the accumulator structure, the natural frequency of tangential mode is larger than the range of the second harmonic frequency of rotational speed. The actual improvement effect can be verified by testing the vibrations of the accumulator before and after the optimization, as shown in Figure 12. It can be seen that the vibration of the accumulator is significantly reduced by using the improved tank-stay, and the average vibration amplitude is reduced to about 52% at multiple rotating speeds, which has achieved good vibration reduction effect.

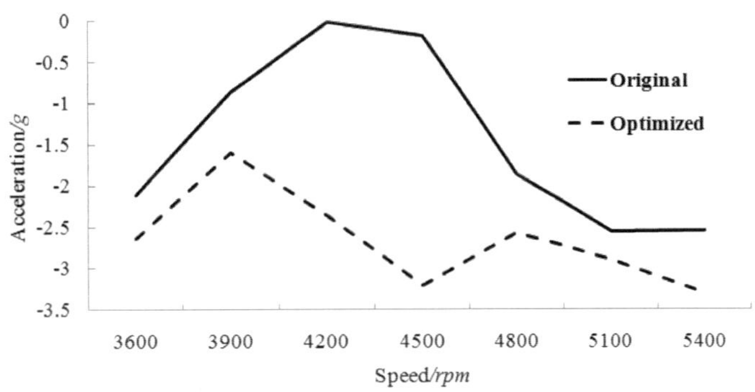

Figure 12: Schematic view experimental result of tangential vibration of the accumulator

The optimal solution was selected for trial production, and the noise test was carried out to verify the results as shown in Figure 13. Compared with the original compressor, it can be seen that the noise power of the compressor decreases by 8.9dB on average in the frequency range of 500-800Hz and 6.1dB in 160Hz, and the noise reduction effect is obvious.

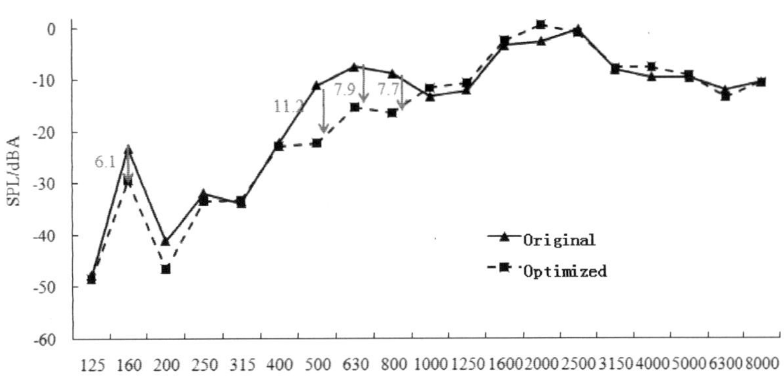

Figure 13: The schematic view noise pressure

6. CONCLUSIONS

In this paper, the high noise of the newly developed variable-speed compressor with a capacity of more than 80CC is studied. The location and generation mechanism of the main noise sources within 1000Hz of the compressor are confirmed by noise experiment, the location of noise source, impact testing and numerical analysis. On the basis of the analysis, the numerical analysis method of the structure of the accumulator and muffler is established, and the structure optimization and evaluation of the two models are carried out. The final optimization scheme test shows that the average vibration amplitude is reduced to about 52% at multiple rotating speeds, and the compressor can reduce noise by 6.1dB in 160Hz and 8.9dB in the frequency range of 500 to 800Hz, achieving good effect of vibration and noise reduction.

REFERENCES

Qian,Y. Zhang, Li. Zhang, X. Ying, Z,Q.(2018). P-V Diagram Measurement of a Variable Speed Rotary Compressor With Large Displacement. *Proceeding of the 24th International Compressor Engineering Conference* ,(1237).

Zhiwu Shang.(2003). Research on Mechanism and Application of Noise Reduction of Rotary Piston Compressor.

Kwang-Ha Suh and Jin-Dong Kim,(2000) The Analysis of the Discharge Muffler in the Rotary Compressor, *Proceeding of the 15th International Compressor Engineering Conference at Purdue*, Purdue University, West Lafayette, Indiana, 2000.651~656.

Yang, W,B.(2018) Noise analysis of outdoor unit of variable frequency air conditioner. *Mechanical and Electrical Information*. 2018(2):121-123.

Huang, Z,Y. Jiang, W-K, Liu, C-H.(2007). Review and Prospect of Research on Fluid Noise of Rotary Compressor. *Journal of Vibration and Shock*. 2007,26(7):159-163.

LI, X,S. JIA, G. ZHOU, Y,H.(2009) Investigation into Noise Analysis and Reduction for Rolling Rotor Compressors. *Chinese Journal of Construction Machinery*. 2009,7(1):105-108.

LIU, N. WANG, T,Y.(2004). Improvement Study of Muffler of Rotary Compressor. *Compressor Technology*. 2004(6):26-29.

ZHU, B, S. GAO, Q. CHEN, Z,H.(2008).Analysis of Acoustic Characteristics of Accumulator of Rotary Compressor *Proceeding of the 9th Air Conditioners, Refrigerators (Cabinets) and Compressors*.2008.186~191.

Xiang H, S, Ye H, H, Hun H J.(2004) Analysis of noise for accumulator of rotary compressor. *Proceedings of the 4th international symposium on heating, ventilating and air conditioning,* Beijing:2004,1(2):940-944.

Comparative Studies of Scroll and Rotary Compressors for US Market Heat Pumps and Air Conditioners

Song LI[1], James CAROW[1], Daniel BACELLAR[1], Cara MARTIN[1*]

[1]Optimized Thermal Systems, Inc., Beltsville, MD
(cmartin@optimizedthermalsystems.com)

*Corresponding Author

ABSTRACT

Rotary compressors have long been developed and adopted for heating, ventilation, and air-conditioning (HVAC) applications across Asia, primarily due to their simpler mechanism and fewer parts as compared to their counterparts such as scroll compressors. However, rotary compressors in heat pumps (HPs) and air conditioners (ACs) in the US have limited market share and are often confined to systems smaller than 3.0 tons (10.6 kW). This paper consists of two parts; the first is on rotary compressor technology and its advantages and disadvantages from both technical and market standpoints. The review consists of a survey of the literature, as well as a survey from field experts through anonymous interviews. Conventionally, rotary compressors are regarded as having lower efficiency in systems larger than 3.0 tons (10.6 kW), which limits their current application to small packaged systems and automotive ACs. The second part includes an experimental investigation of compressor and system efficiencies using scroll and rotary compressors. The compressors compared were drop-in replaced in typical 2.5 (8.8-kW) and 5.0-ton (17.6-kW) R410A split HP systems. Experimental tests, both in cooling and heating modes, were conducted under AHRI 210/240 Standard operating conditions. The test units were extensively instrumented on both the refrigerant and air-side to measure temperature, humidity, pressure, flow rate, and power consumption, according to ASHRAE Standard 41.2. Indoor and outdoor units were placed in a wind tunnel and in an environmental chamber, respectively. The results showed the 2.5-ton (8.8-kW) unit rotary compressor's isentropic efficiency was 2.6% and 14% higher than the scroll compressor in cooling and heating, respectively. At 5.0 tons (17.6 kW), the isentropic efficiency of the rotary compressor was 5.4% lower in cooling and 6.3% higher in heating. In terms of volumetric efficiency, at 2.5 tons (8.8 kW) the rotary compressor was 1.7% lower in cooling than scroll compressor, and comparable to the scroll compressor at two of three heating mode test points. At 5.0 tons (17.6 kW), the rotary compressor volumetric efficiency was 0.7% and 2.8% higher than the scroll compressor in cooling and heating mode, respectively. The overall system with the rotary compressor had 5.7% higher seasonal energy efficiency ratio (SEER) and 3.0% higher heating seasonal performance factor (HSPF) than the scroll compressor at 2.5 tons (8.8 kW). At 5.0 tons (17.6 kW), the system with the rotary compressor was 2.6% higher in SEER and 0.6% higher in HSPF compared to the system with the scroll compressor.

Keywords: rotary compressor, scroll compressor, split heat pump, experimental test, comparison

1. PART I: REVIEW ON ROTARY COMPRESSORS

Compressors plays a vital role in the HVAC and refrigeration (HVAC&R) industry, and they are the most energy demanding component within a vapor compression cycle (VCC). Their characteristics are important in defining their operation and, as such, they are usually selected for specific applications within a given market. For example, rotary compressors have been widely adopted for HVAC applications throughout Asia, due to their simple mechanisms and fewer moving parts. However, in the US the rotary compressor is limited to small (less than 3 tons or 10.6 kW) packaged systems and other small VCCs such as automotive ACs. To better understand the reason behind the limited application of rotary compressors in the US market, a total of twelve direct interviews were conducted, which were compiled, along with feedback received from a diverse set of representatives from the industry in private conversations: 45% were from compressor manufacturers, 44% from HVAC&R system manufacturers, and 11% from academia (Figure 1a). All feedback and comments received were categorized into different groups, and their frequencies were calculated based on the total number of comments, as shown in Figure 1b. The most frequent comment (almost 14%) is that from a performance or efficiency standpoint; rotary compressors are suitable for small systems (less than 3 tons or 10.6 kW), but do not have a comparative performance in larger capacity systems (threshold around 2-4 tons or 7.0-14.1 kW) versus other technologies such as scroll compressors.

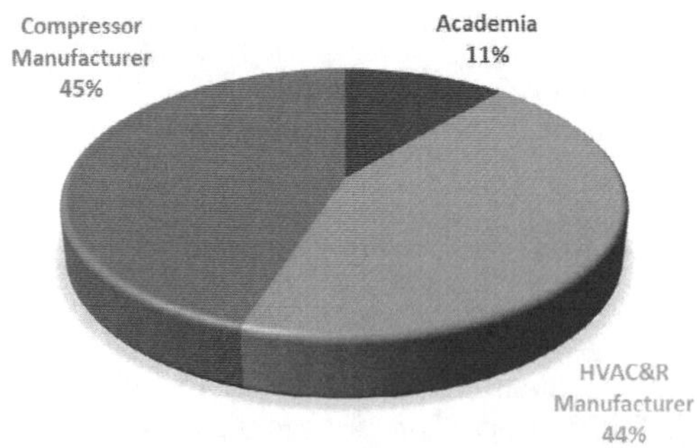

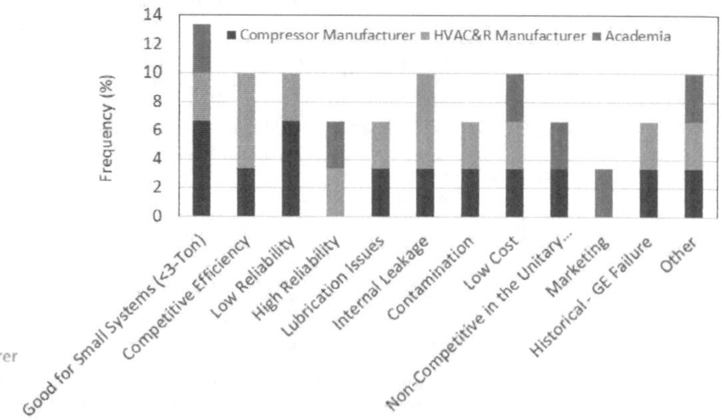

Figure 1: Interview Statistics: a) Source Types; b) Grouped Responses.

Another common response for rotary compressor utilization within the US was the impact of one particular event, namely General Electric's (GE) challenge to introduce rotary compressors in their refrigeration systems during the 1980's (Magaziner and Patinkin, 1989). By 1986, GE built a plant and started producing rotary compressors for their refrigeration products. But in 1988, some compressors from their product portfolio started failing. After investigation, GE discovered improper lubrication led to failing of small parts, and a major recall was announced. Not long after GE's rotary compressor recall, GE sold its rotary compressor manufacturing plant in São Carlos - Brazil to another manufacturer Tecumseh (Form 10-K Tecumseh Products Co. Annual Report, 1994), and then focused its rotary compressor market on recreation vehicle (RV) products instead. Another crucial historical event was Copeland's introduction of their scroll compressor, which has less internal leakage and better performance above 2.5 tons (8.8 kW). From 2014-2018, almost 70% of all US manufacturers' shipments (transfer of ownership) of central ACs and air-source HPs were larger than 2.2 tons (7.7 kW) (Figure **2**) and scroll compressors have been widely used across all brands in AC/HP systems in the US. Both events are better depicted in Figure 3, and seem to have driven the opinion about rotary compressors in the US that, in general, they are not worth the cost of technology development and replacement.

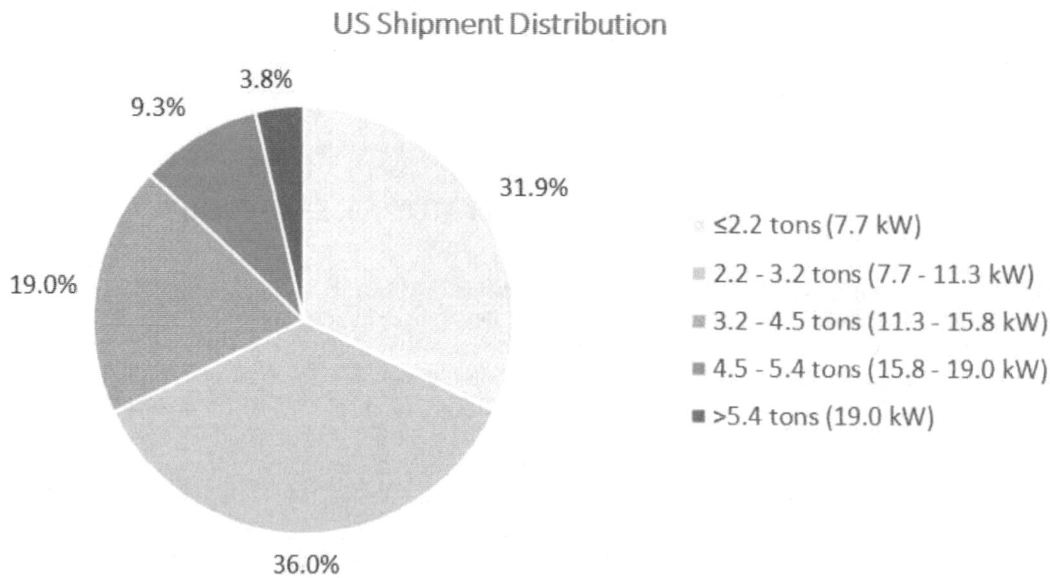

Figure 2: 2014-2018 US Manufacturers' Shipments of Central ACs and Air-source HPs Distribution ("AHRI Statistics.")

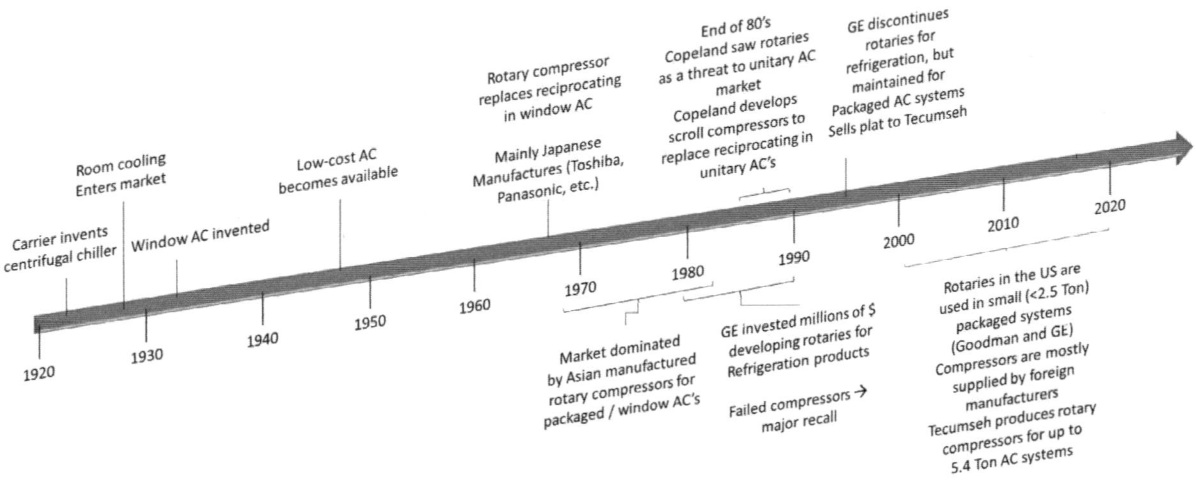

Figure 3: Timeline of Compressors in the US HVAC&R Industry

To better evaluate rotary compressor's potential in the HVAC&R applications, a large amount of research work accomplished. Shi (Shi et al., 2016) replaced the aluminum scroll compressor used in electric vehicle heat pump systems with a steel rotary compressor, which is better at enduring long-term high-pressure application, and studied the impact of different refrigerants on system capacity and COP. An investigation was attempted by another research group, in which several compressor geometric parameters were optimized, and a new rotary compressor model was developed to improve COP and capacity, reaching maximum cooling capacity of 70 HP (52 kW) (Monasry et al. 2018). In contrast, Sathe et al. (2008) developed a prototype miniature rotary compressor with an estimated cooling capacity ranging from 0.046 ton (0.163 kW) to 0.139 ton (0.489 kW), making it ideal for electronic cooling applications.

There are no fundamental aspects of the different compressor technologies, or clear consensus, that suggest which compressor type is better suited for each application. The technical simplicity and low cost of rotary compressors are the main drivers for pursuing a robust technology that can be deployed in any system. In particular, the literature lacks comprehensive compressor investigations for mid-size heat pumps (2-5 tons). The second part of this study consists of a comparative experimental investigation of such systems while using typical scroll compressors and equivalent rotary compressors.

2. PART II: EXPERIMENTAL INVESTIGATION

Instead of only testing rotary compressor performance, the present study conducted a direct comparison between rotary compressor and a baseline scroll compressor and compare how they performed in two split HP systems, one rated at 2.5 tons (8.8 kW) nominal cooling capacity, and the other at 5.0 tons (17.6 kW). Both systems were equipped with a fixed-speed scroll compressor and served as the baseline. With regards to system selection, fixed speed compressors were chosen instead of variable speed compressors; while variable speed systems have their advantages, they currently only account for approximately 20% of the US market (BPA, 2016).

Comprehensive instrumentation was installed on both refrigerant and air-side, and both systems were operated under a series of test conditions. Scroll compressors were then replaced with comparable rotary compressors and test conditions were repeated. Based on the test data, system performance with each of the compressors was compared in both cooling and heating modes.

The two units under test (UUT) were selected to represent a typical split HP. Both UUTs are equipped with a fixed thermal expansion valve (TXV) and run on a single speed scroll compressor that used R410A as the refrigerant. A total of six operating conditions (Table 1) were defined according to AHRI Standard 210/240 (2017), specifically for a single stage air-cooled system.

Before testing, the outdoor unit was placed inside a temperature and humidity-controlled chamber, and the indoor unit was connected to a wind tunnel, in which temperature and humidity were both set to the test conditions. Both UUTs were extensively instrumented on both the refrigerant and air-side to measure temperature, humidity, pressure, flow rate, and power consumption, according to ASHRAE Standard 41.2 (1987). A detailed test schematic is shown in Figure 4. During testing of the 2.5-ton (8.8-kW) unit, refrigerant-side mass flow rate was not recorded continuously and was recorded manually once steady state operation was achieved. For the 5.0-ton (17.6-kW) unit tests, refrigerant flow rate was recorded continuously. The six tests listed in Table 1 were first conducted on both UUTs with their original scroll compressors. Refrigerant charge was tuned to achieve the highest possible energy efficiency ratio (EER), and the test results served as baseline. Then each compressor was replaced by a comparative rotary compressor, and refrigerant charge was again tuned to achieve the highest possible EER. Displacement volumes and refrigerant charges are summarized in Table 2 for all compressors. The series of six tests were then repeated with replacement rotary compressors. For the 5.0-ton (17.6-kW) UUT, a series of indoor units are compatible with the outdoor unit, but only the lowest efficiency indoor unit was available for testing, hence the unit's overall capacity was lower than nominal capacity for both compressors.

Table 1: Test Conditions

Test name	Test mode	Air entering outdoor unit, dry bulb/wet bulb °F (°C)	Air entering indoor unit, dry bulb/wet bulb °F (°C)	Compressor speed	Indoor airflow
A_{Full}	Cooling	95/75 (35/24)	80/67 (27/19)	Full	Full[1]
B_{Full}	Cooling	82/65 (28/18)	80/67 (27/19)	Full	Full[1]
Maximum Operation	Cooling	115/- (46/-)	80/67 (27/19)	Full	Full[1]
$H1_{Full}$	Heating	47/43 (8/6)	70/60 (21/16)	Full	Full[1]
$H2_{Full}$	Heating	35/33 (2/1)	70/60 (21/16)	Full	Full[1,]
$H3_{Full}$	Heating	17/15 (-8/-9)	70/50 (21/10)	Full	Full[1]

1: Manufacturer provided volume flow rate at full speed setting for cooling and heating, respectively.

Table 2: Compressor Specification

Compressor Type	Nominal Capacity ton (kW)	Displacement Volume (cc)	System Charge lb (kg)	Factory Charge lb (kg)	A_{Full} Capacity ton (kW)
Scroll	2.50 (8.8)	22.45	7.72 (3.50)	7.13 (3.23)	2.32 (8.16)
Rotary		23.20	8.16 (3.70)	7.13 (3.23)	2.36 (8.30)
Scroll	5.00 (17.6)	45.43	7.40 (3.36)	10.56 (4.78)	4.37 (15.37)
Rotary		45.50	7.40 (3.36)	10.56 (4.78)	4.43 (15.58)

2.1 TEST RESULTS

All six tests were conducted for rotary and scroll compressors, at both 2.5 (8.8) and 5.0 tons (17.6 kW) capacities, and their performance were compared at both the component (compressor) and system levels. The compressor isentropic efficiencies for each test were calculated based on compressor suction and discharge conditions for both cooling and heating modes (Figure 5), according to Equation (1). At 2.5 tons (8.8 kW), the isentropic efficiency for the rotary compressor was 2.6% higher in cooling and 14% higher in heating than the scroll compressor. Additionally, based on the three test points in cooling mode, the rotary compressor displayed less sensitivity (curvature) than the scroll compressor across a range of condensing temperatures. Similar sensitivity to condensing temperature in cooling mode is observed by Diniz & Deschamps (2016), in which both compressor types were studied under a wider range of condensing temperatures and larger number of test points. In their studies on R22 rotary and scroll compressors at 1.5 tons (5.3 kW), the rotary compressor also displayed less sensitivity/variation to condensing temperatures (Figure 6). At 5.0 tons (17.6 kW), the isentropic efficiency for the rotary compressor was 5.4% lower in cooling and about 6.3% higher in heating mode (Figure 5). For both the 2.5-ton (8.8-kW) and 5-ton (17.6-kW) UUT, in heating mode, the rotary compressor displayed the highest isentropic efficiency at the lowest condensing temperature, while the scroll compressor showed the lowest isentropic efficiency at the lowest condensing temperature. The rotary compressor had higher isentropic efficiency at 2.5 tons (8.8 kW), but lower isentropic efficiency than the scroll compressor at 5.0 tons

(17.6 kW), which corresponds to the feedback from interviews that rotary compressors are more suitable for small systems.

$$\eta_i = \frac{h_{Di} - h_S}{h_D - h_S} \tag{1}$$

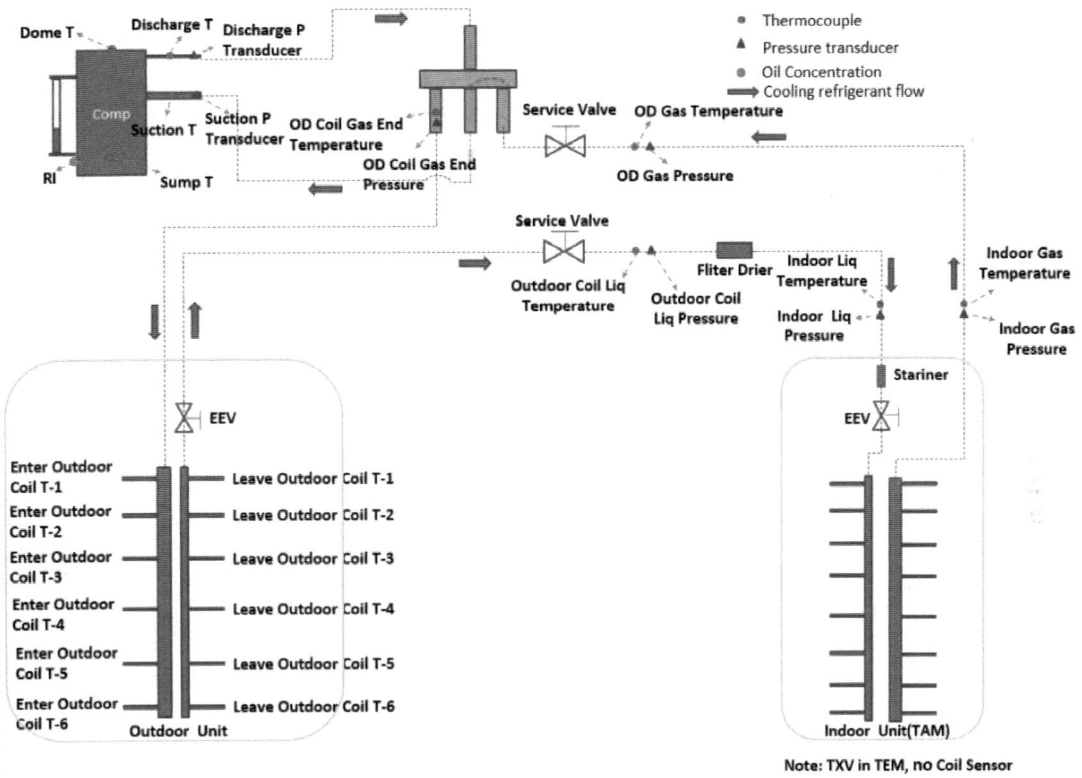

Figure 4: Test Schematic & Instrumentation

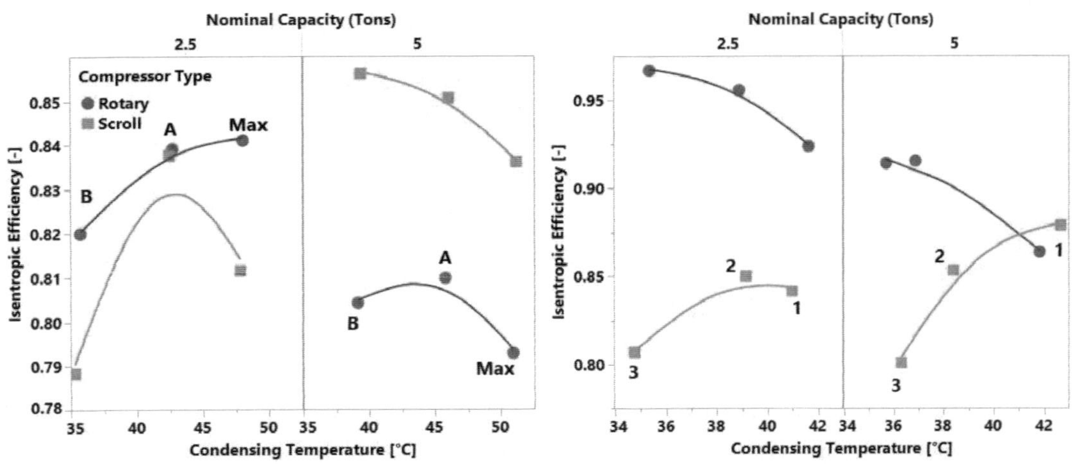

Figure 5: Isentropic Efficiency vs. Condensing Temperature: a) Cooling, b) Heating

Volumetric efficiency (Figure 7), was estimated based on the measured refrigerant flow rate, compressor displacement volume and suction conditions, according to Equation (2). At 2.5 tons (8.8 kW), the volumetric efficiency for the rotary compressor was lower than the scroll compressor by 1.7% in cooling mode (point B, A and Max), and comparable to the scroll compressor at heating mode points 1 and 2. At point 3 (the most extreme heating mode test point), refrigerant flow was unstable during frost accumulation. As mentioned in the previous section, refrigerant flow rate was recorded manually for the 2.5-ton (8.8-kW) UUT, and it may have been recorded outside of steady state, particularly for heating test point 3 for rotary compressor. As a result, volumetric efficiency of the rotary compressor was significantly lower due to lower refrigerant mass flow rate recorded. At 5.0 tons (17.6 kW), volumetric efficiencies for the rotary compressor were higher than the scroll compressor in both cooling and heating mode: 0.7% and 2.8%, respectively. In some cases, the actual volumetric efficiency may differ from what is calculated. As discussed in Jähnig (1999), the true compressor inlet conditions differ from the measured suction conditions due to pressure drop entering the compressor, as well as heat transfer to the refrigerant from the compressor motor and shell. Also, motor slip does occur, and the actual rotational frequency is expected to be less than the supply power's frequency. Moreover, experimental variation and measurement uncertainty can also influence the calculations. When compared with studies by Diniz & Deschamps (2016), both compressors display volumetric efficiency sensitivity to condensing temperatures across a range of capacities, and volumetric efficiencies were observed to decrease at higher condensing temperatures (Figure 8).

$$\eta_v = \frac{\dot{m}_{ref}}{\omega * V * \rho_S} \tag{2}$$

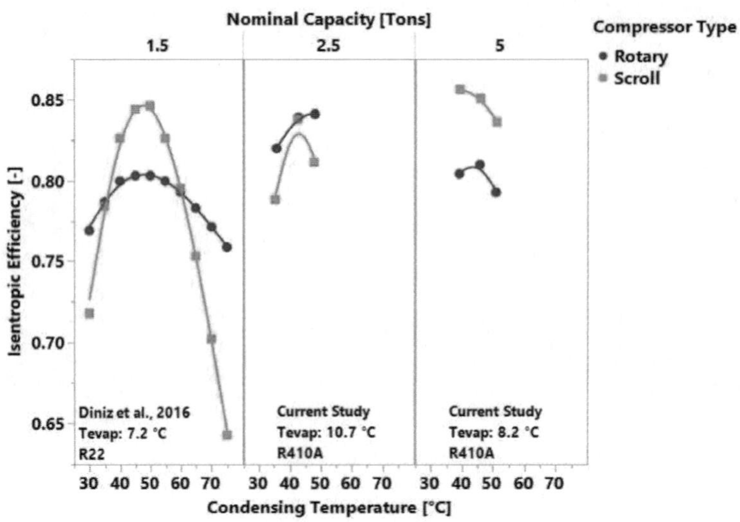

Figure 6: Isentropic Efficiency Comparison in Cooling Mode (Diniz & Deschamps , 2016)

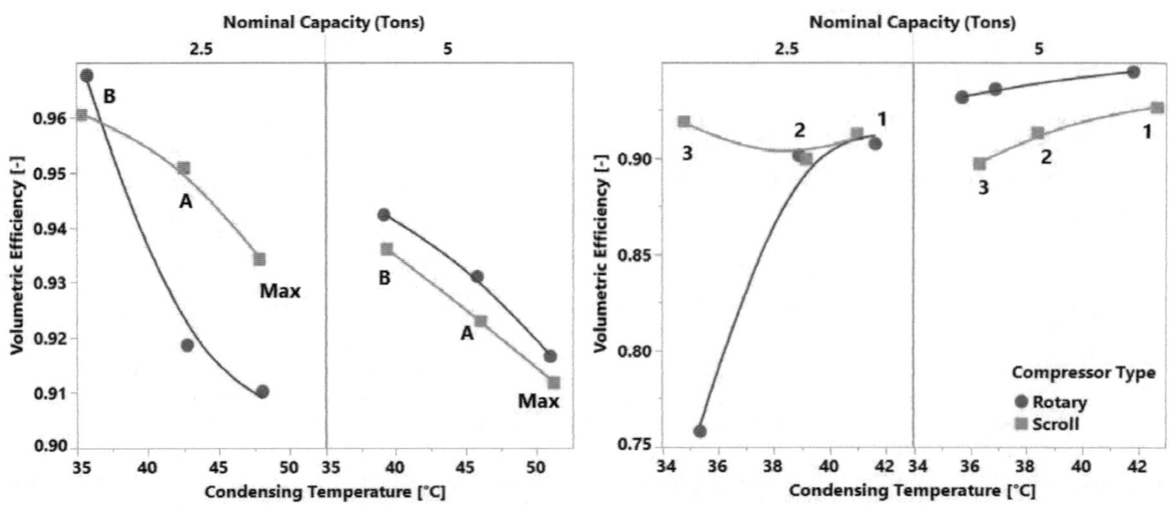

Figure 7: Volumetric Efficiency vs. Condensing Temperature: a) Cooling, b) Heating

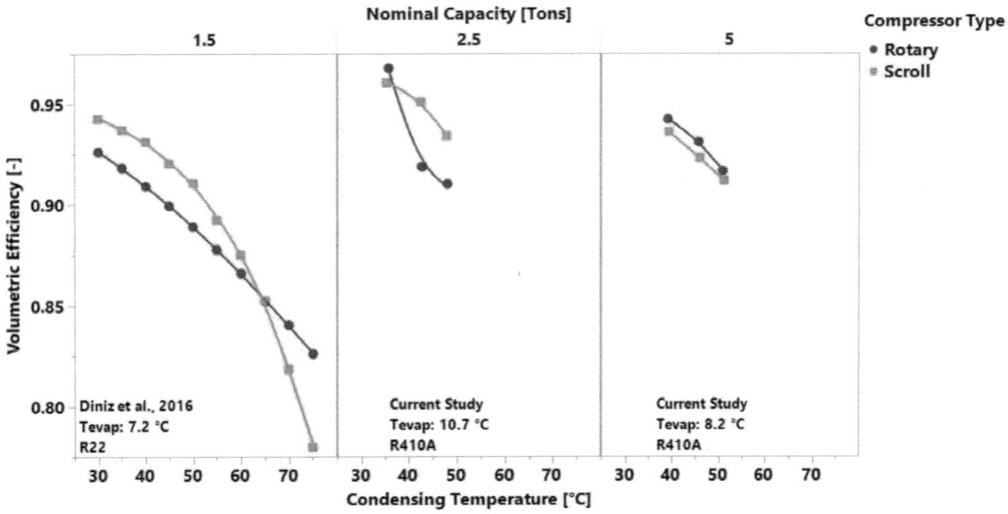

Figure 8: Volumetric Efficiency Comparison in Cooling Mode (Diniz & Deschamps, 2016)

In terms of the overall unit performance, the seasonal energy efficiency ratio (SEER) and heating seasonal performance factor (HSPF) were calculated for cooling and heating modes, respectively. At 2.5 tons (8.8 kW), the rotary compressor had 5.7% higher SEER than the scroll compressor, and 2.6% higher SEER when capacity increased to 5.0 tons (17.6 kW), as shown in Figure 9. Both compressors displayed a higher SEER value at lower capacity. On the other hand, HSPF was calculated for both compressors at the two capacities, across six climatic regions (Region I to VI) defined in AHRI Standard 210/240 (2017). On a regional basis, the 2.5-ton (8.8-kW) unit always had higher HSPF values than the 5.0-ton (17.6-kW). At 2.5 tons (8.8 kW), the UUT with the rotary compressor had, on average, 3% higher HSPF than the UUT with the scroll compressor across six regions, and HSPF was 0.6% higher at 5.0 tons (17.6 kW) (Figure 10). For both capacities, Region I yielded the highest HSPF value, while the lowest HSPF occurred at Region V.

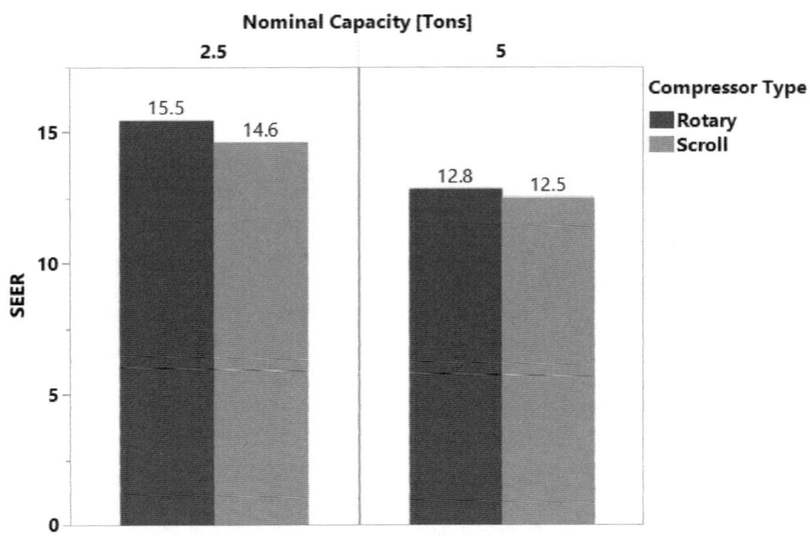

Figure 9: SEER and Nominal Capacity

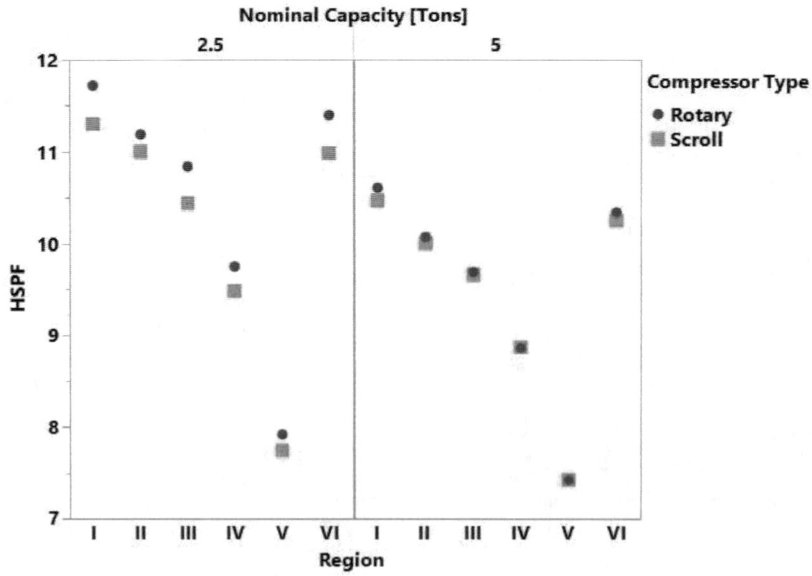

Figure 10: HSPF by Region and Capacity

3. CONCLUSIONS

This work presented: a) a review on rotary compressors; and b) a comparative experimental study between rotary and scroll compressors. The first part included a series of interviews from the industry regarding rotary compressor application as well as supplemental review of the literature. The two most frequent comments received were: limited application of rotary compressors in small capacity systems and the GE failure in using rotary in their refrigeration systems in the 1980's. To evaluate the limitation of rotary compressors in larger capacity systems, a series of cooling and heating tests (AHRI210/240 Standard specified operating conditions) were conducted for two typical US split HPs of 2.5 (8.8) and 5.0-ton (17.6-kW) capacities, using both scroll and rotary compressors. The results showed that at 2.5 tons (8.8 kW), the rotary compressor's isentropic efficiency was 2.6% and 14.0% higher than the scroll compressor, in cooling and heating modes, respectively. For the 5.0 tons (17.6 kW) system, the efficiency of the rotary compressor was 5.4% lower during cooling and 6.3% higher during heating. In terms of volumetric efficiency, at 2.5 tons (8.8 kW) the rotary compressor was lower than the scroll compressor by 1.7% in cooling, and comparable to the scroll compressor at two of the three heating mode test points. At a higher capacity of 5.0 tons (17.6 kW), the rotary compressor was higher than scroll compressor by 0.7% and 2.8% in cooling and heating modes, respectively. When considering system-level efficiency, the 2.5-ton (8.8-kW) system installed with the rotary compressor had 5.7% higher SEER and 3% higher HSPF than with the scroll compressor. For a higher capacity of 5.0 tons (17.6 kW), the unit with the rotary compressor had 2.6% higher SEER and 0.6% higher HSPF as compared to the unit with the scroll compressor.

NOMENCLATURE

AC	Air-conditioner	(-)
AHRI	Air-conditioning, Heating, and Refrigeration Institute	(-)
ASHRAE	American Society of Heating, Refrigerating and Air-Conditioning Engineers	(-)
BPA	Bonneville Power Administration	(-)
EER	Energy efficiency ratio	(Btu/W.h)
GE	General Electric	(-)
h_D	Discharge enthalpy	kJ/kg
h_{Di}	Discharge isentropic enthalpy	kJ/kg
h_S	Suction enthalpy	kJ/kg

HP	Heat pump	(-)
HSPF	Heating Seasonal Performance Factor	(Btu/W.h)
HVAC	Heating, ventilation, and air-conditioning	(-)
HVAC&R	Heating, ventilation, air-conditioning, and refrigeration	(-)
$\dot{m}_{ref}$	Refrigerant mass flow rate	kg/s
η_i	Isentropic efficiency	(-)
η_v	Volumetric efficiency	(-)
ω	Rotational speed	revolution/s
R	Refrigeration	(-)
ρ_S	Suction density	kg/m^3
RV	Recreation vehicle	(-)
SEER	Seasonal Energy Efficiency Ratio	(Btu/W.h)
TXV	Thermal expansion valve	(-)
UUT	Unit under test	(-)
V	Compressor displacement volume	m^3
VCC	Vapor compression cycle	(-)

REFERENCES

AHRI Standard 210/240. (2017). Standard for performance rating of unitary air-conditioning & air-source equipment. Air-conditioning, Heating and Refrigeration Institute.

AHRI, "AHRI Statistics." [Online]. Available: https://www.ahrinet.org/resources/statistics/historical-data/central-air-conditioners-and-air-source-heat-pumps .

ASHRAE Standard 41.2. (1987). Standard methods for laboratory air-flow measurement. American Society of Heating, Refrigeration and Air-Conditioning Engineers, Atlanta.

Bonneville Power Administration, "HVAC MArket Intelligence Booklet," 2016.

Diniz, C., Deschamps, C.J. (2016). "Comparative analysis of two types of positive displacement compressors for air conditioning applications". 23rd International Compressor Engineering Conference at Purdue. West Lafayette, IN.

Form 10-K Tecumseh Products Co. Annual Report. (1994). Securities and Exchange Commission. Washington, DC.

Jähnig, D. (1999). A semi-empirical method for modeling reciprocating compressors in residential refrigerators and freezers. ASHRAE Report RP-870. University of Wisconsin-Madison.

Magaziner, I., Patinkin, M. (1989). Cold Competition: GE wages the refrigeration war. Harvard Business Review, April.

Monasry, J., Aoki, T., Shida, S., Hatayama, M., Hirayama, T. and Okada, M. (2018). "Development of large capacity and high efficiency rotary compressor". 24th International Compressor Engineering Conference at Purdue. West Lafayette, IN.

Sathe, A., Groll, E., Garimella, S. (2008). "Experimental evaluation of a miniature rotary compressor for application in electronics cooling". 19th International Compressor Engineering Conference at Purdue. West Lafayette, IN.

Shi, J., Gao T., Lu, B, Li, W., Zhang, Z., Cai, B., Pan, L., Wang, T. and Yan, R. (2016). "Research on heat pump system using rotary compressor in electric vehicle". 16th International Refrigeration and Air Conditioning Conference at Purdue, West Lafayette, IN.

ACKNOWLEDGEMENT

Test data was provided by Shanghai Highly Electric Appliances Corporation (SHEC).

Study on the Performance of CO_2 Two-stage Rotary Compressor in Freezing Conditions

Yusheng Hu[1,3], Peng Zou[3*], Huijun Wei[1,2,3], Jia Xu[1,2,3], Liping Ren[1,2,3], Pengkai Wan[1,2,3]

[1]State Key Laboratory of Air-conditioning Equipment and System Energy Conservation
Zhuhai, Guangdong, China

[2]Guangdong Key Laboratory of Refrigeration Equipment and Energy Conservation Technology
Zhuhai, Guangdong, China

[3]Gree Electric Appliances, Inc.
Zhuhai, Guangdong, China
E-mail: 18023035429@163.com

* Corresponding Author

ABSTRACT

This paper describes a new type CO_2 two-stage rotary compressor for cold storage and freezing of food. A two-stage compression form with an upper cylinder (first-stage) and a lower cylinder (second-stage), unique oil road structures and technical parameters have been used in the rotary compressor to increase the performance. The results indicating that the optimized CO_2 two-stage rotary compressor has a significant performance advantage, which the coefficient of performance (COP) increases by 4.4% ~ 6.7%.

1. INTRODUCTION

At present, domestic research and application of CO_2 are mainly focus on three aspects: automobile air conditioning, heat pump water heaters, cold storage and freezing of food. Among them, cold storage and freezing of food have great application potential. Due to the design storage temperature and cooling capacity is according to the application occasions, the refrigeration system of cold storage and freezing of food needs different types of compressor. In the range of small cooling capacity at medium and high temperatures (the evaporation temperature range is -15°C～10°C), the type of rotary compressor is one of the mainly used .

Jinghong Ning et al. (2011) used thermodynamic method to calculate the performance parameters of different type refrigeration cycle of CO_2 trans-critical two-stage compression for cold storage and freezing of food. By analyzing and comparing the performance parameters of the refrigeration cycles, they concluded that the CO_2 trans-critical refrigeration cycle with two-stage throttling and complete cooling in middle has some advantages, such as lower energy consumption, better running performance, lower initial investment and so on. Yefeng Liu et al. (2019) combined two-stage compression with CO_2 trans-critical cycle technology and designed a set of refrigerating cabinets which use a small CO_2 two-stage compression refrigeration cycle. Experimental comparison between the refrigeration system with a single-stage compression and a CO_2 two-stage compression proved that the use of CO_2 two-stage compression refrigeration system could improve system performance. The CO_2 two-stage rotary compressors applied in heat pump water heaters and achieved good results by some research organizations, such as SANYO Industrial Co. Ltd. and the University of Maryland et al. However, there are few studies on the application of CO_2 two-stage rotary compressors for cold storage and freezing of food at present.

This paper discusses a new type CO_2 two-stage rotary compressor which can be used in small cooling capacity at medium and high temperatures (the evaporation temperature range is -15°C～10°C) in cold storage and freezing of food, experimentally analyzes the variation of the performance of two compressor schemes and clarifies their advantages of the performance in freezing conditions.

2. COMPRESSOR STRUCTURES AND KEY PARAMETERS

2.1 Compressor Structures

Ying Wang et al. (2010) studied the characteristics of the channel roughness on flow resistance, and found that there will be greater losses when the fluid through a smaller cross-section and the inner wall of the flow passage coarser. As shows in Figure 1, due to the limit of the size, structures and processing conditions of the rotary compressor, there is a similar flow channel (red frame position) inside the pump of conventional CO_2 two-stage rotary compressor.

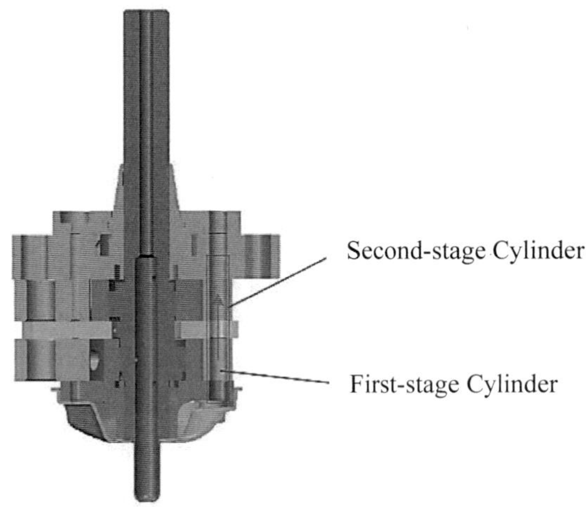

Figure 1: Internal flow channel of conventional pump

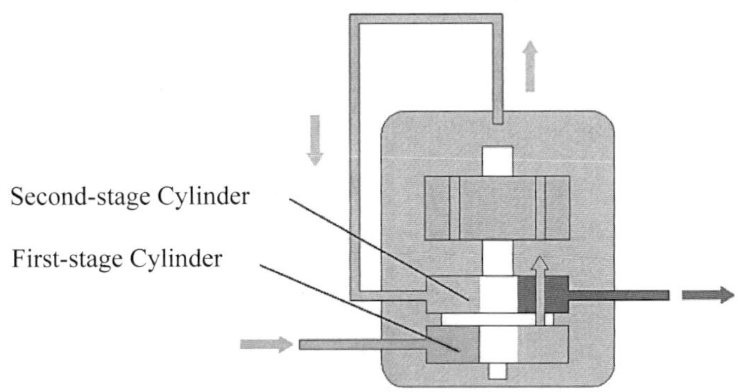

(a): The structure of conventional CO_2 two-stage rotary compressor

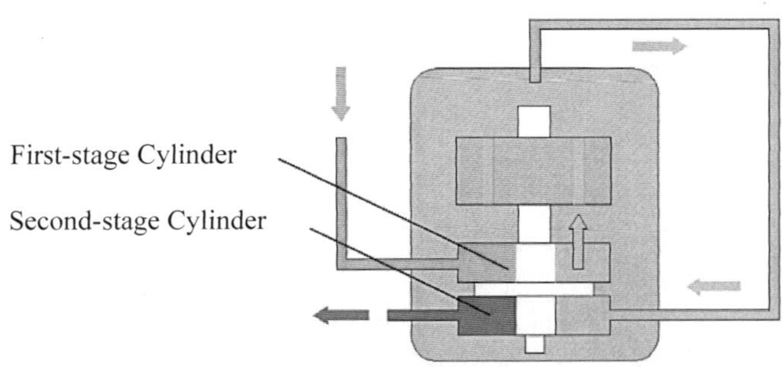

(b): The structure of new type CO_2 two-stage rotary compressor

Figure 2: The structures of the conventional and new type CO_2 two-stage rotary compressors

Figure 2 shows the internal structures of conventional and new type CO_2 two-stage rotary compressors, both of which are two-stage compression structures. The difference between the conventional compressor and the new compressor is the using of the second-stage cylinder that be arranged below the first-stage cylinder. The refrigerant is compressed by the first-stage cylinder and discharges into the airtight housing directly. Compared with the conventional compressors, the new type CO_2 two-stage rotary compressor can avoid flowing through the relatively rough small channels inside the pump and the flow loss can be reduced. Table 1 shows the specifications of the new type CO_2 two-stage rotary compressor.

Table 1: Specifications of the new type CO_2 two-stage rotary compressor

Work volume(cc)	19
Rated power (w)	7500
Motor type	DC inverted frequency
Back pressure type	Medium back pressure
Compression type	Two-stage
Oil type	PAG

2.2 Oil Road Structures

Kun Yang *et al.* (2016) analyzed the compatibility of PAG and CO_2 used in this paper and concluded that they have good mutual solubility. The oil dissolved in CO_2 will enter the compressor pump to each friction pair. However, the pressure of the housing is the first-stage discharge pressure and the pressure of the second-stage cylinder is the condensing pressure which is higher than the first-stage discharge pressure. Under the action of a larger pressure difference, the oil is difficult for the lubricant in the housing to enter the second-stage cylinder for conventional oil road structures.

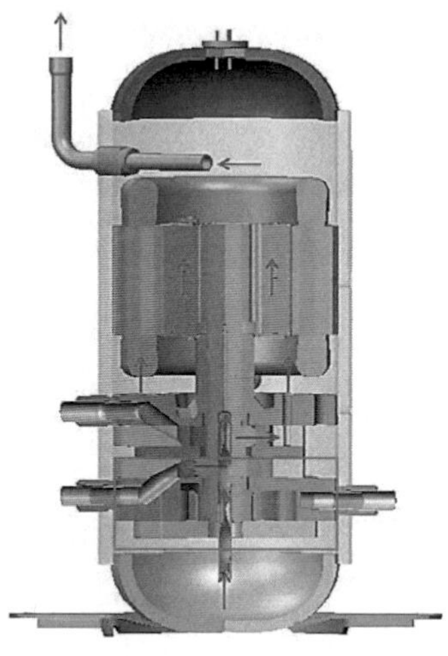

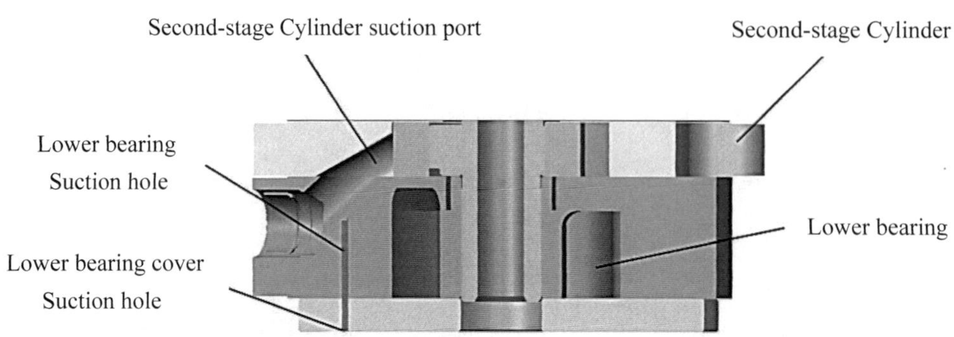

Figure 3: The oil road structures of new type CO_2 two-stage rotary compressor

Figure 3 shows the difference between the oil road structures of the conventional and the new type compressor. In the new type compressor, the oil suction hole is set in the lower bearing and the lower bearing cover, and the top of the second-stage cylinder will be covered with oil. Due to the periodic pulsation of the inlet pressure of the second-stage cylinder, the pressure in the housing is intermittently higher than the second-stage cylinder suction port. Therefore, with the effects of the pressure difference, the oil will be suck into the suction port of the second-stage cylinder and enter the interior of the second-stage cylinder.

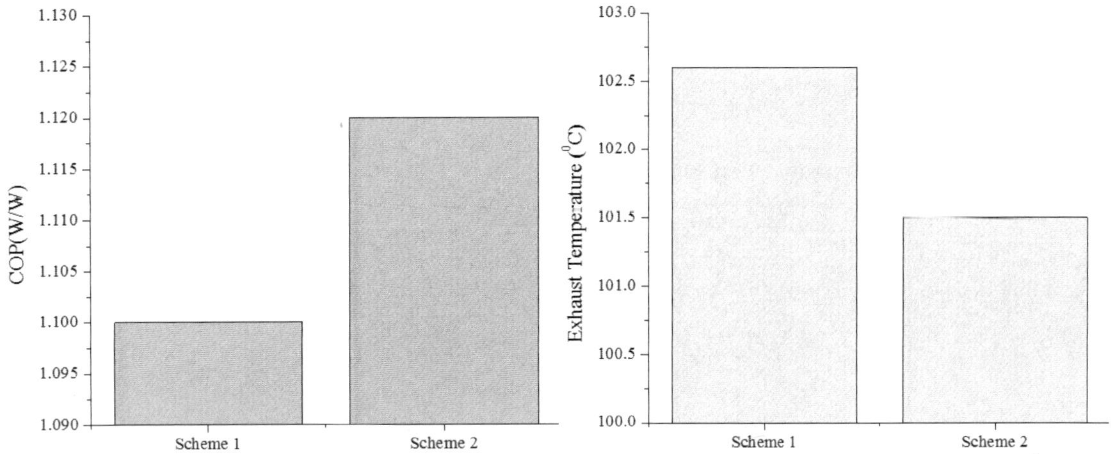

Figure 4: COP and discharge temperature for both schemes

Table 2 is the oil road schemes. As shows in Figure 4, the comparison between the without oil suction hole and with oil suction hole, the test results show that the COP increased by 1.8% and the discharge temperature reduced by 1.1°C. This phenomenon indicates that the second-stage cylinder lubricated effectively when the suction hole are added to the lower bearing and lower bearing cover.

Table 2: Oil road schemes

Oil road schemes	**Lower bearing suction hole**	**Lower bearing cover suction hole**
Scheme 1	Without	Without
Scheme 2	With	With

2.3 Volume Ratio

The volume ratio is the ratio of the working volume of the second-stage and first-stage cylinders of the two-stage compressor. It is one of the key parameters affecting the performance of two-stage compression.

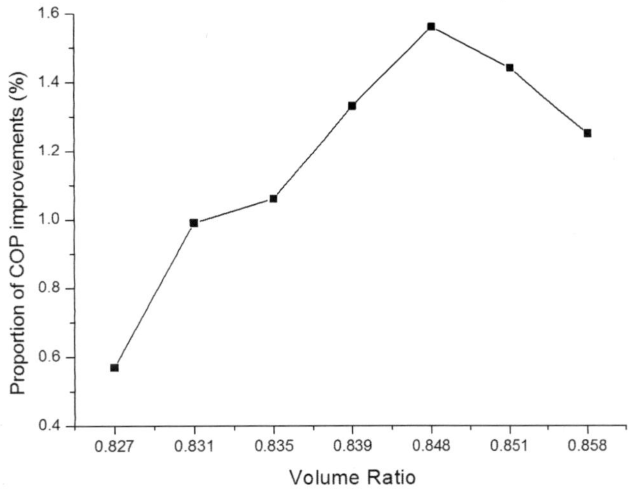

Figure 5: Variation curve of COP improvements proportion with volume ratio under freezing conditions

Qi Liu *et al.* (2014) theoretically calculated the performance parameters of the R410A two-stage compressor and the results showed that when the evaporation temperature range is 3°C~9°C, the condensing temperature range is 35°C~50°C, the range of volume ratio is 3/4~4/5 and the cycle refrigeration coefficient is closer to the maximum refrigeration coefficient. According to the partial calculation method of volume ratio described in the paper, the theoretical analysis of the two-stage throttling intermediate incomplete cooling cycle applied to the CO_2 two-stage rotary compressor. The result shows that the volume ratio of the optimal COP of the compressor corresponding to different operating conditions is not the same. In the freezing conditions, as shows in Figure 5, the proportion of COP increases first and decreases with the increasing of the volume ratio, it shows a parabolic trend. The optimal volume ratio range is 0.839～0.851 and the best COP of volume ratio is 0.848.

3. TEST SCHEMES AND RESULTS

3.1 Compressor Performance Test Facility

As show in Figure 6 to 7, the experiments are carried out on the compressor performance test facility in GREE, which is based on GB/T 29030-2012《Volume Type CO_2 Refrigeration Compressor (Group)》. The main parameters which need to be measured include suction pressure / temperature, discharge pressure / temperature, mass flow and power etc. The range of cooling capacity is from 0.8kW～8 kW.

Figure 6: Compressor performance test facility

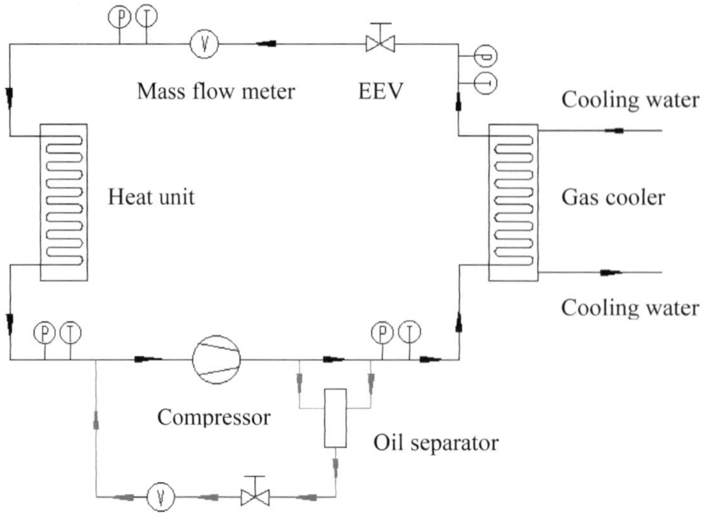

Figure 7: Simplified schematics of test facility

3.2 Test Schemes

As shows in Table 3, there are two schemes for comparative test of two-stage CO_2 compressors from GREE. The test conditions are freezing conditions with evaporation temperature of -10°C, suction temperature of 0°C and discharge pressure of 9.8 MPa.

Table 3: Compressor schemes

Compressor schemes	Compression type	First-stage cylinder	Second-stage cylinder	New oil road structures	Volume ratio
Conventional type	Two-stage	Down	Up	Without	0.750
New type	Two-stage	Up	Down	With	0.848

3.3 Results and Discussions

Figures 8 to 11 show the variation curves of the cooling capacity, power, COP and discharge temperature with the change of frequency of conventional type and new type compressors in freezing conditions. In the entire range of frequency, the cooling capacity, power and discharge temperature of two schemes will increase with frequency, while the COP slightly decreases, and the overall trend is similar.

The comparison result between the two schemes shows that the cooling capacity increase by 1.1%~3.6%, the power reduces by 0.7%~2.9%, the discharge temperature reduces by 3°C~6°C and the COP increases by 4.4%~6.7%, which indicates the new type CO_2 two-stage compressor has significant performance advantage.

The reasons for the above mentioned advantages are the new type CO_2 two-stage rotary compressor has much less flow resistance of CO_2 and the second-stage cylinder has been well lubricated by oil, so the friction loss is reduced, the mechanical efficiency will be improved and the discharge temperature will be lower. Moreover, the leakage channel of the pump also can be well sealed by oil, the leakage of CO_2 under large pressure difference can be reduced and the volumetric efficiency will be improved effectively. Meanwhile, the adoption of the suitable volume ratio for freezing conditions can reasonably distribute the pressure ratio of the first-stage cylinder and second-stage cylinder for the compression efficiency. So the COP of new type CO_2 two-stage rotary compressor improves obviously.

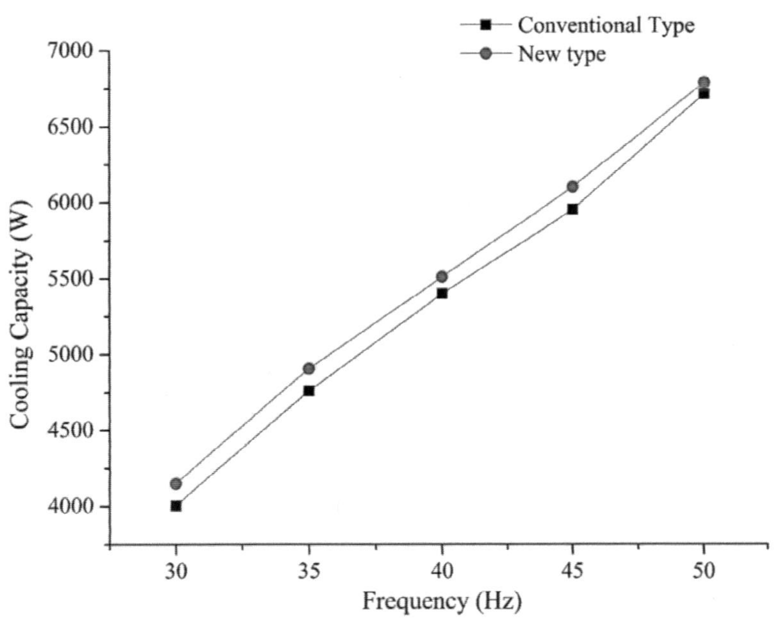

Figure 8: Variation curves of cooling capacity of two schemes with the changing of frequency

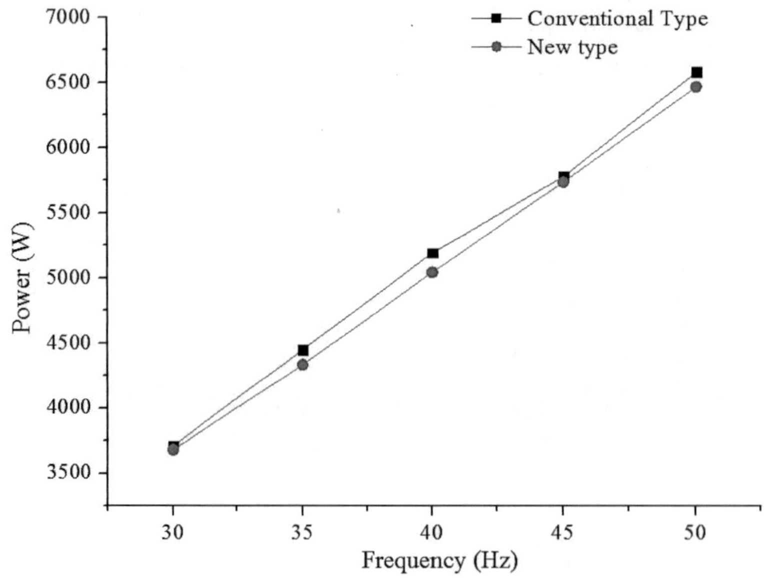

Figure 9: Variation curves of power of two schemes with the changing of frequency

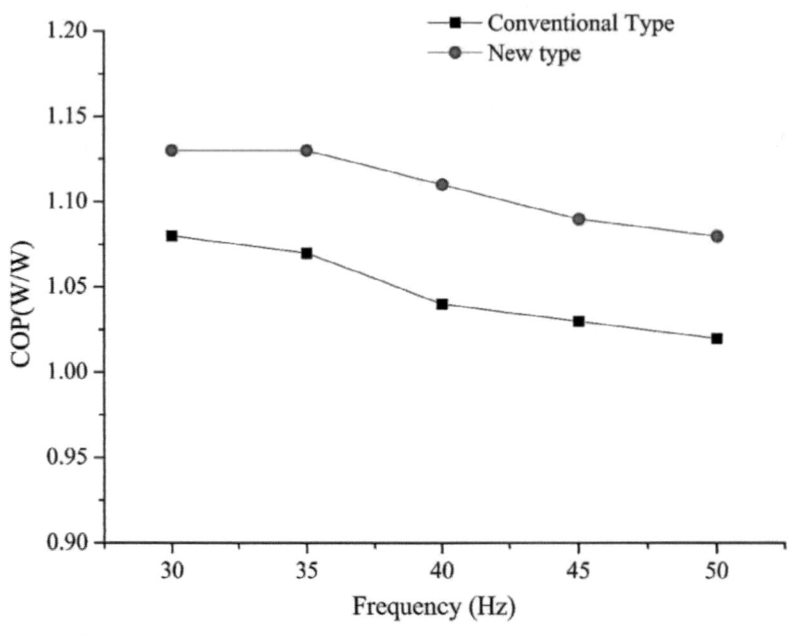

Figure 10: Variation curves of COP of two schemes with the changing of frequency

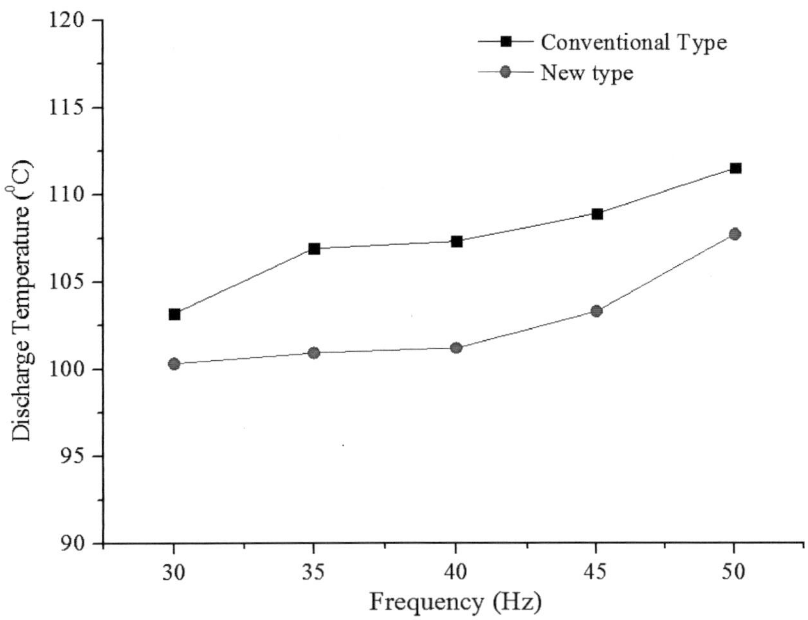

Figure 11: Variation curves of discharge temperature of two schemes with the changing of frequency

4. CONCLUSIONS

This paper describes the technical characteristics of new type CO_2 two-stage rotary compressor and compares the test results with conventional CO_2 two-stage rotary compressor in freezing conditions. The results show that:
(1) Adding oil suction holes in the lower bearing and the lower bearing cover can reduce the discharge temperature and improve COP of the compressor effectively.
(2) The two-stage compression form with an upper cylinder (first-stage) and a lower cylinder (second-stage), including oil suction holes and using the best volume ratio can improve COP of the compressor significantly.

NOMENCLATURE

COP	Coefficient of performance	(W/W)
CO_2	Carbon dioxide	
EEV	Electronic expansion valve	
P	Pressure of the CO_2	(MPa)
T	Temperature of the CO_2	(℃)
m	Volumetric flow rate	(m^3/min)
Rpm	Speed	(Revolutions/min)
V	Specific volume	(m^3)
ΔP	Pressure difference	(MPa)
η_{vol}	Volumetric efficiency	

REFERENCES

M. M. Wang, G. Y. Ma &S. X. Xu. (2014). CO_2 refrigeration technology and its application into freeze and cold

storage. *Refrigeration and Air Conditioning*, 14(04), 54-61.
Z. C. Zhou. (2011). The future application of refrigerant CO_2 in cold storage and freezing of food. *Refrigeration*. 30(01), 24-33.
M. Shi, L. Jia & X. P. Zhang. (2016). Applicability research on CO_2 application in Chinese industrial and Commercial refrigeration industry. *Journal of Refrigeration*. 37(06), 97-103.
Y. Sun, C. X. Ren, H. Zhang, etc. (2014). Research progress of CO_2 Compressor in Refrigeration System. *Refrigeration Technology,* 34(05), 67-71.
J. W.Qiu, L.Zhao, B. M. Wang, etc. (2010). Research and application of CO_2 compressor for food freezing and refrigerating. Chinese Refrigeration Association. (Eds.), *Proceedings of the Seventh National Food Refrigeration Chain Conference,* (201-206).
J. H. NING, H. Y. Li, M. Peng, etc. (2011). Performance analysis of CO_2 refrigeration system for food frozen storage. *Refrigeration and Air conditioning,* 11(05), 118-121.
Y. F. Liu & D. P. Tang. (2019). Meng Deren. Application test of CO_2 two-stage compression refrigeration system in refrigerator. *Fluid Machinery,* 14(04), 76-80.
N. Dreiman, R. Bunch, Y. H. Hwang, etc. (2004). Two-Stage rolling piston carbon dioxide compressor. *Int. Compressor Eng. Conf. at Purdue*, C010.
N. Dreiman, R. Bunch. (2006). Development of carbon dioxide rotary compressor. *Int. Compressor Eng. Conf. at Purdue*, C014.
Y. Wang, Y. H Diao, Y. H. Zhao. (2010). Resistance characteristics of fluid dynamics in micro channels. *Journal of Engineering Thermo physics,* 21(06), 998-1000.
K. Yang, X. P. Zhang, R. J. Wang. (2016). Study on the compatibility and material compatibility of CO_2 refrigerants / lubricants. *Refrigeration Technology,* 44(05), 75-80.
H. F. Lu, L. C. Xie. (2014). Research on suction pressure pulsation of rolling rotor compressors. *Refrigeration and Air conditioning,* 14(02), 106-108.
Q. Liu, G. Y. Ma, S. X. Xu. (2014). Discussion on suitable volume ratio of two-stage rotary compressor. *Refrigeration and Air conditioning,* 14(02), 102-105.

Development of Large-and-Small Volume-Switching Rotary Compressor For Residential Multi-Connected Air-Conditioner

Yusheng Hu[1,3], Huijun Wei [1,2,3], JiaXu[1,2,3], Peizhen Que[1,2,3*] , Yanjun Hu[3]

[1]State Key Laboratory of Air-conditioning Equipment and System Energy Conservation
519070,Zhuhai, Guangdong, China

[2]Guangdon Key Laboratory of Refrigeration Equipment and Energy Conservation Technology
519070,Zhuhai, Guangdong, China

[3]Gree Electric Appliances, Inc.
519070, Zhuhai, Guangdong, China
(Phone: +86-756-8974897, E-mail: que106@163.com)

* Corresponding Author

ABSTRACT

Residential Multi-Connected Air-Conditioner(RMAC) are widely used in urban residential in China, but they have low energy efficiency problems in daily use. They are mainly due to the air-conditioning habits of Chinese residents in "part space and part time", Which makes RMAC most of the time only 1-2 indoor units are turned on, and 60% of the operating time is at a low load rate of less than 30%. During low load operation, the compressor's efficiency drops sharply, and there is also the problem of frequent start and stop when the minimum output is too large. In order to solve this problem, a large-and-small volume switching compressor suitable for RMAC is developed. The compressor can choose single-cylinder small-volume operation or twin-cylinder large-volume operation according to the change of system load. The new compressor technology for RMAC improves the energy efficiency by 124% under 10% load, and avoids the problem of frequent start-stops at low loads, making RMAC more energy efficient and more comfortable.

1. INTRODUCTION

Residential Multi-Connected Air-Conditioner (RMAC) refers to Multi-Connected Air-Conditioner used in residences. It consists of one outdoor unit connected to several indoor units. It has the characteristics of small installation space and beautiful decoration, as shown in Figure 1. In recent years, its market share in China has grown rapidly. According to statistical data (HVAC Information), China's RMAC has a compound annual growth rate of 25% in recent years, accounted for more than 27% of the entire central air-conditioning market share. Therefore, its energy consumption level has an increasingly greater impact on building energy efficiency in China, and reducing the actual energy consumption of RMAC is of great significance.

RMAC is designed and selected according to the heat load of the whole house, that is, it has higher energy efficiency when used in "full space". However, due to Chinese residents' "part space, part time" air-conditioning usage habits, most of the time of RMAC is in low load operation. According to big data statistics, see Cheng, J.(2017), RMAC only operates one indoor unit for 60% of the time, which results in 60% of the operating time in the low load range of less than 30%， as shown in Figure 2 and Figure 3.It is mainly because the bedroom area of

Chinese urban residential is small, which makes the heat load small. The demand for one-bedroom cooling load at night in the air-conditioning cooling season is mainly concentrated in 0.8kW ~ 2.4kW. For most RMAC rated capacity of 12-18kW, its minimum output needs to be reduced to about 5% of the rated output. Because most conventional RMAC use variable-frequency rotor compressors, during low-load operation, the operating frequency of the compressor needs to be reduced to reduce the output. However, due to the decrease of the compressor frequency, its efficiency will be greatly reduced, and due to the minimum frequency of the compressor limitation, when the load demand is below 20%, the minimum output of the compressor is too large, which is prone to frequent start and stop, resulting in high power consumption.

In summary, at present, RMAC uses conventional compressor technology, which has the problems of too large minimum output and low energy efficiency at low load operation, resulting in high energy consumption of RMAC in actual use. Therefore, a new large-and-small volume switching compressor technology was developed, which can significantly improve the energy efficiency of compressors operating at low loads below 30% and reduce the minimum output, achieving significant energy saving effects for RMAC.

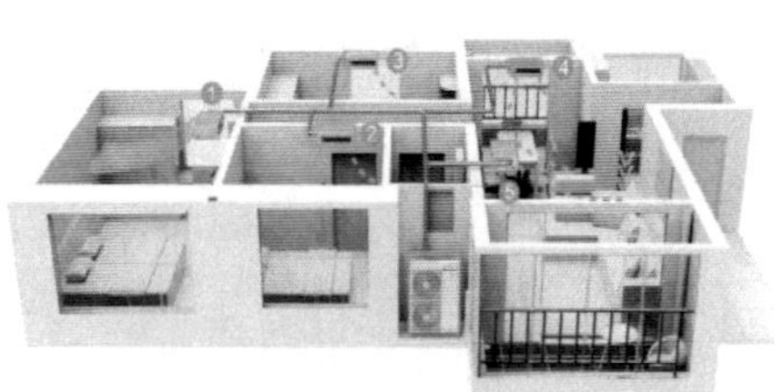

Figure 1: RMAC installation diagram

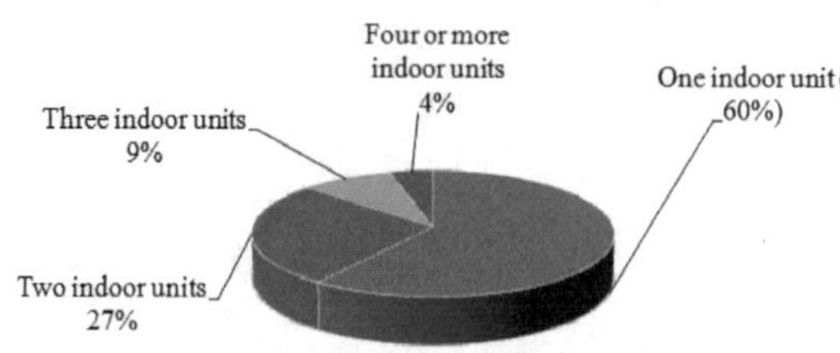

Figure 2: Proportion of indoor unit being turned on

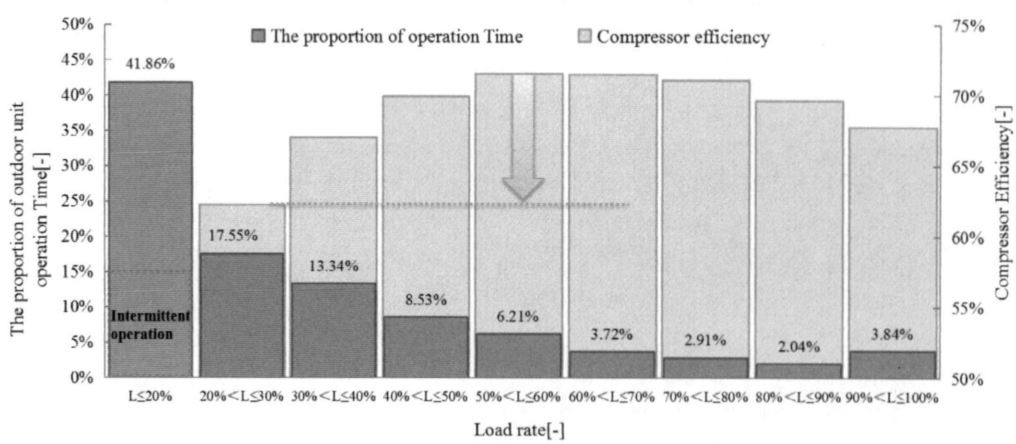

Figure 3: RMAC operation time ratio and corresponding compressor efficiency at different load rate

2. NEW DUAL-EFFICIENCY OPERATING MODE FOR COMPRESSORS

In order to solve the above two major problems of RMAC, a compressor with two modes of small-volume operation and large-volume operation is proposed, which can switch between the two operation modes according to the system load change, as shown in Figure 4 . (a)Small-volume operation(single-cylinder mode) : When RMAC only turns on 1-2 indoor units and is in low-load operation, the compressor operates only with a small-volume cylinder, and a large-volume cylinder is unloaded, and the working volume is reduced by more than

50%. Combined with the reduction of the frequency, the minimum capacity output is greatly reduced, which avoids the frequent start-stop problem caused by the compressor capacity output being too large during low load operation. In addition, due to the reduced operating volume, the compressor's operating frequency at low loads has increased significantly, and the compressor efficiency can be significantly improved. (b)Large-volume operation (Twin-cylinder mode): When multiple indoor units of RMAC are turned on, during medium-high load operation, the small-volume cylinder and large-volume cylinder of the compressor run at the same time to meet the RMAC's high-capacity output and high efficiency demand.

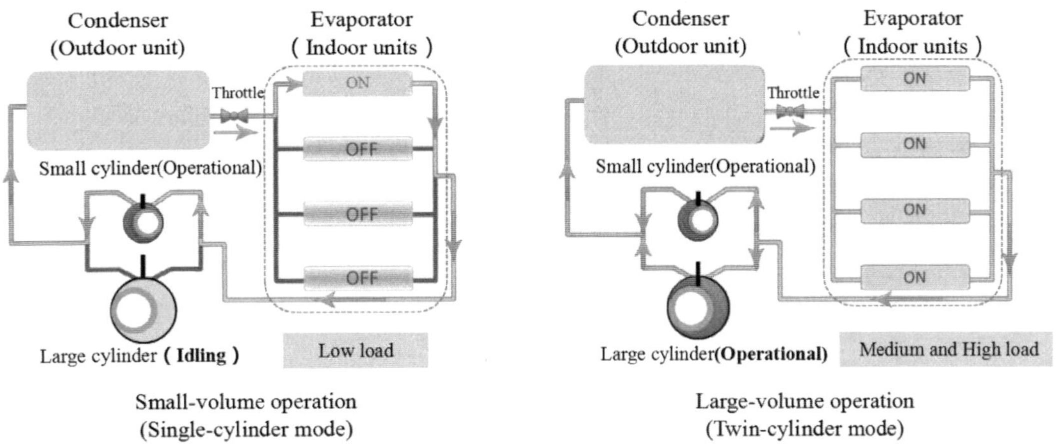

Figure 4: New dual-efficiency operating mode for compressors

3. DEVELOPMENT OF LARGE-AND-SMALL VOLUME SWITCHING COMPRESSOR

3.1 Structure of large-and-small volume switching compressor

The Figure 5 shows a comparison of the compression structure of a conventional twin-rotary compressor and a large-and-small volume switching compressor. Conventional twin-rotary compressor operates with two cylinders of equal volume simultaneously, with a narrow output range. The large-and-small volume switching compressor includes a small-volume cylinder with a volume of V_1, and a large-volume cylinder with a volume of V_2, that is, $V_1 < V_2$. The large-volume cylinder can be switched between operating state and idling state by a vane lock mechanism. That is, a notch is provided on the vane, and a locking mechanism is installed in the lower bearing below the vane. The locking mechanism includes a pin that can move up and down, and a spring at the bottom of the pin.

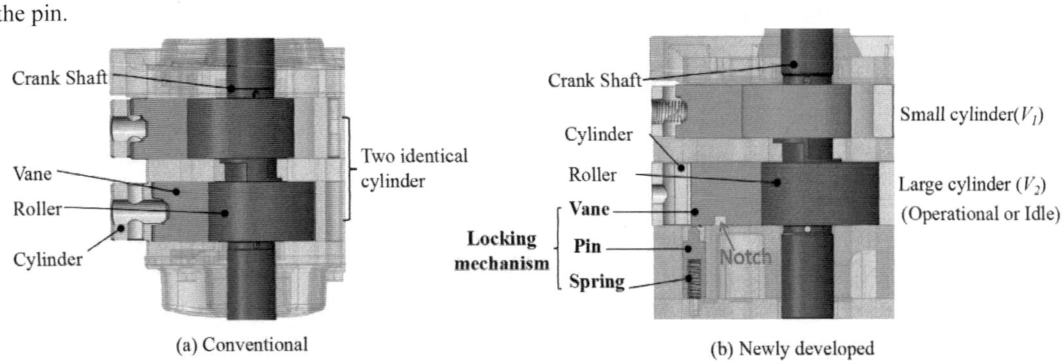

Figure 5: Comparison of compressor structure

3.2 Large and small volume switching principle

In order to achieve the switch between operating state and idling state of the large-volume cylinder, it is realized through the linkage control of the pressure control system and the vane lock mechanism, as shown in the Figure 6.A low-pressure solenoid valve A and a high-pressure solenoid valve B are respectively arranged between the suction port and the exhaust port of the compressor. The two solenoid valves are connected to the sealed cavity at the upper end of the pin through a pressure storage tank, and the lower part of the pin is always connected to the suction low pressure. By opening and closing the two solenoid valves separately, high pressure or low pressure can be applied to the upper end of the pin. The switching process is: (a) the low-pressure solenoid valve A is opened, the high-pressure solenoid valve B is closed, the upper and lower ends of the pin are both low pressure, and the pressure is balanced. The pin is moved upward by the force of a spring force, and the pin locks the vane to separate the vane from the roller. The large-volume cylinder does not compress and realizes idling. (b) The low-pressure solenoid valve A is closed and the high-pressure solenoid valve B is opened. The upper end of the pin is connected with high pressure, and the upper end pressure is higher than the lower end. The pin moves downward against the spring force by the pressure difference between the upper and lower ends, the vane is unlocked, and the sliding blade and the roller fit together, and large-volume cylinder is converted into compression operation.

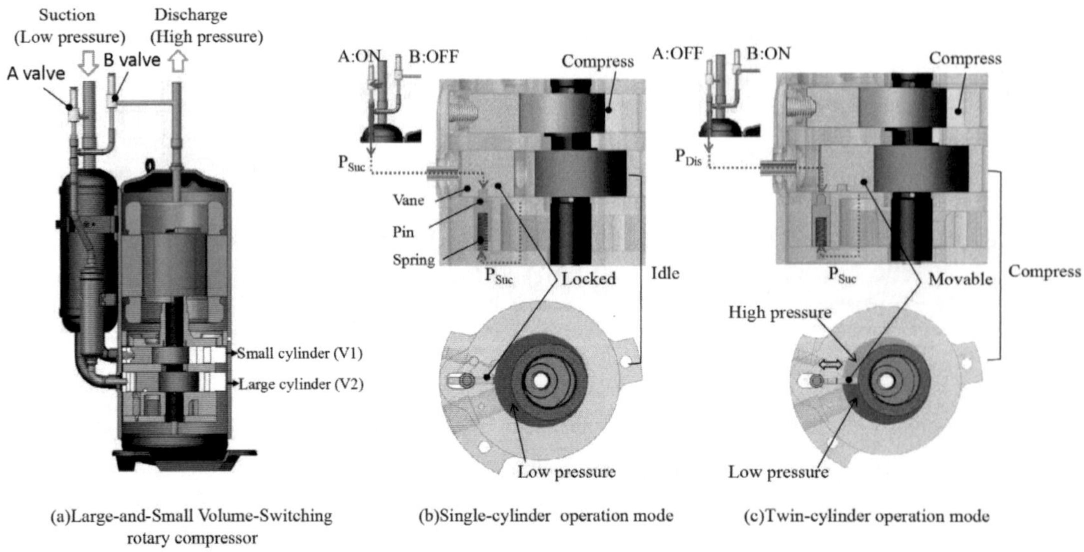

Figure 6: Large-small volume switching principle

3.3 Design of volume distribution ratio of large and small cylinders

Define K as the volume ratio of the small volume cylinder(V_1) to the large volume cylinder(V_2), that is, $K = V_1/V_2$. Through experimental research and theoretical calculations, when the total working volume $(V_1 + V_2)$ is determined, the K value has a very significant impact on the minimum cooling capacity, compressor efficiency and operating stability. Therefore, the design of volume distribution ratio is very important.

3.3.1 Impact of the k value on efficiency and minimum output

As mentioned above, 60% of the time for households to be online is to operate at a low load below 30%, and the minimum output needs to be reduced to 5% of the rated output. Therefore, improving the comprehensive energy efficiency of the compressor under 30% low-load operation and reducing the minimum output is the key to ensuring energy saving for RMAC. Through simulation calculations and experimental studies, the efficiency curves of small-volume operation of compressors with different volume distribution ratios K are obtained, as shown in Figure 7.In the figure , when $K = 1$, the load rate corresponding to the highest point of the compressor efficiency is about 40%. Above or below this load rate, the compressor efficiency will drop significantly. The K value decreases, and the compressor's highest efficiency point also decreases. The peak of the efficiency curve moves toward the load reduction direction, and the capacity output range becomes narrower, and the overall efficiency also decreases. This is mainly because when the value of K is decreased, the useless power consumption caused by the large-volume cylinder (V_2) is increased, resulting in a decrease in the compressor efficiency. In order to ensure the overall energy saving effect, it is necessary to focus on ensuring that the compressor in the low-load area with a load rate within 30% has the highest comprehensive operating efficiency, and also needs to reduce the minimum capacity output. As shown in Figure 8, According to the weighted calculation of compressor efficiency and operating time under low load, within the range of volume distribution ratio K= [0.5, 0.7], the comprehensive efficiency of single-cylinder low load compressor can be guaranteed to reach more than 97% of the highest efficiency value, and the minimum output reaches less than 5% of the rated output.

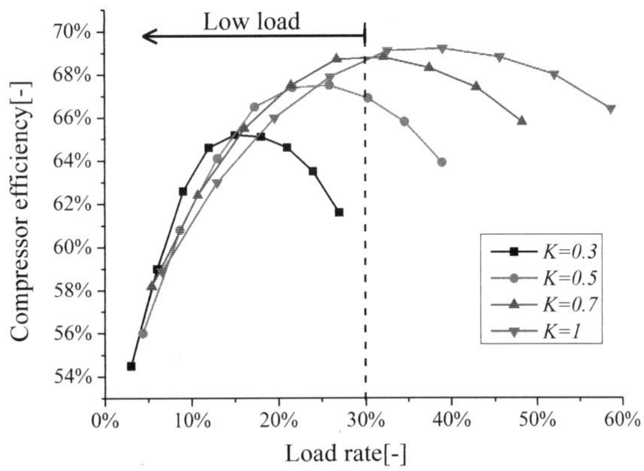

Figure 7: Low-load efficiency of compressor single-cylinder operation with different K values

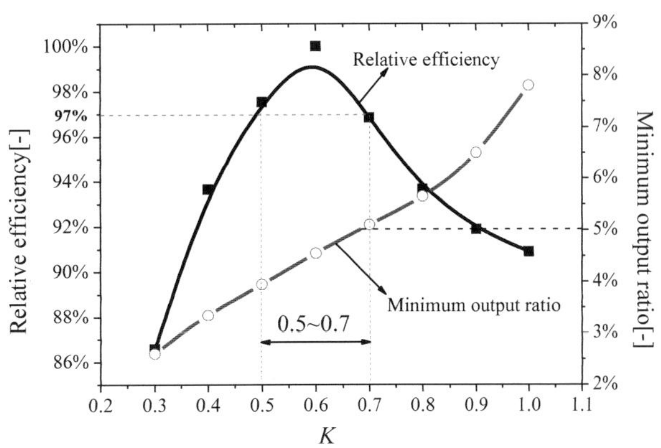

Figure 8: Compressor low load relative efficiency and minimum output ratio at different K values

3.3.2 Impact of the K value on operation stability

For a rolling-rotor compressor, the load torque fluctuates periodically with the rotation angle, and the non-uniformity of torque is the main reason for the low-frequency vibration of the compressor. Because the two cylinders of the large-to-small-volume switching compressor have different volumes, torque fluctuations are large.

Therefore, it is necessary to study the vibration characteristics of different volume distribution ratios K in single-cylinder and twin-cylinder modes in order to design a suitable K value to ensure the requirements for the stability of the compressor operation. The torque when the compressor is running can be calculated as listed in equation(1)-(3), see Ma et al.（2001）and Miao et al.(2000) . According to the compressor force analysis Figure 9, and the speed fluctuation can be calculated from the equations (4)-(5), see Shen, Y. (2004).

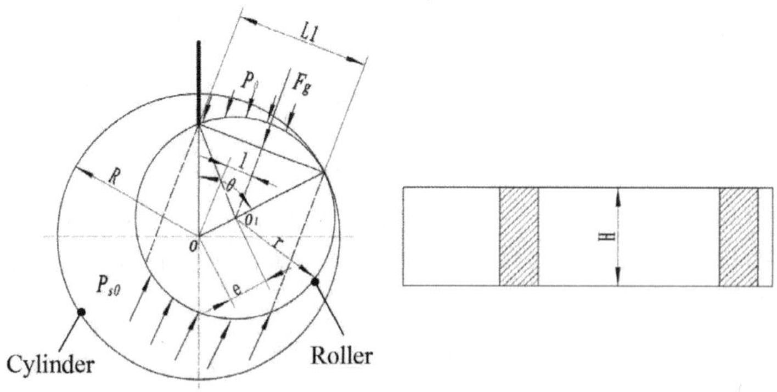

Figure 9 : Compressor force analysis

$$\tau = \frac{e}{R} \tag{1}$$

$$F_g = RH(1-\tau)(p_\theta - p_{s0})\sqrt{2(1-\cos\theta)+\frac{\tau}{1-\tau}(1-\cos 2\theta)} \tag{2}$$

$$M_g = F_g l = \frac{1}{2}R^2 H\tau(1-\tau)(p_\theta - p_{s0})\sqrt{2(1-\cos\theta)+\frac{\tau}{1-\tau}(1-\cos 2\theta)} \tag{3}$$

$$\omega = \sqrt{\omega_0^2 + \frac{2}{J}\left(\int_0^{360} M_d d\varphi - \int_0^{360} M_g d\varphi\right)} \tag{4}$$

$$\xi = \frac{\omega_{\max} - \omega_{\min}}{\omega_m} \tag{5}$$

The calculated load curves for low-frequency operation of the compressor in two-cylinder mode and single-cylinder mode under different volume distribution ratios are shown in Figure 10.From the calculation results: in the single-cylinder mode, only one small-volume cylinder is running. As the volume distribution ratio K increases, the volume of the small-volume cylinder increases, and the compressor torque fluctuation gradually increases. In the two-cylinder mode, when two cylinders are running at the same time, the K value increases, and the smaller the volume difference between the two cylinders, the smaller the compressor load fluctuations. Therefore, when the single-cylinder mode is operated, the smaller the K value is, the more stable the compressor runs; when the twin-cylinder mode is operated, the larger the K value, the more stable the operation. According to the calculations of equations (4)-(5), as shown in Figure 11, selecting $K = [0.5, 0.7]$ can ensure that the unevenness coefficient of the rotation speed at low frequency is less than 5%, which can meet the low frequency stability of single-cylinder and twin-cylinder modes operational requirements.

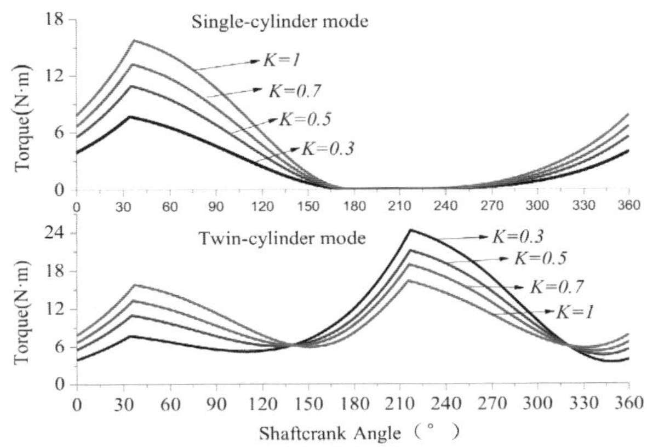

Figure 10: Torque fluctuation with different *K* values

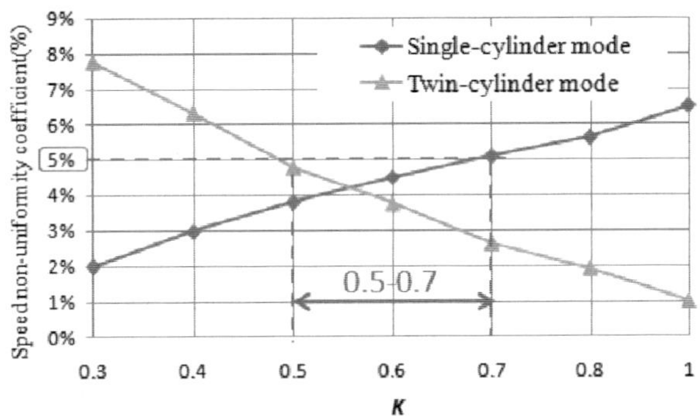

Figure 11: Speed non-uniformity coefficient

Based on the study of the effects of the above-mentioned volume distribution ratio on the compressor's overall efficiency, minimum cooling capacity, and operational stability, it is a better choice to choose a range of compressor volume distribution ratio $K = [0.5, 0.7]$, which can greatly improve the Energy efficiency level of the compressor at low load operation, and can meet the requirements of minimum cooling capacity and operating stability at the same time.

4. COMPRESSOR TEST RESULTS AND APPLICATION IN RMAC

Through the above technology, a 6HP large-and-small volume switching compressor has been developed, and the operating volume of the compressor can be switched between 18cc and 44cc. The compressor energy efficiency test conditions are shown in Table 1. The comparison model is a conventional twin-rotary compressor, which uses a fixed volume (V = 42cc) to operate. The test results are shown in the Figure 12. The conventional compressor maintains high energy efficiency at medium and high capacity output, but when the capacity output decreases below 30% of the rated capacity, the compressor frequency must be continuously reduced, energy efficiency drops sharply. The newly developed compressor uses the single-cylinder small-volume operation mode ($V1$ = 18cc) to reduce the minimum capacity output by 52%, and the new compressor's energy efficiency is increased by 45% during low load operation. This is mainly because the compressor speed can be greatly increased by using a small volume operation at low load, thereby achieving the purpose of increasing the compressor motor efficiency and volume efficiency. In addition, when high-capacity output is required, the new compressor uses the twin-cylinder operating mode (($V2$ = 44cc)). The compressor efficiency has also increased by 5%, mainly due to the compressor's optimized compression structure and efficient new motor.

When the 6HP large-and-small volume switching compressor is applied in a 16kW RMAC, the EER value under a 10% load is increased by 124% compared to the use of a conventional compressor. Mainly at low load operation, the energy efficiency of the compressor is greatly improved, and the minimum output of the compressor is reduced, which avoids frequently start and stop of the compressor, and the power consumption of the compressor is greatly reduced. This new compressor technology enables RMAC to achieve extremely significant energy savings.

Table 1: Energy efficiency test conditions for compressor

Compressor	**Conventional compressor**	**Newly developed compressor (single-cylinder mode)**	**Newly developed compressor (twin-cylinder mode)**
Refrigerant	R410A		
Evaporation temp.	7.2℃		
Suction temp.	35℃		
Condensation temp.	54.4℃		
Subcooling temp.	46.1℃		
Ambient temp.	35℃		
working volume	42cc	18cc	44cc
Rotating Speed	10rps-68rps	15rps-70rps	25rps-65rps

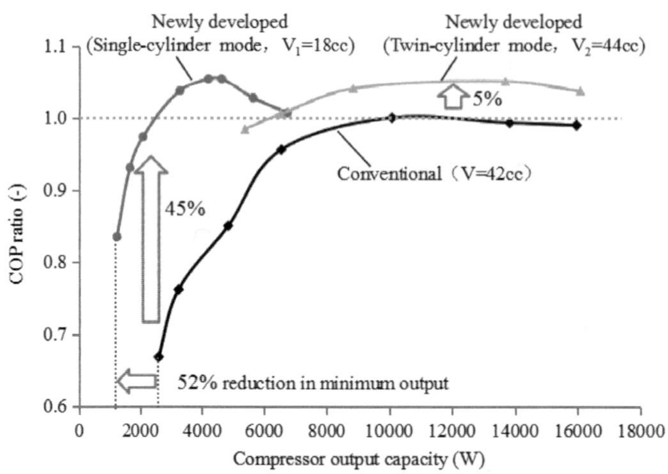

Figure 12: Comparison of compressor energy efficiency

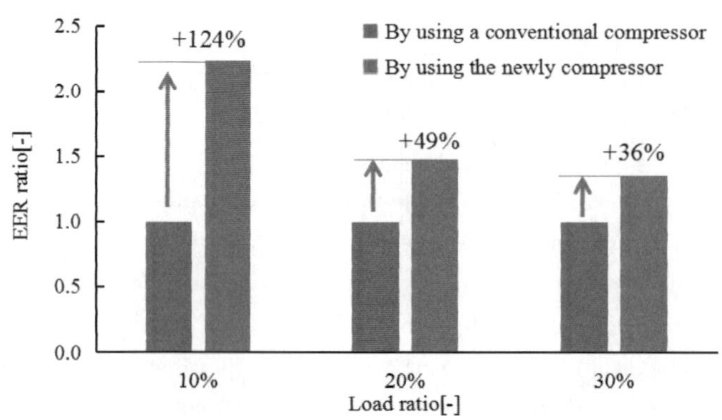

Figure 13: Comparison of energy efficiency at low load with different compressors in 16kW RMAC

5. CONCLUSION

Based on the research on the actual operating load characteristics of RMAC in this paper, it is analyzed that domestic RMAC have problems of excessive minimum cooling capacity and low load energy efficiency when using traditional compressor technology. In order to solve this problem, a high-efficiency large-to-small-volume switching compressor suitable for Chinese households with multiple online connections has been developed.

(1) A new dual high-efficiency operation mode of the compressor is proposed. The compressor can switch the operating volume according to the change of air-conditioning load.

(2) Developed a large-small volume switching compressor, and realized the free switching of two modes of small volume operation and large volume operation in one compressor.

(3) In order to meet the comprehensive requirements of low load energy efficiency, minimum output, and operational stability of the compressor, an optimal volume distribution ratio range of [0.5-0.7] is designed

(4) Compared with traditional compressors, the new compressors developed can improve the low-load energy efficiency by 45%. When it is applied to a 16kW RMAC, the EER value under a 10% load is increased by 124%.

NOMENCLATURE

F_g	gas force	(N)
R	cylinder radius	(m)
L	cylinder height	(m)
E	crankshaft eccentricity	(m)
τ	ratio of the crankshaft eccentricity	(-)
P_θ	compression chamber pressure	(Pa)
P_{s0}	suction pressure	(Pa)
θ	crank angle	(°)
M_g	resistance moment	(N.m)
ω	angular velocity	(rad/s)
ω_0	angular velocity when the crank angle is 0	(rad/s)
M_d	motor output torque	(N.m)
l	arm of force	(m)
J	rotational inertia	(kg·m²)
ξ	speed non-uniformity coefficient	(-)
ω_{max}	maximum angular velocity	(rad/s)
ω_{min}	minimum angular velocity	(rad/s)
ω_m	average angular velocity.	(rad/s)

REFERENCES

HVAC Information. *2016 China Central Air Conditioning Market Development Report* (public version).

Cheng, J. (2017). *Research Report on the Actual Operation Status of Refrigeration and Air Conditioning in China.* Beijing, China Standard Press.

Ma, G., Li, H.(2001).*Rotary compressor* . Beijing, Mechanical Industry Press.

Miao, D. Wu, Y. (2000). *Refrigeration compressor.* Beijing, Mechanical Industry Press.

Shen, Y. (2004). *Mechanical principle tutorial.* Beijing, Tsinghua University Press.

Parametric Studies On A New Schukey Type Rotary Compressor

Bin CUI[1*], Martin GOTTSCHLICH[1], Ulrich LÜDERSEN[1], Stephan KABELAC[2]

[1]Hochschule Hannover, Institute for Process & Energy Engineering and Climate Protection, Hannover, Germany
Phone: +49 (0)511-9296-1652, E-Mail: bin.cui@hs-hannover.de, martin.gottschlich@hs-hannover.de, ulrich.luedersen@hs-hannover.de

[2]Gottfried Wilhelm Leibniz Universität Hannover, Institute for Thermodynamics, Hannover, Germany
Phone: +49 (0)511-762-2277, E-Mail: kabelac@ift.uni-hannover.de

*Corresponding Author

ABSTRACT

Efficient compressors are generally important for applications in air conditioning and refrigeration systems. Therefore, innovative technologies are needed to improve their performance and expand their area of application. A new type of rotary compressor, called “Schukey compressor”, has been investigated both experimentally and theoretically. The Schukey-technology goes back to its founder Jürgen Schukey in 1987, who invented a special gear that enables two rotors to rotate out of phase in the same direction within the compressor housing. Each rotor consists of four rectangular shaped piston wings that interlock and perform non-uniform rotary movements, leading to a periodic change of the working chamber volume. During one revolution of the main shaft, 32 compression cycles are performed. Except for our own research work, no further development on this new device is known to us.

Initial studies have shown that changes in the geometrical design of the machine, especially the position of the suction and discharge ports, have a direct influence on thermodynamical variables e.g. technical work, mass flow or pressure ratio. In this contribution, a parametric study is presented to determine the performance of the machine and to show optimization potentials with regard to the volumetric efficiency. The impact of the annular gap and port size variations is examined by using numerical CFD analysis. It is shown that the annular gap size does has a big impact on the mass flow rate and volumetric efficiency. In order to maintain a high efficiency, the gap size should be kept at least below 0.1 mm, where an efficiency of 85.3% is given. An enlargement of the discharge ports leads to an improvement of the volumetric efficiency up from 85.3% to almost 92% but also leads to a decrease of the pressure ratio at the same time since the compression phase ends off earlier. An enlargement of the inlet ports shows no significant effect in terms of pressure ratio and volumetric efficiency. By using the determined surface plots, the size of the ports can be estimated and adjusted according to the intended application of the compressor.

1. INTRODUCTION

In the past, several types of positive displacement rotary compressors have been used in various applications e.g. air conditioning, refrigeration systems and air-compression. These cover lobe-type, screw-type, vane-type and liquid ring-type machines which work with one or more rotating components that either interlock like screws and lobes or pressure the fluid against the chamber walls. The complex components of screw- and rolling piston-type compressors require costly maintenance and manufacture. Unlike reciprocating compressors, rotary compressors in general do not use valves to control the in- and outflow of the gas.

A new rotary type compressor called “Schukey compressor” was first invented and named after its inventor Jürgen Schukey (Schukey, 1987). He originally designed the machine to be operated as a compression or expansion machine for refrigeration applications. Since then, the interest of the machine was fading. Further research on the machine was only done by Meyer (2019), who experimentally determined the heat transfer coefficient within a working chamber when the machine was operated as an expansion machine in a Rankine power cycle. Except for our own research, no investigations as a compressor have yet been carried out.

Numerical methods as well as the increasing computational power became more and more relevant for the development of compressors. A very important tool over the last years is computational fluid dynamics (CFD), where a lot of research work was done in the past. Shiva Prasad (2004) discussed the benefits and risks for using CFD in the development process of positive displacement compressors. Besides, Fagotti and Possamai (2000) have shown the difficulties of using CFD in the compressor design process, where success highly depends on proper boundary conditions, fine meshing, high computing time and choosing the right modelling approaches. Rodrigues (2014), Rovaris and Deschamps (2006) have investigated several approaches for turbulence modelling in reciprocating compressors considering large-eddy simulations (LES) and RANS-equations, finding out that turbulence modelling is just slightly dependent for the accuracy compared to experimental data. Moreover, Farzaneh-Gord et al. (2015) investigated real and ideal gas models for reciprocating compressors, showing that differences in indicated work and mass flow rates do exist. In addition, Branch (2014) compared two real gas equation-methods with the ideal gas equation in a scroll compressor, determining deviations of mass flow rates and an energy efficiency ratio (EER) up to 14%. Due to the high computational effort of CFD-simulations, other numerical approaches like 0D- or 1D-models have been developed in the past and are still used today, mostly on the basis of the 1st law of thermodynamics to model fluid and heat transfer flows for reciprocating compressors (Tuhovcak et al., 2015). Moreover, the coupling of 0D/1D- with 3D-tools was done by Lang et al. (2008) to simulate the suction line of a hermetic reciprocating compressor.

However, most of the work above only consider the simulation of reciprocating compressors. For rotary compressors, such approaches are rare. Especially for the Schukey compressor no numerical calculations were done yet. In this paper, the design, working principle and possible advantages of the Schukey compressor are discussed and presented.

2. DESIGN AND WORKING PRINCIPLE OF THE MACHINE

2.1 Design of the Schukey compressor

The main components of the examined machine are two interlocking rotors with four wings each, that create eight moving chambers within the housing. The rotational movement of each rotor is specified through its own gearbox with special toothed gears. This allows the change of rotational speeds relative to each other known as the characteristic oscillating "Schukey movement" of the compressor. Both rotors rotate in the same direction, with one rotor accelerating and the other rotor retarding vice versa while the main shaft rotates constantly. As a result, the chambers fluctuate in volume, which lead to its suction, compression and discharge movement. Since the compressor is working without suction and discharge valves, the suction, compression and discharge phases are controlled depending on the position of the inlet and outlet ports. A total of 32 compression cycles are performed during one revolution, which leads to a high-volume turnover of the machine. Compared to screw- and rolling piston-type compressors, possible advantages of the Schukey compressor are fast maintenance and low-cost potential due to the simple construction of the rectangular piston wing geometry. Figure 1 is showing the main components of the Schukey compressor in an exploded view.

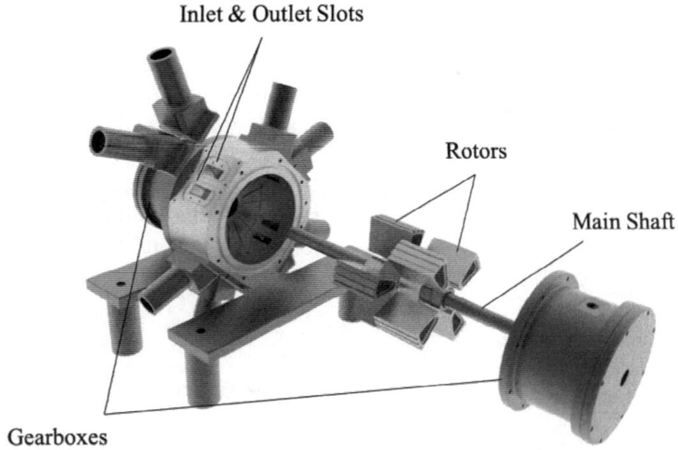

Figure 1: Exploded view of the Schukey compressor.

2.2 Compression cycle

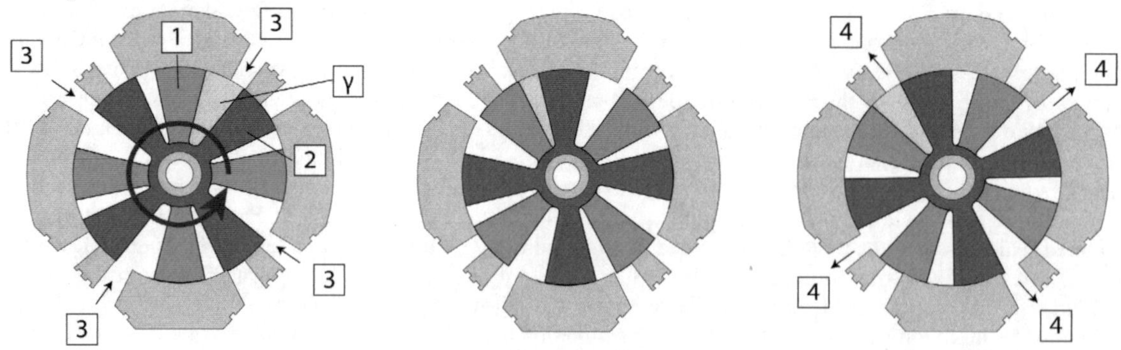

1. Suction 2. Compression 3. Discharge

Figure 2: Compression cycle, Rotor 1 (green), Rotor 2 (blue), Chamber (yellow).

In Figure 2 the compression cycle is shown in three steps. Each cycle begins with the filling of the fluid into the working chamber γ through the inlet port (3). During the suction process, the leading rotor (1) is in its accelerating phase of its movement pattern while the complementary rotor (2) is retarding. In that case, the working chamber increases in volume and the fluid is sucked in. This step ends, when rotor blade (2) closes the inlet port during the counter-clockwise rotation. In that moment, the chamber volume is at its maximum. After closing the inlet, rotor blade (2) starts accelerating while rotor blade (1) starts retarding and thus the chamber volume begins to decrease. Since the working chamber is not connected to any inlet- or outlet port, the compression takes place between rotor blade (1) and (2). After reaching its minimum volume, rotor blade (1) passes the outlet port (4) and the fluid will be discharged. From then on, rotor blade (1) starts to accelerate while rotor blade (2) starts to retard again. The discharge process ends when rotor blade (2) passes and closes the outlet port. From then on, the compression cycle starts again.

The cycle takes places in all chambers simultaneously. Therefore, one revolution of the main shaft leads to 32 compression cycles (8 chambers x 4 sections). The compressor is working without active valves and thus does not need any active control mechanism. The oscillating specific movement pattern for each rotor is determined by its own special toothed gears, which are installed with an offset of 45° to each other and connected to the main shaft.

2.3 Kinematics of the Schukey Compressor

The main shaft rotates at a constant speed, driven by an electric motor. As mentioned, the angular velocities of the rotors are not constant, but oscillating and rotating faster or slower than the main shaft. This results in an accelerating and retarding movement. The relative changes in angular velocity of the rotors are shown in Figure 3, where each rotor passes four minima and four maxima during one revolution of 360° of the main shaft. At the moment, when maxima and minima occur, the chamber angle is in its neutral position of 21°. If the angular velocity curves intersect, the chamber volume reaches its minima or maxima, depending on which chamber is actually considered. This is when the compression or the suction ends.

The angular velocity function of each rotor is determined through incremental measurements of the resulting motion of the gearbox. A non-linear regression with the Levenberg-Marquardt-Algorithm was used to determine the coefficients a, b, c and d of eq. (1) and (2). ω_θ is the (constant) angular velocity of the main shaft. With θ as the crank angle, ω_N being the nominal angular velocity and k_b as an additional constant, the angular velocity of each rotor can be calculated with eq. (1) and (2). The resulting chamber angle- and volume function $\gamma(t)$ and $V_C(\gamma)$ can be derived from the angular velocities in eq. (3) and (4), showing a sine-like behavior illustrated in Figure 3. The geometrical dimensions are represented in Table 2.

Angular Velocity Function:

$$\omega_{R1}(\theta) = \left\{-a \cdot \left[1 - cos\left(\frac{\pi}{45^\circ} \cdot \theta + b\right) + \frac{c}{2} \cdot sin^2\left(\frac{\pi}{45^\circ} \cdot \theta + b\right)\right] + d\right\} \cdot k_b \cdot \frac{\omega_\theta}{\omega_N} + \omega_\theta \tag{1}$$

$$\omega_{R2}(\theta) = \left\{-a \cdot \left[1 - cos\left(\frac{\pi}{45^\circ} \cdot \theta + b + \pi\right) + \frac{c}{2} \cdot sin^2\left(\frac{\pi}{45^\circ} \cdot \theta + b + \pi\right)\right] + d\right\} \cdot k_b \cdot \frac{\omega_\theta}{\omega_N} + \omega_\theta \tag{2}$$

Table 1: Determined coefficients of the angular velocity function

Coefficients	*a*	*b*	*c*	*d*
Values	0.09316	6.28992	0.77746	0.11127

Chamber Angle Function

$$\gamma(t) = \gamma(t - dt) + \int \omega_{\mathrm{R}1,2} - \omega_{\mathrm{R}2,1}\, dt \quad (3)$$

Volume Chamber Function

$$V_C(\gamma) = l \cdot \frac{\pi}{360°} \cdot \gamma \cdot (R_{\mathrm{a}}^2 - R_{\mathrm{i}}^2) \quad (4)$$

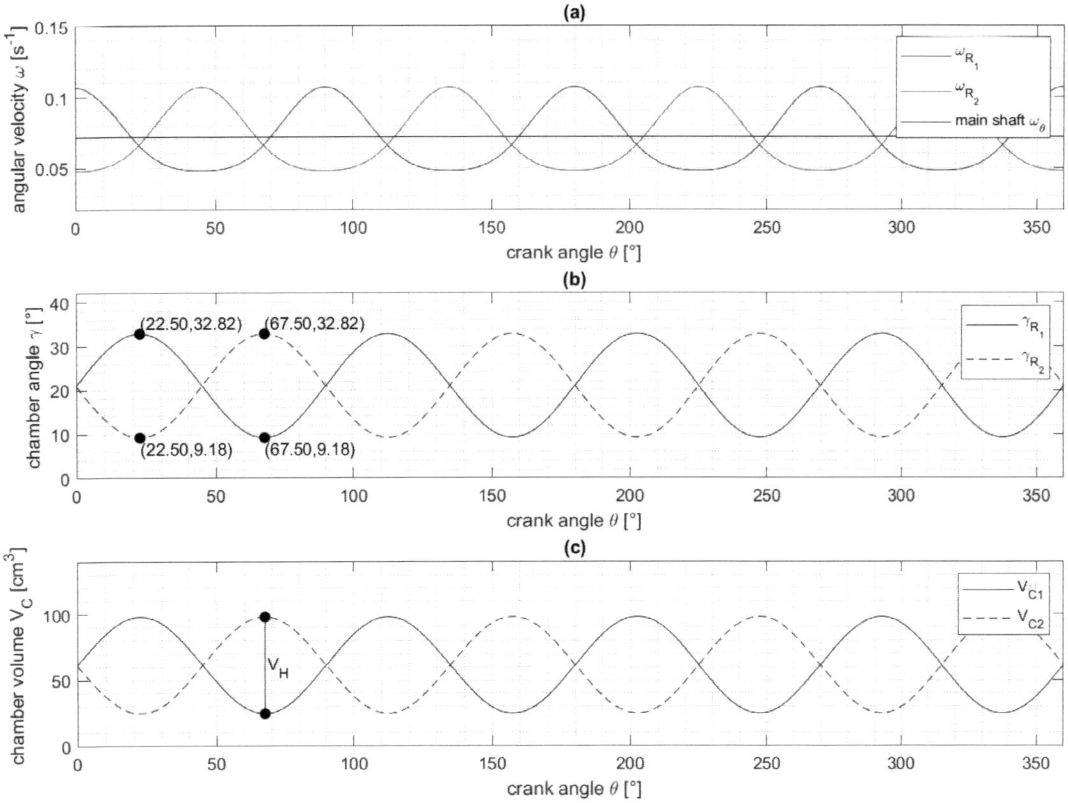

Figure 3: Function of **(a)** angular velocity ω of rotors and main shaft at $\omega_0 = \omega_{\mathrm{N}} = 0.717\ s^{-1}$ =const., **(b)** chamber angle γ with $\gamma_{\max} = 32.82\ °$ and $\gamma_{\min} = 9.18\ °$, **(c)** chamber volume V_{C} during one revolution of the main shaft.

Table 2: Compressor geometry parameters.

Compression Chamber			
Min. chamber angle $\gamma_{\min}$	9.18°	Max. chamber angle $\gamma_{\max}$	32.82°
Min. chamber volume $V_{\min}$	28.46 cm³	Max. chamber volume $V_{\max}$	101.82 cm³
Stroke volume V_{H}	73.36 cm³	Total volume turnover per revolution	2.35 l
Rotor dimensions		**Suction and discharge Ports**	
Outer rotor radius R_{a}	75 mm	Suction port	8°
Inner rotor radius R_{i}	22.75 mm	Discharge port	4°
Rotor length l	69.6 mm		
Rotor blade angle	24.0°		

3. SIMULATION MODEL AND STUDY DESIGN

3.1 Preliminary investigations from experimental measurements

The preliminary experimental investigations from the testing facility of Meyer (2019) provide a basis for the simulation model. The mass flow rate has been measured in the testing facility at a rotational speed of 100 rev/min, showing an average mass flow rate of $\dot{m} = 100\frac{kg}{h}$ (see Figure 4). Further experimental investigations on a new Schukey prototype will be carried out in the near future.

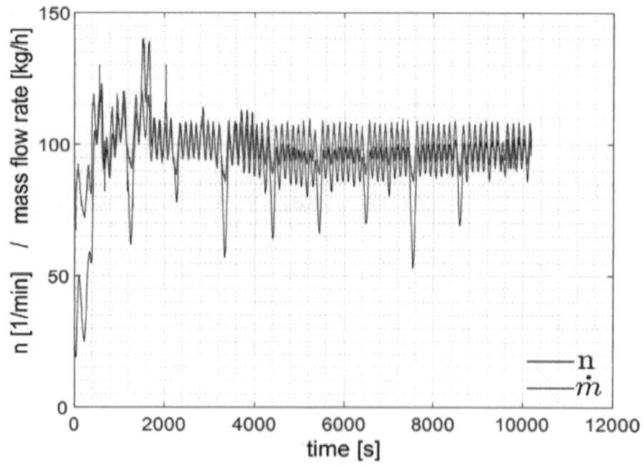

(a) (b)

Figure 4: Experimental mass flow rate measurements by Meyer (2019). **(a)** Machine in its testing facility, **(b)** Measurement of mass flow rate and rotational speed of 100 rev/min.

3.2 Equations of State

A very simple model used in CFD-simulations is the ideal gas model, which represents an idealized state neglecting all intermolecular interactions. The results are sufficient accurate as long as the pressure level is low and the occurring temperatures are high. However, in processes with higher densities like in compression cycles or with gaseous states close to its dew line, the model can lead to significant deviations in fluid flow behavior. In order to obtain the most accurate results, real gas equation-models are required. One method to integrate real gas equations in the used CFD-solver ANSYS CFX is the implementation of the NIST Refprop database (Version 9) with RGP-Tables, which contains access to the real gas properties of the chosen fluid. Another approach is the use of real gas models like the Aungier-Redlich-Kwong (ARK) equation of state. In section 4, the ideal and real gas models are compared to each other in a pressure-volume diagram.

3.3 Model

In order to solve the fluid flow of the refrigerant through the compressor, the commercial CFD solver ANSYS CFX Version 2020 R2, based on finite volume method has been used.

One method to simulate the chamber volumes in positive displacement machines is the immersed boundary method (IBM), first introduced by Peskin (2002). Since then, several efforts have been made to enhance the range of applicability and accuracy (Ghias et al., 2007; Kang et al., 2009). For this method, a background mesh and a mesh for the solid rotor body is required. At the overlaps of both meshes, momentum sources are applied to the fluid, representing the rotor movement. Since there are several restrictions for the IBM among other things its usage only for single-phase incompressible fluids, this method is not eligible to model the compressibility effects inside the Schukey compressor. Another way to model the fluctuating volumes of the chambers is the use of remeshing methods, where a new mesh is generated once the deformed mesh is of poor quality. However, each remeshing iteration leads to high computational time. Therefore, a moving mesh model has been used for the hexa-meshed working chambers. For those, a given mesh stiffness was defined which enables the mesh to be deformed but still maintain a high quality. During the compression cycles, the orthogonal angle criteria >80° is still met and varies between 84° and 90°.

In order to keep the computational effort low, the developed simulation model has been split into rotor- and stator domains with different element types, connected through a GGI (Generalized grid interface) that consider the conversation of timestep dependent connection of the fluxes (see Figure 5). The inlet, outlet and annular gap are defined as the stator domain, while rotor 1 and 2 are defined as rotor domains. The rotors are rotating in counterclockwise direction with fluctuating volumes. Since the design of the machine is symmetrical, periodic boundary conditions are set to take the fluid behavior from neighboring domains into account. Therefore, only ¼ of the whole engine was modelled to save computational time. In total, the mesh contains 11,238,409 nodes and 13,121,510 elements.

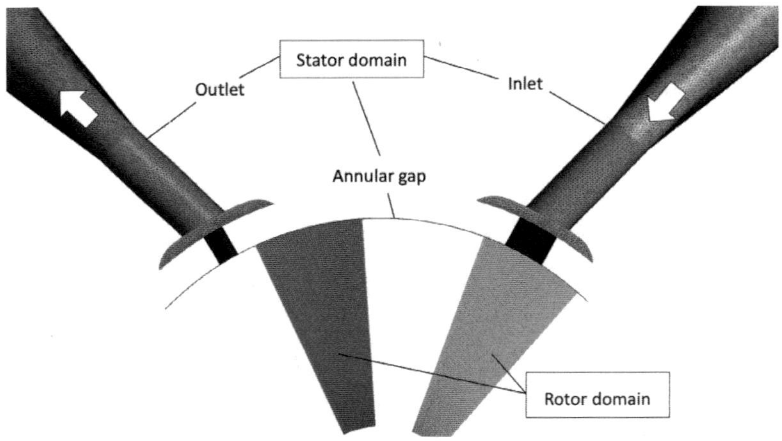

(a)

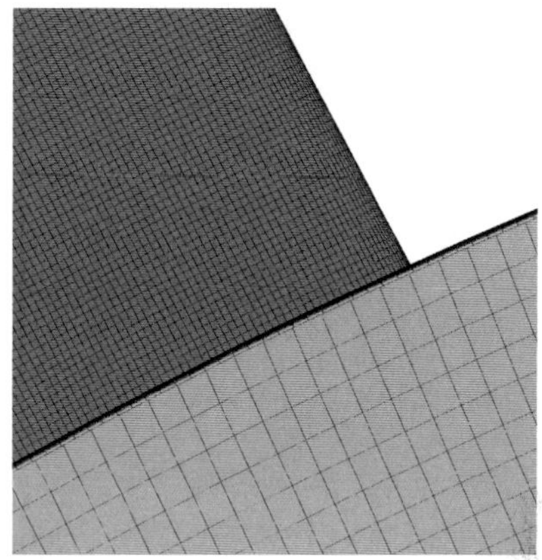

(b)

Figure 5: Simulation model, **(a)** Rotor and stator domains, **(b)** Mesh grid interfaces.

3.4 Boundary conditions

In order to simulate reliable numerical results, a highly fine meshed model and an adequate setup of boundary conditions are required. In this work, different boundary conditions are assumed. Some remain constant during all calculations according to Table 3 and some vary for the parametric study, listed in Table 4. The pressure and temperature at inlet and outlet are specified as opening conditions that enable fluid flows in both directions in and out of the domain. Both, inlet and outlet pressure remain constant. The rotational speed of the main shaft is set constant with 1200 rpm since it has shown very good results in initial studies. To describe the characteristic movement of the rotor wings, specified displacement functions of the left and right rotor walls are set in CFX Expression Language (CEL). Since there were only preliminary measurements of the heat transfer coefficient by Meyer (2019), it is assumed to be constant on the outer walls with 150 W/m²K with an ambient temperature of 35 °C. For turbulence modelling, the Reynolds Averaged Navier Stokes Equations (RANS) are chosen, which solve the average flow behavior separately from unsteady contributions. The Shear Stress Transport (SST) model from Menter (1994) is used since it combines the advantages of the k-ε and the k-ω model and has proven to be suitable for most engineering flows, providing good results and a robust scheme. The simulation considers two full revolutions, in total eight full compression cycles for each chamber, to obtain enough time for flow development inside the machine.

3.5 Parametric study design

Since the machine is operating without active valve mechanisms, a good sealing of the working chambers needs to be ensured. The resulting leakage flow out of the working chamber is modelled by the circumferential annular gap, which varies between 0 mm and 0.2 mm for the parametric study in order to examine the impact of leakage regarding mass flow rate, pressure ratio and volumetric efficiency.

Furthermore, past investigations have shown that the size of the suction and discharge ports can also have a big impact on the mentioned values. Therefore, the correlation between the size of the ports in terms of volumetric efficiency and pressure ratio is determined without permitting a parasitic flow as a result of the direct connection of both openings. The annular gap size in this case was kept constant with 0.1 mm.

It is also of interest to estimate the difference between the ideal gas and real gas models, since the computing time of real gas models is considerably longer. Because of that, the results from the ideal gas-, ARK- and RGP-model are compared to each other in a pressure-volume diagram.

Table 3: Boundary conditions used in simulation.

Boundary condition	Value
Suction pressure	1 bar
Discharge pressure	1.8 bar
Inlet temperature	300 K
Wall heat transfer coeff.	150 W/m²K
Outside temperature	308 K
Rotational speed main shaft	1200 rpm

Table 4: Parametric values.

Parametric variables	Values
Fluid	Air ideal gas, Air ARK-model, Air RGP-Table
Annular gap	0 mm – 0.2 mm
Suction port size	8-16°
Discharge port size	4-14°

4. SIMULATION RESULTS

To compare the results of the parametric studies, pressure-volume diagrams and 3-dimensional surface plots have been created and are shown in Figure 6 and Table 5. Since there is no experimental data available, only simulation results are compared to each other.

The pressure-volume diagram in Figure 6 (a) presents the comparison between the ideal gas-, ARK- and RGP model. The graph only shows small deviations between the compared models, which can also be seen at the results in Table 5. The volumetric efficiency merely deviates about $\pm$ 1% and the internal pressure ratio only differs by $\pm$ 0.03. Therefore, the impact of using real gas models over the ideal gas model is small at these pressure levels and can be neglected to save computational effort.

Figure 6 (b) shows the results of the annular gap study, which do have a more significant impact on the performance, especially at higher pressure areas in this machine. Comparing the gap sizes 0 mm and 0.2 mm, the internal pressure ratio decreases by 0.3 due to the leakage of the fluid. The leakage also leads to a decreased mass flow rate of the machine, which on the other hand, results in a decreasing volumetric efficiency as seen in Table 5 and Figure 6 (c). Without considering an annular gap at all, the volumetric efficiency is nearly ideal with 99.7%. With a growing gap size, the efficiency starts to decrease slowly until a gap size of 0.05 mm with a volumetric efficiency of 97.8% is reached. From then on, the efficiency drops significantly to 85,3% at a gap size of 0.1 mm and 68.5% at the gap size of 0.15 mm. At 0.2 mm the mass flow rate is lowered by nearly 50% which should be avoided for the upcoming prototype of this machine. In order to maintain a high volumetric efficiency >85%, the annular gap size should be kept at least below 0.1 mm, which can be feasible with high manufacturing quality and a good sealing concept.

Lastly, the results of the parametric study on different suction and discharge port sizes are shown in Figure 6 (d-f). The mass flow rate (d) shows an analogous behavior to the volumetric efficiency (e), where a growing suction port size results in no significant improvement of the volumetric efficiency nor the internal pressure ratio. An enlargement of the discharge ports, however, leads to an increase of the volumetric efficiency up to 92% but comes, on the other hand, with a reduction of the internal pressure ratio from its maximum 2.73 to 1.95. This behavior is to be expected and can be explained by the fact, that when the discharge port size is enlarged, the compression phase breaks off earlier and a relatively longer ejection process takes place. The chamber volume has not reached its small volume yet to build up a high pressure. Therefore, it should be prioritized whether a higher pressure ratio or a higher volumetric efficiency is more important. For the intended application, the optimal size of the ports can then be estimated based on the determined surface plots.

Table 5: Results of the parametric studies for real- and ideal gas comparison and annular gap size.

Variable	Internal pressure ratio	MFR	Volumetric Efficiency
	Π	$\dot{m}$	λ
Unit	–	kg/s	%
Air ideal	2.82	0.0506	89.4
Air ARK	2.79	0.0513	90.6
Air RGP	2.84	0.0503	88.8
0 mm	2.81	0.0563	99.7
0.05 mm	2.78	0.0544	97.8
0.1mm	2.73	0.0473	85.3
0.15 mm	2.65	0.0379	68.5
0.2 mm	2.55	0.0274	49.6

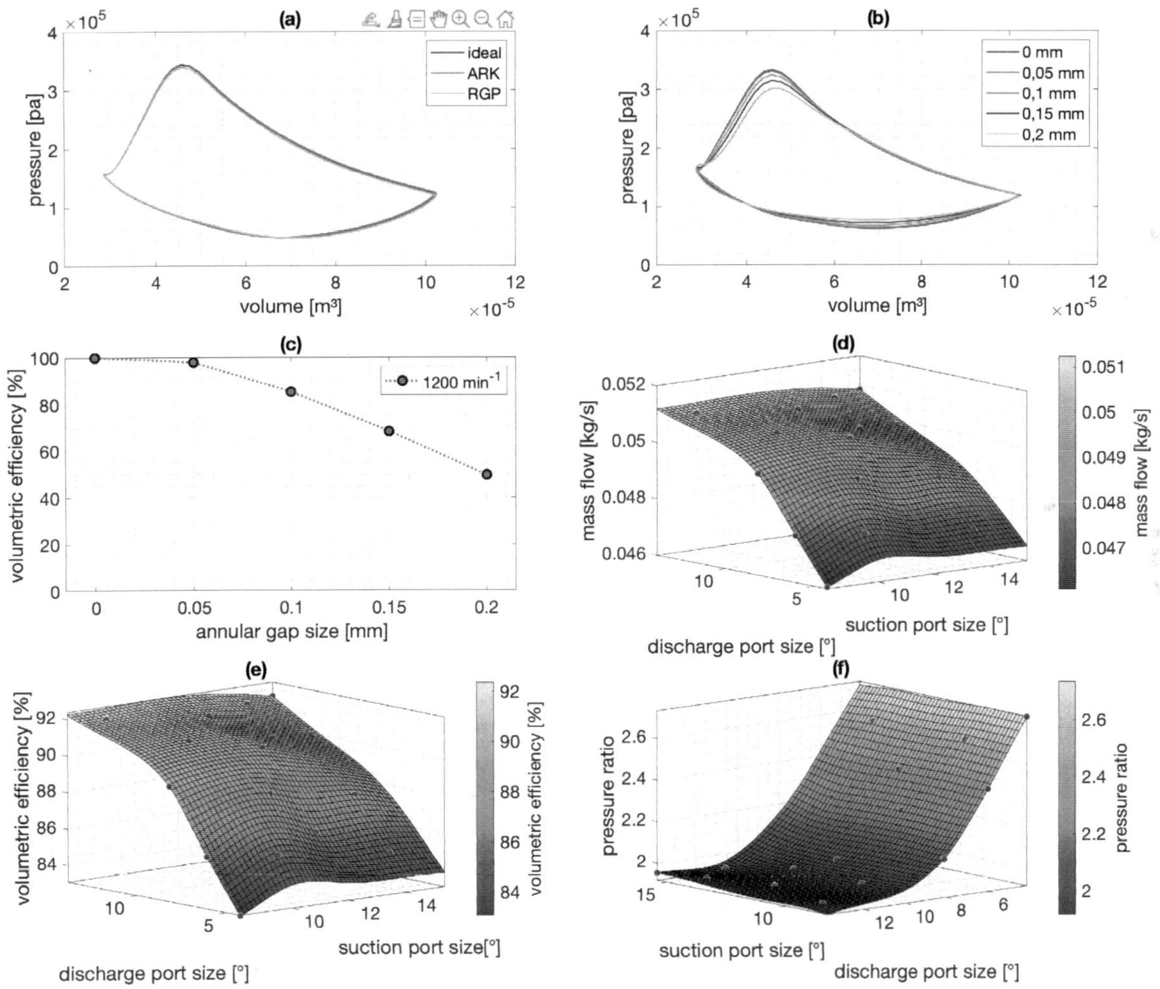

Figure 6: Results of parametric studies at 1200 rev/min: **(a)** Pressure-volume diagram for air ideal gas-, ARK- and RGP model, **(b)** Pressure-volume diagram for annular gap size variation, **(c)** Volumetric efficiency for annular gap size variation, **(d)** Mass flow rates for suction/discharge port size variation, **(e)** Volumetric efficiency for suction/discharge port size variation, **(f)** Internal pressure ratio for suction/discharge port size variation.

5. CONCLUSIONS

A new rotary type compressor named Schukey compressor has been introduced. The working principle is based on two phase-shifted rotors, which are realized by special toothed gears. The construction and working principle are relatively easy due to the simple shape of the rectangular piston wings. Compared to other rotary compressors, this new compressor has a great potential for the application in refrigeration systems because of its high volume turnover. Due to the operation without oil and active valves systems, the compressor requires low maintenance making it more sustainable for the environment after all.

It was shown that the annular gap size does have a high impact regarding mass flow rate and volumetric efficiency. In order to maintain a high efficiency, the gap size should be kept below 0.1 mm, where an efficiency of 85.3% is given. At a gap size of 0.05 mm, the efficiency increases to 97.8% which should be aimed for in the manufacturing process and its sealing concept. Furthermore, this paper showed the differences between using ideal gas and real gas equations. The examined real gas models ARK and RGP only showed small deviations by ± 1% in volumetric efficiency compared to the ideal gas model and therefore can be neglected for the existing pressure range. Lastly, the relation between different suction and discharge port sizes were examined, where an enlargement of the suction port size led to no significant difference in internal pressure ratio and volumetric efficiency. In spite of that, an enlargement of the discharge port size has led to a higher volumetric efficiency from 85.3% to nearly 92%, but comes, on the other hand, with a lower pressure ratio since the compression phase breaks off earlier. Depending on the application, it should be prioritized whether a higher pressure ratio or a higher volumetric efficiency is needed. By using the determined surface plots, the size of the ports can then be estimated and adapted accordingly to its use case.

The machine, however, is still in its development phase where further optimization possibilities exist like finding the optimal wing geometry or changing the kinematic function of the gear. Further investigations are concentrated on these points. Even though the results show high potential in use for air refrigeration processes, the methodology still requires to be fully validated with experimental data before extending analyses should be done.

NOMENCLATURE

ARK	Aungier-Redlich-Kwong	(–)
CFD	Computational Fluid Dynamics	(–)
CEL	CFX Expression Language	(–)
EER	Energy-efficiency-ratio	(–)
EOS	Equation of State	(–)
FSI	Fluid-Structure Interaction	(–)
GGI	Generalized grid interface	(–)
IBM	Immersed boundary method	(–)
LES	Large Eddy Simulation	(–)
MFR	Mass flow rate	$(kg \cdot s^{-1})$
NIST	National Institute of Standards and Technology	(–)
RANS	Reynolds Averaged Navier Stokes	(–)
RGP	Real gas properties	(–)
SST	Shear Stress Transport	(–)
λ	Volumetric Efficiency	(%)
ω	Angular velocity	(s^{-1})
θ	Crank angle	(°)
γ	Chamber angle	(°)
Π	Internal pressure ratio	(–)

Subscript

1,2	Rotor number
θ	Main shaft
c	Chamber
N	nominal

6. ACKNOWLEDGEMENT

This work was supported by the compute cluster, which is funded by the Leibniz Universität Hannover, the Lower Saxony Ministry of Science and Culture (MWK) and the German Research Association (DFG).

7. REFERENCES

Branch, S. (2014). Methods of fluid properties for compressible refrigerant CFD analysis. In International Compressor Engineering Conference (Ed.), *22nd International Compressor Engineering Conference at Purdue 2014: West Lafayette, Indiana, USA, 14-17 July 2014* (pp. 155–165). Curran Associates Inc. https://docs.lib.purdue.edu/icec/2289/

Fagotti, F., & Possamai, F.C. (2000). Using Computational Fluid Dynamics as a Compressor Design Tool: Paper 1377. In International Compressor Engineering Conference (Chair), *International Compressor Engineering Conference,* Purdue.

Farzaneh-Gord, M., Niazmand, A., Deymi-Dashtebayaz, M., & Rahbari, H. R. (2015). Thermodynamic analysis of natural gas reciprocating compressors based on real and ideal gas models. *International Journal of Refrigeration, 56*, 186–197. https://doi.org/10.1016/j.ijrefrig.2014.11.008

Ghias, R., Mittal, R., & Dong, H. (2007). A sharp interface immersed boundary method for compressible viscous flows. *Journal of Computational Physics*, *225*(1), 528–553. https://doi.org/10.1016/j.jcp.2006.12.007

Kang, S., Iaccarino, G., & Ham, F. (2009). DNS of buoyancy-dominated turbulent flows on a bluff body using the immersed boundary method. *Journal of Computational Physics*, *228*(9), 3189–3208. https://doi.org/10.1016/j.jcp.2008.12.037

Lang, W., Almbauer, R., Burgstaller, A., & Nagy, D. (2008). Coupling of 0-,1- and 3-d Tool for the Simulation of the Suction Line of a Hermetic Reciprocating Compressor: Paper 1881. In International Compressor Engineering Conference (Chair), *International Compressor Engineering Conference 2008,* Purdue.

Menter, F. R. (1994). Two-equation eddy-viscosity turbulence models for engineering applications. *AIAA Journal*, *32*(8), 1598–1605. https://doi.org/10.2514/3.12149

Meyer, J. (2019). *Experimentelle Bestimmung und Modellierung des Wandwärmeübergangs innerhalb eines Drehkammerschwingsystems* [Dissertation]. Gottfried Wilhelm Leibniz Universität, Hannover.

Peskin, C. S. (2002). The immersed boundary method. *Acta Numerica*, *11*, 479–517. https://doi.org/10.1017/S0962492902000077

Rodrigues, T. T. (2014). Tubulence Modelling Evaluation for Reciprocating Compressor Simulation: Paper 2297. In International Compressor Engineering Conference (Ed.), *22nd International Compressor Engineering Conference at Purdue 2014: West Lafayette, Indiana, USA, 14-17 July 2014.* Curran Associates Inc.

Rovaris, J. B., & Deschamps, C. J. (2006). Large eddy simulation applied to reciprocating compressors. *Journal of the Brazilian Society of Mechanical Sciences and Engineering*, *28*(2), 208–215. https://doi.org/10.1590/S1678-58782006000200010

Schukey, J. (1987). European Patent Office : Drehkolbenmaschine(0316346). Germany.

Shiva Prasad, B. G. (2004). CFD for Positive Displacement Compressors: Paper 1689. In International Compressor Engineering Conference (Chair), *International Compressor Engineering Conference at Purdue,* Purdue. https://docs.lib.purdue.edu/icec/1689

Tuhovcak, J., Hejcik, J., & Jicha, M. (2015). Modelling of fluid flow and heat transfer in a reciprocating compressor. *IOP Conference Series: Materials Science and Engineering*, *90*, 12018. https://doi.org/10.1088/1757-899X/90/1/012018

Research on Oil Circulation Rate of CO_2 Inverter Rotary Compressor

Li LI[1], Min SUN[2]*

[1]Xi'an Qing'an Refrigeration Equipment Co., Ltd,
Gaoxin 2nd Road, Xi'an, China
E-mail: lili070905@163.com

[2]Xi'an Qing'an Refrigeration Equipment Co., Ltd,
Gaoxin 2nd Road, Xi'an, China
E-mail: sunmin0929@126.com

* Corresponding Author

Abstract

The oil circulation rate(OCR) of a CO_2 inverter rotary compressor is higher than expected. The computational fluid dynamics(CFD) method is used to build a flow field simulation model of the compressor. Boundary conditions are determined according to the real running conditions of the compressor, involving the pressure, the temperature, the mass flow rate and other necessary parameters. There are four paths for refrigeration gas and oil flowing across: the motor rotor channel, the air gap between the motor rotor and the motor stator, the winding gap and the motor stator core cut. The CO_2 mass flow rates of these four paths are obtained and compared. The pressure field in the compressor is obtained and analyzed, which determines the gas and oil flow pattern and path. The velocity is also investigated. The research shows: the CO_2 mass flow rate through motor rotor channel and winding gap have a great influence on the OCR of CO_2 inverter rotary compressors. The simulation result agrees well with the test result.

1. INTRODUCTION

CO_2 is an environmental-friendly refrigerant. However, the OCR of typical CO_2 compressors, a critical parameter for deciding compressor's efficiency, is still high compared to conventional compressors. Analysis of OCR by the flow field simulation method is convenient, visualized and time-saving than the traditional test method.
Guomang YANG et al.(2015) analyze the gas-oil flow field in a rotary compressor by numerical simulation. Some structural changes making the oil returning channel obvious in the compressor are made, so OCR during the whole frequency range is controlled within 0.5%. Lichang XIE et al.(2017) use the CFD method to simulate the circulation of oil and gas in one rolling piston-type compressor, and propose the optimal scheme that effectively reduces the oil rate of the compressor. Tetsuhide Yokoyama et al.(2012) explored an improvement in the oil separation by using CFD analysis, identified two effective measures, and experiments with an actual compressor to verify that oil separation is improved to the target value. But which parameter has the biggest influence on the OCR of CO_2 compressors, the papers did not concerned.
In this paper, a flow field simulation model is built. Also, the flow field of different structures of the compressor is calculated. The relationship between the mass flow rate of CO_2 gas through the four paths were analyzed in detail, which has the biggest influence on the OCR of the compressor according to results.

2. CALCULATION MODEL AND CFD METHODS

2.1 Governing Equations

In this paper, the basic equation of fluid and the multiphase flow model are applied.
The mass conservation equation of incompressible fluid is:

$$\frac{\partial \rho}{\partial t}+u\frac{\partial(\rho u)}{\partial x}+v\frac{\partial(\rho v)}{\partial y}+w\frac{\partial(\rho w)}{\partial z}=0 \quad (1)$$

The momentum conservation equation of incompressible fluid is:

$$\frac{\partial}{\partial t}(\rho\vec{v})+\nabla\cdot(\rho\vec{v}\vec{v})=-\nabla p+\nabla\cdot\left[\mu\left(\nabla\vec{v}+\nabla\vec{v}^{T}\right)\right]+\rho\vec{g}+\vec{F} \tag{2}$$

The volume fraction equation of the multiphase flow model is as follows:
The main phase:

$$\sum_{q=1}^{n}\alpha_q=1 \tag{3}$$

The q phase:

$$\frac{\partial\alpha_q}{\partial t}+v_q\cdot\nabla\alpha_q=\frac{S_{\alpha_q}}{\rho_q}+\frac{1}{\rho_q}\sum_{p=1}^{n}\left(\dot{m}_{pq}-\dot{m}_{qp}\right) \tag{4}$$

$\vec{v}$ ——absolute speed;
u, v, w —— the component of the velocity in the x, y, and z directions;
$\dot{m}_{pq}$ —— mass transport from p phase to q phase;
$\dot{m}_{qp}$ —— mass transport from q phase to p phase.

2.2 Assumptions

(1) The mass flow rate of CO_2 discharged from the compression chamber does not change with time;
(2) The mass flow rates of oil discharged from the compression chamber and eccentric shaft oil hole do not change with time.
(3) The exhaust temperature and the exhaust pressure are constant in the compressor.

2.3 Calculation Model

Figure 1 shows three geometric models of the compressor. The distance between the rotating cap and the motor rotor of Model 1 is 0.022m. The distance between the rotating cap and the motor rotor of Model 2 is 0.0095m. The distance between the rotating cap and the motor rotor of Model 3 is 0.0095m, and there is an additional lower balance weight shell in model 3.

Figure 2 shows the mesh model of model 1. Tetrahedron grids are used to mesh the fluid volume. The grid number is approximately 4,300,000. The mesh method used in Model 2 and Model 3 is the same as in Model 1.

The calculation rotating speed is 60r/s. The boundary conditions are listed in Table 1.

Table 1: Boundary conditions

Boundary conditions	Position
Mass flow inlet	Muffler discharge hole
Pressure outlet	Compressor discharge pipe
Rotary body	Motor rotor and shaft
Static body	Bodies except for rotary body
Interface	Faces on which rotary body and static body coincide

2.4 CFD method

The two-phase model is used to include the mutual interaction of CO_2 gas and oil droplets. CO_2 gas is the continuous phase, and oil droplets are the discrete phase. Four effects are considered when oil droplets colliding on the wall: escaped, wall-jet, reflect and trap. The escaped effect is utilized to the inlet and outlet boundary. The wall-jet effect is utilized to the static wall boundary. The reflect effect is utilized to the rotary wall boundary. The trap effect is utilized to the wall boundary near the oil level. The reference frame model is used to represent the rotation. The rotary body mesh is rotating all the time totally, so it is not needed to mesh again when calculating. The geometric faces are rotating with the rotary body meshes also.

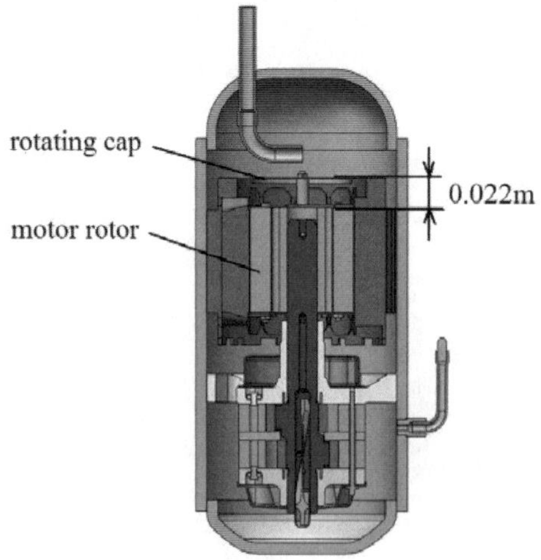

(a) Model 1

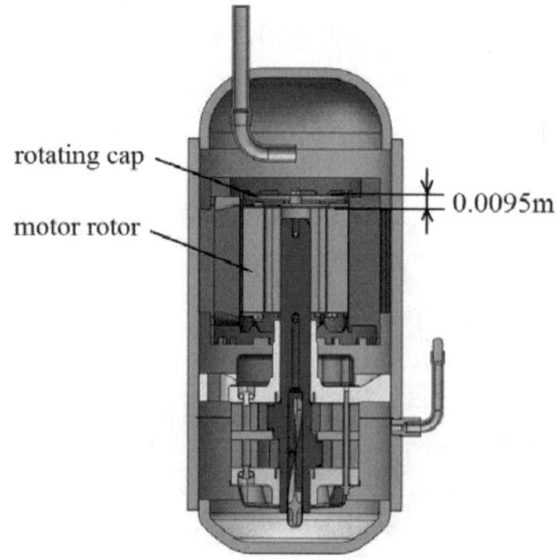

(b) Model 2

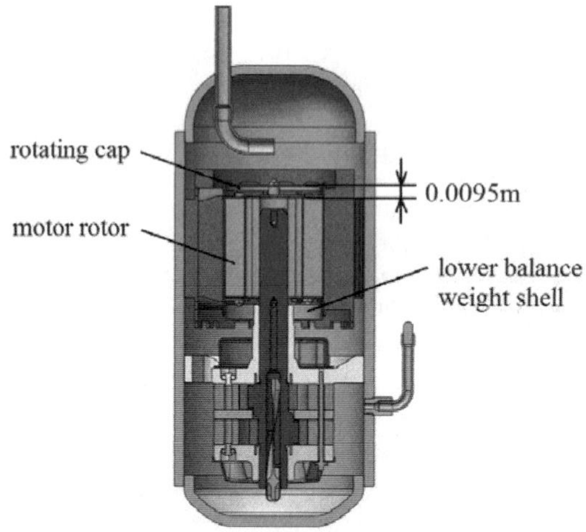

(c) Model 3

Figure 1: Geometric models

2.5 CO_2 gas flow path

There are four paths for CO_2 gas and oil droplets flowing through, either from the space below the motor to the space above the motor, or from the space above the motor to the space below the motor. They are the motor rotor channel, the air gap between the motor rotor and the motor stator, the winding gap and the motor stator core cut respectively, and listed in Table 2. Figure 3 shows the location of the four paths. There are six channels for Path 1, nine channels for Path 3, and nine channels for Path 4.

Table 2: The relationship between path number and location

Path number	Location
Path 1	Motor rotor channel
Path 2	Air gap between the motor rotor and the motor stator
Path 3	Winding gap
Path 4	Motor stator core cut

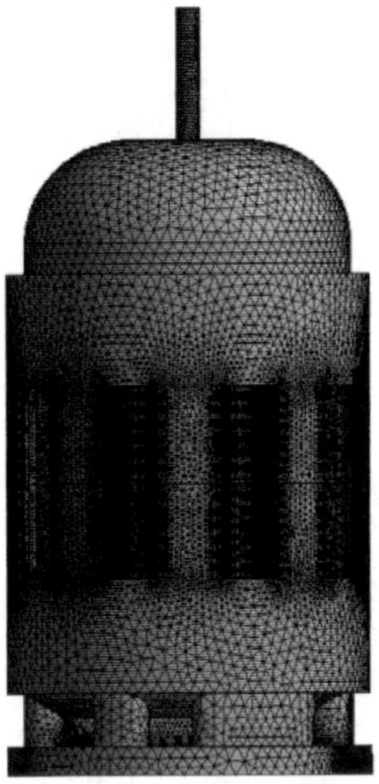

Figure 2: Mesh model of Model 1

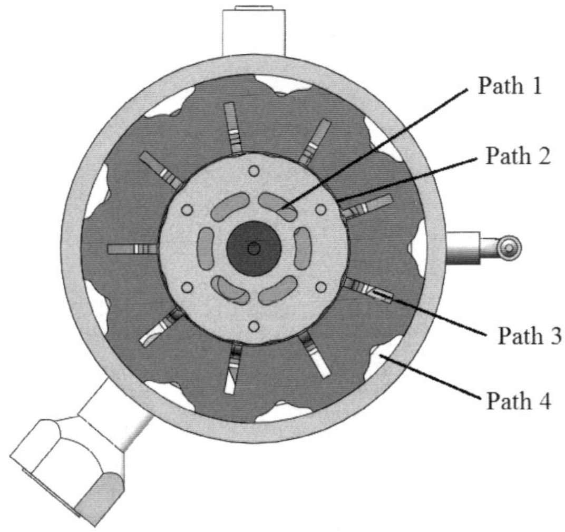

Figure 3: Location of flow paths

3. EXPERIMENT

The experiment was finished in a CO_2 compressor calorimeter. The experimental conditions are listed in Table 3.

Table 3: The experimental conditions

Item	Suction pressure(Pa)	Discharge pressure(Pa)	Rotation speed(r/s)
Data	8e6	2.4e6	45/60

The OCR of 45r/s and 60r/s are tested by oil separation method. Oil in the discharge gas mixture was separated and stored in a transparent graduated container. The mass flow rate of oil was calculated, and the mass flow rate of CO_2 gas was displayed, so the OCR could be calculated.

4. CFD RESULTS AND ANALYSIS

4.1 CO_2 Gas Flow Rate Analysis

CO_2 gas flow rate in the four paths are listed in Table 4. A positive number means that the gas is flowing from the bottom up, and a negative number means that the gas is flowing from the top down.

Table 4: CO_2 gas flow rate in the four paths

Model	Path 1(kg/s)	Path 2(kg/s)	Path 3(kg/s)	Path 4(kg/s)
Model 1	0.392	-0.005	-0.080	-0.206
Model 2	0.259	-0.005	-0.006	-0.147
Model 3	0.092	-0.002	0.054	-0.043

Because CO_2 mass flow rate of muffler discharge hole is 0.101kg/s, so

$$m_1 + m_2 + m_3 + m_4 = 0.101 \quad (5)$$

m_1, m_2, m_3, m_4 —— CO_2 mass flow rate in Path 1, Path 2, Path 3 and Path 4.

According to the data listed in Table 4, Model 1, Model 2 and Model 3 all satisfy Equation (5), consistent with hydrodynamics theory. For Model 3, despite the CO_2 gas flow rate through the motor rotor channel is very low, the CO_2 gas flow direction through the winding gap is from the bottom up, different from model 1 and model 2.
CO_2 gas flow rate from the bottom up and from the top down are calculated, and listed in Table 5. CO_2 gas flow rate from the bottom up of these three models have following relation:

$$m_{model3\_up} < m_{model2\_up} < m_{model1\_up} \tag{6}$$

m_{model1_up}, m_{model2_up}, m_{model3_up} —— CO_2 mass flow rate from the bottom up of Model 1, Model 2 and Model 3.

Table 5: CO_2 gas flow rate from the bottom up and from the top down

Model	From the bottom up (kg/s)	From the top down (kg/s)
Model 1	0.392	-0.291
Model 2	0.259	-0.158
Model 3	0.146	-0.045

Because the motor rotor channels are rotating all the time when the compressor is running, the path lines of gas and oil mixture through Path 1 are upward spirals. In the motor rotor channel, the lower the mass flow rate of the gas, the lower the velocity of the gas, the more turns the gas rotates when it rises to a certain height, and the easier it is for oil droplets to be separated by centrifugal force. So oil discharging from the outlet is less, OCR of the compressor should be lower as well.
Table 6 shows the OCR tested in the experiment. The OCR of Model 2 is lower significantly than Model 1, which is consistent with the analysis above. Model 1 compared to Model 2, the reduction ratio of OCR at 60r/s is greater than the reduction ratio of OCR at 45r/s.

Table 6: OCR tested in the experiment

Model	OCR(%)	
	45(r/s)	60 (r/s)
Model 1	0.30	0.94
Model 2	0.14	0.22
Model 3	0.18	0.61

The OCR of Mode 3 is also lower than Model 1, but higher than Model 2. That seems not consistent with the analysis above. But it is clear that the direction of gas flowing in Path 3 of Model 2 and Model 3 is different. The flow direction in Path 3 of Model 3 is from the bottom up, so the gas and oil mixture has not been subjected to centrifugal force compared to the gas and oil mixture that flows through the motor rotor channels . As a result, the oil droplets in the mixture that flows through Path 3 of Model 3are barely separated, resulting in an increase of OCR.

4.2 Pressure Field Analysis

Figure 4 shows the pressure field of three models. In the motor rotor channels, gas flows from the bottom up, so the bottom pressure is higher than the top pressure. The pressure difference between the motor rotor channel inlet and outlet is showed in Figure 4. The pressure difference in Model 1 is about twice the pressure difference in Model 2. The pressure difference in Model 2 is about 2.5 times the pressure difference in Model 3. The smaller the pressure difference, the lower CO_2 gas flow rate in the channels, which is consistent with the result in Section 4.1. The pressure field below the motor rotor of Model 3 is rearranged, so pressure data from the rotating center to the outside area in the space below and above the motor rotor is uniformly distributed, reducing the pressure difference between the motor rotor channel inlet and outlet significantly. For Model 3, the highest pressure value is located in the outermost part of the area within the cover of the lower balance weight shell below the motor rotor, which leads to the outflow of gas from this area and upward flow through the winding gap. In the above flow, the risk of increased OCR is significant because the oil droplets in the gas mixture are not subjected to centrifugal force. So the highest pressure value location of Model 3 is not good for reducing OCR, and the test results confirm this.

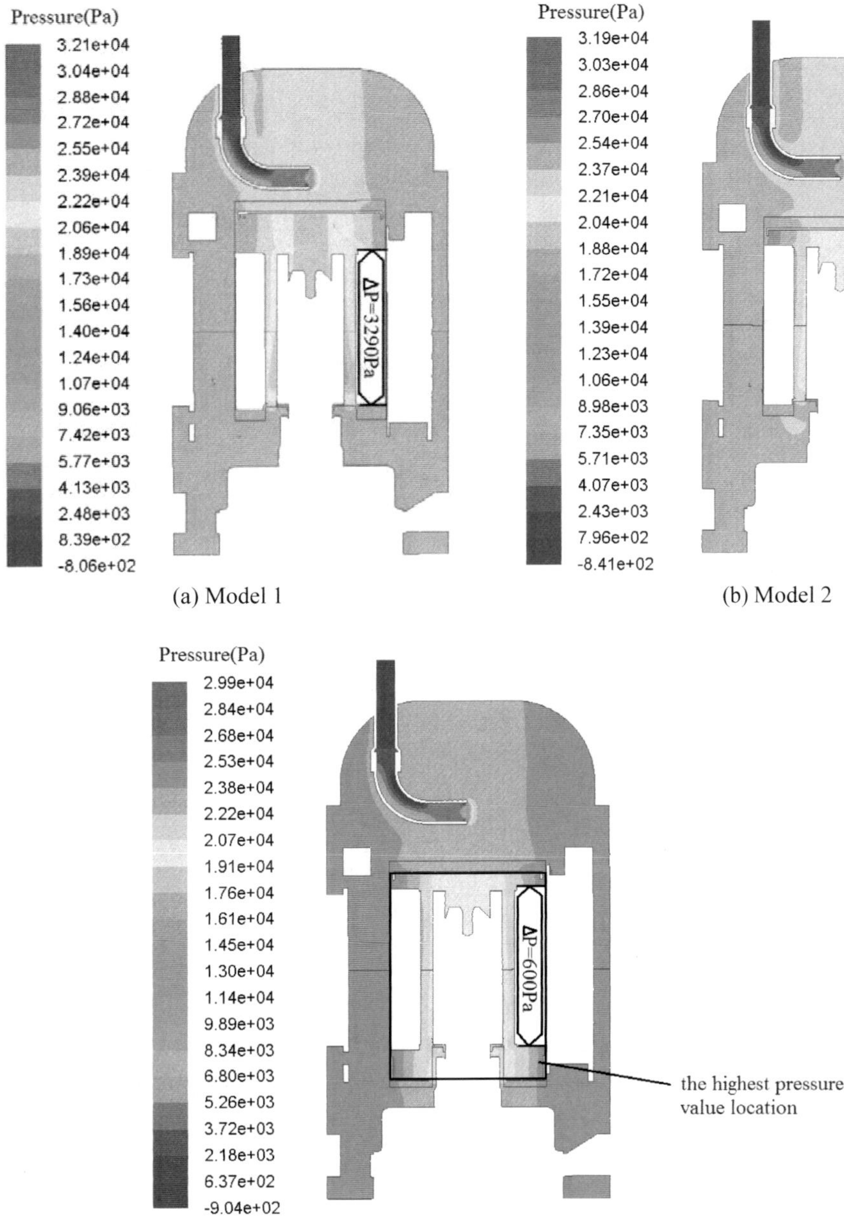

(a) Model 1

(b) Model 2

(c) Model 3

Figure 4: Pressure field

To obtain the lower pressure difference between the motor rotor channel inlet and outlet, the pressure field should be designed. The pressure field determines gas flowing mode and direction, which greatly affects the OCR of the compressor. So not only the compressor parts need to be designed, but also the pressure field in the compressor need to be designed.

4.3 Velocity Field Analysis

Table 7 lists the vertical gas velocity in the motor rotor channels. They have the same relationship with the CO_2 gas mass flow rate illustrated in Equation (6). The gas vertical velocity of Model 3 is smaller than Model 2, and the gas vertical velocity of Model 2 is smaller than Model 1. Lower the velocity, and more oil droplets tend to be separated from the gas, so OCR would be smaller.

Table 7: Gas vertical velocity in the motor rotor channels

Model	Gas vertical velocity (m/s)
Model 1	4.9
Model 2	3.2
Model 3	1.2

5. CONCLUSIONS

(1) By simulation and experiment, it is confirmed that the CO_2 mass flow rate through the motor rotor channel and winding gap have a great influence on the OCR of the compressor.
(2) If the gas flows from up down through the winding gap, OCR has the same changing trend with CO_2 gas flow rate in the motor rotor channels, pressure difference between the motor rotor channel inlet and outlet, and vertical gas velocity in the motor rotor channels. Lower CO_2 mass flow rate, pressure difference and gas vertical velocity in the motor rotor channels, smaller OCR, and vice versa.
(3) If the gas flows from the bottom up through the winding gap, OCR is much higher than the estimated value by the CO_2 mass flow rate in the motor rotor channels.
(4) The pressure field should be designed to obtain small OCR data. The OCR can be reduced from 0.94% to 0.22% at 60r/s when the pressure difference between the inlet and the outlet of the motor rotor channels reduces about 1600Pa.

REFERENCES

[1] Liying, D., Shebing, L., Guomang, Y. (2015). Optimization analysis of oil circulation rate in rotary compressor base on CFD. Refrigeration. 34(3), 84-90.
[2] Ke, C., Haofu, L., Lichang, X. (2017). Simulation and optimization of oil rate in rolling piston type compressor. Refrigeration and air-conditioning. 17(2), 83-87.
[3] Keisuke, S., Shin, S., Tetsuhide, Y. (2012). CFD analysis inside a CO_2 rotary compressor shell to improve oil separation and reduce the shell size. International Compressor Engineering Conference at Purdue(1230,1-10). West Lafayette, America.

Development of 3-Cylinder Rotary Compressor for Large Capacity

Tomohide Shishimoto [1]*, Toshimasa Aoki[1], Koji Hirano[1],
Shogo Shida[1], Takuya Hirayama[2]

[1]Toshiba Carrier Corporation, Compressor Design Department,
Tadehara 336, Fuji City, Shizuoka Prefecture, Japan 416-8521

[2] Toshiba Carrier Corporation, Core Technology Center,
Tadehara 336, Fuji City, Shizuoka Prefecture, Japan 416-8521

* Tomohide Shishimoto (tomohide1.shishimoto@glb.toshiba.co.jp)

ABSTRACT

The authors have developed a large capacity rotary compressor with compact size and low vibration. Recently, we have strong requirements for larger capacity without changing the compressor shell diameter. Considering for larger capacity in conventional 2-cylinder rotary compressor inverter type, there are issues such as high vibration, big sound. As the solution for these problems, we have developed 3-cylinder rotary compressor inverter type which have different compression process of 120 ° phase per revolution for each cylinder. The developed compressor achieves a capacity target, proves one-quarter vibration level comparing with 2-cylinder rotary compressor and ensure the improvement of reliability result. We have been manufacturing the world largest capacity of new rotary compressor and have been launching into the market since September 2020. This new model has successfully attained more large capacity with same shell diameter as our previous large capacity of 2-cylinder rotary compressor inverter type. This new model adopt some new technologies such as the 3-cylinder design for balancing rotation torque, the multi discharge valve structure, the integrated bearing with partition plate and the new wide-range motor. By these technologies, the outdoor units of multiple air conditioner for buildings (Variable Refrigerant Flow-VRF) can achieve a large capacity 20 HP (56kW) with applying one compressor unit only.

1. INTRODUCTION

From global environmental protection point of view, energy saving, resource saving and down-sizing of refrigeration air conditioning system become important issues. Moreover, refrigeration air conditioning system have a tendency to increase the capacity in recent years. However in compressor design, it need to encounter the contradiction between increasing a capacity and its compactness.

In case of adapting the same long-experienced shell with having high pressure resistance safety and economically in design, 2-cylinder rotary compressor within two compression process as known as Twin Rotary Compressor will have an excessive shaft load, low reliability result, high sound and more big vibration when try to enlarge its capacity over than our conventional larger model 100cc. As a solution of those problems, we developed a 3-cylinder rotary compressor inverter type with three compression process inside that is called Triplet Rotary Compressor. This compressor achieve more large capacity and reduce its high vibration. And also, magnifying 20% of its displacement volume from 100cc to 120cc make it as the world's largest one (*) in rotary compressor types. Furthermore, the vibration of developed 3-cylinder compressor reduce to one-third comparing to the 100cc capacity of conventional 2-cylinder rotary compressor type (*According to TCC internal investigation, July 2020)

This developed model has been installing in outdoor units of multiple air conditioner for buildings "SMMS-u" with R410A refrigerant and has been launching into market since September 2020. In this paper, we introduce 3-cylinder design, the multi discharge valve structure design and new wide range motor.

2. THE ELEMENTAL TECHNOLOGIES FOR LARGE CAPACITY

2.1 3-cylinder design

In the large capacity of rotary compressor, from the viewpoint of partial load performance, vibration, and sound, currently the mainstream is a 2-cylinder rotary structure. Generally in conventional 2-cylinder rotary compressor, the eccentric crankshaft arrange 180 ° in opposite to each other so suction and compression are repeated alternately. In the case of increasing capacity, if cylinder height is increased the compression load fluctuation ratio is same but the reliability become worse due to load increasing at bearing. In addition, operating vibration including an increased swing due to increasing moment refer to the shaft rigidity, there is a concern about noise deterioration according to an increasing of pressure pulsation. On the other hand, the torque fluctuation ratio can be reduced by arranging the eccentric of crankshaft at an equal distribution of 120 ° in the developed 3-cylinder rotary compressor.

The eccentric crankshaft at an equal distribution of conventional 2-cylinder rotary structure and developed 3 cylinder rotary structure are shown in figure 1 and 2. Comparison of the torque fluctuation ratios of both model is shown in figure 3 in the rated cooling condition.

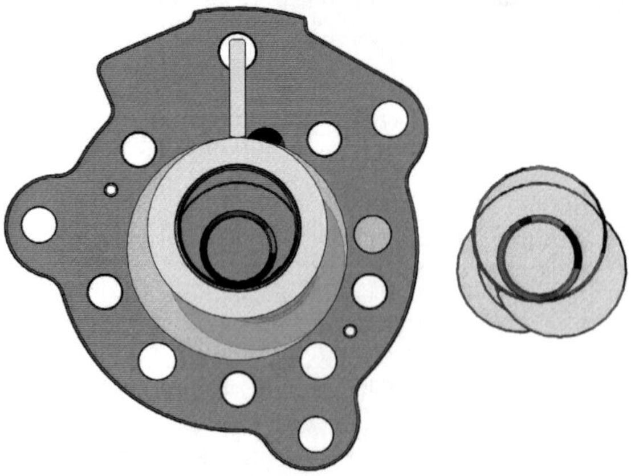

Figure 1 : The eccentric crankshaft at an equal distribution of 120° (Developed 3-cylinder rotary structure)

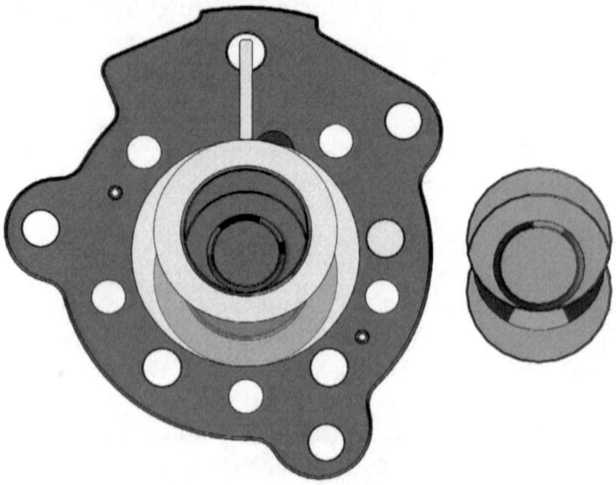

Figure 2 : The eccentric crankshaft at an equal distribution of 180° (Conventional 2-cylinder rotary structure)

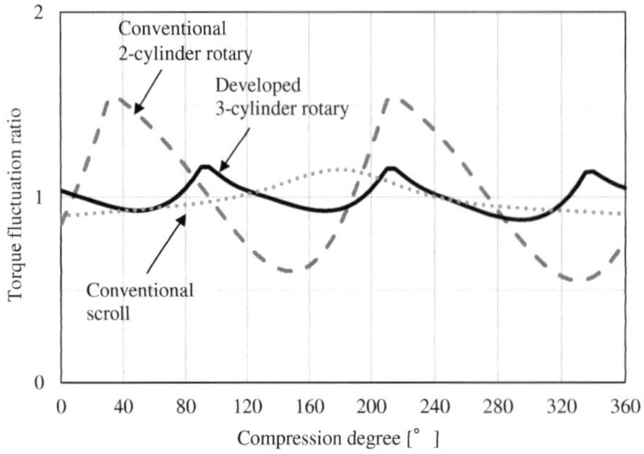

Figure 3 : Torque fluctuation ratio

Compared to the conventional 2 cylinder rotary structure 100cc, developed 3 cylinder rotary structure 120cc has increased the eccentric weight and extend the distance length between the compression chambers, therefore, it is be-comes necessary to increase the balancer weight to balance the moments. On the other hand, increasing the balancer weight lead to more crankshaft deflection, and worsen the reliability of the main bearing particularly. Hence, the deflection of crankshaft is reduced by rotor with lower gravity center and also counter bore at lower side of rotor. In addition of reducing torque fluctuations, a new structure has been adopted to new balancer which is attached at top of rotor and at sub bearing side as maximum as possible to perform a low vibration. Comparison of the vibration ratios of both model is shown in figure 4.The developed model 120cc achieve 78% more low vibration comparing with the conventional 100cc.

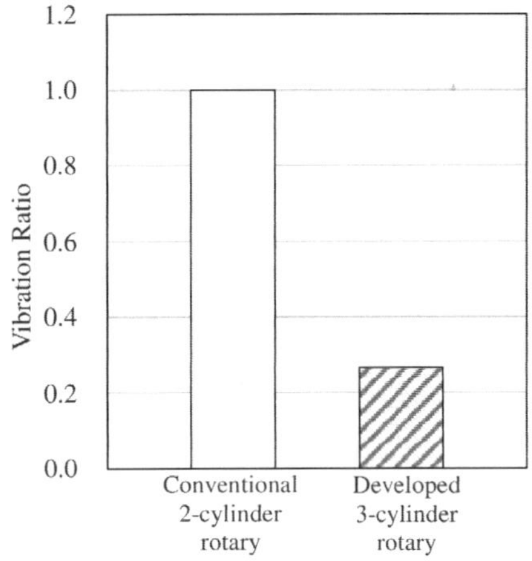

Figure 4 : Vibration ratio

On the other hand, compared to 2-cylinder rotary structure, 3-cylinder rotary structure increases the compression load and the bearing load due to longer distance between main bearing and sub bearing. Therefore, new third bearing is installed, in addition to conventional technology such as providing tapered shape on the upper end of main bearing and counter bore for the center of gravity of the rotor is lowered and installing a ring groove on each bearing's end face. New third bearing which has both the functions of partitioning between each chambers and bearings, is installed to get good reliability and keep the shaft diameter unchanged.
Figure 5 shows the compressor structure of conventional 2-cylinder rotary and developed 3-cylinder rotary.

Bearing reliability index is in figure 6 shows bearing load ratio of cooling load capacity condition for refrigeration 410A environment. Comparing to capacity enlargement 20% of 3 cylinder without new third bearing, bearing load of main bearing and sub bearing with new third bearing of 3 cylinder rotary has improved about half.

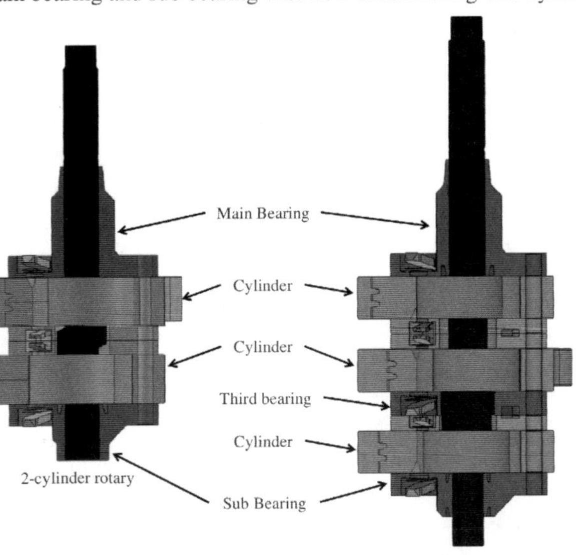

Figure 5 : The compressor structure of the 2-cylinder and 3-cylinder rotary

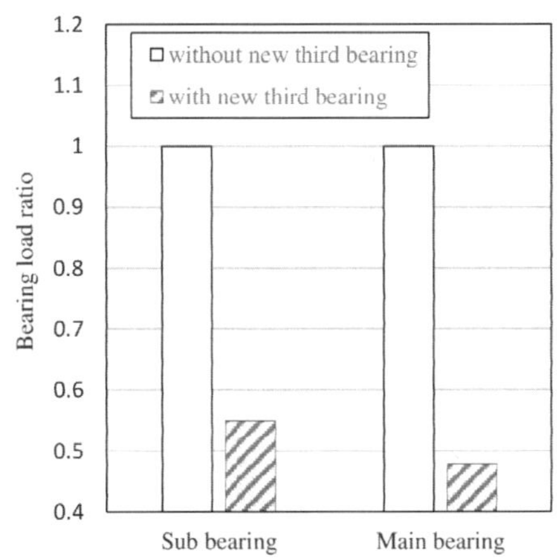

Figure 6 : the ratio of the bearing load under the cooling overload condition with R410A refrigerant

2.2 The multi discharge valve structure design

It is necessary to reduce the flow path loss due to increasing the refrigerant flow rate for maximum performance of larger capacity compressor. The discharge structure of 3-cylinder rotary compressor uses a reed valve structure. In order to suppress the flow path loss by the reed valve structure, it is necessary to increase the cross-sectional area of the discharge port, but it is difficult due to radial restrictions. Therefore, 2 discharge ports are installed in upper and lower side of each compression chamber, and it is tried that the discharge flow loss is reduced.

In addition to the reliability of the valve, the multi discharge valve structure is examined along following 3 focused items. In a general rotary compressor, the discharge port on the bearing side can obtain a sufficient muffler volume for suppressing the pressure pulsation of the discharge gas. While, if the discharge port on the partition plate side between the compression chambers has the same muffler volume, the distance between the eccentric of crankshaft is long and the crankshaft reliability is worse. In the 3-cylinder rotary structure, the eccentric crank shaft are arranged at an equal distribution of 120 °, so the ability to respond of the reed valves and the interference between the reed valves affect the compressor performance compared to the 2-cylinder rotary structure with a 180 ° phase. Therefore, there is a possibility to affect a performance by the discharge gas flow path loss and flow confluence loss from each compression chamber.

In consideration of these, the multi valve structure having a plurality of discharge valve and a muffler structure in one compression chamber is examined. The multi valve structure has a discharge valve on the partition plate side in addition to the bearing side. In order to suppress the pressure pulsation of the discharge gas, the discharge port diameter on the bearing side, which can secure a sufficient muffler volume, is large, and the discharge port diameter on the partition plate side, which limits the space volume of the muffler, is small. The required total cross-sectional area was secured by arranging two discharge ports in one compression chamber.

By the discharge valve and the discharge gas path in the upper and lower compression chambers are symmetrically located in the partition plate side, it is aimed to secure the muffler volume and suppress pressure pulsation and passage loss.

In addition, the discharge valve is designed with a spring constant suitable for each discharge port diameter, and high performance is achieved by effectively opening and closing each valve in the operating range of low rotation to high rotation (low load to high load).

Figure 7 shows the arrangement of multi discharge valve structure, figure 8 and 9 show discharge gas flow path of the discharge valve on bearing and on partition plate side, and figure 10 shows the performance results. As shown in figure 10, the 3-cylinder rotary compressor has achieved a large capacity of over 20HP (56kW), and in the comparison of 56kW efficiency, the efficiency of 3-cylinder rotary compressor is 3.5% higher than 2-cylinder rotary compressor equipped with VRF products of current model.

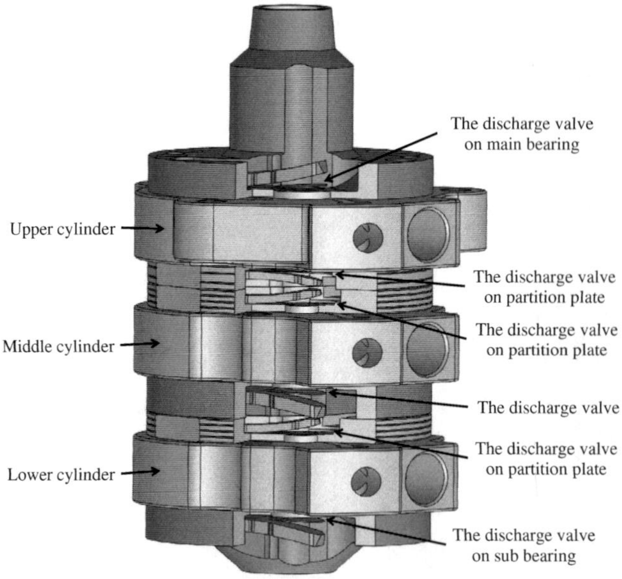

Figure 7 : The arrangement of multi discharge valve structure

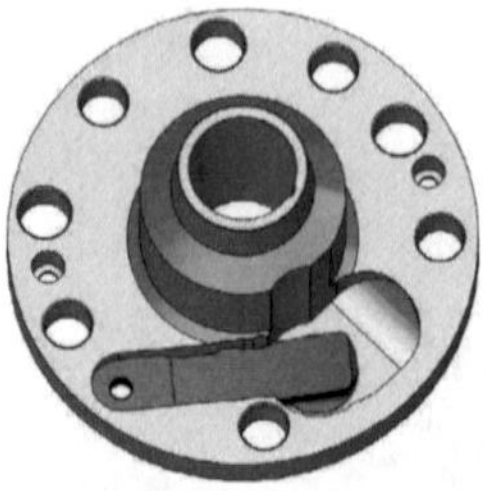

Figure 8 : The discharge gas flow path of the discharge valve on bearing

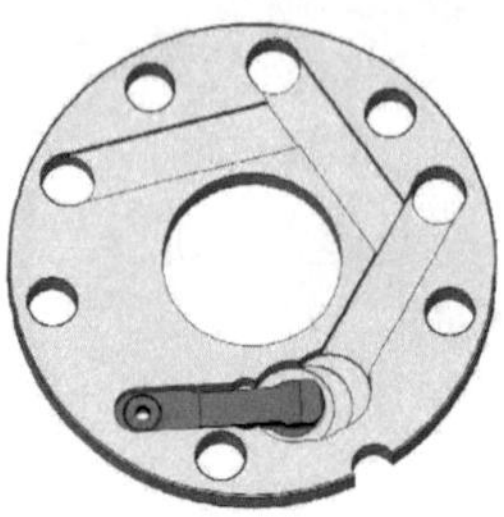

Figure 9 : The discharge gas flow path of the discharge valve on partition plate

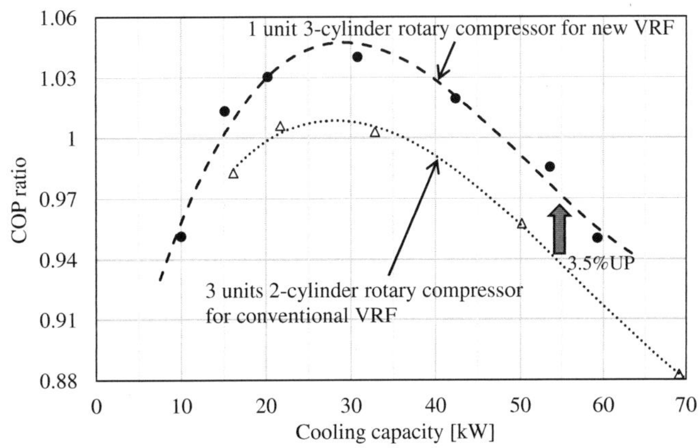

Figure 10 : The performance results between 2 and 3-cylinder rotary compressor

The behavior of the discharge valve (reed valve) installed in each compression chamber is measured by inductive displacement sensor installed in each discharge valve holder. The vertical line shows the maximum lift amount of each valve and the horizontal one the shaft rotating machine angle in figure 11.The operating condition is 90 [1/sec] equivalent to the rated cooling condition.

In each compression chamber discharge valve, the trend can be confirmed that the discharge valve on partition plate side starts to open first and then the discharge valve on bearing side starts to open. In addition, the discharge valve on partition plate side opens and closes twice while the discharge valve on bearing side opens and closes once.

This is designed to make the spring constant of the discharge valve on partition plate side smaller than that of the discharge valve on bearing side, making it easier to open for preventing over-compression and reducing the residual gas in the compression chamber, that is, reducing the re-expansion loss.

On the other hand, in the case of the operation of each the discharge valve on partition plate side is compared, the discharge valve on partition plate side operation of the lower compression chamber tends to be different from that of the middle and upper compression chambers. The reason is presumed that the pressure in the lower flow path is slightly higher than other flow paths due to the gas flow paths in the lower compression chamber is longer than the gas flow path in the middle and upper compression chambers. It can also be confirmed that the discharge valve on partition plate side with a small spring constant functions effectively for the interference of each valve and achieve highly responsive behavior without closing delay.

While, figure 12 shows the valve behavior results in 30[1/sec] condition in which the gas flow rate was reduced under the pressure condition equivalent to the rated cooling condition. In this case, the discharge valve on partition plate side functions as the main discharge flow path instead of the discharge valve on bearing side, and the discharge valve on partition plate side contributes to the performance improvement by suppressing over compression.

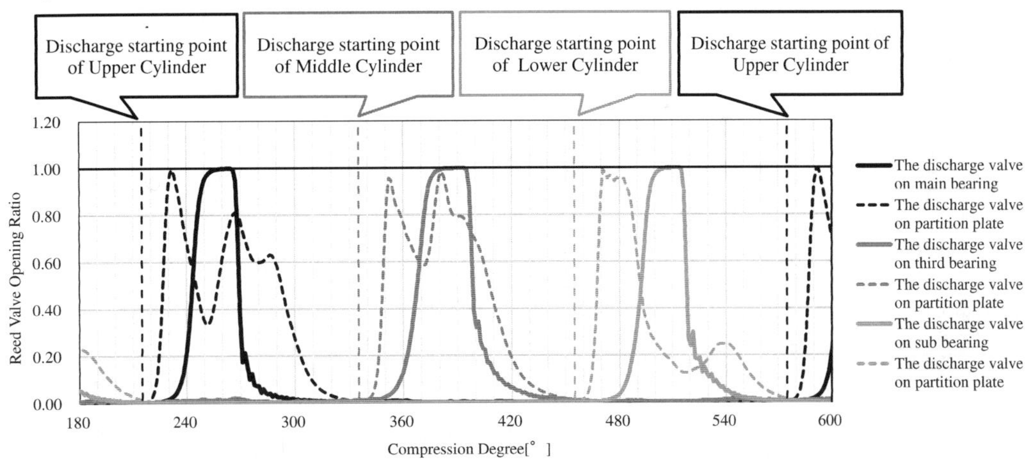

Figure 11 : The behavior of discharge valve in three compression chambers (at rotation speed 90[1/sec])

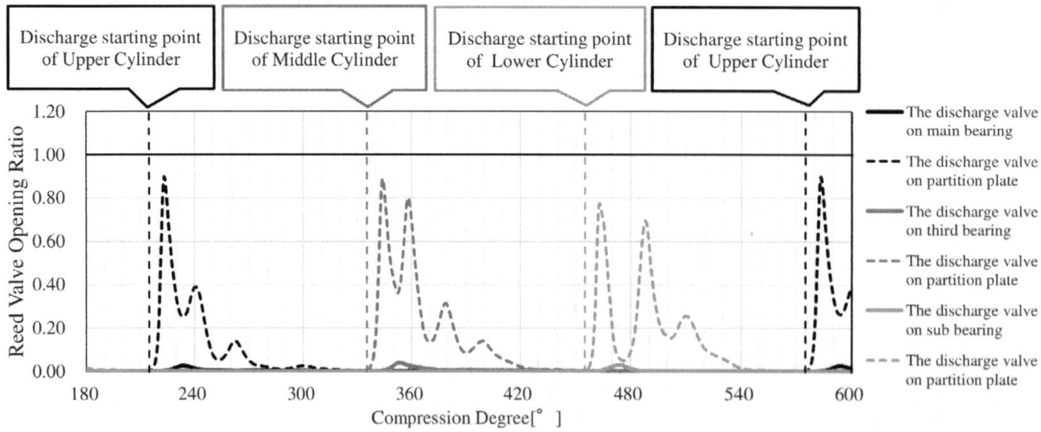

Figure 12 : The behavior of discharge valve in three compression chambers (at rotation speed 30[1/sec])

2.3 New wide range motor (Open winding motor)

In the actually operating air conditioner, the compressor has been operating under the low load operating region for a long time, and the performance improvement in the low load operating region leads to the reduction of the seasonal power consumption and the annual electricity bill. In a general DC inverter motor, the motor efficiency in the low load operating region is improved by reducing the wire diameter and increasing the number of the winding. In this method, the current increases in a high load operating region, so-called field weakening control is used to realize operation under the high speed rotation, but the motor efficiency decreases. Therefore, we have developed the world's first open winding motor (*) as a compressor for air conditioners. (* According to internal research as of July 2020)

Open winding motor is used with dual-state inverters that was developed at the same time as a drive controller.

In addition, open winding motor has a three-phase alternating current with the neutral points separated.

This system is operated with star winding motor drive during load operation below the middle level, and is operated by switching the open winding motor drive during high load operation.

In the case of operating with a star winding motor drive, the motor is driven by one inverter, and operating with an open winding motor drive, the winding is switched and the motor is driven by two inverters.

At the time of open connection, it is possible to apply a voltage about 1.7 times of a standard inverter, and it is possible to operate with a high winding motor at high load operation without decreasing efficiency. Figure 13 shows open winding motor, and figure 14 shows the applied voltage of the open winding motor.

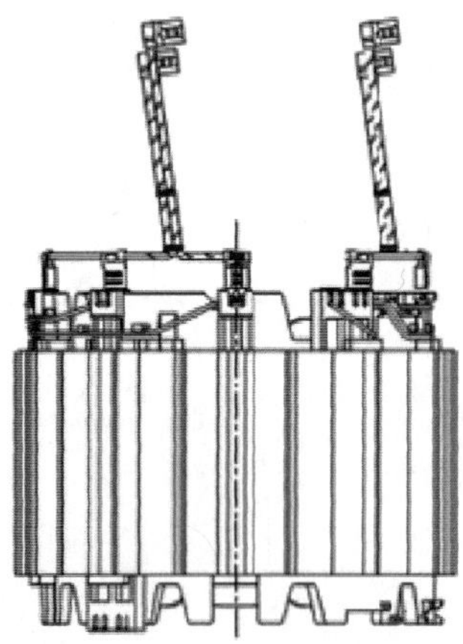

Figure 13 : New wide-range motor

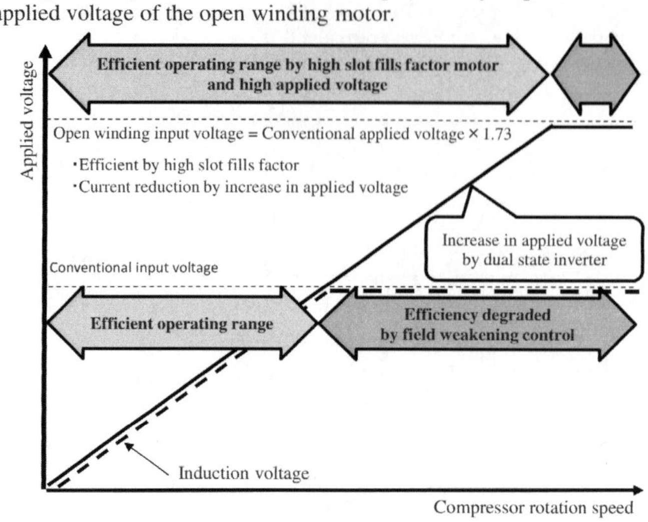

Figure 14 : Relationship between DC-voltage and rotation speed

As shown in figure 15, dual state inverter is composed of two inverters. In the case of the high load operation, the two inverters are used to operate the open winding motor drive. On the other hand, as shown in figure 16, in the case of the low load operation, the inverter loss is reduced and high-efficiency operation is maintained by driving one inverter and stopping the other inverter with star winding motor drive.

Figure 17 shows a performance comparison between open winding and star winding motor drive. Star winding motor drive is superior in the low load operation region, and open winding motor drive is superior in the 20HP (56kW) rated operation region. High-efficiency operation is realized in wide range operation region by selecting the optimal winding according to the operating conditions.

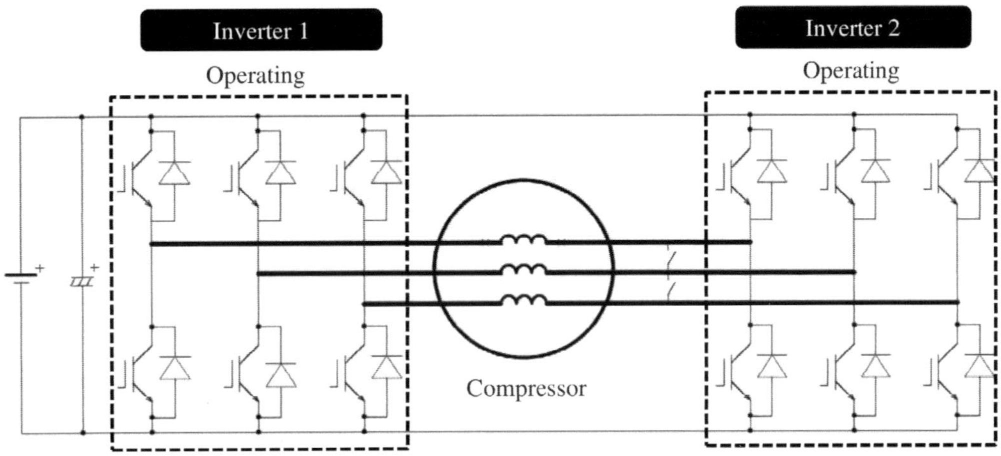

Figure 15 : Operating with open winding

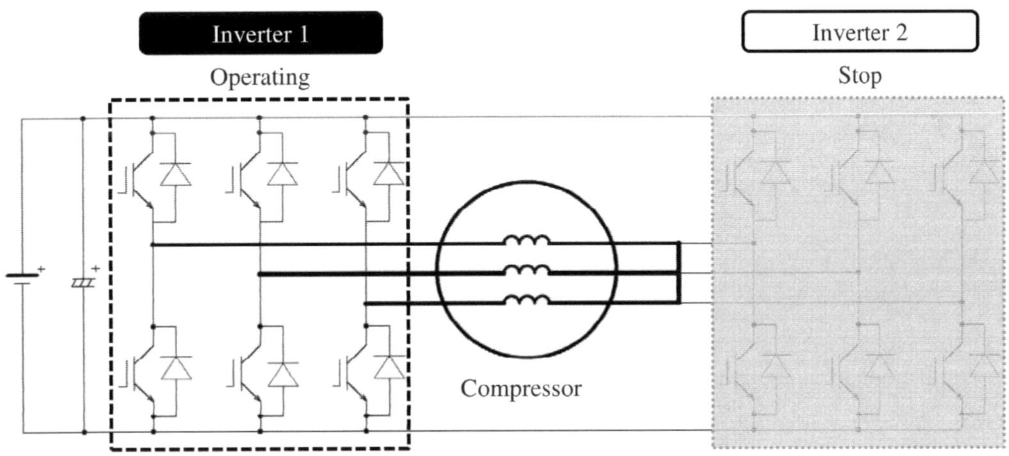

Figure 16 : Operating with star winding

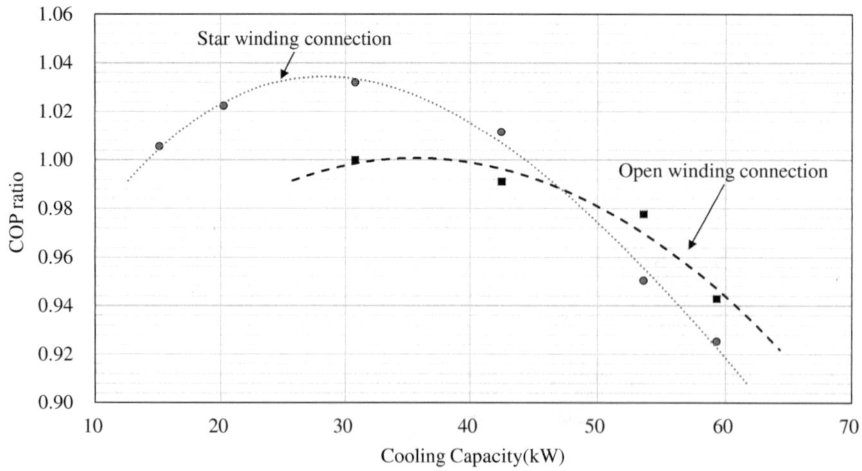

Figure 17 : Performance comparison between open winding and star winding.

3. CONCLUSION

This time, we achieved the development of 3-cylinder rotary compressor with 3 compression chambers and a shaft eccentric angle of 120 °. 3-cylinder rotary compressor is a compact compressor that can obtain a large capacity based on 2-cylinder rotary compressor. 3-cylinder rotary compressor not only has low vibration and small sound, but also realizes high efficiency and wide range of operation from low load to high load by open winding motor and a dual state inverter. In particular, the vibration of 3-cylinder rotary compressor achieve 78% lower comparing with 2-cylinder rotary compressor. For this reason, the large air-conditioning products that used multiple compressors can install one compressor. And, it can contribute to weight reduction, resource saving, and space saving of air-conditioning products.

In the future, using the 3-cylinder rotary compressor as a platform, we will make efforts to reduce the environmental footprint by promoting to lower GWP of the refrigerant and expanding it to various product applications.

REFERENCES

[1] Hirayama. T., Aoki. T., Shida. S., et al., (2017). Study on Large Capacity and High Efficiency of Rotary Compressor. *Trans. of the JASRAE, 34*(3), P261-271.

[2] Kimura. S., Hirayama. T., Aoki. T., et al., (2017). Development of High Efficiency and Large Capacity Rotary Compressor. Proceedings of 2017 JASRAE Annual Conference, Tokyo, B213, JASRAE.

[3] Shida. S., Shishimoto. T., Toda. H., (2020). Development of Large Capacity Rotary Compressor for VRF. *Toshiba Review.*

[4] Hirayama. T., "Rotary Compressor and Refrigeration Cycle Device" (in Japanese), 2011, Japanese Patent 6022247

Development of a New Dual-Cylinder Rolling Piston Compressor and System

Li ZHANG*, Chunhui LIU, Ruixiong WANG, Lei ZHANG

Shanghai Highly Electrical Appliances CO., LTD.
Shanghai, China
(86-2150554560, zhangli@shec.com.cn)

* Corresponding Author

ABSTRACT

In recent years, the requirement of energy efficiency for refrigeration and air condition systems becomes more and more higher while the cost needs to be more and more lower, at the same time some new applications of refrigeration and air condition systems are emerging rapidly. All of these demands need a new compressor to meet. In this paper, a new dual-cylinder rolling piston compressor with two independent suction ports and two independent discharge ports was proposed. The structure and working principal of this compressor were described, and the performance was tested. Then simulation model was built to research the performance of the system using this kind of compressor compared with the traditional system. Simulation results showed that the COP of the system using this kind of compressor is 0.5%~4.4% higher than that of the traditional system in cooling condition, and 2.1%~3.9% higher than that of the traditional system in the heating condition.

1. INTRODUCTION

With the improvement of the living standard, the requirement of the temperature and humidity in the place of residence are also improved. The air conditioners, adjusting the temperature and humidity of the living place, are being used more and more frequently. The energy consumption of air conditioners increases quickly. At present, the proportion of building energy consumption in the social terminal energy consumption has reached 30% (Weihua Zhang, 2017). The proportion of energy consumption of air conditioners in buildings is more than 50% (Weifeng Zhu and YoudiJiang, 2013). Therefore, improving the efficiency of air conditioners is crucial to reduce the energy consumption of buildings.

The energy loss of the air condition system mainly comes from two aspects. Firstly, it is the non-isentropic compressing process in the compressor. Secondly, there is a temperature difference between the refrigerant and the air or water. Therefore, there are mainly two ways to improve the efficiency of the air condition system. One is to reduce the irreversible losses associated with the compressing process. Another is to reduce the temperature difference between the refrigerant and the air or water.

In order to reduce the temperature difference between the refrigerant and the air or water, Caoxiang et al. (2017) proposed a stepped pressures cycle. In this cycle, two or three independent compressors were used. The experimental results showed that the COP increase of dual and triple-subcycle system were achieved by 12.3% and 18.7%.But the cost of this system was higher because of the installation of more than one compressor.

In order to reduce the cost and improve the COP, this paper proposed a new dual-cylinder rolling piston compressor with two independent suction ports and two independent discharge ports.

2. THEORY OF AIR CONDITION SYSTEM

For the reverse Carnot cycle, the efficiency only depends on the temperature of the hot reservoir and cold reservoir. In the actual air condition system, the temperature of the air or water is changeable along with the heat transfer process. In this case the most efficient cycle is Lorenz Cycle. As shown in Figure 1, Lorenz cycle can be regarded as the combination of numerous reverse Carnot Cycles [3]. The COP is only determined by the mean temperatures of the hot reservoir and the cold reservoir, it can be calculated by the Equation (1).

$$COP = \frac{T_L}{T_H - T_L} \tag{1}$$

T_L is the mean temperature of the cold reservoir. T_H is the mean temperature of the hot reservoir.

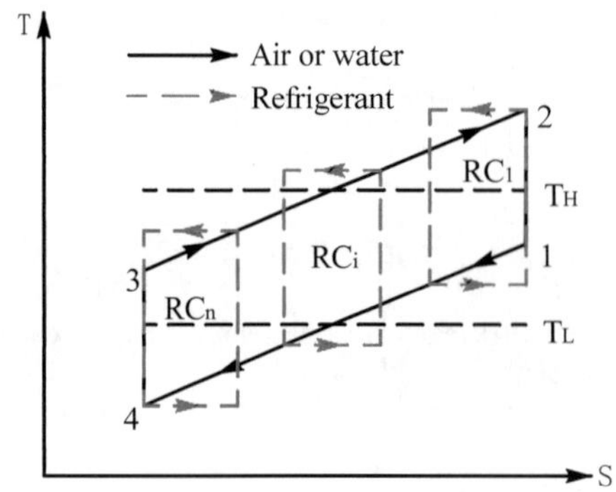

Figure 1: Lorenz Cycle

In order to approach the COP of the Lorenz Cycle, the actual cycle should be split into numerous cycles (Caoxiang et al., 2017). In fact, it could not be realized. The actual cycle could only be split into several cycles. Based on this theory, this paper proposed a new dual-cylinder rolling piston compressor with two independent suction ports and two independent discharge ports. This compressor can split an actual cycle into two subcycles to improve the COP.

3. A NEW ROLLING PISTON COMPRESSOR

A new rolling piston compressor, with two independent suction ports and two independent discharge ports, was proposed. The structure of this compressor is showed in Figure 2. This new compressor has two independent compression units. Both units have their independent suction and exhaust ports. Two units are separated by a middle partition, so the refrigerant in different units will not mix when the compressor works. A common crankshaft drives the two compression units.

The first compression unit consists of the first suction port, the first cylinder, the first exhaust port and so on. The refrigerant, sucked from the first suction port, is compressed in the first cylinder and enters into the first muffler. Then the refrigerant flows into the internal volume of the shell. After cooling the motor, it is discharged out of the compressor from the first discharge port. The second compression unit consists of the second suction port, the second cylinder, the second exhaust port and so on. The refrigerant, sucked from the second suction port, is compressed in the second cylinder then discharged into a sealed cavity, after that it is discharged out of the compressor directly from the second discharge port.

When the compressor works, the lubricating oil in the compressor could leak into the cylinders through the clearances between moving parts. For the first cylinder, the refrigerant carries the oil droplets into the interior of the compressor and those oil droplets finally sink to the bottom of the compressor due to the gravity and separation effect coming from the motor and the cavity above the motor. But for the second cylinder, the oil droplets are carried into the sealed cavity and then directly discharged from the second discharge port. When the compressor works for some time, more and more oil droplets will leave the compressor, which would lead to mechanical failures due to the lack of the lubricating oil. In this paper a new scheme was proposed to solve this problem shown in Figure3. An oil separator was used in this scheme. The oil separator has three pipes. The inlet pipe connects the second discharge port of the compressor, and the outlet pipe connects the condenser. The oil return pipe connects the first suction port of the compressor. In order to control the mass flow rate through the oil pipe, a throttling capillary is installed in the oil return pipe. This oil separator can separate the oil from the refrigerant coming from the second compression unit. Driven by the pressure difference, the oil could flow back to the first suction port. Then, the oil droplets finally return back to the compressor. By using this method, the problem of lacking lubricating oil could be solved.

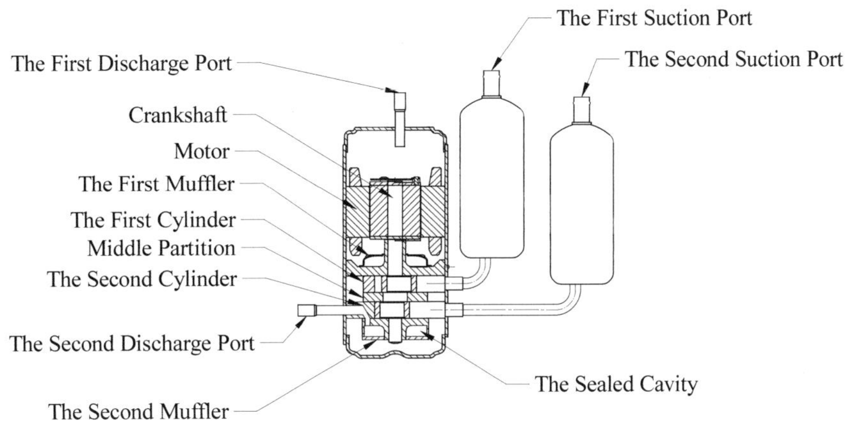

Figure 2: The structure of the new rolling piston compressor

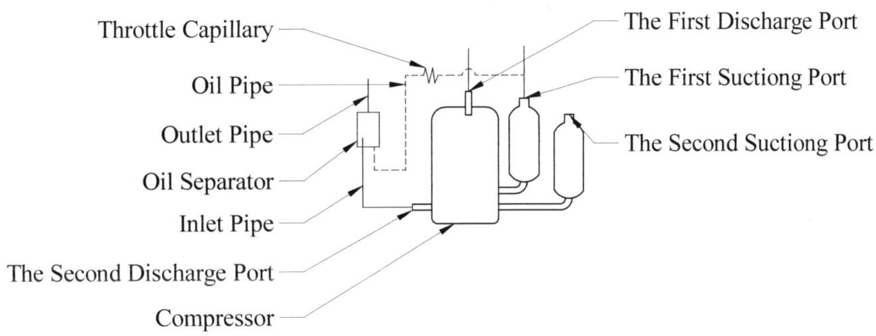

Figure 3: A schematic view applying the new compressor

4. THE SIMULATION OF THE DUAL-LOOP SYSTEM AND TRADITIONAL SYSTEM

In order to know the performance of a dual-loop air condition system applying this new compressor, a system simulation model was built. At the same time, a traditional air condition system simulation model was built too as a comparison.

4.1 The Performance Model of the Compressor

To build a system simulation model, a compressor performance model for the new compressor and traditional compressor is needed. So the performance at different operating conditions of this new compressor and a traditional dual-cylinder compressor with the same displacement was tested at a compressor performance test rig (calorimeter). Based on the test data, a mathematical model using AHRI polynomial equation was built.

4.2 The Simulation Model of the Dual-loop System and Traditional System

In order to simulate the performance of a air conditioner system, a parameterized mathematical model of the air conditioner was built, which mainly includes:

- The indoor and outdoor heat exchangers of the air conditioner were modeled, and the appropriate heat exchange correlations were used.

  The C.C.Wang model was used for heat exchange of the air-side. The Shah-Tang model was used for the heat exchange of the condensation side. The Kattan-Thome-Favrat mode（Kattan, N., Thome, J.R. and Favrat et al.1998）was used for boiling heat exchange of the evaporation side.

- The mathematical models of fan and EEV（electronic expansion valve）were established. The fan adopts efficiency model, and the orifice valve model was used for the EEV, which includes the physical and mathematical models according to the two-phase flow characteristics of refrigerant.

Based on this model, the performance of a dual-loop system using the new compressor was simulated. Figure 4 shows the schematic of the simulation model which consists of the first loop and the second loop. The first suction port and the first discharge port connect an evaporator and a condenser separately, forming the first loop. The second suction port and the second discharge port connect another evaporator and condenser separately, forming the second loop. For the condenser side, the air flow firstly enters into the condenser in the second loop, and it is heated by the refrigerant. Then it enters into the condenser in the first loop. For the evaporator side, the air flow firstly enters into the evaporator in the first loop, and it is cooled by the refrigerant. Then it enters into the evaporator in the second loop.

As a comparison, a simulation model was also built for a traditional air conditioner. Figure 5 shows the schematic of the simulation model. In this model, the heat transfer area of the condenser and the evaporator, the volume flow of the air and the expansion valve are same as the dual-loop system, the compressor is a traditional dual-cylinder compressor with the same displacement as the new compressor, and its performance model was built at section 4.1.

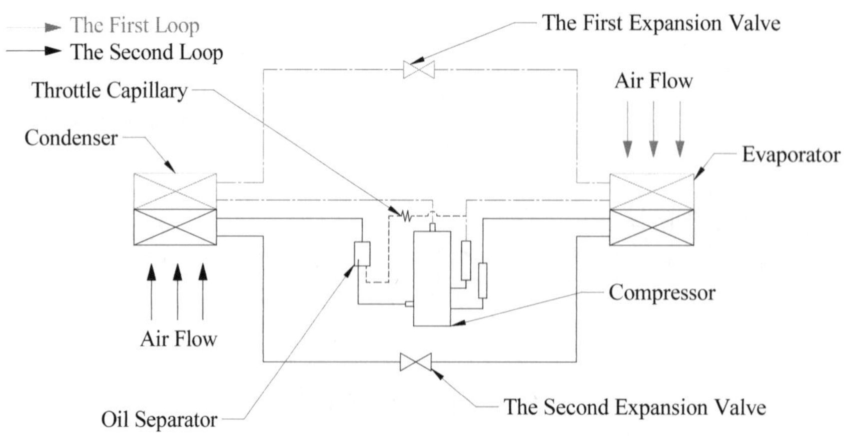

Figure 4: The schematic of the simulation model applying the new compressor

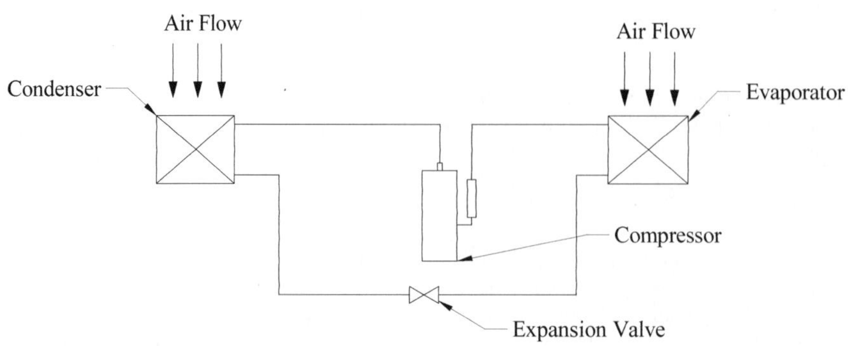

Figure 5: The schematic of simulation system applying the traditional compressor

4.3 Operating Conditions

In order to compare the performance of the dual-loop system and the traditional system, different operating condition need to be set for the simulation models. Table 2 shows the detailed parameters of the operating conditions.

Table 1: Simulation conditions of the system

Conditions	Indoor temperature (dry bulb / wet bulb) (℃)	Outdoor Temperature (℃)	Relative humidity
Heating	20/15	-3, 2, 7, 12, 17	87%
Cooling	27/19	30, 32.5, 35, 37.5, 40	40%

Both the new compressor and the traditional compressor are variable frequency compressors. But during the simulation, the rotational speed is set as constant. Table 2 shows some detail parameters of compressors.

Table 2: Parameters of those compressors

Content	Cooling condition	Heating condition
Refrigerant	R410A	
Rotational speed(rpm)	3000	4800

5. SIMULATION RESULTS AND DISCUSSION

5.1 Suction Pressure and Discharge Pressure

Figure 6 and Figure 7 show the variation of suction pressure and discharge pressure of the dual-loop system compared with the traditional system under cooling and heating condition. With the increasing of outdoor temperature, the suction pressure and the discharge pressure of the dual-loop system and traditional system increase in cooling and heating condition. Both the suction pressure and the discharge pressure of the traditional system are lower than the dual-loop average pressure. The suction pressure and discharge pressure of the traditional system are between that of the first loop and the second loop. The reason is that the mean temperature of air is different between the dual-loop system and the traditional system. Higher mean temperature of the air could result in higher pressure of the refrigerant. For the evaporator, the air temperature decrease along with the heat transfer process. The air firstly enters into the evaporator in the first loop, and is cooled by the refrigerant. Then it enters into the evaporator in the second loop. Therefore, the mean temperature of the first loop is higher than that of the traditional system, while the mean temperature of the second loop is lower than that of the traditional system. For the condenser, the air temperature increase along with the heat transfer process. The air firstly enters into the condenser in the second loop and is heated by the refrigerant. Then it enters into the condenser in the first loop. Therefore, the mean temperature of the first loop is higher than that of the traditional system, while the mean temperature of the second loop is lower than that of the traditional system.

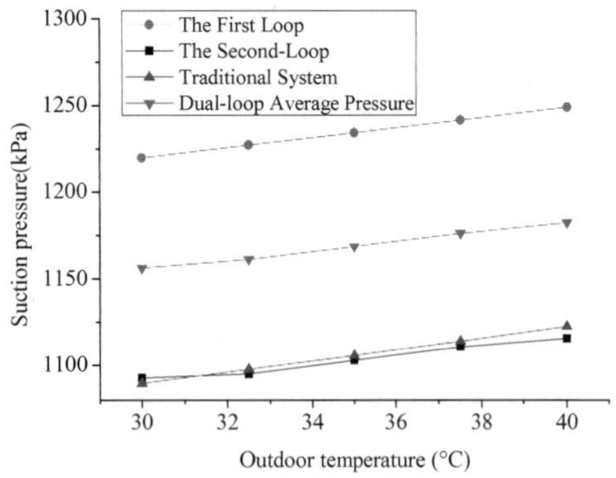

(a) Suction Pressure

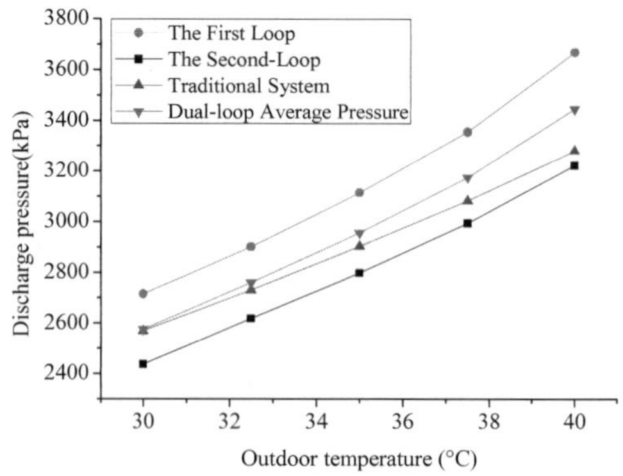

(b) Discharge Pressure

Figure 6: Cooling condition

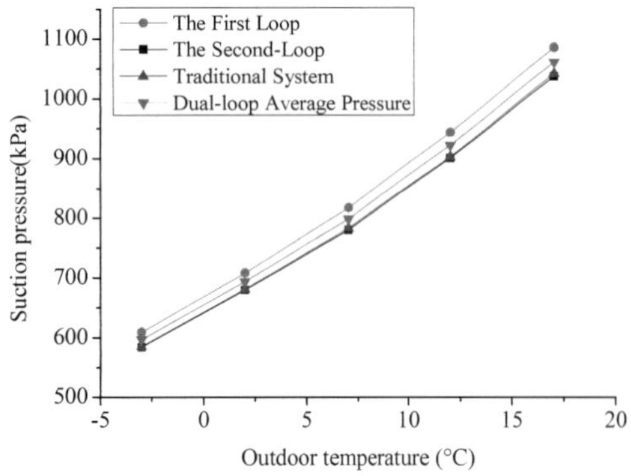

(a) Suction Pressure

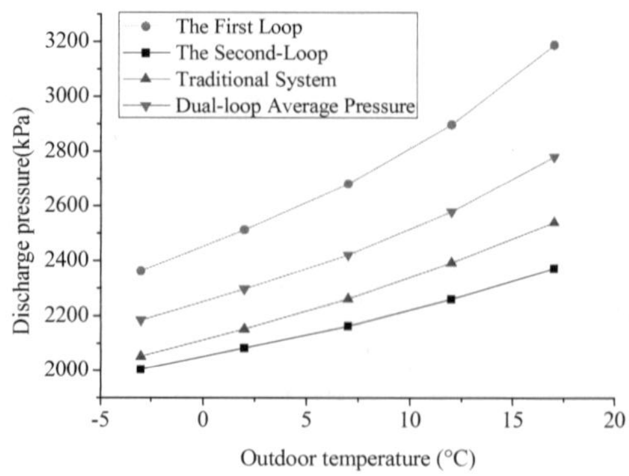

(b) Discharge Pressure

Figure 7: Heating condition

5.2 Mass Flow Rate

Figure 8 shows the variation of mass flow rate of the dual-loop system compared with the traditional system under cooling condition and heating condition. With the increasing of the outdoor temperature, the mass flow rate of the dual-loop system and traditional system increase in both cooling condition and heating condition. The mass flow rate of the dual-loop system is higher than that of the traditional system in both cooling condition and heating condition. The reason is that the average suction pressure of the dual-loop system is higher than that of the traditional system, while the displacement of the two type compressor is same. The suction pressure determines the refrigerant density in the suction port. The higher of the suction pressure results in the higher of the refrigerant density. Therefore, the mass flow rate of the dual-loop system is higher than that of the traditional system. The mass flow rate of the dual-loop system is 5.4% ~8.0% higher than that of the traditional system in cooling condition and1.4%~2.5% higher than that of the traditional system in heating condition.

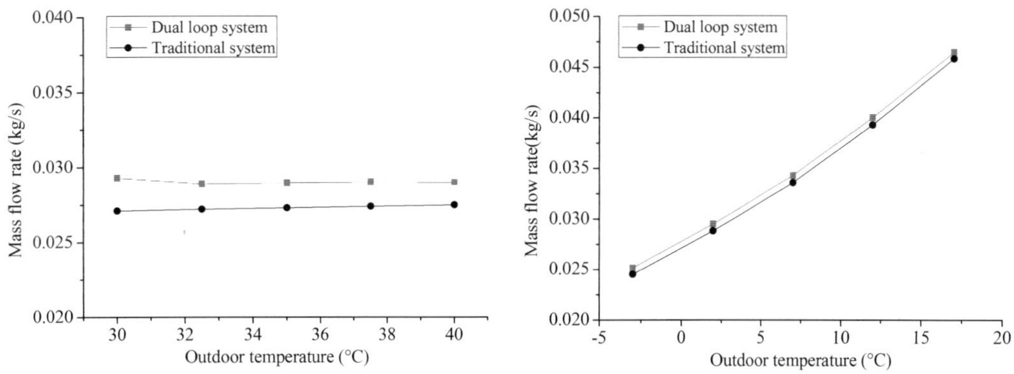

(a) Cooling condition (b) Heating condition

Figure 8: Mass flow rate

5.3 Power Consumption

Figure 9 shows the variation of power consumption of the dual-loop system compared with the traditional system under cooling condition and heating condition. With the increasing of the outdoor temperature, the power consumption increase in both cooling condition and heating condition. The power consumption of the dual-loop system is higher than that of the traditional system in both cooling condition and heating condition. The reason is that mass flow rate of the dual loop system is higher than that of the traditional system. Compressing more refrigerant means the compressor consumes more power. The power consumption of the dual-loop system is 0.5~3.8% higher than that of the traditional system in cooling condition and 4.8%~7.1% higher than that of the traditional system in heating condition.

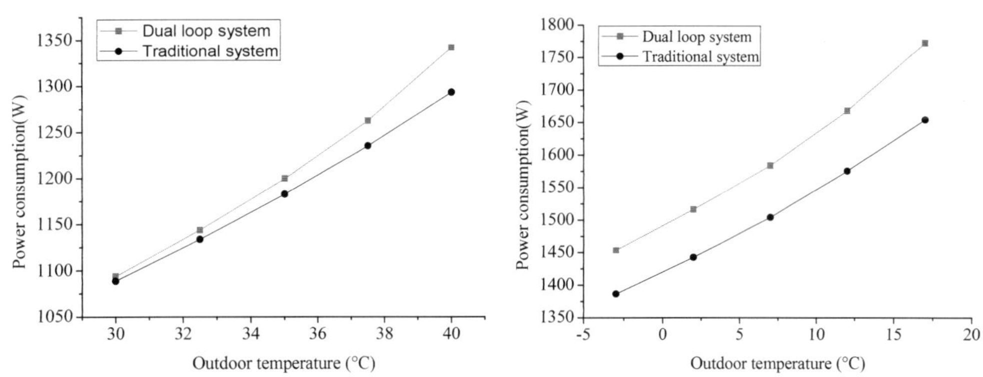

(a) Cooling condition (b) Heating condition

Figure 9: Power consumption

5.4 Cooling and Heating Capacity

Figure 10 shows the variation of cooling capacity and heating capacity of the dual-loop system compared with the traditional system under cooling condition and heating condition. With the increasing of the outdoor temperature, the cooling capacity decrease, while the heating capacity increase. Both the cooling capacity and heating capacity of the dual-loop system is higher than that of the traditional system in cooling condition and heating condition. The reason is that mass flow rate of the dual loop system is higher than that of the traditional system. The cooling capacity of the dual-loop system is 4.3~4.8% higher than that of the traditional system and the heating capacity of the dual-loop system is 7.0~11.3% higher than that of the traditional system.

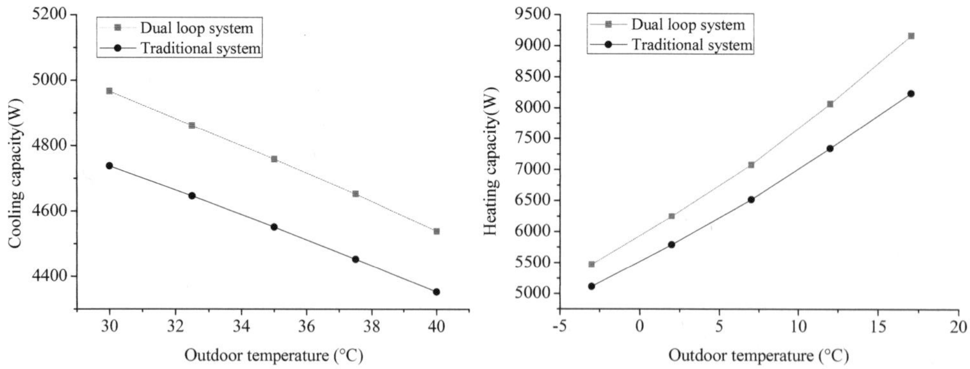

(a) Cooling condition (b) Heating condition

Figure 10: Cooling capacity and heating capacity

5.5 COP

Figure 11 shows the variation of COP of the dual-loop system compared with the traditional system under cooling condition and heating condition. With the increasing of the outdoor temperature, the COP decreases in cooling condition, while the COP increases in heating condition. The COP of the dual-loop system is higher than that of the traditional system in cooling condition and heating condition. It shows that the dual-loop system is more efficient than the traditional system. The COP of the dual-loop system is 0.5%~4.4% higher than that of the traditional system in cooling condition and 2.1%~3.9% higher than that of the traditional system in heating condition.

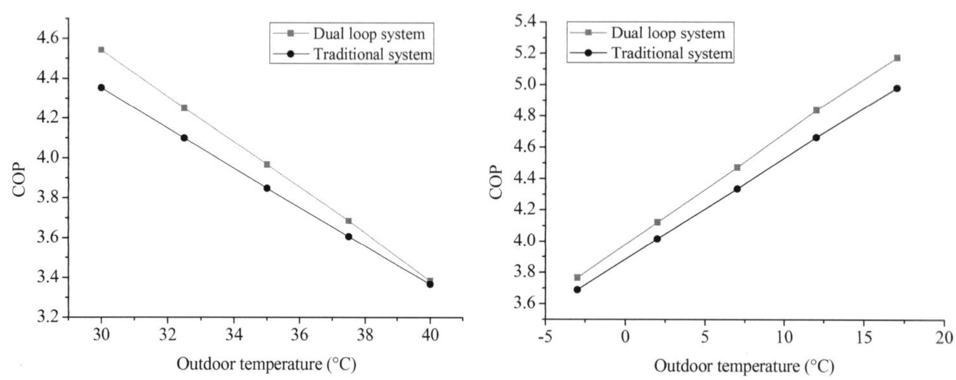

(a) Cooling condition (b) Heating condition

Figure 11: Cooling capacity and heating capacity

6. CONCLUSIONS

A new dual-cylinder rolling piston compressor with two independent suction ports and two independent discharge ports was proposed, and this compressor was applied in a dual-loop system. Through building the simulation model of the dual-loop system and the traditional system, this paper came to the following main conclusions. The COP of the dual-loop system is 0.5%~4.4% higher than that of the traditional system in cooling condition and 2.1%~3.9% higher than that of the traditional system in heating condition.

NOMENCLATURE

The nomenclature of the equation during the text:

T	temperature	(K)

Subscript

H	High
L	Low

REFERENCES

Weihua Zhang.(2017)Basic research on standard system of residential building energy consumption index classification [D]. Xian University of Architecture and Technology, 1-134.

Weifeng Zhu & YoudiJiang (2013). Item-based energy consumption reference indicators of large-scale public buildings in Shanghai [J].Heating Ventilating and Air Conditioning , 43(11),3-7.

Xiang Cao, Chunlu Zhang & ZiYang Zhang.(2017) Stepped pressure cycle-A new approach to Lorenz cycle[J]. International Journal of Refrigeration, 74, 283-294.

Wang C C,(2001). Parametric study of the air - side performance of slit fin - and – tube heat exchangers in wet conditions[J] . Journal of Mechanical Engineering Science , 215(9) ,1111 - 11221

Wang C. C,(1999). A heat transfer and friction correlation for wavy fin-and-tube heat exchangers. Int. J. of Heat and Mass Transfer , 42(10) ,1919～1924.

Shah, M. M,(1982). Chart Correlation for Saturated Boiling Heat Transfer: Equations and Further Study, ASHRAE Trans, Vol. 88, Part 1, 185-196.

Tang,(1997). Empirical Study of New Refrigerant Flow Condensation inside Horizontal Smooth and Micro-Fin Tubes, Ph.D. Thesis, University of Maryland.

Kattan, N., Thome, J.R. and Favrat, D,(1998). Flow Boiling in Horizontal Tubes. Part 1:Development of a Diabatic Two-Phase Flow Pattern Map, J. Heat Transfer, Vol. 120, No. 1, 140-147.

Kattan, N., Thome, J.R. and Favrat, D,(1998). Flow Boiling in Horizontal Tubes. Part 2: New Heat Transfer Data for Five Refrigerants, J. Heat Transfer, Vol. 120, No. 1, 148-155.

Kattan, N., Thome, J.R. and Favrat, D,(1998). Flow Boiling in Horizontal Tubes. Part 3:Development of a New Heat Transfer Model Based on Flow Patterns, J. Heat Transfer, Vol. 120,No. 1, 156-165.

Study on PMF Rotary Compressor for Unitary Air Conditioners in N.A. Market

Bin Gao*, Qiang Gao, Huaming Li, Jianqiang Huang

GMCC, R&D Department, Shunde, Foshan, Guangdong Province, China

ABSTRACT

Unitary air conditioner in N.A. market takes the non-bleed TXV as its throttle valve. When the system turns off, the TXV automatically shut down to maintain the pressure in heat exchangers. This is benefit for SEER and comfort in use. However, pressure maintaining resulted in a fast-restart issue to fix-speed rotary compressor applying in this kind of system. To solve this problem, this paper firstly gave two concepts and principles of passive pressure balance and active pressure balance to conclude the means to achieve pressure balance in rotary compressor. Then, designed the technical scheme of PMF (Pressure Maintaining and Fast-restart) solution and analyzed the working process based on the two pressure balance principles. Furthermore, verified the PMF solution by compressor and system performance tests, PMF function test and reliability test. The results show that rotary compressor with PMF solution has good performance, stable PMF function and high reliability, which is a suitable product for unitary air conditioners.

1. INTRODUCTION

The US style ducted air conditioner as known as unitary air conditioner is the major product form for the detached house in North America market. Currently, 95% of this products are using fix speed scroll compressors. Along with the developing of rotary compressor, there is a trend that rotary compressor will replace the scroll type for the unitary ACs which capacity range between 1.5 ton to 3 ton.

Today, there are two types of throttling valves applying to fix speed unitary air conditioner, including Non-Bleed TXV and piston. The TXV system is more popular due to its advantages in energy efficiency and comfort. Since the TXV closes automatically during down time, the heat exchangers have pressure maintaining effect, which helps the system back to cycle balance after the system restart, hence also good for customer's experience.

However, because of this pressure maintaining function, the pressure difference between condenser and evaporator remains high level after the system shutdown. This pressure difference brings the restart issue for traditional rotary compressor when it wants to restart within 3 minutes after last shutdown.

To solve this problem, some of the system manufactures applying some assistive devices calls Hard Start Kit to increase the starting torque of the compressor motor to improve the starting ability. This can improve but cannot ensure a successful restart when the pressure difference is at a high level. Another way out is to develop a new compressor with fast restart function when system keeps on high pressure difference.

2. Technical Principle

The definition of pressure balance is that the pressure among components of the system reaches equilibrium state. In this case, there is no pressure difference or the pressure difference is insufficient for the fluid to flow any further. Rotary compressor can be started when the pressure between suction line and discharge port reaches pressure balance.

2.1 Passive pressure balance principle

The passive pressure balance means the spontaneous action happened in the system that makes the system trend to be balanced, such as the leakage under a pressure difference between the compressor shell and the accumulator.

Base on the structure of rotary compressor, there are several fit clearances in the compressing pump. When the non-bleed TXV system with a conventional rotary compressor stopped from running state, refrigerant in condenser leaks to evaporator through those clearances due to the pressure difference ΔP.

As shown in Figure 1, the compressing pump leakage paths include: (1) piston end clearance; (2) piston outer diameter clearance; (3) slider side clearance; (4) slider end clearance and head clearance. Under the effect of pressure difference ΔP, the fluid on the high-pressure side of the system flows to the low-pressure side through these gaps gradually until pressure balance is reached. This process usually takes more than twenty minutes because the volume of heat exchangers is too large compare to those micrometer level clearances as the routes.

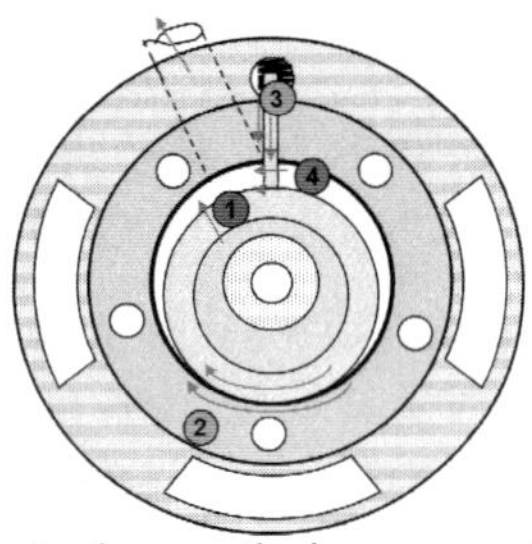

Figure 1: Leakage paths in compressing pump

Depending on the height of the oil surface after compressor stopped, the fluid leaking from the gap may be oil or gaseous refrigerant. Oil leaks when oil level is higher than the cylinder. Because of its high viscosity, the balance time is much longer than gas leaking.

According to the flow calculation equations, equation (1) and (2), shown below. The mass of the fluid flowing through a gap can be obtained:

$$m = \frac{\pi \times h^3}{6 \times \mu \times \ln(R_2 / R_1)} \times \Delta P \tag{1}$$

$$m = \frac{b \times h^3}{12 \times \mu \times L} \times \Delta P \tag{2}$$

Equation (1) is used to calculate the flow between two parallel ring gaps, which is applicable to the leakage analysis of piston end clearance. Equation (2) is used to calculate flow between parallel plate gaps, which is applicable to the leakage analysis of slide clearance.

The inlet pressure of the gap is the same as the pressure in compressor shell, while the outlet pressure of the gap is the same as the pressure in suction hole as well as the accumulator.

2.2 Active Pressure balance principle

The active pressure balance using additional mechanism that set in the system and triggered in time to relief pressure to reach faster pressure balance.

The passive pressure balance process has strong correlation to the ΔP valve, when the ΔP decrease, the flow velocity through the gap drops rapidly that makes pressure balance time extends. The active pressure balance scheme is designed to help shorten this process at low ΔP state.

In this paper, an active pressure balance mechanism was designed by referring to the direct acting relief valve which has maturely been used in the hydraulic system. As shown in figure 2, 1 is the valve body, 2 is the valve core, 3 is the pressure regulating compression spring, 4 is the pressure regulating rod, port P is the pressure fluid inlet, T is the overflow port.

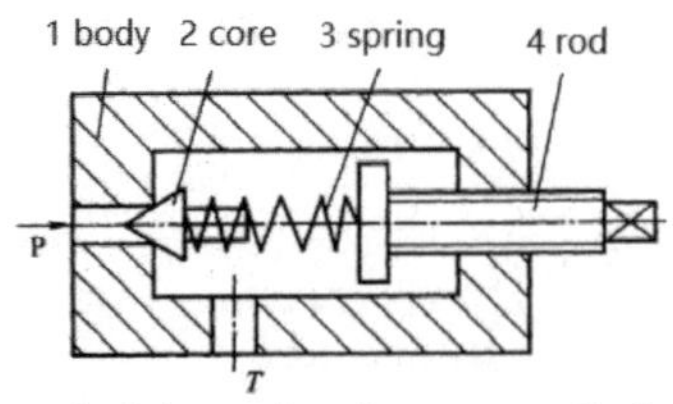

Figure 2: Schematic of pressure relief valve

This kind of relief valve using a simple but reliable spring-pressure adaption structure, automatically open and close with the change of ΔP between ports P to T. When the differential pressure force is greater than the spring force, pressure is relieved from port P to T. When the differential pressure is less than the spring force, the valve closes.

3. Design of PMF solution

3.1 Technical scheme

The design purpose is separated to two aspects for simplification: after the system shutdown, the ΔP of two heat exchangers can be maintained before the next startup, at the same time, the compressor should successfully restart in 3 minutes.

To keep the pressure difference between the heat exchangers, the direct way is to cut off the flow path between them. Since the non-bleed TXV automatically closed after shutdown, the only path between the heat exchangers is flowing through the compressor.

A check valve is added on the both suction and discharge side of the rotary compressor. When compressor is running, the check valves are in the open position that without affecting the operation. When compressor stopped, the check valves closed automatically that divide the system to condenser side with higher pressure, compressor and evaporator side with lower pressure, then the pressure difference between two heat exchangers is maintained.

Due to the function of the two check valves, the range of pressure balance is redefined to the internal pressure balance of the compressor. Compare to the heat exchangers, the inner volume of compressor is much lesser so that the pressure balance is much faster.

Because of the effect of passive pressure balance, the ΔP of the compressor suction and discharge sides decreased gradually until a threshold value is reached. Under this threshold value, the leakage velocity through the gaps slowdown very fast so that the complete balance time will much longer than 3 minutes.

A pressure relief valve is a solution to this problem. Opening and closing of this valve is automatically acted by differential pressure ΔP between suction and discharge sides of compressor by the design opening differential pressure ΔPr of the relief valve. The valve is set to connect with the inner cavity of the exhaust muffler, and the outlet is connected with the suction hole of the cylinder.

During normal operation, the valve is shutdown that has no impact on compressor operation. When compressor stopped and after passive pressure balance process, the value of ΔP is less than ΔPr, the relief valve automatically opened to achieve absolutely pressure balance.

The solution of Fast-restart Solution with Differential Pressure in system, as call PMF as an abbreviation. This PMF technical schematic is as shown in Figure 3. It consists of discharge check valve Vd, suction check valve Vs and pressure relief valve Vr.

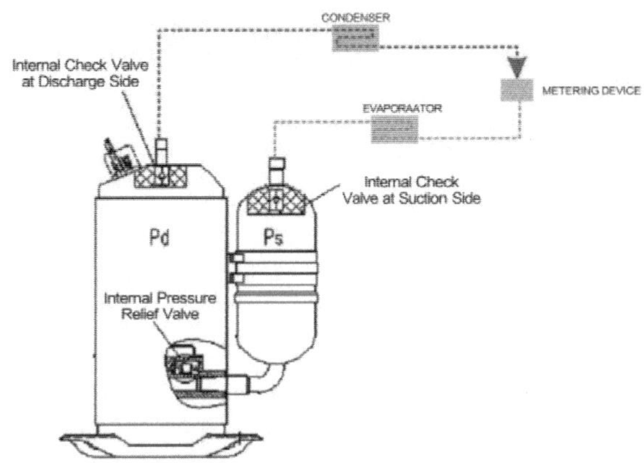

Figure 3: PMF technical schematic

To improve the reliability, the valves are all designed in the compressor. Vd is set up in the inlets of discharge pipe on the upper case. Vs is located in the accumulator as well as suction flow path. Vr is in the compressing pump, inlets and outlets connect discharge volume and suction hole respectively. The layout of PMF valves in compressor as shown in figure 4.

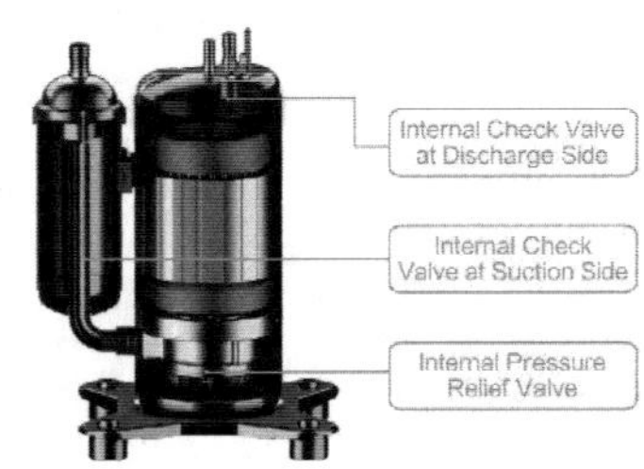

Figure 4: PMF valves in compressor

3.2 Working Process

To understand the working process, The whole cycle, from running to shutdown to pressure maintaining and then restart, is divided into 5 steps, as shown in figure 5, where Pe and Pc represent pressure in evaporator and condenser respectively. During the process, the PMF valves can automatically open and close along with the pressure change around the valves.

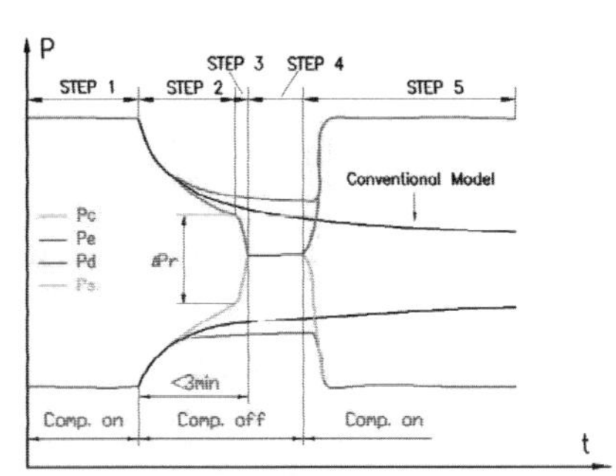

Figure 5: working process of FDSP solution

In which:

Step 1: During the system operation, the check valve Vd and Vs are on open state, the pressure relief valve Vr is on off state while ΔP=Pd-Ps>ΔPr. The compressor is no different from the conventional rotary compressor;

Step 2: After the compressor shutdown, the non-bleed TXV automatically closed after a while. In the compressor, the check valve Vd and Vs are automatically closed, the pressure relief valve Vr is keep on off state because ΔP=Pd-Ps>ΔPr. At this period, the passive pressure balance working makes ΔP decrease gradually.

Step 3: The compressor still in shutdown state, the check valve Vd and Vs are keeping closed. The pressure relief valve Vr automatically opens when ΔP=Pd-Ps<Δ Pr because of the passive pressure balance. Pd and Ps quickly balanced to Pm that achieves absolutely pressure balance.

Step 4: The compressor still in shutdown state. The check value Vd and Vs are keeping closed and the pressure relief value Vr is keeping open. At this period, the pressures in compressor Pd=Ps=Pm, then the compressor can be restarted anytime. Because of the check valves, the pressure Pe<Pm<Pc, that means the unitary system can maintain the pressure difference between the condenser and evaporator.

Step 5: When the compressor restarts, the Pd rises and Ps falls quickly, when Pd>Pc and Ps<Pe, the check valve Vd and Vs automatically open. At the same time, when ΔP=Pd-Ps>ΔPr, Vr closes automatically. This starting process is the same as that of conventional rotary compressor, and then the circle accomplished and back to the step1 state.

Therefore, with the application of PMF,-The pressure in compressor achieve balanced state within 3 minutes by optimize the setting of ΔPr of the relief valve. At the same time, maintaining pressure difference between condenser and evaporator during system shutdown period.

4. Verification of PMF

4.1 Optimization of valve structure

To improve reliability and decrease affection of valves on compressor performance, the FEM and CFD tools are used to analyze and optimize the structure relative to reliability and performance, including the strength, flow resistance and operational stability of the valves. The results are summarized and shown in Table1.

Table 1: Results of structure optimization

Method	Content	Result
FEM	Strength of valve plate and valve base	Conformed material and dimension
	Strength of relief valve	Conformed location and dimension
CFD	Flow resistance of check valve	Structure of flow path is optimized
	Operational stability optimization	Stroke of the relief valve is optimized

With FEM and CFD simulation, the original valve structure was designed and the compressor test with the valves was accomplished.

4.2 Experimental analysis

4.2.1 Leakage principle verification

Firstly, the effect of passive pressure balance was confirmed by adding two check valves on both compressor suction and discharge sides. The tested ΔP curve is shown in Figure 6. It was witnessed that the ΔP decreased gradually as a result of the check valves closes automatically, which was in accordance with the expectation.

According to the ΔP curve, at the beginning stage of compressor shutdown, ΔP value drops rapidly, along with the leakage, ΔP value decrease makes the balancing velocity slow down, especially at the lower ΔP level like less than 0.5MPa.

With passive pressure balance, the ΔP between discharge and suction side of the compressor drops significantly after 3 minutes from shutdown compared to the original system. However, if the oil level is higher than upper surface of cylinder when compressor stopped, the leakage fluid of passive pressure balance is lubricants with high viscosity. Sometimes there is still a considerable ΔP exist after shutdown for 3 minutes. In this case, the compressor will not restart properly.

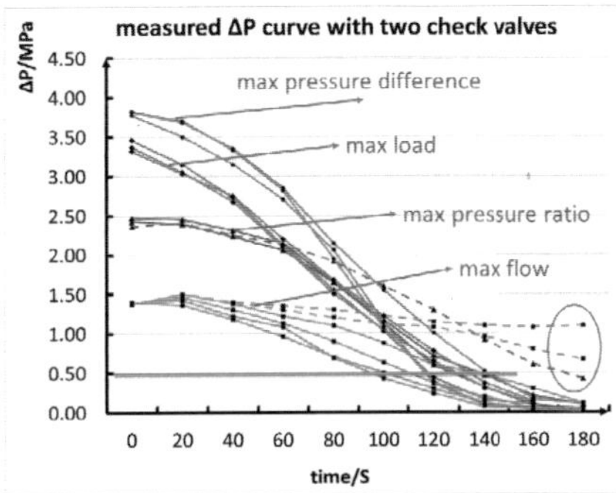

Figure 6: ΔP curve by passive pressure balance

The effect of active pressure balance with the relief valve also conformed base on the two check valves compressor. The results show that the deviation between the tested ΔPr and the designed ΔPr is less than 0.05MPa. With the relief valve, the pressure between discharge and suction sides reaches absolute pressure balance within 10s after the valve is opened, that makes the balance time of all working conditions can be within 3min when the ΔPr value is properly designed. That indicates a very good feasibility of this design.

In summary, the PMF solution that include two check valves and single relief valve have confirmed by several tests with different working conditions.

4.2.2 Performance Impact verification

Because the suction and discharge check valves are set on the compressor flow path, it is inevitable that compressor performance will be impacted because of the pressure loss of the check valves.

Regarding to the relief valve, at normal working conditions, the running ΔP is much higher than the ΔPr makes the relief valve keep closing during operation, so that the relief valve has no effect on the performance.

The compressor performance comparison of samples with and without PMF solution are confirmed. Under the typical working condition A, the COP drops because of the two check valves is less than 0.6 percent relatively. Meanwhile, the noise with and without valves is at the same level.

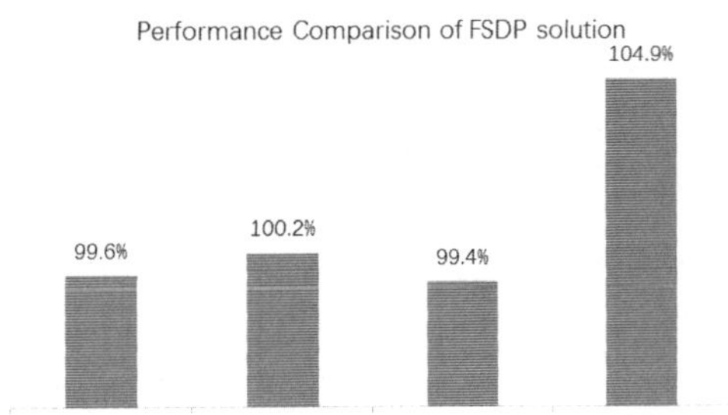

Figure 7: Performance Comparison of PMF solution

In addition, the system performance is also confirmed by testing. The test applies to two samples with and with PMF solution on the same unitary system to carry out the SEER comparative test. Both the compressor and system test results are shown in Figure 7.

According to the result, although the sample with PMF has lower compressor performance, but the result of system SEER is 4.9% higher. This is because the PMF helps the system get a much lower Cd coefficient during SEER test which has significant effect on SEER calculation.

4.2.3 PMF function test
PMF function tests are used mainly to confirm pressure maintaining and compressor restart function after shutdown from operation.

The compressor ASG240N with PMF solution was applied on a 2.5 ton unitary system. The pressure data of condenser Pc, evaporator Pe, compressor shell pressure Pd, and accumulator pressure Ps were measured respectively on the test required by AHRI standard 210/240.

The typical pressure change curve is shown in figure 8. Compare to the curve in figure 5, the process of the test record has very good alignment with the expectation.

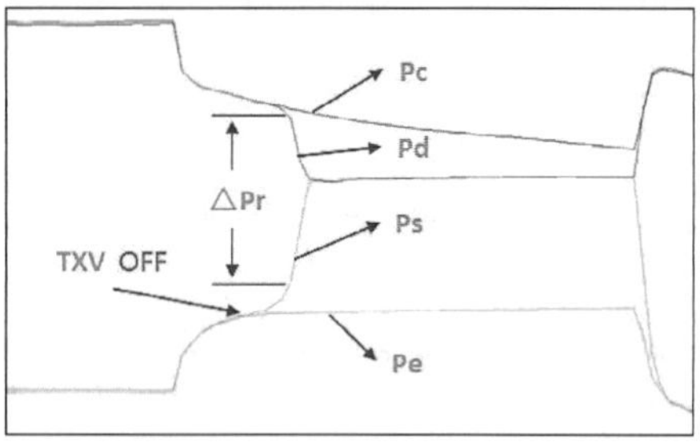

Figure 8: Typical pressure curve of PMF solution

When the system stopped from operation, the check valves closed immediately, while the TXV still on at this time, Pd equal to Pc and Pe equal to Ps. They move toward each other because of the fast flow through TXV. As the TXV closed, Pc and Pe starts the maintaining stage while Pd and Ps changed by the passive pressure balance. When the ΔP in the compressor decreased to ΔPr, the relief valve opened made the Pd and Ps achieved absolutely balance within a short time.

According to the system test, the balancing time of all the test conditions were within 3 minutes, typically within 2.5 minutes. The compressor could restart immediately as long as the balance state was reached. If the compressor was not restarted, the pressure difference between Pc and Pe could be maintained for several hours.

4.2.4 Reliability Analysis
Reliability is the priority of PMF compressor development. For this typical model ASG240N, detailed analyses using TRS tool were conducted to help make the reliability test scheme. The list and results are shown in Table 2.

Table 2: Reliability test scheme and results

Object	Targets	How to make it	Results
Compressor	Long life time	Oil, structure and material optimization	Accomplished
		Multiple reliability test	Approved
	Large gas charge amount	Oil level and liquid level test	Approved
valves	On/off time	270 thousand times test	Approved
	Noise	No unmoral noise	Approved
	leakage	Structure and material optimization	Accomplished
	Strength	Pressure tight test	Approved

Currently, all the reliability tests are accomplished, and the results are good according to standards.

5. CONCLUSIONS

In this paper, a new rotary compressor with PMF solution is proposed, with these results accomplished:

- The decline in compressor performance is within 1%, but the system SEER improved significantly.
- After system shutdown, the pressure difference between heat exchangers can be maintained for several hours, while the compressor can be restarted by 3 minutes.
- The reliability of both compressor and PMF valves are optimized and verified to ensure a reliable product for N.A. market.

Based on those researches and developments, the rotary compressor with PMF solution is a product of potential for unitary air conditioners for N.A. market.

NOMENCLATURE

m	leakage flow	(g)
h	gap width	(m)
R1	inner diameter of piston	(m)
R2	outer diameter of piston	(m)
μ	leakage hydrodynamic viscosity	(Pa·s)
ΔP	pressure difference between inlet and outlet of gap	(Pa)
L	Length of leakage channel	(m)

Subscript

ΔPr	designed open pressure difference of relief valve
PMF	Pressure Maintaining and Fast-restart
COP	Coefficient of Performance
Cd	Degradation Coefficient

REFERENCES

[1]Gang. Y, Jianhua. W, & Yezheng. W. (2005). Investigation of the Radial Clearance Leakage in Rolling Piston Compressor. *FLUID MACHINERY, Vol.33, No.9*, page72-75.

[2]Biye. Y, Ruisong. S & Jingxu. Y. (2019). Research on influence of axial clearance of rotary compressor on performance. Chinese Joural of Refrigeration Technology. Vol.39, No.1, page 51-54.

[3]Zhi. Z. (2020). Improvement and Research on Cylinder Structure of Rolling Piston Compressor. PhD Thesis. Shengyang Construction University.

[4]Kai. W. (2018). Research on Internal Flow Principle and Leakage Property of Scroll Compressor. PhD Thesis. Xi'an Technology University.

[5]Limei. Y. (2018). Analysis of the Effect of Gap Leakage on Lubricant System of Scroll Compressor. PhD Thesis. Langzhou Technology University.

Performance Comparison between Rolling Piston and Coupled Vane Compressor

International Compressor Engineering Conference at Purdue 2020

Q.J. Chen[1], K.T. Ooi[2]

School of Mechanical and Aerospace Engineering. Nanyang Technological University, 50 Nanyang Avenue, 639798 Singapore, Singapore

[1]chen1179@e.ntu.edu.sg, [2]mktooi@ntu.edu.sg

ABSTRACT

In this paper, the performance of the rolling piston compressor and the coupled vane compressor are compared, and differences or similarities are presented and discussed. The comparison is based on one of the available sizes of the rolling piston compressor as used in a room air-conditioner. The extent of the theoretical comparison includes their respective design features, relative sizes, individual flow and frictional energy losses, details of PV diagrams, valve flow analysis and other performance parameters such as "compressor" COP. The coupled vane compressor is a novel compressor recently design with the objective to save raw materials in the production process while matching the performance of current compressors out in the market. It does this through its unique design allows for rotor size to not be limited by the stator size, thus increasing the working chamber to cylinder volume ratio. Through arbitrarily chosen values for their geometrical parameters, the comparison of the two compressors shows that the coupled vane compressor has potential to replace current compressors, thus saving a significant amount of raw materials and cost for compressor manufacturing and creating a more sustainable environment in the future.

NOMENCLATURE

a	R_r/R_c	[-]
A	area	[m^2]
b	distance between the rotor centre and the cylinder centre	[m]
C	specific heat capacity	[$J\ kg^{-1}\ K^{-1}$]
c	damping coefficient	[-]
D	diameter	[m]
E	Young's modulus	[$N\ m^{-2}$]
e	eccentric distance	[m]
F	force	[N]
f	motor operating frequency	[Hz]
g	acceleration due to gravity	[$m\ s^{-2}$]
h	specific enthalpy	[$J\ kg^{-1}$]
h_b	vane height	[m]
I	moment of inertia	[m^4]
k	thermal conductivity	[$W\ m^{-1}\ K^{-1}$]
l	length	[m]
m	mass	[kg]
N	number of items	[-]
O	origin	[-]
P	power	[W]
p	pressure	[Pa]
Q	heat	[J]
q	specific heat	[$J\ kg^{-1}$]
R	radius	[m]
Re	Reynolds number	[-]
r	radial coordinate	[m]
s	entropy	[$J\ kg^{-1}\ K^{-1}$]
T	temperature	[K]
t (t_b)	thickness (vane thickness)	[m (m)]
u	specific internal energy	[$J\ kg^{-1}$]
V	volume	[m^3]
v	velocity	[$m\ s^{-1}$]
v_s	specific volume	[$m^3\ kg^{-1}$]
W	work	[J]
w	width	[m]

Greek symbols

α	contact angle at the leading vane	[rad]
β	contact angle at the trailing vane	[rad]
ε	eccentricity ratio	[-]
η	efficiency	[-]
λ	friction factor	[-]
μ	dynamic viscosity	[Pa·s]
ρ	density	[$kg\ m^{-3}$]
θ	rotation angle	[rad]
ω	angular speed; natural frequency	[$rad\ s^{-1}$; Hz]
ζ	damping ratio	[-]

Subscripts

i	inlet (suction) conditions
o	outlet (discharge) conditions
1	upstream conditions
2	downstream conditions
c	cylinder chamber conditions

1. INTRODUCTION

Positive displacement compressors in general are mostly applied in small and medium cooling capacity applications. Rolling piston compressor is the most widely used positive displacement rotary compressor in the world today. It is particularly popular in household air-conditioners and those small and medium size cooling load applications, generally from fractional kW up to 15 kW of cooling capacity. Its annual production rate has been increasing, and for the past few years, it has been hovering around 170 to nearly 200 million pieces annually, making it the most produced compressor in the market. As compared to other types of positive displacement compressors, it has the inherent characteristics of a simple geometry, having few parts, low vibration and being compact.

Due to the large annual production volume of the rotary compressors, large amounts of material go towards making the compressor and most of these materials are metallic materials. If the general weight of a compressor is taken as 20 kg, 3-4 billion kg of metal will be used annually just for fabricating the rotary compressors alone. In the interest of saving materials, one of the newly introduced rotary compressor design, namely coupled-vane compressor (CVC) was born. (Ooi and Shakya, 2018a) For the same cooling capacity, this new compressor needs about half, if not less, of the material in fabricating it, and yet has about the same number of parts, if not fewer.

As a result of the innovative design feature, this CVC allows a rotor to cylinder ratio of less than 0.4, instead of greater 0.7 for all the existing rotary compressors, making it the most compact rotary compressor design available.

This report aims to study and compare the differences in the design and working principles of the rolling piston compressor with that of the CVC. The mathematical model used in the simulation programme to predict the performance of the compressors will also be briefly presented and discussed. The results of the simulation programme following certain operating parameters are compare and presented, including the frictional energy losses, PV diagrams, valve flow analysis, and other performance parameters.

2. ROLLING PISTON COMPRESSOR

Rolling piston compressors are known for having high efficiency and producing low noise. Because of this, they are widely found in domestic refrigeration and air-conditioning units (Matsuzaka and Nagatomo, 1982).

2.1 Design

A rolling piston consists of a rotor which is being housed in a cylinder, with its centre being eccentric relative to the cylinder centre. The compressor operates through the rotation of the rotor along the cylinder walls which moves the vane, causing it to reciprocate within its slot found along the cylinder walls. This vane separates the working chamber into its suction and compressor chambers.

These design factors of the rolling piston result it in having low vibrational problems as its parts are well-balanced throughout each compression cycle. However, there are multiple rubbing components and leakage paths which might lead to losses in the overall compressor efficiency (Teh and Ooi, 2009).

2.2 Mathematical Model

A mathematical model can be formulated by including heat transfer, valve dynamics, and the thermodynamics and flow aspects. To go more in-depth, stress and mechanical losses can also be considered. Similar to a previous study done (Ooi, 2004), the mathematical model referred to in this project does not include heat transfer for ease of computation. Results produced in that simulation through the omission of heat transfer has been shown to be comparable to that of actual measured results and is thus concluded to be suitable for use in this evaluation.

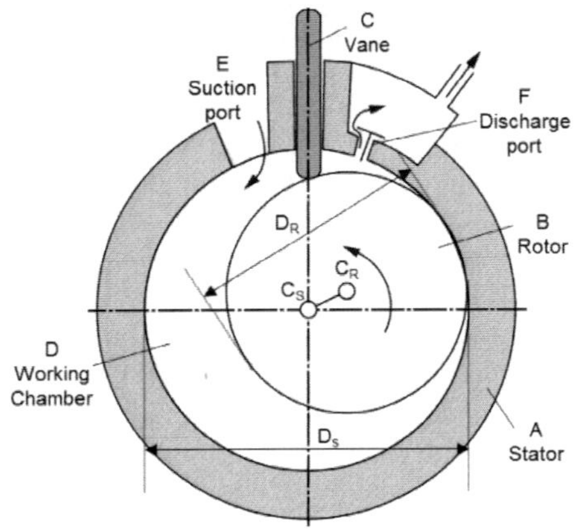

Figure 1: Rolling Piston Compressor Schematics (Soedel, 2007)

2.2.1 Geometrical Model: Figure 1 shows the schematics of the rolling piston compressor. Its working volume is split into the compression and suction chambers. The total working chamber volume is the volume found between the stator, rotor, and piston, bound by the two end plates of the compressor. This can be calculated as follows:

$$V(\theta) = \frac{IR_c^{\,2}}{2}\Bigg[(1-a^2)\theta - \frac{(1-a^2)}{2}\sin 2\theta - a^2\sin^{-1}\left(\left\{\frac{1}{a}-1\right\}\sin\theta\right) - a(1-a)\sin\theta\sqrt{1-\left(\frac{1}{a}-1\right)^2\sin^2\theta}\Bigg] - \frac{t_b h_b}{2}R_c\left[1-(1-a)\cos\theta - \sqrt{(1-a)^2\cos^2\theta + 2a - 1}\right] \quad (1)$$

2.2.2 Thermodynamic Processes: The First Law of thermodynamics can be used to determine the variation of pressure, mass, and temperature of the working fluid through the suction, expansion, compression, and discharge processes. Gas relations properties and equations for the conservation of mass are also used. The above variables can be connected through these equations.

$$\frac{dT_c}{dt} = \frac{\frac{dv_s}{dt}\left[\left(\frac{\partial h_c}{\partial v_s}\right)_T - v_c\left(\frac{\partial p_c}{\partial v_s}\right)_T\right] - \frac{1}{m_c}\left[\frac{dQ}{dt} + \sum\frac{dm_i}{dt}(h_i - h_c) - \sum\frac{dm_o}{dt}(h_o - h_c)\right]}{\left[v_c\left(\frac{\partial p_c}{\partial T_c}\right)_{v_s} - \left(\frac{\partial h_c}{\partial T_c}\right)_{v_s}\right]} \quad (2)$$

$$\frac{dp_c}{dt} = \frac{dT_c}{dt}\left(\frac{\partial p_c}{\partial T_c}\right)_{v_s} + \frac{dv_s}{dt}\left(\frac{\partial p_c}{\partial v_s}\right)_T \quad (3)$$

$$\sum\frac{dm_i}{dt} - \sum\frac{dm_o}{dt} = \frac{dm_c}{dt} \quad (4)$$

In-cylinder convection analysis can be conducted using convective heat transfer model to determine the value of Q.

3. COUPLED VANE COMPRESSOR

The coupled vane compressor is a novel compressor designed to improve current compressor designs through the reduction in the material used for its manufacturing and increasing the working volume for a certain size of compressor but reducing the need for the rotors to be of a minimum size as compared to the cylinders (Ooi and Shakya, 2018b).

3.1 Design

The coupled vane compressor is a novel compressor designed with a twin-vane system which run diametrically through the rotors, and slides in and out of the rotor and along each other during the operation of the compressor. Pressure from the discharge chamber will balance with the centrifugal force of the vane's rotation and the pressure from the suction chamber to keep the vane tips in contact with the wall.

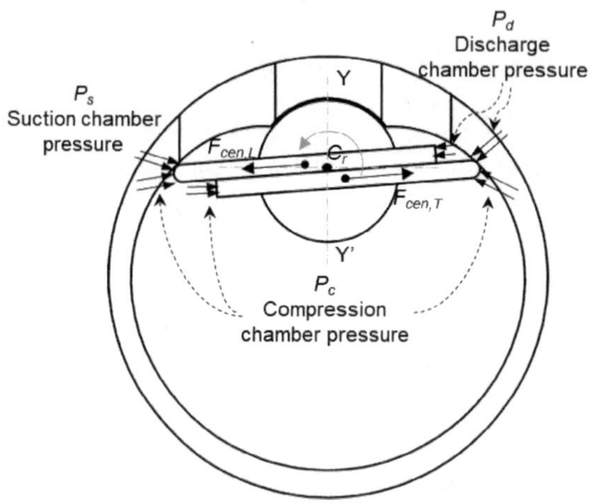

Figure 2: Coupled Vane Compressor Schematics (Ooi and Shakya, 2018c)

3.2 Mathematical Model

Similar to the simulation of the rolling piston compressor, the simulation for the CVC will include a mathematical model through derivations of the working chamber geometry, working fluid thermodynamics, suction and discharge port primary flows, and secondary leakage flow occurring at the internal clearances. Heat transfer is also excluded in this case for simple computations. Details the mathematical model and its derivation have been done by Ooi and Shakya (2018).

3.2.1 Geometrical Model: Figure 2 shows the schematics of the CVC. Similar to the rolling piston compressor, its working volume can be split into the compression and suction chambers. The total working volume, however, defers from the rolling piston due to their different design geometries. For the CVC, the working volume can be calculated as follows:

$$V_c(\theta_r) = \frac{l_c}{2}\int_0^{\theta_r} \left(r(\theta_r)\right)^2 d\theta_r - l_c\left(\pi R_r^2 \times \frac{\theta_r}{2\pi}\right)$$

$$= \frac{l_c}{2}\left[R_c^2\theta_r + \frac{b^2}{2} sin(2\theta_r) - bsin\theta_r\sqrt{R_c^2 - (b\sin\theta_r)^2} - R_c^2\tan^{-1}\left(\frac{bsin\theta_r}{\sqrt{R_c^2 - (b\sin\theta_r)^2}}\right)\right] - l_c\left(\pi R_r^2 \times \frac{\theta_r}{2\pi}\right) \quad (5)$$

3.2.2 Thermodynamic Processes: The complete working cycle of the CVC includes the suction, compression, and discharge processes. A control volume is assumed for the prediction of the thermodynamic properties of the working fluid. The following equations are used to determine the thermodynamic properties:

$$\sum \dot{m}_{in}(u + pv)_{in} - \sum \dot{m}_{out}(u + pv)_{out} + \dot{Q} - \dot{W} = [\dot{m}(u) + m(\dot{u})]_{system} \quad (6)$$

$$\frac{dp}{dt} = \left(\frac{\partial p}{\partial T}\right)_{\rho} \frac{dT}{dt} + \left(\frac{\partial p}{\partial \rho}\right)_{T} \frac{d\rho}{dt} \quad (7)$$

$$\dot{m}_{cv} = \dot{m}_{in} + \dot{m}_{leak,in} - \dot{m}_{out} - \dot{m}_{leak,out} \quad (8)$$

4. SIMULATION RESULT

The simulation is carried out with the AHRI Standard 540, 2020 Standard for Performance Rating of Positive Displacement Refrigerant Compressors. Details of which are briefly covered as follows:

Table 1: Compressor Testing Standards

Compression Cycle			Low Side		High Side – Subcritical	
Application	Rating Test Point	Cycle Type	Suction Dew Point Temperature, °C	Superheat, K (or Return Gas Temperature), °C	Discharge Dew Point Temperature, °C	Condenser Exit Sub Cooling , K
Refrigeration	Medium	Subcritical	-10	10 (20)	45	0

The above ratings are tested at 35°C ambient temperature surrounding the compressor. Further details of the testing standards can be found in the AHRI Standard 540 – 2020.

4.1 Volume Tested

The volume tested for each compressor is equal to bring them towards a fairer comparison. This is done by ensuring that the maximum working chamber volume per cycle is the same for each compressor. This maximum volume occurs at theta equal to 360° for the rolling piston compressor and 270° for the CVC.

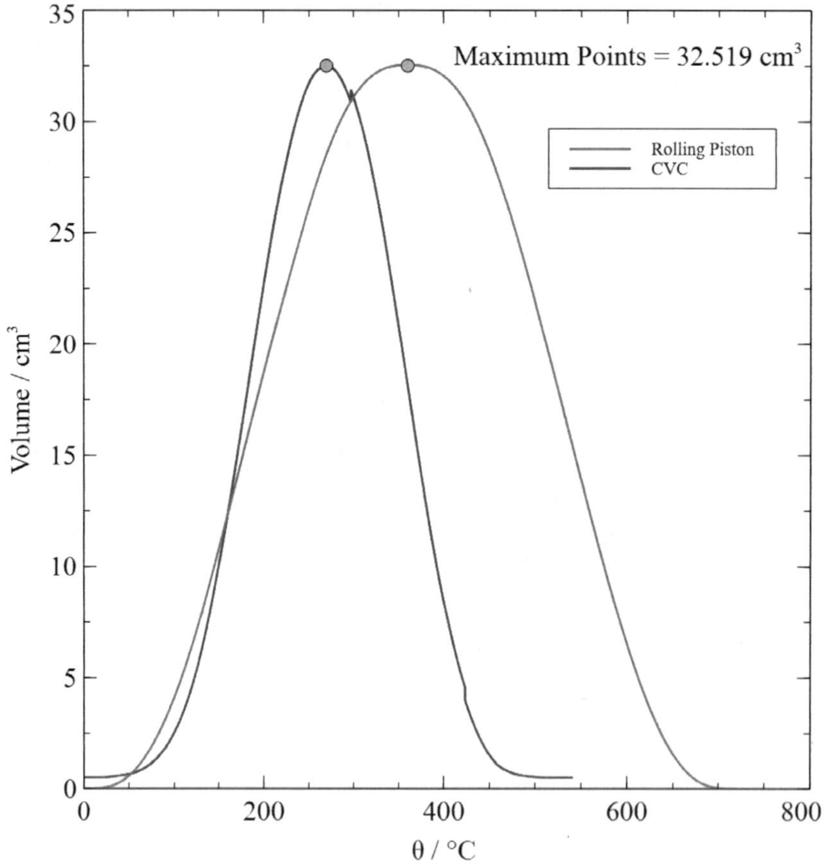

Figure 2: Volume of Working Chamber for Rolling Piston Compressor and CVC

The above can be achieved by ensuring that the geometric parameters for both compressors results in the same volume.

Table 2: Geometric Parameters of Rolling Piston Compressor

Geometric Parameter (RP)	Value
Cylinder Radius / *mm*	58.020
Rotor Radius / *mm*	46.780
Working Chamber Height / *mm*	35.150
Bearing L/S / *mm*	22.250
Suction Port Diameter / *mm*	16.0
Discharge Port Diameter / *mm*	7.0
Working Chamber Maximum Vol. / cm^3	32.519

Table 3: Geometric Parameters of Coupled Vane Compressor

Geometric Parameter (CVC)	Value
Cylinder Radius / mm	32.500
Rotor Radius / mm	20.250
Working Chamber Height / mm	18.759
Suction Port Diameter / mm	16.0
Discharge Port Diameter / mm	7.0
Working Chamber Maximum Vol. / cm^3	32.519

The cross-sectional dimensions being the main variable factors which affect their performance. Suction and discharge valve sizes in both the rolling piston and CVC have the same value.

The rolling piston compressor is running at 3000 rpm while the CVC is running at 4000 rpm. This is due to each compression cycle of the CVC going through 1.333 times the angle of a full compressor cycle of the rolling piston compressor. This variation in the rotation rate is to ensure that both compressors compress the same amount of air per unit time.

4.2 Fluid Tested

R134a is selected for testing with both compressors through the simulation programme.

5. RESULTS AND DISCUSSIONS

Results from the simulation studies are presented below. It is worth noting that the simulation programmes, although based on the mathematical models of the working principles of the compressors, is not able to completely encompass all the details of a compressor's operation. They have been compared with results of actual physical testing and have shown a reasonable degree of accuracy for them to be used as an estimate of the performance of the compressor. Thus, results obtained should be used as a gauge and not of the actual performance of the compressor.

5.1 Results

The following shows the results for testing of the compressor with R134a as the working fluid:

Table 4: Mechanical Analysis of Rolling Piston Compressor using R134a

Frictional power losses	W	Contribution (%)
Loss due to vane side reactions	29.952	29.955
Loss due to vane tip roller force	6.432	6.433
Loss due to roller and eccentric friction	31.863	31.866
Loss due to roller and cylinder head friction	0.019	0.019
Loss due to eccentric and cylinder head faces	0.750	0.750
Loss due to bearing S and shaft	11.308	11.309
Loss due to bearing L and shaft	19.666	19.668
Total friction power	99.990	100%

As can be seen from the rolling piston, a large portion (around 60%) of the losses occur as a result of roller and eccentric friction.

Table 5: Mechanical Analysis of CVC using R134a

Frictional energy losses (180deg)	J	Contribution (%)
Loss due to vane tip 1	1.174	32.062
Loss due to vane tip 2	1.620	44.242
Loss between vane and rotor	0.182	4.957
Loss between vane faces	0.097	2.641
Loss due to rotor endface fluid	0.191	5.218
Loss due to upper bearing	0.182	4.960
Loss due to lower bearing	0.217	5.919
Total friction power	3.663	100%

Simulation studies done on the CVC running at 3000rpm and using stainless steel 17-7 PH vanes resulted in the main contribution for frictional loss being between the vane tip and the cylinder, that is almost 80% of total frictional loss.

The P-V diagrams for both simulations carried out are as shown:

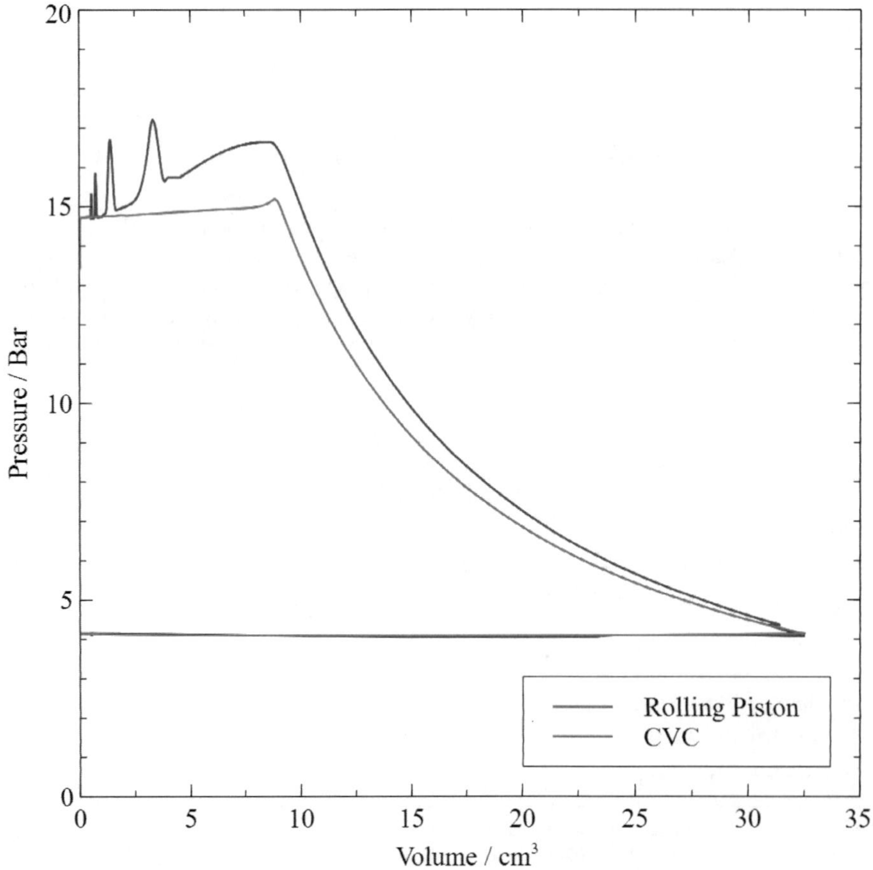

Figure 3: P-V Diagrams for Rolling Piston Compressor and CVC

As can be seen from the diagrams, both compressors undergo similar variations in their pressure and volume throughout their compressor cycles. Because of the difference in rotation rates, the CVC goes through more of such compression cycles per unit time.

Additional performance summary of both the compressors are presented in the table below:

Table 6: Performance Summary of each Compressor

Compressor Type	COP	Mechanical Efficiency	Volumetric Efficiency
Rolling Piston	3.297	89.78%	97.92%
Coupled Vane	2.351	67.29%	95.68%

5.2 Discussion

There are further improvements that can be done to ensure a better comparison between these two compressors. In the simulation studies done, cylinder radius, rotor radius, and vane height were the main geometric parameters varied. In future studies done, more parameter can be altered such as the vane thickness, length, and clearance values.

In addition, rotation rates of each compressor can be equal to ensure a fairer comparison. This can be done by changing the working chamber volume of the compressors with resulting overall volume compressed per unit time remaining the same. The simulations can also be repeated with other refrigerant such as R32 or R1234yf or other fluids to test the consistency of the results.

Actual experiments can also be conducted to ensure that the results from the simulation can be replicated and that actual results do not suffer too large a difference when compared to the results of each compressor's mathematical model.

6. CONCLUSIONS

It is difficult to compare two different compressors due to their unique design and the differing required operating parameters for each of them. The overall results show that the dimensions and thus raw material required for the CVC is smaller than that of the rolling piston compressor for the same amount of volume being compressed per unit time. This is mainly due to the relatively smaller size of the rotor compared to the cylinder for the CVC, in this case 32.50mm for cylinder size. The rolling piston compressor has a rotor to cylinder radius ratio of 58.02mm.

The CVC suffers in some of its overall performance parameters, with a lower COP, mechanical efficiency, and volumetric efficiency as compared to the rolling piston compressor, which means that more power is required by the CVC to carry out the same compression process. The most significant difference lies in the mechanical efficiency, having a value of 67.29% for the CVC, which pales in comparison to the 89.78% for the rolling piston compressor as shown in Table 6. This is likely a result of the significant frictional losses between the vane tips and inner cylinder walls. To counter this, the vane dimensions, or angular velocity of the compressor can be varied to ensure that the force balance at the vane tip is more equal. If successful, less reaction force will be required by the cylinder inner walls to counter the centrifugal forces of the vanes, resulting in a reduction of the frictional forces between them as well. In addition to this, a more suitable lubrication method can be chosen to further reduce friction. With less frictional losses from the vane tips, improvements to the mechanical efficiency of the CVC will be observed.

It is worth noting that the CVC is a novel compressor and is therefore still in its early phase of development. Many improvements can still be made to its design and thus overall efficiency - a department where it is found lacking when compared to the rolling piston. For the CVC, the absence of too large a restriction on its rotor size allows for a larger working chamber volume per cylinder size. This proves a significant advantage in its design and possible future potential.

It is possible to conduct optimisation studies on each compressor and comparing their performance thereafter to ensure a fairer comparison of the two as opposed to arbitrarily chosen values for their geometrical parameters.

REFERENCES

[1] Ooi K.T. and Shakya P. (2018). A New Compact Rotary Compressor: Coupled Vane compressor. *The 24th International Compressor Engineering Conference, Purdue.*

[2] T. Matsuzaka and S. Nagatomo (1982). Rolling Piston Type Rotary Compressor Performance Analysis. *International Compressor Engineering Conference.*

[3] Teh Y. L. and Ooi K. T. (2009). Theoretical study of a novel refrigeration compressor-Part III: Leakage loss of the revolving vane (RV) compressor and a comparison with that of the rolling piston type. *international journal of refrigeration*. 32(5): p. 945-952.

[4] Soedel W. (2007). Sound and vibrations of positive displacement compressors. *Boca Raton, FL : CRC Press*.

[5] AHRI STANDARD 540-2020, Standard for Performance Rating of Positive Displacement Refrigerant Compressors.

ACKNOWLEDGEMENT

The author would like to show his appreciation to previous work done by other researchers in creating the mathematical models and simulation programmes of the rolling piston as referenced. Dr. Shakya readily provided the simulation and design details done for the CVC despite short notice. Much gratitude should also be shown to the academic mentors, friends, and family who aided in this project and showed their confidence in the author's abilities.

CFD Analysis of a Twin Screw Compressor with Oil Injection

Junfeng WANG[1*], Hui DING[1], Sham RANE[2], Ahmed KOVACEVIC[2]

[1] Simerics Inc.,
1750 112th AVE NE Ste C205, Bellevue, WA 98004, USA
jw@simerics.com

[2] Centre for Compressor Technology, City, University of London,
London, UK
sham.rane@city.ac.uk, A.Kovacevic@city.ac.uk

* Corresponding Author

ABSTRACT

In this paper, a full 3D transient Computational Fluid Dynamics (CFD) model of a twin screw compressor with oil injection is described in detail. Volume of Fluid (VOF) two phase flow model is used for gas and liquid phases. The rotor meshes are generated by *SCORG*^{TM} grid generator and read into the CFD solver at each time step. The simulation runs efficiently and results are obtained in about 12 hours on 24 CPU cores. The CFD prediction of air mass flow rate and indicated power has a good agreement with the experimental data measured at City University of London. The compression pocket is tracked using a customized post-processing technique and the pressure-volume diagram within it is also obtained and analyzed. Simulations demonstrate that the approaches used in this paper are robust and fast, and can be readily applied to industrial compressor systems for rapid design iterations and improvements.

1. INTRODUCTION

A rotary twin screw compressor is commonly used in a wide variety of industrial applications due to its reliability, compact structure and wide range of operating pressures and flow rates. Comparing to the dry twin screw compressor, the added oil in oil flooded twin screw compressor offers several advantages. Most importantly, the oil can significantly reduce the gas temperature in the compression chamber as well as the rotor temperature to prevent seizure. Secondly, the oil can seal clearance gaps to improve volumetric efficiency. Thirdly, the oil lubricates rotors and bearings. It is also essential to optimize the oil injection to reduce the power loss associated with the oil for a better performance. However, it is extremely difficult to experimentally determine oil distribution.

Recently, 3D CFD simulation gradually gains its popularity because CFD simulations can deliver deep insights into the performance of a twin screw compressor and provide detailed flow pattern of oil and gas. However, 3D CFD transient simulations faced several challenges. In earlier days, the major challenge was the complex deforming fluid domain in the screw compressor chamber, which required a robust CFD grid to deal with. This difficulty prevented the broad application of CFD simulations to twin screw compressors. A breakthrough was achieved by Kovacevic (Kovacevic, 2002, 2005) who applied the analytical rack generation method of Stosic (Stosic, 1998) to generate an algebraic, adaptive block-structured grid. Screw Compressor Rotor Grid Generator (SCORG) was then developed to generate grids and pre-processing scripts for other CFD software to use. Rane (S. Rane, 2015) proposed a new approach to generate a conformal interface between the rotor domains to improve the accuracy of indicated power and flow rate. The rotor meshes generated by the conformal way are employed in this study. Another challenge is to solve two phase flow with complex moving parts and small leakage gaps. The high density ratio between liquid and gas, sophisticated interaction among the phases, and the interface tracking with complex shape makes the problem difficult to solve. In such case, gas phase has to be treated as compressible, heat transfer effects have to be included, and interface tracking has to be done inside moving, deforming volumes. The major issues for such multiphase CFD simulations are poor convergence, very long simulation time, and unsatisfactory mass/energy conservation. Ding (Ding & Jiang, 2017) performed a numerical simulation of an oil flooded twin screw compressor using CFD package Simerics-MP+ to show the cooling and sealing effects of oil. The simulation demonstrated good convergence, fast calculation speed, and excellent conservation of mass and energy. In this paper, a screw compressor with oil injection model is investigated and validated with test data. The CFD model will be described in detail. Simulation results of indicated power and gas flow rate will be compared with the experimental data. Oil distribution and temperature distribution will be visualized and analyzed.

2. CFD SOLVER AND GOVERNING EQUATIONS

2.1 Conservation Equations for Gas Liquid Mixture

The CFD package Simerics-MP+ is employed to perform the numerical simulations. In Simerics-MP+, the conservation of mass, momentum, and energy of a compressible fluid are solved using finite volume method. These conservation laws can be written in integral representation as

$$\frac{\partial}{\partial t}\int_{\Omega(t)} \rho d\Omega + \int_{\sigma} \rho(\mathbf{v}-\mathbf{v}_{\boldsymbol{\sigma}})\cdot \mathbf{n} d\sigma = 0 \tag{1}$$

$$\frac{\partial}{\partial t}\int_{\Omega(t)} \rho \mathbf{v} d\Omega + \int_{\sigma} \rho((\mathbf{v}-\mathbf{v}_{\boldsymbol{\sigma}})\cdot \mathbf{n})\mathbf{v} d\sigma = \int_{\sigma} \boldsymbol{\tau}\cdot \mathbf{n} d\sigma - \int_{\sigma} p\mathbf{n} d\sigma + \int_{\Omega} \mathbf{f} d\Omega \tag{2}$$

$$\frac{\partial}{\partial t}\int_{\Omega(t)} \rho E d\Omega + \int_{\sigma} \rho((\mathbf{v}-\mathbf{v}_{\boldsymbol{\sigma}})\cdot \mathbf{n})E d\sigma = \int_{\sigma} \kappa\nabla T\cdot\mathbf{n} d\sigma - \int_{\sigma} p\mathbf{v}\cdot \mathbf{n} d\sigma + \int_{\sigma} (\mathbf{v}\cdot\boldsymbol{\tau})\cdot \mathbf{n} d\sigma + \int_{\Omega} \mathbf{f}\cdot \mathbf{v} d\Omega \tag{3}$$

in which $\Omega(t)$ is the volume of the computational domain or control volume, σ is the surface of $\Omega(t)$, $\mathbf{n}$ is the surface normal of σ pointed outwards, ρ is the fluid density, p is the pressure, $\mathbf{f}$ is the body force, $\mathbf{v}$ is the fluid velocity and $\mathbf{v}_{\boldsymbol{\sigma}}$ is surface motion velocity. The shear stress tensor $\boldsymbol{\tau}$ is a function of the fluid viscosity μ and the velocity gradient.

The standard $k-\varepsilon$ two-equation model (Launder & Spalding, 1974) is used to account for turbulence,

$$\frac{\partial}{\partial t}\int_{\Omega(t)} \rho\kappa d\Omega + \int_{\sigma} \rho((\mathbf{v}-\mathbf{v}_{\boldsymbol{\sigma}})\cdot\mathbf{n})\kappa d\sigma = \int_{\sigma} (\mu + \frac{\mu_t}{\sigma_\kappa})(\nabla\kappa\cdot\mathbf{n}) d\sigma + \int_{\Omega} (G_t - \rho\varepsilon) d\Omega \tag{4}$$

$$\frac{\partial}{\partial t}\int_{\Omega(t)} \rho\varepsilon d\Omega + \int_{\sigma} \rho((\mathbf{v}-\mathbf{v}_{\boldsymbol{\sigma}})\cdot \mathbf{n})\varepsilon d\sigma = \int_{\sigma} (\mu + \frac{\mu_t}{\sigma_\varepsilon})(\nabla\varepsilon\cdot\mathbf{n}) d\sigma + \int_{\Omega} (c_1 G_t \frac{\varepsilon}{\kappa} - c_2\rho\frac{\varepsilon^2}{\kappa}) d\Omega \tag{5}$$

Together with equation of state, where density is a function of temperature and pressure, to form a closed system:

$$\rho = f(p, T) \tag{6}$$

2.2 VOF Model

VOF models are widely used in simulation of two phase flow (Hirt & Nichols, 1981)]. VOF solves a set of scalar transport equations representing the volume fraction of each fluid component occupies in every computational cell. The transport equation of the volume fraction for each fluid component can be written as:

$$\frac{\partial}{\partial t}\int_{\Omega(t)} \rho_i F_i d\Omega + \int_{\sigma} \rho_i(v - v_\sigma)\cdot \mathbf{n} F_i d\sigma = 0 \tag{7}$$

where F_i is the volume fraction of the ith fluid component, and ρ_i is the local density of ith fluid component. The weighted mixture density of the fluid in equation (1) to (5) are then calculated as:

$$\rho = \Sigma \rho_i F_i \tag{8}$$

Both implicit and explicit methods can be used to solve the equation. High Resolution Interface Capturing (HRIC) scheme can be used for the convective term in the transport equation.

3. CASE STUDY

The test compressor used for this study is an oil flooded twin screw compressor with a 4/5 lobe arrangement and 'N' profiles. The nominal interlobe, radial and axial leakage gaps are 60 μm. The main rotor diameter is 127.0 mm and the wrap angle is 306°. The tests with different rotor speeds and discharge pressures were carried out at City University of London (A. Rane S.R. Kovacevic & Stosic, 2016) and the results have been used in this paper for comparison with those from Simerics-MP+.

3.1 CFD Model

Figure 1 and Figure 2 show the simulation fluid domain and the corresponding meshes in 3D view. The rotor part of the twin screw is meshed as a single domain using a grid generation software SCORG which creates a series of mesh files for the rotor at different rotation angles. The rotor mesh files are read into the solver via volume remesh approach. The suction port and discharge port of fluid volumes are meshed using Simerics-MP+ binary tree unstructured mesher. The oil injection is through four injection holes that are connected to the rotor domain. All the fluid volumes are connected together using Mis-matched Grid Interface (MGI). The total number of cells is about 0.795 million. Simulations under three operating conditions have been carried out, listed in Table 1.

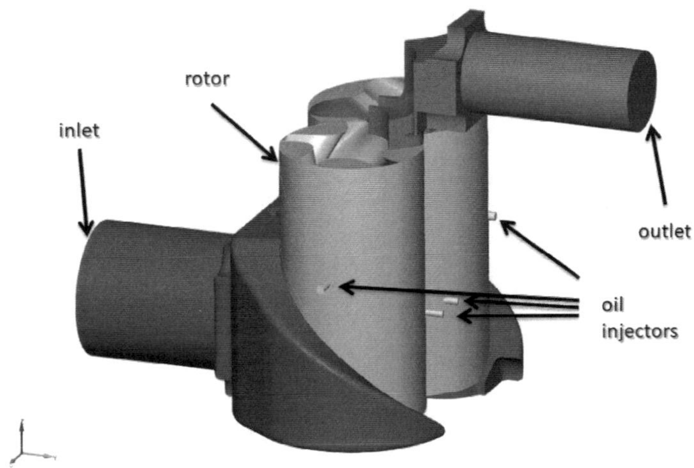

Figure 1: Fluid volume

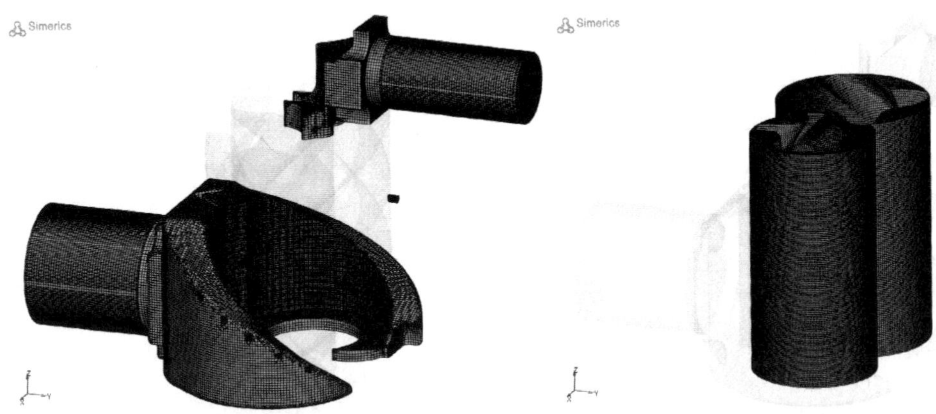

(a)

(b)

Figure 2: Ports and rotor meshes

The axial end clearances can be included in the CFD model without causing divergence or sacrificing running speed.

Table 1: CFD model cases and operating conditions

Case	Male Rotor Speed (rpm)	Suction Pressure (bar)	Suction Temperature (K)	Discharge Pressure (bar)	Oil Injection Pressure (bar)	Oil Injection Temperature (K)
1	3000	1.00	298.0	6.0	5.5	323.0
2	3000	1.00	298.0	8.0	7.5	323.0
3	6000	1.00	298.0	8.0	7.5	323.0

However, they are not included for simplification purpose. The air and oil properties are summarized in Table 2. The air phase is defined as an ideal gas state with molar mass of 28.96 kg/kmol.

Table 2: Gas and oil properties

	Gas	Oil
Density	Air (using ideal gas law)	800 kg/m^3
Specific heat capacity	1004.4 J/kgK	1800 J/kgK
Dynamic viscosity	1.831^{-5} $Pa \cdot s$	0.0088 $Pa \cdot s$
Thermal conductivity	0.0261 W/mK	0.18 W/mK

The discharge pressure is gradually increased to the designed pressure within 4 full male rotor rotations. Results becomes periodic within 15 revolutions for all simulations. Mass imbalance for air is below 3% and for oil phase is within 1.5%. The simulations are carried out at Dell PowerEdge computing nodes @ 2.2GHz. The simulation time is about 2.4 hours per revolution using 6 cores and is reduced to 0.84 hours per revolution on 24 cores.

3.2 Comparison of the Air Flow and Indicated Power with Experimental Results

Table 3 shows both predicted CFD results and experimental results of the air mass flow rate and the indicated power for three cases. It can be seen that for Case1 and Case2 at 3000 rpm the predicted air mass flow rates are below 3% of the experimental values. The air mass flow rate prediction in Case3 is 4% higher than the measured value. Comparing to the flow rate of Case2 at 3000 rpm, the flow rate of Case3 at 6000 rpm has increased by 87% which is slightly less than the ideal increasement factor of 1 due to the leakages and drop in suction air density. The indicated powers predicted by CFD simulations in Case1 and Case2 are 2% and 6% higher than the measured data respectively, while that in Case3 is above 7%. The indicated power in experiments is obtained by multiplying the measured shaft power with an estimated mechanical efficiency of 70%. This could underestimate the indicated power in experiments and is also the possible reason that CFD simulation predict higher powers than those in experiments.

Table 3: Comparison of the flow and indicated power with experimental results

	Air Mass Flow Rate (kg/min)			Indicated Power (kW)		
Case	Experimental result	CFD result	Difference	Experimental result	CFD result	Difference
1	4.73	4.58	-3%	16.31	16.65	2%
2	4.67	4.52	-3%	19.15	20.32	6%
3	8.75	9.10	4%	39.22	42.07	7%

3.3 Pressure-Volume Diagram

A user defined script is employed in the software to track the compression pocket as well as to calculate its volume and averaged pressure. Figure 3 shows the pressure contour of the compression pocket at different male rotor angles for Case2. The color map ranges from 1.0 bar to 8.0 bar. It can be seen that the pressure in the pocket is low when it is initially formed near the suction port and then the pressure increases with the pocket volume being reduced. The compression pocket averaged pressure and volume diagram is also shown in Figure 4. It is observed that for three cases the pressure during the compression process before connecting to the discharge port aligns closely. After the compression pocket connects to the discharge port, it is noted that Case3 with 6000 rpm and 8.0 bar discharge pressure has the highest maximum pressure pulsation value of 10.5 bar, while Case1 with 3000 rpm and 6.0 bar discharge pressure has the lowest pressure pulsation.

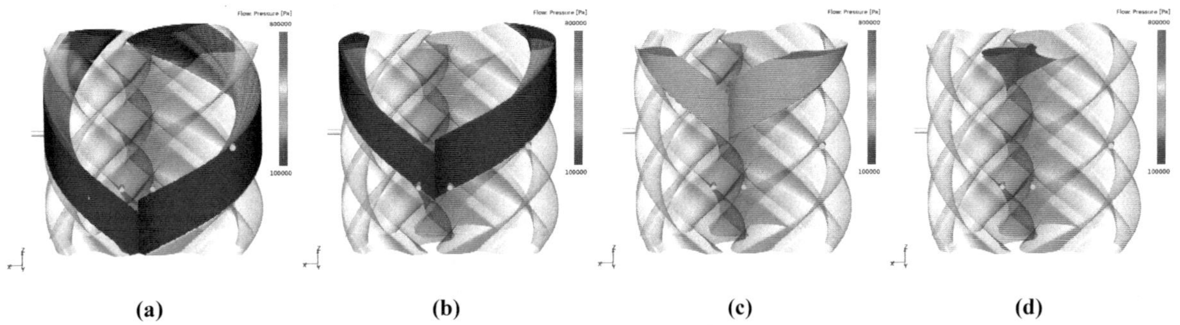

Figure 3: Compression pocket and pocket pressure contour for Case2 at different male rotor angles: (a) 45° (b) 135° (c) 225° (d) 315°

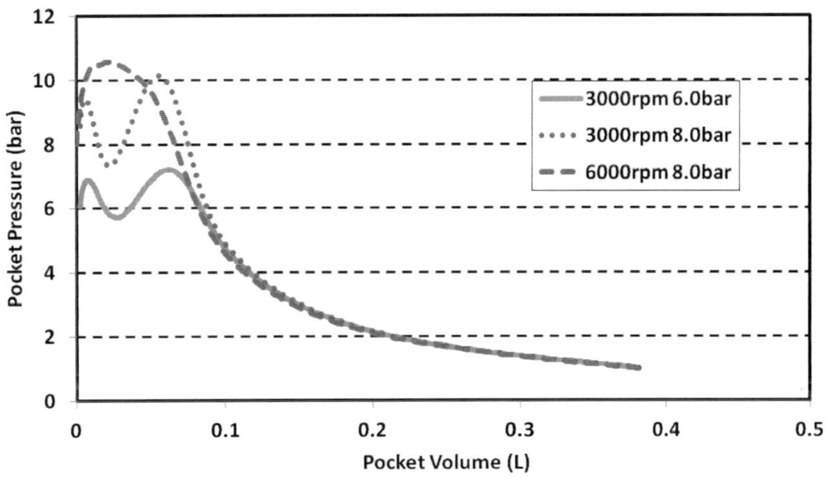

Figure 4: Averaged pressure and volume diagram of compression pocket

3.4 Oil Distribution

Figure 5 shows oil concentration contours at four different male rotor angles for Case2. The color map ranges from 0% to 15% volume fraction of oil with magenta representing high concentration and blue representing low concentration. Oil has higher concentration close to the injection and then been carried away by the movement of pocket. It is also observed that oil is accumulated in the rotor tips and discharge port. Figure 6 plots instantaneous oil flow rate at the discharge port for one complete male rotor cycle after solution becomes periodic. The oil flow rate for Case1 with 3000 rpm and 6.0 bar is lower than that of Case2 with same rotation speed but higher pressure 8.0 bar. Case3 with 6000 rpm and 8.0 bar exhibits bigger oscillation with higher maximum value and lower minimum value comparing to other Cases. This oil distribution variation has a direct impact on the temperature distribution, which will be discussed in detail in the next section.

3.5 Temperature Distribution

Figure 7 shows temperature contour for Case2 at four different male rotor angles. The color map ranges from 290 K to 350 K with magenta representing high temperature and blue representing low temperature. The temperature inside each pocket is highly non-uniform due to the cooling effects of injected oil. By comparing the temperature contour in Figure 7 with the oil distribution contour in Figure 5, it can been seen that typically higher oil concentration areas have lower temperature and lower oil concentration areas have higher temperature. As shown in Figure 8, the averaged temperature of mixture (both oil and air) at the discharge is periodic. Case3 with 6000 rpm and 8.0 bar has the maximum

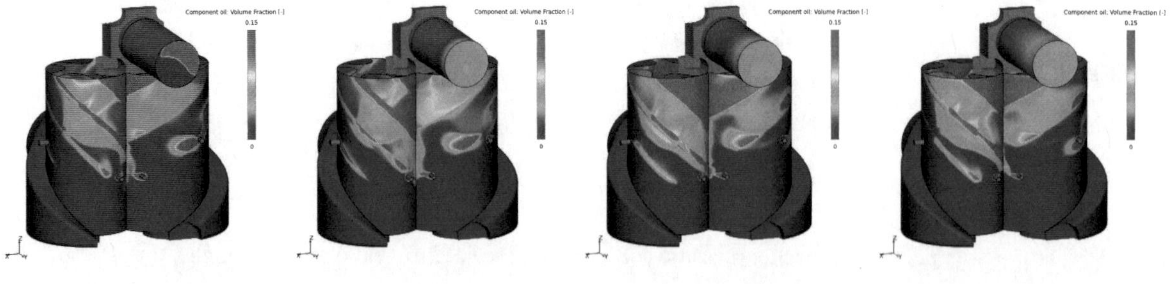

(a) **(b)** **(c)** **(d)**

Figure 5: Oil distribution for Case2 at different male rotor angles: (a) 45° (b) 67.5^o (c) 90° (d) 112.5^o

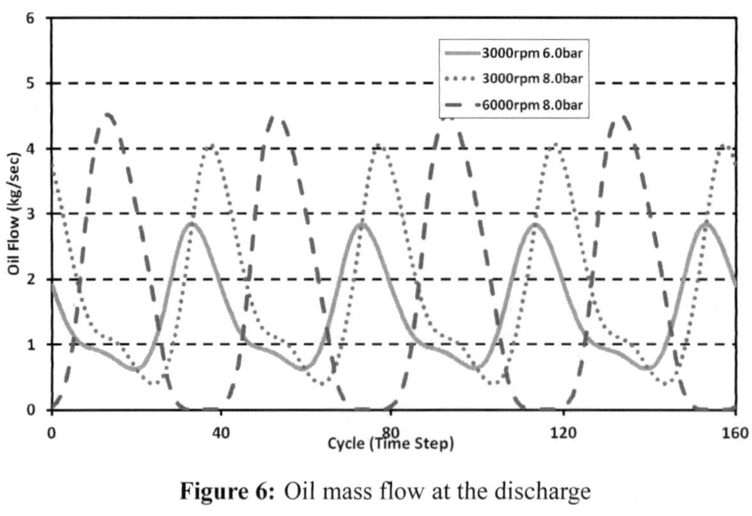

Figure 6: Oil mass flow at the discharge

peak temperature of 370 K around time steps at 0,40,80,120 and 160, which is resulted from the close to zero oil flow at the same time steps as shown in Figure 5. Other than these time steps, the temperature is lower since the oil flow is higher. Case 1 and Case2 exhibit the similar correlation between the oil distribution and temperature distribution, which has demonstrated the oil cooling effects.

There are three parameters that significantly impact the temperature in the compression chamber and discharge; a) Discharge Pressure, b) Rotational speed and c) Oil injection. With increase in both discharge pressure and rotational speed, the gas temperature increases due to higher compression work and lower residence time available for injected oil heat transfer. With increase in oil injection mass the gas temperature decreases. Both these effects have been observed from Figure 6,7 and 8.

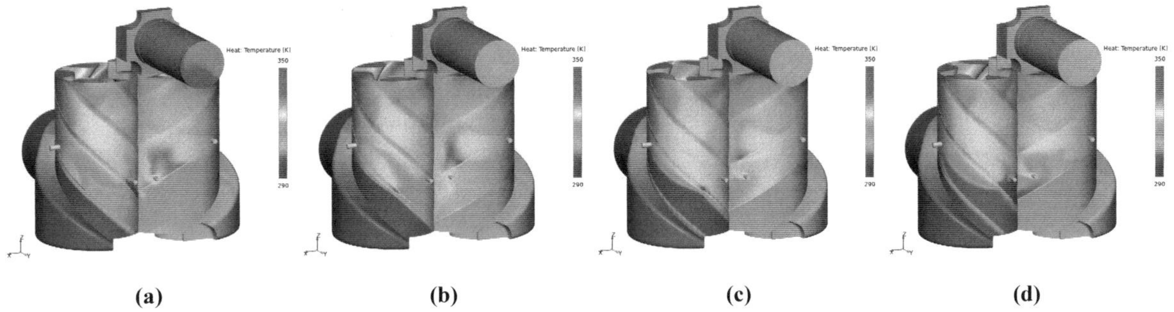

Figure 7: Temperature distribution for Case2 at different male rotor angle: (a) 45° (b) 67.5^o (c) 90° (d) 112.5^o

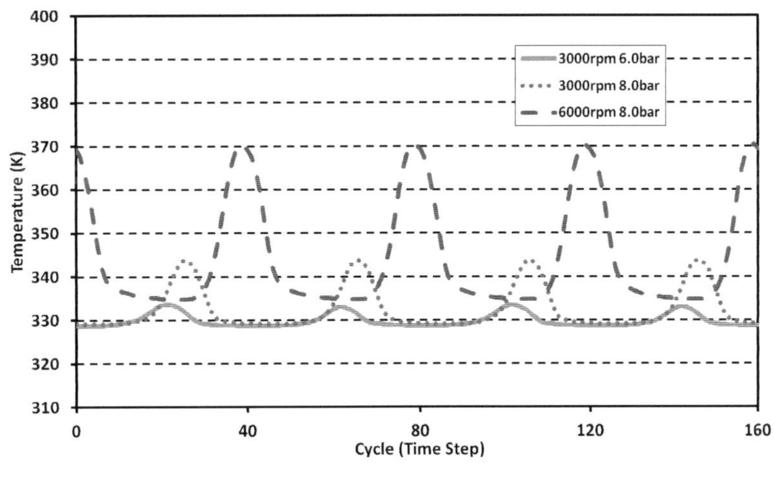

Figure 8: Averaged temperature at the discharge

4. CONCLUSIONS

This paper reports the CFD simulations for a twin screw compressor with oil injection using VOF multiphase model. The rotor mesh is generated by $SCORG^{TM}$ and read into the solver at each time step. Three simulations at different rotor speeds and discharge pressures are performed. The results of gas flow rate and indicated power are compared with the experimental measurements in City, University of London.

- Gas mass flow rates in numerical simulations have a good agreement with the test values in three cases.
- The indicated powers obtained by numerical simulation show a well match with the measured shaft power.
- The averaged pressure in compression pocket is tracked and pressure-volume diagram is also plotted and compared for three cases.

- Oil distribution clearly exhibits the cooling effect on the temperature distribution.
- The solver provides a very efficient tool for numerical simulation and to better understand and improve the design.

NOMENCLATURE

Ω	control volume	(m^3)
σ	surface of control volume	(m^2)
$\mathbf{n}$	surface normal	(-)
ρ	fluid density	(kg/m^3)
p	pressure	(Pa)
$\mathbf{f}$	body force	(N)
$\mathbf{v}$	fluid velocity	(m/s)
$\mathbf{v}_\sigma$	surface motion velocity	(m/s)
$\boldsymbol{\tau}$	shear stress tensor	(-)
μ	fluid viscosity	$(Pa \cdot s)$
F	volume fraction	(-)

Subscript

i	fluid component

REFERENCES

Ding, H., & Jiang, Y. (2017). CFD simulation of a screw compressor with oil injection. *IOP Conference Series: Materials Science and Engineering*, *232*, 012020. Retrieved from https://doi.org/10.1088%2F1757-899x%2F232%2F1%2F012020 doi: 10.1088/1757-899x/232/1/012020

Hirt, C., & Nichols, B. (1981). Volume of fluid (vof) method for the dynamics of free boundaries. *Journal of Computational Physics*, *39*, 201-225.

Kovacevic, A. (2002). Three-dimensional numerical analysis for flow prediction in positive displacement screw machines. *Doctoral Thesis*.

Kovacevic, A. (2005). Boundary adaptation in grid generation for cfd analysis of screw compressors. *International Journal for Numerical Methods in Engineering*, *64*(3), 401-426.

Launder, B., & Spalding, D. (1974). The numerical computation of turbulent flows. *Computer Methods in Applied Mechanics and Engineering*, *3*, 269-289.

Rane, A., S.R. Kovacevic, & Stosic, N. (2016). Cfd analysis of oil flooded twin screw compressor. *International Compressor Engineering Conference*.

Rane, S. (2015). Grid generation and cfd analysis of variable geometry screw machines. *Doctoral Thesis*.

Stosic, N. (1998). On gearing of helical screw compressor rotors. *Proceedings of the Institution of Mechanical Engineers, Part C: Journal of Mechanical Engineering Science*, *212*, 587-594.

Application of Computational Fluid Dynamics to the Lubrication Study of an Oil-injected Screw Compressor

Sheng-Hung Hsieh*, Jin-Wei Zhang, Chi-Shun Huang

Hanbell Precise Machinery Co. Ltd., Taiwan
Taoyuan City, Taiwan (R.O.C.)

* shhsieh@hanbell.com

ABSTRACT

Computational fluid dynamics and mesh generation tools have been well developed, and could be used to calculate the performance of an oil-injected screw compressor. Designers could now effectively obtain reliable results of efficiency, temperature, force and torque. Some physical phenomena inside a screw compressor are not easy to be observed through experiments, such as the oil distribution and the oil film thickness on the rotor surface. Under the ideal lubrication condition, the oil film of proper thickness should be maintained on the local surfaces which are going to contact with each other. This could be numerically explored by computational fluid dynamics. The first case in this study showed the effect of centrifugal force on a thin free surface flow on the rotor surface. Designers could graphically understand how the oil film flows on the rotor surface when rotors separately rotate without meshing with each other and doing the compression work. The second case in this study was the rotor lubrication. The clearance distribution between rotors in the actual contact area was designed by the minimum film thickness and was 1 μm. The pressure gradient on the rotor surface was used to see if the hydrodynamic pressure appeared. Designers could optimize the design of oil injection not only by considering efficiencies, but also by analyzing the pressure gradient and the oil film distribution on rotor surfaces.

1. INTRODUCTION

The gas is compressed by a pair of rotors in an oil-injected screw compressor. During the compression process, the proper quantity of oil is injected into the compressor to reduce the leakage flow rate, and to cool the compressed gas temperature. The rotors are also lubricated by the oil for the long-time operation. The designer could effectively predict the performance of a screw compressor by using the well-developed thermodynamic model and computational fluid dynamics (CFD) tools. The pressure curve, efficiencies, the gas power consumption, the gas force, and the gas torque could be accurately and efficiently calculated. However, there were few recent studies that focused on the rotor lubrication in the screw compressor. In the previous studies of the bearing lubrication, there were rich references to the theoretical model and the experimental verification.

The rotors in the screw compressor could be considered as two cylinders which were in rotational contact with each other. Hydrodynamic lubrication or Elastohydrodynamic lubrication (EHD or EHL) would appear in the contact area. The pressure of oil rises rapidly. The pressure gradient in the direction of rotation on the cylinder surface would be significant positive followed by negative (Zhang *et al*., 2019). The pressure distribution in the contact area would match the elastic Hertzian contact pressure distribution (Khonsari and Booser, 2008). Under the condition of flooded EHD, Dowson (1970) gave the equations to estimate the minimum film thickness. When the insufficient quantity of oil was in the contact area, the starved EHD appeared. Kostal *et al*. (2015, 2017) established the equipment to accurately measure the meniscus position and the oil film thickness. They experimentally examined the parameters in the equation of film thickness reduction. The parameters were presented by the study of point contact lubrication (Chevalier et al., 1998), and the study of elliptical contact lubrication (Damiens et al., 2004). Compared with the meniscus position as the criterion for determining the starvation level, the film thickness in the inlet area could be applied to identify the starvation level more clearly in the severely starved case, which film thickness reduction was less than 0.5 and the meniscus position was very close to the Hertzian contact circle. Kostal *et al*. (2017) presented that the inlet film thickness was close to the central film thickness under severely starved lubrication. When the starvation level was improved, the film thickness reduction curves of different contact types would grow divergently. In the studies of the effect of centrifugal force on the liquid movement, Zamfirescu et al. (2004) numerically analyzed the liquid drop in the compression chamber. The liquid drop was affected by the centrifugal force and the gas flow, and filed away from the rotor center to stay on the housing and rotors' tips. Zoelen et al. (2008) studied the grease film on raceways of the taper roller bearing and the spherical roller bearing by experimental measurement and numerical analysis. Under the influence of centrifugal force, the transient film thickness distribution curves on these two raceways were different. These studies showed that the surface film thickness was a key parameter of the rotor lubrication. In order to achieve the good lubrication on rotors which were in rotational contact with each other, the oil might first fly to the rotor surface and then formed an oil film. The oil film driven by the centrifugal force, the air flow velocity, and the pressure difference would flow on the rotor surface. The oil film of proper thickness should be maintained on the local surfaces which were going to contact with each other. Finally, the hydrodynamic lubrication or EHD would appear in the contact area.

The goal of this study was to apply CFD to the lubrication study of an oil-injected screw compressor. In the grid model, the minimum film thickness calculated by the equation provided by Dowson (1970) was used to be the clearance between rotors in the actual contact area. TwinMesh was used to generate the dynamic grid model of the compression chamber. The built-in multi-phase flow model and free surface model in ANSYS CFX were used to calculate the oil distribution on rotor surfaces. Two cases were presented in this study. In the first case, rotors rotated separately. The transient oil distributions on rotor surfaces under the influence of centrifugal force were analyzed. In the second case, the six CFD models with different positions of oil injection ports were analyzed. The time-average axial distribution curves of the average volume fraction of oil on rotor surfaces were presented. The pressure gradient on rotor surface was used to see if the hydrodynamic pressure appeared, and would be one of the design objectives of the oil injection.

2. CFD MODEL AND GRID MODEL

The CFD model used in this study was based on an oil-injected screw compressor, which displacement is 260 m^3/hr at 3,000 rpm. Figure 1 (a) shows the solid model of the compressor which is constructed by the inlet part, a pair of "5x6" rotors, the outlet part, and the oil port(s). In order to reduce the physical calculation time, the simplified model of the compressor was adapted, as shown in Figure 1 (b). The continuous equations, momentum equations, and energy equations for the turbulence flow were solved by ANSYS CFX. The built-in inhomogeneous model and free surface model in ANSYS CFX were used to describe the multi-phase flow, and to present the oil distribution on rotor surfaces. The selected models and schemes in ANSYS CFX are listed in Table 1 (Rane *et al.*, 2016 and Basha *et al.*, 2018). The flow paths were divided into the dynamic and the static grid models. The dynamic grid model of compression chambers is shown in Figure 2 (a), and was generated by TwinMesh. The edge size of elements on the casing was about 2 mm. The edge size of elements on the interface between rotors was about 1 mm. The radial element number was 6. The first radial element height close to the rotor surface was about 0.1 mm. The axial element size was about 2 mm. There were about 310,000 elements in the dynamic grid model. The static grid models of the inlet part, the outlet part, and the oil port(s) are shown in Figure 2 (b). There were about 130,000 elements in the static grid models.

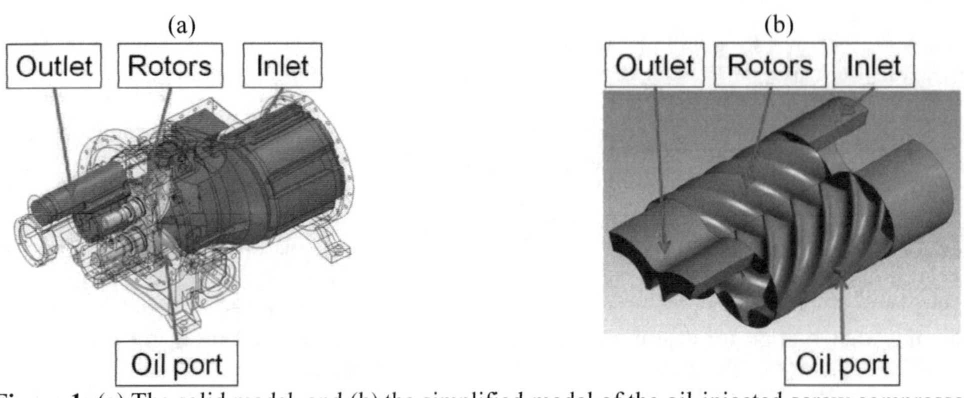

Figure 1: (a) The solid model, and (b) the simplified model of the oil-injected screw compressor

Table 1. The selected models and schemes in ANSYS CFX

Models and Schemes	Selection
Turbulence Model	SST
Multiphase Model	Inhomogeneous Model Free Surface Model
Advection Scheme	High Resolution
Turbulence Scheme	First Order Upwind
Transient Scheme	Second Order Backward Euler
Convergence Criterion	10^{-3} Max Residual Level

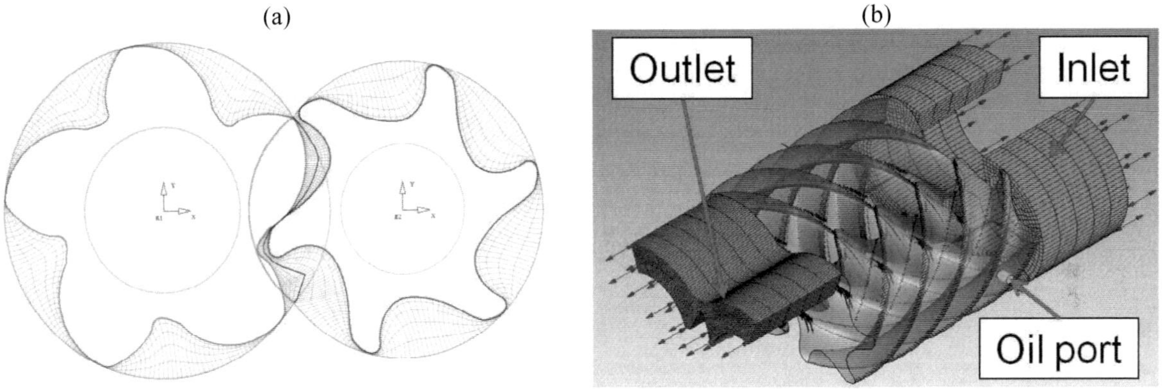

Figure 2: (a) The dynamic grid model, and (b) the static grid models of flow paths

The non-contact rotor profiles were needed to generate the dynamic grid model. Figure 3 (a) shows the concept design of clearance distribution between rotors. The proper clearance distribution should be designed on the rotor profiles to ensure that rotors rotated smoothly. The clearance was 0 μm on the parts of rotor profiles which were in contact with each other, and was 16 μm on the parts of rotor profiles which were not in contact with each other. When the rotors were lubricated by the oil, there was a thin oil film in the actual contact area. The minimum oil film thickness could be calculated by Equation (1) (Dowson, 1970). The material parameter, the speed parameter, and the load parameter in Equation (1) were all listed in Table 2 and 3. The minimum oil film thickness was affected by the relative speed between rotors, material properties of oil and rotors, curvatures of surfaces, and the condition of load. In this study, the actual contact area was design on the profiles where the relative velocities between rotors were between 3 and 5 m/s. The clearance between rotors in the actual contact area was 1 μm. This ideal design of clearance distribution between rotors is shown in Figure 3 (b). In addition, the dynamic viscosity would be changed rapidly by pressure, and was given in Equation (2).

$$h_o/R = 2.65G^{0.54}U^{0.7}/W^{0.13} \tag{1}$$

$$G = \alpha E'$$

$$U = (\mu_0 u)/(E'R)$$

$$W = w/(E'LR)$$

$$\mu = \mu_o \exp(\alpha \cdot P) \tag{2}$$

μ_o : the dynamic viscosity at atmospheric (essentially zero) pressure

α : the pressure-viscosity coefficient

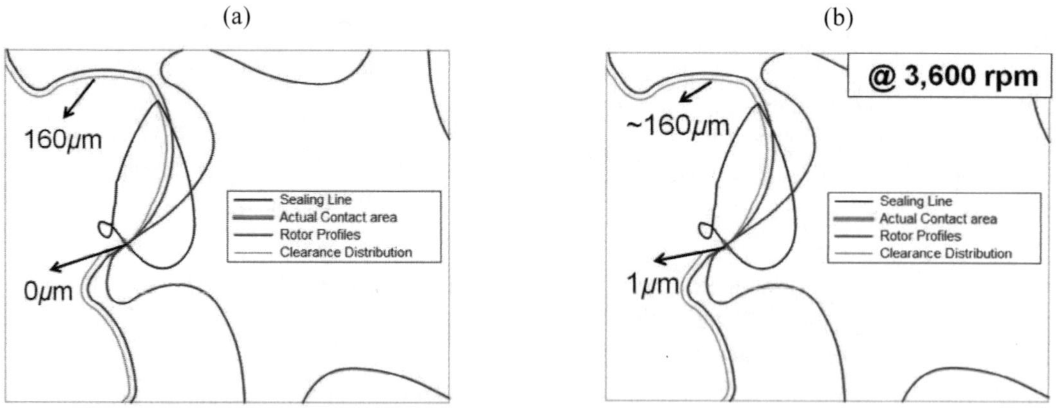

Figure 3: (a) The concept design, and (b) the ideal design of clearance distributions between rotors

Table 2. Material properties of fluids

Material Properties	**Unit**	**Value**
Primary fluid	--	R134a (ideal gas)
Secondary fluid	--	Oil
Density of oil	g/cc	0.92
Dynamic viscosity of oil (μ_o)	cP	36.13
Surface tension of oil	N/m	0.03
Contact angle of oil	--	45°
Roelands' pressure-viscosity index (Z)	--	0.63
Pressure-viscosity coefficient (α)	Gpa^{-1}	0.02

Table 3. The minimum film thickness and the contact area width

Parameter	**Unit**	**Value @ 3,600 rpm**		
Relative speed	m/s	1~3	3~5	5~7
Minimum Film Thickness(h_o)	mm	0.0006	0.0010	0.0015
Contact area width ($2b$)	mm	0.138	0.138	0.151
Max. Hertzian contact pressure (P_o)	MPa	241.2	226.5	212.1
h_o/R	$x10^5$	3.869	6.118	7.448
Material Parameter (G)		4,454.2	4,454.2	4,454.2
Pressure-viscosity coefficient (α)	Gpa^{-1}	0.020	0.020	0.020
Young's modulus (E)	N/mm^2	202,000	202,000	202,000
Poisson's ratio (γ)		0.280	0.280	0.280
E'	N/mm^2	219,184	219,184	219,184
Speed Parameter (*U*)	$x10^{11}$	2.121	3.988	5.154
Dynamic viscosity (μ_o)	cP	36.13	36.13	36.13
u	mm/s	2,024.0	4,047.1	6,095.6
R_1	mm	30.3	30.3	30.3
R_2	mm	32.7	37.4	54.8
R	mm	15.7	16.7	19.5
Load Parameter (*W*)	$x10^6$	7.609	6.712	5.883
w	N	426.8	440.5	455.8
L	mm	16.3	17.9	18.1

3. RESULT AND DISCUSSION

There were two cases discussed in this section. In the first case, rotors rotated separately. The transient distribution of the volume fraction of oil on the rotor surface under the influence of the centrifugal force was analyzed. In the second case, six CFD models with different oil injection positions were analyzed by considering the CFD model and grid models presented in Section 2. The time-average axial distribution curves of the average volume fraction of oil on rotor surfaces were presented. The pressure gradient on the rotor surface was used to see if the hydrodynamic pressure appeared. The detailed calculation conditions and results of these two cases were explained and discussed in following sub-sections.

3.1 Effect of Centrifugal Force on Oil Film Flow on Rotor Surfaces

In this case, there was the thin film of uniform thickness on the rotor surface before the rotor rotation. Figure 4 shows the oil film domain and the gas domain in the CFD model. The oil film domain was full of oil to be the 50 μm thick thin film as the initial condition. The rotational wall (A) described the rotor surface in the CFD model. The opening type boundary condition (B) described the other surfaces in the CFD model. The rotational speed of the male rotor was 3,600 rpm, and the one of the female rotor was 3,000 rpm. Figure 5 shows the transient distribution of the volume fraction of oil on rotor surfaces when rotors have separately rotated 0, 1/8, 1/4, 1/2, and 1 turn. The oil film moved on the rotor surface toward the tip of rotor and left the rotor surface. Figure 6 shows the axial distribution curves of the average volume fraction of oil on rotor surfaces. These axial distribution curves were considered to be flat, and decreased significantly near the outlet end of the male rotor and the inlet end of the female rotor. Although the rotational speed of the male rotor was faster than the one of the female rotor, the average value of the distribution curve on the male rotor decreased a little slower than the one on the female rotor.

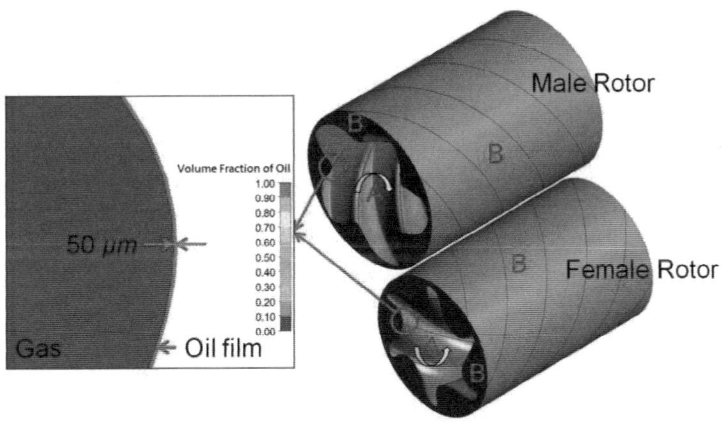

Figure 4: The initial condition and boundary conditions are used in the analysis of the effect of centrifugal force on oil film flow on rotor surfaces. "A" is the rotational wall. "B" is the opening type boundary condition.

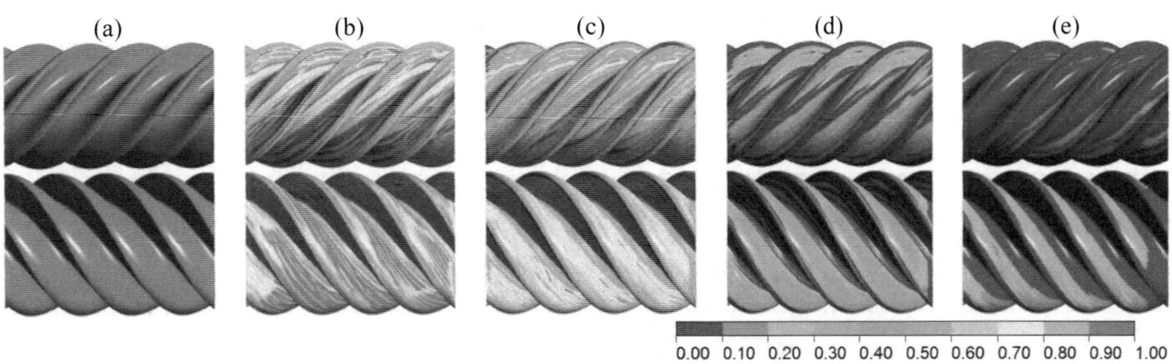

Figure 5: These contour graphs show the transient distributions of the volume fraction of oil on rotor surfaces when rotors have separately rotated (a) 0 turn, (b) 1/8 turns, (c) 1/4 turns, (d) 1/2 turns and (e) 1 turn.

(a) (b)

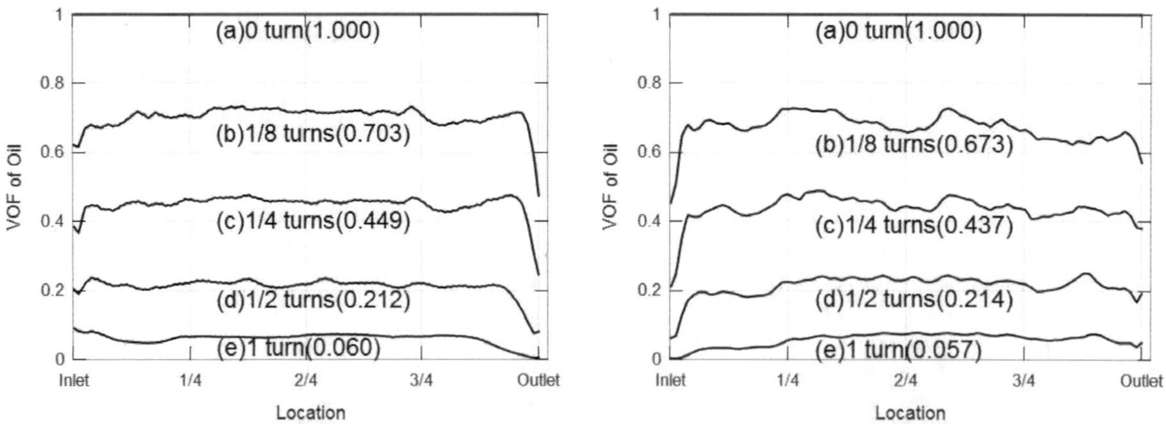

Figure 6: The axial distribution curves of the average volume fraction of oil on (a) the male rotor surface and (b) the female rotor surface

Figure 7 shows the distribution curves of the volume fraction of oil on tooth profiles in the middle of rotors. The wavy curves occurred when rotors rotated. Figure 7 (a) shows the distribution curves on the tooth profile in the middle of the male rotor. The average absolute radius gradient along the short side of the male rotor profile was 0.71, and the one along the long side of the male rotor profile was 0.41. The amplitude of the wavy curve on the short side was larger than the one on the long side. The values of distribution curves on the short side decreased faster than the ones on the long side. Figure 7 (b) shows the distribution curves on the tooth profile in the middle of the female rotor. The average absolute radius gradient along the short side of the female rotor profile was 0.76 mm, and the one along the long side of the female rotor profile was 0.55 mm. The amplitudes of the wavy curve on both sides of the female rotor profile were quite the same. The values of distribution curves on the short side decreased a little faster than the ones on the long side. When rotors have rotated one turn, the values of distribution curves on tooth flanks of both rotors were higher than the ones on roots and tips of both rotors.

(a) (b)

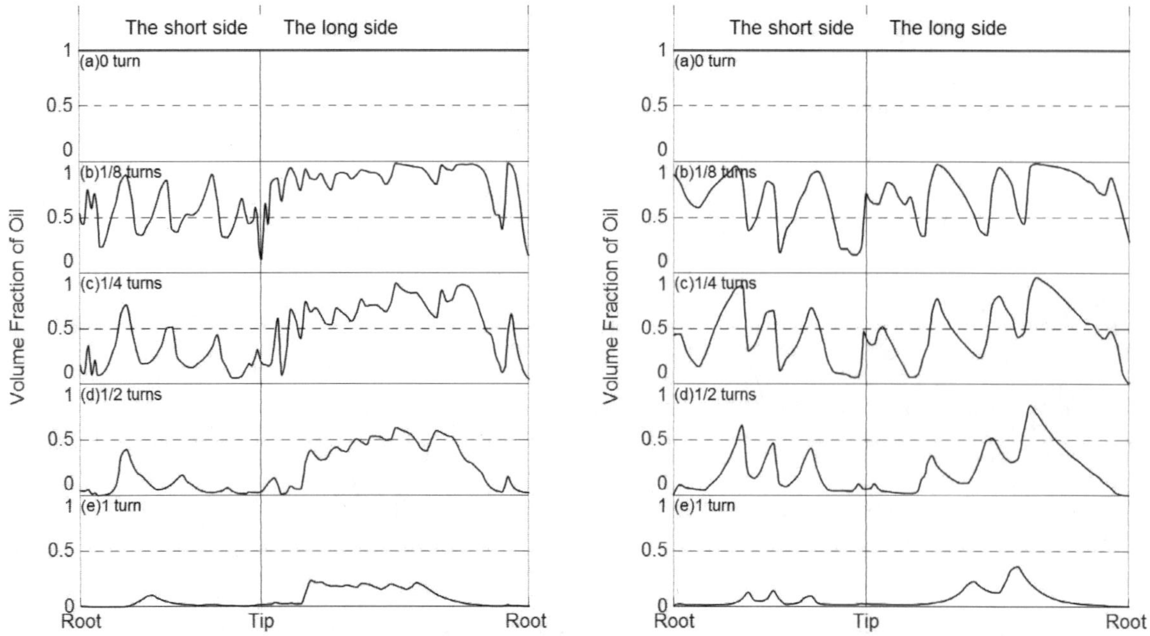

Figure 7: The distribution curves of the volume fraction of oil on tooth profiles in the middle of (a) the male rotor and (b) the female rotor

3.2 Oil Film Distribution and Pressure Gradient on Rotor Surfaces

In this case, the different oil injection ports were considered in six CFD models. The inlet pressure was 3.6 bar. The outlet pressure was 9.5 bar. The inlet temperature was 15 °C. The outlet temperature was 65 °C. Figure 8 (a) shows the positions of oil injection ports and six CFD models. The injection ports named M1 and F1 were located on the 1/4 length of rotor from the inlet end. The injection ports named M2 and F2 were located on the middle of rotors. The injection ports named M3 and F3 were located on the 1/4 length of rotor from the outlet end. The names of CFD models meant the oil was only injected through the corresponded named port(s). The oil flow rate of each injection port was 20 Litre/min. Figure 9 shows the contour graphs of the volume fraction of oil on rotor surfaces. After the oil was injected into the compression chamber, it would be collided with rotor surfaces. The oil would spread on surfaces of both rotors and be transported toward the outlet end. The oil on surfaces of both rotors would be transferred to each other in the meshing area. The oil distribution also changed for the centrifugal force, the gas flow, and the gas pressure difference. In the CFD model named M2, the oil was injected through the port named M2 on the side of the male rotor. Part of oil flowed to and was transferred to the female rotor surface. In the CFD model named F2, the oil was injected through the port named F2 on the side of the female rotor. Less of oil flowed to and was transferred to the male rotor surface.

The time-average axial distribution curves of the average volume fraction of oil on the entire rotor surface are shown in Figure 10 (a), and the ones on the local surface in the rotor meshing area are shown in Figure 10 (b). The rotor meshing area was between the cusps of the casing, where were 56° and 123° in the angle coordinate system shown in Figure 8 (b). Figure 10 shows that low values of distribution curves were between the oil injection port and the inlet end. All peak values of distribution curves appeared near the positions of oil injection ports. There was more oil collided with the male rotor surface than with the female rotor surface. Between the oil injection port and the outlet end, the distribution curves gradually dropped from peak points to become flat curves. Figure 10 (b) shows that there were noticeable distribution curves of the CFD model named M1F1. The positions of oil injection ports in this CFD model were closer to the inlet end than the ones in the other CFD models. The oil was widely distributed on the local surface in the rotor meshing area.

CFD Model	Oil Port	
	M	F
M2	2	X
F2	X	2
M1F1	1	1
M2F2	2	2
M3F3	3	3
M1F3	1	3

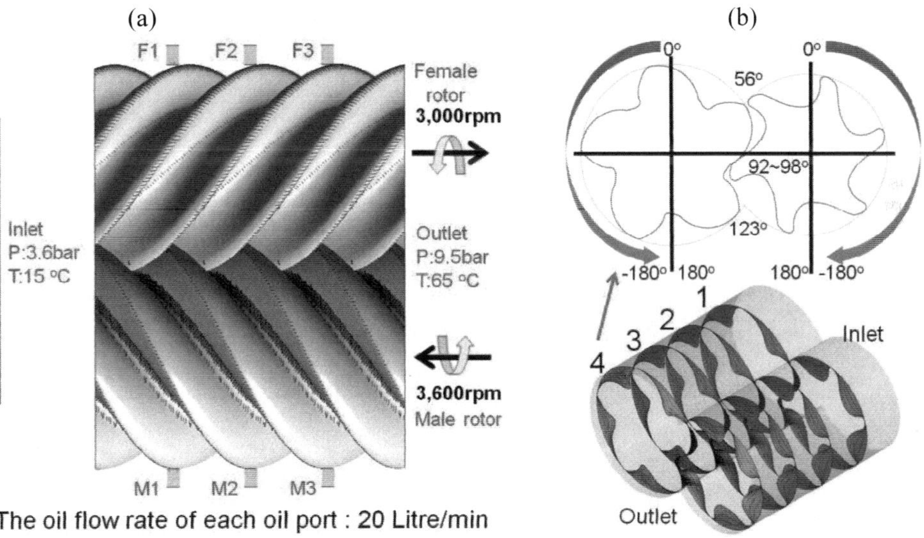

Figure 8: (a) Positions of oil injection ports and six CFD models, and (b) positions of four cross sections and the angular coordinate system

The pressure gradient along the rotor profile in the rotation direction on the rotor surface could be analyzed to understand if a good lubrication condition was in the actual contact area between rotors. Figure 11 shows the pressure gradient curves on male rotor surfaces of six CFD models at the same calculation time steps. These curves were obtained on four cross sections where rotors were actually in rotational contact with each other. The positions of these four cross sections and the angular coordinate system are shown in Figure 8 (b). The cross section 1 was near the oil injection ports named M1 and F1. There were the same correspondences for the cross sections 2 and 3 to the oil injection ports. The cross section 4 was near the outlet end. The actual contact area was between 92° and 98°.

The minimum clearance between rotors on these four cross sections all appeared at 96.21°. The values of pressure gradient curves shown in Figure 11 slightly changed between 84° and 94°. Peak values of most of pressure gradient curves appeared between 94° and 96°. The pressure would grow to the peak value when the pressure gradient curve drops from the peak to the zero. Most of these zero points of pressure gradient curves shown in Figure 11 appeared near 96°. However, the peak values of pressure did not match the Hertzian pressure. These curves only showed the phenomenon of hydrodynamic lubrication. Phenomenon of EHD did not appear in the actual contact area. One of the possible reasons was that the proper quantity of oil was not maintained in the actual contact area. Another one was that the edge size of element was larger than the actual contact area in the grid model. The spatial resolution was not good enough to show the pressure variation of EHD. Peak values of pressure gradient curves recorded in Figure 11 could be used to identify if the hydrodynamic pressure appeared near the actual contact area. In CFD models named M2, F2, M2F2 and M3F3, there were no obvious hydrodynamic pressure appeared on the cross section 1s. In CFD models named M1F1 and M1F3, the obvious hydrodynamic pressures appeared on four cross sections. The above analysis showed that the designer could check if the proper quantity of oil was maintained in the actual contact area based on the physical phenomenon of lubrication.

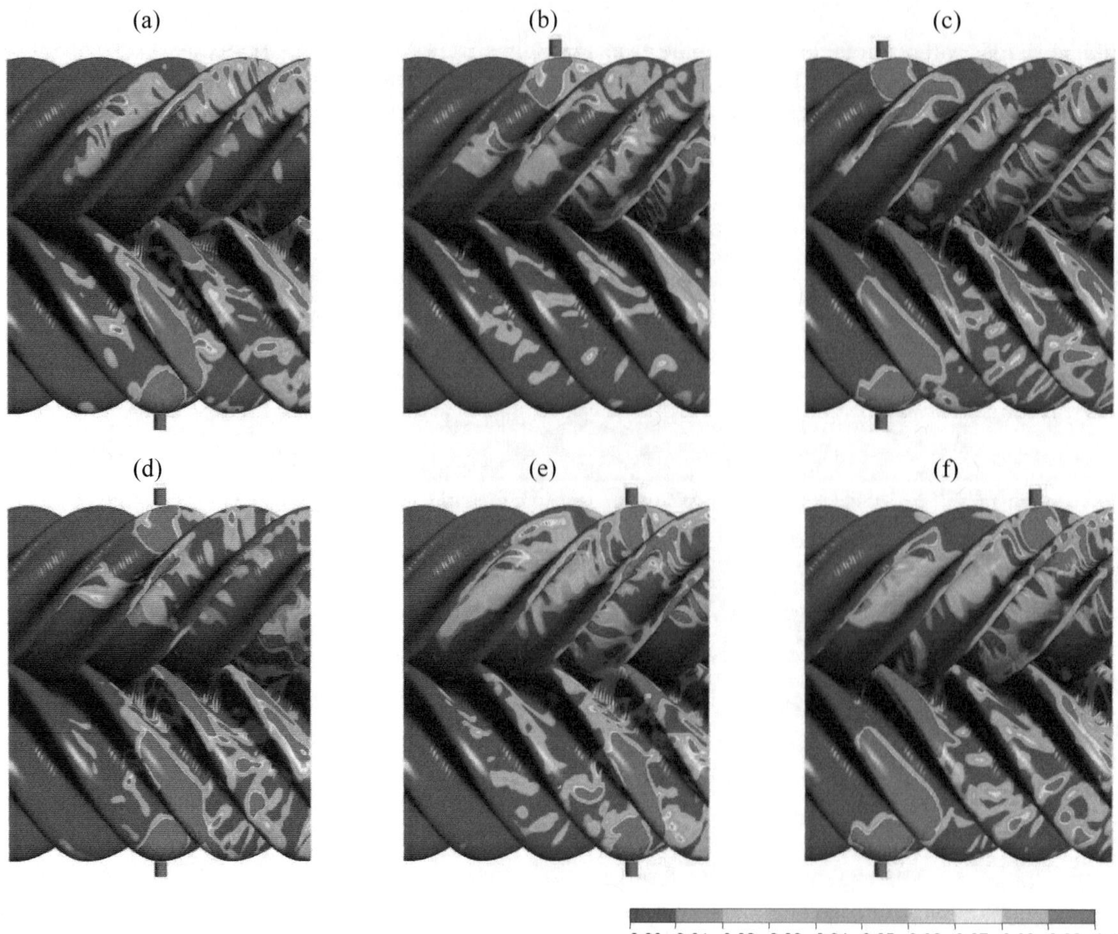

Figure 9: The contour graphs show the distributions of volume fraction of oil on rotor surfaces of CFD models named (a) M2, (b) F2, (c) M1F1, (d) M2F2, (e) M3F3, and (f) M1F3.

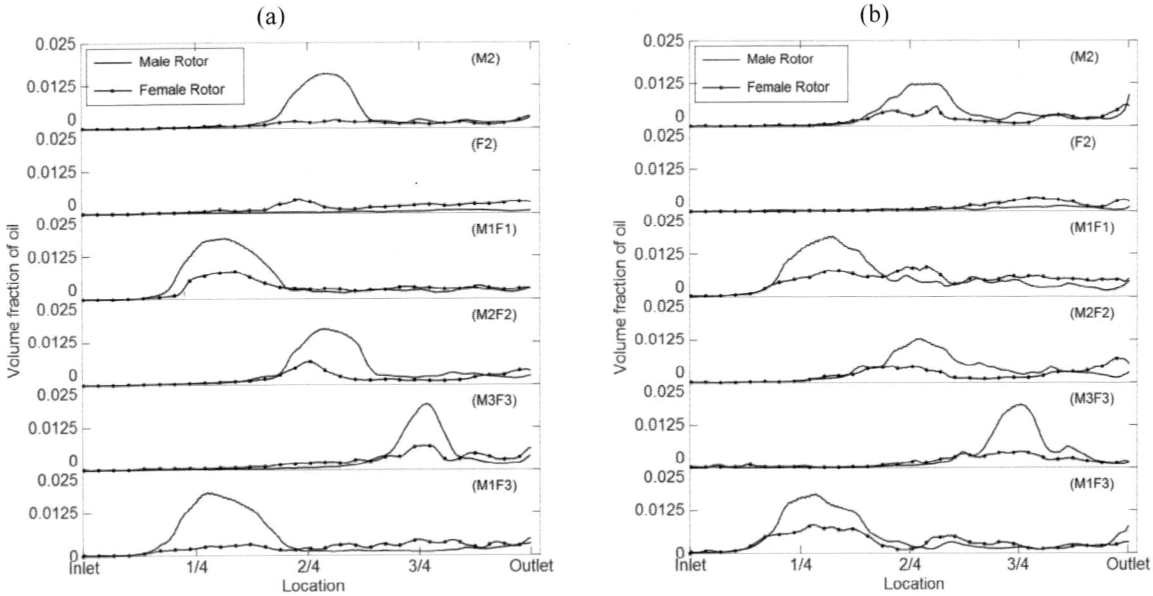

Figure 10: The time-average axial distribution curves of the average volume fraction of oil on (a) the entire rotor surface, and (b) the local surface in the rotor meshing area

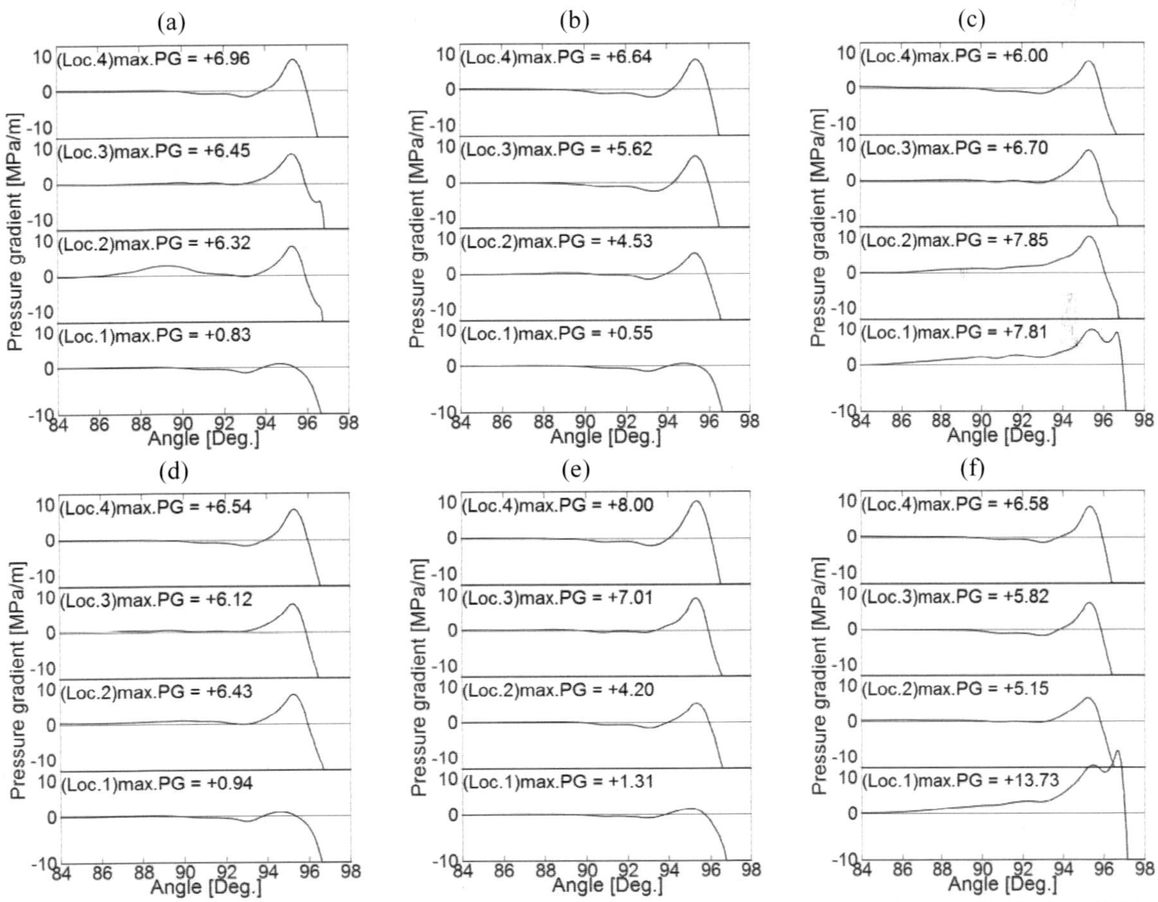

Figure 11: The pressure gradient curves on male rotor surfaces of CFD models named (a) M2, (b) F2, (c) M1F1, (d) M2F2, (e) M3F3, and (f) M1F3.

4. CONCLUSIONS

The ideal CFD model of an oil-injected screw compressor was applied to the study of the rotor lubrication. In the dynamic grid model, the clearance distribution between rotors in the actual contact area was designed by the minimum film thickness, and was 1 μm. The value of this clearance would be affected by the relative speed between rotors, material properties of oil and rotors, curvatures of surfaces, and the condition of load.

The surface oil moved toward the rotor tip and left the rotor surface as the rotor rotation. Although the rotational speed of the male rotor is faster than the one of the female rotor, the average volume fraction of oil on the male rotor surface decreased a little slower than the one on the female rotor surface. The wavy distribution appeared on the part of profile of large radius gradient. The distribution curves on the short side of the tooth profile decreased faster than the one on the long side of the tooth profile. When rotors have rotated one turn, the values of distribution curves on tooth flanks were higher than the ones on roots and tips of rotors.

The axial distribution curves of the volume fraction of oil on rotor surfaces were shown in this study. The pressure gradient on the rotor surface was used to see if the hydrodynamic pressure appeared. In order to achieve a good rotor lubrication, the proper quantity of oil should be maintained on the local surfaces which were going to contact with each other. The pressure gradient and the oil distribution on rotor surfaces would be design objectives of the oil injection, and could be effectively obtained by CFD.

In the future studies of rotor lubrication in the oil-injected screw compressor, the central oil film thickness could be considered in the actual contact area of rotors. The grid model should be improved to obtain a fine spatial resolution in the actual contact area to present the pressure variation of EHD, and to be applied to address the sensitivity of the clearance distribution to the rotor lubrication.

REFERENCES

Basha, N., Rane, S., & Kovacevic, A. (2018). Multiphase Flow Analysis in an Oil-injected Twin Screw Compressor. *Proceedings of the 3rd World Congress on Momentum, Heat and Mass Transfer. doi: 10.11159/icmfht18.132.*

Chevalier, F., Lubrecht, A. A., Cann, P. M. E., Colin, F., & Dalmaz, G. (1998). Film Thickness in Starved EHL Point Contacts. *Journal of Tribology, 120(1)*, 126–133.

Damiens, B., Venner, C. H., Cann, P. M. E., & Lubrecht, A. A. (2004). Starved Lubrication of Elliptical EHD Contacts. *Journal of Tribology, 126(1)*, 105–111.

Dowson, D. (1969). Elastohydrodynamic Lubrication. In *Interdisciplinary approach to the lubrication of concentrated contacts: proceedings of a NASA-sponsored symposium held July 15-17, 1969, in Troy, New York*. Washington: Scientific and Technical Information Division, National Aeronautics and Space Administration.

Khonsari, M. M., & Booser, E. R. (2008). *Applied Tribology: Bearing Design and Lubrication, Second Edition*. Chichester, England: Wiley.

Kostal, D., Sperka, P., Svoboda, P., Krupka, I., & Hartl, M. (2015). Experimental Observation Of Elastohydrodynamically Lubricated Contacts Replenishment. *MM Science Journal, 2015(03)*, 640–644.

Kostal, D., Sperka, P., Svoboda, P., Krupka, I., & Hartl, M. (2017). Influence of Lubricant Inlet Film Thickness on Elastohydrodynamically Lubricated Contact Starvation. *Journal of Tribology*, *139*(5).

Rane, S., Kovačević, A., & Stošić, N. (2016). CFD Analysis of Oil Flooded Twin Screw Compressors. In *the 23rd International Compressor Engineering Conference at Purdue*.

Zamfirescu, C., Nannan, N., Marin, M. & Infante Ferreira, C.A., (2004). Oil free two phase ammonia (water) compressor. *Final report of NOVEM project BSE-NEO 0268.02.03.03.0002, Delft University of Technology, The Netherlands [report K-336]*.

Zoelen, M. T. V., Venner, C. H., & Lugt, P. M. (2008). Free Surface Thin Layer Flow on Bearing Raceways. *Journal of Tribology*, *130*(2).

A CFD Investigation of the Suction Process in a Screw Compressor With Variations in Port Shape and Rotor Speed

Matt CAMBIO[1*], Scott BRANCH[2], Alberto SCALA[3], Jon HARRISON[4]

[1*]Trane Technologies, Compressor Development,
La Crosse, WI, US,
608-787-3571, mcambio@trane.com

[2]Trane Technologies, Compressor Development,
La Crosse, WI, US,
608-787-3881, Sbranch@trane.com

[3]Trane Technologies, Compressor Development,
La Crosse, WI, US,
608-787-2717, Alberto.Scala@tranetechnologies.com

[4]Gamma Technologies LLC, Staff Engineer - Business Development,
Westmont, IL, US,
630-207-8603, j.harrsion@gtisoft.com

* Corresponding Author

ABSTRACT

Computational fluid dynamics (CFD) has been widely applied to screw compressors. CFD has allowed engineers to gain a better understanding of the complex flows that occur inside the screw compressor. In many cases, the details revealed are only possible with CFD compared to 1-D simulations. Both CFD and 1D can reduce development costs and as in this case, guide decisions on the choice of hardware to be tested. This paper presents the flow details of the suction process in a screw compressor and the effects on performance. The cases presented are a twin screw rotor set applied to two different suction port designs. One suction port matches the helix angle of the rotors and closes at the maximum suction volume. The other suction port stays open for 20 degrees of rotation past the maximum suction volume and does not match the rotor helix angle. The paper presents the effects of the geometry difference at various rotor rotational speeds.

1. INTRODUCTION

Simulation of compressors has long been a method of reducing development time through experimentation. Schwerzler and Hamilton (1972) demonstrated a technique of modeling compressor valves as a systems of orifices and concluded that the ability to do parametric studies based on geometry would save weeks of testing. Sauls (1988) concluded through computer simulation the engineer could optimize screw compressor slide valve geometry before the compressor is built, saving costly hardware iterations. In addition, Sauls (2011) presented the use of high fidelity simulations such as computation fluid dynamics (CFD) and finite element analysis (FEA) to calibrate and improve 1-D simulations in order to maintain speed and flexibility of the design process.

A CFD simulation was performed to determine the impact of a delayed suction port closure vs. a design without suction delay which closed completely when maximum pocket volume was reached. The expected result of the delayed suction port closure was a loss in volumetric efficiency, but the CFD simulation contradicted expectations. The CFD showed flow continued to enter the pocket after the volume was decreasing resulting in improved volumetric efficiency. A review of literature showed that this effect is understood in positive displacement compressors. It is referred to as inertial charging.

Inertial charging was described by Redfield (1912) for reciprocating air compressors as an increase in cylinder pressure at the end of an intake stroke due to the inertia of a moving air column. Cui and Sauls (2006) presented a

CFD analysis of a scroll compressor which indicated a pre-compression of the gas during the suction process. Kameya et al. (2011) performed CFD and verified through testing that improvements to a screw compressor's volumetric efficiency can be achieved by delaying the closing of the suction.

Inertial charging is also a phenomenon widely seen in IC engines and is a major contributor to volumetric efficiency. This is in fact a requirement to predict inertial charging when using 1-D simulation for analyzing IC engines. Wu, et al (2019) showed optimization of the gas exchange process in a 2 stroke engine, and Sahoo (2014) which showed the influence of intake runner geometry on maximum air trapping in the cylinder.

A new screw compressor rotor set was designed that did not match the existing production rotor housing's helix angle and timing of the suction port closure. This paper presents both a CFD analysis and 1-D analysis of inertial charging in a screw compressor and the effects of speed. This is significant in a high speed compressor design since the effect is magnified.

2. DESIGN STUDY

2.1 Suction Port Geometry

A screw compressor optimized for variable speed as explained by Cambio and Powell (2014) has a nominal speed of 4800 rpm. A new rotor geometry was designed as to increase the volume flow of the compressor. The new rotor geometry included a new rotor profile, lobe count, rotor diameter, and wrap angle. In order to use the same rotor housing casting the rotor length and center distance was left unchanged.

Figure 1 is an overlay of the normalized volume curves of the production and new rotor designs. The axial and radial suction port's relative area are shown as well. The dashed line represents the point at which the suction ports close. It is aligned with the maximum volume of the production rotor set but closes 20 degrees after maximum volume is achieved on the new rotor design. In addition, there is a 20 degree difference in the helix angle of the production rotor compared to the new rotors. The expectation was that volumetric efficiency would decrease because of the mismatched suction port.

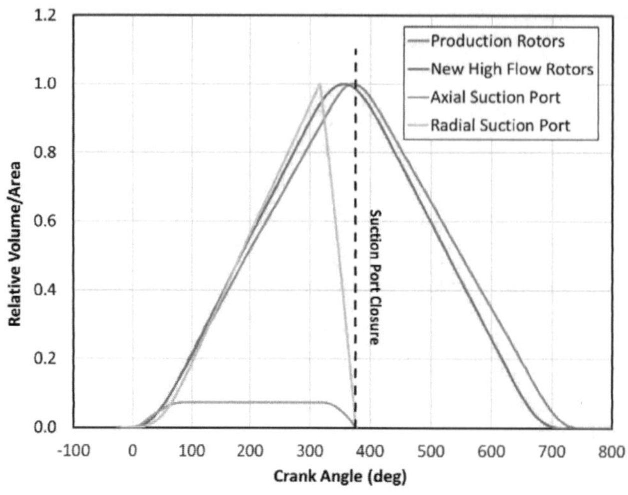

Figure 1: Volume curve comparison of production and new rotors.

2.2 CFD Solution

CFD simulations were performed with the new rotor set and two rotor housings, the production housing with a suction delay, and a modified housing with the suction port aligned with the rotor's helix and suction port closure at the maximum volume. Simulations were done using CFX version 18.2. The simulations were built in three sections. The suction volumes, the compression volumes, and the discharge volumes. Figure 2 is a representation of the solution domains.

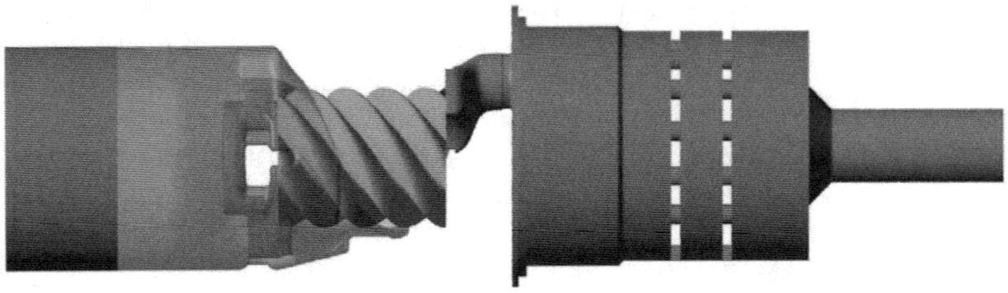

Figure 2: CFD domain consisting of (l to r) suction volume, compression volume, and discharge volume.

The screw rotor's mesh size was 2.27 million nodes and the total mesh size was 8.92 million nodes. Large screw rotor meshes are needed to achieve mesh independence and match actual compressor mass flow rates. An SST turbulence model and a total energy heat transfer model with viscous work terms were applied. All the walls were adiabatic. The fluid was a refrigerant gas without the presence of oil. A lightly applied inertial loss was applied to an isotropic loss model to approximate the oil restriction in the mesh clearance and the tip gaps. An isotropic loss model is a momentum loss through an isotropic process. It is used to slow the flow of the gas in the regions of interest. The simulations were conducted using around 100 cores.

2.3 1-D Solution

A 1-D model was created in GT-SUITE v2020 to capture the inertial charging effect seen in the CFD model. The code solves the conservation equations for mass, energy and momentum as shown in equation (1) from Gamma Technologies (2019).

$$\frac{d\dot{m}}{dt} = \frac{\overbrace{dpA}^{\text{Pressure Gradient}} + \overbrace{\sum_{boundaries}(\dot{m}u)}^{\text{Momentum Terms}} - \overbrace{4C_f \frac{\rho u|u|}{2}\frac{dxA}{D}}^{\text{Friction Loss}} - \overbrace{K_p\left(\frac{1}{2}\rho u|u|\right)A}^{\text{Pressure Drop}}}{dx} \qquad (1)$$

The equation has terms to account for the momentum of the fluid as well as the pressure gradient of the flow field. The term dpA will become negative in the presence of an adverse pressure gradient that occurs as the pocket volume decreases prior to the completion of the suction process. Because equation (1) attempts to conserve momentum it is possible to capture inertial charging.

3. RESULTS

3.1 CFD Results

Figure 3(a) shows the CFD solution of the new high flow rotors with the production rotor housing. The suction port is the grey translucent area at the bottom of the picture. The rotors are color contoured by pressure. The rotors are approximately in their position of maximum pocket volume. It can be seen that the rotor tips are within the suction port area all along the observable rotor helix such that the pocket is still in communication with the suction flow path. The difference is on average 20 degrees.

Figure 3(b) is the new high flow rotors in the same position as the rotors in 3(a) with a suction port that is designed to close at maximum pocket volume. Notice the small gap on the left hand side of the picture between the suction area and the male rotor tip. It demonstrates that the pocket is no longer in communication with the suction port at the same rotor position.

The solution shown in Figure 3 is at 4800 rpm. Comparing the pressure contour colors reveals that the pocket pressure with the suction delay is higher than without the suction delay. The higher pressure in the delayed suction case (Figure 3a) is due to more mass flow entering the rotor pocket. This is the effect described as inertial charging.

The CFD analysis was done at speeds of 1200, 2400, and 4800 rpm. Figure 4 is a plot of mass in the rotor pocket as a function of crank angle. The dashed line at 360 degrees indicates the maximum pocket volume and suction port closure of a new rotor housing design. The dashed line at 394 degrees is where the production suction port closes. Figure 4(a) is at a rotor speed of 4800 rpm. It is shown that with the delayed suction porting the mass in the pocket continues to increase past the point of maximum pocket volume. There is a small decrease in the pocket mass just prior to the suction port closing but it is not enough to offset the extra mass in the pocket. After the suction process is complete there is a 4.3% difference in mass in the pocket. At a rotor speed of 2400 rpm the difference is only 1.2% and at 1200 rpm the difference is -1.1%. As the speed is reduced there is less momentum in the gas and the open port allows gas to escape.

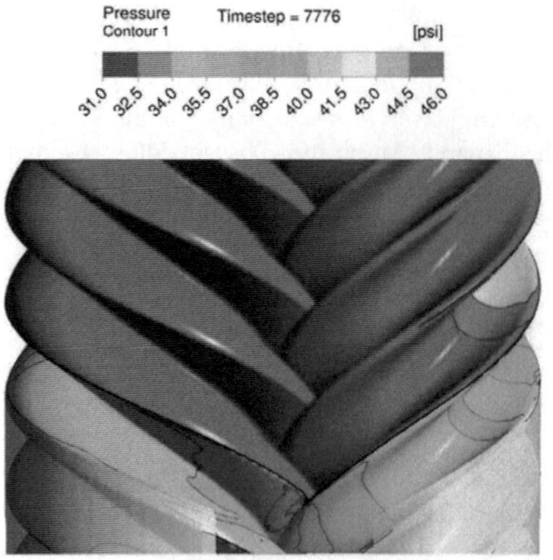

(a) High flow rotor with suction delay

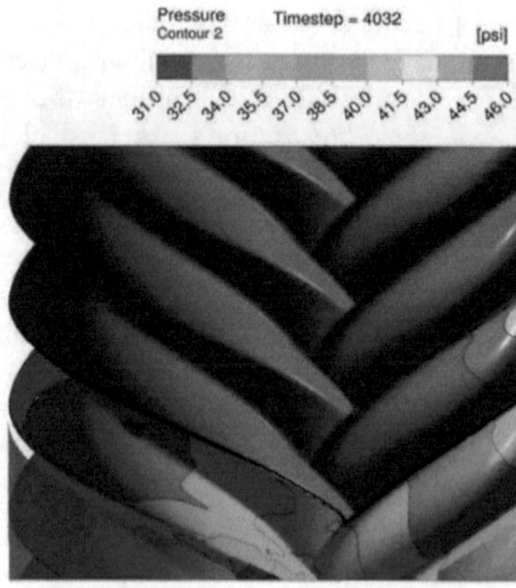

(b) High flow rotors without suction delay

Figure 3: Suction port geometry with and without delay

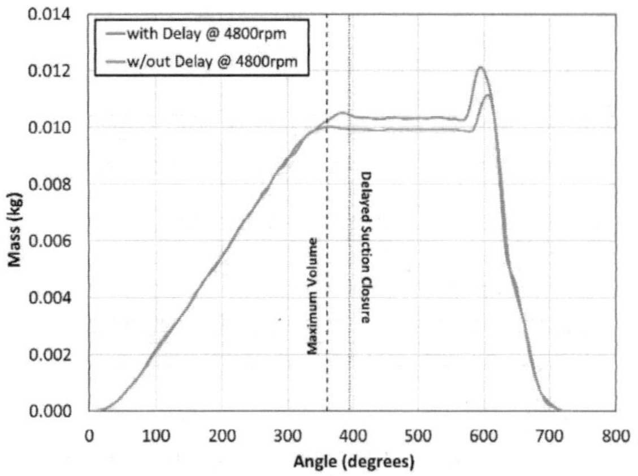

(a) 4800 rpm

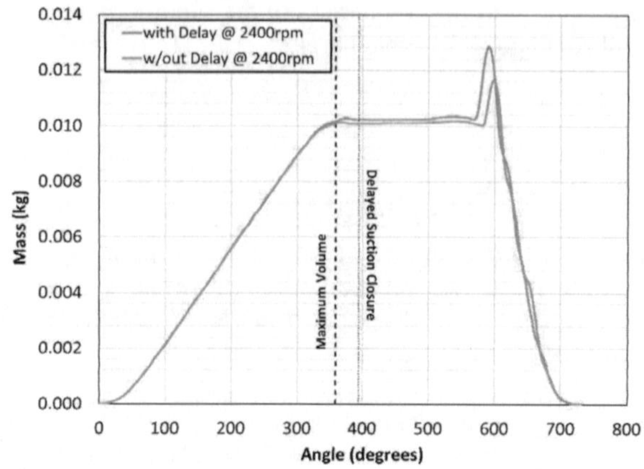

(b) 2400 rpm

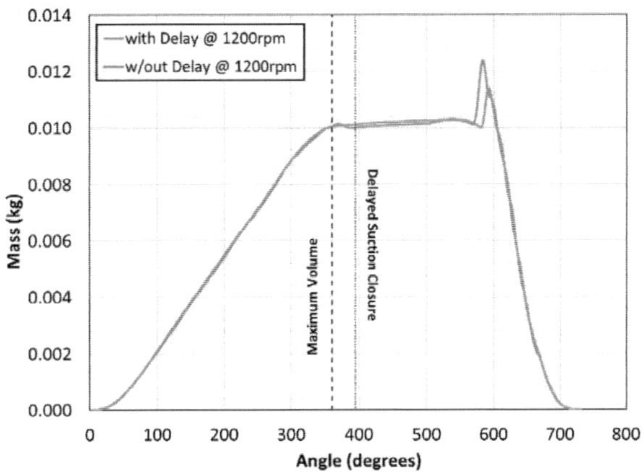

(c) 1200 rpm

Figure 4: CFD pocket mass vs. crank angle for various speeds

The CFD analysis has been done using the axial discharge port. Because of the different rotor designs the axial volume ratio is not the same on the two analysis. This is evident in the amount of back flow between the two curves. It should be noted that the blue curve shows the discharge port opens completely about 10-15 degrees earlier. We believe this fact did not cause any effect on how the chamber is filled on the suction side and the resulting volumetric efficiency.

Figure 5 is a plot of volumetric efficiency vs. speed. The black curve is the volumetric efficiency as a result of testing of the current production design. The orange line is the CFD calculation for the new rotors without suction port delay. There is no reason to believe that the new rotor design would have a significantly lower volumetric efficiency than the production design. Therefore it can be said the CFD is under predicting volumetric efficiency and over predicting leakage. This is reasonable to expect since oil is not present in the solution. The CFD also shows a larger decrease in volumetric efficiency at lower speeds than what is seen in the current production compressor.

Comparing the two CFD solutions the suction delay improved the volumetric efficiency by 4.6% at 4800 rpm, 0.6% at 2400 rpm, and -2.7% at 1200 rpm. The lines cross at about 2100 rpm. This characteristic is not only beneficial at full load but part load as well since a lower flow will be achieved at the compressor's minimum speed. This will create a wider modulation range. Figure 6 is a plot of relative mass flow vs. speed. The suction delay allows for 1.5% less suction flow improving the minimum flow operation.

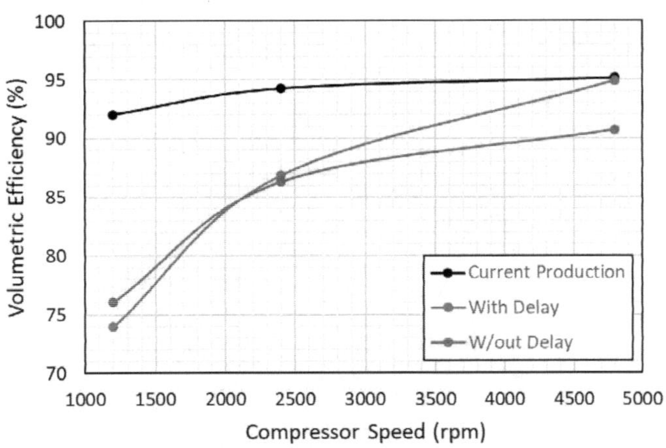

Figure 5: Volumetric efficiency vs. speed

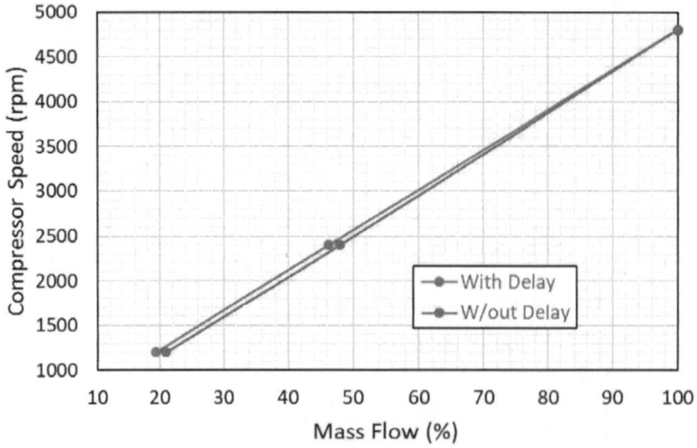

Figure 6: Speed vs. percent mass flow

3.2 1-D Results

In the 1-D model particular attention was placed on discretization of the suction port volume, as spatial discretization is required to capture momentum of the fluid. Figure 7 shows three views of the suction flow path. The flow path was divided into five paths to the rotor suction ports through the bypass areas around the bearing pedestal. Once the flow passes through the bypass channels it can enter the rotors through axial or radial ports.

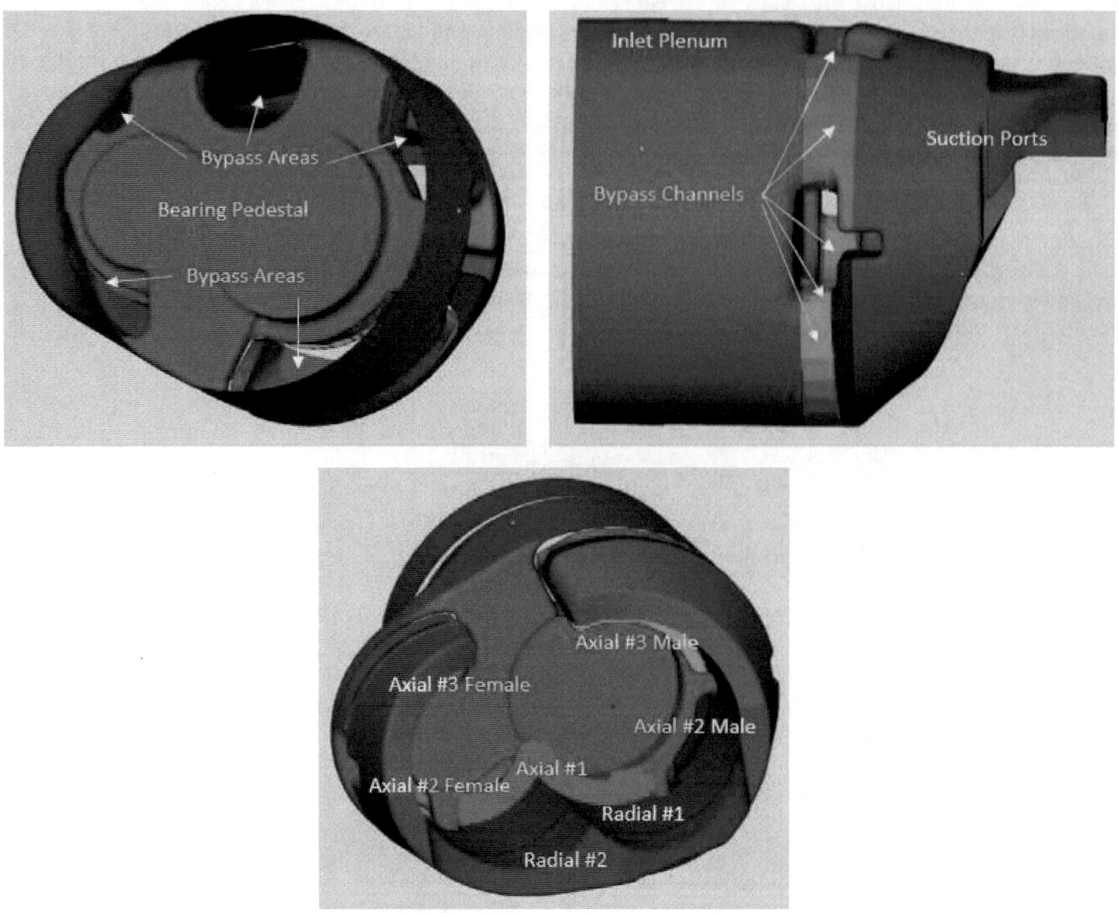

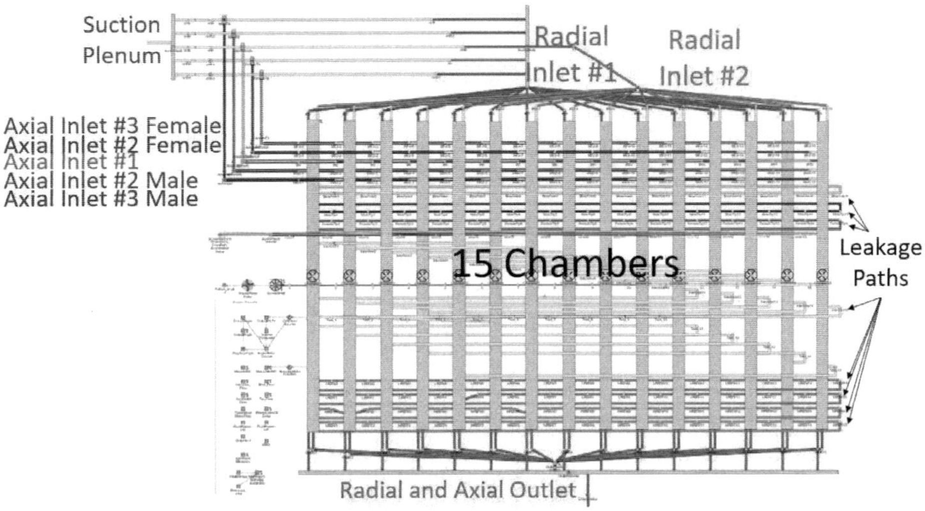

Figure 7: 3-D and 1-D suction flow path volumes and ports.

The model schematic in Figure 7 shows the 1-D flow paths from the suction plenum, into the five axial and two radial suction ports, into the chambers with different leakage paths, and finally out through the radial and axial outlet ports. The suction area profiles for with and without delayed closing are also shown in Figure 8. These profiles were generated using GT-Converge which performs automated volume and area profile extraction using Boolean operations from a 3D CAD model of the flow path.

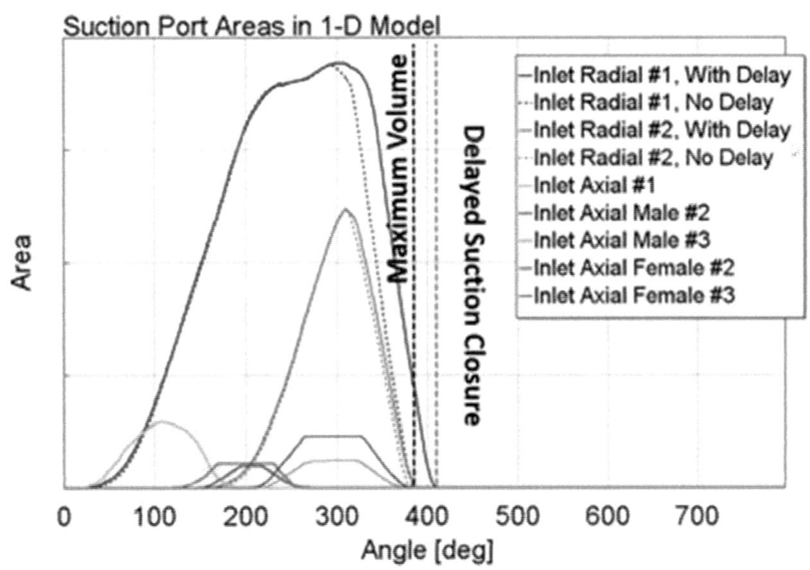

Figure 8: 1-D suction port area profiles obtained by GT-Converge for delayed and no-delayed closing.

In Figure 9 the mass in the rotor pocket is shown from the 1-D simulation. Pocket mass is shown for the two cases, with and without suction delay. There are two vertical lines, the heavy dashed line is the point at which maximum pocket volume is achieved. The second light dashed line is the point at which the delayed suction closes. The three plots are for 4800, 2400, and 1200 rpm.

In comparison to Figure 4 the 1-D simulation results show a similar trend to CFD where the amount of mass trapped in the chamber increases not only with speed, but also when a delayed closing is used at the highest speed case of 4800 rpm, there is an increased amount of mass trapped compared to no delayed closing. The 1-D result shows a slightly lower magnitude of increase using a delayed closing at 4800 rpm of roughly 1% compared to the CFD prediction of 4.6%. The trend in the 1-D result is also very similar to CFD where the delayed closure has less trapped mass at low speed compared to no delayed closure.

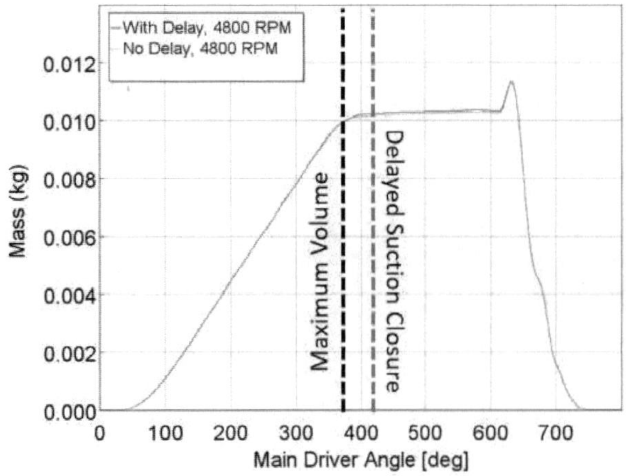

(a) 4800 rpm

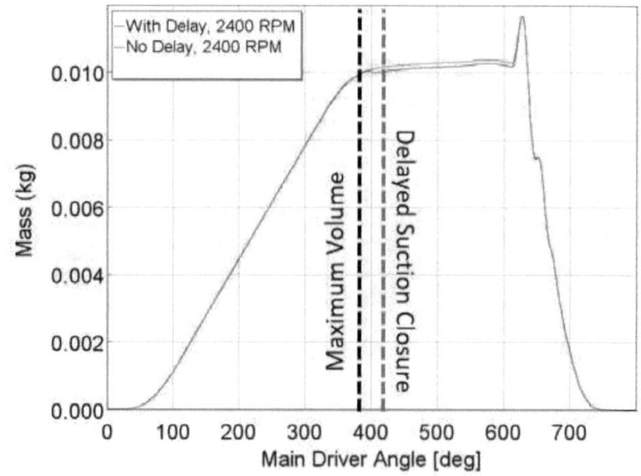

(b) 2400 rpm

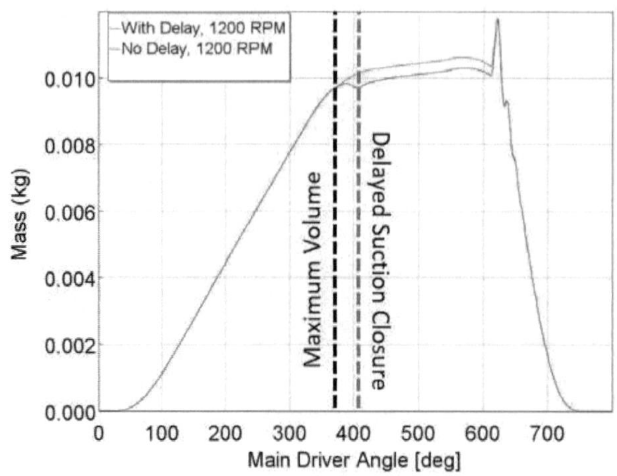

(c) 1200 rpm

Figure 9: 1-D pocket mass vs. crank angle for various speeds

Figure 10 is a comparison of the volumetric efficiency between CFD (solid lines) and 1-D (dashed lines) along with the current production compressor. It is important to note that the production volumetric efficiency is for a completely different rotor set and would be expected to have a different volumetric efficiency compared to the simulations. However, the intent is to show the production characteristic as a function of speed. In general both simulations predict a more dramatic drop in volumetric efficiency with a reduction in speed. Neither of the simulations include oil effects which could explain this behavior.

Focusing on 4800rpm, both CFD and 1-D predict a higher volumetric efficiency with delay compared to without delay which illustrates the inertial charging effect.

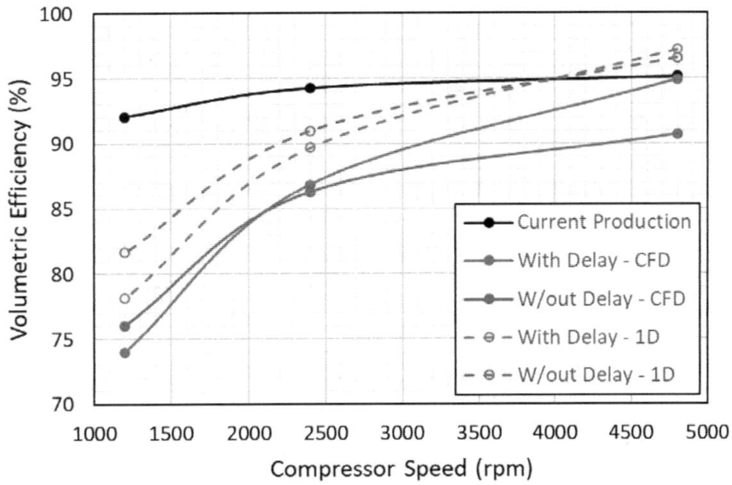

Figure 10: Comparison of 1-D to CFD of volumetric efficiency vs. speed

The CFD and 1-D predictions for the case with delay is interesting in the fact that the curvature of the characteristic is very different from the others. It shows a greater decrease in volumetric efficiency with decreasing speed such that at 2400 rpm it is roughly equivalent to the case without delay and less than it at 1200 rpm.

The speed at which the delayed closing provides more benefit than no delayed closing is around 2400 rpm in the CFD prediction, while the 1-D prediction occurs at a higher speed of roughly 3200 rpm.

4. CONCLUSIONS

An analysis has been presented of a new screw rotor pair in a production rotor housing where the suction port does not match the rotors in terms of the suction port closing at maximum volume and referred to as delayed suction closure. The analysis was done to explore the impact of choosing to use the production casting rather than applying resources to design a new casting. The CFD analysis shows that the momentum of the gas continues to fill the rotor pocket after maximum volume is reached and the pocket volume begins to decrease. The delayed suction closure improves volumetric efficiency by 4.6% at 4800 rpm when compared to a suction port that closes at maximum volume. The effect decreased as a function of speed such that at 2400 rpm there was 0.6% improvement and at 1200 rpm the volumetric efficiency decreased by 2.6%. The 1-D analysis also shows that the momentum of the gas continues to fill the rotor pocket after maximum volume is reached and the pocket volume begins to decrease at the high speed case. The delayed suction closure improves volumetric efficiency by 0.7% at 4800 rpm when compared to a suction port that closes at maximum volume. The effect decreased as a function of speed such that at 2400 rpm there was 1.3% reduction and at 1200 rpm the volumetric efficiency decreased by 3.5%.

Unfortunately, the test plan does not include testing both configurations with and without delay to validate the CFD and 1-D simulations. The performance effects shown by CFD and 1-D are believed to be accurate for engineering purposes and have provided the guidance to suggest that using the production rotor housing is acceptable. Because of this conclusion the analysis preserved the project's initial estimates of cost and schedule.

For compressors that modulate flow with speed and are designed to operate at tip speeds above 40 m/sec inertial charging is an advantage. Improving volumetric efficiency at full speed and reducing volumetric efficiency at low speed will give the compressor a wider modulation range. Although not shown in this study it is assumed that there will be a point at which too much of a suction delay would produce diminishing returns and cause the compressor to lose volumetric efficiency. For this reason it is important to have design tools show this effect. Specifically a 1-D tool that would be better suited for performing design optimization.

NOMENCLATURE

A	cross sectional area	m^2
C_f	Fanning friction factor	
D	diameter	m
dx	length of mass element in flow direction	m
K_P	pressure loss coefficient	
$\dot{m}$	boundary mass flux into volume	kg/sec
p	pressure	Pa
ρ	density	kg/m^3
u	velocity at the boundary	m/sec

REFERENCES

Cambio, M., Powell, G. (2014). The Selection of Screw Rotor Geometry with Compressor Speed as a Design Variable. *Proceedings of the 22nd International Compressor Engineering Conference at Purdue*. West Lafayette, IN. Herrick Laboratories.

Cui, M. M., Sauls, J. (2006). Investigation of Suction Process of Scroll Compressors. *Proceedings of the International Compressor Engineering Conference at Purdue*. West Lafayette, IN. Herrick Laboratories.

Gamma Technologies (2019). *GT SUITE Flow Theory Manual*. Westmont, IL: Gamma Technologies.

Kameya, H., Ishikawa, M., Saito, T. (2011). Improvement of Volumetric Efficiency for Screw Compressors Using Inertial Charging. *Proceedings of the 7th International Conference of Compressors and Their Systems.* London, UK. Institute of Mechanical Engineers.

Redfield, S. B. (1912). Inertia of Air Compressor Intake. *Power*, 35, 162.

Sauls, J. (1988). Slide Valve Design Using a Screw Compressor Performance Simulation, *Proceedings of the 1988 SRM Compressor Conference*. Stockholm, Sweden. Svenska Rotor Maskiner AB (SRM)

Sauls, J. (2011). Use of Finite Element and Computational Fluid Dynamics Analyses in the Development of Positive Displacement Compressor Simulations. *Proceedings of the International Conference on Compressors and Their Systems.* London, UK. Institute of Mechanical Engineers.

Sahoo, D. (2014). Study of Intake System Parameters for Incylinder Charging Efficiency – 1D CFD Method. *GT-SUITE India Users Conference*. Bangalore, IN. https://www.gtisoft.com/wp-content/uploads/publication/RNTBCI_GTConferenceIndia2014.pdf

Schwerzler, D., D., Hamilton, J., F. (1972). An Analytical Method for Determining Effective Flow and Force Areas for Refrigeration Valving Systems. *Proceedings of the 1972 Purdue Compressor Technology Conference*. West Lafayette, IN. Herrick Laboratories.

Wu, G., Li, W., and Xu, Z. (2019). Optimizing Gas Exchange for a 2-stroke Free Piston Engine. *GT-SUITE North American Users Conference*. Plymouth, MI. https://www.gtisoft.com/wp-content/uploads/2019/11/Optimizing-Gas-Exchange-for-a-2-Stroke-Free-Piston-Engine-Guangquan-Wu-Weichai.pdf

AKNOWLEDGEMENTS

We would like to thank Trane Technologies for encouraging the publication of our work and the time and resources to do so and Gamma Technologies for their participation in this work as well.

Numerical Analysis on a Perforated Muffler Applied in the Discharge Chamber of a Twin Screw Refrigeration Compressor Based on Fluid-Acoustic Coupling Method

Shizhong Sun[1], Ziwen Xing[1*], Wenqing Chen[2], Minglong Zhou[2], Yu Hua Huang[3], Po Chia Su[3]

[1]School of Energy and Power Engineering, Xi'an Jiaotong University
Xi'an, Shannxi, People's Republic of China
(Phone: 113709188502, E-mail: zwxing@mail.xjtu.edu.cn)

[2]Suzhou Academy, Xi'an Jiaotong University
Suzhou, People's Republic of China

[3]R&D Department,Hanbell Precise Machinery CO., LTD
Taiwan

* Corresponding Author

ABSTRACT

The twin screw compressor has been widely used in the refrigeration systems due to advantages such as compact structure, stable operation, high efficiency and good adaptability. Intermittent gas flow generates gas pulsation that cause serious problems such as structural vibration and noise in the twin screw refrigeration compressor. Because the mechanical noise can be controlled well with the improvement of machining and assembly accuracy, the aerodynamic noise induced by gas pulsation even has become the main noise source of the twin screw refrigeration compressor. In order to reduce the pressure pulsation, a broadband perforated panel muffler applied in the discharge chamber of the twin screw refrigeration compressor is proposed based on the noise spectrum and flow characteristics of the compressor. In order to obtain the noise spectrum of the twin screw refrigeration compressor, the pressure fluctuation in discharge chamber based on a three-dimensional CFD simulation model is calculated, and the acoustical model is established based on fluid-acoustic coupling method. Then the impacts of different structural parameters on the performance of a perforated panel muffler are investigated, including perforation rate, perforation diameter and panel thickness. Through the optimization of the perforated muffler, a better reduction effect of broadband noise can be achieved. Results of fluid-acoustic coupled analysis can provide guidance on the design and optimization of the perforated muffler and noise reduction of the twin screw refrigeration compressor.

1. INTRODUCTION

Twin-screw compressors have the advantage of a simple structure, less easily damaged parts, excellent running stability, and strong adaptability, and have been widely used in refrigeration and air conditioning fields. Intermittent gas flow generates gas pulsation that cause serious problems such as structural vibration and noise in the twin screw refrigeration compressor. Because the mechanical noise can be controlled well with the improvement of machining and assembly accuracy, the aerodynamic noise induced by gas pulsation even has become the main noise source of the twin screw refrigeration compressor. Fujiwara *et al.* (1986) conducted the first comprehensive measurement of the vibration and noise of a screw compressor. Holmes (2003) analyzed the noise of screw compressors and proposed a method to solve mechanical noise. Mujic *et al.* (2011) conducted a comprehensive analysis of the noise of the screw compressor, and concluded that the gas pulsation is the main cause of screw compressor noise. Therefore, in order to reduce the vibration noise of refrigeration screw compressor, it is necessary to first analyze and attenuate the gas pulsation during operation. J Hauser (2013) predicted the pressure pulsation in the discharge port of a screw compressor using Computational Fluid Dynamics model, which shows great accuracy compared with experimental investigations. Wu *et al.* (2017) constructs a simplified dynamic model of a screw rotor system that uses multibody dynamics to predict the compressor's dynamic response and to determine the reasons for structural vibration. The stimulated vibration frequencies of the proposed model fit well with those measured vibration frequencies of the experimental model. K Matsuo (2013) studied noise and vibration reduction in medium-size oil-flooded twin-screw compressor using the helix modification. Noise reduction by 3dB was achieved under full load conditions. Applications of muffles are effective ways of reducing gas pulsation and absorbing aerodynamic noise of compressors. Kim *et al.* (2016) proposed a suction muffler applied in hermetic reciprocating compressor, which improved both flow performance and acoustic performance. Oh *et al.* (2016a) studied the topology optimization of a suction muffler in a fluid machine to maximize energy efficiency and minimize broadband noise. Oh *et al.* (2016b) also applied a suction muffler on a linear compressor to study the variation of acoustical resonance frequency of duct with orifice depending on periodic motion. However, little literature about the applications and optimization of the perforated panel muffler in twin-screw compressor is available. A fluid-acoustic coupling model of the screw compressor is established in this paper, which provides guidance on the applications and optimization of perforated panel muffler in twin screw compressor to reduce the aerodynamic noise.

2. INTRODUCTION OF FLUID-ACOUSTIC COUPLED MODEL

In order to obtain the aerodynamic noise spectrum of the screw refrigeration compressor, a fluid-acoustic coupling model is established. The pressure pulsation of the discharge section is obtained by calculating the unsteady flow of the screw compressor, which is coupled with the acoustic model of discharge section to calculate the aerodynamic noise of the screw compressor during operation.

2.1 CFD Model

The CFD simulation in this paper is based on the following assumptions:

(1) The influence of lubricating oil on the refrigerant is not considered

(2) The discharge process is adiabatic, and the heat exchange between gas and wall, wall and ambient is not considered.

Due to the complex geometry of the compressor's working chamber, the internal fluid domain and the clearance changes continuously with the rotation of the rotor, which made the requirements of the grid extremely demanding. The generated grids of rotors fluid domain are shown in Figure 1.

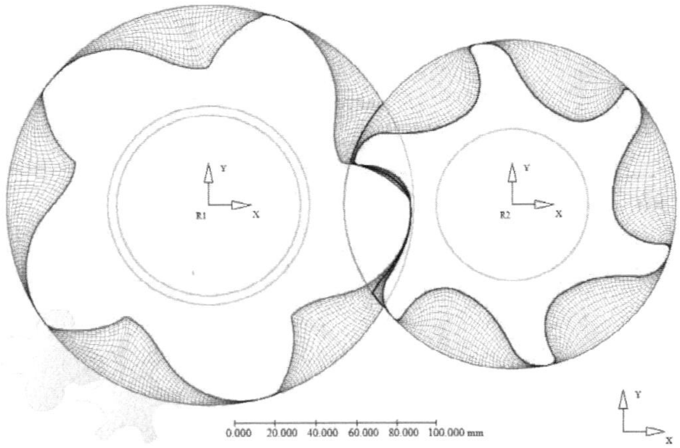

Figure 1: Two-dimensional mesh of rotors fluid domain

In order to ensure the accuracy of simulation of discharge process flow, the corresponding discharge flow channel model is established according to the actual compressor discharge end-seat structure. The simplified suction and discharge basin grid are generated by Ansys Meshing as shown in Figure 2.

Figure 2: Suction and discharge port fluid domain grid

The refrigerant is R134a, and its physical properties are given by the physical property data fitted by the Peng–Robinson equation in the database. As shown in Table 2, the inlet and outlet temperature and pressure are used as the boundary conditions for the CFD simulation.

Table 2: Boundary conditions for CFD simulation

Parameters	Values
Rotation speed of the male rotor, rpm	3000
Suction pressure, bar	3.50
Suction temperature, K	288.15
Discharge pressure, bar	9.63

2.2 Acoustic Model

The numerical model of acoustic field is a basic control equation which establishes the relationship between sound pressure, density variation and particle velocity, and then solves other acoustic parameters.

$$\begin{cases} \rho_0 \dfrac{\partial v}{\partial t} = -\dfrac{\partial p}{\partial x} \\ \rho_0 \dfrac{\partial v}{\partial x} = -\dfrac{\partial \rho'}{\partial t} \\ p = c_0^2 \rho' \end{cases} \tag{1}$$

where ρ_0 is the static density, ρ' is the density increment, p is the sound pressure, v is the particle velocity, x is the particle displacement and c_0 is the speed of sound.
Assuming that the refrigerant is an ideal fluid, the three-dimensional wave Equation (2) of small amplitude sound wave in a homogeneous ideal fluid medium can be derived from the basic governing Equation (1) of acoustic field under the assumption of static, uniform and no loss of the medium, thus the finite element Equation (3) can be obtained:

$$\nabla^2 p = \frac{1}{c_0^2} \frac{\partial^2 p}{\partial t^2} \tag{2}$$

$$\left[K + \omega C - \omega^2 M\right] p = F_A \tag{3}$$

where K is the stiffness matrix, C is the damping matrix, M is the mass matrix, and F_A is the normal velocity.
The transmission loss (TL) of the muffler can be calculated by Eq. (4) based on the sound pressure nephogram of sound field inside muffler obtained by Equation (3).

$$TL = 10\lg(\frac{W_{in}}{W_{out}}) = 10\lg(\frac{p_{in}^2 S_{in}}{p_{out}^2 S_{out}}) \tag{4}$$

where W_{in} and W_{out} are input and output sound power of the muffler, p_{in} and p_{out} are the incident and transmitted sound pressure of the muffler, and S_{in} and S_{out} are the area of the inlet and outlet of the muffler.
Generally, it is assumed that there are 6 acoustic units in the minimum wavelength of sound wave, that is, the side length of the largest element should be less than 1/6 of the shortest wavelength, as the expression shown:

$$L \le \frac{c_0}{6 f_{\max}} \tag{5}$$

where c_0 is the propagation velocity of sound wave in fluid medium and L is the linear element length.
In order to improve the simulation accuracy and quality of the grid, the maximum side length of the whole grid cell can be controlled below 1/12 of the shortest wavelength. The surface boundary of the acoustic model is set as a rigid wall, and the boundary conditions of constant velocity excitation at the entrance and nonreflection at the exit are used for the acoustic finite element calculation.

2.3 Fluid-Acoustic Coupled Model

In order to further study the internal relationship between air flow pulsation and aerodynamic noise and mechanism of aerodynamic noise induced by air flow pulsation, the unsteady CFD results are mapped to the acoustic field model. The fluid-acoustic coupling model of the screw compressor discharge section is established, as shown in Figure 3. There are two noise monitors at a distance of 100mm from the discharge pipe outlets.

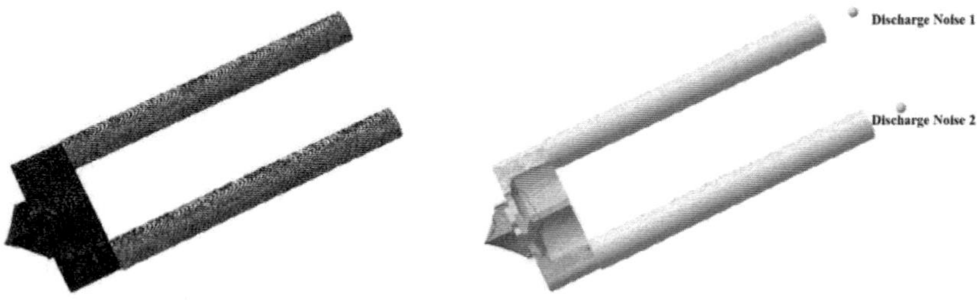

(a) Unsteady CFD results (b) Fluid-acoustic coupling model

Figure 3: Fluid-acoustic coupling model of the screw compressor discharge section

3. RESULTS OF FLUID-ACOUSTIC COUPLED SIMULATION

3.1 Results of CFD Simulation

Figure 4 shows the pressure distribution and temperature distribution of the suction and discharge port fluid domain and the surface of the rotor when the discharge ports are connected, which are obtained by CFD simulation. The rotor surface pressure and temperature gradually increase from the suction end to the discharge end.

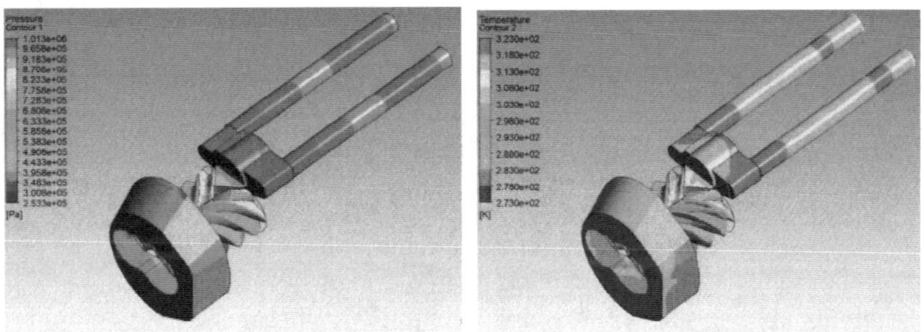

Figure 4: Pressure and temperature distribution when the discharge port is connected

Figure 5 shows the pressure pulsation monitored by the monitoring point in the working chamber of the compressor. From the low-pressure suction section to the high-pressure discharge section, the pulsation energy increases sharply, especially in the final compression chamber. The data shows that the final compression pressure when the radial discharge port is connected is about 1153kPa, which is much higher than the 963kPa set as the discharge pressure. After the axial discharge port opens, the compression pressure continues to decrease to 767kPa. There is an intense pressure pulsation in the compression chamber of the discharge section, while the pressure pulsation is much lower in the discharge pipe due to the large volume which provides buffer and attenuation functions.

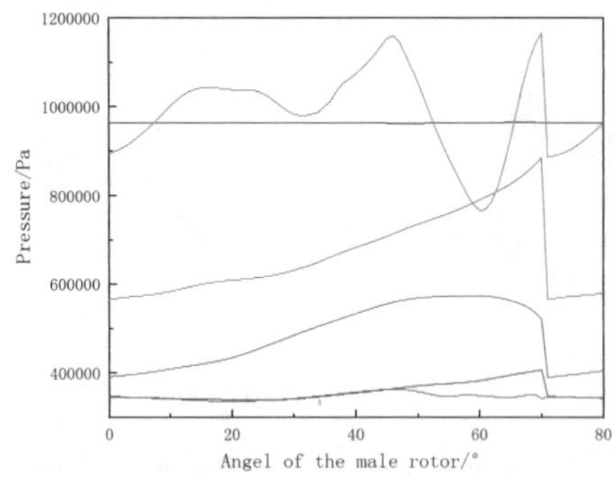

Figure 5: Pressure pulsation in the working chamber

3.2 Noise Characteristics of the compressor

Figure 6 shows the calculation results of two monitoring points of discharge aerodynamic noise of screw refrigeration compressor. The discharge aerodynamic noise of the screw refrigeration compressor has significant harmonic characteristics. The noise values at the fundamental frequency, the second frequency and the third frequency of the engagement frequency of the male and female rotors are particularly significant, which need to be improved. Therefore, the perforated panel muffler is designed to reduce the discharge noise aimed at the noise of the fundamental frequency, double frequency and triple frequency of the screw compressor.

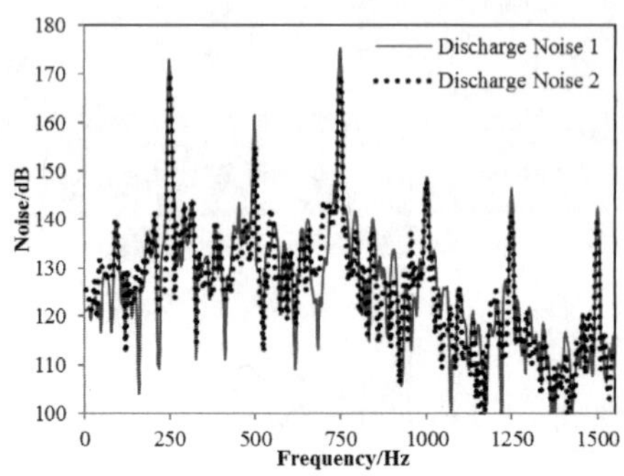

Figure 6: Discharge noise spectrum of screw compressor at the speed of 3000rpm

4. DESIGN AND OPTIMIZATION OF PERFORATED MUFFLER

In order to reduce the noise of screw refrigeration compressor, a perforated panel muffler suitable for screw refrigeration compressor is proposed. Its typical structure is shown in Figure 7. The perforated panel muffler is mainly composed of the perforated panel and the cavity behind it. l is the panel thickness, a is the perforation radius, d is the distance between the centers of the neighboring holes, and D is the depth of the cavity behind the perforated panel. There are many factors that affect the performance of the perforated muffler, such as the perforation rate, perforation diameter, panel thickness and resonator volume. In the application of screw compressor noise reduction, the volume of perforated panel muffler is not easy to change and the depth of cavity is limited due to the compact structure of compressor, so the perforation rate, perforation diameter and panel thickness are particularly important for the optimization of acoustic performance of perforated panel muffler.

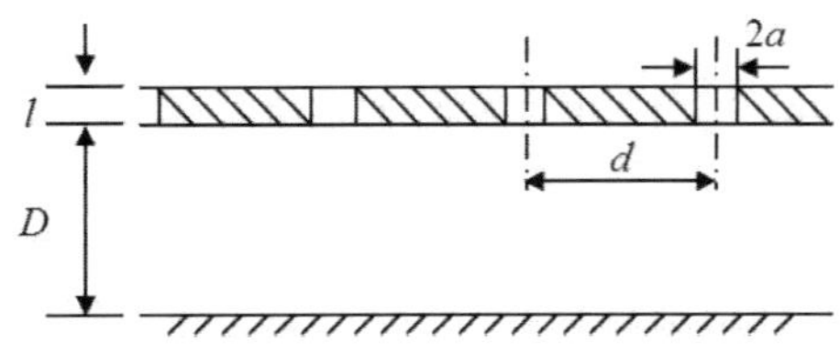

Figure 7: Typical structure of a perforated panel muffler

4.1 Analysis of Structural Characteristic

4.1.1 Perforation diameters: The transmission loss of the perforated panel muffler with different perforation diameters is shown in Figure 8, when cavity depth of the perforated panel is 16.65mm, the number of perforations is 36, and the thickness of the perforated panel is 2mm. As the perforation diameter decreases, the resonance peak of the perforated tube muffler shifts to low frequency, and the sound absorption effect of the resonance peak increases first and then decreases. With the decrease of the perforation diameter, the damping and friction effect of the perforation wall on the acoustic wave is enhanced, the acoustic energy of the acoustic wave is attenuated, the acoustic energy of the acoustic wave is reduced, and the sound absorption performance of the muffler is improved. When the perforation diameter continues to decrease, the number of acoustic waves passing through the perforation diameter is reduced, resulting in the decrease of the acoustic wave energy of the attenuation, and the noise reduction effect is reduced accordingly. As a result, the resonant frequency of the perforated panel muffler can be reduced by reducing the perforation diameter properly, and the sound absorption performance of the perforated panel muffler can be improved at the same time.

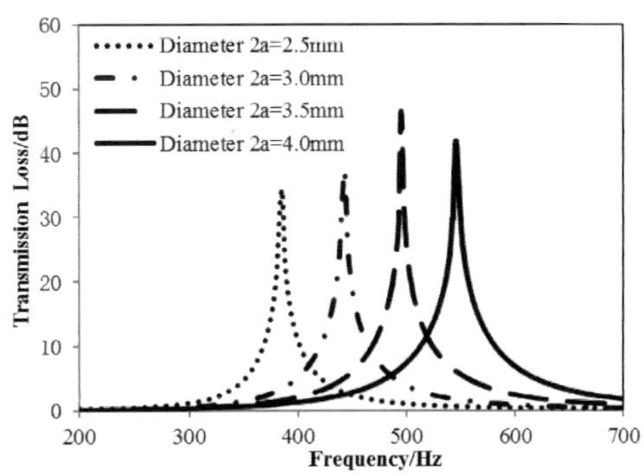

Figure 8: Acoustic absorption performance with different perforation diameters

4.1.2 Perforation rate: The acoustic absorption performance of the perforated panel muffler with different perforation rate is shown in Figure 9, when cavity depth of the perforated panel is 16.65mm, the perforation diameter is 3.5mm, and the thickness of the perforated panel is 2mm. With the increase of perforation rate, the resonance peak of the perforated panel muffler shifts to high frequency, and the absorption performance of perforated plate muffler increases first and then decreases. Although the damping and friction effect of each perforation wall on acoustic wave is unchanged, with the increase of perforation rate of perforated panel and the number of perforations, the total damping and friction effect of all perforation walls increases, and the total energy of acoustic wave attenuation increases, so the sound absorption performance of the perforated panel muffler is improved. When the number of perforations continues to increase, due to the decrease of the spacing between neighboring holes, they will interfere with each other and affect the final sound absorption performance. As a result, with the increase of the perforation rate of the perforated panel muffler, the resonant frequency gradually shifts to the high frequency, and the absorption performance of the formant increases first and then decreases.

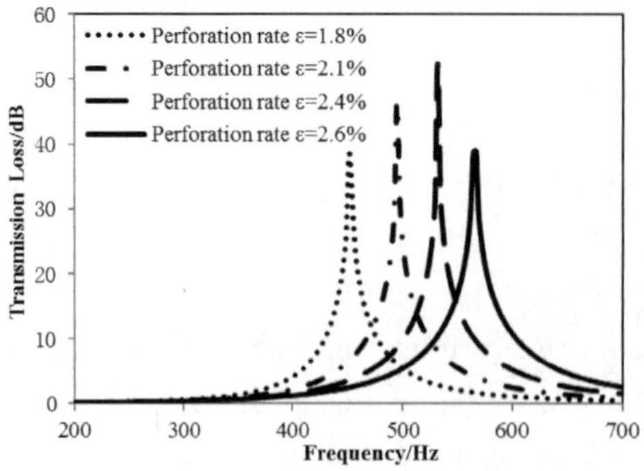

Figure 9: Acoustic absorption performance with different perforation rate

4.1.3 Panel thickness: The acoustic absorption performance of the perforated panel muffler with different panel thickness is shown in Figure 10, when cavity depth of the perforated panel is 16.65mm, the perforation diameter is 3.5mm, and the number of perforations is 36. With the increase of panel thickness, the resonant frequency of perforated panel muffler shifts to low frequency, and the absorption performance of formant increases first and then decreases. With the increase of the panel thickness, the depth of the acoustic wave passing through the perforation wall increases, and the damping and friction effect in the propagation process increases. Therefore, the energy of the acoustic wave attenuated on the wall increases, which leads to the improvement of the sound absorption performance of the perforated panel muffler. When the panel depth continues to increase, the increasing the damping and friction of the wall may reduce the number of sound waves passing through the perforated panel, thus affecting the sound absorption performance. As a result, the resonant frequency of the perforated panel can be reduced and the acoustic absorption performance of the perforated panel muffler can be improved by properly increasing the panel thickness.

To sum up, the structure parameters of perforated panel muffler such as perforation rate, perforation diameter and panel thickness have significant impact on the resonance frequency and absorption performance, and the analysis of structure parameters can provide guidance on the design and optimization of the perforated muffler.

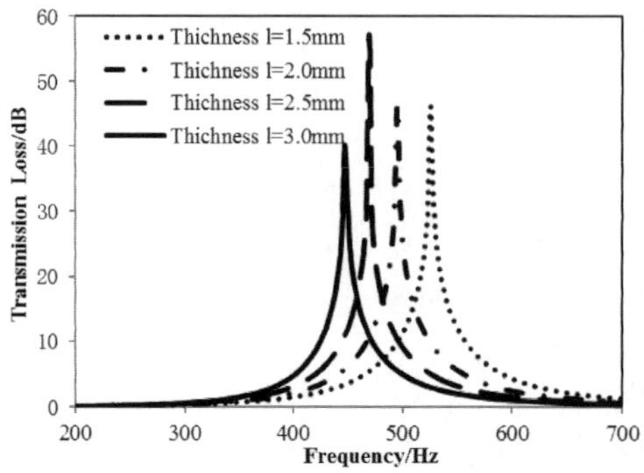

Figure 10: Acoustic absorption performance with different panel thickness

4.2 Optimization of Acoustic Characteristic

Based on the noise characteristics of screw refrigeration compressor, it can be seen that the first three times frequency noise of the engagement fundamental frequency of the male and female rotors of screw compressor is particularly significant, which needs to be improved. In order to reduce the noise of the screw refrigeration compressor, based on the influence of the structural parameters of the perforated panel muffler on the performance of the muffler, a broadband perforated panel muffler is designed in series with multiple chambers, and then its structure is optimized to improve its performance, as shown in Figure 11.

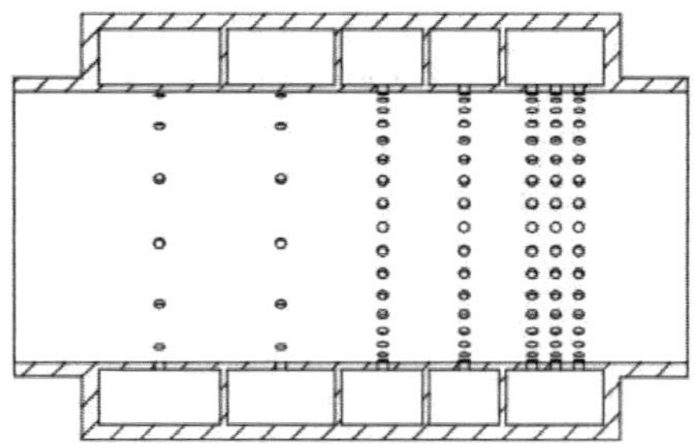

Figure 11: Sectional view of a multi-chamber broadband perforated panel muffler

Figure 12 shows the acoustic absorption performance of the optimized multi-chamber broadband perforated panel muffler. As shown in the Figure 12, the perforated panel muffler has remarkable acoustic absorption performance at the fundamental frequency, second frequency and third frequency of the discharge noise of the screw refrigeration compressor, which can effectively reduce the discharge noise of the compressor.

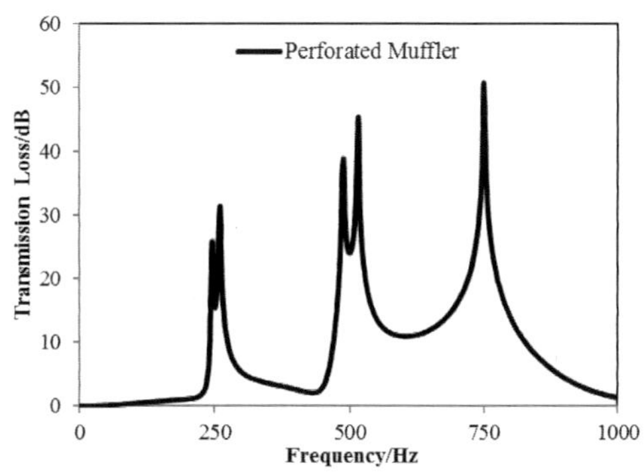

Figure 12: Acoustic absorption performance of the perforated panel muffler

5. CONCLUSIONS

A fluid-acoustic coupling model of a twin screw refrigeration compressor is established. The pressure pulsation and noise spectrum of the discharge section are obtained. In order to reduce the noise of screw refrigeration compressor, a perforated panel muffler suitable for screw refrigeration compressor is proposed. The impacts of different structural parameters on the performance of a perforated panel muffler are investigated, including perforation rate, perforation diameter and panel thickness. The conclusions are as follows:

- Results of CFD simulation show that there is an intense pressure pulsation in the compression chamber of the discharge section, while the pressure pulsation is much lower in the discharge pipe due to the large volume which provides buffer and attenuation functions.
- The resonant frequency of the perforated panel muffler can be reduced by reducing the perforation diameter properly, and the sound absorption performance of the perforated panel muffler can be improved with proper diameter.
- With the increase of the perforation rate of the perforated panel muffler, the resonant frequency gradually shifts to the high frequency, and the absorption performance of the formant increases first and then decreases.
- The resonant frequency of the perforated panel can be reduced and the acoustic absorption performance of the perforated panel muffler can be improved by properly increasing the panel thickness.
- The optimized multi-chamber broadband perforated panel muffler has remarkable acoustic absorption performance at the fundamental frequency, second frequency and third frequency of the discharge noise of the screw refrigeration compressor

REFERENCES

Fujiwara, A & Sakurai, N. (1986). Experimental analysis of screw compressor noise and vibration. *Journal of Sound & Vibration, 28*(3), 375–401.

Holmes, CS. (2003). Inspection of screw rotors for the prediction of compressor performance, reliability and noise. *In Proceedings of the 4th International Conference on compressor and refrigeration,* 82-96.

J, Hauser & M, Beinert. (2013). CFD analysis of pressure pulsation in screw compressors – combine theory with practice. *8th International Conference on Compressors and their Systems*, 625-633.

Matsuo, K & Holmes, Christopher. (2013). Experimental study of noise and vibration reduction in a medium-size oil-flooded twin-screw compressor by the application of helix relief. *8th International Conference on Compressors and Their Systems,* 363-371.

Mujic, E & Kovacevic, A & Stosic, N & Smith, I, K. (2011). Noise generation and suppression in twin-screw compressors. *Proceedings of the Institution of Mechanical Engineers. Part E, Journal of Process Mechanical Engineering, 225*(2), 127-148.

Oh, Seungjae & Wang, Semyung & Cho, Sungman. (2016). Topology optimization of a suction muffler in a fluid machine to maximize energy efficiency and minimize broadband noise. *Journal of Sound and Vibration*, *366*, 27-43.

Oh, Seungjae & Wang, Semyung & Kim, Jongnam. (2016). Experimental study on variation of acoustical resonance frequency of duct with orifice depending on periodic motion. *Applied Acoustics, 111*, 96-104.

Sanghyeon, Kim & Cheong, Cheolung & Park, Jaeseong & Kim, Haeseung & Lee, Hyojae. (2016). Investigation of flow and acoustic performances of suction mufflers in hermetic reciprocating compressor. *International Journal of Refrigeration*, *69*, 74-84.

Wu, Xiaokun & Xing, Ziwen & Chen, Wenqing & Wang, Xiaolin. (2018). Performance investigation of a pressure pulsation dampener applied in the discharge chamber of a twin screw refrigeration compressor. *International Journal of Refrigeration, 85*, 74-84.

The Study on Dynamic Characteristics of Twin-Screw Compressor Rotor

Baoshun XIONG[1]*, Huagen Wu[2], Kanlong LIN[3], Beiyu ZHANG[4]

[1]Xi'an Jiaotong University, Department or Equivalent,
Xi'an, Shanxi, China
xiongbaoshun98@stu.xjtu.edu.cn

[2]Xi'an Jiaotong University, Department or Equivalent,
Xi'an, Shanxi, China
hgwu@xjtu.edu.cn

[3]Xi'an Jiaotong University, Department or Equivalent,
Xi'an, Shanxi, China
longerme@stu.xjtu.edu.cn

[4]Xi'an Jiaotong University, Department or Equivalent,
Xi'an, Shanxi, China
zby555544444@stu.xjtu.edu.cn

* xiongbaoshun98@stu.xjtu.edu.cn

ABSTRACT

In the working process of twin-screw compressor, the rotors are subjected to multiple physical effects of the gas temperature, pressure and force, and presents a periodic change. In this paper, three-dimensional Computational Fluid Dynamics (CFD) simulation of screw compressor is carried out, and the characteristics of temperature distribution, pressure distribution and gas force distribution on the rotors' surface are studied. Firstly, the rotor domain, suction and exhaust face grids are generated by TwinMesh, and then they are imported into the CFX calculation model. Through the fluid-structure interaction analysis and calculation, the strain shape of the screw rotor under the multiple alternating physical action of gas temperature, pressure and force is analyzed, and the deformation of the rotor structure caused by the pressure and temperature in the working process is obtained, which plays a good guiding role in the design of the screw rotor and the improvement of the performance of the compressor.

1. INTRODUCTION

Screw compressors can be divided into two types: single-screw compressors and twin-screw compressors. Twin-screw compressors are composed of a pair of female and male rotors that mesh with each other in parallel. Screw compressors are widely used in industrial production and chemical industries due to their compact structure, stable operation, convenient maintenance, and long life. In recent years, screw compressors have developed rapidly: on the one hand, screw profiles and structural designs have Made great progress. On the other hand, the introduction of a screw rotor special milling machine, especially a grinding machine, has improved the processing accuracy and efficiency of the key parts, and effectively improved the performance of the screw compressor.

In recent years, many scholars at home and abroad have done a lot of research on screw compressor. N. Stosic et al. presented some results of mathematical modelling and experimental investigation of the influence of oil injection upon the screw compressor working process. Xing analyzed the screw rotor meshing conditions and gave a general method for the design of the female and male rotor profile. N. Stosic et al. introduced a suitable procedure for optimization of the screw compressor shape, size, dimension and operating parameters, which results in the most

appropriate design for a given compressor application and fluid. Cai et al. provided a simplified method for calculating the contact line shape and length of screw compressor. D. Zaytsev et al. deduced a forming method of rotor profile of twin-screw compressor. LIU et al. simulated the lubrication problem between the rotor and the bearing. An studied the influence of rotor speed on the selection of screw compressor. Li et al. verified the speed transmission ratio of screw rotor. Zhao et al. made a dynamic simulation study on the working condition of a certain type of screw compressor by using virtual prototyping technology. Seshaish. N. et al. studied the influence of internal pressure and leakage of oil injected compressor. Wu et al. presented a design method for the rotor profile of twin-screw compressor from an arbitrary sealing line. Ahmed El Shorbagy designed the rotor tooth profile of twin-screw compressor, and analyzed the three-dimensional flow field of the tooth profile by using CFX and TwinMesh. Feng et al. proposed the mean pressure model and sector pressure model to calculate the axial force on the rotor of the twin-screw refrigeration compressor. Rané et al. studied a twin-screw compressor and found that the refinement of the grid along the circumferential direction of the rotor profile directly affected the prediction of mass flow. Li et al. studied the rotor structural characteristics of twin-screw compressor under fluid-solid coupling.
In summary, at present, there are many studies on the rotor profile and speed of screw compressor, but there are few studies on the dynamic characteristics of the rotor. In this paper, through the computer simulation of the twin-screw refrigeration compressor, the force and deformation of the screw rotor are studied and analyzed.

2. METHODS

In this paper, the twin-screw refrigeration compressor is simulated by CFD, and the heat flux module of CFX in ANSYS Workbench is used for calculation. The rotor segment is meshed by TwinMesh software, and the inlet and outlet part is modeled by SolidWorks and imported into ANSYS Mesh (ICEM) for meshing. The simulation solution is solved by 12-core parallel operation, which reduces the time needed to solve the problem.
Based on the actual working conditions of twin-screw refrigeration compressors, a fluid-solid thermal coupling model is established in this paper. CFX is used to calculate the flow field distribution of the twin-screw compressor during operation, and the temperature distribution of the female and male rotors during operation are calculated by fluid-solid thermal coupling. Then the pressure field distribution calculated in the CFX flow field is imported as a result file into the ANSYS statics module to calculate the force and deformation characteristics of the female and male rotors in the working process.
Table 1 shows the detailed structural parameters of the twin screw refrigeration compressor. The ratio of the rotor length to the outside diameter of the male rotor (L/D) was 1.09742. The theoretical volume flow rate of the compressor was 3.797 $m^3 \cdot min^{-1}$.

Table 1: Structural parameters of the screw compressor

Parameters	Male Rotor	Female Rotor
Number of Teeth	5	6
Diameter of Rotor D/mm	138.507	109.76
Length of Rotor L/mm	152	152
Wrap Angle of Male Rotor	300°	250°
Screw Lead of Male Rotor T/mm	181.761	218.113

3. SIMULATION SETTINGS

3.1 Boundary Condition Settings

The turbulence model in this paper was the SST k-ω model, which combines the advantages of the k-ω model in the near-wall region calculation and the advantages of the k- ε model in the far-field calculation. The refrigerant was R134a in twin-screw compressor. The inlet and outlet temperatures and pressures is shown in Table 2. To simplify the calculation, the heat transfer of the oil was not considered in the CFD simulation. The rotor speed was 3000 rpm.

Table 2: Boundary conditions of computational fluid dynamics (CFD) simulation

Parameters	Value
Inlet Pressure	0.285 MPa
Inlet Temperature	274 K
Discharge Pressure	1.505 MPa
Discharge Temperature	355 K
Volume flow	3.254 $m^3 \cdot min^{-1}$

In the simulation, it is also necessary to set the pressure of the initial flow field in the twin-screw compressor. In this paper, the pressure of the initial flow field is set as the inlet pressure of the compressor, and the reference pressure is 0 bar. All wall surfaces of screw compressors are non-slip wall surfaces.

3.2 Meshing

As the structure of the female and male rotor is relatively complicated, this paper focuses on the deformation and force of the rotor under the condition of heat and force. Therefore, the model of the rotor is simplified. The grid of the rotor segment needs to establish a structural grid with the same number of grid points, so TwinMesh is used to mesh the rotor segment. The rotor grid is divided as shown in Figure 1. The total number of the female and male rotor grids is 1344401.

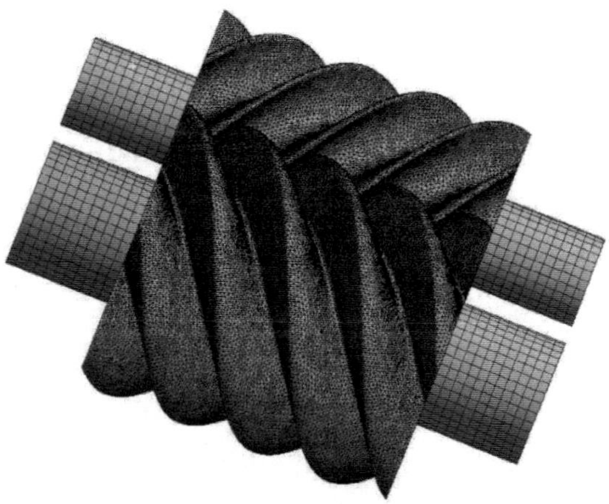

Figure 1: The rotor meshing

4. RESULTS

The effect of the pressure and temperature distribution inside the compressor on the efficiency and performance of the compressor is a topic for engineers in the compressor field. The distribution of pressure and temperature in the compressor was calculated by the CFD software. Figure 2 shows the distribution of pressure and temperature with the male rotor and female rotor. In the process of working, the pressure on the rotor surface is greatest near the exhaust orifice.

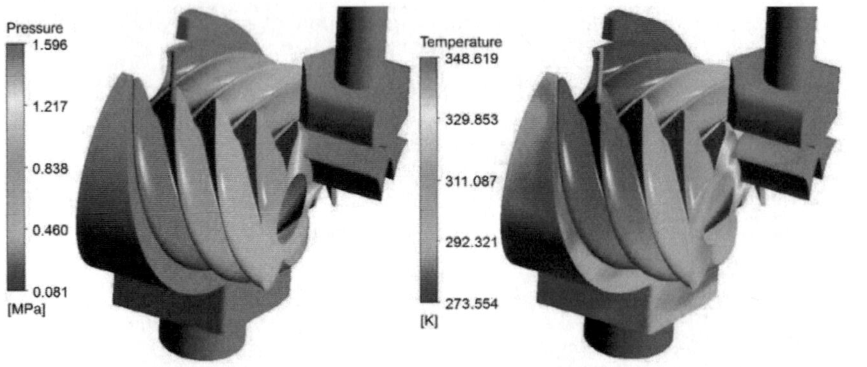

Figure 2: The distribution of pressure and temperature inside the compressor

During the working process of the screw compressor, the rotor is deformed due to the change of temperature and rotor force. The deformation of the twin-screw compressor rotor under three different conditions: pressure field, temperature field and pressure temperature field were analyzed in this paper.

4.1 Pressure Field

In the working process of twin-screw compressor, the pressure distribution in different tooth grooves is constantly changing, so the forces on the male and female rotors of the compressor are constantly changing.

The pressure field distribution calculated in the CFX flow field is imported into the ANSYS statics module as a result file. Figure 3 is the force diagram of the male rotor. In the working process, the axial force on the rotor, the radial force on the suction end and the radial force on the discharge end all show periodic changes. Figure 4 is the force diagram of the female rotor. The force change of female rotor is similar to that of male rotor, but the overall force is smaller than that of male rotor.

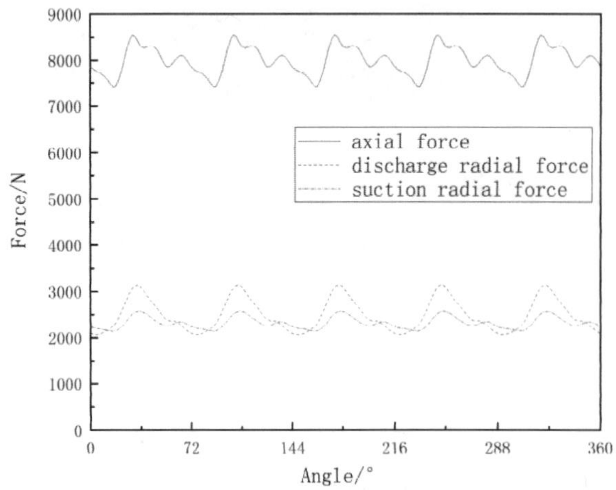

Figure 3: Male rotor force

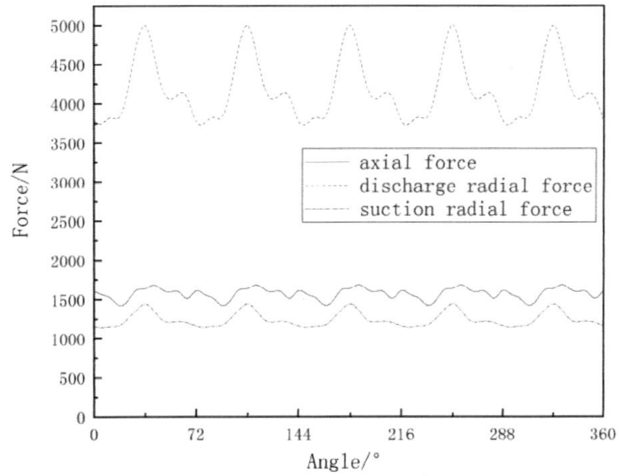

Figure 4: Female rotor force

Figure 5 shows that the internal torque of the rotor varies with the angle of the male rotor. And the power change of the rotor is shown in Figure 6.

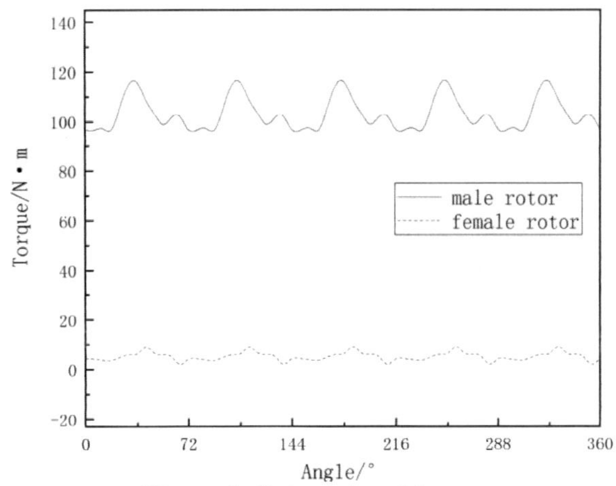

Figure 5: Rotor internal torque

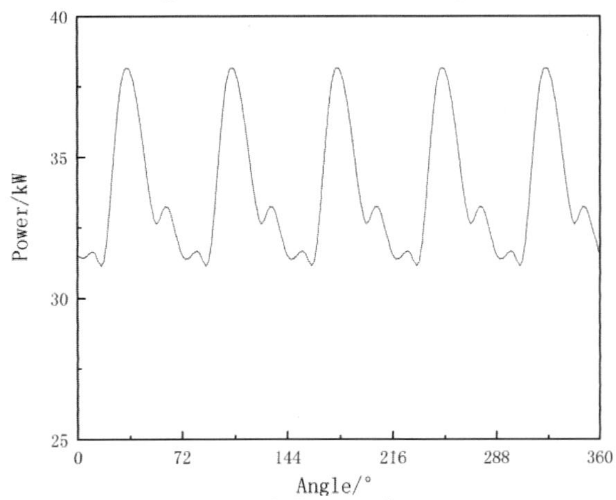

Figure 6: Power change

The forces on the female and male rotor have similar changes. With the increase of the rotation angle of the male rotor, the forces of female and male rotor change little. Among the axial force of female and male rotor and the radial force of suction and discharge end, the axial force of male rotor is the largest, and its maximum value can reach 8548N.For the internal torque of the male rotor, with the increase of the angle of rotation of the male rotor, its value gradually increases and then decreases. For the internal torque of the female rotor, its value basically does not change with the increase of the rotation angle of the male rotor. In addition, the internal gas torques of the female and male rotors are both positive value, which indicates that both torques are gas resistance moments. The power also increases at first and then decreases, and the maximum value reaches 38.16kW.
The following uses the rotor at four different angles of 12°, 24, 36°and 60° as examples to calculate the deformation of the rotor as shown in Figure 7.

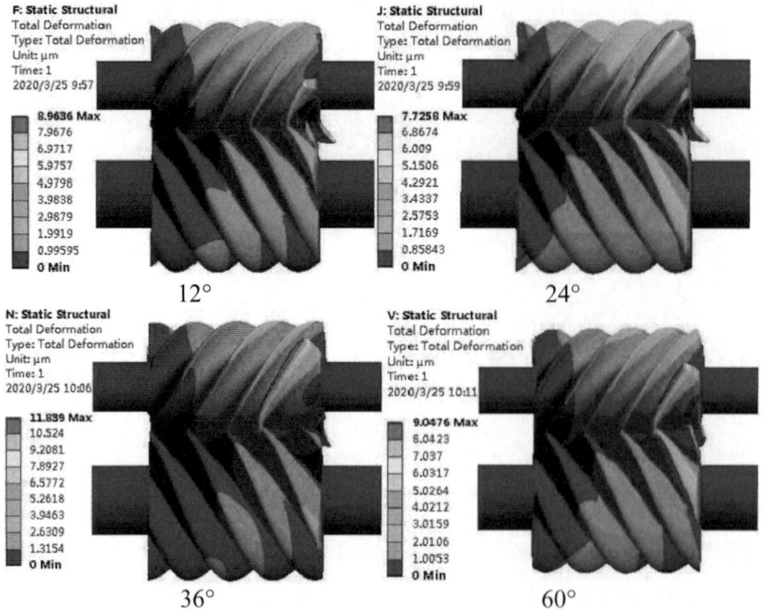

Figure 7: Deformation of the rotor at different angles

It can be seen from Figure 7 that the deformation of the female rotor is greater than that of the male rotor at different angles. This is because the teeth of the male rotor are thicker than the teeth of the female rotor, and the tooth stiffness of the male rotor is better. Moreover, the deformation of the rotor near the discharge end is larger than that of the suction end. This is because the compressor tooth groove has a higher pressure near the discharge end.
Through fluid-solid coupling calculation, the maximum deformation of the rotor under the pressure field varies with the angle of the male rotor as shown in Figure 8.

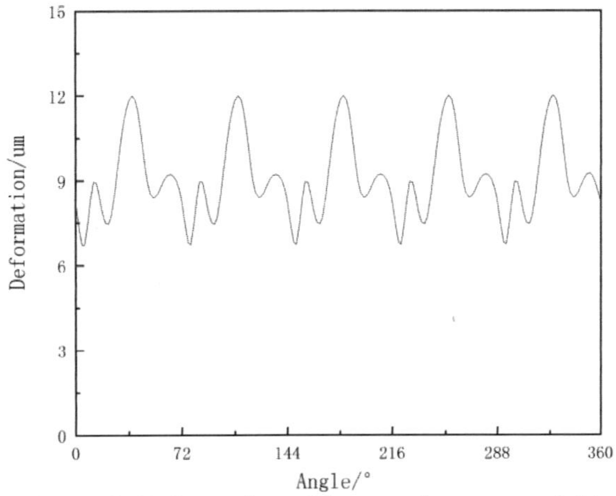

Figure 8: Deformation of the rotor in pressure field

It can be seen from Figure 8 that the maximum deformation of the rotor changes continuously with the change of the angle of the male rotor. When the angle of the male rotor reaches 36°, the deformation of the rotor reaches the maximum of 11.84μm.

4.2 Temperature Field

The change of temperature will lead to thermal stress in the working process of rotor. The female and male rotor will have thermal deformation under the temperature field.

When only affected by the temperature field, the deformation of the rotor of the twin-screw compressor at different angles is shown in Figure 9. And the maximum deformation of the rotor is obtained by fluid-solid coupling calculation as shown in Figure 10.

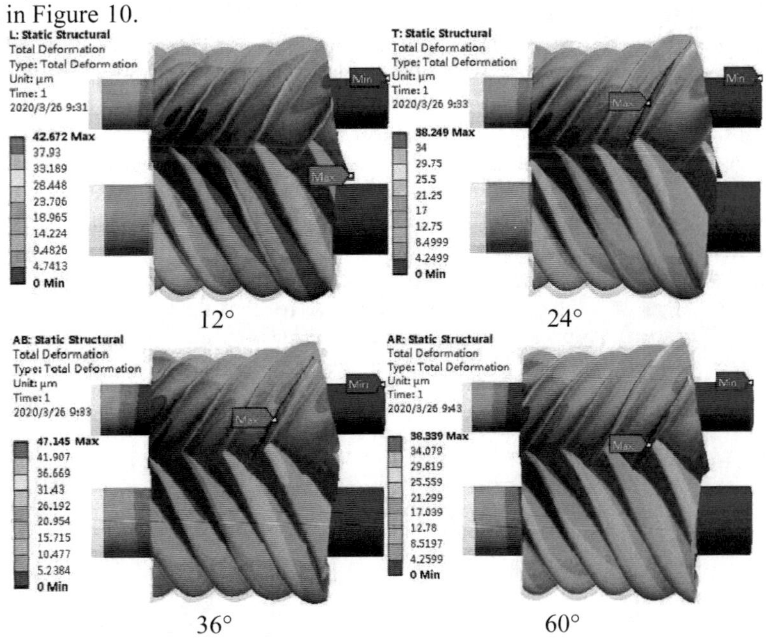

Figure 9: Deformation of the rotor at different angles

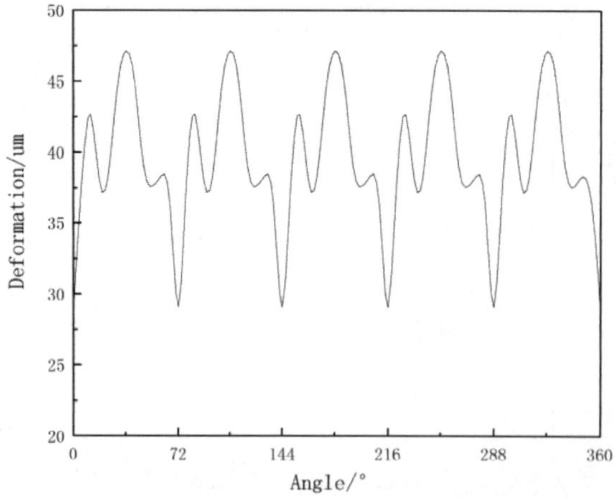

Figure 10: Deformation of the rotor in the temperature field

As a whole, under the temperature field, the thermal expansion and deformation of the male rotor near the discharge end is smaller than that of the female rotor, and the shrinkage deformation of the male rotor near the suction end is larger than that of the female rotor. It can be seen from Figure 10 that the maximum deformation of the rotor changes periodically. When the angle of the male rotor reaches 36 ° ,the deformation of the rotor reaches the maximum of 47.15μm.

4.3 Pressure and Temperature Field

In the working process of the twin-screw compressor, the force and deformation of the female and male rotor are affected by the temperature field and pressure field. The research on the rotor under the joint action of the two physical fields is closer to the actual situation of the rotor in the working process.

The deformation of the rotor of twin-screw compressor at different angles under the combined action of temperature and pressure fields is shown in Figure 11. As a whole, when the temperature and pressure fields are combined, the deformation of the female rotor is still larger than that of the male rotor. The maximum deformation position of the rotor at different angles is about 36°. This is due to the end of the compressor's external compression process at this time, the discharge temperature and discharge pressure have reached the maximum, resulting in the largest deformation there. The maximum deformation of the rotor is located on the female rotor, and the maximum deformation position is around the discharge orifice, so the deformation of the rotor is mainly affected by the discharge pressure and discharge temperature.

The maximum deformation of the rotor is obtained by fluid-solid coupling calculation as shown in Figure 12. When the angle of the male rotor reaches 36°, the deformation of the rotor reaches the maximum of 56.69μm.

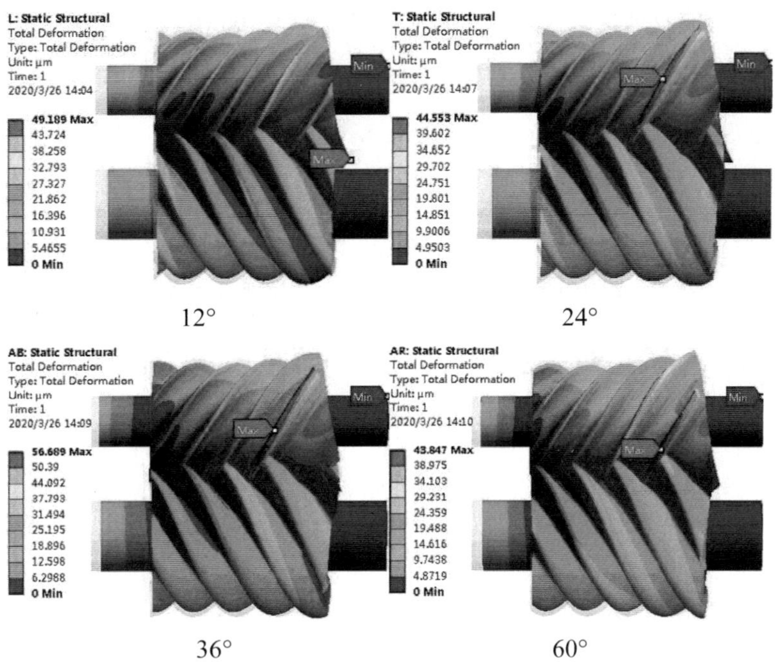

Figure 11: Deformation of the rotor at different angles

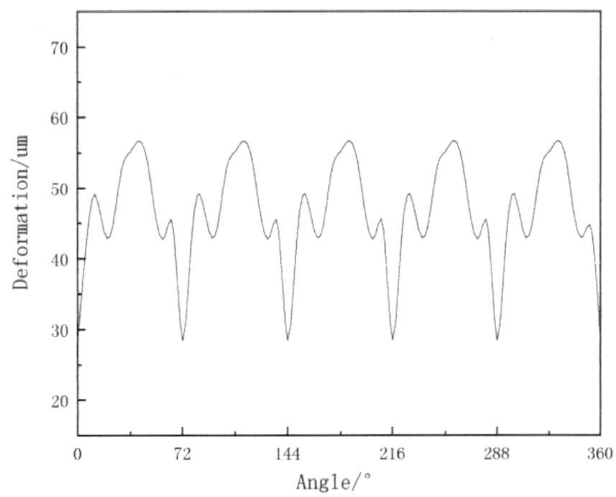

Figure 12: Deformation of the rotor in the pressure and temperature field

5. CONCLUSIONS

In this paper, the strain shape of the screw rotor under the multiple alternating physical action of gas temperature and pressure is analyzed, and the deformation of the rotor structure caused by the pressure and temperature in the working process is obtained. Our results can be concluded as follows.

Under the action of gas pressure, the force of the female and male rotor shaft of the twin-screw compressor is obtained by calculation. Under this condition, the maximum axial force on the male rotor is about 8500N, the maximum radial force on the discharge end is about 3100N, and the maximum radial force on the suction end is about 2500N. The radial force on the discharge end of the female rotor is the largest, and the maximum value is about 5000N. The maximum axial force is about 1700N and the maximum radial force at the suction end is about 1400N.

Through the analysis of the force and deformation of the rotor of the compressor under the combined action of pressure field and temperature field, it can be seen that the deformation of the rotor is mainly affected by the internal temperature distribution of the compressor and tends to expand at the discharge end and shrink at the suction end. Under the combined action of the two physical fields, the maximum deformation of the rotor is 56.69 μm. In this paper, the research on the force and deformation of the rotor provides a theoretical basis for the bearing selection and minimum clearance design of the twin-screw compressor, and the study has important guiding significance for improving the performance and structural safety of the twin-screw compressor.

REFERENCES

1. N. Stosic, Lj. Milutinovic, K. Hanjalic. (1992). Investigation of the influence of oil injection upon the screw compressor working process. *International Journal of Refrigeration,15*(4), 206-220.
2. XING Ziwen. (2000). Screw compressors-theory, design and application. *Beijing: Machinery Industry Press*, 26-28.
3. N. Stosic, I. K. Smith, & A. Kovacevic. (2003). Optimization of screw compressors. *Applied Thermal Engineering.23*(10), 1177-1195.
4. CAI Hong & YANG Wei. (2019). Calculation Method for Contact Line Shape and Length of Screw Compressors. *Compressor Technology,* (4),28-30.
5. ZAYTSEV D, FERREIRA, & INFANTE C A. (2005). Profile generation method for twin screw compressor rotors based on the meshing line. *International Journal of Refrigeration,* 28, 744-755.
6. Liu H, Xu H, & Ellison PJ. (2010) Application of Computational Fluid Dynamics and Fluid-structure Interaction Method to the Lubrication Study of a Rotor-bearing System. *Tribology Letters,* 38(3), 325-336.
7. AN Shaopei. (2015). Discussion of the Screw Compressor Selection Based on Rotor Speed. *Compressor Technology,* (3),45-47,56.
8. LI Hailin, ZHANG Zhengwu, & ZHOU Bin. (2016). Motion Simulation Analysis of the Screw Rotor System Based on ADAMS. *Mechanical Research and Application ,29*(1),77-79.
9. ZHAO Jialei, & LI Zhifeng. (2019). Study on dynamics performance of twin screw compressor. *Journal of Shaanxi University of Technology,* (04) ,28-32.
10. Seshaish, N., K. Ghosh, S. K., Sahoo, R. K. and Sarangi, S. K., (2006). Mathematical modeling of the working cycle of oil injected rotary twin screw compressor, *Applied Thermal Engineering, 27*, 145-155.
11. Yuren Wu, & Zhanghua Fong, (2008). Improved rotor profiling based on the arbitrary sealing line for twin-screw compressors, *Mechanism and Machine Theory,43*(6), 695-711.
12. Ahmed El Shorbagy, Auslegung, Konstruktion und numerische(2014). Simulation eines trockenlaufenden Schraubenverdichters und Vergleich der Simulationsergebnisse mit den Entwurfsanforderungen, *Technischen Universität Berlin.*
13. Feng Hou, Zhaorui Zhao, Zhiqiang Yu, et al.(2017). Experimental study of the axial force on the rotors in a twin-screw refrigeration compressor, *International Journal of Refrigeration,* (75),155-163.
14. Rane. S, Kovacevic. A., Stosic. N, Kethidi, M. (2014) Deforming grid generation and CFD analysis of variable geometry screw compressors. *Computers Fluids, 99*, 124–141.
15. LI Tuo-lei, WANG Jun-li, LEI shuai, et al. (2019). Research on Rotor Structure Characteristics of Twin-screw Compressor with Fluid-solid Coupling. *Compressor Technology,* (5),5-11.

Mechanistic Force Analysis of Single-Screw Compressor with 6-11 Geometry Configuration

Haotian LIU[1]*, Ashvin DHUNPUT[2], Harry MIYAMURA[2], Terry YOUNG[2], Dominic LAFFORD[2], Eckhard A. GROLL[1], Davide ZIVIANI[1]

[1] Ray W. Herrick Laboratories, School of Mechanical Engineering, Purdue University
West Lafayette, 47907-2099, USA
liu1460@purdue.edu; groll@purdue.edu; dziviani@purdue.edu

[2] ADC (UK) – Daikin Group
Unit 41 Pearson Way, Questor Estate, Dartford, Kent DA1 1JN, U.K.
ashvin.dhunput@jehall.co.uk, harry.miyamura@jehall.co.uk, twt.young@jehall.co.uk, dominic.lafford@jehall.co.uk

* Corresponding Author

ABSTRACT

In commercial and industrial Heating, Ventilating, Air Conditioning & Refrigeration (HVAC&R) applications, single-screw compressors have been widely used and proven to be efficient, reliable and economical in various operating conditions and capacity ranges. The efficiency of single-screw compressor depends on several factors, including the pressure ratio, volume ratio, and leakage flows as well as frictional, heat and mechanical losses. Thus, estimating forces and moments play an important role with respect to reducing mechanical losses. It is also critical to the reliability and safe operation of the machine. In this paper, a detailed mechanistic model of a 6-11 configuration single-screw compressor is described. The model is used to investigate the impact of design changes on compressor performance. The governing equations of conservation of mass and energy across the multi chambers of the compressor as well as the overall energy balance is developed to solve the thermodynamic model to capture the pressure and volume variation at different crank angles. The energy efficiency, mass flow rate and other compressor performance indices are also calculated. The simulation results are validated using experimental data obtained with a 6-11 screw compressor with R134a as the working fluid under several different operating conditions. The mechanistic model is achieved by implementation force and momentum balances on both the main rotor and starwheels to analyze the bearing forces, friction forces and rotor forces. The mechanistic model is integrated with the thermodynamic model and implemented in the Python programming language in the open source simulation package PDSim.

1. INTRODUCTION

Single-screw compressors are widely used and proven to be efficient, reliable and economical in vapor compression systems. The concept of single-screw compressor was firstly introduced in 1960s (Vrinat, 1984) and the first working prototype of single-screw compressor was built in 1963 (Zimmern and Patel). After the first prototype, the single-screw compressors have been widely used and largely developed in the following decades. There are several different possible positions for the central rotors which leads to different configurations for the shapes and locations of the starwheels (Masuda et la. 2013) The most common single-screw configuration is the Cylindrical-Plate type, which uses the meshing between a cylindrical central rotor and two planar starwheels. The central rotor is usually the prime mover with 6 grooves and driven by an electric motor. Two star rotors are equipped with 11 teeth and driven by the central rotor. This is usually referred as 6-11 configuration. The screw-like grooves are formed by the meshing, which gather vapors from the suction port, trap them in the pockets between the grooves and compressor housing. With the rotation of the rotor, the vapors are forced to the discharge port with decreasing in volume and increasing in pressure.

In compressor design process, detailed first-principle compressor simulation models are extremely useful in accurately predicting the compressor performance and identifying improved compressor designs. There are several modeling work focusing on single-screw compressor (van Male, 1978; Chan et al., 1981; Bein and Hamilton, 1982). More recently, model for single screw machine using an open source generalized simulation framework for positive displacement (PDSim) has been employed to predict the expander perfomance (Ziviani et al., 2019). The geometry analysis from the expander can be also used in the compressor.

In this paper, a detailed model of a single-screw compressor based on geometric parameters and a set of differential equations employing PDSim framework is developed and validated based on the previously developed expander model (Ziviani et al., 2019). Based on the model, a detailed mechanical forces analysis is performed to capture the pressure change, bearing force, reaction force and frictions of the central rotor and starwheels, which is important to many benefits in operation including long working life, high volumetric efficiency, low leakage, low noise and low vibration.

2. SINGLE SCREW COMPRESSOR MODEL

The single-screw compressor modeled in this paper is a 6-11 configuration with a sliding valve for variable volume ratio (VVR slide). The working fluid is R134a. The compressor model developed in the study is derived from the well-known compressor model established by Bein and Hamilton (1982). The model was improved by Ziviani et al. (2018) to have a more comprehensive description of the machine with increased complicity in chamber model, heat transfer and energy balance. There are two sub-models: single-screw geometry model that calculates the volume change of the control volume and thermodynamic model that handles the thermodynamic relation, heat transfer and energy balance. The general model descriptions and specific model modifications and improvements are briefly described in this section.

2.1 Geometry model

The main geometric characteristics can be found in Figure 1as well as the 3D CAD model of the single screw compressor. The complete formulation of the geometry model can be found in Ziviani et al. (2014). The control volumes and porting for each chamber are calculated by polygon description of single-screw geometry. The groove volume and surface are calculated by using the Green's theorem on area and volumes on the intersections of central rotor polygon and tooth polygon. The portion of tooth engaged with the groove can be described analytically by defining the coordinates of the four corner points, which represents the limits of the tooth engaged with the respective groove, as shown in Figure 1. The discrete increment of the groove volume is given by the product of the engaging tooth area A and the distance of its centroid from the main rotor axis at each rotation angle. The total groove volume is calculated as a sum of each discrete volume increment as

$$V=\sum_i \Delta V=\sum_i A_i c_r \Delta\theta \tag{1}$$

Figure 1: 3D CAD model of the single compressor assembly and schematic of the single engaged tooth into the mating groove

Note that this only work with increasing of the volume from 0 to 180° of rotation. The decrease of groove volume is done by subtracting the volume of the chamber on the other side of the rotor. As a tooth is disengaged, the groove is engaged on the opposite side of the rotor and as the rotation proceeds the groove decreases its volume on one side and simultaneously increases on the other side. With this approach, both starwheels are considered so that a complete evolution of the groove volume over one rotation of the rotor can be obtained.

Polygon approach is also used to determine the suction and discharge ports. The central rotor and the groove are unwrapped into a 2D plane. The suction ports are determined by the through the polygon intersection by the shape of the end of the groove. The length of leading and trailing edges can be determined at each crank angle. An improvement has been made on the discharge ports to add the function of sliding valve. The intersection of the unwrapped groove with the discharge port polygon determines the discharge port. Instead using a fixed port, a four-point discharging port achieved from the sliding valve of the CAD file is used in the current model. When the valve slides, the corresponding coordinates in the (Θ,z) reference system will change and create the new discharging port polygon accordingly. As shown in Figure 2, it can be observed at the same rotation angle. with the increased opening of the sliding valve, the discharging port polygon changes accordingly with a larger area. The discharging port which is the intersecting area of the two polygons also increases. This also influences the starting and ending location of the discharge process, which varies the volume ratio of the compressor. The PDSim model also has the capacity of inputting a set of (x,y) coordinates of the true geometry of the sliding valve to achieve higher accuracy, which will be necessary in the case of an irregular sliding valve geometry design. The (x, y) coordinates are the relative location in the unwrapped 2D plane with respect to the center of the screw rotor.

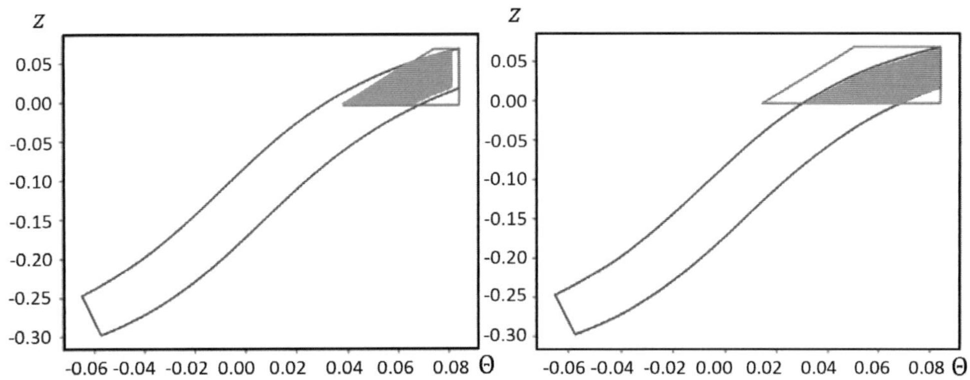

Figure 2: Groove polygon intersecting with sliding valve polygon in two different sliding valve locations

2.2 Thermodynamic Model

Thermodynamic model needs to be coupled with the previously described geometry model to solve the overall energy balance. In order to accurately capture the compressor performance, the thermodynamic model includes calculation flow paths, leakage flows, heat transfer and overall energy balance.

Flow paths are required to connect the different control volumes or inlet or outlet of control volumes within the compressor. In a flow path, the upstream state is fully specified and, in general, the downstream pressure is known. The flow path model is then used to predict the mass flow rate through the flow path. There is a wide range of flow models that can be employed for a flow path. Flow models are used to describe suction and discharge processes as well as leakage flows. The flow model selected for a particular flow path depends on the geometry, flow type, and other parameters such if the flow can be treated compressible flow or not. The flows associated with the suction and discharges are assumed to be isentropic with a contraction flow coefficient. More detailed description of the flow models and correlations can be found in Bell et al. (2013). The leakage flows are also calculated by the corresponding flow models and correlations.

There are nine main leakage paths in a single-screw compressor as identified by Ziviani et al. (2019) and they can be classified based on the starwheel or central rotor locations:

- Starwheel: tip; leading flank; leading flank blowhole; trailing flank; trailing flank blowhole; window.
- Central rotor: leading thread land; trailing thread land; discharge end band

The leakage flows are related to the geometry of the starwheel and rotor assembly, such as the contact lines between the tooth and groove, the gap between the tooth and housing and the gap between the end rotor surface and the wall. The gap sizes are either measured from the 3D CAD file or assumed to be zero due to the lack of data.

Heat transfer happens during the compression process due to the temperature difference in the control volumes. However, since the structure of a positive displacement machine can be complicated, the temperature distribution inside the shell is subjected to spatial and temporal variations. Studying the overall heat exchange network inside a compressor or expander is of particular interest in order to accurately predict the performance of the machine. Moreover, some of heat transfer phenomena such as heating up of the suction gas in a compressor can be critical with respect to the decrease in volumetric efficiency. In order to simplify the overall heat transfer analysis, the spatial variations of temperatures within the different compressor elements are neglected which implies assuming that thermal gradients inside each solid or oil are relatively small compared to the average temperatures. It follows that temperatures associated with each compressor elements are uniform at any instant during the transient process. Such approach is usually referred to as the lumped capacitance method. By using the lumped capacitance method, a lumped temperature is assigned to each of the compressor components.

Both single-lumped temperature and multi-lumped temperature are adapted here. The single lumped temperature is a simplified model that assumes a uniform temperature for all the compressor elements. It is effective in solving the energy balance with reasonable accuracy. The accounted heat transfers in the single-lumped temperature are summarized here:

- Between compressed gas and lumped mass
- Between lumped mass and the ambient

After solving the model using single-lumped temperature model, the multi-lumped temperature model can be applied to have a better understanding of the thermal loss and heat transfer inside the compressor. Each main component in the compressor is considered as a lumped mass with a uniform temperature. The higher the number of elements identified, the more complex the thermal network model becomes. For example, if considering the rotor housing, the gas in the shell and the shell as three lumped mass, the corresponding heat transfers are shown in Figure 3 as thermal resistance network. The thermal resistance and heat transfer correlation calculations has been thoroughly discussed in Ziviani et al. (2018). Same method has been used in this paper.

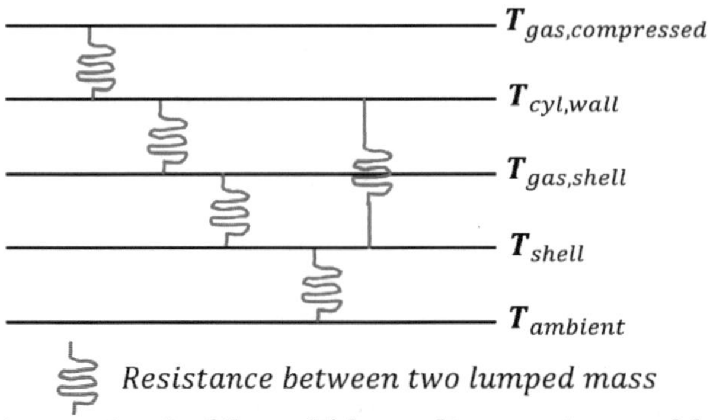

Figure 3: Thermal resistance network of the multi-lumped temperature model single-screw compressor.

Inside the compressor, in addition to the previously discussed heat transfer, there are also other mechanical or electrical energy converted into thermal energy such as friction losses, motor losses etc. A steady-state lumped energy for each of the lumped temperature can be obtained:

$$\sum_i (\dot{Q}_{lump,in,i} - \dot{Q}_{lump,out,i}) + \dot{Q}_{lump,gen} + \sum_j \dot{W}_{loss,j} = 0 \quad (2)$$

where $\dot{Q}_{lump,gen}$ is the heat generated rate from the lumped mass (e.g., chemical reactions), and $\dot{Q}_{lump,in,i}$ is the *i-th* heat transfer rate term calculated according to Figure 3. In all The thermodynamic properties of the working fluid are provided using the CoolProp routines (Bell et al., 2014). The resulting system of equations is solved at the end of each working cycle by employing a modified Broyden method (Bradie, 2006) to ensure the overall energy balance.

3. MECHANISTIC FORCE ANALYSIS

The single-screw compressor under investigation has a 6-11 configuration which has symmetrical starwheel location with balance hole connecting both ends of the screw rotor. The main mechanical force that needs to be overcome by the compressor driving mechanism is the rolling friction from the bearings that support the central rotor. Theoretically, there is no axial force and radial force on the central rotor which is one of the major characteristics of the single screw compressor. However, the weight of the screw rotor and overall shaft configuration (e.g., location of the motor) needs to be balanced, but it is out of the scope of this work. Such favorable mechanical configuration of the single-screw compressor yields to an efficient, quiet and reliable operation. The main challenges are associated with the wear of the tooth engaging profile. In Ziviani et al. (2014), the mechanical analysis was performed by a simplified model considering the friction torque. In the present work, a detailed mechanistic force analysis is proposed and implemented within the PDSim framework, which can accurately calculate the forces, torques and moments.

For the central rotor, the only existing force is the tangential force which is determined by the starwheel tooth meshing area in the groove and the corresponding gas pressure. It can be calculated as:

$$F_{gi} = P_{gas,i} \bullet A_{mesh,i} \tag{3}$$

$P_{gas,i}$ is the pressure at the corresponding tooth achieved from the calculation results and $A_{mesh,i}$ is the meshed area of the same tooth. The tangential force direction is opposite to the direction of the pressure force. For each tooth with meshing face, the pressure forces act on the centroid of the meshing area. The torque of the tangential force can be calculated by multiplying the force with the distance from the centroid to the axial of the rotor, which is the work needed by the compressor.

The starwheel force analysis plays an important role single screw compressor design and manufacture. The wear on the tooth surface increases the gap distance between the groove and tooth, which will lower the discharge volume and efficiency. The reaction force and corresponding friction forces on tooth surface determines the wear and losses. In order to investigate the surface wear and friction losses of the tooth, the detailed starwheel analysis is performed.

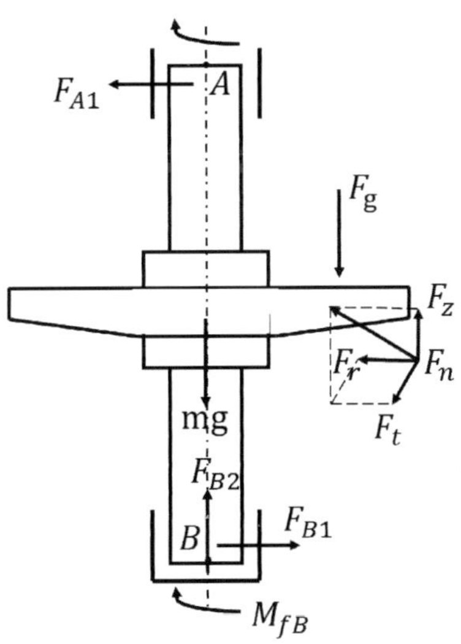

Figure 4: Free body diagram for the starwheel system including the axis AB and bearing

The starwheel system (including the axis and bearing) forces and moments balance is shown in Figure 4. The gas force F_g equals to the gas pressure on all meshed teeth.

$$F_g = \sum_{i=1}^{3} F_{gi} \tag{4}$$

where $F_{g,i}$ can be calculated in Equation 3. The normal force on the starwheel can be calculated as:

$$F_n = \sum_{i=1}^{3} F_{ni} \tag{5}$$

where F_{ni} has its three orthogonal decomposition as the tangential force F_{ti}, radial force F_{ri} and axial force F_{zi}.

$$\begin{cases} F_{ti} = 2T_{2i} / d_2 \\ F_{ri} = F_{ti} \tan\alpha \\ F_{zi} = 2T_{1i} / d_1 \end{cases} \tag{6}$$

where $T_{1,i}$, $T_{2,i}$ are the torque acting on the screw rotor and starwheel; α is the helix angle on the meshing point; d_1, d_2 are the diameter of the screw rotor and starwheel. The torque on the screw rotor and starwheel is related by the transmission efficiency:

$$\begin{gathered} T_2 = T_1 i \eta_m \\ T_1 = \sum_{i=1}^{3} T_{1i} \\ T_2 = \sum_{i=1}^{3} T_{2i} \end{gathered} \tag{7}$$

where T_1 is the total torque on the screw rotor and T_2 is the total torque of the starwheel; i is the transmission ratio and η_m is the transmission efficiency assumed to be 0.87. Friction between the starwheel and the groove has a significant effect on the tooth wear, which can be calculated with the normal force on the surface and friction coefficient at the surface:

$$F_f = \mu \sum_{i=1}^{3} F_{ni} \tag{8}$$

When the starwheels operate in steady state, the force and moment balance must be achieved:

$$\begin{cases} T_2 = M_{fA} + M_{fB} \\ -F_{Ai} - F_r + F_{B1} = 0 \\ F_{B2} + F_z - F_g - mg = 0 \end{cases} \tag{9}$$

$$\text{with} \begin{cases} F_r = \left| \sum_{i=1}^{3} F_{ri} \right| \\ F_z = \left| \sum_{i=1}^{3} F_{zi} \right| \end{cases}, \begin{cases} M_{fA} = F_{fA} r_A \\ M_{fB} = F_{fB} r_B + \frac{2}{3} r_{B2} \mu_{B2} F_{B2} \end{cases} \text{and} \begin{cases} F_{fA} = F_{A1} \mu_A \\ F_{fB} = F_{B1} \mu_{B1} \end{cases} \tag{10}$$

where F_{fA} and F_{fB} are the friction forces at the bearings and M_{fA} and M_{fB} are the corresponding friction torques at the bearings; r_A and r_B are the radius of the bearing; F_{A1}, F_{B1}, F_{C1} are the reaction forces at the bearings; μ_{A1}, μ_{B1}, μ_{C1} are the friction coefficients at the corresponding locations; m is the mass of the starwheel system.

By referring to Figure 4, if point A is considered to balance the moments (B can also be selected to run the calculation), the moment balance of the system is given by:

$$F_{B1} L_{AB} + \frac{F_z d_2}{2} - F_g r_c - F_r L_A = 0 \tag{11}$$

where L_{AB} is the distance between A and B; L_A is the distance from A to the top tooth surface of the starwheel; r_c is the radius of the circle where the centroid of the meshed face is. The gas pressure at each rotation angle needs to be calculated firstly as input to the mechanistic analysis. Equations (3) to (11) represent a system of equations with equal number of variables and equations. Thus, the system is determined and can be solved.

4. RESULTS AND DISCUSSION

4.1 Model Validation

A semi-hermetic single-screw compressor with a 6-11 configuration and sliding valves was tested at various working conditions including different pressure ratios and volume ratios. The working fluid of the compressor is R134a and the experimental operating condition is reported in Table 1. Note that at these different testing conditions, the pressure ratio varies in the range 1.14-6.91, indicating that the sliding valve location is changed to have proper volume ratios to minimize over-compression losses.. In the experiment, the superheat and subcooling are also controlled. The mass flow rate, volumetric efficiency and isentropic efficiency are measured and calculated. The actual values of these are not reported in the paper but the normalized value will be reported for validation purpose..

Table 1: Experimental testing data for the given single screw compressor with R134a as working fluid

Test case:		A	B	C	D	E
Suction Temp.	**°C**	1.7	4	5	5	0
Discharge Temp.	**°C**	51.7	50	38	16.9	68
Suction Pressure	**kPa**	311.5	337.9	349.9	349.9	293
Discharge Pressure	**kPa**	1376	1319	963.7	519.2	2025
Pressure Ratio	**-**	4.42	3.90	2.75	1.48	6.91

The experimental data was used to validate the mechanistic model. The input parameters were set as suction temperature, suction temperature superheat and discharge pressure. The mass flow rate, volumetric efficiency, and overall isentropic efficiency have been predicted and compared with the experimental values. After the initial results with no tuning done on the model was achieved. The model was tuned by modifying the parameters that are guessed, including gap sizes, friction coefficients, etc. A summary of the results is shown in Table 2.

Table 2: Error in simulation results of mass flow rate, volumetric efficiency, and overall isentropic efficiency

	Case:	A	B	C	D	E
No tuning	Mass flow	0.46%	0.28%	6.48%	-0.09%	13.83%
	Volumetric eff.	3.39%	3.18%	2.92%	2.77%	17.06%
	Isentropic eff.	12.10%	10.51%	2.27%	-44.92%	53.87%
With Tuning	Mass flow	-3.11%	-2.07%	6.74%	2.31%	3.09%
	Volumetric eff.	-1.30%	-0.25%	2.14%	4.17%	4.94%
	Isentropic eff.	-9.76%	-1.56%	7.40%	-17.57%	-0.78%

The corresponding percentage error plots of the mass flow rate, volumetric efficiency and overall isentropic efficiency are shown in Figure 5. And the parity plot of the normalized mass flow rate is shown in Figure 7. It can be observed that without tuning, the model can predict the compressor performance under five different operating conditions relatively accurately. For all three parameters, the model over-predicted the values with only one under predicted data point in the prediction of isentropic efficiency because not all the possible losses and leakages were considered and calculated with given information. In the untuned model, when assumptions were used to achieve values such as gap size, friction coefficients, among others, conservative numbers were selected based on the experience. The errors also vary largely with different cases. For instance, test conditions A, B and C were predicted more accurately than cases D and E. The isentropic efficiency in case D and E had a significant error. The mass flow rate and volumetric efficiency of case E also had a relatively large error. The main reason is the extreme pressure ratio applied in the experiment, where the sliding valve has been modified to adjust the volume ratio achieved by the compressor. In order to address these errors, we further tuned the model to capture the actual compression process. The leakage paths, gap sizes are modified. More importantly, the slider valve openings are adjusted to achieve the reported volume ratio in the test for each operating conditions. The tuned model predicted all five conditions more accurately. It can be seen in Figure 5, the tuned results showed good consistence in predicting all three parameters. Most of the error is within 5% with only a few outliers. Nevertheless, the validation indicates that the model can predict the perfomance at different operating conditions. To improve the accuracy of the model, future tuning can be performed along with gathering additional physical information about the compressor.

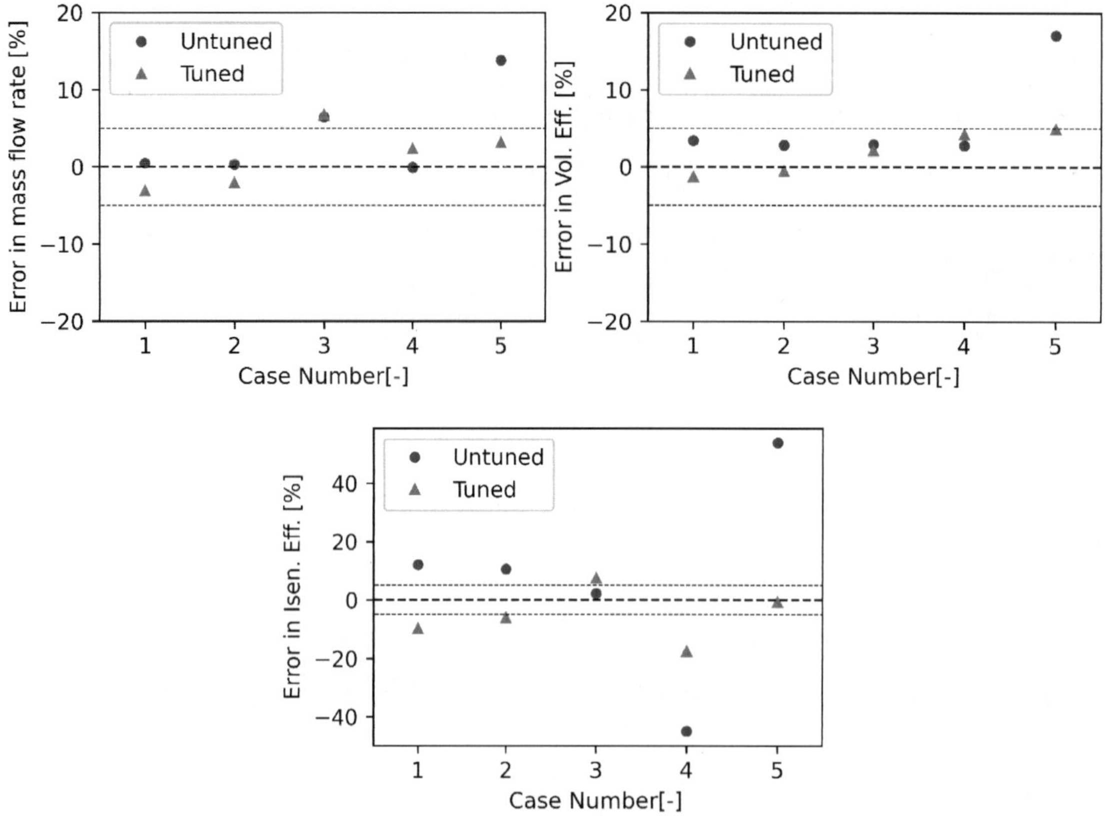

Figure 5: Percentage error of the mass flow rate, volumetric efficiency and isentropic efficiency of the results

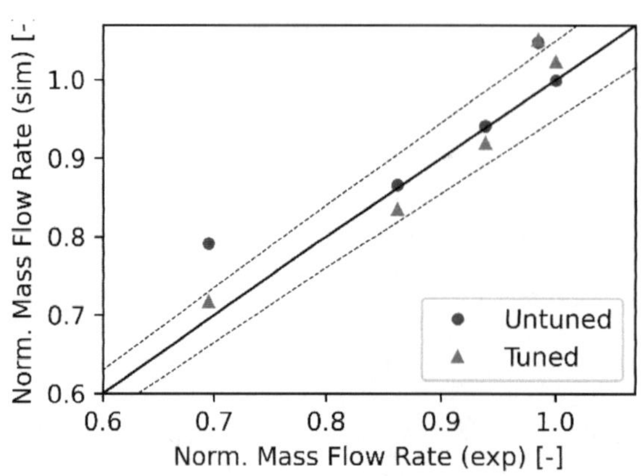

Figure 6: Parity plots of the normalized mass flow rate

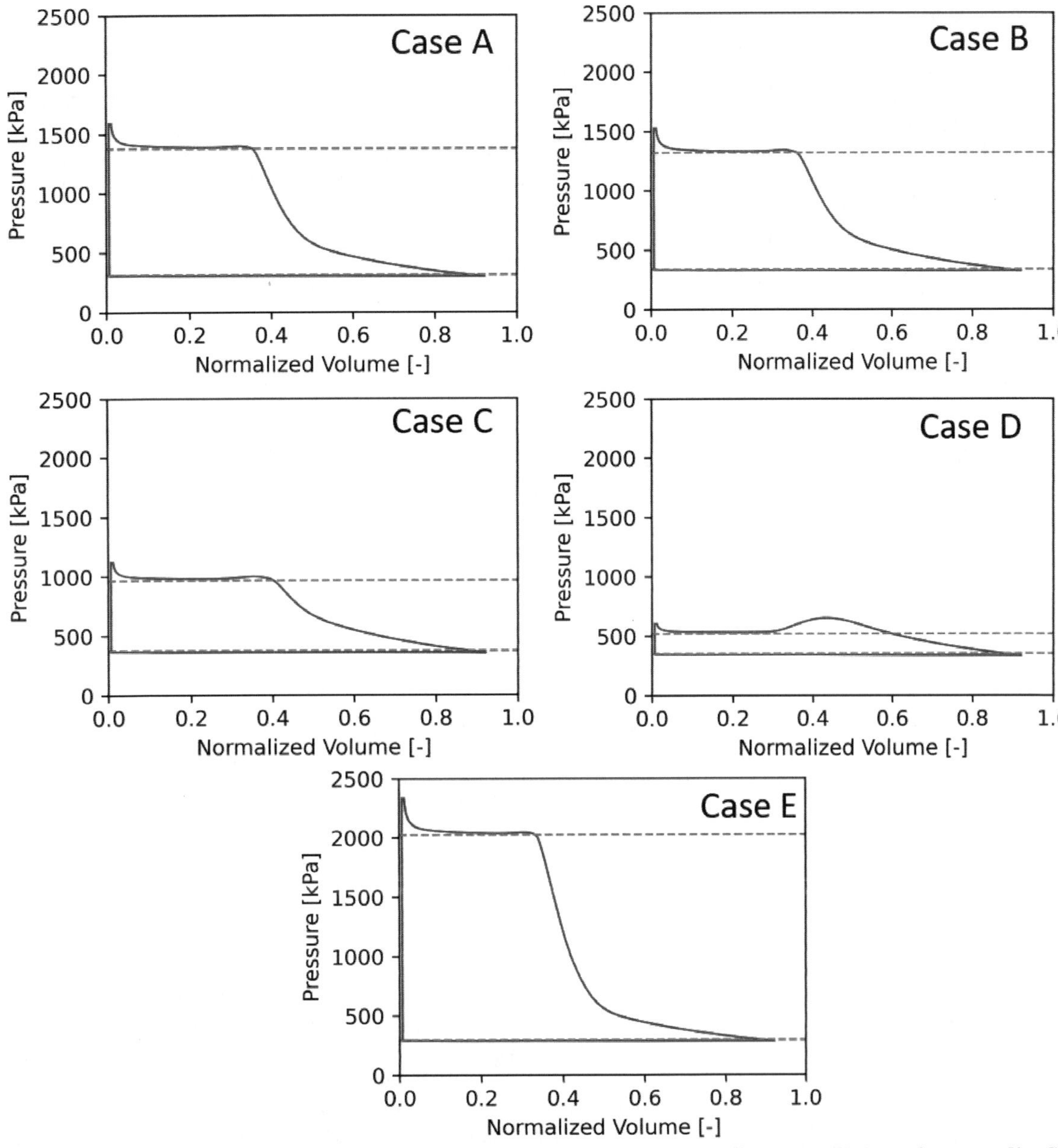

Figure 7 Transit results: P-V diagram of the PDSim simulation in all five cases (Volume is normalized)

4.2 Transient Results with Force Analysis

The model can also predict the transient trends of the compression process and instantaneous force variations with respect to the rotation angle. These transient results can provide helpful insights on the compression process, which helps the design process and balance the loads. The calculated indicated diagrams are shown in Figure 7. The discharge and suction pressure of each operating conditions are indicated by dash lines accordingly. The indicated diagrams provide information regarding the position of the slider and possible compression losses. The suction process starts at a pressure slightly lower than the suction pressure and continues until the groove volume achieves its maximum. After suction closure, as the screw starts to rotate, the groove volume decreases and the pressure increases, until it reaches the discharge port. It can be seen in the different cases that the in-chamber pressure reaches discharging pressure at different volumes, which indicates the sliding valve is needed to adjust the volume ratio according to the pressure ratio. The discharge process starts once the compression chamber is exposed to the discharge port, where the pressure will balance while the volume continues to decrease, until the discharging process finish. This completes a whole compression process for one chamber and for 6-11 compressor. However, six chambers will go through same compression process in one complete rotation of the screw rotor. The model is able to track all the chambers for a complete working cycle. Most of the models reported in the literature are limited to the closed compression process. By analyzing the P-V plots, there are two pressure overshoots that can be observed. The first one happens the moment preceding the opening of the compression chamber to the discharge cavity and the amplitude of the overshot is dictated by the location of the sliding valve that can create a mismatch between pressure ratio and volume ratio. The second overshot happens at the end of the compression process where the groove is reaching the minimum volume that represents a dead volume in which the gas is being compressed before opening to suction on the other side of the screw rotor. Since the model can capture these phenomena, it can be used to further optimize the discharge port and also the timing of the compression process at different operating conditions by adjusting the slider with a feedback controller or model-based controller.

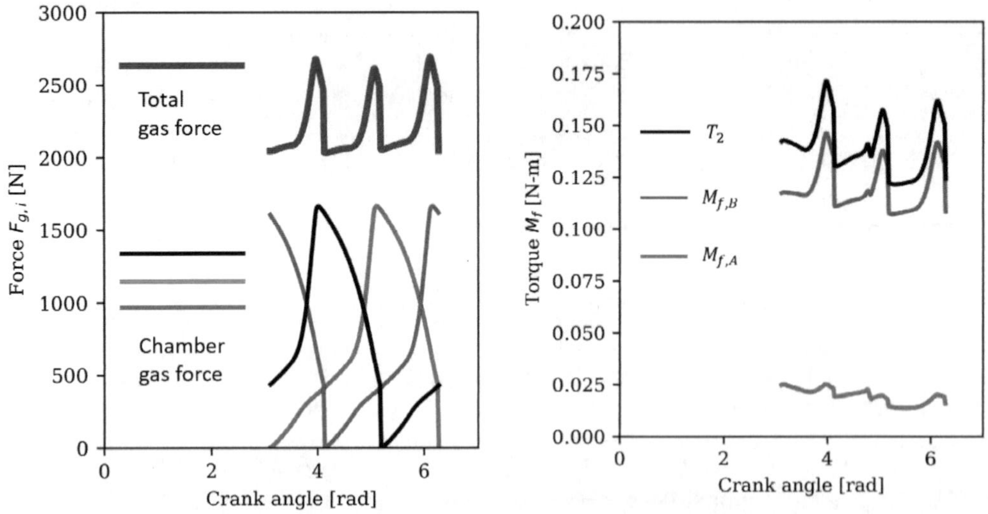

Figure 8 Transient results: chamber gas forces and total gas force change with crank angle and starwheel torque forces and friction torque forces at end A and end B

The forces of the single screw compressor have also been analyzed. Due to the similarity in different operating conditions, case A was selected as an example to show the detailed force analysis. Figure 8 shows the gas forces for each chamber and total gas force acting on one side of the screw rotor with one star-wheel. The other side of the screw rotor has same results according to symmetry. With solved gas forces, the reaction forces, torques, friction forces and friction torque can also be calculated. The torques at the starwheel as well as the friction torques at both ends are shown in Figure 8. It can be seen that the friction torque at end B follows the change with gas force while the friction torque at end A stays relatively constant. The total torque at the starwheel equals the sum of the friction torques. This is consistent with the analysis as indicated in Figure 4. The reaction force at end B balances the gas force along z-axis while reaction force at end A is not directly determined by the gas force. The estimated torque forces can guide the design of the bearings and material selection. The bearing at the shaft end B should be stronger than the bearing at the shaft end A to resist the higher friction torque with larger variation.

5. CONCLUSIONS

In this paper, a comprehensive model of a 6-11 configuration single-screw compressor has been presented and validated. By using the polygon approach, a geometry model has been built with variable volume ratio (VVR slide) enable with sliding valve. A detailed mechanical force analysis is performed focusing on the force and moment balance on the starwheel system. The model is validated with experimental data obtained at five different operating conditions. The model predictions show good agreement (within 20%) for all the performance indices. Further investigation of the detailed measurements taken with the studied compressor at different sliding valve locations can further improve the model accuracy. The force and torques measurement should also be considered for the validation of the mechanical force analysis.

NOMENCLATURE

A	area	m^2
C_r	centroid distance from axis	m
F	force	N
g	acceleration of gravity	m/s^2
i	transmission ratio	-
L	length	m
m	mass	kg
M	moment	Nm
P	pressure	kPa
$\dot{Q}$	heat transfer rate	kW
r	radius	m
T	torque	Nm
$\boldsymbol{T}$	temperature	K
V	groove volume	m^3
$\dot{W}$	work transfer rate	kW
z	z direction along rotor axis	m^3
α	angle	rad
Θ	unwrapped groove angle	rad
η	efficiency	-
μ	friction coefficient	-
ΔV	incremental groove volume	m^3
$\Delta\theta$	incremental rotation angle	rad

Subscript

1	screw rotor
2	star wheel
cyl	cylinder
f	friction
g	gas
gen	generation
i	index of summary
$lump$	lumped
m	mechanical transmission
mesh$i$	meshing area
n	normal direction
r	tangential direction

REFERENCES

Bein, T. W., & Hamilton, J. F. (1982). Computer modeling of an oil flooded single screw air compressor. In International compressor engineering conference. (Paper 383)

Bell, I. H., Groll, E. A., Braun, J. E., and Horton, W. T. (2013). A computationally efficient hybrid leakage model for positive displacement compressors and expanders. Int. J. Refrigeration, 36:1965--1973.

Bell, I. H., Wronski, J., Sylvain, Q., and Lemort (2014). Pseudo-pure fluid thermophysical property evaluation and the open-source thermophysical property library Coolprop. Ind. Eng. Chem. Res., 53:2498--2508.

Bradie, B. (2006). A friendly introduction to numerical analysis. Pearson Prentice Hall, USA.

Masuda M.; Ueno H.; Inoue T.; Hori K.; Hossain M. (2013) Effact of variable volume index on performance of single screw compressor. Proceedings of the 8th International Conference on Compressors and their Systems, City University of London, London, pp. 257-264.

Vrinat G. (1984). Helical or screw compressors. Int. J. Refrig., 7 (2), 107-114

Ziviani, D; Bell, I.; Zhang, X.; Lemort, V.; De Paepe, M.; Braun, J. E.; and Groll, E. A., PDSim: Demonstrating the capabilities of an open-source simulation framework for positive displacement compressors and expanders, International Journal of Refrigeration (2019)

Ziviani, D; De Paepe, M.; Braun, J. E.; and Groll, E. A., "Detailed Thermal Model of an Open-Drive Single-Screw Expander for ORC Applications" (2018). International Compressor Engineering Conference. Paper n. 2617, pp. 1-11.
Ziviani, D., Groll, E. A., Braun, J. E., & De Paepe, M. (2018). Review and Update on the Geometry Modeling of Single-Screw Machines with Emphasis on Expander. International Journal of Refrigeration, Accepted.
Ziviani, D., Groll, E. A., Braun, J. E., Horton, W. T., De Paepe, M., & van den Broek, M. (2017). Non-symmetric approach to single-screw expander and compressor modeling. In IOP Conf. Ser.: Mater. Sci. Eng. (Vol. 232).
Zimmern, B. and Patel, G. C. (1972). Design and operating characteristics of the Zimmern single screw compressor. In International Compressor Engineering Conference. Paper 16.

ACKNOWLEDGEMENT

The authors would like to thank Daikin Applied Development Center for sponsoring this research project.

Selection of Twin Screw Compressor Economizer Port Location to Optimize Unit Efficiency

Engineering Conference at Purdue 2020

Tasha Williams[1]*, Matthew Cambio[2]

[1] Michigan State University, Mechanical Engineering,
East Lansing, MI, USA
E-mail: will3218@msu.edu

[2] Trane Technologies, Compressor Development,
La Crosse, WI, USA
E-mail: mcambio@trane.com

ABSTRACT

Twin screw compressors are widely used in many applications. The performance limits of screw compressors are continuously being pushed to meet increased demand for energy efficiency. Several design adjustments can be made to the compressor to create a more efficient unit, such as the proper design of an economizer. The economizer is a port where refrigerant vapor is (re-)injected. General compressor effects of using an economizer are increase in compressor power, increase in discharge mass flow, and reduction of isentropic efficiency. However, the unit experiences an increase in refrigeration capacity and efficiency. The use of economizers for screw compressors is a developing application. This paper examines the addition of an economizer to a water chiller system with the intent of achieving a unit efficiency improvement. The economizer port size and location impacts the compressor performance. For this study, the port size location is varied to understand how it affects the unit performance. A useful procedure is developed to optimize the unit efficiency using internal compressor and system modeling tools is discussed within this paper. The selected port location will be used to create a new screw compressor design that is anticipated to achieve a 2% to 3% improvement in the unit coefficient of performance.

1. INTRODUCTION

Refrigeration and Air-conditioning systems have been around since the 1900's implementing the basic vapor compression cycle. The incorporation of these systems have become vital to daily life. However, there are several challenges that are met with the use of refrigeration and air-conditioning systems. One major challenge is global warming which stems from poor refrigerants and a high electrical demand. Over the years, many adjustments have been made to these systems to address these challenges. In order to improve environmental conditions, enhancements can be made to the vapor compression cycle or the components within the vapor compression cycle (Moon et.al, 2008). The vapor compression cycle consists of compressor, condenser, expansion valve, and evaporator. The compressor component has a large impact on the systems overall performance thusly the compressor will be the component of focus within this paper specifically the screw compressor. Screw compressors have been used for several years for a wide variety of applications. In addition to altering the compressor component, the cycle is also being modified to incorporate vapor injection. In the recent years, the advantages of vapor injection or economizers in compressor types has been explored and is known to have a positive impact on the overall performance of the cycle. The addition of an economizer within a twin screw compressor has two key benefits; increase in coefficient of performance (COP) and capacity (Wang et al., 2009). This paper observes the impact the economizer port size and location have on a water chiller system with the intent of achieving a unit efficiency improvement.

2. VAPOR COMPRESSION CYCLE

The vapor compression cycle in Figure 1 is commonly used in refrigeration and air-conditioning systems. During the vapor compression cycle the refrigerant goes through four stages. The refrigerant is compressed, after compression the vapor is condensed to a pure liquid. The pure liquid is then expanded reducing in pressure and temperature. The quality of the refrigerant changes as it goes through the expansion process and consist of two phase flow. Lastly, the refrigerant goes through the evaporator and leaves as a gas. Useful cooling and boiling is obtained using the liquid within the evaporator however the vapor is energy wasted. This is referred to as the lost refrigeration effect. The total loss varies with system pressure differential and refrigerant being used. However, this inefficiency has been known to contribute to over 30% of loss using R134a (Cambio, 2015).

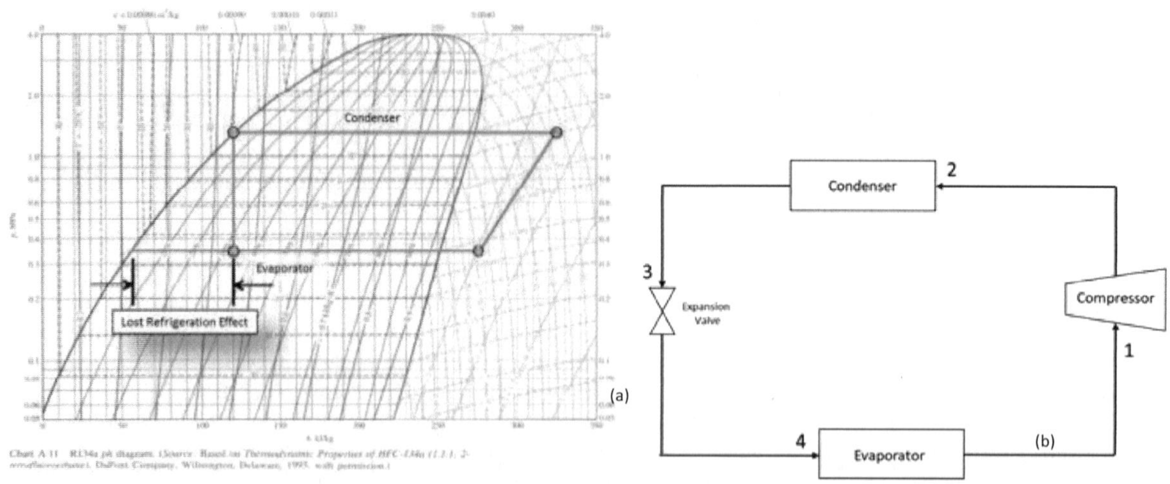

Figure 1: (a) p-h chart of vapor compression cycle (b) schematic diagram of vapor compression

One modification that can be implemented to improve the lost refrigeration effect within the vapor compression cycle is the inclusion of vapor injection also known as an economizer. The vapor compression with vapor injection cycle includes basic components with the addition of a flash-tank and modified compressor that contains an economizer port to allow for vapor injection. The vapor compression cycle with vapor injection (seen in Figure 2) is slightly different from the basic cycle. The refrigerant vapor enters the condenser (point 4). From the condenser (point 5) the refrigerant is expanded to an intermediate pressure and enters the flash-tank (point 6). The liquid (point 7) and vapor (point 9) are separated from each other within the flash-tank. This is done by using both velocity reduction and gravity effect which allows the liquid portion to exit the bottom of the tank. The liquid portion is further expanded and continues on to the evaporator (point 8). The refrigerant enters at suction of the compressor (point 1) from the evaporator which starts the 1st stage of compression. The vapor portion in the flash tank is injected into the compressor through the economizer port (point 9) and mixes with the refrigerant in the compression chamber (point 2) and this starts the 2nd stage of compression (point 3). This is often referred to as flash-tank vapor injection. There are several benefits of using a flash-tank cycle, namely; lower compressor discharge temperature, increases capacity, and reduces power consumption (Xu et al., 2010). In order to obtain the efficiency benefits, the economizer port size and location are important parameters that should be optimized.

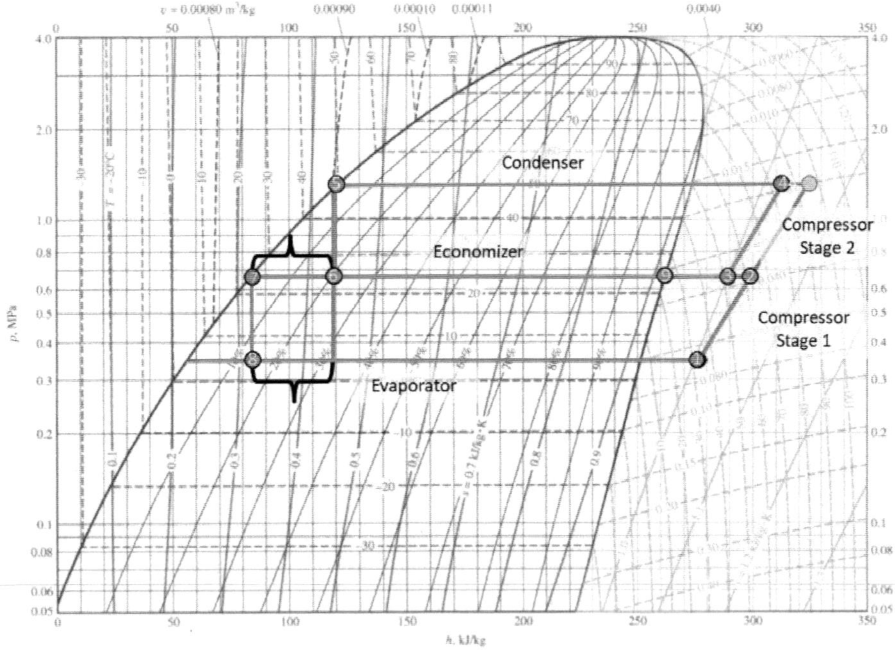

Chart A-11 R134a ph diagram. (Source: Based on Thermodynamic Properties of HFC-134a (1,1,1, 2-tetrafluoroethane). DuPont Company, Wilmington, Delaware, 1993, with permission.)

Figure 2: Vapor compression cycle with vapor injection

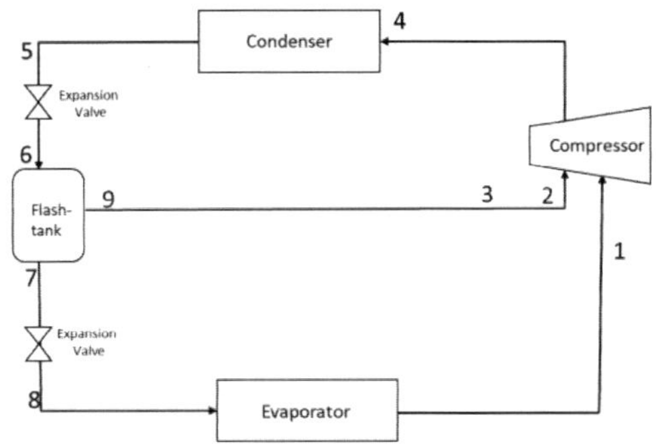

Figure 3: Schematic diagram of vapor compression cycle with vapor injection

3. ECONOMIZER PORT OPTIMIZATION

The inclusion of an economizer increases the overall unit performance. Two critical design parameters of compressors with economizer are port size and location. In-house analytical tools were used to determine a location and size that maximizes the overall unit performance. Each parameter affects the performance differently and the prime combination is desired. The final results are compared to the unit without an economizer port with COP, power, and capacity at 5.89, 317.43 kW, and 413.14 tons respectfully. R134a is the working fluid for this study. Theoretically the flash-tank cycle was modeled to understand how COP changes with respect to the economizer injection pressure while assuming a 100% compressor isentropic efficiency. The economizer injection pressure is calculated using Equation (1) where evaporator and condenser saturation temperature is 40°F and 100°F respectfully. The pressure split, r, is defined as the percentage of pressure between condenser and evaporator and is varied from 10% to 80% with 10% being closer to the pressure of the evaporator and 80% being closer to the condenser. As seen in Figure 4, the maximum COP is found around an injection pressure of 40% of the compressor total pressure rise. Furthermore, this shows us that the COP has a strong dependence on the economizer pressure in the vapor compression cycle.

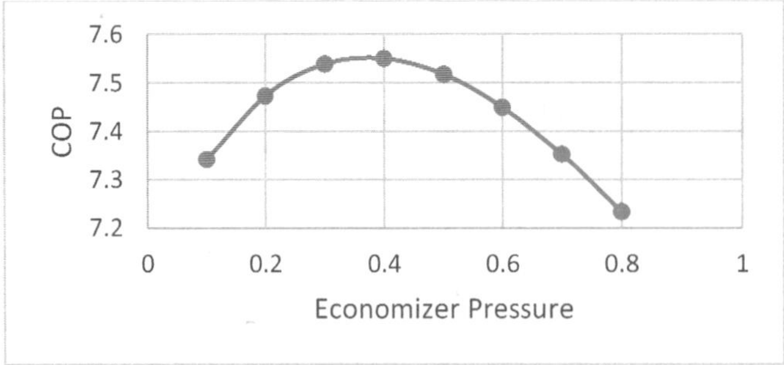

Figure 4: COP vs. Economizer Pressure

Applying this understanding, the economizer port size and location parameters are evaluated using a screw compressor model with the addition of an economizer port. The analysis was completed with an in-house 1-D compressor model along with a unit model. The 1-D screw compressors model is a Trane internal design code. The model follows conservation of mass and energy by modeling a single compression chamber. Leakages from the upstream and downstream chambers are accounted for. The model is comprised of an inner loop which use commercial differential equation solver for the compression process and an outer loop which takes care of parasitic losses such as motor and bearing losses, heat transfer, etc. Inputs to the compressor model include chamber volume and port areas vs. crank angle, as well as motor efficiency and mechanical loss models. In this case a flash tank economizer was used to simplify the analysis. Other inputs include suction, discharge and economizer saturation conditions. However, when conducting a study of economizer port geometry and attempting to capture the unit effects it was necessary to link the compressor model to a Trane internal unit modeling tool. The unit model has detailed models of the heat exchangers and economizer so that the effects on the unit of changing mass flow are accounted for. This was done by transferring mass flows and power from the compressor model to the unit model, solving the unit model and transferring saturation conditions back to the compressor model in an iterative method until mass flows and power was converged. This modeling tool produced unit capacity, coefficient of performance (COP), and other unit values. Suction flow, economizer flow, and power, are calculated by the compressor model using Equations (2) - (6).

$$p_{econ} = r\left(p_{cond} - p_{evap}\right) + p_{evap} \tag{1}$$

$$\dot{m}_{suc} = \frac{V * N * \omega}{v_{suc} * 2\pi} \tag{2}$$

$$\dot{m}_{econ} = \int_{t_1}^{t_2} \frac{dm_{econ}}{dt} dt \tag{3}$$

$$\frac{dm_{econ}}{dt} = A_{econ} * C_d * \left(\rho_{econ} \sqrt{\frac{2 * k * p_{econ}}{(k-1)\rho_{econ}}} * \sqrt{\left(\frac{p_{pocket}}{p_{econ}}\right)^{\frac{2}{k}} - \left(\frac{p_{pocket}}{p_{econ}}\right)^{\frac{k+1}{k}}} \right) \tag{4}$$

$$P_{shaft} = \dot{m}_1(h_2 - h_1) + \dot{m}_9(h_2 - h_9) \tag{5}$$

$$P_{kw} = \frac{P_{shaft}}{\eta_{motor}} \tag{6}$$

3.1 Size

For this study, the port size ranges from 0.2 in. to 1.2 in. while keeping the port location at 25 degrees of the male rotor rotation. This position is just after closure of the suction port. The evaporator and condenser saturation temperatures are 43°F (6.11° C) and 96°F (35.56° C). When the size goes below 0.2, the injection flow decreases substantially and there is no reason to reduce the port further in size. Figure 5 shows the change in performance due to port size. It is seen that as the port decreases in size, the unit performance increases. This is partly due to the decrease in leakage effect. The increase in COP is mostly due to the economizer injection pressure moving closer to the middle of the cycle. The middle of the cycle is defined using the condenser and evaporator pressures. Likewise, the economizer port size and injection pressure affect the economizer injection flow which in turn affects the pocket pressure and fixed volume ratio effects. Additionally, as the economizer port gets smaller in size the leakage effect from the high-pressure pocket to the low pressure pocket decreases. Observing the trend, there is no maxima reached and this is likely due to the placement of the economizer port being so close to suction and the large reduction in economizer flow. As the economizer flow decreases, the condenser flow and saturation temperature decreases as well which leads to a lower condenser pressure and an increase in cycle performance.

Table 1: Size vs Unit Performance

Case #	D_{econ} in (cm)	Angle (Deg)	Econ Flow CFM (m^3/s)	COP	% COP Improvement	Power (kW)	Capacity Tons
1	1.2 (305)	25	79.48 (0.038)	6.09	3.3%	264.72	458.18
2	1 (254)		69.57 (0.033)	6.10	3.6%	262.18	455.05
3	0.8 (203)		54.02 (0.026)	6.12	4.0%	258.10	449.35
4	0.6 (152)		33.34 (0.016)	6.14	4.3%	251.95	440.03
5	0.4 (.02)		15.10 (0.007)	6.17	4.7%	244.85	429.51
6	0.2 (50.8)		3.80 (0.002)	6.18	5.0%	240.36	422.58

Table 2: Running conditions per case with varying size

Case #	D_{econ} in (cm)	Angle (Deg)	Econ Pressure Psi (MPa)	Evap Sat Temp °F (°C)	Cond Sat Temp °F (°C)	Normalized Compressor Efficiency
1	1.2 (305)	25	76.32 (0.526)	43.40 (6.33)	96.39 (35.77)	0.957
2	1 (254)		79.40 (0.547)	43.40 (6.33)	96.37 (35.76)	0.959
3	0.8 (203)		84.80 (0.585)	43.41 (6.34)	96.23 (35.68)	0.963
4	0.6 (152)		92.79 (0.640)	43.43 (6.35)	96.01 (35.56)	0.957
5	0.4 (102)		95.68 (0.660)	43.46 (6.37)	95.75 (35.42)	0.988
6	0.2 (50.8)		96.66	43.47 (6.37)	95.58 (35.32)	1.000

3.2 Location

The location of the economizer port can be chosen to maximize capacity, performance, or something in between. The scope of improvement has a huge impact on the placement of the economizer port. Higher COP is found when the vapor is injected at an optimum pressure between the suction and discharge pressures of the compressor. There are several things to consider when placing the injection port. Injecting the vapor too close to suction can cause the vapor to enter the suction. The effects of vapor entering at the suction decreases the unit performance (Wang et.al, 2009). The injected vapor displaces flow from the evaporator and reduces suction capacity. Likewise, placing the injection port too close to the discharge will have a negative impact on economizer effectiveness due to discharge gas reversing flow through the economizer. Due to the helical nature of the rotors, the angular position of the male rotor translates to and axial distance along the axis of rotation of the rotor (seen in Figures 5 and 6).

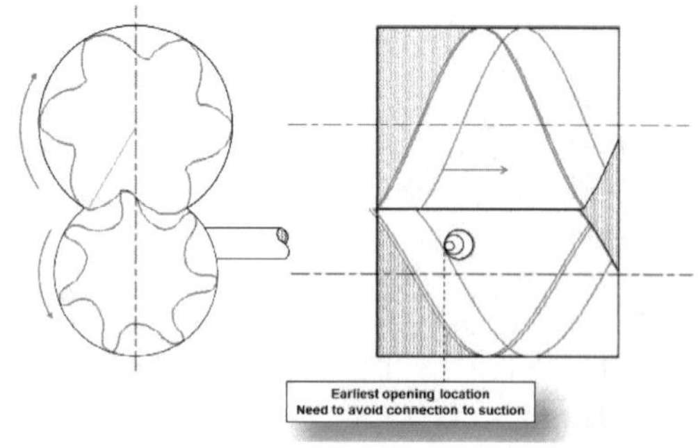

Figure 5: Economizer port location representation at the low-pressure end of rotors

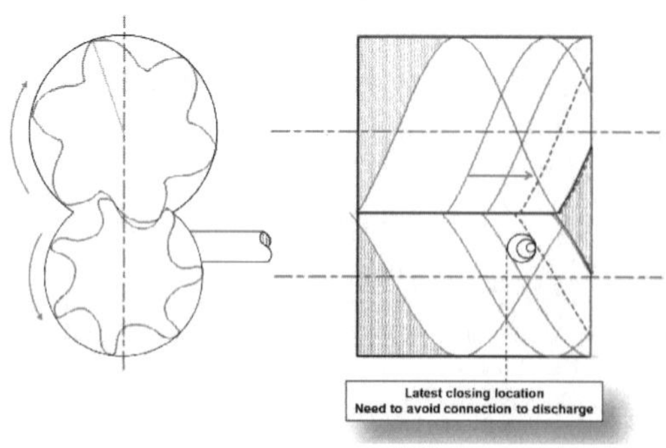

Figure 6: Economizer port location representation at the high-pressure end of rotors

For this study, the location ranges from 25 to 120 degrees of the male rotor rotation with the evaporator and condenser saturation temperatures roughly at 43°F (6.11° C) and 96°F (35.56° C). The maximum location was determined by the opening of the discharge port. Once the economizer port overlapped the discharge process, the compressor model would no longer run. Table 2 shows the change in performance and capacity due to port location. As the location moves further from suction the unit performance increases. When the port was placed at an intermediate pressure with the economizer pressure at roughly 95.52 psi (0.659MPa), the positive improvement on COP began to decrease. The intermediate pressure is approximately 55% of the difference of the evaporator and condenser pressures. From this information, it is seen that the unit performance has a strong dependence on the intermediate pressure. As the location of the port moves to a higher pressure in the compression process, the performance increases. In Figure 5, it is seen that the location has a greater influence on the COP than size does. It is seen in Tables 1 and 2 that the size and location of the port has a large influence on economizer pressure. The COP trend agrees with the original theoretical analysis shown in Figure 4 with a variation in the location of the peak. In theory, for the ideal case the maximum COP would be located around 40% while in this study it was located around 50%. Most likely the discrepancy between the theoretical analysis and the compressor model are changes in efficiency due to volume ratio effects. From this analysis, it was seen that when decreasing the size and moving the location further from suction an optimum improvement can be found. Both alterations affect the economizer injection pressure and overall system performance. Additionally, the compressor evaporator to condenser saturation temperatures ratio is designed for 40°F (4.44° C) to 100°F (37.78° C) while the conditions for this analysis are a lower pressure ratio which causes over compression. Over compression is increased with economizer injection which explains why the maximum compressor efficiency occurs with the smallest economizer flow.

Table 3: Location (angle) vs. Unit Performance

Case #	D_{econ} in (cm)	Angle (Deg)	Econ Flow CFM (m^3/s)	COP	% COP Improvement	Power (kW)	Capacity Tons
1	0.8 (203.2)	25	54.22 (0.026)	6.12	3.90%	258.23	449.42
2		60	42.17 (0.02)	6.17	4.82%	253.62	445.24
3		70	37.69 (0.018)	6.19	5.07%	251.88	443.21
4		85	30.75 (0.015)	6.21	5.39%	249.10	439.67
5		95	25.61 (0.012)	6.21	5.51%	247.13	436.68
6		109	17.92 (0.008)	6.22	5.52%	244.39	431.90
7		120	11.82 (0.006)	6.21	5.49%	242.25	427.99

Table 4: Running conditions per case with varying location (angle)

Case #	D_{econ} in (cm)	Angle (Deg)	Econ Pressure Psi (MPa)	Evap Sat Temp °F (°C)	Cond Sat Temp °F (°C)	Normalized Compressor Efficiency
1	0.8 (203.2)	25	84.72 (0.584)	43.41 (6.34)	96.23248 (35.68)	0.965
2		60	89.47 (0.617)	43.42 (6.35)	96.12066 (35.62)	0.976
3		70	91.20 (0.629)	43.43 (6.35)	96.06919 (35.59)	0.980
4		85	93.56 (0.645)	43.44 (6.36)	95.98098 (35.54)	0.986
5		95	94.62 (0.652)	43.44 (6.36)	95.90901 (35.51)	0.990
6		109	95.52 (0.659)	43.45 (6.36)	95.79605 (35.44)	0.995
7		120	95.84 (0.661)	43.46 (6.37)	95.70431 (35.39)	1.000

Figure 5: Graph of Unit Performance showing difference between size and location

4. CONCLUSION

The basic refrigeration cycle improvement in HVAC systems are continuing to improve. The exploration of economizers continues to increase COP within various application thusly reducing the impact on the environment. An optimized screw compressor economizer port location and size was analytically discovered in this paper to increase the unit efficiency within a water-chiller system. The design produced a predicted COP improvement of 5.52% which exceeded the 2-3% target. Ultimately, the location proved to have a larger impact on the COP in comparison to size. The change in overall performance with varying size did not follow the theoretical trend due to the influence of other parameters such as compressor efficiency, and saturation temperatures. Further investigation is required to determine size impacts when placed a significant distant past suction. Moreover, test will be conducted to determine the accuracy of these results. Based on the analytical models used, the results indicate:

- Moving the injection port location further away from suction increases the performance as a result of the intermediate pressure becoming closer to the optimum economizer injection pressure
- The size of the port caused an increase in performance due to a decreased leakage effect, reduced economizer flow, and movement of the economizer pressure towards the middle of the cycle.
- Optimal improvement can be found when adjusting both port size and location

NOMENCLATURE

The nomenclature should be located at the end of the text using the following format:

A	Area	(m^2)
C_d	Injection Port Flow Constant	(–)
COP	Coefficient of Performance	(–)
D	Diameter	(cm)
g	Gravitational Constant	(–)
h	Enthalpy	(kJ/kg)
k	Isentropic Exponent	
$\dot{m}$	Mass Flow	(kg/s)
m	Mass	(kg)
N	Number of Male Lobes	
η	Efficiency	(–)
p	Pressure	(Pa)
P	Power	(kW)
ρ	Density	(kg/m^3)

v	Specific volume	(kg/m^3)
V	Suction Chamber Volume	(m^3)
ω	Shaft angular velocity	(rad/sec)
X	vapor quality	(–)

Subscript

cond	Condenser
econ	Economizer port
evap	Evaporator
kW	Electirical
pocket	Compression Chamber
shaft	Shaft
suc	Suction

REFERENCES

Cambio, M. (2015). Model for Screw Compressors with Economizers. *Trane Internal Technical Note*. La Crosse, WI.

Moon, C.-U., Choi, K.-H., Yoon, J.-I., Kim, Y.-B., Son, C.-H., Ha, S.-J., Jeon, M.-J., An, S.-Y., & Lee, J.-H. (2018). Experimental study on the performance of the vapor injection refrigeration system with an economizer for intermediate pressures. *Heat and Mass Transfer*, *54*(10), 3059–3069.

Qiao, H., Aute, V., & Radermacher, R. (2015). Transient modeling of a flash tank vapor injection heat pump system – Part I: Model development. International Journal of Refrigeration, 49, 169–182.
Wang, X., Hwang, Y., & Radermacher, R. (2008). "Performance Investigation of Refrigerant Vapor-Injection Technique for Residential Heat Pump Systems" *International Refrigeration and Air Conditioning Conference*.

Wang, X., Hwang, Y., & Radermacher, R. (2008). Performance investigation of refrigerant vapor-injection technique for residential heat pump systems.

Wang, B., Shi, W., Han, L., & Li, X. (2009). Optimization of refrigeration system with gas-injected scroll compressor. *International Journal of Refrigeration*, *32*(7), 1544–1554.

Xu, X., Hwang, Y., & Radermacher, R. (2011). Refrigerant injection for heat pumping/air conditioning systems: Literature review and challenges discussions. *International Journal of Refrigeration*, *34*(2), 402–415.

ACKNOWLEDGEMENT

The authors would like to thank Trane Technologies and everyone who provided guidance for this research.

A Prediction Method for Slug Limit of Scroll Compressor using CFD Two-Phases Simulation

Kang ZHENG*, Jiangbo LIN, Liu WANG

Danfoss (Tianjin) Ltd., Tianjin Key Laboratory of Refrigeration and Air-conditioning Compressor Technology, Tianjin, P.R. China

Phone: +86 18622525618, Fax: +86 22 8219 6862, zhengkang@danfoss.com

* Corresponding Author

ABSTRACT

Slug test is used to simulate a real application work condition in lab. During operation, a compressor eventually gets to endure dynamic slugging which is characterized by the arrival of a liquid refrigerant at the suction side when the compressor is already running. Because of this dynamic, the liquid is able to reach the compression side. The presence of incompressible liquid in the scrolls (volumetric compression) involves mechanical strength quickly becoming unbearable for the compressor involute breakage or Oldham coupling breakage are the consequences suffered by the compressor following a severe slugging.

In this paper, a prediction method of slug limit is introduced which is based on simulation result numerical fitting to experimental result. In laboratory, engineers use dynamometer to monitor the image of compressor current intensity which shows a steep peak when more and more liquid refrigerant injected. Once the current peak ratio exceeds our goals, we would take the corresponding liquid quantity as slug limit of the compressor.

And this liquid injection process is transient simulated by CFD method using two-phases flow model – Eulerian multiphases model which allows for the modeling of multiple separate yet interacting phases. The phases can be liquids, gases, or solids in nearly any combination. Liquid refrigerant is thought as granular phase and calculated. Mass flow rate of each phase at outlets is recorded during calculation process and divided by normal mass flow rate (only gas suction). The maximum ratio of mass flow rate is used to predict slug limit comparing to experimental criterion. Using this method, it's easier to comparing with different parts structure which one is better for slug performance and which one is worse, saving time and cost and more and more helpful for engineering design.

1. INTRODUCTION

As we know, test can tell the slug limitation of compressor, but it cost too much, for example machining and assembling samples. It costs not only a lot of money but also quite a few of time. Using simulation tool is a reasonable solution to save many traditional tests. Following detailed introduces why we need slug test and how we did it in lab. We use current peak ratio as a criterion to judge the slug limitation of compressor rather than mass flow rate or pressure or temperature which are used in CFD code.

In this paper, we use CFD tool to simulate mass flow rate of liquid flow into scroll, which has similar changing tendency with current peak ratio. In other words, the current fluctuation happens when there is slug flow into compressor because motor load raises up suddenly which is also the result of a big mass flow into scroll. But we have to make some assumptions, for example the liquid mass particles are rigid body other than phase changeable liquid refrigerant particles. We must balance the calculation time and model complexity. If the calculation model was over-complex and cost too much time, it violates our original intension.

From our experience, the criterion of current peak ratio of the max load and no load of slug is set down for different type of compressors. But it is a different number for mass flow rate ratio. We simulate many types of Danfoss compressors and get a simulation judgement method by curve fitting to test result. Using this method, engineer gets many conveniences when design or modify a part which is related to slug performance, improve flooded start ability, application of different manifolding system and so on. The aim of slug simulation is not necessarily getting a specific number. To quantized judge the improvement level is more important for engineers.

2. TEST PROCEDURE AND RESULT ANALYSIS

2.1 Test Procedure

At the beginning of every test cycle, the system insures there is enough liquid refrigerant as test standard in a tank. Then open the valve inject the liquid into compressor in a very short time and record the current fluctuation during the injection process. As the current peak get the standard value, we think it get the limitation of this type of compressor.

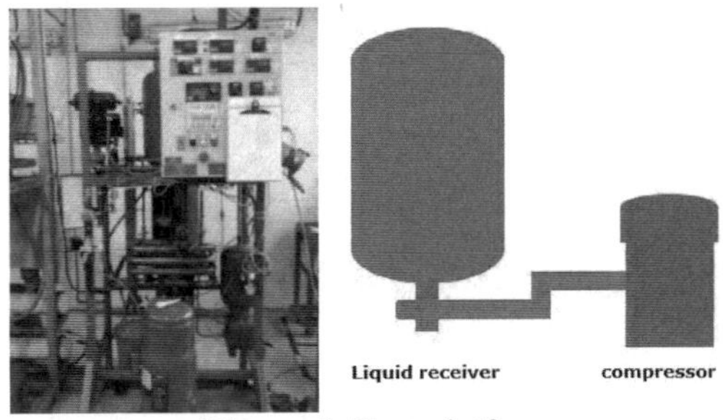

Figure 1. Test platform

2.2 Test Result Analysis

After each cycle, laboratory technician saves the screen shot of amperemeter, and calculates the ratio of max current peak and normal current value. There is an example of test result analysis shown as figure 2. The title is different quantity of liquid injection each time. When the liquid quantity is increasing, the current of compressor is also increasing at the same time. We set a standard of evaluation in advance, to judge the performance of compressor facing to this slug phenomenon in lab.

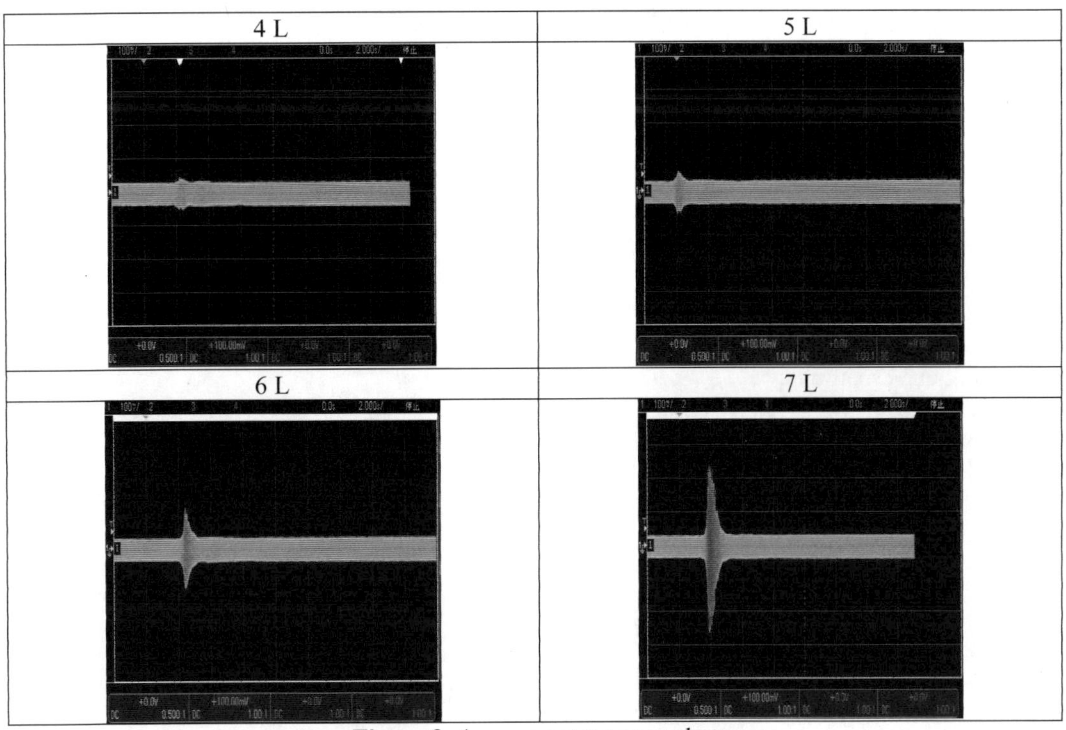

Figure 2. Amperemeter screen shots

In figure 2, injection quantity is varying from 4 to 7 liters. The current peak is getting steeper as the increasing liquid. Meanwhile, it can be record numerically as following figure 3.

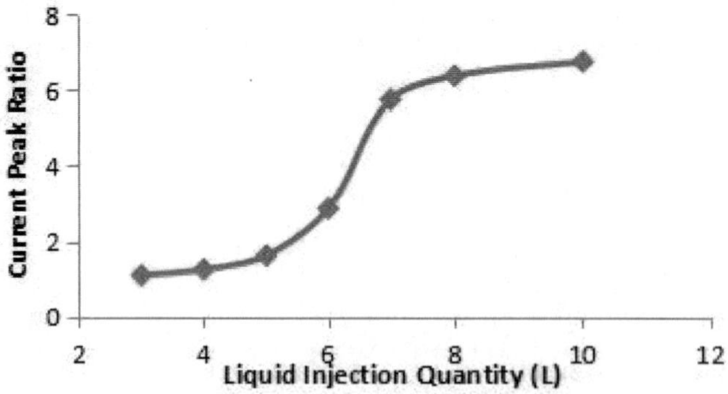

Figure 3. Current peak ratio by test

3. NUMERICAL SIMULATION

3.1 Calculation model

The calculation model is only about tested compressor and doesn't take the test system into account which is shown in figure 4. The inlet is compressor suction fitting and outlet is close to the scroll involute shown in figure 5. Although the area should be moving when it was running but we assumed a fixed vent as the calculation outlet. The whole domain is called low pressure side where pressure is close to suction pressure.

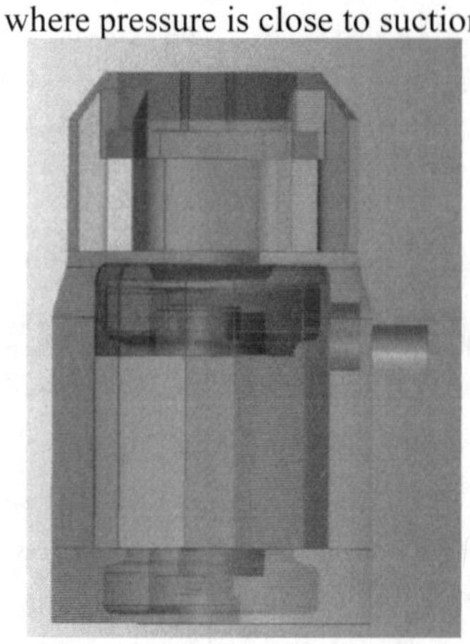

Figure 4. Calculation model

The calculation is to check the flow path of liquid refrigerant and velocity at the suction port of involute in transient process. Here are some hypotheses:

- Not considered geometry moving. We simulated dynamic process, but the model didn't move at all. Only boundary was related to time;
- Considered two phases flow. Used Eulerian multiphase model;
- Not considered heat transfer. The properties (density, viscosity) of material were constant.

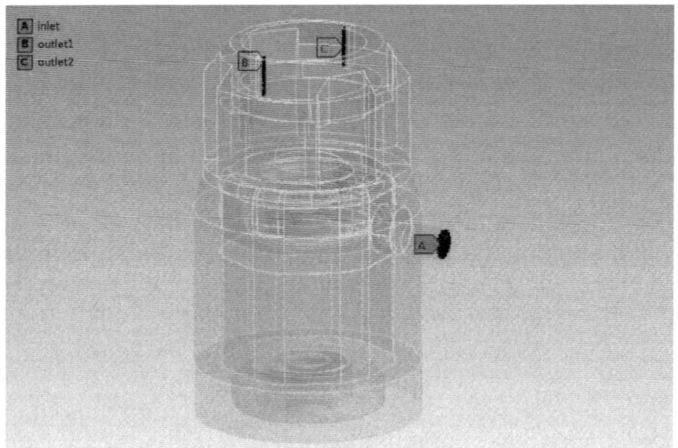

Figure 5. Boundary set

In initialization, whole domain is refrigerant gas and boundary condition for inlet is mass flow rate with all liquid. We monitored pressure changing during liquid injection process in test, and then record mass flow rate changing process, for instance 2 liters liquid injected into compressor in 2 seconds. We assumed the mass flow rate is linear decreasing as calculation input. Meanwhile we monitored mass flow rate at the outlets as the simulation output comparing with current ratio by test and got a simulated slug limit which could save many tests.

3.2 Calculation result

The calculation is unsteady and monitored result is varying with time shown in figure 6.

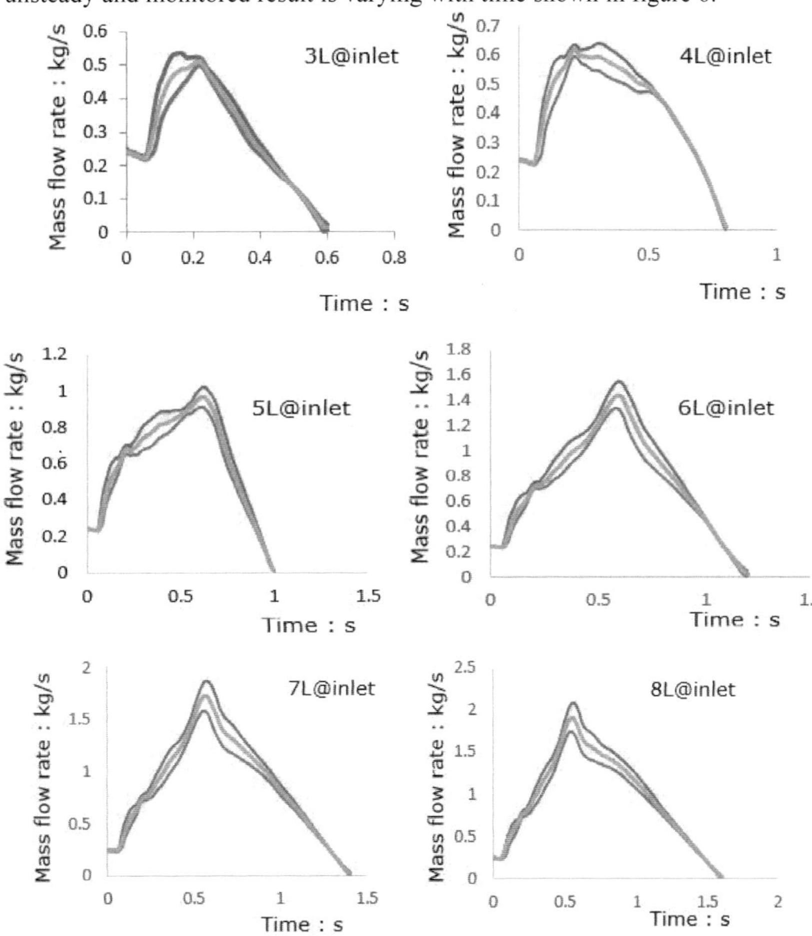

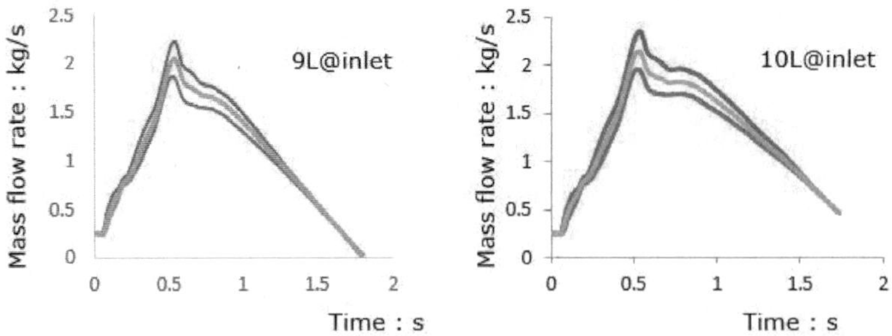

Figure 6. Mass flow rate at outlets with different liquid quantity

The figure 6 is liquid mass flow rate at two outlets (red line is outlet1, blue line is outlet2 and green line is the average of two outlets). We simulate different liquid quantity as input and record the max number during the procedure. The trends are very close which is half rising and half declining. But they are different time when the max mass flow rate come. When we use 3 liters liquid injection, it come to max at 0.25 seconds. When we use 4 liters liquid injection, it is almost 0.4 seconds. However, when injection liquid is bigger than 5 liters, the max happen over 0.5 seconds (around 0.6s). Then we also record the total max number (both liquid and gas) which is shown below.

Table1: Max total mass flow rate at different liquid injection

Liquid Injection Quantity Unit: L	Max total mass flow rate Unit: kg/s
3	1.04
4	1.35
5	2.03
6	2.93
7	3.66
8	4.01
9	4.28
10	4.48

Based on the 10 liters liquid injection case, the velocity distribution at different moments are visualized in figure 7 by contour of velocity magnitude located on center section surface. At the outlets, velocity keeps increasing from 0 to 0.5 seconds, and decreasing from 0.5 to 2 seconds.

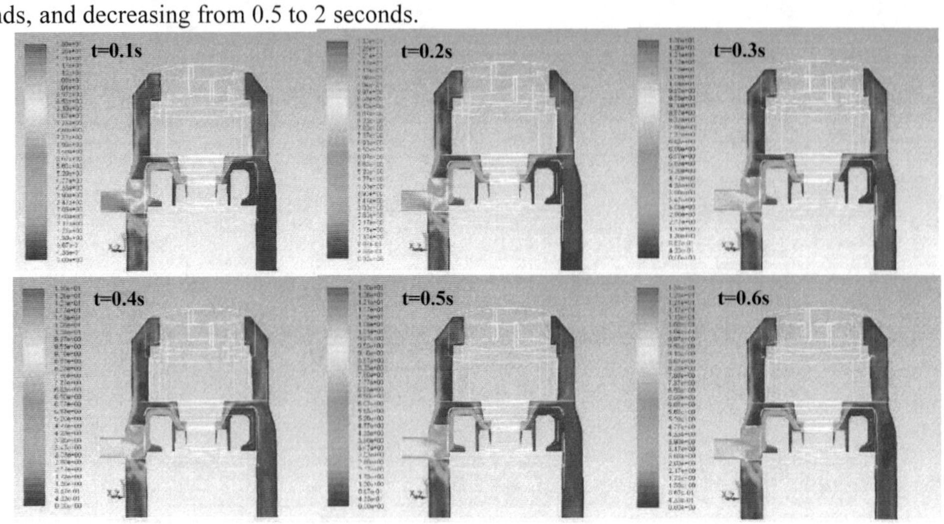

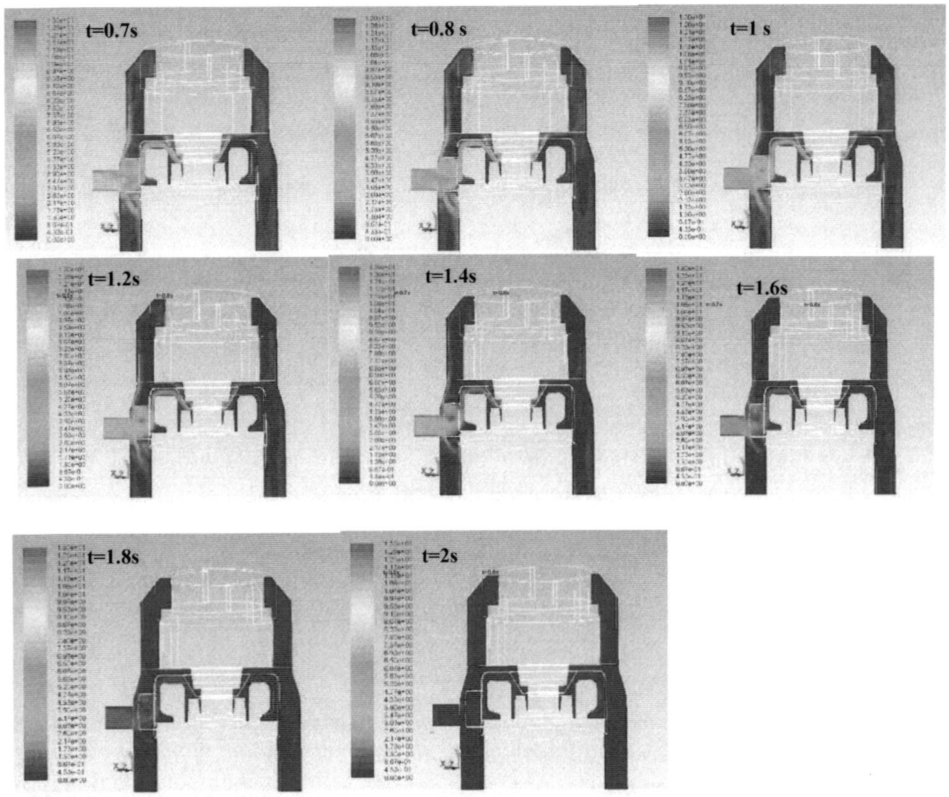

Figure 7. Velocity contour of 10 liter liquid injection

3.3 Result analysis and discussion

Use max total mass flow rate divided by theoretical mass flow rate only suction gas to get a ratio of mass flow rate:

$$\lambda = \frac{Q_t}{Q_g}$$

λ : Ratio of mass flow rate.

Q_t : Total mass flow rate including gas and liquid in slug work condition

Q_g : Mass flow rate only suction gas in normal work condition.

The aim of this calculation is to get a non-dimensional value to compare with test result which is also a non-dimensional ratio value of current peak. The figure 8 shows simulation result and test result tendency, blue line is tested current peak ratio and red line is simulated max mass flow rate ratio. The tendency has a good agreement. The liquid injection procedure basically can be divided into three stages. Firstly, little liquid injected (far less than slug limit, for instance less than 4 liters liquid), compressor managed to overcome the impact of liquid and wouldn't bring big current fluctuation. Secondly, more liquid injected (close to slug limit, beyond the limit, for instance in the range of 4 liters-8 liters), more and more liquid sucked into scroll brought a huge motor load and obvious current peak occurred. Thirdly, much more liquid injected (over 8 liters in this example), the scroll can not take more liquid away so mass flow rate basically got the limitation and the motor showed same phenomenon-the current peak wouldn't get larger anymore. If the quantity continued to be increased, the compressor could be broken down.

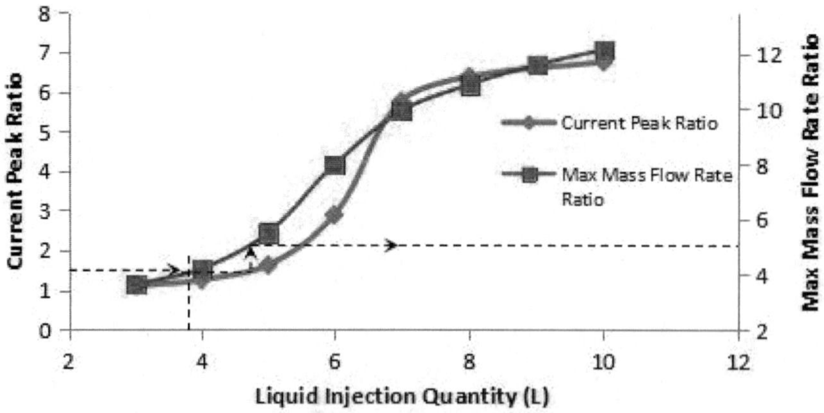

Figure 8. Comparison of test and simulation

Based on test criterion we get slug limit for a compressor and following along the vertical dotted line we get the relevant max mass flow rate ratio with same liquid injection quantity. We use the relevant value as the simulated slug limit. If a compressor from same platform has a larger ratio than the simulated slug limit, we think it take risk of failure when liquid flow back.

4. CONCLUSION

This paper shows a CFD approach to simulate a kind of scroll compressor reliability test – we called it slug limit test. For Danfoss scroll compressor, there are many different types and slug limit criteria. We choose a typical one as the research target, make a slug limit test, record the current peak ratio and get the quantity of limitation. Then we built a numerical model, make simplifications, assumptions, and calculations. From calculation result, we know max mass flow rate has increasing and decreasing process in one slugging cycle. The time of getting the peak is later and later until the total liquid injection is close to limit. After making a curve of different liquid injection quantity, we know the whole process can be divided into three stages which is including the first steady stage, the second sharp rising stage and the third relevant steady stage which is almost the compressor limitation of taking liquid ability. After that we use simulated result curve fit to test curve and get another criterion which is in same level with test criterion. The simulated max mass flow rate ratio has a very good agreement with tested current peak ratio. It has same tendency which help to set a series of criteria for other types of compressor. By this criterion, engineer can tell the difference of new design with the former one or other ideas when he just has simulation result. Based on the simulation accordance to experiment, a lot of time and money can be saved in the product development procedure.

REFERENCES

Ding, Hui & Jiang, Yu. (2016). CFD Simulation of An Oil Flooded Scroll Compressor Using VOF Approach. Proceedings of the 23nd International Compressor Engineering Conference at Purdue (Paper 1616). West Lafayette, USA: Purdue University.

Tadeu Tonheiro RODRIGUES. (2018). Refrigerant Liquid Slugging in The Suction System of Reciprocating Compressor. Proceedings of the 24th International Compressor Engineering Conference at Purdue (Paper 1137). West Lafayette, USA: Purdue University.

A Study of "OVAL" Scroll Compressor for Capacity Increase

Wataru IWATAKE[1]*, Raito KAWAMURA[2], Keisuke NARUMI[3], Daisuke KUDO[4],

[1, 2] Mitsubishi Electric Corporation, Advanced Technology R&D Center,
8-1-1 Tsukaguchi-Honmachi, Amagasaki, Japan
Phone: +81-6-6497-3499, Fax: +81-6-6497-7285,
[1] Email: Iwatake.Wataru@bk.MitsubishiElectric.co.jp
[2] Email: Kawamura.Raito@cj.MitsubishiElectric.co.jp

[3] Mitsubishi Electric Corporation, Air-Conditioning & Refrigeration Systems Works,
6-5-66 Tebira, Wakayama, Japan
Phone: +81-73-436-2121, Fax: +81-73-428-9022,
Email: Narumi.Keisuke@aj.MitsubishiElectric.co.jp

[4] Mitsubishi Electric Corporation, Shizuoka Works,
3-18-1 Oshika, Suruga-ku, Shizuoka, Japan
Phone: +81-54-287-3258, Fax: +81-54-287-3127,
Email: Kudo.Daisuke@ah.MitsubishiElectric.co.jp

ABSTRACT

In recent years, there has been demand for a large-capacity scroll compressor to reduce cost and improve performance. To increase the capacity of the scroll compressor, technology to expand the stroke volume without increasing the shell size is required. To expand the stroke volume, expanding the wrap height in the axial direction is most effective. However, the higher wrap height raises the stress on the wrap and causes the orbiting scroll to overturn. So, we started to study a new shape of wrap to expand the stroke volume in the horizontal direction. The base plate of an orbiting scroll is usually circular. However, the trajectory of the fixed scroll on the base plate of the orbiting scroll is not circular, so wasted space that is not useful for compression exists on the base plate. We devised an oval wrap to make the trajectory of the fixed scroll on the base plate circular and reduce this space. The radius of basic circle was constant regardless of the involute angle in the conventional wrap. On the other hand, we can make the wrap shape "oval" by changing the radius of the basic circle sinusoidally (every 180 degrees). In this conference, we will present the geometric theory of our new oval wrap and the experimental result of the prototype. The results are follows: (1) It is possible to increase the stroke volume by more than 20% with the oval wrap. (2) A prototype with the oval wrap operates stably, and compressor efficiency is equivalent to the conventional wrap.

1. INTRODUCTION

In scroll compressors, technologies to increase capacity without increasing compressor size are required toward improving performance and reducing cost. Technologies to increase capacity are classified into two types, technologies that increase the rotation speed and technologies that expand the stroke volume. However, increasing the rotation speed may increase sliding loss, and may deteriorate the compressor performance. For expanding the stroke volume, increasing the scroll wrap height is most effective. However, higher wrap height raises the stress at the scroll root, and may deteriorate the compressor performance. To cope with this problem, many studies have been performed with regard to the scroll shape in the horizontal direction, including that described in a paper by John P. Elson et al. (2008). We defined again geometric functions of scrolls forming the compression chamber, expanded scrolls in the horizontal direction, and developed "oval scrolls" to expand the stroke volume. In this paper, we will introduce the geometric theory and effect of oval scrolls, and explain the experimental result of prototypes.

2. SCROLL GEOMETRIC THEORY

Figure 2.1 shows the conventional scroll shape, and Figure 2.2 shows the oval scroll shape. The base plate of an orbiting scroll is usually circular. However, the trajectory given when an orbiting scroll performs oscillation movements is not circular, but is oval in the winding end direction. As a result, wasted space that is not useful for stroke volume is generated partially on the base plate of the orbiting scroll as shown in the figure. We thought it desirable for expanding the stroke volume to effectively utilize the space on the base plate of an orbiting scroll by making the scroll shape oval and moving the base circle center from the compressor center, and devised an oval scroll. Figure 2.3 shows an example of the trajectory of the conventional scroll, and Figure 2.4 shows an example of the trajectory of the oval scroll we devised. In these figures, an orbiting wrap is fixed and a fixed wrap is moved. The trajectory of the oval scroll is close to a circular shape as shown in the figure, and the space on the base plate of an orbiting scroll can be utilized effectively.

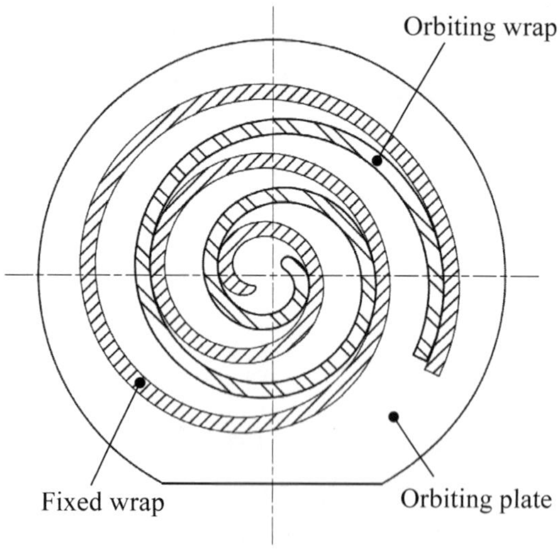

Figure 2.1: Conventional scroll shape

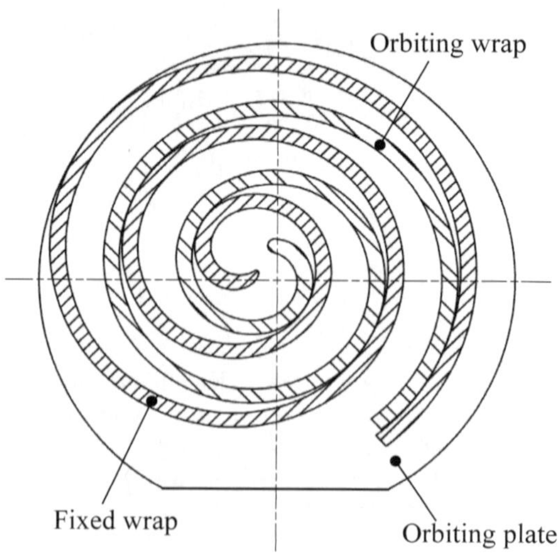

Figure 2.2: Oval scroll shape

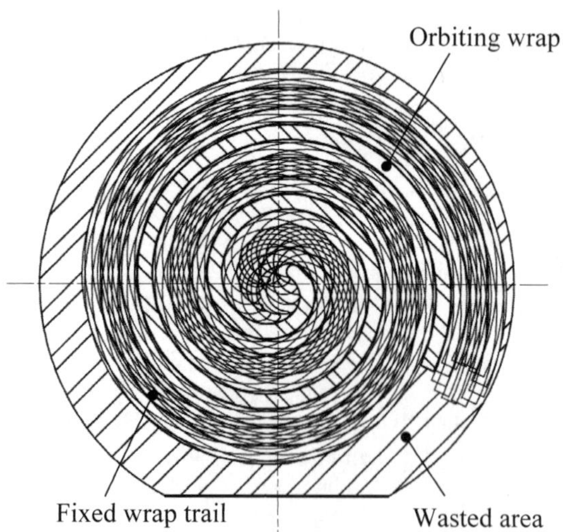

Figure 2.3: Trajectory of conventional scroll

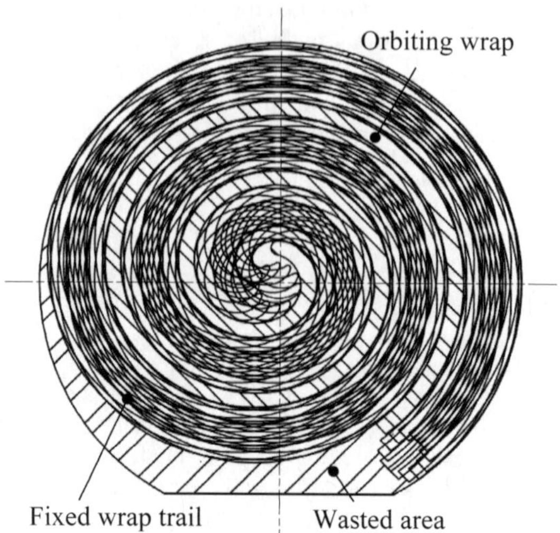

Figure 2.4: Trajectory of oval scroll

The geometric theory for the outward face of the oval scroll will be explained first. Formula (2.1) and formula (2.2) indicate the x-coordinate and y-coordinate of the conventional scroll, respectively. The conventional scroll is defined by the involute curve of a circle whose normal line is a plane curve always in contact with a constant base circle radius.

$$x = a\left(cos(\theta - \zeta) + \theta sin(\theta - \zeta)\right) \tag{2.1}$$

$$y = a\left(sin(\theta - \zeta) - \theta cos(\theta - \zeta)\right) \tag{2.2}$$

Formulas (2.3) and (2.4) indicate the x-coordinate and y-coordinate of the newly devised scroll, respectively. Here, "$a(\theta)$" is the base circle radius, and "θ" is the involute angle. Different from the conventional scroll, the base circle radius is given by a function of the involute angle "θ". Formula (2.5) indicates the base circle radius of the oval scroll, and Figure 2.5 shows changes in the base circle radius "$a(\theta)$" against the involute angle "θ". For forming an oval scroll, it is necessary to change the base circle radius "$a(\theta)$" sinusoidally against the involute angle "θ" in the cycle of "π rad". In this way, the scroll shape is made into an oval whose long diameter direction is the direction in which the base circle radius "$a(\theta)$" becomes the maximum.

$$x = a(\theta)\left(cos(\theta - \zeta) + \theta sin(\theta - \zeta)\right) \tag{2.3}$$

$$y = a(\theta)\left(sin(\theta - \zeta) - \theta cos(\theta - \zeta)\right) \tag{2.4}$$

$$a(\theta) = a_{min}(1 + \alpha \sin^2(\theta - \xi)) \tag{2.5}$$

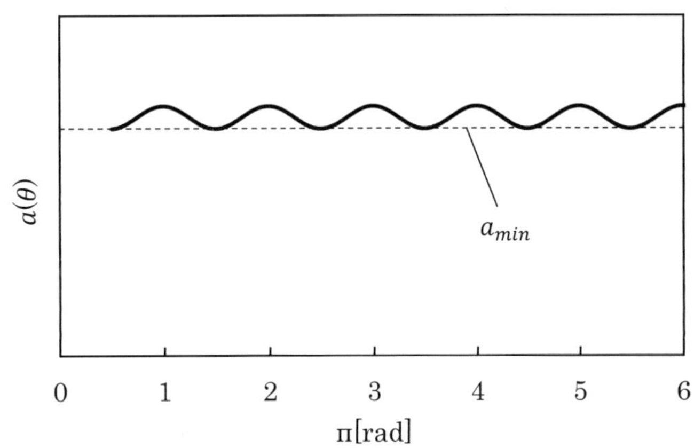

Figure 2.5: Base circle radius "$a(\theta)$" of oval scroll

Figure 2.6 shows an example of the curve of an oval scroll. The flatness "α" is 0.1 in (a), and 0.4 in (b). As the flatness "α" is larger, changes in the base circle radius "$a(\theta)$" in one cycle become larger, and the scroll becomes closer to oval.

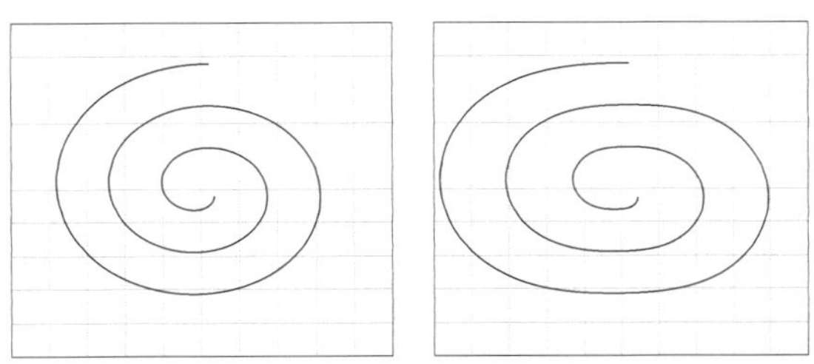

(a) Flatness "α" = 0.1 (b) Flatness "α" = 0.4

Figure 2.6: Curve of oval scroll

The geometric theory for the inward face of the oval scroll will be explained next. Figure 2.7 shows the inward face drawing procedure. For forming the compression chamber without leakage, it is necessary that the outward face of the orbiting scroll is always in contact with the inward face of the fixed scroll at one point, and that the outward face of the fixed scroll is always in contact with the inward face of the orbiting scroll at one point while the orbiting scroll is performing oscillation movements. To meet this necessity, the inward face must be the outer envelope of a circle group whose center is located on a curve made by inverting the outward face by "π rad". When the fixed scroll and orbiting scroll have a same average wrap thickness for any π rad section, the inward face and outward face of the two scrolls have symmetrical shapes. The next section will describe prototypes of scrolls configured based on this theory.

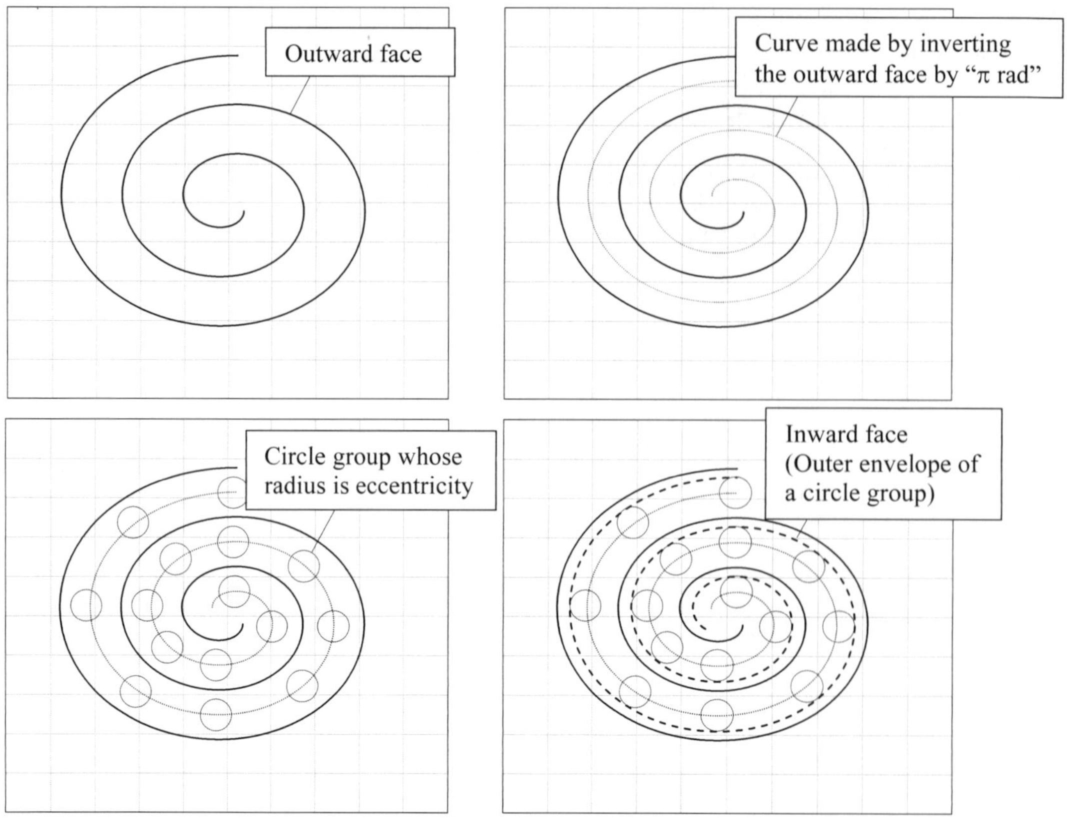

Figure 2.7: Oval scroll inward face drawing procedure

3. SPECIFICATION OF THE PROTOTYPE

3.1 Compressor Configuration

Figure 3.1 shows the entire configuration of the prototype we made, and Table 3.1 shows the basic specifications of the prototype. Used refrigerant is R290, and the stroke volume of the compressor is 36 cm^3. In the frame compliant method we adopted, the orbiting scroll floats together with the guide frame, and the compression chamber is sealed in the axial direction. We adopted a high-pressure shell whose inside is filled with high-pressure refrigerant after compression.

Refrigerant flows as follows: Low-pressure refrigerant is drawn from the suction tube into the compression mechanism. Refrigerant is compressed from low pressure to high pressure by the compression mechanism that consists of the fixed scroll and orbiting scroll, and discharged to the upper area of the compressor. After that, refrigerant flows out of the compressor from the discharge tube by way of the gap between the shell and the stator.

Table 3.1: Prototype basic specifications

Refrigerant	R290 (propane)
Stroke Volume	36 cm^3
Shell type	High pressure type

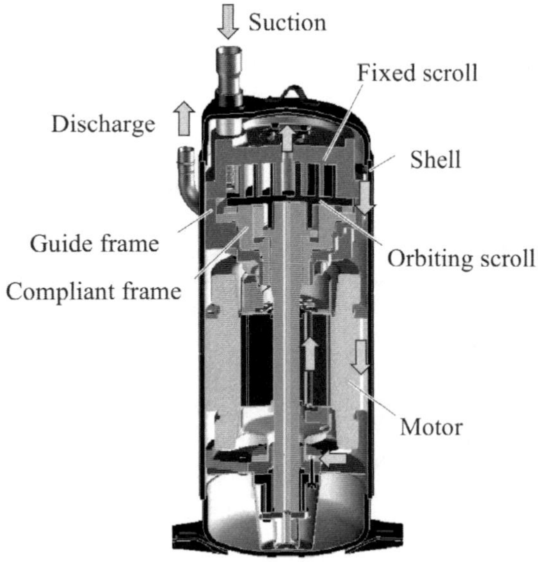

Figure 3.1: Prototype entire configuration

3.2 Scroll Shape

Figure 3.2 shows the shape of prototype scrolls, and Table 3.2 shows the specifications of prototype scrolls. It is held constant between the three prototype scrolls except for flatness, wrap height, and base circle center. (a) shows the base specifications of the conventional scroll, (b) shows an oval scroll whose flatness "α" is 0.025, and (c) shows an oval scroll whose flatness "α" is 0.050. When the wrap height is constant, the stroke volume is largest in (c). The oval scroll gives a stronger effect on expansion of the stroke volume when the base circle center is moved from the compressor center. In (b) and (c), the base circle center is moved by -2.868 mm in the x-axis direction, and by +2.233 mm in the y-axis direction. Figure 3.3 shows prototype oval scrolls whose flatness "α" is 0.050. (a) shows a fixed scroll, and (b) shows an orbiting scroll. Both the fixed scroll and orbiting scroll are made of FCD450, and machined by a general NC machine tool.

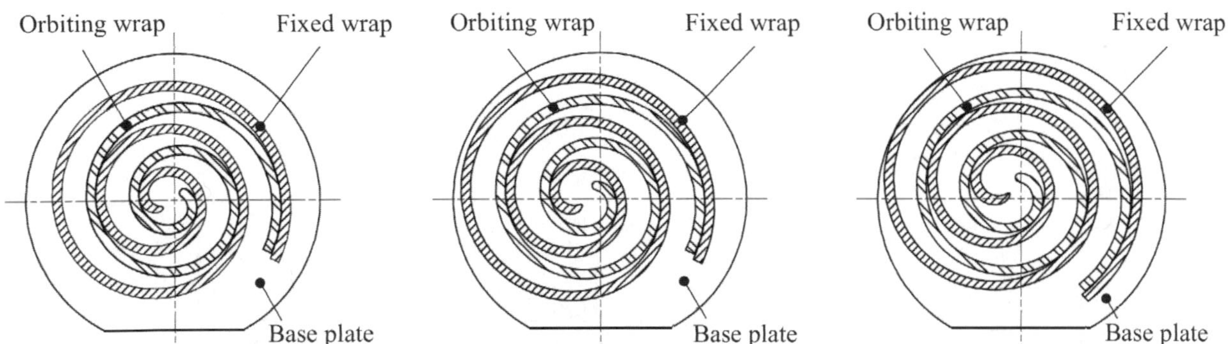

(a) Conventional scroll (b) Oval scroll ("α" = 0.025) (c) Oval scroll ("α" = 0.050)

Figure 3.2: Prototype scroll shapes

Table 3.2: Prototype scroll specifications

	(a) Conventional scroll	(b) Oval scroll	(c) Oval scroll
ρ (−)	2.30	2.30	2.30
e (mm)	4.15	4.15	4.15
t_{ave} (mm)	3.0	3.0	3.0
h (mm)	25.8	23.3	21.7
α (−)	-	0.025	0.050
(x_c, y_c) (mm)	(0, 0)	(-2.868, 2.233)	(-2.268, 2.233)

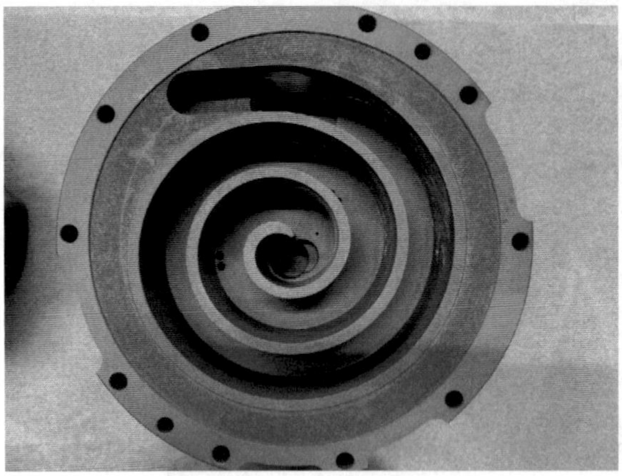

(a) Fixed scroll

(b) Orbiting scroll

Figure 3.3: Prototype oval scrolls ("α" = 0.050)

Figure 3.4 shows the stroke volume ratio of oval scrolls. The ratio of the margin area against the base plate area is 26.0% in the conventional scroll, 17.8% in the oval scroll ("α" = 0.025), and 10.1% in the oval scroll ("α" = 0.050). By effectively utilizing the space on the base plate of the orbiting scroll, the stroke volume can be expanded by 12.0% in the oval scroll ("α" = 0.025) and by 20.0% in the oval scroll ("α" = 0.050) compared with the conventional scroll. This means that the stroke volume can be expanded while the wrap height is unchanged. However, to compare the performance together, the wrap height is designed lower in the prototypes because the scroll is extended in the horizontal direction.

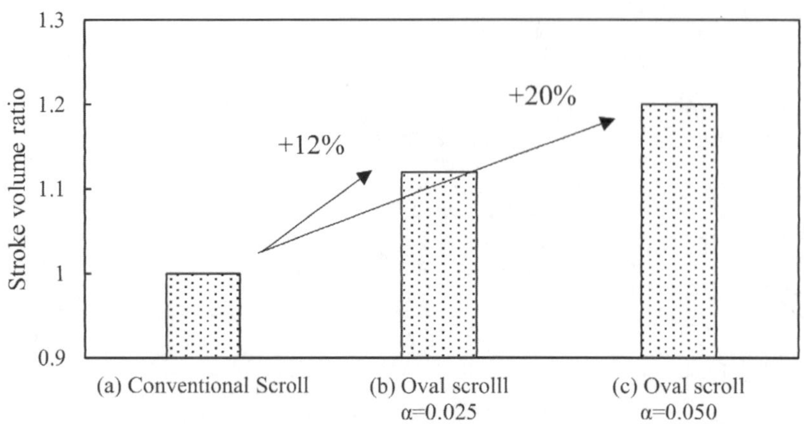

Figure 3.4: Stroke volume ratio of oval scrolls

4. TEST RESULTS

Table 4.1 shows the prototype operation condition, and Figure 4.1 shows the performance evaluation result of the prototype oval scrolls compared with the conventional scroll. Formula (2.6) indicated secondary COP.

The secondary COP (coefficient of performance) is improved by +0.6% in the oval scroll ("α" = 0.025) and by +1.7% in the oval scroll ("α" = 0.050) compared with the conventional scroll. It is estimated that the wrap height becomes lower because the oval scrolls are extended in the horizontal direction, leakage loss from the compression chamber is reduced, and then the performance is improved.

$$Secondary\ COP = \frac{W_{ref}}{W_{in}} \tag{2.6}$$

Figure 4.2 shows gas load theoretical values of the conventional scroll and the oval scroll ("α" = 0.050). It is calculated by ideal chamber pressure and geometric pressure-receiving area which change from time to time. It is assumed that gas refrigerant is mixed instantaneously when the innermost chamber is interconnected. Values calculated from REFPROP 9.1 are used for the refrigerant physical properties. Though the oval scroll is extended in the horizontal direction, conversely, the wrap height is reduced. Its gas load "$F_{g\theta}$" in the tangential direction and gas load "F_{gr}" in the radial direction are similar to those of the conventional scroll, and its sliding loss is similar to that of the conventional scroll.

Figure 4.3 shows the scrolls after operation for 50 hours. (a) shows the conventional scroll, and (b) shows the oval scroll ("α" = 0.050). It is proved that the oval scroll can operate stably without significant sliding marks in the same way as the conventional scroll.

Table 4.1: Operation condition

CT (℃)	54.4
ET (℃)	7.2
SH (K)	11.1
SC (K)	8.3
N (rps)	60

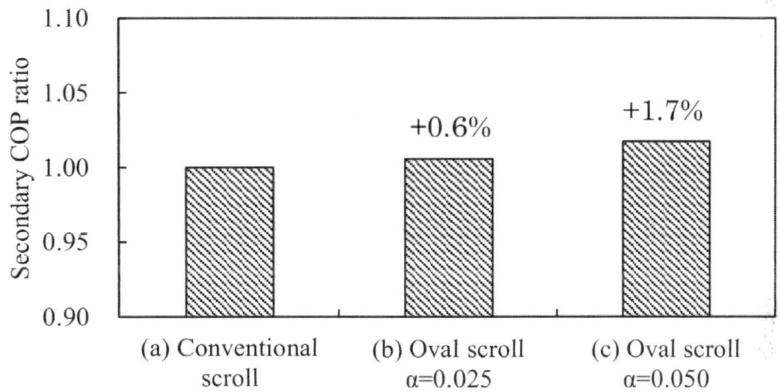

Figure 4.1: Performance evaluation result

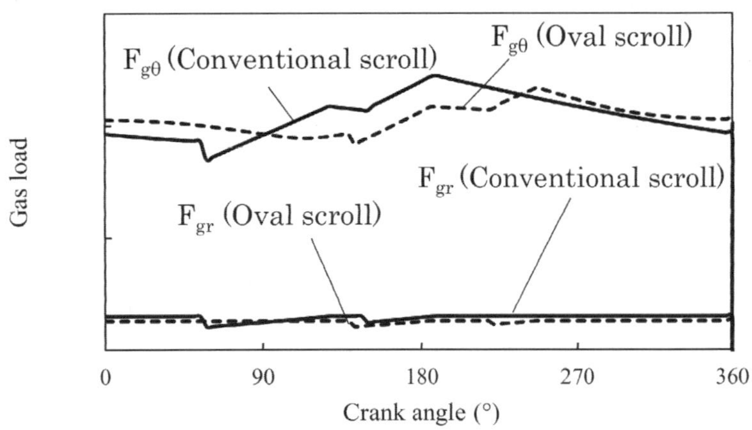

Figure 4.2: Gas load theoretical values

(a) Conventional scroll

(b) Oval scroll (α=0.050)

Figure 4.3: Scrolls after operation for 50 hours

5. CONCLUSIONS

In this paper, we have proposed an oval scroll compressor, introduced the geometric theory, and presented the performance evaluation result. We have obtained the following conclusions:

• Oval scrolls can be realized by making the base circle radius "$a(\theta)$" as a curve changing sinusoidally in the cycle of "π rad".

• Oval scrolls can expand the stroke volume by +20% compared with the conventional scroll.

• When the stroke volume (performance) is constant, oval scrolls operate stably in the same way as the conventional scroll, the secondary COP is improved by +1.7% because the wrap height is lower, and the leakage loss is reduced.

NOMENCLATURE

x	x-coordinate	(mm)
y	y-coordinate	(mm)
a	base circle radius	(mm)

θ	involute angle	(rad)
ζ	set rotation angle	(rad)
$a(\theta)$	base circle radius (oval)	(mm)
α	flatness	(-)
ξ	angle of flat direction	(rad)
ρ	build-in volume ratio	(-)
e	eccentricity	(mm)
t	wrap width	(mm)
h	wrap height	(mm)
CT	condensing pressure	(℃)
ET	evaporating pressure	(℃)
SH	degree of superheating	(K)
SC	degree of subcooling	(K)
F_g	gas force	(N)
W	work	(W)

Subscript

ave	average
min	minimum
c	base circle radius center
θ	rotation direction
r	radial direction
ref	refrigeration
in	compressor input

REFERENCES

John ELSON, Norbert KAEMMER, Simon WANG, Michael PEREVOZCHIKOV. (2008). Scroll Technology: An Overview of Past, Present and Future Developments. *International Compressor Engineering Conference at Purdue* (1204).

E.W.Lemmon, M.L.Huber, and M.O.McLimden. Reference Fluid Thermodynamic and Transport Properties-REFPROP, Version9.1, NIST Standard Reference Database 23, 2013.

Horizontal scroll compressor for refrigeration applications

Mikhail ANTIMONOV*, Michael PEREVOZCHIKOV, Kirill IGNATIEV

Emerson Climate Technologies,
Sidney, Ohio, USA
Mikhail.Antimonov@Emerson.com, Michael.Perevozchikov@Emerson.com,
Kirill.Ignatiev@Emerson.com

* Corresponding Author

ABSTRACT

New efficiency regulations, as well as trends in applications of new low GWP refrigerants, force compressors to be re-designed, especially for A3 refrigerants, because of the regulatory limitations of the refrigerant charge. The paper addresses compressor design allowing for more efficient use of the refrigerant charge.

An oil sump retains significant amount of refrigerant diluted in the oil. Elimination of the oil sump allows to use the limited refrigerant charge more effectively. Furthermore, it reduces the size of the compressor and simplifies the compressor structure. Another benefit of this type of compressor is its orientation flexibility- the same compressor can be installed vertically or horizontally. These design features are beneficial in close-coupled systems, as well as transport (truck/trailer) refrigeration systems.

The sump-less design requires a revision of friction pairs in the compressor: radial bearings, thrust bearings, tip-to-base contact and others. Existing refrigeration compressors rely on the continuous oil flow for journal bearing lubrication, while the lubrication mechanism in the sump-less compressor relies on the oil mist. This design leverages the approach utilized in the automotive scroll compressors, but the requirements and conditions for the compressor operation and reliability are different. The paper discusses the major design steps and results of preliminary reliability and system tests with an experimental determination of the optimal oil charge.

1. INTRODUCTION

HFC (hydrofluorocarbon) refrigerants are to be phased out soon because of their high GWP (Global Warming Potential). Most of their replacements are A2 or A3 refrigerants. Application of those usually restricts the refrigerant charge of the systems, since flammable fluid possess safety risk. For instance, the maximum charge of the system with A3 refrigerant is 150 grams, according to level specified by ASHRAE-15. The problem of minimization of the refrigerant charge was considered previously, for example, in Padilla et al., (2014), Poggi et al., (2008), Fernando et al., (2004), Bamigbetan et al., (2018). Most of the refrigerant is in the heat exchangers during the system operation (Hoehne and Hrnjak, (2004)). Thus, the approach was usually in optimization of the heat exchangers and their replacement with micro-channel heat exchangers (Padilla et al., (2014), Fernando et al., (2004) and Butrymowicz et al., (2016)). However, a large portion of the refrigerant is also staying in the compressor sump being dissolved in the oil (Hoehne and Hrnjak, (2004), Padilla et al., (2014)). Reduction of the oil charge will be beneficial for the reduction of the refrigerant charge (Fernando et al., (2004)).

In the refrigeration system, the oil is required for lubrication of the friction pairs inside the compressor. In a scroll compressor, the oil is pumped from the oil sump via the hole through the shaft. The oil charge in the oil sump is optimized for each compressor for sufficient lubrication of friction pair and to minimize the oil circulation rate in the system. Thus, the problem of oil charge reduction becomes a problem of the oil sump elimination and revision of the oil management in the compressor and in the whole system. Elimination of the oil sump results in a significant re-design of a compressor. The friction pairs of this compressor should still be lubricated, but all this oil continuously travels with the refrigerant through the system. Oil mist lubrication is used as a lubrication method. These changes lead to compressor designs similar to compressors in automotive AC systems.

2. COMPRESSOR DESIGN

A successful sumpless design must be both cost effective and efficient compared to existing compressors. Utilizing radial and dual-force axial compliance will help to achieve this challenge.

Axial compliance: Design of axial compliance mechanism was chosen as a first step in the compressor design in this work. Floating fixed scroll design used in production scroll compressors includes thrust bearing requiring hydro-dynamic lubrication which functions with continuous oil supply to the friction pairs. Even though it was demonstrated that thrust surface is viable if the oil mist with the suction gas is directed to the thrust surface, a floating orbit scroll design was chosen for this application.

Shell pressure: Both high side and low side shell options were considered and tested. Finally, a high side design was chosen because of the simpler construction, easier variable volume ratio valves application, discharge pulse and shell oil entrapment mitigation.

The cross-section of the designed compressor was described in Perevozchikov et al. (2020) and Ignatiev et al. (2020). It is shown in Figure 1. The compressor has direct suction. The oil moves with the suction flow through the compression mechanism. The compressor can have both horizontal and vertical orientation.

Bearings: The journal bearings, typically used in the scroll compressors, require continuous oil supply and cannot rely on an oil mist lubrication. Therefore, rolling element bearings are selected in proposed sump-less compressor design.

The rolling element bearings must be sized properly in dependence on the application that the compressor is designed for. Each application can be represented as a set of weighted compressor conditions, where weight is a time period while the compressor is operating at this condition. The forces acting on the bearings are calculated and are used in equation (1) to calculate the minimum required load rating of the bearing for a given life requirement:

$$C_{r\,min} = \sqrt[p]{\frac{60L_{10}\sum t_i n_i F_i^p}{10^6}} \tag{1}$$

where C_r is the dynamic load rating, i denotes different load conditions defined by t_i as the weight of the condition, n_i as the speed (rpm), F_i as the load, L_{10} is the required bearing life (hours), p is 3 for ball bearings and 10/3 for roller bearings.

Ball bearings are chosen as the main and lower bearings and needle bearing as a drive bearing.

Flow management: The scroll set may have discharge ports on both fixed and orbit scrolls. If a discharge valve is required for efficiency of high-pressure ratio conditions, then it can be installed on both scrolls (see Ignatiev et al. (2020)).

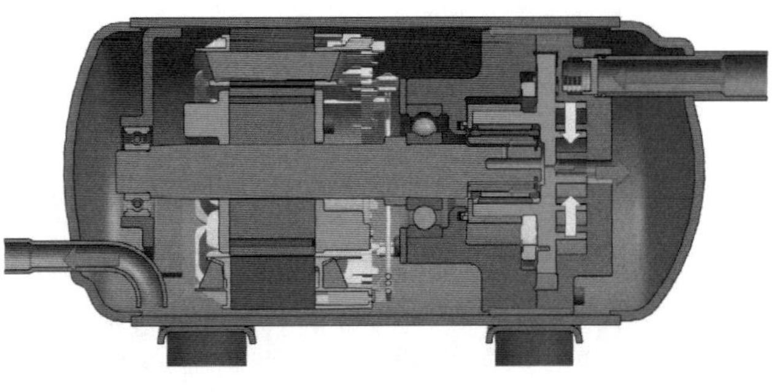

Figure 1: Refrigerant flow in the sump-less compressor.

One of the tested 3-ton sump-less compressor had a discharge port in the orbit scroll only. The cross-section of this compressor is shown in Figure 2. Total flow of the compressed refrigerant together with the oil mist passes through the radial compliance mechanism and drive bearing, providing the best lubrication to the friction pairs.

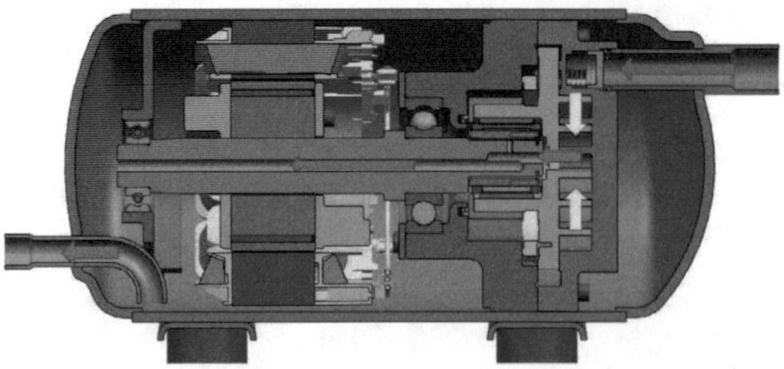

Figure 2: Sump-less compressor with discharge port in the orbit scroll

Radial compliance mechanism: A mechanism comprising unloader and eccentric shaft pin with flat surfaces was chosen as a primary radial compliance mechanism (see Caillat et al., (1993)).

3. COMPRESSOR PERFORMANCE

The performance of the compressor was measured within the operating envelope of a typical AC compressor including rating points according to ASHRAE Standard 540-2019 and at the rating conditions corresponding to the system test points according to AHRI Standard 210/240-2017. The tests were taken in calorimeter. The compressor efficiency was equivalent to the efficiency of 3-ton production scroll compressor. Its volumetric efficiency was slightly higher because of the direct suction. However, the mechanical losses were higher resulting in overall efficiency equivalent to the efficiency of the production scroll compressor. The test results at specific point are presented in Table 1.

Table 1: Test results for a compressor prototype

Tevap F (C)	Tcond F (C)	Speed RPM	Efficiency %
45 (7.2)	130 (54.4)	3600	70
50 (10)	115 (46.1)	Nominal	71

4. RELIABILITY TESTS

For the preliminary assessment of the new design reliability, we chose three reliability tests that allow to conduct accelerated preliminary tests with an emphasis on the compressor components that may have major reliability concerns. The tested prototypes passed these tests without any major failure modes.

4.1 High-Load Test

High-load test was performed to demonstrate the ability of the bearings to operate for the required life of the compressor. This test was performed for both horizontal and vertical compressor orientations. The compressor ran at the condition in the top-right corner of the envelope at the maximum designed speed. We assumed that the test was successfully passed if the compressor could operate with about the same efficiency in a predefined time period and the bearing did not have signs of failure.

Two sump-less compressors were tested at high-load conditions. Both compressors successfully passed the minimum predefined test period and did not show bearing failure modes in the extended tests.

4.2 Defrost Test

The compressor is designed for refrigeration application. In this application, the compressor can operate at a condition when the liquid refrigerant from the evaporator returns to the compressor suction in defrost cycles. Reliability of the sump-less compressor at this condition was assessed. The focus in this test was on the drive bearing and scroll set. The liquid refrigerant could wash the oil from the surfaces of the scroll resulting in accelerated wear of the scroll surfaces. The compressor prototype with the cross-section shown in Figure 2 was chosen for this test because it was less favorable for the drive bearing lubrication. The refrigerant was discharged through the port in the orbit scroll onto the drive bearing, the liquid refrigerant might not be fully evaporated in the scroll and might reach the drive bearing washing the oil. It might take some time for oil to relubricate the compressor components. The drive bearing would be starving during this time.

The compressor was running at the condition simulating an ice maker. The liquid refrigerant was dumped into the compressor suction line periodically.

Figure 3 shows suction gas temperature and the gas temperature measured just below the drive bearing. The latter temperature represents the discharge gas temperature. This temperature drops close to the suction temperature and remains almost constant for the period of liquid refrigerant dumping. This shows that some liquid refrigerant might reach the drive bearing for several second during each cycle.

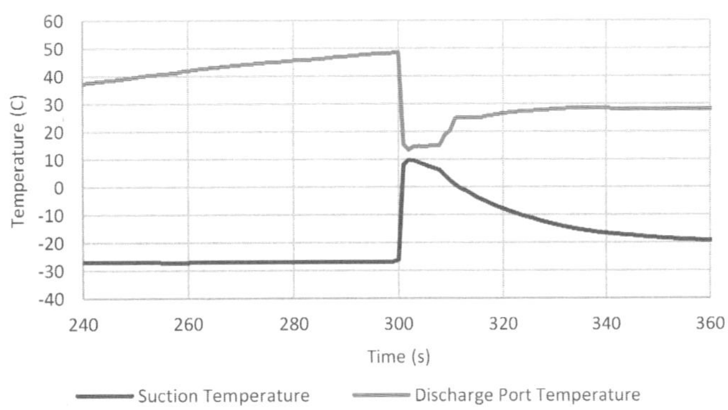

Figure 3: Suction gas temperature and temperature of the gas next to the discharge port.

After the test, the bearings did not have signs of damage due to the oil starvation.

4.3 Start/Stop Test

Another compressor failure mode can relate to compressor start moment. When the bearing suddenly starts spinning, hydrodynamic layer may be not formed resulting in boundary lubrication between rolling elements and races and the rolling elements may skid along the bearing races.

Start/stop test was done with a sump-less compressor. The compressor was cycled with running and idle periods. The condition was set to be close to ARI condition. The compressor passed through the number of cycles equivalent to the compressor life period.

No significant wear was observed on the bearings after the test.

5. SYSTEM TESTS WITH SUMP-LESS COMPRESSOR

The sump-less compressor was tested in refrigeration and air-conditioning systems: bottle cooler, two-door freezer and AC split system.

At first, a sump-less compressor was tested in a bottle cooler with R290 and R513A as refrigerants. It was demonstrated that the system with a sump-less compressor ran efficiently with only a 2-oz oil charge which corresponded to less than 1% oil circulation rate. This allowed to reduce by several times the refrigerant amount dissolved in the oil.

Air-conditioning split system performance was evaluated at different oil charge to determine the optimal oil charge. The original compressor in an off-the-shelf system was replaced with a sump-less compressor. A weighted average system efficiency was measured with using A_2, B_2, B_1 and F_1 conditions in AHRI Standard 210/240-2017. The dependence of the system efficiency on the oil charge is shown in Figure 4. The efficiency increased in the beginning, when the oil charge was increased from minimum amount, because the friction losses in the compressor were decreasing when we added oil, but the efficiency of heat exchangers did not change. Then, when we continued increasing the oil charge, the efficiency of the heat exchangers dropped resulting in higher pressure ratio condition and, as a result, in lower system efficiency.

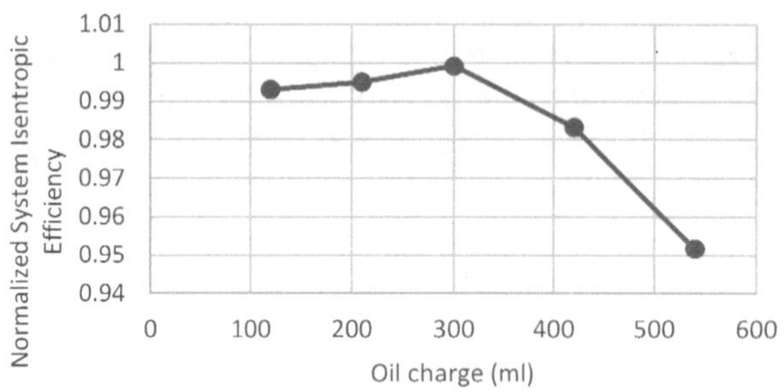

Figure 4: Normalized weighted average system isentropic efficiency in dependence on the oil charge.

As the result from the series of tests, an initial relation of oil to refrigerant charge of air-conditioning split system has been established.

6. CONCLUSIONS

Sump-less design reduces amount of oil needed in the system and thus can make even larger systems capable to run with 150g of flammable refrigerant

Use of rolling element bearings instead of journal bearings allows to simplify the compressor envelope and to minimize the effect of speed on the oil circulation rate.

Sump-less compressor passed preliminary reliability tests. There were no major failure modes after the tests.

Optimal oil charge was determined for stand-alone refrigeration units charged with propane and a residential AC split system charged with R410A refrigerant.

REFERENCES

AHRI Standard 210/240-2017, Performance Rating of Unitary Air-conditioning & Air-source Heat Pump Equipment, 2017, Air-Conditioning, Heating, and Refrigeration Institute, 2111 Wilson Boulevard, Suite 500, Arlington, VA 22201, U.S.A.

Padilla , Y., Elbel, S., & Hrnjak, P.S. (2014). Extremely Low Refrigerant Charge Beverage Display Cooler Technology using Propane. *Proceedings of the 15th International Refrigeration and Air Conditioning Conference at Purdue,* West Lafayette, Indiana, USA

Butrymowicz, D., Gagan, J., Skiepko, T., et al. (2016) Experimental Investigations of Propane Minichannel Condenser and Evaporator. *Proceedings of the 16th International Refrigeration and Air Conditioning Conference at Purdue,* West Lafayette, Indiana, USA

Caillat, J-L. M., Weatherston, R.C., Bush, J.W. (1993) Scroll-type machine having a lubricated drive bushing. US Patent N 5,197,868

Poggi, F., Macchi-Tejeda,H., Leducq D., Bontemps A. (2008), Refrigerant charge in refrigerating systems and strategies of charge reduction, *International Journal of Refrigeration*, 31(3), 353-370.

Bamigbetan, O., Eikevik, T.M., Nekså, P., et al. (2018) Theoretical analysis for charge reduction in a 200 KW hydrocarbon high temperature heat pump. *13th IIR Gustav Lorentzen Conference on Natural Refrigerants (GL2018). Proceedings*. Valencia, Spain.

Fernando P., Palm B., Lundqvist P., Granryd E. (2004). Propane heat pump with low refrigerant charge: design and laboratory tests, *International Journal of Refrigeration* 27, 761–773.

Hoehne, M., Hrnjak, P.S. (2004) Charge minimization in systems and components using hydrocarbons as a refrigerant, Air Conditioning and Refrigeration Center, University of Illinois, Technical Report 224

Perevozchikov, M.M., Gehret, K.J. & Ignatiev, K.M. (2020) Oil flow through the bearings of a scroll compressor. US Patent N 10,801,495

Ignatiev, K.M., Perevozchikov, M.M. & Gehret, K.J. (2020) Compressor discharge valve assembly. US Patent N 10,753,352

ACKNOWLEDGEMENT

The authors would like to acknowledge the support of management of the Emerson Commercial and Residential solutions for their support of the research project and preparation of paper.

Experimental Comparison Of Seasonal Performance In R410A Chiller Using Single Speed And Two Stage Compressor

Sugun Tej INAMPUDI[1], Francesco BOTTICELLA[2], Stefan ELBEL[1,2]*

[1]University of Illinois at Urbana-Champaign, Department of Mechanical Science and Engineering, Air Conditioning and Refrigeration Center, 1206 West Green Street, Urbana, IL 61801, USA

[2]Creative Thermal Solutions, Inc., 2209 North Willow Road, Urbana, IL 61802, USA

* Corresponding Author: elbel@illinois.edu

ABSTRACT

Rising cooling needs for the residential and commercial air conditioning sectors and the requirement of higher SEER values increased the focus on energy-efficient part load performance. Although several studies exist on the topic of capacity modulation, there is no comprehensive study that compares different modulation strategies in the same experimental facility. In a first step, this paper aims to experimentally compare two different capacity controls for scroll compressors (single speed vs two-stage compressor) using the same R410A water ethylene glycol (WEG) chiller having a nominal cooling capacity of 8kW. The tests were conducted according to AHRI Standard 551/591 (2018). The Integrated Part Load Value (IPLV.SI) was used to fairly compare all these capacity modulation strategies. Based on the experimental results, the difference in the order of 8% in IPLV.SI is found. While both compressors showed comparable capacities and COPs at the higher modulation conditions, two stage compressor has better COP at the lower condition because of its ability to operate at a lower stage and better meet the required load.

1. INTRODUCTION

Almost all HVAC&R systems need to be designed to handle cooling loads below the target or design condition. These load variations are often caused by ambient weather variations, different levels of product loading, or the fact that systems are overdesigned to achieve quick pull-down. The range of load variations can be very substantial, in some cases even down to 30% or less of the design capacity. Different capacity control strategies employ different methods so that they can be implemented into the compressor or outside the compressor, in which case the system needs to provide adequate provisions. The modulation target is the same across the different techniques employed: the (average) refrigerant flow rate across the evaporator is reduced to adjust for the different levels of cooling capacity needed. There are many different methods of how to achieve this goal.

The main shortcoming of any literature-based comparison to the best of our knowledge is the fact that different data sets were obtained with different systems using different refrigerants, evaporation and condensation temperatures, secondary flow rates, and different levels of system controls. It is therefore nearly impossible to derive meaningful comparisons between the different studies when it comes to efficiency.

As a first step, this paper aims to experimentally compare two capacity modulation strategies, a single speed scroll compressor and a two stage scroll compressor using the same R410A WEG chiller system. These compressors have the comparable nominal cooling capacity. The tests are conducted according to AHRI Standard 551/591 (2018) and the comparison is done using a single figure of merit called IPLV.SI (Integrated Part Load Value).

2. MODULATION STRATEGIES BEING CONSIDERED

The following are the different compressor modulation strategies being considered as part of this study. As a first step, this paper compares a single speed and two stage compressor. All the compressors are scroll and have similar capacity and belong to the same generation.

2.1 Single speed compressor

A single speed compressor modulates capacity by cycling i.e. turning on and off. The on/off capacity control is the simplest method of adjusting the predetermined temperature (setpoint) using a thermostat. After the temperature reaches the setpoint, the thermostat stops the compressor and circulating refrigerant in the cycle. Since the secondary fluid continues to circulate, the temperature of the water or air gradually raises. When the thermostat detects, this rise turns the compressor on (Ekren, 2017).

2.2 Two stage compressor

This is a commercially available two stage compressor. It is different from conventional two stage compressors that use a low-pressure stage and a high-pressure stage. Instead, it has two internal bypass ports which enable the system to run at 67%-part load capacity. If the compressor is operated at the same condenser and evaporator water temperature, then the capacity at the 67% stage is approximately 67% of the capacity at the 100% stage. This compressor can be operated either at a high stage (100% capacity) or a low stage (67% capacity). The capacity is reduced by bypassing a portion of the gas in the scroll back to suction. These bypass ports are opened or closed using a solenoid valve present in the scroll chamber (Wang *et al.*, 2012).

2.3 Variable speed compressor

This compressor modulates its capacity by varying the speed of the motor (Qureshi and Tassou, 1996). The focus will be on the inverter driver variable speed compressor with a brushless DC motor. The selected compressor can be operated between 20 Hz and 100 Hz.

2.4 Digital scroll compressor

In this commercially available compressor, the modulation is achieved through rapid engagement/disengagement of the upper scroll (using a piston and solenoid valve). The scrolls are separated in a periodic cycle to obtain a time-averaged capacity based on the ratio of loading and unloading times. This provides a modulation range from 10% to 100% (Wu and Wu, 2004).

2.5 Tandem compressors

In this configuration multiple compressors operate in parallel. The modulation is achieved by turning a different combination of compressors on and off (Cecchinato, 2010). This study will cover a tandem combination of two compressors. The study will include combinations of a single speed compressor with a variable speed, digital scroll, two stage, and another single speed compressor.

3. DETAILS OF THE EXPERIMENTAL SETUP

Figure 1 shows the schematic of R410A Water Ethylene Glycol (WEG) chiller. A mixture of 20 % water and ethylene glycol is used as the secondary fluid. Two closed WEG loops are connected to the evaporator and condenser. Variable speed pumps and electric heaters are used to control the WEG flow rate and the inlet temperature of the condenser and the outlet temperature of the evaporator. An additional heat exchanger with chilled water flowing through is included in the condenser WEG loop to reject the heat from the condenser WEG loop. All the heat exchangers used in the facility are brazed plate heat exchangers. A 0.9 L receiver is connected between the condenser and the subcooler. Superheat is controlled by an Electronic Expansion Valve (EEV) while the subcooling is a function of the charge. The geometric dimensions of the evaporator, condenser, and subcooler can be found in Table 1. Both the compressors used are scroll compressors with a nominal speed of 1800 min^{-1}.

Type-T thermocouples, absolute and differential pressure transducers, and Coriolis-type mass flow meters are used to obtain the refrigerant side measurements while type-T thermocouples, differential pressure transducers, and Coriolis-type mass flow meters are used to obtain WEG measurements. Data is collected at steady-state conditions at 5s intervals for 20 consecutive minutes, and the data is averaged over the collection period. REFPROP 10.0 was used to calculate the WEG and R410A properties (Lemmon *et al.*, 2012).

The uncertainties of the calculated properties associated with the instrument uncertainties of measured properties are calculated by uncertainty propagation as shown in Equation (1):

$$U_Y = \sqrt{\sum_i (\partial Y/\partial X_i)^2 U_{X_i}^2} \tag{1}$$

where U_Y is the uncertainty of calculated property and U_{X_i} is the uncertainty of measured property. The uncertainty of the other sensors used in the experimental facility is presented in Table 2. This uncertainty estimation does not include the uncertainty in thermophysical properties. However, the uncertainty in enthalpy difference can be approximated as the uncertainty in specific heat which is around ±0.5% (Lemmon *et al.*, 2012).

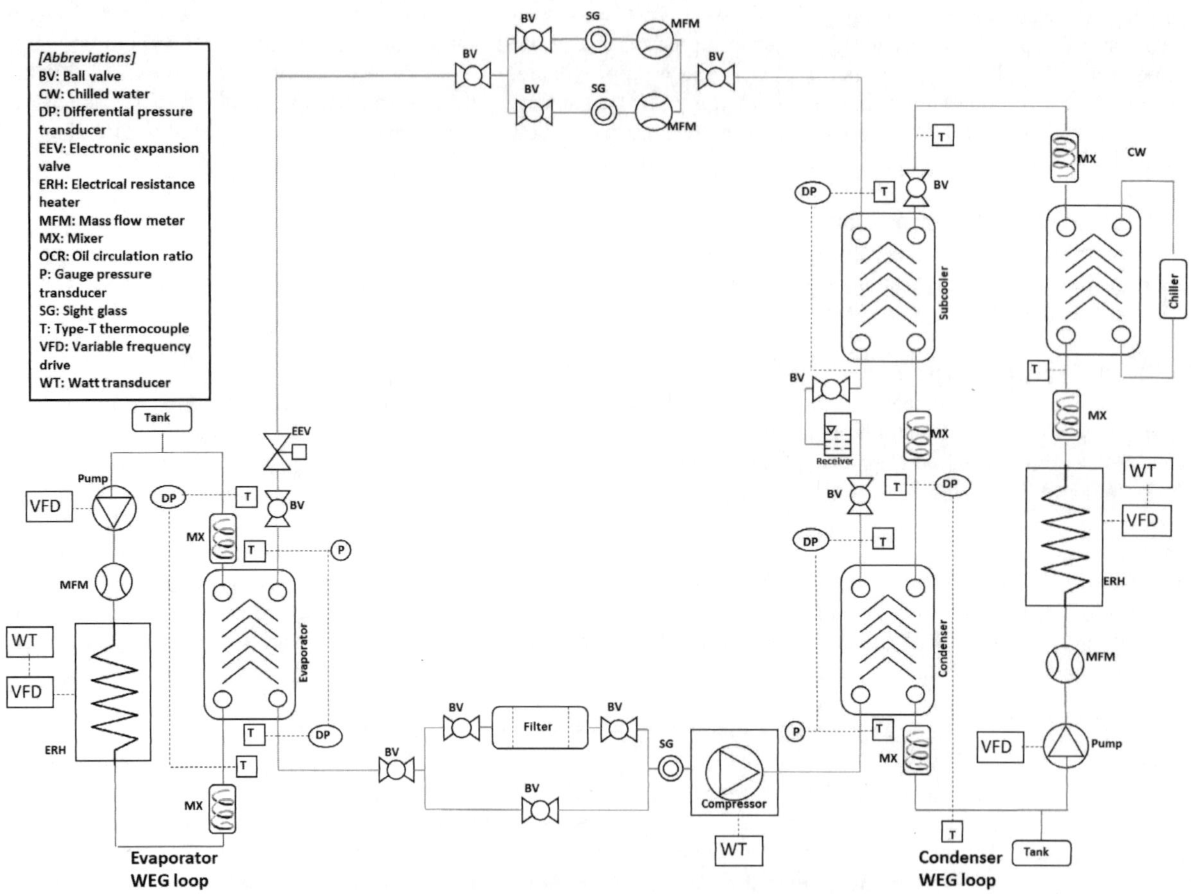

Figure 1: Schematic of the R410A experimental facility

Table 1: Dimensions of the brazed plate heat exchangers

Heat exchanger	Length (mm)	Width (mm)	Number of plates
Evaporator	311	111	28
Condenser	311	111	14
Subcooler	207	77	14

Table 2: Summary of measured and calculated property uncertainties

Instrument	Thermocouple (°C)	Pressure transducer (kPa)	Mass flow meter (g/s)	Wattmeter (kW)	Capacity (kW)	COP (-)
Uncertainty	±0.1	±0.2%	±0.2%	±0.5%	±1.5%	±1.6%

Capacity is calculated on the refrigerant side and WEG side. For the WEG side, mass flow rate, temperature, and specific heat are used to calculate capacity as shown in Equation (2). For the refrigerant side, temperature and pressure are used to calculate the enthalpy which is then used with the mass flow rate to calculate the capacity as shown in Equation (3). The capacity reported is the average of the refrigerant side and WEG side capacity given by Equation (4). The difference between the two capacities is indicated by the error given by Equation (5). This error is always less than 3% for the part load rating tests. Power consumed by the compressor is measured using a Wattmeter. The ratio of the average capacity and power consumed by the compressor is used to calculate the COP_{test} as shown in Equation (6).

$$\dot{Q}_{ev,WEG} = \dot{m}_{WEG} C_p \Delta T \tag{2}$$

$$\dot{Q}_{ev,ref} = \dot{m}_{ref} \Delta h \tag{3}$$

$$\dot{Q}_{ev,avg} = \frac{(\dot{Q}_{ev,WEG} + \dot{Q}_{ev,ref})}{2} \tag{4}$$

$$\varepsilon_{Qev} = \frac{(\dot{Q}_{ev,avg} - \dot{Q}_{ev,WEG}) \cdot 100}{\dot{Q}_{ev,avg}} \tag{5}$$

$$COP_{test} = \dot{Q}_{ev,avg} / \dot{W}_{cp} \tag{6}$$

4. AHRI 551/591 (2018) STANDARD

AHRI 551/591 is used for the determination of the part-load performance of water chillers. The standard defines a single number part-load efficiency figure of merit called Integrated Part Load Value (IPLV.SI) calculated at part load rating conditions. These part load rating conditions are shown below in Table 3. IPLV.SI is the weighted average of the COP_R measured at these standard rating conditions as shown in Equation (7). These factors in Equation (7) are based on the weighted average of the most common building types and operations using average weather in 29 U.S cities.

$$IPLV.SI = 0.01 \cdot A + 0.42 \cdot B + 0.45 \cdot C + 0.12 \cdot D \tag{7}$$

$$A = COP_R \text{ at } 100\%$$
$$B = COP_R \text{ at } 75\%$$
$$C = COP_R \text{ at } 50\%$$
$$D = COP_R \text{ at } 25\%$$

If a compressor cannot be unloaded to 25%, 50%, or 75% load point, then the compressor is run at the minimum step of unloading at the condenser entering water shown in Table 3 for 25%, 50%, or 75% capacity points as required. Once the COP_{test} is calculated at these conditions using Equation (6), it is degraded to COP_R using the Equation (8), (9), (10), and (11).

$$COP_R = COP_{test} / C_D \tag{8}$$

$$C_D = (-0.13 \cdot LF) + 1.13 \tag{9}$$

$$LF = \frac{(\%Load)(Q_{ev\ 100\%})}{(Q_{ev\ min\%Load})} \quad (10)$$

$$\%Load = {(Part\ load\ net\ capacity)}/{(Full\ load\ rated\ net\ capacity)} \quad (11)$$

Table 3: AHRI 551/591 part load conditions for IPLV.SI

Condition	Part load ratio (%)	Condenser Inlet/Outlet (°C)	Evaporator Inlet/Outlet (°C)
A	100	30/35	12/7
B	75	24.5/*	*/7
C	50	19/*	*/7
D	25	19/*	*/7

Table 3 shows the part load conditions. As seen in the table, condenser inlet, outlet, and evaporator inlet, outlet are mentioned for A condition while for the B, C, and D conditions, only the condenser inlet and evaporator outlet temperature are mentioned. Standard required that the condenser and evaporator WEG flow rate used for the A condition be used for the B, C, and D conditions.

5. RESULTS

Before testing the compressor at the part load rating conditions, charge optimization was done at the A condition i.e. condenser WEG inlet temperature of 30°C for both the compressors to determine the optimum charge for the compressor. After that, the compressors are tested at the part load rating conditions at their respective optimum charge. The next section shows the charge optimization of a single speed compressor.

5.1 Charge optimization of a single speed compressor

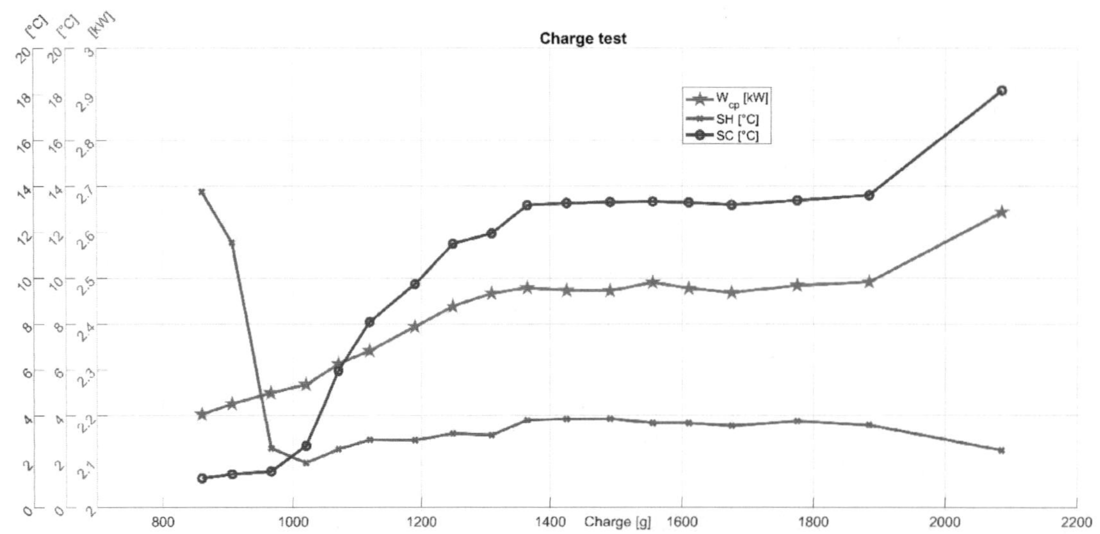

Figure 2: Variation of W_{cp}, superheat and subcooling with charge

Figure 2 shows the variation of the compressor work, superheat, and subcooling with the charge. As the charge increases, subcooling increases until it reaches a plateau. Superheat remains the same after the first few points because the EEV maintains the same superheat. The high superheat at the initial charge is because of the undercharged system. Figure 3 shows the variation of COP, average capacity, and the compressor discharge pressure with the charge. There is a range of charge from 1300 g to 1800 g where the curves remain almost constant. This range exists because of the receiver present after the condenser. As the charge increases in this range, the receiver gets filled and the amount of charge present in the remaining system remains almost constant. If the charge is increased beyond this charge, the receiver can no longer hold the charge and liquid starts backing up in the condenser. This is seen by an increase in the compressor discharge pressure and increase in the subcooling. A charge of 1500 g is selected as the optimum charge.

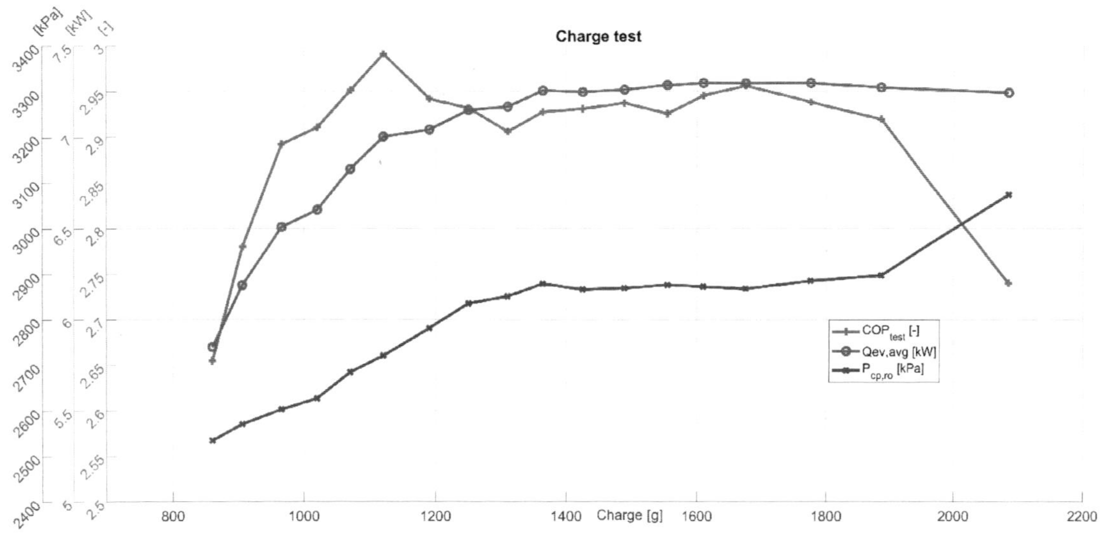

Figure 3: Variation of COP_{test}, $Q_{ev,avg}$, and $P_{cp,ro}$ with charge

5.2 Single speed compressor part load performance

As explained before, as single speed compressor does not have any modulation, it can operate only at 60 Hz. The variation of the capacity with part load rating conditions is shown in Figure 4. The load is a linear function of the part load rating conditions.

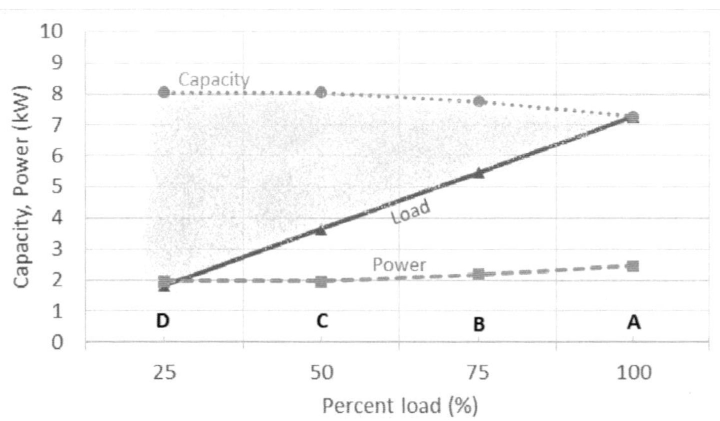

Figure 4: Variation of capacity, power with percent load for single speed compressor

As seen in Figure 4, as the percent load changes from 100% to 25%, the standard requires a reduced condenser WEG inlet temperature, this reduces the condensing temperature which in turn causes the capacity to increase. The shaded region between the capacity curve and the load curve is an indication of how much the obtained capacity differs from the required load and is a representation of the cycling losses. The capacity remains constant for the C and D conditions because the condenser WEG entering temperature is maintained the same as required by Table 3, only the percent load changes. Since a single speed compressor only operates at one stage, and thus the performance of the compressor at these percent loads is the same. As expected, the compressor power reduces as the percent load reduces because of reduced condensing temperature and compressor discharge pressure.

5.3 Two stage compressor part load performance

A two stage compressor can be operated at either high stage (100%) or low stage (67%). The performance of this compressor at different part load conditions is shown in Figure 5.

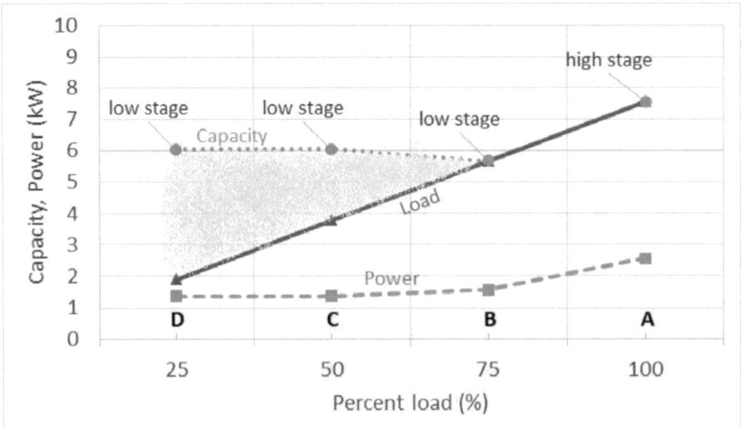

Figure 5: Variation of capacity, power with percent load for two stage compressor

For the A condition, the compressor is operated at a high stage. For the B, C, and D conditions, as the required load reduces, the compressor is operated at the low stage. The capacity is not a continuous line between A and B conditions because this compressor does not provide a continuous modulation, it provides a step modulation. At the B condition when the compressor is operated at a low stage, it provides 75% capacity which matches the required load. However, at the C and D conditions, the low stage cannot match the capacity and it behaves similarly to a single speed compressor and hence there will be cycling losses. The shaded region is a representation of the cycling losses. By comparing Figures 4 and 5, you can see that the shaded region i.e. cycling losses are higher in a single speed compressor compared to a two stage compressor.

5.4 Comparison between single speed and two stage compressor

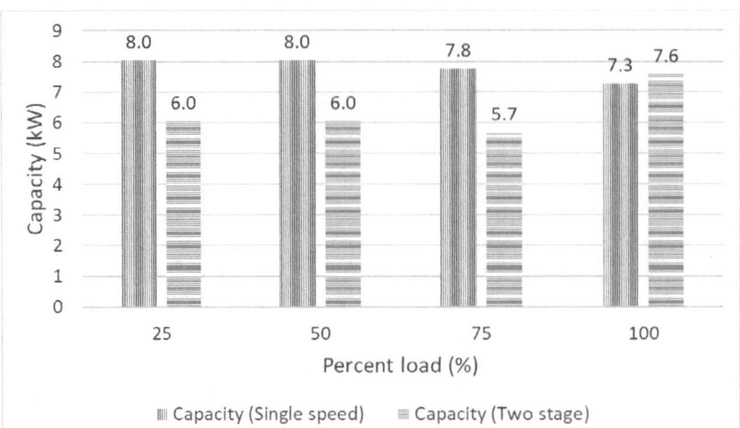

Figure 6: Comparison of capacity between single speed and two stage compressor

Figure 6 compares the capacities between a single speed and two stage compressor. At the 100% load, both the compressors have a similar capacity which indicates that both the compressors are comparable. However, for the remaining conditions, the single speed compressor has higher capacities than a two stage compressor because the two stage compressor can operate at a low stage.

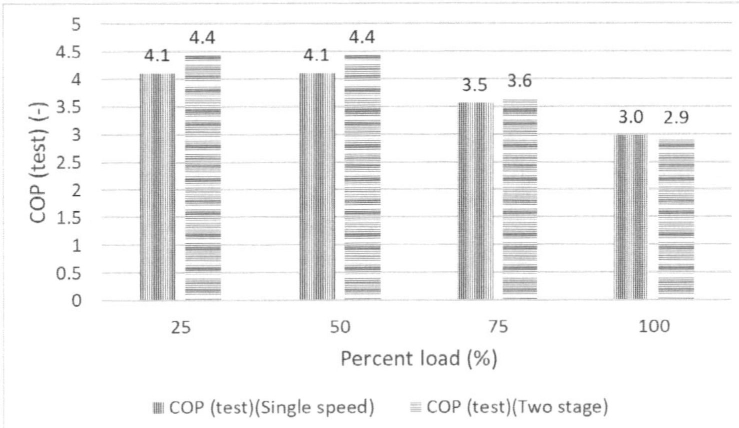

Figure 7: Comparison of COP_{test} between single speed and two stage compressor

Figure 7 shows the COP_{test} given by Equation (6). Both the compressors have similar capacity and power consumption at the 100% load, so it is not surprising that the COP_{test} are around the same. It is interesting to note that at the 75% load, though the two stage compressor is operating at the low stage and the single speed compressor is operating at its highest (only) stage, the COP_{test} are still comparable. However, the COP_{test} for the two stage is significantly higher at the lower percent loads because of reduced capacity and reduced compressor power. The COP_{test} for each compressor remains the same for the 50% and 25% load because the part load operating conditions are the same for these points.

Figure 8 shows the COP_R given by Equation (8). Except for the 100% load when both the compressors have similar performance, the two stage compressor has a higher COP_R than a single speed compressor. This is because of its ability to operate at lower stage. This trend can also be graphically explained by the cycling losses represented in Figures 4 and 5. Though the COP_{test} is the same for each compressor at the 25% and 50% load, COP_R at the 25% load is lower than 50% load because higher cycling losses reduce the COP_{test} by greater value as shown in Equation (9).

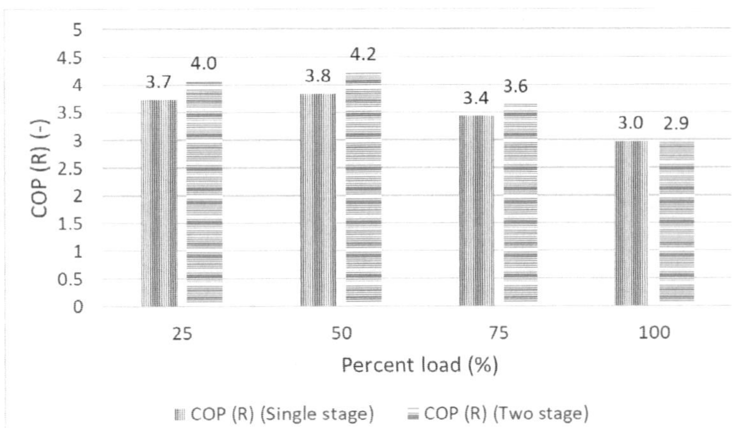

Figure 8: Comparison of COP_R between single speed and two stage compressor

IPLV.SI is calculated using the COP_R from Figure 8 and Equation (7). A single speed compressor has an IPLV.SI of 3.6 while the two stage compressor has an IPLV.SI of 3.9. This 7% difference can be explained by the higher COP_R of the two stage compressor at 50% and 75% load which have the highest weighting factors in Equation (7). A two stage compressor does not have any degradation coefficient at the 75% load and though it does incur some cycling losses at the 50% load, it is still operating at the low stage and these losses are lower than that incurred by the single speed compressor.

6. CONCLUSIONS

This study experimentally compared a single speed and a two stage compressor in the same R410A WEG chiller. Both the compressors have similar capacity and they belong to the same generation. Charge optimization was done to determine the optimum charge for both the compressor and AHRI 551/591 was used to determine the IPLV.SI for both the compressors at their respective optimum charge. Both compressors showed comparable capacities and COP_{test} at the higher modulation conditions. However, two stage compressor has a higher COP_R at the lower condition because of its ability to operate at a lower stage and better meet the required load, this reduces the C_D. Single speed compressor has an IPLV.SI of 3.6 while the two stage compressor has an IPLV.SI of 3.9.

NOMENCLATURE

C_d	degradation coefficient [-]	m	mass flow rate [kg/s]
COP	coefficient of performance [-]	Q	heat transfer rate [kW]
C_p	specific heat [kJ/kg-K]	SC	subcooling [°C]
ε	error [-]	SH	superheat [°C]
IPLV.SI	integrated part load value [-]	T	temperature [°C]
h	enthalpy [kJ/kg]	W	power [kW]
LF	load factor [-]	WEG	water ethylene glycol

Subscript

avg	average
cp	compressor
ev	evaporator
ref	refrigerant
ro	refrigerant, outlet
weg	water ethylene glycol mixture

REFERENCES

1. A. H. R. I. (2018). AHRI Standard 551/591 (SI): 2018 Standard for performance rating of water-chilling and heat pump water-heating packages using the vapor compression cycle. *Air-Conditioning, Heating, Refrigeration Institute, Arlington, VA, USA*.
2. Cecchinato, L. (2010). Part load efficiency of packaged air-cooled water chillers with inverter driven scroll compressors. *Energy Conversion and Management*, *51*(7), 1500-1509.
3. Ekren, O. (2017). Refrigeration System: Capacity Modulation Methods. *IntechOpen, London, UK,* 119-143.\
4. Lemmon, E. W., Bell, I. H., Huber, M. L., & McLinden, M. O. (2018). NIST Standard Reference Database 23: Reference Fluid Thermodynamic and Transport Properties-REFPROP, Version 10.0, National Institute of Standards and Technology. *Standard Reference Data Program, Gaithersburg.*
5. Qureshi, T. Q., & Tassou, S. A. (1996). Variable-speed capacity control in refrigeration systems. *Applied Thermal Engineering*, *16*(2), 103-113.
6. Wang, B., Han, L., Shi, W., & Li, X. (2012). Modulation method of scroll compressor based on suction gas bypass. *Applied Thermal Engineering*, *37*, 183-189.
7. Wu, G. H., & Wu, K. Q. (2004). Design and Application Research on Digital Scroll Compressor in Air Conditioning System, *Proc. of the International Compressor Engineering Conf,* Paper 1672, West Lafayette, IN, USA

ACKNOWLEDGEMENT

The authors would like to thank the member companies of the Air Conditioning and Refrigeration Center at the University of Illinois at Urbana-Champaign for their funding to support this project, Emerson / Copeland for compressor samples, Danfoss for brazed plate heat exchangers, and Creative Thermal Solutions for the technical support.

A critical analysis of the characterization of scroll compressors energy consumption

Javier MARCHANTE-AVELLANEDA[1], José M. CORBERAN[1]*,
Emilio NAVARRO-PERIS[1], Som S. SHRESTHA[2]

[1]Universitat Politècnica de València, Institute for Energy Engineering,
Valencia, Spain
Contact Information (corberan@iie.upv.es)

[2]Energy and Transportation Science Division, Oak Ridge National Laboratory,
Oak Ridge, TN 37831-6070, USA
Contact Information (shresthass@ornl.gov)

* Corresponding Author

ABSTRACT

This paper presents the analysis of the energy consumption of scroll type compressors. The study has included the data of several AHRI reports: (especially AHRI-11 and AHRI-21) as well as data from other source. A total of 8 different scroll compressors, of different size, some of them tested with various refrigerants (R134a, R32, R410A, R404A…) have been considered in the study.
The values of the compressor consumption, and of the corresponding compressor efficiency, and the shape of the corresponding response surfaces, for all the studied compressors and refrigerants, have been analyzed with the objective of understanding better the dependence of the energy compressor consumption on the operating conditions and the refrigerant. The analyzed data include tests following different superheat control, i.e. constant superheat or constant return temperature, so the effect of the inlet temperature on the energy consumption and efficiency are also discussed.
The paper includes the analysis of the compressor consumption as dependent of the temperatures of the tested points, and alternatively as dependent of the corresponding pressures, and as a result it will be shown that the representation as a function of pressures is more universal than the one made with temperatures.
Two simple correlation polynomials, based on suction and discharge pressures, are presented, which require less empirical information and have better interpolation-extrapolation characteristics than the AHRI standard correlation.

1. INTRODUCTION

Nowadays, the use of mathematical models allows the estimation of the systems performance and therefore are very useful in assisting the systems design, analysis and control. The heart of any refrigeration or HP system is the compressor; therefore, the estimation of its performance is of particular importance in the evaluation of the system performance. Numerous models have been proposed in the Literature to estimate the compressor performance. There could be classified as: theoretical, based in some way on the modelling of the thermodynamics of the involved processes across the compressor, or fully empirical, i.e. based on functionals (for instance polynomials) which are fitted to performance data. Theoretical models can be purely theoretical, based only on physical principles, or on the integration of the corresponding conservation equations, or theoretical but adjusted in some way by some empirical coefficients, which help to fit the results to performance data, adjusting for effects which have not been represented, adequately enough, in their formulation. We will call this last type of models semi-empirical.
A thorough review of compressor models has been included in several recent papers/reports, for instance in Byrne et al., 2014 and Hermes et al., 2019
Although a good number of semi-empirical models have been proposed over the years, fully empirical models are still in use, and it is the way that most of the compressors manufacturers report their compressor performance. In fact, as

reported in Cheung & Wang, 2018 when semi-empirical models have been compared with empirical models, in the cases where a large number of experimental data points are available all across the compressor envelope, the fully empirical models show better agreement in the representation of the compressor performance. The reason for this is that the semi-empirical models employ a pre-defined functional, with some coefficients which need to be adjusted to experimental data, but with dependences on the input variables which are implicit to the functional and therefore, cannot be changed along the fitting. Whereas, fully empirical models, involve no physics so that they require of experimental data to be fitted to, but are more flexible to adapt their shape to the actual compressor performance surface if enough experimental data is available.

Semi-empirical models are, in contrast, able to catch up the influence of the most important variables, for instance pressure ratio, in the performance and hence, are able to give a reasonable good estimation of the compressor performance with a reduced amount of experimental data, see for instance Jähnig et al., 2000, Cuevas & Lebrun, 2009, Navarro-Peris et al., 2013, and the recent paper by Hermes et al., 2019, but cannot reproduce the actual compressor performance with high accuracy.

The classical fully empirical model employ to characterize the compressor performance is the 10 coefficients third degree AHRI polynomial (ANSI/AHRI 540, 2015). These polynomials are able to provide a very accurate prediction of the compressor performance: refrigerant mass flow-rate and compressor energy consumption, across its entire working envelope by fitting the 10 coefficients. There has always been a discussion about, how many experimental data points and where to place them, are necessary in order to reach a reasonable good accuracy all across the compressor envelope. This topic was recently researched by Aute et al., 2015, Aute & Martin, 2016 and Cheung & Wang, 2018.

A few authors have proposed other empirical models, with the objective of reducing the amount of experimental data points required for their fitting, and maybe additionally improving the interpolation and extrapolation capabilities of the functionals in comparison with the AHRI polynomial. Among this, it should be first mentioned the more compact 2nd degree polynomial proposed by Shao (in Shao et al., 2004), the functional proposed by Aute in (Aute et al., 2014) and the proposed by Navarro (Navarro-Peris et al., 2013).

This paper presents an analysis of the energy consumption of scroll compressors, in order to better understand its shape and dependence on the operating parameters, with the objective of developing a functional for its characterization better than the 10 coefficients AHRI polynomial in the sense of requiring a lower number of experimental points for an accurate characterization of the performance and also maybe better extrapolation capabilities.

2. COMPRESSOR PERFORMANCE DATA

A few years ago, AHRI disclosed a series of results of performance of different compressors, scroll and piston, with conventional and new refrigerants and mixtures. These experimental results are included in several reports within the AHRI Low-GWP Alternative Refrigerants Evaluation Program. This study has considered all those AHRI reports containing scroll compressor tests: AHRI-11 (Shrestha, Mahderekal, et al., 2013), AHRI-21 (Shrestha, Sharma, et al., 2013), AHRI-24 (Rajendran & Nicholson, 2013), AHRI-33 (Shrestha et al., 2014), AHRI-34 (Rajendran & Nicholson, 2014a), AHRI-36 (Rajendran & Nicholson, 2014b), AHRI-38 (Rajendran & Nicholson, 2014c), AHRI-39 (Rajendran & Nicholson, 2014d), AHRI-58 (Rajendran et al., 2016a), AHRI-65 (Rajendran et al., 2016b), and AHRI-66 (Suindykov et al., 2016), and additionally the performance data published in Cuevas & Lebrun, 2009; totaling 8 different scroll compressors, and a huge variety of refrigerants: R134a, R32, R410A, R404A, R447A, R454B, DR5, DR7, L40, L41a, L41b, ARM31a, D2Y65.

From the analysis of all the compressor consumption data, we have found two different behaviors depending on the application range: at moderate-high evaporation temperatures M-HT, the compressor consumption is quite independent of the evaporation temperature, sometimes showing a slight increase, sometimes a slight decrease. While, at low evaporation temperatures LT, the compressor consumption increases significantly with the decrease of the evaporation temperature with a certain hyperbolic behavior. AHRI -11 is characteristic of the first kind: M-HT, while AHRI -21 is characteristic of the second kind: LT. Those reports have been selected for discussion in this paper because they include many experimental test points, covering the entire operating domain of the respective compressors. The compressor studied in Cuevas (Cuevas & Lebrun, 2009) belong the M-HT category.

Table 1 summarizes the main characteristics of the compressors tested in AHRI-11, AHRI-21, and Cuevas & Lebrun, 2009. Finally, the Table 2 shows the Mass% compositions of the tested refrigerants' mixtures. The thermophysical properties of the mixtures analyzed has been obtained with the NIST's Refprop software package (Lemmon et al., 2018), evaporation and condensation temperatures considered at dew point.

Table 1. Main compressor characteristics and tested refrigerants

Source	Model	Manufac.	Disp. (freq.) (cm^3) (Hz)	Refrigerants tested	Test points	Conditions by refrigerant test
AHRI 21	ZS21KAE-PFV	Copeland	50.96 (60)	R404A/ARM31a/ D2Y65/L40/(R32 +R134a)	191/186/ 183/173/ 133	SH=11°K SH=22°K SC=6°K Tsuc=18°C
AHRI 11	ZP21K5E-PFV	Copeland	20.32 (60)	R410A/R32/DR5	196/166/ 189	
Cuevas(2009)	-	-	54.25 (50)	R134a	18	SH=6.8K

New refrigerants composition (mass %)

Source	Name	Composition
AHRI 21	ARM-31a	R-32/R-134a/R-1234yf (28/21/51)
	D2Y-65	R-32/R-1234yf (35/65)
	L-40	R-32/R-152a/R-1234yf/R-1234ze(E) (40/10/20/30)
	R-32/R-134a	R-32/R-134a (50/50)
AHRI 11	DR-5	R-32/R-1234yf (72.5/27.5)

2. COMPRESSOR PERFORMANCE ANALYSIS

The analyzed data includes tests following different superheat control, e.g. constant superheat or constant return temperature, so the effect of the inlet temperature on the compressor consumption and efficiency are also discussed One of the possibilities to characterize the compressor consumption has always been to characterize the compressor efficiency, η_c. This presents the advantage of being a non-dimensional parameter quite independent of the size of the compressor and mainly dependent on the pressure ratio.

$$\eta_c = \frac{\dot{m} \cdot \Delta h_{is}}{\dot{W}_c}$$

In fact, many authors have employed this approach to characterize the compressor. Moreover, both compressor and volumetric efficiencies are quite characteristic of the compressor and they can even be used to estimate the compressor performance with a different refrigerant. Some authors have proposed non-dimensional parameters, similar to the efficiencies, which are even more general and provide a slightly better estimation of the compressor performance (Pierre, 1982; and Navarro-Peris et al., 2013)

Figure 1, shows the compressor efficiency of compressor ZS21KAE-PFV versus pressure ratio, for all the tests points included in AHRI 21 for the reference refrigerant R404A. Three sets of data were measured, corresponding to three different conditions at the suction: constant superheat of 11.11 K (SH11), constant superheat of 22.22 K (SH22), and constant return temperature 18 °C (T18). Figure 2 shows the same experimental data separating in different boxes the tests performed at the same evaporation temperature (left side of the figure), or separating in different boxes the tests performed at the same condensation temperature (right side of the figure).

Figure 2 shows that for a constant superheat, and if one performs the tests at a constant evaporation temperature and varies only the condensation temperature, then one obtains a clear curve with pressure ratio with a maximum at a certain pressure ratio. This is not the case, if the tests are performed the other way around, keeping constant the condensation temperature and varying the evaporation temperature (right side).

Compressor efficiency has the typical shape as a function of the pressure ratio (with a maximum at a given pressure ratio) only for compressors working at a more or less constant evaporation temperature (see figure 2), which is just the case of compressors for air conditioning or chillers.

Figure 1 also shows that the suction temperature clearly affects compressor efficiency, with a trend to increase with the increase of the superheat. The highest efficiencies are the ones corresponding to SH22, with exception of some points at low evaporation temperatures where at return temperature 18°C the superheat is higher than 22.

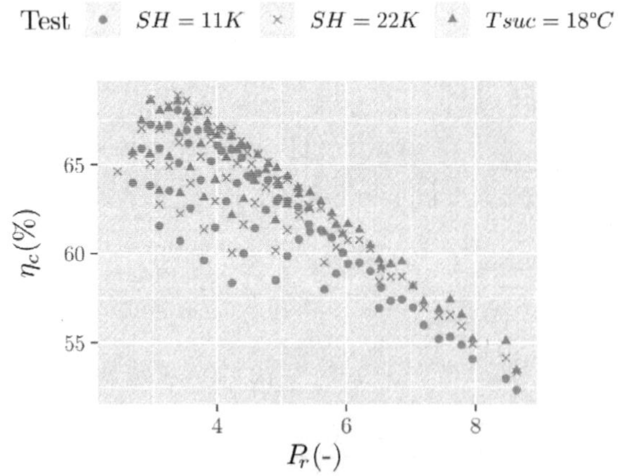

Figure 1: Compressor efficiency versus pressure ratio of compressor ZS21KAE-PFV, for all the tests points included in AHRI 21 for the reference refrigerant R404A

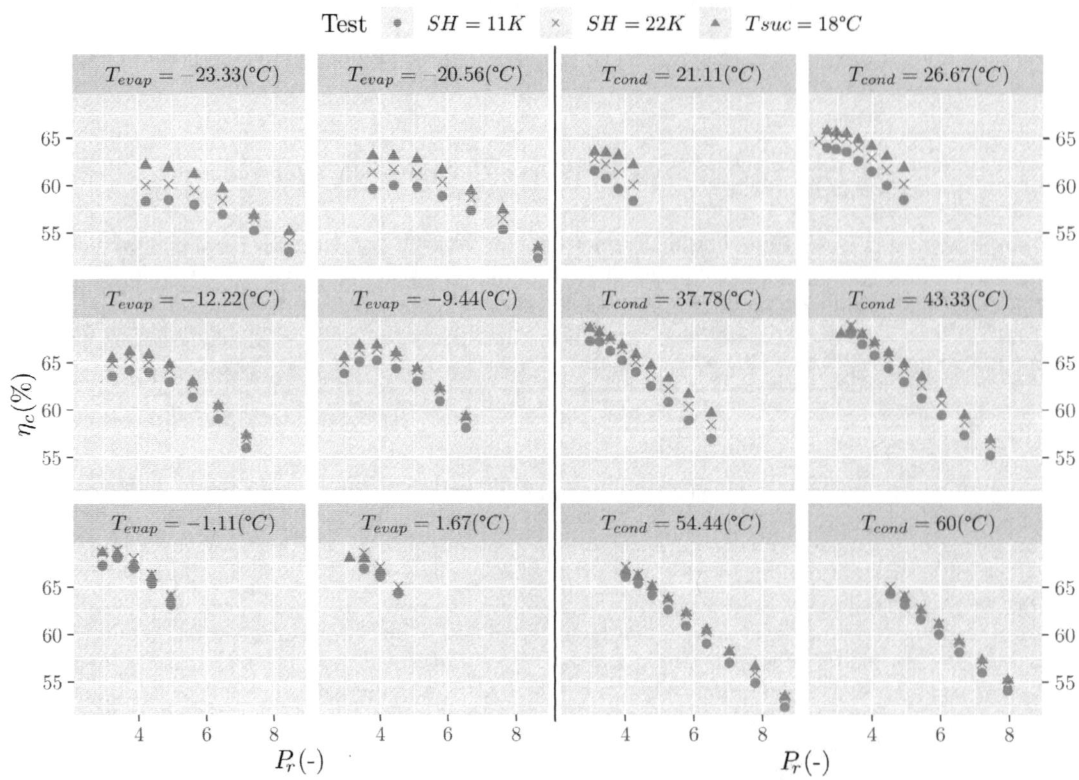

Figure 2: Compressor efficiency versus pressure ratio of compressor ZS21KAE-PFV with refrigerant R404A. Left side: compressor efficiency at given evaporation temperatures. Right side: compressor efficiency at given condensation temperatures.

In other words, compressor efficiency is a complex function of evaporation and condensation conditions plus the superheat. In contrast, if one represents the compressor consumption versus the evaporation and condensation temperatures (see figure 3) it is hard to see any influence of the superheat. Figure 3 shows the AHRI-21 compressor consumption vs. condensation temperature at various constant evaporation temperatures at the left side, and the compressor consumption vs. evaporation temperature at various constant condensation temperatures at the right side.

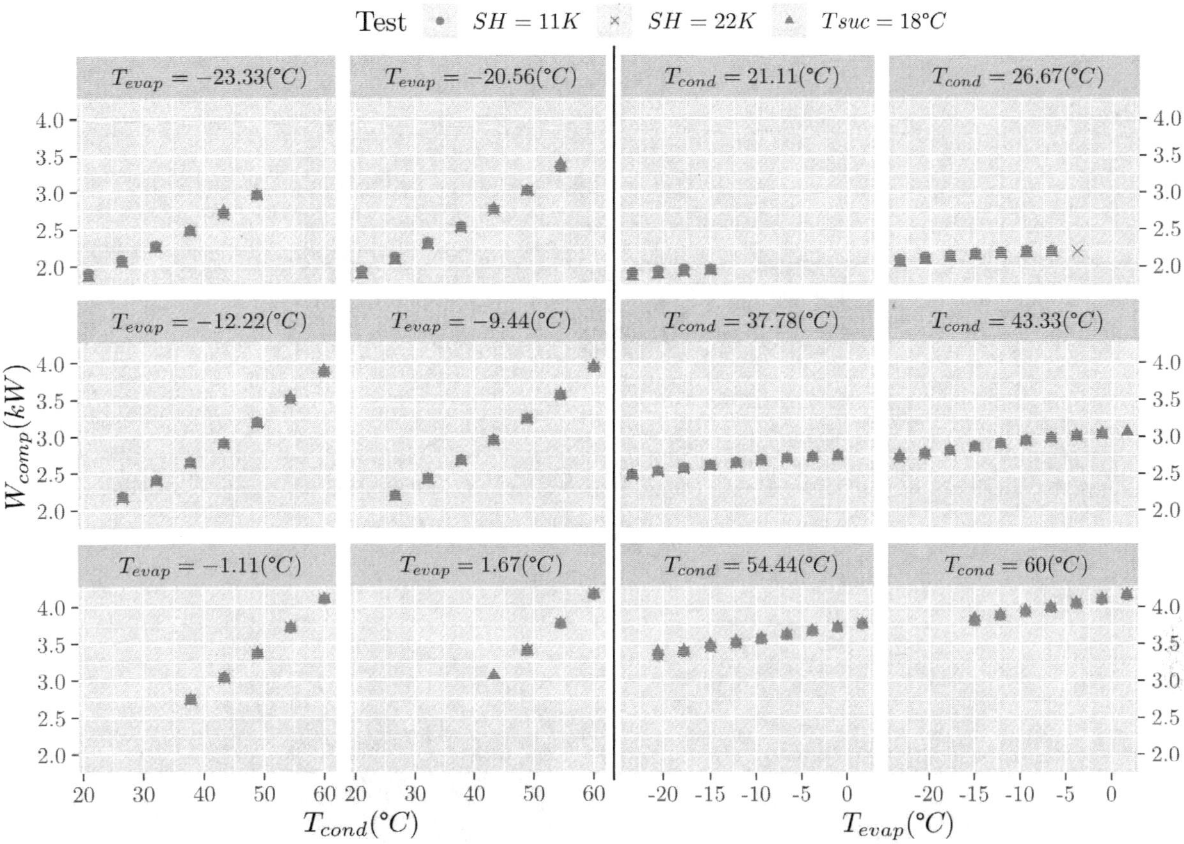

Figure 3: Compressor consumption versus evaporation or condensation temperatures of compressor ZS21KAE-PFV with refrigerant R404A. Left side: compressor efficiency at given evaporation temperatures. Right side: compressor efficiency at given condensation temperatures.

Figure 4 shows the energy consumption of compressor ZS21KAE-PFV with refrigerant R404A in a 3D plot as a function of evaporation and condensation temperatures. As it can be observed the compressor consumption surface is quite smooth. Figures 3 and 4 show that the energy consumption of scroll compressors is mainly dependent on the condensation temperature, increasing with it, with a slight dependence on the evaporation temperature.
Figure 5 shows the compressor consumption maps of compressors ZS21KAE-PFV (AHRI-21) and ZP21K5E-PFV (AHRI-11) for their reference refrigerant, R404A and R410A respectively.
As can be seen, and as mentioned above, the energy consumption of scroll compressors mainly depends on the condensation temperature with a slight dependence on the evaporation temperature. AHRI-21 compressor, which has been tested at low temperatures, show a decreasing trend of the consumption with the evaporation temperature, while AHRI-11 compressor shows a slight increasing trend with the evaporation temperature.

All M-HT analyzed scroll compressors of the referenced AHRI reports, and the one tested by Cuevas & Lebrun, 2009, show this slight decrease of the compressor consumption, almost linear, with the increase of the evaporation temperature. While the compressors of reports AHRI-21, AHRI-34 and AHRI-36 show the consumption decreasing trend with the evaporation temperature. It should be pointed out that the compressors of reports AHRI-21, AHRI-34 are for LT applications and in fact they are Liquid injection type. Therefore, it can be concluded that the dependence of the scroll compressors consumption with the evaporation temperature is weak and it depends on the application range, slightly hyperbolic decreasing for LT applications while it is slightly linear increasing for MT-HT applications.

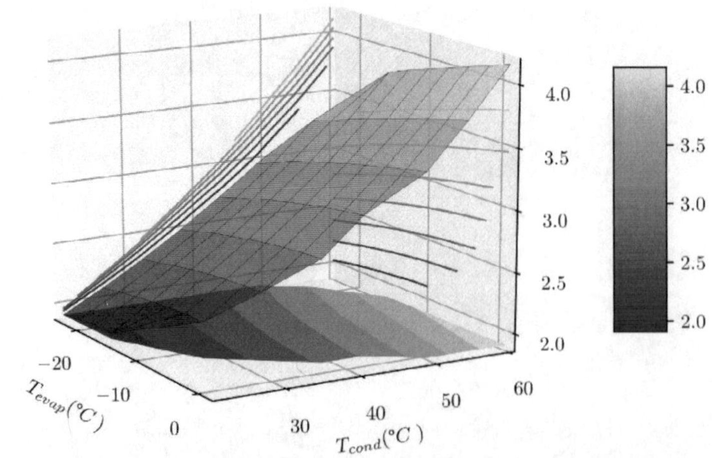

Figure 4: 3D plot of compressor consumption versus evaporation and condensation temperatures of compressor ZS21KAE-PFV with refrigerant R404A (SH=11K).

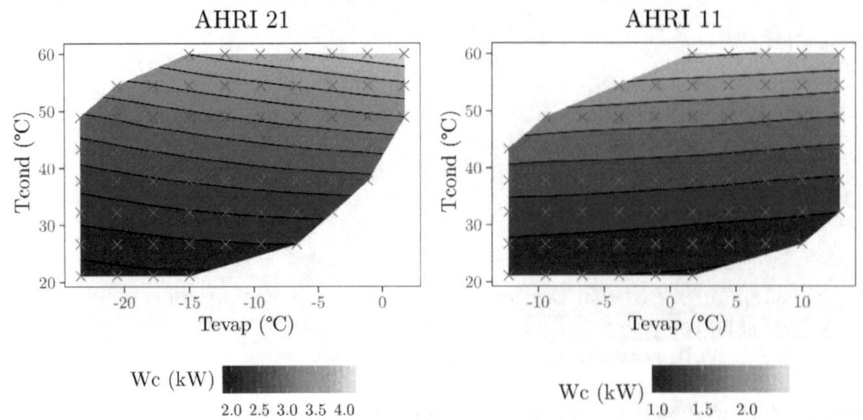

Figure 5: Compressor consumption maps of compressors ZS21KAE-PFV (AHRI-21) and ZP21K5E-PFV (AHRI-11) for their reference refrigerant, R404A and R410A respectively (SH=11K).

3. COMPRESSOR CONSUMPTION CORRELATION

3.1 Correlations employed

If one observes the surface representing the compressor consumption versus the condensation and evaporation temperatures shown in figure 4, it is easy to understand why the 10 coefficients AHRI polynomial (ANSI/AHRI 540, 2015) is able to reproduce the response surface so well when enough experimental data points are available for the fitting and they are well distributed all across the operation domain. In fact, the authors have employed the more compact polynomial proposed by Shao et al., 2004 and have found that it provides the same ability to represent the surface with only 6 coefficients. The polynomial proposed by Shao employs only the main terms of the AHRI polynomial and in the experience of the authors, it is able to represent very well the energy consumption of scroll and rotary compressors.

$$W_c = b_1 T_c^2 + b_2 T_c + b_3 T_c T_e + b_4 T_e^2 + b_5 T_e + b_6$$

In the mentioned paper, Shao proposes a correlation for variable speed compressors based on the described functional at the nominal speed and a correction with the speed, which also works very well.

However, when one plots the compressor consumption for different refrigerants the surfaces show different levels depending on their respective refrigerant properties. The authors have found that if alternatively, the consumption is plotted as a function of the refrigerant pressure, instead of the temperatures, it turns out that the surfaces are much more similar with each other. Figure 6 shows the compressor consumption for the compressor ZS21KAE-PFV of AHRI-21, for 4 different refrigerants. From figure 6, one can see that the representation versus the pressures is much more universal than versus the temperatures.

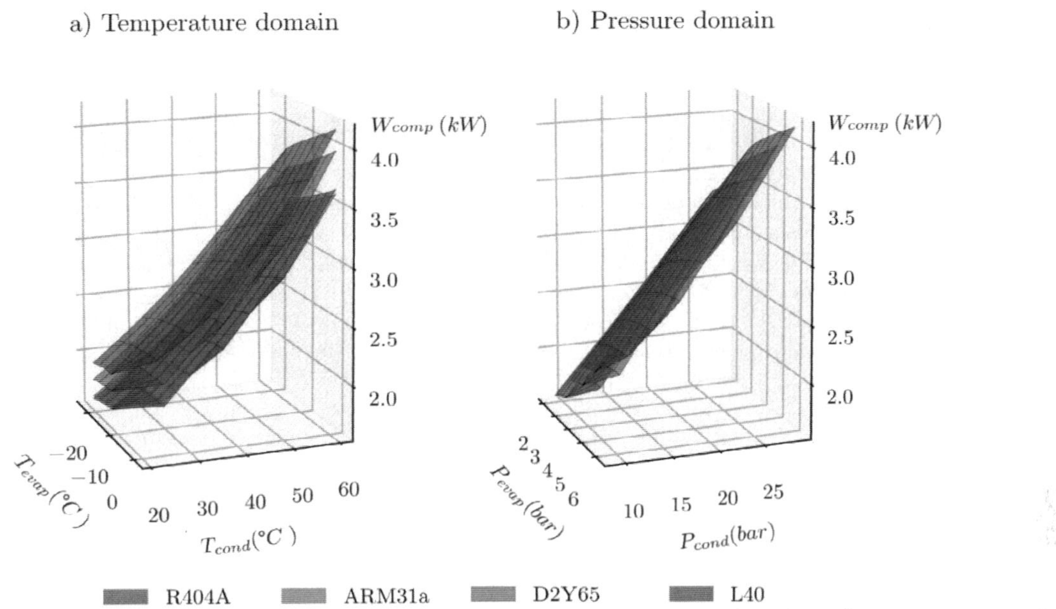

Figure 6: Left side: 3D plot of compressor consumption versus evaporation and condensation temperatures of compressor ZS21KAE-PFV for 4 different refrigerants. Right side: 3D plot of compressor consumption versus evaporation and condensation pressures of compressor ZS21KAE-PFV for 4 different refrigerants (SH=11K).

The representation of figure 6 vs. evaporation and condensation pressures have been performed for the wide range of refrigerants included in the referenced reports and has proven that that representation of the compressor consumption is much more independent of the refrigerant hence more representative of the compressor. Furthermore, the authors have found that the correlation of polynomials based on the condensation and evaporation pressures is as effective as the one based on the dew temperatures. Therefore, we propose to correlate the performance as a function of the evaporation and condensation pressures.

As it can be observed in figure 6 right side, the compressor consumption is a quite flat surface when represented versus the evaporation and condensation pressures. The authors have found that a simple linear polynomial containing linear terms on both evaporation and condensation pressures and one cross-term with their product leads to a very robust correlation with very decent accuracy for all the analyzed compressors and refrigerants.

Correlation 1: $W_c = C_0 + C_1P_e + C_2P_c + C_3P_eP_c$

This polynomial will be referred as to Correlation 1 in the following comparison of results. If one wants to increase the accuracy of the correlation, one should add first at second order dependence on the condensation pressure, since it has most of the influence on the consumption, and it becomes a bit quadratic at high condensation pressures. Also, the authors have found that a second order on the evaporation temperature is good for LT compressors. Adding those two terms to Correlation 1, one gets the second correlation proposed by the authors:

Correlation 2: $W_c = C_0 + C_1P_e + C_2P_c + C_3P_e^2 + C_4P_c^2 + C_5P_eP_c$

Finally, we will compare the results with the same correlation but employing the dew temperatures instead of the pressures. This correlation is exactly the one proposed by Shao et al., 2004. We will name this one, correlation 3.

Correlation 3: $W_c = C_0 + C_1T_c^2 + C_2T_c + C_3T_e^2 + C_4T_e + C_5T_eT_c$

3.2 Comparison of correlations

The described correlations, 1 2 and 3, were fitted to the compressor consumption results included in all the available AHRI reports with scroll compressors, mentioned in the first section, and to the set of the test points of Cuevas & Lebrun, 2009 corresponding to 50 Hz constant compressor frequency.

The results of the fitting were very good for all the analyzed compressors and refrigerants. We also did the fitting to the original 10 coefficients AHRI polynomial but the results did not improve, and a big portion of the coefficients did not have significance enough. Table 2 shows a summary of the correlation results for compressors ZS21KAE-PFV (AHRI-21) and ZS21KAE-PFV (AHRI-11) for 4 different refrigerants each, and for the compressor tested by Cuevas & Lebrun, 2009, for the three correlations defined above. The Table includes the values of the coefficients (estimates) for correlation 1, 2 and 3, as well as the maximum relative error (MRE) in (%) and the Root Mean Square Error (RMSE) in W. For each compressor and refrigerant, the correlations are fitted to all available test points, including all different suction conditions. The coefficients are meant to provide the compressor consumption in kW with temperatures expressed in °C and pressures in bar.

As can be seen in table 2, both MRE and RMSE are very low with practically all the analyzed correlations, providing a very good representation of the compressor consumption across the entire envelope. The highest accuracy is reached with Correlation 2, proving that the correlation with pressures is better than with temperatures, and as discussed above, less dependent on the employed refrigerant. The fact of adding coefficients allows for a better fitting of the experimental results, however, different terms could lose significance.

The compressor consumption results of Cuevas & Lebrun, 2009 are not so many as the other two considered compressors, however the polynomials show slightly higher RMSE when fitted to those results. The authors have plotted the surface and it is very similar to the other MT-HT compressors. The origin for the errors could then be caused by a higher experimental uncertainty. Anyhow, all the proposed correlations, were able to describe those results with still high accuracy.

4. CONCLUSIONS

A thorough analysis of the energy consumption characteristics of scroll compressors have been performed. The study has included all scroll compressor results included in the AHRI reports corresponding to the AHRI Low-GWP Alternative Refrigerants Evaluation Program. The following main conclusions can be drawn from the performed study.

- The first conclusion is that when the compressor is measured in a wide range of operating conditions, inside its envelope, the compressor efficiency shows a complex shape, and it is clearly sensitive to the suction conditions (superheat). In contrast, the compressor consumption is a smooth surface when plotted versus the evaporation and condensation temperatures (or pressures), and it shows very little dependence on the superheat. Therefore, compressor consumption is much easier to characterize by fitting a polynomial than compressor efficiency.
- For scroll compressors it is not necessary to employ a 10 coefficients polynomial, as proposed in ANSI/AHRI 540, 2015, to characterize the compressor consumption. The much compact expression proposed by Shao et al., 2004 is accurate enough and requires many less test points to be fitted to.
- The authors have found that if the compressor consumption is correlated versus the condensation and evaporation pressures, the correlation results are better and moreover, it is more universal, being more independent of the employed refrigerant.
- The energy consumption of scroll compressors is a quite plane and smooth surface. A simple correlation with linear terms on the condensation and evaporation pressures together with a cross-term with their product, requires only 4 coefficients and provides a very simple and robust representation. If more accuracy is required, a 6 coefficients polynomial including a quadratic term for each pressure provides an excellent accuracy all across the compressor envelope.

Table 2: Correlations results

	Correlation 1			Correlation 2			Correlation 3				
Coeff	Estimate	MRE (%)	RMSE (W)	Estimate	MRE (%)	RMSE (W)	Estimate	MRE (%)	RMSE (W)	Source	Fluid
C0	7.650E-01			5.876E-01			1.364E+00				
C1	-2.086E-03			1.100E-01			4.177E-04				
C2	8.803E-02	2.34	21.50	8.301E-02	1.54	13.44	2.082E-02	2.19	14.04	AHRI21	R404A
C3	4.986E-03			-1.782E-02			-2.083E-04				
C4	-			-8.686E-05			-1.109E-02				
C5	-			7.071E-03			4.832E-04				
C0	6.718E-01			5.179E-01			1.165E+00				
C1	1.598E-02			1.197E-01			3.626E-04				
C2	8.459E-02	1.89	17.42	8.342E-02	1.29	9.40	1.830E-02	1.65	9.73	AHRI21	ARM31a
C3	5.617E-03			-2.242E-02			-1.556E-04				
C4	-			-2.473E-04			-8.665E-03				
C5	-			8.578E-03			4.503E-04				
C0	7.080E-01			5.574E-01			1.260E+00				
C1	1.104E-03			1.028E-01			4.285E-04				
C2	8.712E-02	2.61	18.41	8.351E-02	1.61	11.01	1.873E-02	1.85	11.98	AHRI21	D2Y65
C3	5.247E-03			-1.886E-02			-1.699E-04				
C4	-			-1.473E-04			-1.027E-02				
C5	-			7.683E-03			4.824E-04				
C0	6.125E-01			5.073E-01			1.164E+00				
C1	1.549E-02			1.149E-01			4.079E-04				
C2	9.143E-02	2.27	15.79	8.526E-02	1.50	9.04	1.627E-02	1.29	9.12	AHRI21	L40
C3	5.172E-03			-2.182E-02			-1.650E-04				
C4	-			-1.149E-04			-8.714E-03				
C5	-			8.141E-03			4.321E-04				
C0	6.063E-01			5.289E-01			1.306E+00				
C1	-5.939E-03			1.076E-01			5.739E-04				
C2	1.025E-01	1.46	15.51	8.770E-02	0.91	8.89	1.395E-02	1.28	13.62	AHRI21	R32/R134a
C3	4.996E-03			-2.107E-02			-2.337E-04				
C4	-			9.159E-05			-1.140E-02				
C5	-			7.826E-03			4.759E-04				
C0	2.758E-01			3.242E-01			7.719E-01				
C1	-2.954E-02			-1.080E-02			4.470E-04				
C2	5.846E-02	2.57	12.86	4.813E-02	0.72	4.95	1.002E-03	1.21	7.26	AHRI11	R410A
C3	4.317E-04			-2.900E-04			-7.612E-05				
C4	-			2.913E-04			-3.930E-03				
C5	-			-1.346E-04			-2.073E-05				
C0	4.033E-01			3.962E-01			7.952E-01				
C1	-4.484E-02			-1.635E-02			4.836E-04				
C2	5.235E-02	2.71	16.88	4.307E-02	3.19	12.53	6.100E-04	3.12	14.43	AHRI11	R32
C3	1.305E-03			-6.050E-04			-7.108E-05				
C4	-			3.195E-04			-6.163E-03				
C5	-			5.302E-04			6.753E-05				
C0	2.880E-01			3.566E-01			7.815E-01				
C1	-3.288E-02			-1.259E-02			4.626E-04				
C2	5.778E-02	4.95	17.94	4.479E-02	4.19	10.61	-1.732E-03	4.62	13.00	AHRI11	DR5
C3	6.835E-04			3.952E-05			-4.084E-05				
C4	-			4.213E-04			-2.703E-03				
C5	-			-2.444E-04			-3.607E-05				
C0	2.959E-01			3.021E-01			6.949E-01				
C1	-3.105E-02			-7.373E-03			3.922E-04				
C2	5.503E-02	2.29	10.48	4.661E-02	1.02	4.64	1.556E-03	1.25	6.33	AHRI11	L41a
C3	7.344E-04			-8.129E-04			-8.312E-05				
C4	-			2.678E-04			-4.315E-03				
C5	-			2.042E-04			2.430E-05				
C0	9.187E-02			1.779E-01			1.383E+00				
C1	-3.761E-02			-2.548E-03			7.654E-04				
C2	1.549E-01	3.16	60.75	1.321E-01	2.58	51.17	-3.100E-02	3.05	54.48	Cuevas, Lebrun	R134a
C3	-6.756E-04			-9.498E-04			-2.660E-04				
C4	-			5.152E-04			1.959E-02				
C5	-			-1.101E-03			-1.686E-04				

NOMENCLATURE

SH	Superheat (K)	$\eta_c = \frac{\dot{m} \cdot \Delta h_{is}}{\dot{W}_c}$	Compressor efficiency (%)
SC	Subcooling (K)	$\dot{m}$	Refrigerant mass flowrate (kg/s)
T_{suc}	Suction temperature (°C)	$\Delta h_{is} = h_{2s} - h_1$	Enthalpy difference (isentropic compression) (kJ/kg)
T_{evap}	Evaporation temperature at dew point (°C)	P_r	Pressure ratio (-)
T_{cond}	Condensation temperature at dew point (°C)	$\dot{W}_c$	Compressor power input (kW)
P_{evap}	Evaporation pressure (bar)	MRE	Maximum Relative Error (%)
P_{cond}	Condensation pressure (bar)	$RMSE$	Root Mean Square Error

REFERENCES

ANSI/AHRI 540. (2015). *AHRI 540 - Standard for performance rating of positive displacement refrigerant compressors and compressor units*. Air-Conditioning, Heating, and Refrigeration Institute (AHRI).

Aute, V., & Martin, C. (2016). A Comprehensive Evaluation of Regression Uncertainty and the Effect of Sample Size on the AHRI-540 Method of Compressor Performance Representation. *In: Proceedings of the Twenty-Third International Compressor Engineering Conference at Purdue*, 1–9.

Aute, V., Martin, C., & Radermacher, R. (2015). AHRI Project 8013 : A Study of Methods to Represent Compressor Performance Data over an Operating Envelope Based on a Finite Set of Test Data. *Air-Conditioning, Heating, and Refrigeration Institute*.

Aute, V., Qiao, H., Kwon, L., Radermacher, R., Hall, G. M., & Park, C. (2014). Transient modeling of a multi-evaporator air conditioning system and control method investigation. *11th IEA Heat Pump Conference 2014, May 12-16 2014, Montréal (Québec) Canada*.

Byrne, P., Ghoubali, R., & Miriel, J. (2014). Scroll compressor modelling for heat pumps using hydrocarbons as refrigerants. *International Journal of Refrigeration*, *41*, 1–13. https://doi.org/10.1016/j.ijrefrig.2013.06.003

Cheung, H., & Wang, S. (2018). A comparison of the effect of empirical and physical modeling approaches to extrapolation capability of compressor models by uncertainty analysis: A case study with common semi-empirical compressor mass flow rate models. *International Journal of Refrigeration*, *86*, 331–343. https://doi.org/10.1016/j.ijrefrig.2017.11.020

Cuevas, C., & Lebrun, J. (2009). Testing and modelling of a variable speed scroll compressor. *Applied Thermal Engineering*, *29*(2–3), 469–478. https://doi.org/10.1016/J.APPLTHERMALENG.2008.03.016

Hermes, C. J. L., Santos, G. Z., & Ronzoni, A. F. (2019). Performance characterization of small variable-capacity reciprocating compressors using a minimal dataset. *International Journal of Refrigeration*, *107*, 191–201. https://doi.org/10.1016/j.ijrefrig.2019.07.014

Jähnig, D. I., Reindl, D. T., & Klein, S. A. (2000). *A Semi-empirical Method for Representing Domestic Refrigerator/Freezer Compressor Calorimeter Test Data*.

Lemmon, E. W., Bell, I. H., Huber, M. L., & McLinden, M. O. (2018). *NIST Standard Reference Database 23: Reference Fluid Thermodynamic and Transport Properties-REFPROP, Version 10.0, National Institute of Standards and Technology*. https://doi.org/https://doi.org/10.18434/T4/1502528

Navarro-Peris, E., Corberán, J. M., Falco, L., & Martínez-Galván, I. O. (2013). New non-dimensional performance parameters for the characterization of refrigeration compressors. *International Journal of Refrigeration*, *36*(7), 1951–1964. https://doi.org/10.1016/j.ijrefrig.2013.07.007

Pierre, B. (1982). *Kylteknik, Allmän Kurs*. Inst. Mekanisk Värmeteori och Kylteknik. Kungl Tekniska Högskolan.

Rajendran, R., & Nicholson, A. (2013). *TEST REPORT #24*.

Rajendran, R., & Nicholson, A. (2014a). *TEST REPORT #34*.

Rajendran, R., & Nicholson, A. (2014b). *TEST REPORT #36*.

Rajendran, R., & Nicholson, A. (2014c). *TEST REPORT #38*.

Rajendran, R., & Nicholson, A. (2014d). *TEST REPORT #39.*
Rajendran, R., Pham, H., Bella, B., & Skillen, T. (2016a). *TEST REPORT #58.*
Rajendran, R., Pham, H., Bella, B., & Skillen, T. (2016b). *TEST REPORT #65.*
Shao, S., Shi, W., Li, X., & Chen, H. (2004). Performance representation of variable-speed compressor for inverter air conditioners based on experimental data. *International Journal of Refrigeration*, *27*, 805–8015. https://doi.org/10.1016/j.ijrefrig.2004.02.008
Shrestha, S., Mahderekal, I., Sharma, V., & Abdelaziz, O. (2013). *TEST REPORT #11.*
Shrestha, S., Sharma, V., & Abdelaziz, O. (2013). *TEST REPORT #21.*
Shrestha, S., Sharma, V., & Abdelaziz, O. (2014). *TEST REPORT #33.*
Suindykov, S., Zhang, L., & Gernemann, A. (2016). *TEST REPORT #66.*

ACKNOWLEDGEMENT

The present work has been partially funded by the Ministerio de Educación, Cultura y Deporte through the 'Formación de Profesorado Universitario' programme ref. FPU15/03476. The authors also want to acknowledge the financial support provided by the project "ENE2017-83665-C2-1-P" funded by the "Ministerio de Ciencia, Innovación y Universidades" of Spain.

This research used resources at the Building Technologies Research and Integration Center, a DOE Office of Science User Facility operated by the Oak Ridge National Laboratory

Theoretical Analysis of Rigid-Body Vibration in Swing Compressors

Daiki Kikutake*, Tatsuya Katayama, Hideki Matsuura

Technology and Innovation Center, Daikin Industries, Ltd.,
1-1, Nishi-Hitotsuya, Settsu, Osaka, Japan
Phone: +81-6-6195-7074,
E-mail: daiki.kikutake@daikin.co.jp (Daiki Kikutake),
tatsuya.katayama@daikin.co.jp (Tatsuya Katayama),
hideki.matsuura@daikin.co.jp (Hideki Matsuura)

* Corresponding Author

ABSTRACT

This study aims to theoretically analyze the rigid-body vibration in swing compressors. Currently, low noise and vibration (NV) are qualities recognized as a desired value for an air conditioner by end users. Because the compressor is one of the main sources of NV in an air conditioner, low NV compressors are needed. Designing a low NV compressor requires the understanding of the physical phenomena occurring in a compressor that contribute to NV. In this study, we theoretically analyzed rigid-body vibration, which has been one of the main phenomena causing NV in a swing compressor. Main factors causing rigid-body vibration in the swing compressor were classified into three types of excitation forces. By formulating them in a Fourier series representation, the primary and secondary components in the frequency domain were calculated for each excitation force without any numerical calculation method, including FFT. Furthermore, by multiplying the mass and inertial tensor of the compressor to this, we calculated both the amplitude and phase of the rigid-body vibration generated by each excitation force. We then confirmed that the calculated vibration was consistent with the experimental results. The contribution of each excitation force to the radial and tangential vibrations of typical swing compressors was theoretically analyzed. The results of the analysis clearly indicated that vibration was effectively reduced by not only simply reducing the magnitudes of the excitation forces but also by balancing the excitation forces.

1. INTRODUCTION

Low noise and vibration (NV) are qualities recognized as a desired value for an air conditioner by end users. Low NV of an air conditioner is very important for reliability and user comfort. A compressor is one of the main sources of NV in an air conditioner. Therefore, for an air conditioner to provide high comfort and reliability, a compressor with low NV is required.

NV of a compressor arise from various phenomena that occur in the compressor. Designing a low NV compressor requires a theoretical understanding of the physical phenomena occurring in the compressor that contribute to NV. This theoretical understanding of the phenomena enables us to predict and analyze compressor NV and is very useful for the effective design of a low NV compressor before trial production, troubleshooting of compressors with large NV, and developing new techniques for a low NV compressor.

Rigid-body vibration of a compressor is one of the main phenomena causing NV of an air conditioner. Rigid-body vibration of the compressor, especially the primary and secondary components in the frequency domain, causes sound to not only be radiated from the compressor but also causes high stress for pipes and solid-borne sound to the outdoor unit. Therefore, it is very important to understand the mechanism of the rigid-body vibration of the compressor and to reduce that vibration. As for rotary compressors, which are mainly used in domestic air conditioners, some types of excitation forces are known to cause rigid-body vibration (Imaichi et al., 1982). In order to design a low NV rotary compressor, all the excitation forces should be considered, and the manner by which each excitation force affects the rigid-body vibration of the compressor must be understood.

In this study, we conducted a theoretical analysis of rigid-body vibration for a swing compressor, which is a rotary-type compressor developed by Daikin Industries, Ltd. Firstly, we formulated the rigid-body vibration of a swing compressor. The main factors causing rigid-body vibration in a swing compressor were classified into three types of excitation forces and, by formulation in a Fourier series representation, we obtained the primary and secondary components in the frequency domain without any numerical calculation method, including FFT. Furthermore, by multiplying the mass and inertial tensor of the compressor to this, both amplitude and phase were formulated for the rigid-body vibration generated by each excitation force. Secondly, we measured the rigid-body vibrations of several typical swing compressors, including both single-cylinder type and twin-cylinder type. The calculated vibrations were confirmed to be consistent with the measured ones. Thirdly, for the rigid-body vibrations of each single-cylinder and twin-cylinder compressor, we conducted a theoretical analysis on the contribution of each excitation force to the radial and the tangential vibrations for the primary and secondary components in the frequency domain.

2. THEORY

In this section, we present theoretical equations of rigid-body vibration of a swing compressor. This study classified the main factors causing rigid-body vibration in a swing compressor into three types of excitation forces: centrifugal force of the rotor system, inertial force of the piston, and torque which is generated when refrigerant is compressed. In subsection 2.1, based on an equation of motion for a rigid body, we formulate an arbitrary frequency order for components in the frequency domain for vibration caused by an excitation force. In subsections 2.2 – 2.4, the excitation forces are formulated in a Fourier series representation.

2.1 Equation of Motion for Rigid-Body Vibration

A motion of a rigid body is composed of translational and rotational motions having 3 degrees of freedom, respectively. An equation of motion for a rigid body is expressed as

$$m\ddot{\boldsymbol{r}} + C_r\dot{\boldsymbol{r}} + K_r\boldsymbol{r} = \boldsymbol{F},\ I\ddot{\boldsymbol{\Theta}} + C_\Theta\dot{\boldsymbol{\Theta}} + K_\Theta\boldsymbol{\Theta} = \boldsymbol{T}, \tag{1}$$

where $\boldsymbol{r} = (x, y, z)^{\mathrm{T}}$, $\boldsymbol{\Theta} = (\Theta_x, \Theta_y, \Theta_z)^{\mathrm{T}}$, $\boldsymbol{F} = (F_x, F_y, F_z)^{\mathrm{T}}$, $\boldsymbol{T} = (T_x, T_y, T_z)^{\mathrm{T}}$, m, I, $C_{r,\Theta}$, $K_{r,\Theta}$, are the translational displacement, the rotational displacement, the force, the torque, the mass, the inertial tensor, viscosity matrix, and stiffness matrix, respectively. As for the rigid-body vibration of a compressor, $C_{r,\Theta}$ and $K_{r,\Theta}$ represent the influence of pipes and supporting foot rubbers. This study ignores the influence of the pipes and the rubbers to analyze the compressor's inherent vibration:

$$m\ddot{\boldsymbol{r}} = \boldsymbol{F},\ I\ddot{\boldsymbol{\Theta}} = \boldsymbol{T}. \tag{2}$$

When the rotational speed of the compressor is N, the excitation forces of the vibration become a periodic function with the period of $1/N$. Therefore, $\boldsymbol{F}$ and $\boldsymbol{T}$ are expressed in a Fourier series representation with a fundamental angular frequency of $\omega = 2\pi N$:

$$\begin{aligned}\boldsymbol{F} &= \boldsymbol{F}_0 + \textstyle\sum_j \left(F_{jx}\cos\left(j\omega t + \phi_{F_{jx}}\right), F_{jy}\cos\left(j\omega t + \phi_{F_{jy}}\right), F_{jz}\cos\left(j\omega t + \phi_{F_{jz}}\right)\right)^{\mathrm{T}},\\ \boldsymbol{T} &= \boldsymbol{T}_0 + \textstyle\sum_j \left(T_{jx}\cos\left(j\omega t + \phi_{T_{jx}}\right), T_{jy}\cos\left(j\omega t + \phi_{T_{jy}}\right), T_{jz}\cos\left(j\omega t + \phi_{T_{jz}}\right)\right)^{\mathrm{T}},\end{aligned} \tag{3}$$

where $\boldsymbol{F}_j = (F_{jx}, F_{jy}, F_{jz})^{\mathrm{T}}$, $\boldsymbol{\phi}_{F_j} = \left(\phi_{F_{jx}}, \phi_{F_{jy}}, \phi_{F_{jz}}\right)^{\mathrm{T}}$, $\boldsymbol{T}_j = (T_{jx}, T_{jy}, T_{jz})^{\mathrm{T}}$, $\boldsymbol{\phi}_{T_j} = \left(\phi_{T_{jx}}, \phi_{T_{jy}}, \phi_{T_{jz}}\right)^{\mathrm{T}}$ are the jth component in the frequency domain of the amplitude and the phase of the force and the torque. With Equations (2) and (3), jth components in the frequency domain of the translational and rotational vibration acceleration (jn translational and rotational vibration) in the compressor are expressed as

$$\ddot{\boldsymbol{r}}_j = m^{-1}\left(F_{jx}\cos\left(j\omega t + \phi_{F_{jx}}\right), F_{jy}\cos\left(j\omega t + \phi_{F_{jy}}\right), F_{jz}\cos\left(j\omega t + \phi_{F_{jz}}\right)\right)^{\mathrm{T}},$$
$$\ddot{\boldsymbol{\Theta}}_j = I^{-1}\left(T_{jx}\cos\left(j\omega t + \phi_{T_{jx}}\right), T_{jy}\cos\left(j\omega t + \phi_{T_{jy}}\right), T_{jz}\cos\left(j\omega t + \phi_{T_{jz}}\right)\right)^{\mathrm{T}}. \tag{4}$$

When you want to calculate the jn translational vibration at a position $\boldsymbol{R} = (X, Y, Z)^{\mathrm{T}}$ on the compressor $\boldsymbol{a}(\boldsymbol{R})_j$, it is calculated with $\ddot{\boldsymbol{r}}_j$ and $\ddot{\boldsymbol{\Theta}}_j$ as

$$\boldsymbol{a}(\boldsymbol{R})_j = \ddot{\boldsymbol{r}}_j + \ddot{\boldsymbol{\Theta}}_j \times \boldsymbol{R}. \tag{5}$$

Rigid-body vibration of a compressor has multiple excitation forces. Total jn vibration of the compressor is expressed as the sum of the vibration caused by each excitation force:

$$\ddot{\boldsymbol{r}}_{j\mathrm{total}} = \Sigma_{\mathrm{force}}\, \ddot{\boldsymbol{r}}_j,\ \ddot{\boldsymbol{\Theta}}_{j\mathrm{total}} = \Sigma_{\mathrm{force}}\, \ddot{\boldsymbol{\Theta}}_j,\ \boldsymbol{a}(\boldsymbol{R})_{j\mathrm{total}} = \Sigma_{\mathrm{force}}\, \boldsymbol{a}(\boldsymbol{R})_j. \tag{6}$$

This means that if each excitation force is formulated in a Fourier series representation, we can obtain the contribution of the excitation force to the jn vibration separately.

In this study, two types of orthogonal coordinates are defined as shown in Fig. 1. The x, y, z coordinate, which is shown in Fig. 1 (a), has its origin at the center of the cylindrical casing. The z axis coincides with the central axis of the casing and the x axis passes through the central axis of the accumulator. The $x_{\mathrm{cyl}}, y_{\mathrm{cyl}}, z_{\mathrm{cyl}}$ coordinate, which is shown in Fig. 1 (b), has its origin at the center of the cylinder. The z_{cyl} axis coincides with the central axis of the crankshaft (and also with the z axis) and the y_{cyl} axis passes through the center of the swing bushes.

For each excitation force, $\boldsymbol{F} = \left(F_x, F_y, F_z\right)^{\mathrm{T}}$ and $\boldsymbol{T} = \left(T_x, T_y, T_z\right)^{\mathrm{T}}$ are formulated as the following procedure:

1. Formulate $\boldsymbol{F}_{\mathrm{cyl}} = \left(F_{x_{\mathrm{cyl}}}, F_{y_{\mathrm{cyl}}}, F_{z_{\mathrm{cyl}}}\right)^{\mathrm{T}}$ on the $x_{\mathrm{cyl}}, y_{\mathrm{cyl}}, z_{\mathrm{cyl}}$ coordinate in a Fourier series representation
2. Formulate $\boldsymbol{F} = \left(F_x, F_y, F_z\right)^{\mathrm{T}}$ on the x, y, z coordinate by using the following equation:

$$\begin{pmatrix} F_x \\ F_y \\ F_z \end{pmatrix} = \begin{pmatrix} \sin\theta_{\mathrm{suc}} & \cos\theta_{\mathrm{suc}} & 0 \\ -\cos\theta_{\mathrm{suc}} & \sin\theta_{\mathrm{suc}} & 0 \\ 0 & 0 & 1 \end{pmatrix} \begin{pmatrix} F_{x_{\mathrm{cyl}}} \\ F_{y_{\mathrm{cyl}}} \\ F_{z_{\mathrm{cyl}}} \end{pmatrix} \tag{7}$$

3. Formulate $\boldsymbol{T} = \left(T_x, T_y, T_z\right)^{\mathrm{T}}$ on the x, y, z coordinate by using the following equation:

$$\boldsymbol{T} = \boldsymbol{r} \times \boldsymbol{F}, \tag{8}$$

where $\boldsymbol{r} = (x, y, z)^{\mathrm{T}}$ is the position at which $\boldsymbol{F} = \left(F_x, F_y, F_z\right)^{\mathrm{T}}$ applies

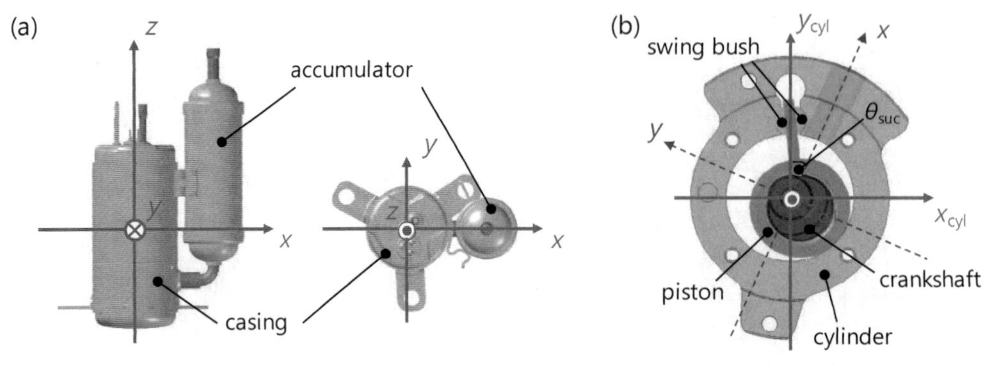

Figure 1: Orthogonal coordinates. (a) x, y, z coordinate. (b) $x_{\mathrm{cyl}}, y_{\mathrm{cyl}}, z_{\mathrm{cyl}}$ coordinate.

2.2 Centrifugal Force of the Rotor System

We define a system composed of a crankshaft, a motor rotor, and balancers attached on the motor rotor as the rotor system, as shown in Fig. 2 (a). When the centrifugal forces applied to the eccentric parts of the system are unbalanced, the unbalanced force excites the vibration of the compressor via its bearings. The balancers are usually designed so that the centrifugal force of the rotor system and the centrifugal force applied to the cylindrical part of the piston are balanced. However, in the usual design of the balancers, the influence of the deflection of the crankshaft is not often considered. As shown in some previous studies (Huang *et al.*, 2010 and Zhang *et al.*, 2014), the influence of the deflection of the crankshaft on the centrifugal forces is not negligible. Therefore, the deflection of the crankshaft should be calculated at the same time when the centrifugal forces are calculated. The deflection was calculated with a finite element model (FE model) in the previous studies, but this method requires a detailed FE model and takes much calculation time. In this subsection, we formulate the deflection of the crankshaft with a beam deflection based on a material mechanics, and then obtain the unbalanced centrifugal force including the influence of the deflection.

The rotor system was modeled as a beam supported at the upper end of the main bearing (bearing 1) and the lower end of the sub-bearing (bearing 2), as shown in Fig. 2 (b). When the centrifugal forces applied to the upper balancer, the lower balancer, and the motor rotor are denoted as $F_{\mathrm{ubw}}, F_{\mathrm{lbw}}, F_{\mathrm{ro}}$, respectively, and the deflections of the crankshaft at their respective positions as $y_{\mathrm{ubw}}, y_{\mathrm{lbw}}, y_{\mathrm{ro}}$, $\boldsymbol{y} = (y_{\mathrm{ubw}}, y_{\mathrm{lbw}}, y_{\mathrm{ro}})^{\mathrm{T}}$ is expressed with $\boldsymbol{F}_{\mathrm{cent}} = (F_{\mathrm{ubw}}, F_{\mathrm{lbw}}, F_{\mathrm{ro}})^{\mathrm{T}}$ as

$$\boldsymbol{y} = A_1 \boldsymbol{F}_{\mathrm{cent}}, \tag{9}$$

where A_1 is influence coefficient matrix. Also, $\boldsymbol{F}_{\mathrm{cent}}$ is expressed with $\boldsymbol{y}$ as

$$\begin{aligned} \boldsymbol{F}_{\mathrm{cent}} &= (m_{\mathrm{ubw}}(y_{\mathrm{ubw}} + r_{\mathrm{ubw}})\omega^2, m_{\mathrm{lbw}}(y_{\mathrm{lbw}} - r_{\mathrm{lbw}})\omega^2, m_{\mathrm{ro}} y_{\mathrm{ro}} \omega^2)^{\mathrm{T}} \\ &= \mathrm{diag}(m_{\mathrm{ubw}}\omega^2, m_{\mathrm{lbw}}\omega^2, m_{\mathrm{ro}}\omega^2)\boldsymbol{y} + (m_{\mathrm{ubw}} r_{\mathrm{ubw}} \omega^2, -m_{\mathrm{lbw}} r_{\mathrm{lbw}} \omega^2, 0)^{\mathrm{T}} \\ &\equiv A_2 \boldsymbol{y} + \boldsymbol{b}_1 \end{aligned} \tag{10}$$

From Equation (9) and (10), $\boldsymbol{y}$ and $\boldsymbol{F}_{\mathrm{cent}}$ are obtained as follows:

$$\boldsymbol{y} = -(A_1 A_2 - I_3)^{-1} A_1 \boldsymbol{b}_1, \; \boldsymbol{F}_{\mathrm{cent}} = [I_3 - A_2 (A_1 A_2 - I_3)^{-1} A_1] \boldsymbol{b}_1, \tag{11}$$

where I_3 is a unit matrix.

As for a single-cylinder type swing compressor, the forces which the rotor system receives from the bearings F_{bear1} and F_{bear2} satisfy following equilibrium equations:

$$F_{\mathrm{bear1}} + F_{\mathrm{bear2}} = F_{\mathrm{ubw}} + F_{\mathrm{ro}} + F_{\mathrm{lbw}} + F_{\mathrm{ecc}}, \tag{12}$$

$$F_{\mathrm{bear1}} h_{\mathrm{bear1}} + F_{\mathrm{bear2}} h_{\mathrm{bear2}} = F_{\mathrm{ubw}} h_{\mathrm{ubw}} + F_{\mathrm{ro}} h_{\mathrm{ro}} + F_{\mathrm{lbw}} h_{\mathrm{lbw}}, \tag{13}$$

where $F_{\mathrm{ecc}} = m_{\mathrm{ecc}} e \omega^2$ and $h_{\mathrm{bear1,bear2,ubw,lbw,ro}}$ is the height shown in Fig. 2(b). From Equations (12) and (13), F_{bear1} and F_{bear2} are obtained. $\boldsymbol{F}_{\mathrm{cyl}}(\theta) = \left(F_{x_{\mathrm{cyl}}}, F_{y_{\mathrm{cyl}}}, F_{z_{\mathrm{cyl}}}\right)^{\mathrm{T}}$ caused by $F_{\mathrm{bear}i}$ $(i = 1, 2)$ is expressed as

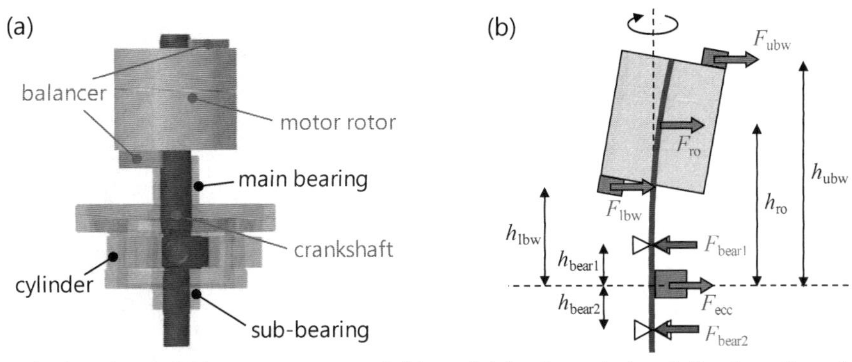

Figure 2: Schematic drawings of (a) rotor system and (b) model for formulating deflection of crankshaft and centrifugal forces

$\boldsymbol{F}_{\text{cyl}}(\theta) = (F_{\text{beari}} \sin\theta, F_{\text{beari}} \cos\theta, 0)^{\text{T}}$. $\boldsymbol{F}(\theta)$ and $\boldsymbol{T}(\theta)$ caused by F_{beari} are then obtained with Equations (7) and (8). $\boldsymbol{F}(\theta)$ and $\boldsymbol{T}(\theta)$ for a twin-cylinder type can be formulated in the same way as those of a single-cylinder type, except that the contribution of the centrifugal force of another eccentric part of the crankshaft is added to Equations (12) and (13).

2.3 Inertial Force of the Piston

Swing motion of the piston is a phenomenon which only exists in a swing compressor, not in a rotary compressor. The swing motion of the piston causes an inertial force, which is one of the excitation forces for rigid-body vibration. The swing motion can be categorized into two types of motion: one is the reciprocating motion in the blade direction and the other is the pendular motion in the swing angle direction, as shown in Fig. 3 (a). The excitation force caused by each motion is formulated in this subsection.

The inertial force caused by the reciprocating motion $F_{\text{iner\_BL}}(\theta)$ is applied to the crankshaft and it excites the vibration of the compressor via its bearings. $F_{\text{iner\_BL}}(\theta)$ is calculated as

$$F_{\text{iner\_BL}}(\theta) = m_{\text{pis}} \frac{\mathrm{d}^2 L_{\text{b}}}{\mathrm{d}t^2} \cos(\pi - \theta - \varphi), \quad (15)$$

where L_{b} is the length shown in Fig 3(b) and φ is the swing angle, which are expressed as

$$\begin{aligned} L_{\text{b}} &= \sqrt{e^2 + L^2 - 2eL\cos\theta} - r - (L - R) \cong e\left(1 - \cos\theta - \frac{e}{4L}\cos 2\theta\right), \\ \varphi &\cong \sin\varphi = \frac{e\sin\theta}{\sqrt{e^2 + L^2 - 2eL\cos\theta}} \cong \frac{e}{L}\sin\theta + \frac{1}{2}\left(\frac{e}{L}\right)^2 \sin 2\theta + \frac{27}{8}\left(\frac{e}{L}\right)^3 \sin 3\theta. \end{aligned} \quad (16)$$

$\boldsymbol{F}_{\text{cyl}}(\theta)$ caused by $F_{\text{iner}_{\text{BL}}}(\theta)$ is expressed as $\boldsymbol{F}_{\text{cyl}}(\theta) = \left(F_{\text{iner\_BL}} \sin\theta, F_{\text{iner\_BL}} \cos\theta, 0\right)^{\text{T}}$. $\boldsymbol{F}(\theta)$ and $\boldsymbol{T}(\theta)$ caused by $F_{\text{iner\_BL}}(\theta)$ are then obtained with Equations (7) and (8).

The inertial force caused by the pendular motion $F_{\text{iner\_phi}}(\theta)$ is applied to the swing bush and it excites the vibration of the compressor via its cylinder. $\boldsymbol{F}(\theta)$ and $\boldsymbol{T}(\theta)$ caused by $F_{\text{iner\_phi}}(\theta)$ are calculated in the same way as those caused by $F_{\text{iner\_BL}}(\theta)$. $F_{\text{iner\_phi}}(\theta)$ is expressed as

$$F_{\text{iner\_phi}}(\theta) = -\frac{I_{\text{pis}}\ddot{\varphi}}{\sqrt{e^2 + L^2 - 2eL\cos\theta}} \cong \frac{I_{\text{pis}} e\omega^2}{L^2}\left[\sin\theta + \frac{5e}{2L}\sin 2\theta + \frac{19}{4}\left(\frac{e}{L}\right)^2 \sin 3\theta\right]. \quad (17)$$

$\boldsymbol{F}_{\text{cyl}}(\theta)$ is calculated as $\boldsymbol{F}_{\text{cyl}}(\theta) = \left(-F_{\text{iner\_phi}} \cos\varphi, -F_{\text{iner\_phi}} \sin\varphi, 0\right)^{\text{T}}$, and then $\boldsymbol{F}(\theta)$ and $\boldsymbol{T}(\theta)$ are obtained with Equations (7) and (8).

As for a twin-cylinder type swing compressor, you can formulate the inertial force caused by both the reciprocating motion and the pendular motion in the same way as described in the following. When the inertial force in one cylinder is expressed as $\boldsymbol{F}_{\text{cyl1}}(\theta)$, which is formulated as Equations (15) or (17), the force in the other cylinder $\boldsymbol{F}_{\text{cyl2}}(\theta)$ is expressed as $\boldsymbol{F}_{\text{cyl2}}(\theta) = \boldsymbol{F}_{\text{cyl1}}(\theta + \pi)$. The total inertial force $\boldsymbol{F}_{\text{cyl\_total}}(\theta)$ is expressed as $\boldsymbol{F}_{\text{cyl\_total}}(\theta) = \boldsymbol{F}_{\text{cyl1}}(\theta) + \boldsymbol{F}_{\text{cyl2}}(\theta)$, and then $\boldsymbol{F}(\theta)$ and $\boldsymbol{T}(\theta)$ are obtained with Equations (7) and (8).

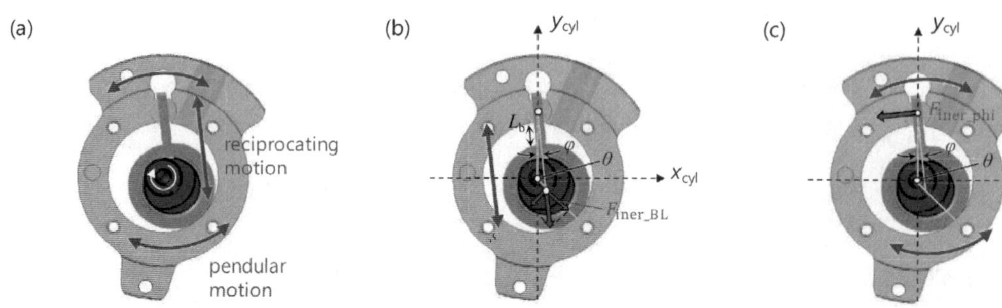

Figure 3: Schematic drawings of (a) swing motion of piston, (b) inertial force caused by reciprocating motion in the blade direction and (c) inertial force caused by pendular motion in the swing angle direction.

2.4 Compression Torque

As shown in Fig. 4 (a), torque caused by the pressure difference between the compression chamber and the suction chamber is applied to the rotor system. The reaction torque, which we call the compression torque, then excites the rigid-body vibration of the compressor. The compression torque has the only component of around z axis $T_{\mathrm{comp\_}z}(\theta)$, which is expressed as

$$T_{\mathrm{comp\_}z}(\theta) = -\Delta P(\theta) V_{SL}(\theta), \tag{18}$$

where $\Delta P(\theta)$ is the pressure difference between the compression chamber and the suction chamber, and $V_{SL}(\theta)$ the product of the area at which $\Delta P(\theta)$ applies and the arm length of the compression torque. When both $\Delta P(\theta)$ and $V_{SL}(\theta)$ are formulated in a Fourier series representation, $T_{\mathrm{comp\_}z}(\theta)$ is also expressed in the representation. In this sub-section, the Fourier series representations of $\Delta P(\theta)$ and $V_{SL}(\theta)$ are formulated.

From the Poisson's law for adiabatic process in thermodynamics, the pressure of the compression chamber at a crank angle θ is expressed as

$$P_{\mathrm{c}}(\theta) = \begin{cases} P_{\mathrm{s}}\left(\dfrac{V_{\mathrm{cc}}}{V_{\mathrm{c}}(\theta)}\right)^{\gamma} & (0 \leq \theta \leq \theta_{\mathrm{d}}) \\ P_{\mathrm{d}} & (\theta_{\mathrm{d}} \leq \theta \leq 2\pi) \end{cases} \tag{19}$$

where $V_{\mathrm{c}}(\theta)$ is geometrically obtained as follows:

$$V_{\mathrm{c}}(\theta) \cong \frac{eH}{2}\left[(R+r)(2\pi - \theta) + \left(R + \frac{r^2}{L}\right)\sin\theta - d(1-\cos\theta)\right]. \tag{20}$$

You can obtain the Fourier coefficients of $P_{\mathrm{c}}(\theta)$ numerically by performing FFT. If you want to obtain the Fourier coefficients without FFT, they can be obtained by the calculation based on the definition of the Fourier coefficients: that is, by calculating the inner product of $P_{\mathrm{c}}(\theta)$ and trigonometric functions. However, Equation (19) is too complex to calculate the inner product. In this study, to make the calculation of the inner product easier, $P_{\mathrm{s}}(V_{\mathrm{cc}}/V_{\mathrm{c}}(\theta))^{\gamma}$ was approximated with a quartic function as $P_{\mathrm{s}}(V_{\mathrm{cc}}/V_{\mathrm{c}}(\theta))^{\gamma} \cong A\theta^4 + B\theta^3 + C\theta^2 + D\theta + E$, where A, B, C, D and E were the coefficients. These coefficients can be formulated by the boundary conditions at $\theta = 0$, $\pi/2$ and π. This approximated function was in good agreement with the original function, as shown in Fig. 4 (b). Once $P_{\mathrm{c}}(\theta)$ is expressed by using a polynomial function, the inner product with trigonometric functions can easily be calculated.

$V_{\mathrm{SL}}(\theta)$ can be geometrically obtained. The detail process of the calculation is omitted in this paper, but it is expressed as

$$V_{SL}(\theta) \cong \frac{1}{2}eH\left\{R + r - \left[2R - \frac{e(R+r)}{L}\right]\cos\theta + e\left(1 - \frac{3R+r}{2L}\right)\cos 2\theta\right\}. \tag{21}$$

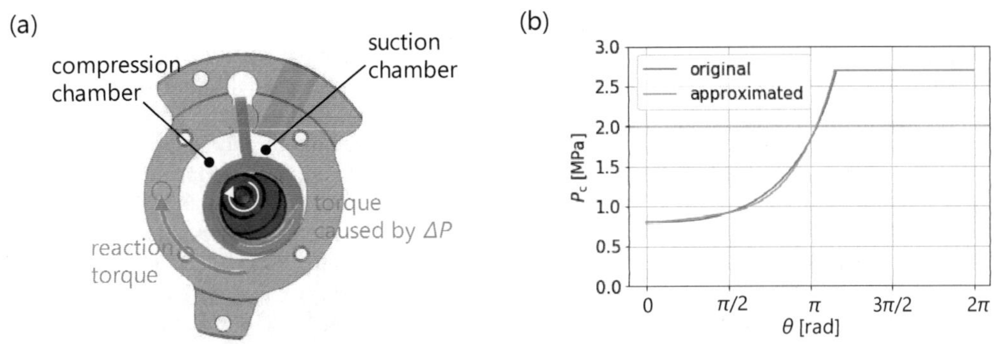

Figure 4: (a) Schematic drawing of compression torque. (b) An example of the comparison between original $P_{\mathrm{c}}(\theta)$ and approximated $P_{\mathrm{c}}(\theta)$.

3. Method

We performed theoretical calculations and experimental measurements for vibration of typical swing compressors listed in Table 1. The conditions for the calculations and experiments of each compressor are also presented in Table 1. Calculations were performed with the theoretical equations presented in the previous section. The experiments were performed on the equipment with the flexibly supporting system developed in the previous study (Wada *et al.*, 2016). The vibration of each compressor listed in Table 1 was measured with a triaxial acceleration sensor attached to the compressor as shown in Fig. 5. The radial and tangential vibrations measured by the sensor were compared with the calculated vibrations.

We carried out detailed theoretical analysis on compressors No. 1 and No. 3. The amplitude and the phase of the vibration caused by each excitation force were calculated separately. The contribution of each excitation force to the vibration was then analyzed.

Table 1: Conditions for the calculations and experiments.

Compressor No.	Type	Condensation Temperature	Evaporation Temperature	Rotational Speed	Refrigerant
1	single cylinder	45 °C	−7 °C	32 - 116 rps	R32
2	single cylinder	55 °C	10 °C	52 - 100 rps	R32
3	twin cylinder	45 °C	−7 °C	32 - 116 rps	R32
4	twin cylinder	45 °C	−5 °C	42 - 118 rps	R32

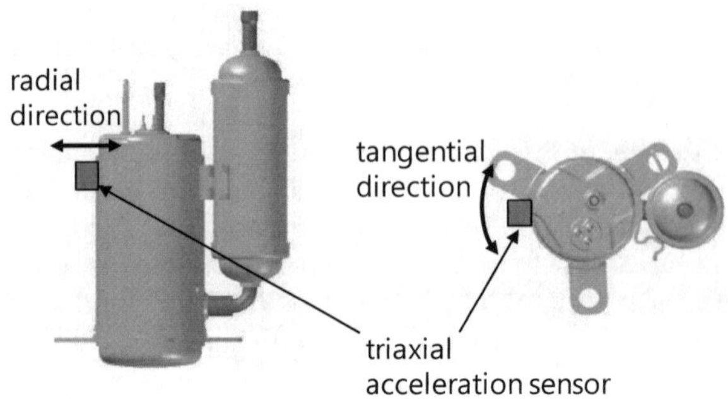

Figure 5: Schematic drawing of the position of the triaxial acceleration sensor.

4. RESULTS AND DISCUSSION

Firstly, Fig. 6 shows the comparison between calculated and measured vibrations. The calculated vibration was in good agreement with the measured one for both $1n$ and $2n$ vibrations of each compressor. This result indicates the validity of the theoretical calculations of the vibrations for both single- and twin-cylinder types of swing compressors.

In the following, we show the results of the theoretical analysis of the contribution of each excitation force for $1n$ and $2n$ vibrations. Figs. 7 - 9 show the results of the detailed analysis of the $1n$ vibration of compressor No. 1, the $1n$ vibration of compressor No. 3, and the $2n$ vibration of compressor No. 1, respectively. The contribution of each excitation force is shown in each figure. (a) and (b) of each figure show the radial vibration, whereas (c) and (d) show the tangential vibration. Also, (a) and (c) of each figure shows the amplitude and phase of the vibration generated by each excitation force at the rotational speed of 100 rps, whereas (b) and (d) of each figure shows the dependence on N of the magnitude of the vibration caused by each excitation force. Here, since the results of the analysis on compressors No. 1 and No. 3 show the typical vibration properties of single- and twin-cylinder type, we call compressors No. 1 and No. 3 simply 1YC (single-cylinder type swing compressor) and 2YC (twin-cylinder type swing compressor). Furthermore, for simplicity of description, we call the centrifugal force of the rotor system, the inertial force caused by the reciprocating motion, the inertial force caused by the pendular motion, and the compression torque F_{rotor}, $F_{\text{pis\_BL}}$, $F_{\text{pis\_phi}}$ and F_{comp}.

As shown in Fig 7 (a) and (b), the radial $1n$ vibration of 1YC is contributed by F_{rotor}, $F_{\text{pis\_BL}}$ and $F_{\text{pis\_phi}}$. The amplitude of the vibration caused by $F_{\text{pis\_phi}}$ is small compared to that by F_{rotor} and $F_{\text{pis\_BL}}$ because the direction of F_{pis_phi} is almost orthogonal to the radial direction as shown in Fig. 3 (c). It should be noted that the vibrations by F_{rotor} and F_{pis_BL} are almost in opposite phase with each other, and the amplitudes are almost the same. This relationship between F_{rotor} and F_{pis_BL} results in the totally small radial $1n$ vibration of 1YC, in spite of the large amplitudes of the vibrations by F_{rotor} and F_{pis_BL}. That is due to the design of balancers. F_{pis_BL} includes the centrifugal force applied to the cylindrical part of the piston. The balancers are designed so that the centrifugal force of the rotor system and the force of the cylindrical part of the piston are balanced. As shown in Fig. 7 (b), and also formulated in Sec. 2, F_{rotor}, F_{pis_BL} and F_{pis_phi} increase in proportion to N^2. Nevertheless, the radial $1n$ vibration of 1YC stays small as long as these excitation forces are balanced.

The tangential $1n$ vibration of 1YC has a completely different property from that of the radial one. In addition to the contribution of F_{rotor}, F_{pis_BL} and F_{pis_phi}, the tangential $1n$ vibration of 1YC is largely contributed by F_{comp}. Although F_{rotor}, F_{pis_BL} and F_{pis_phi} are proportional to N^2, F_{comp} is constant regardless of N, resulting in the large tangential $1n$ vibration of 1YC at any N. Moreover, unlike in the case of the radial $1n$ vibration of 1YC, F_{rotor}, F_{pis_BL} and F_{pis_phi} are not balanced in the tangential one. As a result, the total tangential $1n$ vibration of 1YC increases with N, in addition to the constantly large vibration caused by F_{comp}. The large tangential $1n$ vibration of single-cylinder type rotary compressor, including 1YC, often becomes a problem in the development of the compressor. The results shown in this study indicates that there are two types of solutions for the problem: one is to reduce the amplitude of each excitation force, and the other is to balance the excitation forces with each other.

The radial $1n$ vibration of 2YC has the same property as that of 1YC, except for the smaller amplitudes of the vibrations by F_{rotor}, F_{pis_BL} and F_{pis_phi} in 2YC compared to those in 1YC. 2YC has two eccentric parts in the crankshaft. The centrifugal forces applied to the eccentric parts generally have the same magnitudes and are in opposite directions with each other, resulting in the more balanced, that is, the smaller F_{rotor} of 2YC compared to that of 1YC. The same explanation can be applied to F_{pis_BL} and F_{pis_phi}. The one piston moves in the opposite phase with the other piston, resulting in the smaller F_{pis_BL} and F_{pis_phi}. It is important that the total $1n$ vibration of 2YC is almost the same with that of 1YC although the amplitudes of the vibrations by F_{rotor}, F_{pis_BL} and F_{pis_phi} of 2YC are quite different to those of 1YC. This result indicates that the small total vibration does not necessarily require that the amplitudes of the excitation forces be small. Even though the excitation forces have large amplitude, the total vibration becomes small if these excitation forces are balanced with each other.

The property of the tangential $1n$ vibration of 2YC is not similar to that of 1YC. Whereas the tangential $1n$ vibration of 1YC has the large contribution of F_{comp}, the vibration of 2YC has no contribution of F_{comp}. 2YC has two cylinders and the compression of the refrigerant in one cylinder is performed in the opposite phase with that in the other cylinder. Therefore, the compression torque in one cylinder cancels out that in the other cylinder, resulting in no contribution of F_{comp} to the tangential $1n$ vibration of 2YC. As a result, the property of the tangential $1n$ vibration of 2YC is similar to that of the radial $1n$ vibration of 2YC, except for having a bit larger contribution of F_{pis_phi}.

As for the $2n$ vibration, 1YC and 2YC have the same property. The results of the analysis on $2n$ vibration described below are then applied to both for 1YC and 2YC. A characteristic of $2n$ vibration, which is different from $1n$ vibration, is that F_{rotor} does not contribute to both radial and tangential $2n$ vibrations. This indicates that the balancers cannot be designed for the reduction of $2n$ vibration. The excitation forces which contribute to $2n$ vibration are F_{pis_BL}, F_{pis_phi} and F_{comp}. F_{pis_BL} and F_{pis_phi} contribute to both radial and tangential $2n$ vibration, and F_{comp} contributes to only the tangential $2n$ vibration. The radial $2n$ vibration is very small and does not have any notable property, whereas the tangential $2n$ vibration has an interesting property. As shown in Fig. 9 (c), the phase of the tangential $2n$ vibration caused by F_{comp} is almost opposite to that caused by F_{pis_phi}, meaning that the vibrations caused by F_{comp} are canceled out by F_{pis_phi}. These excitation forces have different dependence on N: F_{comp} is constant regardless of N, whereas F_{pis_phi} increases in proportion to N^2. Therefore, the canceling effect becomes larger with the increase of N. As a result, the total tangential $2n$ vibration decreases with the increase of N, as shown in Fig. 9 (d). This canceling phenomenon is owing to F_{pis_phi}, which does not exist in rotary compressors. It means that this vibration property is one of the distinct properties of swing compressors and is not observed in rotary compressors.

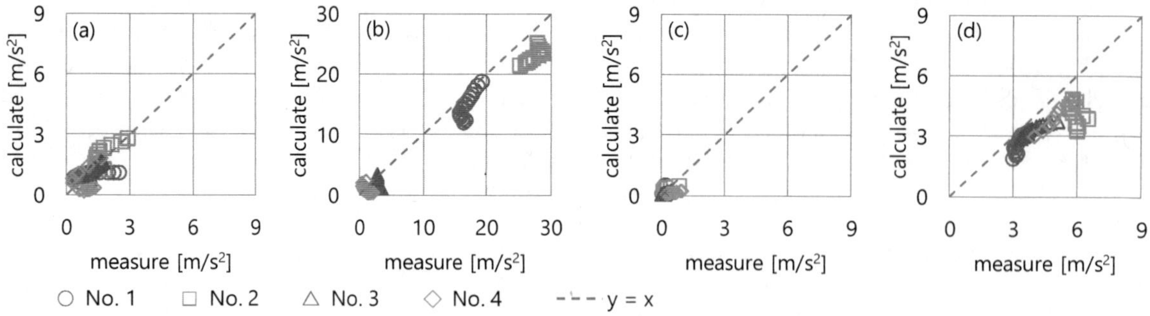

Figure 6: Comparison between the measured vibration and calculated vibration. (a) Radial $1n$ vibration. (b) Tangential $1n$ vibration. (c) Radial $2n$ vibration. (d) Tangential $2n$ vibration.

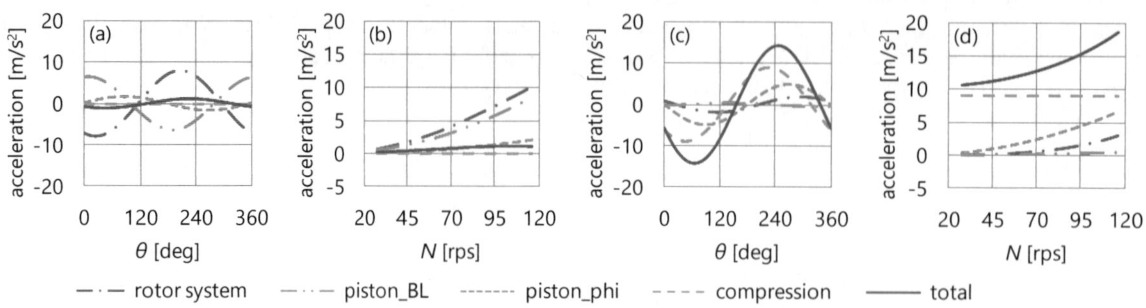

Figure 7: Contribution of each excitation force to $1n$ vibration of 1YC. (a) Amplitude and phase of the radial $1n$ vibration caused by each excitation force at 100 rps. (b) Dependence on N of magnitude of the radial $1n$ vibration caused by each excitation force. (c) Same as (a), but for the tangential $1n$ vibration. (d) Same as (b), but for the tangential $1n$ vibration.

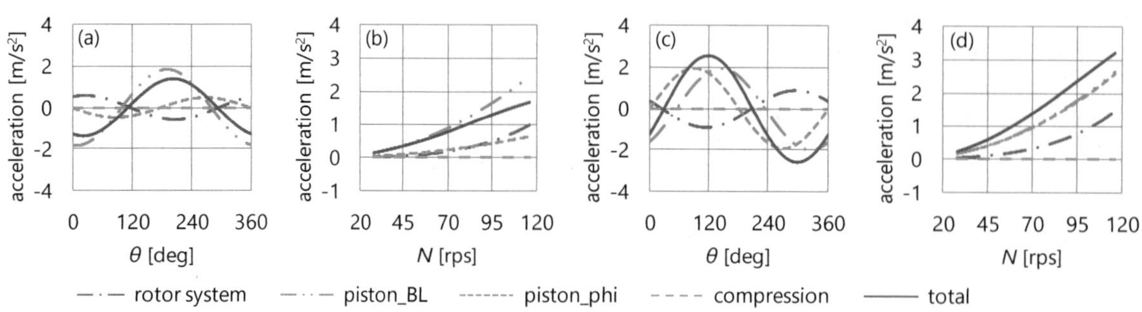

Figure 8: Same as Figure 7, but for $1n$ vibration of 2YC.

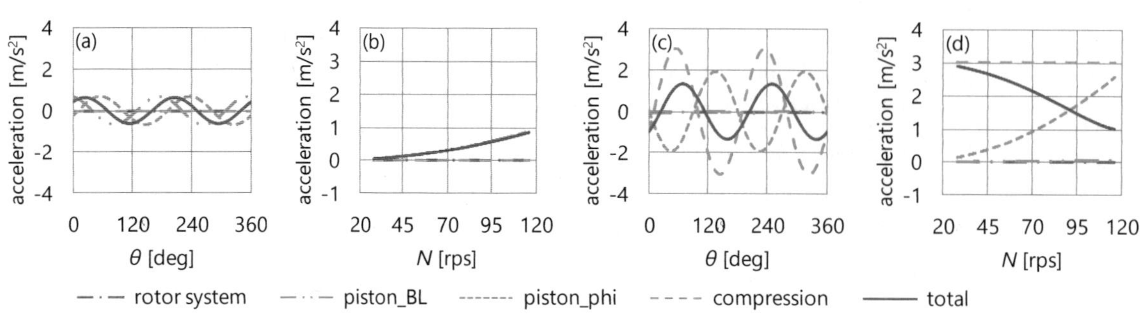

Figure 9: Same as Figure 7, but for $2n$ vibration of 1YC.

5. CONCLUSIONS

This study performed a theoretical analysis of rigid-body vibration in a swing compressor. Firstly, we classified the main factors of the vibration into three type of excitation forces and formulated the vibration caused by each excitation force. Secondly, the rigid-body vibrations of typical swing compressors were measured. The calculated vibrations were confirmed to be consistent with the measured ones. Thirdly, the theoretical analysis of the properties of rigid-body vibrations of both single- and twin-cylinder type swing compressors were performed. The primary and secondary components in the frequency domain of the radial and tangential vibrations were theoretically analyzed from the viewpoint of the contribution of each excitation force. To our knowledge, this is the first time to perform the following:

- Both the amplitude and the phase of the primary and secondary components of the vibration caused by each excitation force in the frequency domain were formulated separately without any other numerical calculation technique, including FFT.
- The contribution of each excitation force to vibration, including the relationship of the phase of each excitation force, was analyzed.

The results of the theoretical analysis clearly indicated that not only was simply reducing the magnitudes of the excitation forces effective but balancing the excitation forces was also very effective for the reduction of the vibration. Balancing the excitation forces is more advanced than reducing the magnitudes of the excitation forces because both amplitude and phase of each excitation force must be obtained to balance the excitation forces. This study theoretically formulated both amplitude and phase of the excitation forces which are the main causes of vibrations, and the validity of the theoretically-calculated vibration is confirmed by a comparison to a measured one. We expect that this study will help the development of a more advanced technique for the reduction of the vibration in a theoretical viewpoint.

NOMENCLATURE

e	eccentricity of piston center
H	cylinder height
I_{pis}	moment of inertia of piston
L	length between center of cylinder and center of swing bushes
$m_{\mathrm{ubw/lbw/ro}}$	mass of upper balancer/lower balancer/motor rotor
m_{ecc}	mass of eccentric part of crankshaft
m_{pis}	mass of piston
P_{c}	pressure of compression chamber
P_{d}	discharge pressure
P_{s}	suction pressure
R	inner radius of cylinder
r	outer radius of piston
$r_{\mathrm{ubw/lbw}}$	distance between center axis of motor rotor and center of gravity of upper/lower balancer
V_{cc}	cylinder volume
V_{c}	volume of compression chamber
γ	polytropic index
θ	rotating angle of crankshaft
θ_{d}	rotating angle of crankshaft at which P_{c} reaches P_{d}

REFERENCES

Huang, J *et al.* (2010). Dynamic Balance Technology of Inverter Controller Rotary Compressor. *Proceedings of the International Compressor Engineering Conference at Purdue* (Paper 1140).

Imaichi, K. *et al.* (1982). Vibration Analysis of Rotary Compressor. *Proceedings of the International Compressor Engineering Conference at Purdue* (Paper 407).

Wada, H. *et al.* (2016). Vertical-Vibration Suppressing Design of Accumulator with New Vibration-Measuring Method. *Proceedings of the International Compressor Engineering Conference at Purdue* (Paper 1158).

Zhang, H. *et al.* (2014). Dynamic behaviors of the crankshafts in single-cylinder and twin-cylinder rotary compressors. *International Journal of Refrigeration*, 47, 36-45.

Evaluation of Compressor Reliability for HFO Refrigerants

Arisa KAWAE[1]*, Hideki MATSUURA[2], Masaru TANAKA[3]

[1]Technology and Innovation Center, Daikin Industries, Ltd.,
1-1, Nishi-Hitotsuya, Settsu, Osaka, 566-8585, Japan
(Phone: +81-6-6195-7074, arisa.kawae@daikin.co.jp)

[2]Technology and Innovation Center, Daikin Industries, Ltd.,
1-1, Nishi-Hitotsuya, Settsu, Osaka, 566-8585, Japan
(Phone: +81-6-6195-7074, hideki.matsuura@daikin.co.jp)

[3]Technology and Innovation Center, Daikin Industries, Ltd.,
1-1, Nishi-Hitotsuya, Settsu, Osaka, 566-8585, Japan
(Phone: +81-6-6195-7074, masaru.tanaka@daikin.co.jp)

*Corresponding Author

ABSTRACT

Responses to an international flamework on climate change, such as the Kigali Agreement in 2016, have been discussed in an industrial field. The phase down regulation of CFC and HCFC refrigerant has been discussed and performed under the Montreal Protocol for refrigerants from the viewpoint of ozone layer protection. From the viewpoint of preventing global warming, HFC reduction plans are currently being proposed by North America, Europe, India, etc. within the framework of the Protocol. A developing country where the phase-out of CFCs and HCFCs began in 2013, are growing trend to switch to HFC or HFO mixed refrigerants with lower GWP than R410A. A revision of the regulations related to HFC are also being promoted in each region.

The GWP upper limit for each equipment application was set together with the HFC phase down in Europe which due to the F-gas regulations. R32, HFO refrigerants such as R1234yf and R1234ze, or a mixed refrigerant of HFC and HFO are proposed as candidates for low GWP refrigerants. The problem of R32 was the higher discharge gas temperature than R410A, but the problem could be solved by temperature control and suitable refrigeration oil.

On the other hand, the problem of refrigerants HFO and mixed HFO is easy to decompose than HFC. When HFO is decompose, it is generated acid such as fluorinated acid and trifluoroacetic acid. These acids make which is reduced the compressor reliability. We selected of suitable refrigeration oil and materials for the HFO and confirmed the compressor reliability

1. INTRODUCTION

Phase down regulations for CFC and HFC refrigerants are being implemented from the perspective of preventing global warming, and F-gas regulations, etc., necessitate equipment development that is compatible to low global warming potential (GWP) refrigerants for each equipment application [1]. While each of the low GWP refrigerant candidates, including R32, HFO, and HFO blends, has its own issues, Daikin Industries has already addressed the issues related to R32 and is currently marketing R32 products [2]. This paper focuses on the HFO and HFO blend refrigerants, which are among other candidates for alternative refrigerants, and reports on results for confirming the reliability of product systems and compressors by selecting refrigeration oil suitable for the single-component refrigerants R1234yf and R1234ze, which are HFO refrigerants, as well as the refrigerants containing them.

2. Issues of HFO and HFO Blend Refrigerants

Tables 1 and 2 show the characteristics of HFO and HFO blend refrigerants and Table 3 shows their issues. HFO refrigerants easily decompose due to their low GWP, and the decomposed product can lead to such problems as deterioration in the stability of refrigeration oil, wear of sliding parts, and corrosion of the expansion valve.

Table 1: Characteristics of R410 Alternative Refrigerants Including HFO Blends

		Current	Alternative refrigerant candidates		
		R410A	R32	R454B	AMOLEA 400X
Refrigerant Composition	R32	50wt%	100wt%	68.9%	
	R125	50wt%			
	R1234yf			31.1%	
	R1123				
Refrigerant Properties	GWP (IPCC4)	2090	675	466	405
	Flammability and Toxicity	A2L	A2L	A2L	A2L

Table 2: Characteristics of R134a Alternative HFO and HFO Blend Refrigerants

		Current	Alternative refrigerant candidate		
		R134a	R1234yf	R1234ze	R513A
Refrigerant Composition	R134a	100wt%			44wt%
	R1234yf		100wt%		56wt%
	R1234ze			100wt%	
Refrigerant Properties	GWP (IPCC4)	1430	4	6	573
	Flammability and Toxicity	A1	A2L	A2L	A1

Table 3: Technical Issues of HFO Refrigerant Air Conditioning Systems

Elements	Technical Issues
Refrigerant	·Chemical Stability
Compressor	·Wear (Corrosive Wear)
Refrigeration Oil	·Miscibility with HFO Refrigerant ·Chemical Stability ·Mixture Viscosity
Motor Materials	·Compatibility with HFO Refrigerants Refrigerant & Lubricant ·Thermal Stability
Refrigeration Cycle	·Corrosion in Expansion Valve ·Capillary Tube Clogging ·Influence of Process Materials ·Decrease in Volume Capacity

3. Selection of Refrigeration Oil Used for HFO and HFO Blend Refrigerants

3.1 Refrigerant stability

To investigate the specifications of refrigeration oil, the stability of single-component refrigerants was confirmed by the shield tube test. Table 2 shows the test conditions and Figure 1 shows the test results. Although there was no decomposition for HFO refrigerants R1234yf and R1234ze under conditions with no air and water or with only water, R1234ze and R1234yf decompose more easily than R32 when air (oxygen) is present. The amount of detected acid follows in order of R22 > R1234yf > R1234ze > R32. As shown in Scheme 1, when R1234ze and R1234yf decompose, strong acids such as hydrofluoric acid and trifluoroacetic acid are generated and cause expansion valve corrosion, capillary clogging, and abnormal wear. Consequently, it is necessary to select refrigeration oil and materials that are resilient to the effects of the strong acid generated by the decomposition of HFO refrigerants.

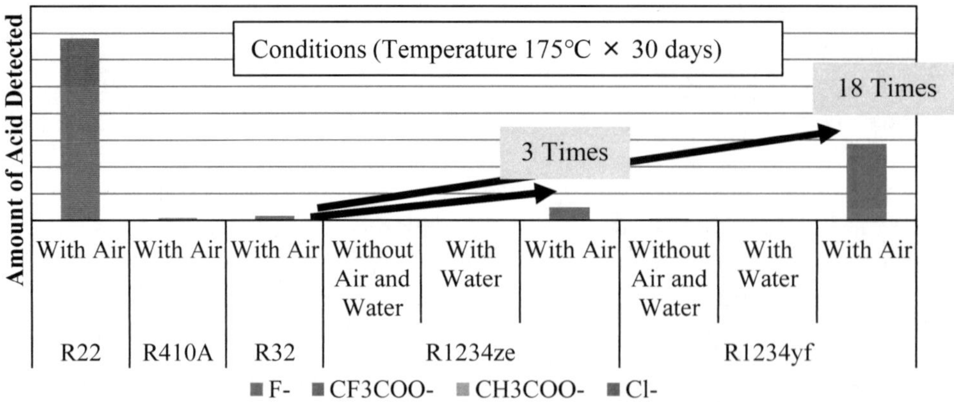

Figure 1: Detected Ratio of Various Acids after Refrigerant Stability Test

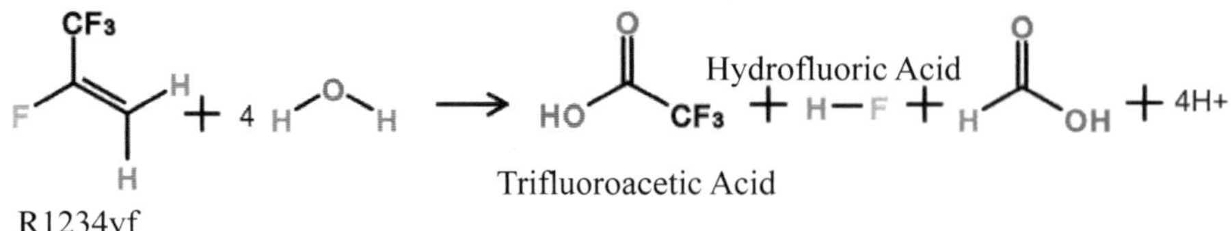

Scheme 1: Decomposition Reaction of R1234yf

3.2 Investigation of refrigeration oil specifications

Stability is an issue for HFO refrigerants. For this reason, it is important to select a base oil with high stability and include an additive mixture that can neutralize the decomposition product of the refrigerants. Because HFO refrigerants have no compatibility problem with refrigeration oil, it is possible to use polyvinyl ether (PVE) and polyol ester (PVE), which are used for conventional HFC refrigerants. PVE, which excels in compatibility with HFO refrigerants, hydrolyzability, and lubricity and is the same oil used for R32, was used as the base oil [3]. Because HFO refrigerants decompose more easily than R32, a combination of additives and a new stabilizer were employed to suppress HFO refrigerant decomposition. Table 4 shows the mixture of additives that was used in the oil for the HFO refrigerants.

Table 4: Refrigeration Oil Additive Specifications for HFO and HFO Blend Refrigerants

	Refrigeration Oil for HFO Refrigerant
Extreme Pressure Agent	Same as that for R32
Antioxidant Agent	Same as that for R32
Acid Catcher Agent	Same as that for R32
New Stabilizer	Added

4. Refrigeration Oil Lubricity Test Results

Lubricity was tested with a ring-on-disk test using the selected refrigeration oil. Table 5 shows the test conditions, whereas Figure 2 and Table 6 show the test results. Evaluation was performed for a test piece with a ring made of cast iron and disks made of cast iron and aluminum alloy. For cast iron, the surface pressure for R454A and Current-PVE, which is used with R410A, was equivalent to that of R410A and Current-PVE, and good lubrication properties were confirmed for R454B and R1234yf using New-PVE, which is used for HFO refrigerants. For the aluminum alloy, lower surface properties were seen for Current-PVE, but the combination of New-PVE and R454B demonstrated surface properties comparable to R410A and Current-PVE. A large amount of fluoride ions was detected on the aluminum alloy surface by EPMA mapping of the aluminum alloy surface after the test, but detection of fluoride ions was suppressed for New-PVE and R454B. For R454B and Current-PVE, we believe that the R1234yf contained in R454B reacted with the aluminum surface to generate aluminum fluoride, and this caused a reduction in lubricity. However, for R454B and New-PVE, the new stabilizer effectively suppressed a reaction between R1234yf and the aluminum surface, thereby preventing the formation of aluminum fluoride, and exhibited lubricity comparable to that of R410A and Current-PVE. Consequently, we selected New-PVE as the refrigeration oil to be used with HFO refrigerants and performed tests to confirm its refrigeration cycle reliability.

Table 5: Conditions of Max Loading Surface Test

Refrigerant	R410A or R454B	
Sliding Velocity	3.5 m/s	
Load	From 0 MPa until the test stops (Up to 10MPa)	
Ring material	Cast iron	
Disk material	Cast iron	Al-Alloy

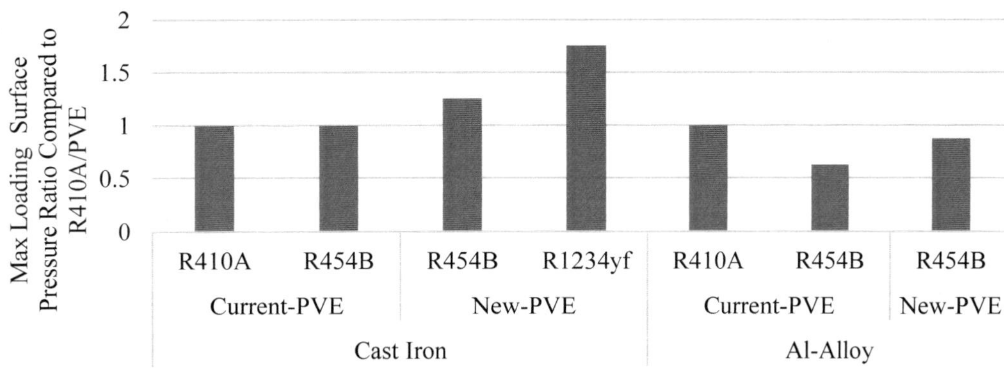

Figure 2: Results of Lubricity Conformation

Table 6: EPMA Mapping Image of Cast Iron and Al-Alloy

	R410A/Current-PVE	R454B/Current-PVE	R454B/New-PVE	scale
Al-Alloy	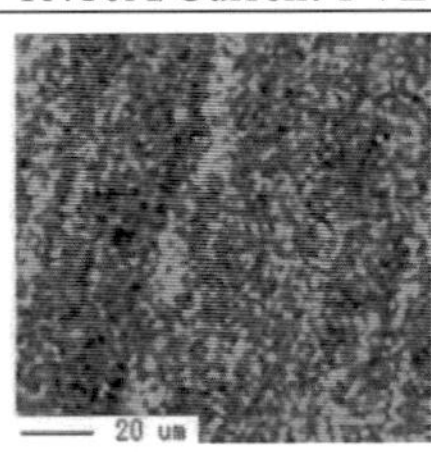	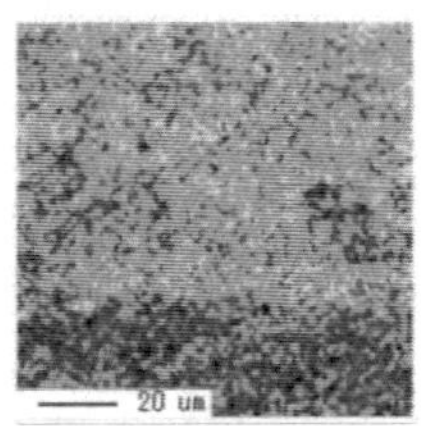	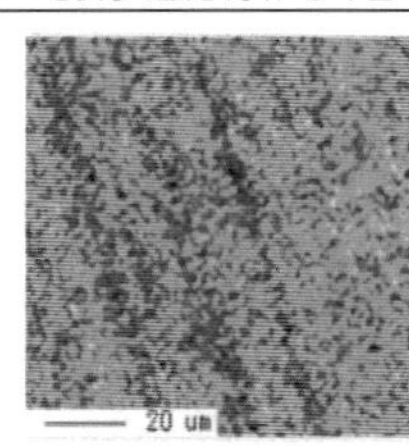	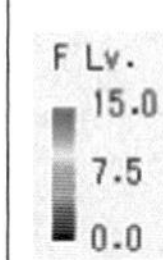

5. Reliability Confirmation Results of HFO and HFO Blend Refrigerants in the Refrigeration Cycle by Drop-in Test

5.1 Confirmation results of system reliability by HFO blend drop-in test (Mini-split Air Conditioner)

To verify applicability to refrigerant systems employing HFO blends, a drop-in test was performed on a 2.8 kW R32 refrigerant system. Table 7 shows test conditions and Table 8 and Table 9 show test results. While discoloration and corrosion were seen in the expansion value for PVE and R410A, discoloration and corrosion were not seen for PVE and HFO refrigerants, and there was also no discoloration in the oil. Additionally, in the test used for PVE and R410A, acid is generated in the refrigeration oil, and the HFO refrigerants are believed to decompose in the system, causing corrosion from the decomposition product. In contrast, because the formation of acid is suppressed in PVE and HFO refrigerants, we believe that this demonstrates the effect of the additive.

Table 7: Conditions of System Drop-in Test

Test No.	No. 1	No. 2
System	2.8 kW System for R32	
Refrigerant	R454B	
Compressor Type	Swing Compressor for R32	
Casing Pressure	High Pressure Dome Casing Type	
Rotational Speed	Variable	
Refrigeration Oil	Current-PVE	New-PVE
Additive	Same as additive used in PVE for R410A	New stabilizer added
Expansion Valve	Electric Expansion Valve (Material: Cu-Zn Alloy (Brass)	
Discharge Gas Temperature	120°C Max.	
Operating Time	1000 hours	

Table 8: Results of System Drop-in Test

Test No.	No. 1	No. 2
Expansion Valve	Corroded	No Problem
Refrigeration Oil		
Acid Number	0.02 mg KOH/g	0.01 mg KOH/g
Fluorine ion	>1 mg/L	>1 mg/L
Trifluoroacetic ion	>1 mg/L	>1 mg/L
Acetic acid ion	2 mg/L	2 mg/L

Table 9: Expansion Needle Valves after System Drop-in Test (Needle Material: Cu-Zn Alloy [Brass])

	No. 1	No. 2
Appearance	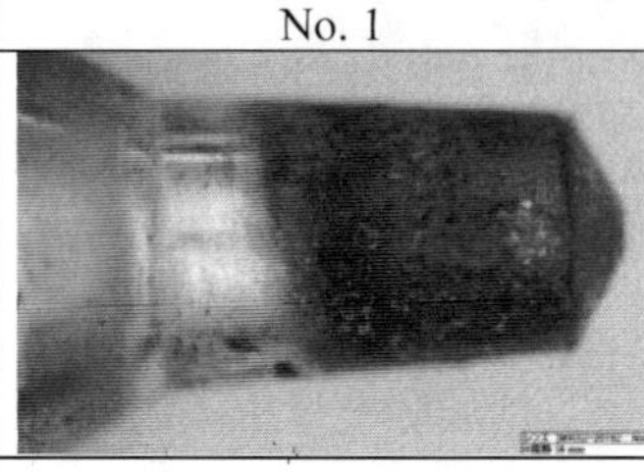	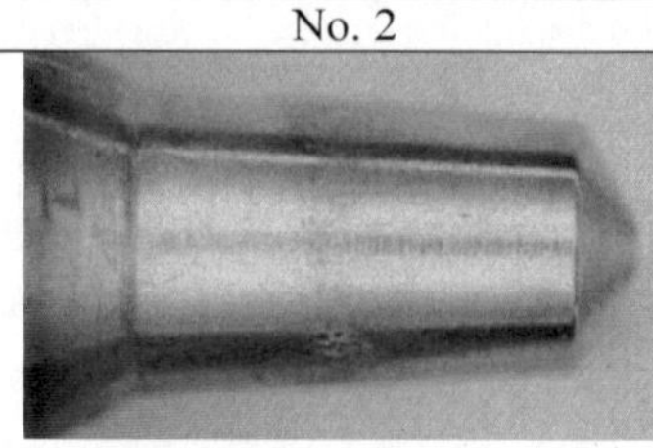
Detected Fluorine by EPMA	16.2 wt%	0.9 wt%

5.2 Confirmation Results for System Reliability by HFO Refrigerant Drop-in Test (Screw Chiller)

To verify applicability to HFO refrigerant systems, a drop-in test was performed on a 100kW R134a screw chiller. Table 10 shows the test conditions, and Table 11, Table 12, and Figure 2 show the test results. No corrosion was observed in the expansion valve, motor, screw rotor, and gate rotor after the test. The oil showed no rise in acid number after the test, and there was no acid formation derived from refrigerant decomposition product.

Table 10: Conditions of System Drop-in Test

Test No.	No. 3
System	Screw Chiller
Refrigerant	R1234ze
Compressor Type	Screw Compressor 100kw
Rotational Speed	Variable
Refrigeration Oil	New-PVE
Additive	New stabilizer added
Expansion Valve	Electric Expansion Valve (Material: Cu-Zn Alloy [Brass])
Discharge Gas Temperature	120 °C MAX
Operating Time	2000 hours

Table 11: Results of System Drop-in Test

Test No.	No. 3
Expansion Valve	No Problem
Compressor	No Problem
Refrigeration Oil	
Acid Number	0.02 mg KOH/g
Fluorine ion	>1 mg/L
Trifluoroacetic ion	>1 mg/L
Acetic Acid ion	>1 mg/L

Table 12: Appearance of Refrigeration Oil after System Drop-in Test

Time	0 hour	957 hours	2000 hours
Appearance of Refrigeration Oil		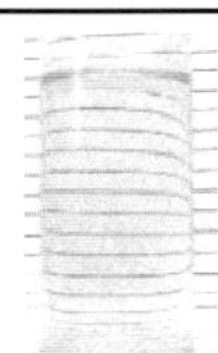	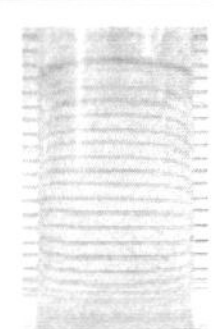

Figure 3: (A) Motor, (B) Expansion valve, (C) Screw rotor, (D) Gate rotor, after System Drop-in Test

5.3 Confirmation Results of Scroll Compressor Reliability by HFO Blend Drop-in Test

To verify compressor reliability for HFO blends, an HFO blend refrigerant was dropped into an R32 scroll compressor, and a durability test was performed. Table 13 shows the test conditions and Table 14 shows the test results. Worsening of shaft surface roughness and abnormal bearing wear were not observed even under conditions of high discharge gas temperature for which refrigerant decomposition is a concern. Also, the oil showed no increase in acid number after the test, and there was no acid formation derived from refrigerant decomposition product.

Table 13: Durability Test Conditions of Scroll Compressor

Test No.	No. 1	No. 2	No. 3
Operating Conditions	High Compression Ratio	Max Speed, Wet	Overload Wet, On-off Action
Operating Time	400 hr	400 hr	400 hr
Refrigerant	R454B		
Refrigeration Oil	New-PVE		
Additive	New stabilizer added		
Discharge Gas Temperature	120 °C	50 °C	65 °C
Discharge Gas Pressure	High	Normal	Low

Table 14: Durability Test Results of Scroll Compressor (Wear Property of Sliding Parts)

Test No.	No. 1	No. 2	No. 3
Shaft (Gray Cast Iron)	Same as R32		
Bearing (PTFE Coating Bronze)	Same as R32		
Oldham (Al)	Same as R32		
Refrigeration Oil			
Appearance of Refrigeration Oil	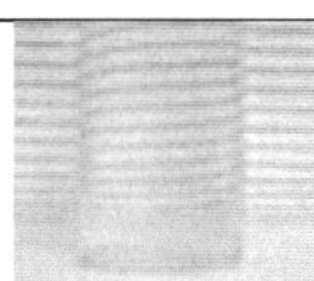	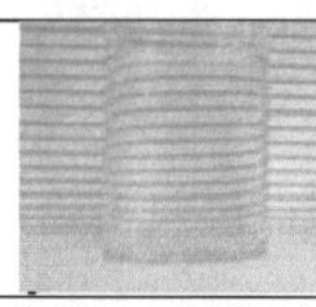	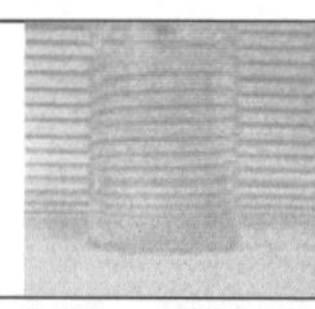
Acid Number	0.01 mg KOH/g	0.01 mg KOH/g	0.01 mg KOH/g
Fluorine ion	>1 mg/L	>1 mg/L	>1 mg/L
Trifluoroacetic ion	>1 mg/L	1 mg/L	>1 mg/L
Acetic Acid ion	>1 mg/L	>1 mg/L	>1 mg/L

5.4 Confirmation Results for Screw Compressor Reliability by HFO Refrigerant Drop-in Test

To verify compressor reliability for HFO refrigerant, an HFO refrigerant was dropped into an R1234yf screw compressor and a durability test was performed. Table 15 shows the test conditions and Table 16 shows the test results. Even at discharge gas temperature of 120°C for which there is concern for refrigerant decomposition, no corrosion of parts was observed after the test. Also, the oil showed no increase in acid number after the test, and there was no acid formation derived from refrigerant decomposition product.

Table 15: Durability Test Conditions of Scroll Compressor

Test No.	No. 4
Operating Conditions	High Compression Ratio
Operating Time	2000 hr
Refrigerant	R1234yf
Refrigeration Oil	New-PVE
Additive	New stabilizer added
Discharge Gas Temperature	120 °C
Discharge Gas Pressure	High

Table 16: Durability Test Results of Screw Compressor (Wear Property of Sliding Parts)

Test No.	No. 4
Shaft	Same as R134a
Bearing	Same as R134a
Refrigeration Oil	
Appearance of Refrigeration Oil	
Acid Number	0.01 mg KOH/g
Fluorine ion	>1 mg/L
Trifluoroacetic ion	>1 mg/L
Acetic acid ion	>1 mg/L

6. CONCLUSIONS

Product systems and compressors using either R32 and R134a refrigerants were charged with HFO and HFO blend refrigerants, and reliability was evaluated by a drop-in test.

- The HFO blend refrigerants containing R1234yf react with the surface of the aluminum alloy and reduces lubricity.
- It was found that HFO and HFO blend refrigerants generate more acid due to deterioration of stability than R32 refrigerant, and the acid that is generated corrodes the expansion valve.
- Adding a new stabilizer is effective to enhancing the stability of HFO and HFO blend refrigerants.
- For practical application, it is necessary to evaluate not only a simple element test as performed during refrigerant conversion in the past but also a reliability test of the actual system. Additionally, it is important to conduct field tests to sufficiently confirm lubrication oil, functional parts, etc., that are likely to affect the system during actual use conditions.

REFERENCES

[1] United Nations, Montreal Protocol celebrates another milestone as agreement to reduce climate-warming gases is set to enter into force in 2019. Available from https://www.unenvironment.org/news-and-stories/press-release/montreal-protocol-celebrates-another-milestone-agreement-reduce

[2] Shigeharu TAIRA etc., "*Development of Air Conditioner with Low GWP Refrigerant R32,*" The International Symposium on New Refrigerants and Environmental technology 2012, JRAIA, November 8-9, 2012

[3] Matsumoto, T., Nakajima, S., and Kawaguchi, Y, "*Evaluations of PVE Lubricants for Refrigeration Systems with the Low GWP Refrigerants*", The International Symposium on New Refrigerants and Environmental Technology 2016, 227-230 (2016)

ACKNOWLEDGMENTS

We would like to thank Idemitsu Kosan Co., Ltd. for providing the oils and cooperating in the oil analyses after the tests and lubricity tests.

Study on Bearing Characteristics of Reciprocating Compressor for Refrigerator

Masanobu GONDO[1]*, Hirotaka KAWABATA[1], Hiroto HAYASHI[1]

[1]Panasonic Appliances Refrigeration Compressor Japan, Kusatsu, Shiga, Japan
Phone: +81-77-566-4813, Fax: +81-77-565-9895, E-mail: gondo.masanobu@jp.panasonic.com

* Corresponding Author

ABSTRACT

In recent years, in order to reduce the energy consumption of refrigerators through reduction of input power and improved refrigeration capacity, reciprocating compressors are gradually converted to inverter type. Correspondingly, we have studied the sliding bearing characteristics of the crankshaft journal at variable speed which is the key to improving the efficiency of reciprocating compressor.

Generally, the characteristics of a sliding bearing are explained with a Stribeck curve. In the fluid lubrication regime, there is sufficient lubricant between metal surfaces. As the rotational speed decreases, the viscous resistance is reduced and it will result in the reduction of friction coefficient. Conversely, it is widely known that further reduction of rotational speed will shift to boundary lubrication regime where the friction coefficient increases due to metal to metal contact.

In the inverter reciprocating compressor, it is presumed that the sliding of crankshaft on bearing at high rotational speed is in fluid lubrication regime. The input power can be reduced with gradual decrease of friction coefficient by lowering rotational speed, but then the sliding characteristic of the crankshaft at each rotational speed was unclear such that when the speed is too low it will progress into the boundary lubrication regime.

This paper explains our experimental study to obtain a Stribeck curve at low load, low speed, and low viscosity oil in inverter operation using a bearing tester that estimate sliding crankshaft on bearing. The results obtained enable the prediction of sliding characteristic of crankshaft and optimization of the bearing property in reciprocating compressor.

1. INTRODUCTION

The Paris Agreement which is an international framework for combating global warming was signed in 2016. Its content is to reduce the emission of CO_2 and other greenhouse gases. Japan has set a goal to reduce greenhouse gas emissions by 26% compared to 2013 by 2030 in the INDC submitted to UNFCCC. On the other hand, EU has pledged to achieve 40% reduction as compared to 1990 by 2030. The INDC clearly showed that countries in the worldwide are setting high targets to achieve the aim of Paris Agreement. In this effort, there are two solutions for reducing greenhouse gases for household refrigerators. Firstly, it is the conversion to hydrocarbon refrigerants with low global warming potential such as R600a. Secondly it is to reduce power consumption by improving the energy efficiency as most of the power consumption is from the compressor. Hamaoka *et al.* (1996, 1997) studied energy saving in refrigerators by optimizing the rotational speed of an inverter reciprocating compressor according to the usage of the refrigerator. In recent years, works to further reduce power consumption were shown through development of improved heat insulation performance of refrigerators and reciprocating compressor that operates at rotational speed below 1200rpm. In order to achieve reduction of power consumption under such operating conditions, it is important to grasp sliding characteristics to predict and minimize sliding loss. This study focus on the plain bearing characteristics of the crankshaft and the bearing, which account for most of the sliding loss in reciprocating compressors. Conventionally, the characteristics of plain bearing under high rotational speed or fluid lubrication regime can be described by Reynolds equation. However, in the low rotational speed, theoretical calculation is complex when metal surface comes into contact. Kobayashi *et al.* (1998) reported the bearing characteristics of a reciprocating compressor for R134a using a journal bearing tester. In this test, the load was larger than that of R600a now widely used in refrigerators. Besides, the test was conducted on induction compressor operating at constant speed of maximum 3500rpm with a higher viscosity grade oil than current mainstream. Thus, it was challenging to accurately predict the sliding loss on inverter compressors. In this paper we discuss the experimental studies on bearing characteristics at low load, low speed, and low viscosity oil under variable speed operation by using a journal bearing tester. As we were able to estimate the reduction of the sliding loss in the reciprocating compressor accurately, the results from this study were applied in compressor development.

2. SLIDING BEARING THEORY

The general plain bearing characteristic is described by a Stribeck diagram as shown in Figure 1. The horizontal axis is the Sommerfeld number (S), a dimensionless quantity for evaluating the sliding state, and the vertical axis shows the friction coefficient. The equation for the Sommerfeld number is shown in equation (1).

$$S = \frac{\eta N}{P} \left(\frac{r}{c}\right)^2 \tag{1}$$

In above formula, η is oil viscosity [Pa·s], N is rotational speed [rps], P is bearing pressure [Pa], r is shaft radius [mm] and c is radial clearance [mm]. The right side of the Stribeck curve with larger Sommerfeld number describes the fluid lubrication regime where adequate oil film separates the metal surfaces to prevent wear and failure. The friction coefficient decreases with reduction of viscous resistance from the effect of lowering rotational speed or oil viscosity. The midsection of the Stribeck curve around the transition point is the mixed lubrication regime in which metal contacts occurs and sliding loss increases as the rotational speed is lowered. The lubrication shifts towards boundary lubrication regime on left side of the transition point as rotational speed is reduced further. The thin oil film is inadequate to prevent metal contact resulting in the increase of friction coefficient. In order to reduce the sliding loss, it is vital to clearly understand factors affecting the friction coefficient property in the mixed lubrication regime.

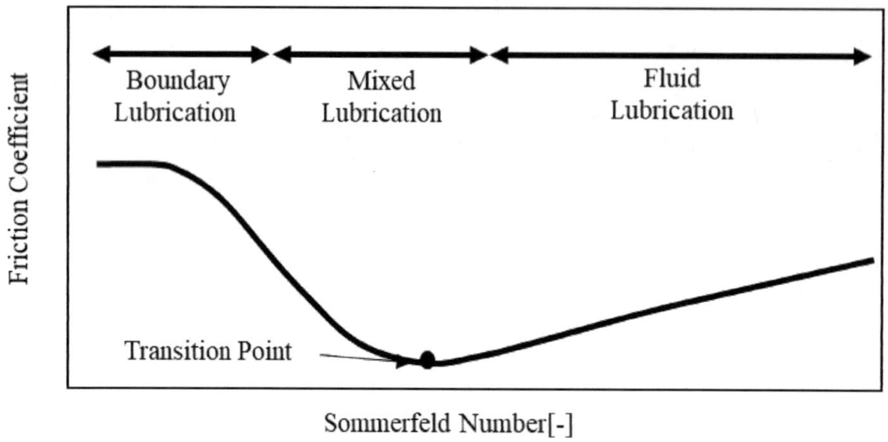

Figure 1: General Stribeck Diagram

3. EXPERIMENTAL STUDIES

3.1 Test Method

Figure 2 is the schematic diagram of the journal bearing tester used in this study. In this tester, the shaft is placed horizontally and driven by an inverter motor. The sliding length is defined by the bearing length. The bearing which is mounted on a housing is replaceable. The calculation of frictional force is explained as follows. A load (F) is applied to the bearing from the bottom of the housing. The rotational force (F_l) of the housing attached with the bearing is measured at a distance (l) from the center of the shaft to the measurement point by load cell as shown in Figure 2. F_l is generated as the shaft rotates and it is used to obtain the frictional force (F_r) in equation (2).

$$F_r = F_l \cdot \frac{l}{r} \tag{2}$$

The friction coefficient (μ) is derived from equation (3) using F_r and F.

$$\mu = \frac{Fr}{F} = \frac{F_l}{F} \cdot \frac{l}{r} \quad (3)$$

The shaft is lubricated by oil supplied to the oil pool on the housing which flows through the oil groove and fills the sliding portion of shaft and bearing.

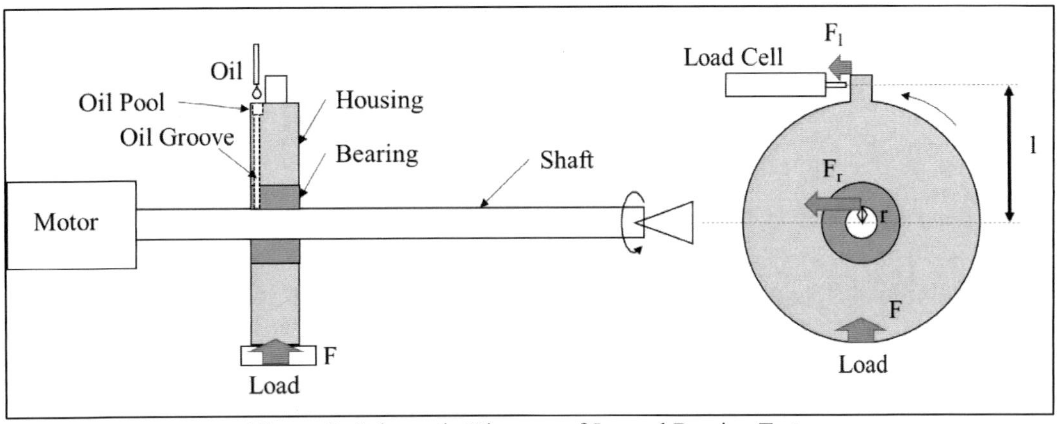

Figure 2: Schematic Diagram of Journal Bearing Tester

3.2 Specimen & Test Condition

The shaft and the bearing is fabricated from gray cast iron material. A surface modification is applied on the shaft by chemical conversion treatment. The shaft has a diameter of 16mm and the bearing length is 13mm. The surface roughness of the shaft and bearing after finishing is Ra 0.16μm and Ra 0.24μm respectively. Table 1 shows the test conditions of the journal bearing tester.

Table 1: Journal Bearing Test Condition

Item	Unit	Value
Load	N	98, 147, 196
Rotational Speed	rpm	100 ~ 2400
Oil Viscosity Grade (Mineral)	-	VG5, VG8

4. RESULTS & DISCUSSIONS

4.1 Effects of Load on Friction Coefficient

Firstly, it is important to verify that the change of friction coefficient can be measured even at low load condition using the journal bearing tester. Figure 3 shows the result for VG8 lubricating oil. The friction coefficient is expressed in ratio after normalized with respect to the coefficient at 196N applied load. The test was conducted at rotational speed of 200rpm and 800rpm. It was confirmed that at higher rotational speed of 800rpm there is fluid lubrication where the friction coefficient ratio increases as the load decreases. The result at low rotational speed of 200rpm shows mixed or boundary lubrication where the friction coefficient ratio decreases with load.

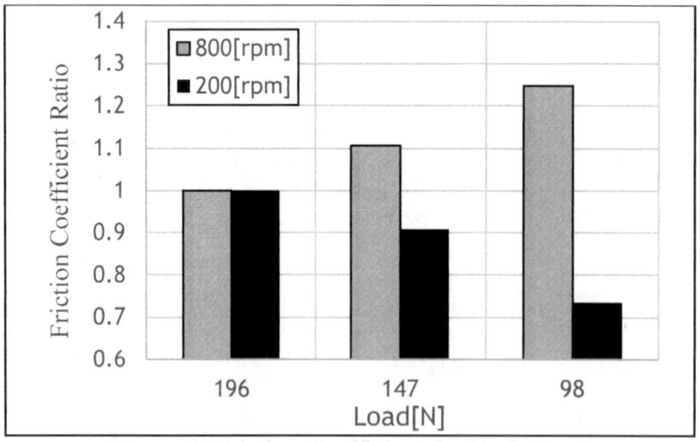

Figure 3: Friction Coefficient for Each Loads

4.2 Effects of Rotational Speed on Friction Coefficient

Next, it was determined whether a friction coefficient change in a low rotational speed regime can be evaluated by the bearing tester. Figure 4 shows the relationship between the rotational speed and the friction coefficient. The rotational speed was varied between 100rpm to 1200rpm. Both the rotational speed and friction coefficient are normalized with respect to the minimum friction coefficient obtained. The effect of rotational speed was investigated at 98N and 196N applied load with VG8 lubricating oil. Result in Figure 4 shows that when the speed ratio is greater than 1, there is the fluid lubrication regime where the friction coefficient ratio increases with rotational speed ratio for both applied load condition. It was also confirmed that as the speed ratio falls below 1, there is the mixed or boundary lubrication regime where the friction coefficient ratio increases with reduction of the rotational speed. Therefore, it was verified that the friction coefficient change in a low rotational speed regime can be measured by using the tester.

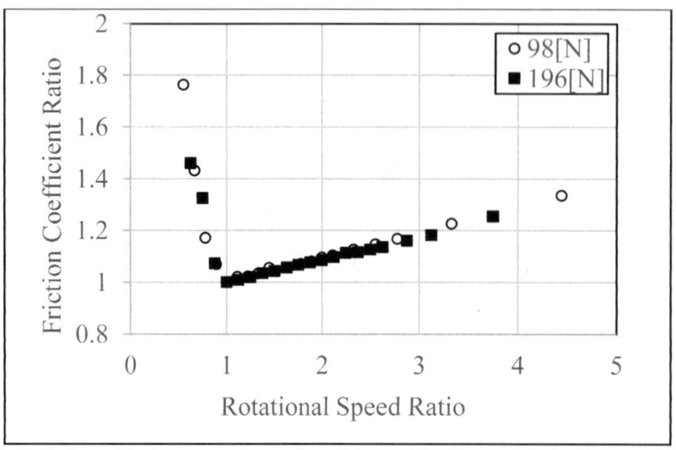

Figure 4: Relationship between Rotational Speed Ratio and Friction Coefficient Ratio

4.3 Effects of Oil Viscosity on Friction Coefficient

Next, we clarify whether a friction coefficient change using lower viscosity oil can be measured using the tester. Figure 5 shows the relationship between the rotational speed ratio and the friction coefficient ratio using lubricating oil viscosity grade VG5 and VG8. The test was performed at minimum 98N applied load and the rotational speed was varied in the range of 100rpm to 1200rpm. The result obtained shows similar trend as previous section, such that as the rotational speed ratio is higher than 1, there is the fluid lubrication regime where friction coefficient ratio increases with speed ratio for both oil. Similarly, as the speed ratio is less than 1, mixed or boundary lubrication regime appears where the friction coefficient ratio increases with reduction of rotational speed for both oil. Thus, it was verified that change of friction coefficient using low viscosity oil can be evaluated by using the tester.

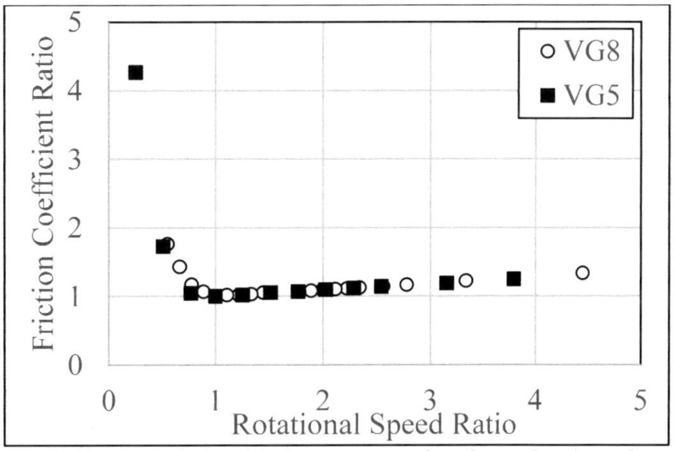

Figure 5: Relationship between rotational speed ratio and friction coefficient ratio due to oil viscosity grade

4.4 Sommerfeld Number

The results from the preceding tests show that the friction coefficient in the low load, low speed, and low viscosity condition can be evaluated with the bearing tester. The friction coefficient was plotted against the Sommerfeld number as shown in Figure 7. A Stribeck curve with a minimum transition point where mixed lubrication regime shifts to boundary lubrication regime is evident. Furthermore, it was able to detect the small variation of the friction coefficient in the region near the transition point. Therefore, this bearing test method is effective for the analysis of bearing characteristics in the low load, low speed, and low viscosity. Using the results from this test, it is possible to predict the sliding loss in the reciprocating compressor by an approximation equation in which the friction coefficient is a function of the Sommerfeld number. The prediction of sliding loss results using a reciprocating compressor are described in the next section.

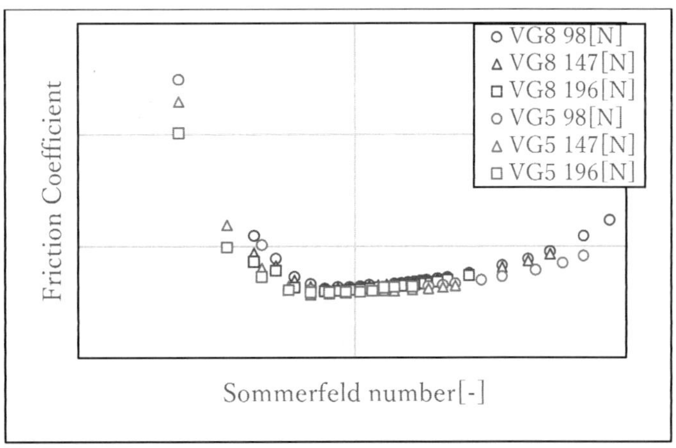

Figure 7 Relationship between Sommerfeld Number and Friction Coefficient

5. VALIDATION TEST ON RECIPROCATING COMPRESSOR

The sliding loss in a reciprocating compressor was calculated from the friction coefficient obtained in the experimental studies, and a validation test was performed. Table 2 shows the compressor test conditions using crankshafts with 4 different sliding length. The result is shown by comparing the actual input with estimated input.

Table 2: Compressor Validation Test Conditions

Item	Unit	Value
Cylinder Capacity	cc	9.1
Refrigerant	-	R600a
Oil Viscosity Grade	-	VG5 (Mineral)
Operating Frequency	rpm	1020, 1320, 1620,2220
Evaporator temperature	°C	-23.3
Condenser temperature	°C	38

Figure 8 displays the relationship between the calculated input value and the actual measured input value normalized to the average input value of each operating frequency. The result has a good correlation based on the coefficient (R) 0.90. This validates that the sliding loss of the reciprocating compressor can be estimated accurately based on experimental results from the journal bearing tester.

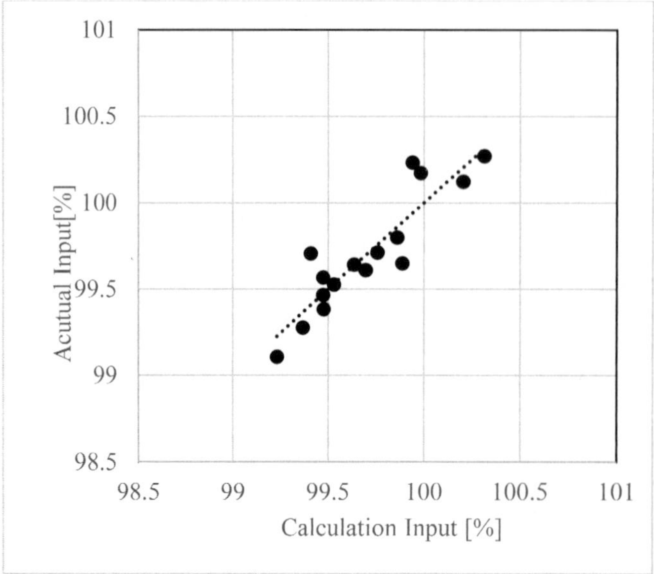

Figure 8 Relationship between predicted input value and measured input value

Using the estimation above, a reciprocating compressor with improved crankshaft design was developed. The reduction of sliding loss at low speed has enable us to achieve improvement in the inverter compressor COP up to 3%.

6. CONCLUSIONS

The following results were obtained from this study.

- We verified the plain bearing characteristics in the low load, low speed, and low viscosity regimes using the journal bearing tester.
- Sliding loss in the operating conditions of reciprocating compressor for household refrigerators could be accurately predicted.
- Using this prediction, reduction in sliding loss and improvement of COP up to 3% were achieved in the development of inverter reciprocating compressor

REFERENCES

Hamaoka, K., Ogawa, K., Ogahara, H., & Nyuji, T. (1996). Development of inverter reciprocal compressor system for refrigerator. *Proceedings of the Japanese Joint Conference on Air-conditioning and Refrigeration*, 30, 1-4.

Hamaoka, K., Ogawa, K., Ogahara, H., & Nyuji, T. (1997). Control for the Starting of an inverter Reciprocating Compressor Using a Brushless DC Motor. *JIPE*, 23(1), 66-71.

Kobayashi, M., Hayashi, A., & Yoshimura, T. (1998). Experimental Study on Journal Bearing Characteristics in Reciprocating Compressors for HFC-134a. *Proceedings of International Compressor Engineering Conference at Purdue*, 1258.

Lubrication Quality of Compressor Rolling Bearings with Oil-Refrigerant Mixtures

Rudolf HAULEITNER[1]*, Guillermo E. MORALES-ESPEJEL[2,3]

[1] SKF Product Development,
Steyr, Austria

[2] SKF Research & Technology Development,
Houten, The Netherlands

[3] Université de Lyon, INSA-Lyon, CNRS,
LaMCoS UMR5259, F69621, France

* Corresponding Author

ABSTRACT

Rolling bearings in refrigerant compressors are lubricated with a mixture of oil and refrigerant. This has always represented a challenge for the bearing lubrication quality estimation (e.g.: kappa or lambda parameters). Even if the mixture rate is known, the exact effect on the lubricant film thickness remains doubtful due to the unknown piezo-viscosity and compressibility of the refrigerant and their variation with pressure and temperature. In this work, existing mixing laws for viscosity and piezo-viscosity are examined and adapted to better represent existing measurements. The results are compared with published Daniel plots showing reasonable agreement. Once this is done a modification to the bearing lubrication quality parameter kappa is proposed to better reflect the effect of the refrigerant on the lubrication quality of compressor rolling bearings. This is a first step in the direction of predicting the bearing life for this challenging application.

Considerations of machine efficiency and environmental impact lead to ever new development of refrigerants and refrigeration oils. The refrigeration industry must adopt quickly to these changes. The capability to predict accurately the film thickness in lubricated contacts and having tools available to perform a reliable bearing selection are prerequisites for a robust machine design. The modelling capability of the lubrication quality does not only support the bearing life prediction but will facilitate the assessment of other fundamental bearing performance ratings such as friction and heat generation. The potential of the lubrication quality prediction is outlined in an example of a screw compressor in an air conditioning application.

1. INTRODUCTION

The estimation of the lubrication quality in rolling bearings of refrigerant compressors has always been a challenge due to many unknowns, e.g.: the lubricant dilution by refrigerant, the bearing temperature, the refrigerant piezo-viscosity and viscosity variation with pressure and temperature, the chemical effects of the refrigerant on the bearing raceways. In rolling bearings, the lubrication quality parameter κ (Greek kappa) is used, while in general machine design the parameter Λ-ratio (Greek lambda) is employed. A discussion and comparison between the two are presented in (Morales-Espejel *et al.*, 2010). However, even if the effective viscosity of the oil-refrigerant mixture is known in an application, the classical way to calculate the bearing lubrication quality parameter κ, as described in the ISO 281:2007 and (SKF, 2018) cannot be applied because this method assumes the lubricating film thickness in the bearing as calculated for an oil. The reduction of the piezo-viscosity in the mixture and the increase of the compressibility with the presence of the refrigerant are not considered. Therefore, in the past several researchers proposed modifications to the original model.

Wardle *et al.* (1992) measured viscosity of mixtures of oils and refrigerants HFC-134a and HCFC-22 as well as film thickness in a ball-on-disc configuration to validate modifications to the classical film thickness calculation equations. Meyers (1998) introduced these modifications into the calculation of the lubrication quality parameter in rolling bearings. Then, he applied this parameter in the L10 life estimation for rolling bearings. But no endurance tests were presented to validate the methodology. He also describes in an internal communication, that Jacobson found that the reduced chlorine of the HCFC-22 refrigerant and the absence of chlorine in the HFC-134a refrigerant significantly increases the viscosity requirements for rolling bearing lubrication. He estimated that two times greater operating viscosity is need for a bearing lubricated by a mixture of HCFC-22 and mineral oil and three times greater viscosity for a mixture of HFC-134a and polyol ester (POE) oil compared to a mineral oil lubricated bearing. For the first time and "engineering" safety factor due to refrigerant chemistry (beyond the simple calculation of film thickness) was introduced in the calculation of the bearing required viscosity. This model is summarized in (Meyers, 1997).

Much later after these developments, Morales-Espejel *et al.* (2018) reviewed the progress in refrigerant compressor technology, where they propose as a complement of the L10 life estimation a life calculation based only on surface distress (primary failure mode in these applications), with modifications to the previously published surface distress model (Morales-Espejel and Brizmer, 2011). While the L10 estimation remains a challenge in this application, it is perhaps an ideal candidate for a bearing life model that explicitly separates the surface from the subsurface, as in (Morales-Espejel *et al.*, 2015).

However, the methodology suggested in (Morales-Espejel *et al.*, 2018) requires a correct film thickness estimation in the contact, which is only possible if the viscosity, piezo-viscosity and compressibility of the mixture can be estimated. Some researchers have tried to develop methodologies for this. Notably the work of Gunsel and Pozebanchuk (1999a) with several mixture measurements reported. Also, Akei *et al.* (1996) and Akei and Mizuhara (1997) contributed with proposed mixture laws for viscosity and piezo-viscosity. As well as the recent experimental work of Bair (2017) and the full elastohydrodynamic lubrication (EHL) numerical calculations of film thickness and pressures for mixtures by Bair *et al.* (2017a).

Despite of the described progress very few publications exist with an engineering methodology on the estimation of oil-refrigerant mixture viscosity, piezo-viscosity, film thickness and required viscosity. The current work describes a methodology to achieve this by independently applying the viscosity variation with pressure and temperature of the oil and of the refrigerant. Then the lubricating properties of the mixture are calculated with the use of modified mixture law equations. From this, Daniel plots can be reproduced and the estimation of the bearing required viscosity and lubrication quality parameter is possible.

2. MIXTURE LAWS

2.1 Literature Models

As discussed before, studies have been carried out in the past to measure and estimate the film thickness in EHL contacts lubricated with oil-refrigerant mixtures. In the calculation of film thickness in EHL contacts lubricated with oil-refrigerant mixtures the estimation of the mixture properties is a key intermediate step. Diverse equations have been suggested to estimate viscosity and piezo-viscosity of oil-refrigerant mixtures. For example, Akei and Mizuhara (1997) use the Eyring theory to derive equations for the piezo-viscosity coefficient and viscosity:

$$\alpha_{mix} = \frac{ms_{ref}\left(\alpha_{ref} - \alpha_{oil}\right)}{s_{ref}(m-1)+1} + \alpha_{oil} \quad (1)$$

$$\ln(\eta_{mix}) = \left(\ln(\eta_{ref}) - \ln(\eta_{oil})\right)\frac{ms_{ref}}{s_{ref}(m-1)+1} + \ln(\eta_{oil}) \quad (2)$$

Where $m = M_{oil}/M_{ref}$, being M the molecular mass of the component. Equations (1) and (2) can be plotted for different refrigerant fractions and molecular mass ratios, as shown in Figure 1 as illustrative example.

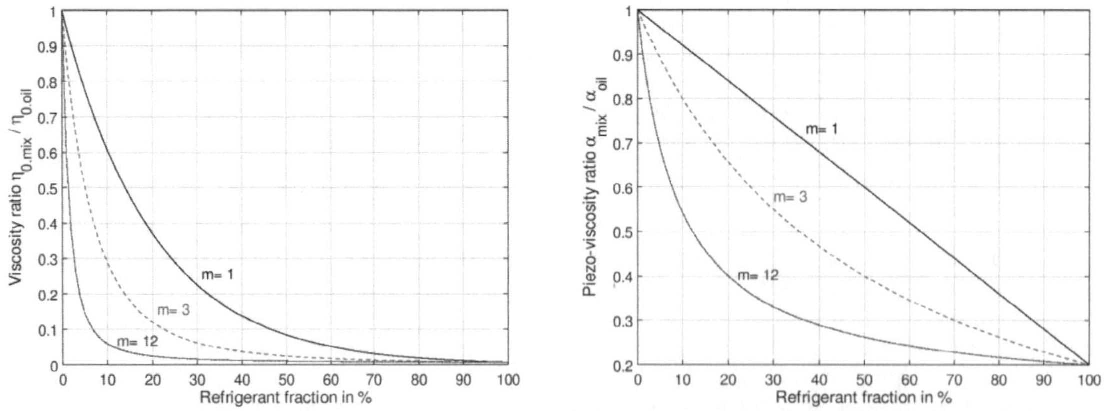

Figure 1: Illustrative example of the variation of viscosity (left) and piezo-viscosity coefficient (right) in a mixture of oil and refrigerant as a function of refrigerant fraction and molecular mass ratio m, following equations (1) and (2).

In (Morales-Espejel and Hauleitner, 2020) the predictions of Equation (2) were confronted with different oil-refrigerant mixtures where experimental data exist. The disagreement between the predictions using Equation (2) together with the molecular mass ratios taken from publications and measurements motivated the modification of the equations to better predict the measurements.

2.2 Adapted Model

Equations (1) and (2) are modified with two constants (k_{al}, k_{et}) multiplying the molecular mass ratio, as follows:

$$\alpha_{mix} = \frac{\boldsymbol{k_{al}} m s_{ref}(\alpha_{ref} - \alpha_{oil})}{s_{ref}(\boldsymbol{k_{al}} m - 1) + 1} + \alpha_{oil} \tag{3}$$

$$\ln(\eta_{mix}) = \left(\ln(\eta_{ref}) - \ln(\eta_{oil})\right) \frac{\boldsymbol{k_{et}} m s_{ref}}{s_{ref}(\boldsymbol{k_{et}} m - 1) + 1} + \ln(\eta_{oil}) \tag{4}$$

These constants are calibrated using viscosity measurements of different oil-refrigerant mixtures, so calibration functions that depend on the oil viscosity and working temperature can replace the originally constant parameters (k_{al}, k_{et}). Thus,

$$k_{al} = f(\eta_{0.oil}, T), \; k_{et} = f(\eta_{0.oil}, T) \tag{5}$$

In (Morales-Espejel and Hauleitner, 2020) it was shown that with the improved model given by equation (4) a much better agreement with measurements of a mixture of refrigerant HFC-134a and POE 100 oil could be obtained. The model was tested with several other combinations of oils and refrigerants at different temperatures and dilutions and the agreement with measurements was always much better than the original model of equations (1) and (2).

In this work the described approach is applied to the mixture of refrigerant HFC-134a and POE 170 oil, a typical combination used in refrigerant screw compressors (Fuchs Schmierstoffe GmbH, 2020). The considered fluid properties are summarized in Table 1. The Daniel plot for this mixture is shown in Figure 2 and was published in the brochure (Fuchs Schmierstoffe GmbH, 2017).

Table 1: Oil and refrigerant properties
(Fuchs Schmierstoffe GmbH, 2017), (Climalife, 2013) and (Morales-Espejel and Hauleitner, 2020)

Fluid name	η_0	α*	M	ρ_0
	mPa·s	1/Pa	g/mol	kg/m³
POE 170	173 (at 40 °C)	20x10^{-9}	1070	972 (at 15 °C)
HFC-134a	0.2 (at 25 °C)	5x10^{-9}	102	1207 (at 25 °C)

Following the approach above allows to reproduce the curves of the Daniel plot based on the properties of the individual fluids. In this example, the data from Table 1 is used to reproduce the viscosity of the POE 170 oil with 10% of refrigerant HFC-134a. Figure 3 shows a good agreement between the calculated data with the data from the Daniel plot.

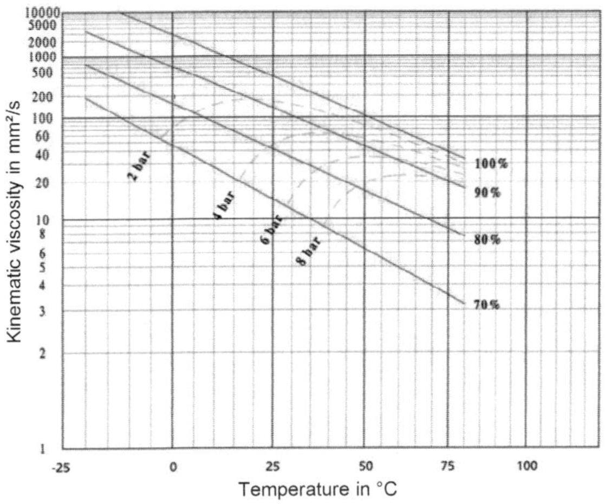

Figure 2: Daniel plot for Fuchs RENISO TRITON SE 170 and R-134a (Fuchs Schmierstoffe GmbH, 2017)

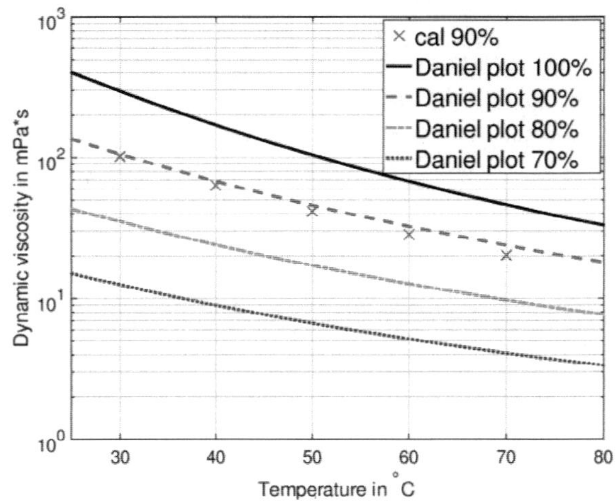

Figure 3: Original data (lines) from (Fuchs Schmierstoffe GmbH, 2017) and reproduced data (red x's) for the case of POE 170 and HFC-134a.

3. LUBRICATION QUALITY IN ROLLING BEARINGS

The central film thickness formula for an EHL contact given by Hamrock and Dowson (1981) is:

$$h_c = 2.69 R_x W^{-0.067} U^{0.67} G^{0.53} (1 - 0.61 e^{-0.73k}) \tag{6}$$

Where:

$$k = 1.03 \left(\frac{R_y}{R_x}\right)^{0.64}, W = \frac{F}{E' R_x}, U = \frac{\eta_0 \bar{u}}{E' R_x}, G = \alpha E' \tag{7}$$

Thus, the ratio of central film thickness for two lubricants with different piezo-viscosity coefficients α, is:

$$\frac{h_{c.A}}{h_{c.B}} = \left(\frac{\alpha_A}{\alpha_B}\right)^{0.53} \tag{8}$$

From (International Standard, 2007b) the relationship between the lubrication quality parameter κ and the lambda parameter (which is proportional to the film thickness) is given by:

$$\kappa \approx \Lambda^{1.3} \tag{9}$$

For two equal bearings using Equation (8) with subscript A for oil-refrigerant mixtures and subscript B for oils and with Equation (9) one can derive the influence factor of piezo-viscosity on the film thickness,

$$f_{piezo} \approx \left(\left(\frac{\alpha_{mix}}{\alpha_{oil}}\right)^{0.53}\right)^{1.3} \approx \left(\frac{\alpha_{mix}}{\alpha_{oil}}\right)^{0.7} \tag{10}$$

Now, for bearings lubricated with oil-refrigerant mixtures the methodology to calculate kappa needs to consider the refrigerant mass fraction, and is as follows:

By knowing the mass fraction of refrigerant in the oil s_{ref} and the molecular mass ratio m, the viscosity and piezo-viscosity of the mixture is calculated by using equations (3) and (4). Since the required viscosity of the bearing ν_1 is only valid for oil (or grease) lubrication and not for oil-refrigerant mixtures, the actual mixture viscosity ν_{mix} (at ambient pressure or saturation pressure of the mixture) in the bearing (so far calculated with Equation (4) or obtained from a Daniel plot if available) needs to be adjusted for the reduced piezo-viscosity introduced by the refrigerant.

Even this adjusted viscosity needs to be corrected by a safety factor, similar to the one introduced by Meyers (1998) (so-called "Jacobson" safety factor) f at a dilution $s_{ref.c}$ which will smoothly become 1 when the dilution of the oil by refrigerant becomes zero. A convenient empirical function reflecting this is:

$$f_s = 1 + (f - 1)\tanh\left(\frac{4s_{ref}}{s_{ref.c}}\right) \tag{11}$$

The final adjusted actual viscosity of an oil-refrigerant mixture for kappa calculation $\left(\kappa = \nu_{adj}/\nu_1\right)$ is:

$$\nu_{adj} = \nu_{mix}\frac{f_{piezo}}{f_s} \tag{12}$$

An adjustment factor for the actual viscosity in rolling bearings as calculated for oil lubrication is then defined as

$$f_{adj} = \frac{f_{piezo}}{f_s} \tag{13}$$

Thus,

$$\kappa_{adj} = \kappa_{oil} \cdot f_{adj} \tag{14}$$

with κ_{oil} being the lubrication quality factor calculated as oil but with the viscosity of the mixture.

An additional compensation factor can be introduced for the compressibility of the refrigerant as described by Morales-Espejel and Hauleitner (2020) and multiplied to the right-hand side of equation (13). Because of the lower piezo-viscosity coefficient, differences in compressibility, in lubricity and chemistry of dissolved liquid refrigerant, the viscosity of the oil-refrigerant mixture cannot be used for calculation of kappa without adjustment. Mathematically the adjustment factor f_{adj} has no limit and can reach mixture rates of 100% refrigerant but of course, all-steel bearings will not survive this, as discussed in (Morales-Espejel *et al.*, 2018). Therefore, it is proposed to limit this compensation technique for practical reasons to a maximum dilution of 30% refrigerant ($s_{ref} < 0.3$).

4. APPLICATION EXAMPLE

In the following, we consider a practical application example and demonstrate the potential of the described methodology and the effect of lubricant viscosity behavior for system calculations. The lubrication quality of bearings in a screw compressor of an air conditioning equipment is analyzed. Table 2 summarizes the operating conditions.

Table 2: Nominal bearing operating conditions

Bearing designation	Rotational speed	Radial load	Axial load	Max. contact pressure	Discharge pressure
	rpm	N	N	MPa	bar
ACBB 7308	3600	0	6600	1500	8
CRB NU 308	3600	5900	0	1350	8

While in the previous sections the behavior of the fluid properties is investigated against one changing input parameter only (e.g.: mixture viscosity versus dilution in Figure 1, or mixture viscosity versus temperature in Figure 3), we want to exemplify how the combined influence of several parameters could affect the lubrication quality of bearings in reality. For this, certain relationships between input parameters are assumed, as explained next.

4.1 Influence of temperature and dilution

The characteristics of the dilution versus pressure and temperature is specific for each oil-refrigerant mixture and this information is included in the Daniel plots. In this example, we will assume that the lubricant conditions are determined by the discharge pressure of the compressor. The resulting oil-refrigerant mixture ratio is defined in the Daniel plot by the intersection of the pressure line (isobar lines) and the temperature. In Figure 4 this information is extracted for two pressures from the Daniel plot given in Figure 2. For the mixture of POE 170 and HFC-134a, the dilution of the oil by refrigerant at 6 and 8 bar reduces with higher temperature. In the observed temperature range and at 8 bar, the mixture contains between 8 and 22% of refrigerant.

The dilution at a given pressure and temperature is an important factor influencing the lubrication quality of the bearing and depends on the actual combination of oil and refrigerant. The refrigerants with low global warming potential (GWP) and zero ozone depletion potential (ODP) often lead to higher dilutions compared with the previous refrigerant generation, like HFC-134a.

Since mixture viscosity drops (as well as piezo-viscosity) with higher dilution (see Figure 1) and viscosity reduces with higher temperature (see Figure 3), following the characteristic of dilution and temperature from Figure 4 will lead to a superposition of these two opposing trends. This leads to an interesting characteristic forming a maximum viscosity value within the considered temperature range and a drop of viscosity towards lower as well as higher temperatures. This effect is also shown in the Daniel plot when following the isobar line. Figure 5 shows the reproduction of the mixture viscosity ν_{mix} at 8 bar according Equation (4) as well as the adjusted (mixture) viscosity ν_{adj} according to Equation (12), including the reduction of viscosity resulting from the piezo-viscosity effect and the proposed safety factor. Throughout this example, a Jacobson safety factor f equal to 1.5 is used.

Practically, it is not possible to distinguish the results between the angular contact ball bearing (ACBB) and the cylindrical roller bearing (CRB) under conditions listed in Table 2. This was already noticed in (Morales-Espejel and Hauleitner, 2020) also giving an example of an ACBB and a CRB. To simplify the graphical representation, in the following figures only the results of the ACBB 7308 are displayed.

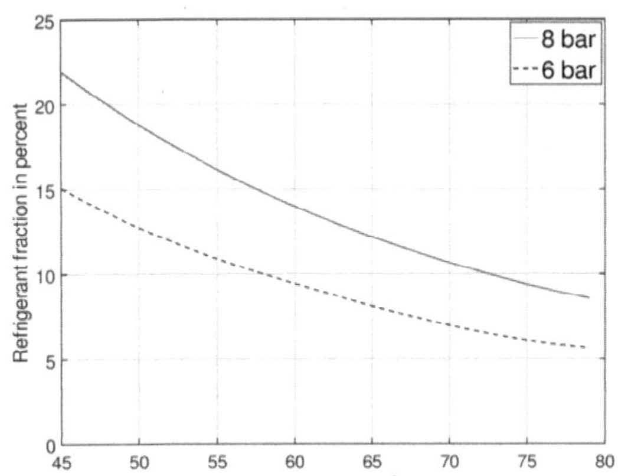

Figure 4: Refrigerant fraction in the oil-refrigerant mixture at 8 bar (red, solid line) and at 6 bar (blue, dashed line) extracted from (Fuchs Schmierstoffe GmbH, 2017).

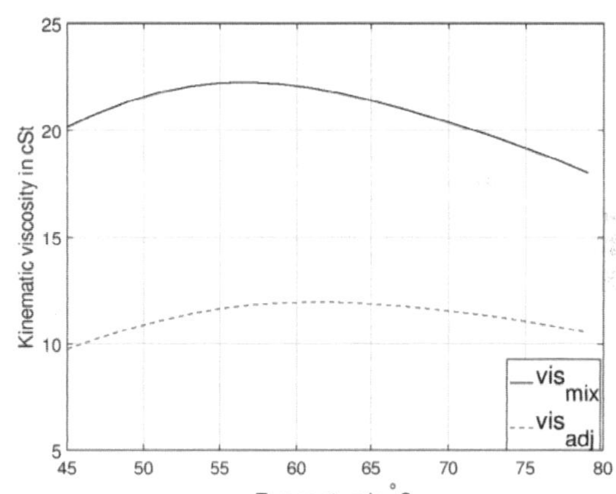

Figure 5: Mixture (black, solid line) and adjusted (red, dashed line) viscosity for the mixture POE 170 with HFC-134a at 8 bar following the relation in Figure 4.

The maximum of the adjusted viscosity ν_{adj} is shifted to a higher temperature compared to the maximum of the mixture viscosity ν_{mix}. The main reason for this effect is that the adjustment factor of Equation (13) is dependent on the dilution.

With the bearing operating conditions from Table 2 the lubrication quality parameter is calculated according Equation (14). The results in Figure 6 (left) show that the lubrication quality significantly reduces from the pure oil to the mixture and that the described additional effects (piezo-viscosity, safety factor) lead to a further severe reduction in kappa. It should be noticed that kappa in this graph is plotted on a logarithmic scale. The adjusted kappa is approximately by a factor 2 smaller than the mixture viscosity (given by common Daniel plots). While for lubrication with pure oil, a higher temperature leads to a reduction in kappa, the behavior for an oil-refrigerant mixture indicates challenging lubrication conditions at the lower and the higher end of the observed temperature range. At the lower end of the temperature axis with increasing temperature the effect of dilution reduction (see Figure 4) dominates, so kappa increases. At the higher temperature end increasing temperature reduces the viscosity of the heavily oil-charged mixture, so kappa is reduced.

4.2 Influence of discharge pressure

Next, we repeat the same evaluation but assume a discharge pressure of 6 bar instead of 8 bar (keeping the same rotational speed as in chapter 4.1). Figure 4 shows the reduction in dilution due to the lower pressure. Within the observed temperature range, the refrigerant content drops to between 6 and 15% when reducing the pressure from 8 to 6 bar.

Figure 6 (right) displays the lubrication quality for the conditions at 6 bar. As indicated by the Daniel plot, the lower pressure results in a lower dilution at the same temperature and, therefore, increases mixture viscosity as well as kappa compared to the 8 bar case.

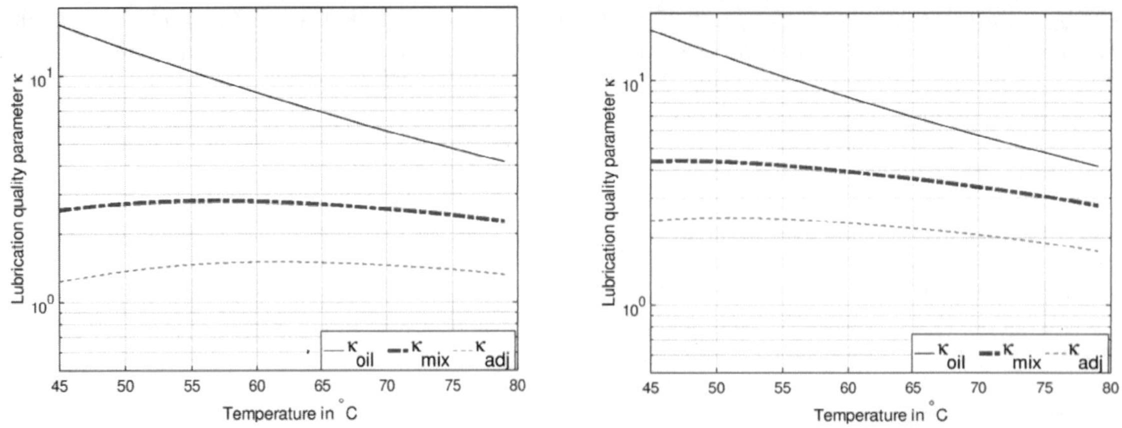

Figure 6: Lubrication quality parameter for pure oil (black, solid line), mixture (thick blue, dashed line) and adjusted mixture (red, dashed line) of POE 170 with HFC-134a at 8 bar (left) and 6 bar (right).

4.3 Influence of rotational speed

With the introduction of variable speed drives (VSD), an additional dimension in the operating range is the compressor rotational speed. The speed is an important parameter for the film thickness in the rolling element bearing (see Equations (6) and (7)). We resume the condition of Chapter 4.1 and plot the data with the additional dimension of speed.

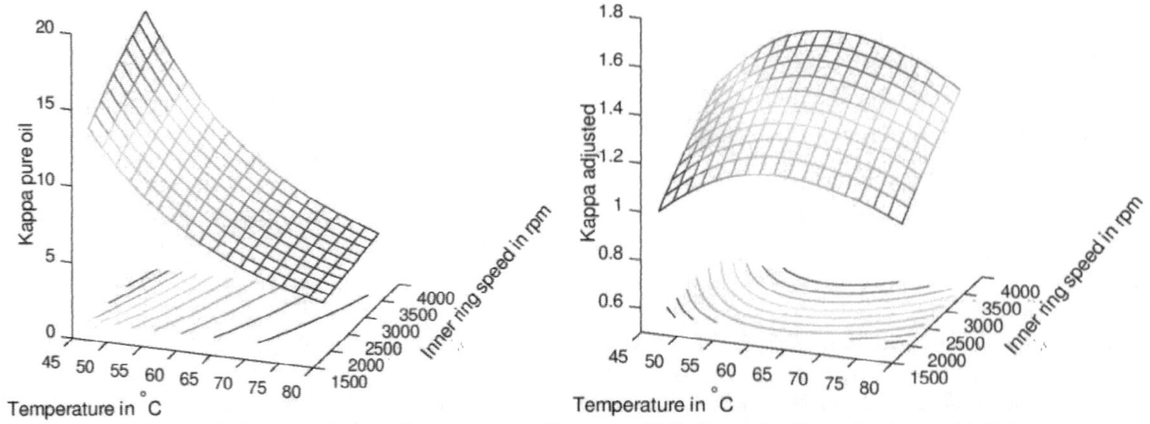

Figure 7: Lubrication quality parameter for pure oil (left), and adjusted mixture (right) of POE 170 with HFC-134a at 8 bar.

Figure 7 compares the characteristics of the lubrication quality parameter of a pure oil (although having an unrealistic high viscosity for the given conditions) against an oil-refrigerant mixture. For the case of pure oil, the regime with lowest lubrication quality is located at high temperature and low speed. The case of oil-refrigerant mixture shows significantly lower kappa values. The characteristic of kappa versus temperature and speed indicates two areas with the lowest lubrication quality, which are both in the low speed range once in the low and secondly in the high temperature area of the observed range. This explains why most of the time the critical bearing position in terms of lubrication quality is on the female - slower rotating - rotor of the twin screw compressor. Ultimately, the effect of reduced speed should be superimposed with the one for reduced discharge pressure.

5. CONCLUSIONS

A complete methodology for the estimation of the rolling bearing lubrication quality parameter κ has been described. First, accurate mixing law equations for viscosity and piezo-viscosity are developed. With these equations, the corresponding Daniel plots can be derived if not available. Then these equations are used to estimate the viscosity of the mixture. Further, the mixture viscosity is adjusted for piezo-viscosity and safety. The methodology derives an adjustment factor for the actual viscosity in the bearing when a mixture of oil and refrigerant is present. This adjustment factor when multiplied with the oil-calculated-method mixture viscosity will reduce its value to a safe level.

As shown in this work the Jacobson safety factor f leaves some freedom to the engineer to adapt the safety of the pursued design, since there are still several unknown effects. One of them already discussed in (Morales-Espejel *et al.*, 2018) is the chemical aggressiveness of the refrigerant towards the bearing steel, reducing its fatigue strength due to corrosion or poor lubricity. For example, refrigerants with very low global warming potential (GWP) and zero ozone depletion potential (ODP) might be considered as highly reactive/corrosive fluids, especially with the presence of moisture. In those cases, engineers are encouraged to use $f \geq 2$, otherwise $1.5 \leq f \leq 2$ would be sufficient.

For a real application, all parameters influencing the lubrication quality of the mixture (e.g.: temperature, pressure, rotational speed) will be interrelated by the operating cycle of the air conditioning unit. Following the proposed approach, it is possible to predict the lubrication quality for the bearings for each point within the operating map. It was shown that the dilution of the oil by refrigerant drastically reduced the lubrication quality and that the behavior gets more complex with this added degree of freedom (mixture ratio between oil and refrigerant).

From this analysis, the following conclusions can be drawn:

1. The oil-refrigerant mixture law equations found in literature need some adaptation functions dependent on lubricant viscosity and temperature to be accurate in the prediction of mixture viscosity and piezo-viscosity. After this Daniel plots can be calculated.

2. The oil-calculated-method mixture viscosity in a rolling bearing still needs to be adjusted for piezo-viscosity and safety before it can be used to calculate the lubrication quality of a bearing working in oil-refrigerant conditions. The factor f_{adj} is introduced for this purpose.

3. The Jacobson safety factor is an adaptable variable for the design of an application, highly reactive refrigerants (especially in the presence of moisture) will require larger values of this safety factor.

NOMENCLATURE

E'	Combined Elasticity modulus	(Pa)
F	Contact force	(N)
f	Jacobson safety factor at maximum dilution	(-)
f_s	Safety multiplication factor for required viscosity in oil-refrigerant mixtures	(-)
G	Dimensionless Dowson-Higginson material parameter	(-)

h_c	Central lubricant film thickness	(m)
L10	Basic Rating Life (90% survival probability)	(h)
m	Molecular mass ratio between oil and refrigerant	(-)
M	Molecular mass of a substance	(g/mol)
p	Pressure in the contact	(Pa)
R_x	Equivalent radius of curvature along rolling direction	(m)
R_y	Equivalent radius of curvature perpendicular to rolling direction	(m)
s	Fraction content in the mixture	(-)
u_1	Contact surface velocity of body 1	(m/s)
u_2	Contact surface velocity of body 2	(m/s)
$\bar{u}$	Entrainment velocity, $\bar{u} = (u_1 + u_2)/2$	(m/s)
U	Dimensionless Dowson-Higginson speed parameter	(-)
W	Dimensionless Dowson-Higginson load parameter	(-)

Greek symbols

α	Piezo-viscosity coefficient of fluid	(Pa^{-1})
α^*	Integrated piezo-viscosity coefficient, $\alpha^* = \left[\int_0^\infty \frac{\eta(p=0)}{\eta(p)} dp\right]^{-1}$	(Pa^{-1})
η	Dynamic viscosity	(Pa·s)
κ	Lubrication quality parameter for rolling bearings, $\kappa = \nu/\nu_1$	(-)
ν	Actual kinematic viscosity in the bearing	(cSt)
ν_1	Required kinematic viscosity of the lubricant	(cSt)
ρ	Liquid density	(kg/m^3)

Subscript

0	Property at atmospheric condition (oil) or at saturated condition (refrigerant)
adj	Adjusted for an oil-refrigerant mixture
mix	Oil-refrigerant mixture
oil	Oil
piezo	Adjusted by piezo-viscosity
ref	Refrigerant

REFERENCES

Akei, M., Mizuhara, K., Taki, T., Yamamoto, T. (1996). Evaluation of Film-Forming Capability of Refrigeration Lubricants in Pressurized Refrigerant Atmosphere. *Wear, vol. 196, pp. 180-187.*

Akei, M., Mizuhara, K. (1997). The Elastohydrodynamic Properties of Lubricants in Refrigerant Environments. *Trib. Trans., vol. 40, 1, pp. 1-10.*

Bair, S. (2017). A New High-Pressure Viscometer for Oil/Refrigerant Solutions and Preliminary Results. *Trib. Trans., Vol. 60(3), pp. 392-398.*

Bair, S., Habchi, W., Baker, M., Pallister, D.M. (2017a). Quantitative Elastohydrodynamic Film-Forming for an Oil/Refrigerant System. *Trans. ASME, J. of Trib., vol. 139, pp. 061501-1-7.*

Bair, S., Baker, M., Pallister, D.M. (2017b). Revisiting the Compressibility of Oil/Refrigerant Lubricants. *Trans. ASME J. of Trib., vol. 139, pp. 024501-1-4.*

Climalife (2013). R-134a technical Data Sheet. *Ref.: FF.58 /04.13/V6/ EN.*

Fuchs Schmierstoffe GmbH. (2017). Produkt Information RENISO TRITON SE 170. *PI 4-1327, PM 4 / 08.16.*

Fuchs Schmierstoffe GmbH. (2020). RENSIO Kaeltemaschinenoele 2020/2021. 930087126 09/2020 0,5

Gunsel, S., Pozebanchuk, M. (1999a). Elastohydrodynamic Lubrication with Polyolester Lubricants and HFC Refrigerants. *Final Report, vol 1, prepared for The Air Conditioning and Refrigeration Technology Institute under ARTI MCLR Project Number 670-54400. DOE/CE/23810-102.*

Gunsel, S., Pozebanchuk, M. (1999b). Elastohydrodynamic Lubrication with Polyolester Lubricants and HFC Refrigerants. *Final Report, vol 2, prepared for The Air Conditioning and Refrigeration Technology Institute under ARTI MCLR Project Number 670-54400. DOE/CE/23810-102.*

Hamrock, B. J., Dowson, D. (1981). Ball Bearing Lubrication – The Elastohydrodynamics of Elliptical Contacts. *Whiley-Interscience, New York.*

International Standard (2007a). ISO - Rolling Bearings – Dynamic Load Rating and Rating Life, *ISO 281.*

International Standard (2007b). ISO - Rolling Bearings – Explanatory Notes on ISO 281 – Part 2. *Technical Report ISO TR/ 1281-2.*

Jacobson, B.O., Morales-Espejel, G.E. (2006). High Pressure Investigation of Refrigerants HFC245fa, R134a and R123. *In Proceedings of the International Compressor Engineering Conference, Purdue University, paper 1789, pp. 1-8.*

Meyers, K.E. (1997). Creating the Right Environment for Compressor Bearings. *SKF Evolution.* http://evolution.skf.com/creating-the-right-environment-for-compressor-bearings/

Meyers, K. E. (1998). New Life Method for Rolling Bearings in Compressors. *International Compressor Engineering Conference. Paper 1209.*

Morales-Espejel, G.E., Gabelli, A., Ioannides, E., Hooke, C.J. (2010). Micro-Geometry Lubrication and Life Ratings of Rolling Bearings. *Proc. IMechE Part C: J. Mechanical Engineering Science, Vol. 224, pp. 2610-2626.*

Morales-Espejel, G.E., Brizmer, V. (2011). Micropitting Modelling in Rolling–Sliding Contacts: Application to Rolling Bearings. *Trib. Trans. Vol. 54, pp. 625-643.*

Morales-Espejel, G.E., Gabelli, A., de Vries, J.C. (2015). A Model for Rolling Bearing Life with Surface and Subsurface Survival—Tribological Effects. *Tribol. Trans., Vol. 58, 5, pp. 894-906.*

Morales-Espejel, G.E., Wallin, H.H., Hauleitner, R., Arvidsson, M. (2017). Progress in Rolling Bearing Technology for Refrigerant Compressors. *Proc. IMechE Part C: J. Mechanical Engineering Science, Vol. 232(16), pp. 2948-2961.*

Morales-Espejel, G.E., Hauleitner, R. (2020). Lubrication Quality of Compressor Rolling Bearings with Oil-Refrigerant Mixtures. *Tribology Online, Vol. 15, No. 1 (2020) pp. 45 - 52.*

SKF (2018). Rolling Bearings. *Catalogue PUB BU/P1 17000 EN.*

Wardle, F., Jacobson, B., Dolfsma, H., Hoglund, E., Jonsson, U. (1992). The Effect of Refrigerants on the Lubrication of Rolling Element Bearings Used in Screw Compressors. *International Compressor Engineering Conference. Paper 843.*

ACKNOWLEDGEMENT

The authors wish to thank Bernie van Leeuwen, Director SKF Research and Technology Development, for his permission to publish this article.

Numerical Analysis and Optimization for Hydrodynamic Lubrication in Journal Bearings of Rotary Compressor

Bing YU[1,2,3*], Ou Xiang YANG[1,2,3], Jia XU[1,2,3], Hui Jun WEI[1,2,3], Hai Hong ZHAO[3]

[1] State Key Laboratory of Air-conditioning Equipment and System Energy Conservation, ZhuHai, Guangdong, China
Tel: +86-756-8587795, E-mail: jvcanjian@163.com

[2] Guangdon Key Laboratory of Refrigeration Equipment and Energy Conservation Technology
Zhuhai, Guangdong, China

[3] Gree Electric Appliances, Inc. of Zhuhai, Zhuhai, Guangdong, China

* Corresponding Author

ABSTRACT

Based on the average Reynolds' equation of hydrodynamic lubrication, the axis locus and the minimum oil film thickness of the journal bearings under the dynamic load were numerically solved in this paper, and the movement and bearing characteristics of the compressor bearings at different rotation speeds were analyzed to evaluate the influence of the friction loss of the journal bearings on the performance of the rotary compressor. The simulation results indicated that the minimum oil film thickness of the sub bearing is smaller than the critical oil film thickness at low rotation speed ($n \leq 1800$ rpm), therefore, the friction power increases significantly with the decrease of rotation speed. The effects of width-diameter ratio, clearance and viscosity of lubricating oil on the improvement of minimum oil film thickness were further analyzed. By optimizing the width-diameter ratio of the bearing, the load carrying capacity of the oil film is improved, and the friction power of sub bearing reduces by more than 80%. The experimental results showed that the performance of rotary compressor can be improved by more than 1% at low rotation speed.

1. INTRODUCTION

The main and sub bearings are the supporting mechanism of the crankshaft of the rotary compressor. Their lubrication and friction have a critical impact on the performance and reliability of the compressor. Therefore, many scholars and engineers have conducted a lot of research on them. Okada and kuyama (1982) solved the motion and lubrication of rolling piston based on the simplified Reynolds' equation, and compared the influence of structural parameters of eccentric bearing on friction power. Hirayama et al. (2006) coupled the modified Reynolds' equation and the elastic contact equation in consideration of surface roughness and elastic deformation of the bearing of the rotary compressor, and analyzed the influence of the surface roughness on the lubrication characteristics of the bearing. Ito et al. (2010) conducted a numerical analysis of the rolling piston rotation based on four kinds of friction forces acting on the rolling piston, in which the oil film viscous force between the piston and the crankshaft was solved according to the Reynolds' equation. Matsui et al. (2010) established an analysis model of mixed lubrication of reciprocating compressor bearings, calculated the loss caused by oil viscosity and solid contact pressure, and designed a new bearing which can reduce its friction loss by 20%. Mi and Meng (2016) numerically analyzed the thermodynamic lubrication of rolling piston and two journal bearings in rotary compressors based on the transient energy equation of viscous fluids, and compared the results with isothermal analysis. The results show that the difference of lubrication performance between the thermal and isothermal cases is relative small, due to the mild temperature rise in the bearing system. Fu et al. (2016) evaluated the lubrication state of the bearings by measuring the electrical resistance between lubricated surfaces. The research shows that the lubrication state in the compressor using high viscosity oil is better than that using low viscosity. Kong et al. (2016) proposed to use contact stress to evaluate the wear condition of the bearings in rotary compressor, and analyzed the influence of bearing height, diameter and clearance on the wear through finite element simulation and accelerated life test.

The performance of compressor at low speed (n ≤ 1800 rpm) has a great influence on the APF of air conditioner, while the operation at low speed often results in poor lubrication of the bearing and high friction power of the bearings. Based on the average Reynolds' equation of hydrodynamic lubrication, the axis locus and the minimum oil film thickness of the compressor bearings under dynamic load are numerically solved. The movement and lubrication characteristics of the bearings under different rotating speeds are analyzed. An improvement scheme is proposed to reduce the friction power of the compressor.

2. ANALYSIS MODELS AND CONDITIONS

2.1 average Reynolds' equation of hydrodynamic lubrication

Based on the Reynolds' equation of average flow model (Pair and Cheng, 1978) and the contact factor (Wu and Cheng, 1989), the basic control equation of hydrodynamic lubrication in journal bearing is expressed as:

$$\frac{\partial}{r^2\partial\theta}\left(\varphi_\theta h^3\frac{\partial p}{\partial\theta}\right)+\frac{\partial}{\partial y}\left(\varphi_y h^3\frac{\partial p}{\partial y}\right)=\frac{6\mu(U_{\mathrm{j}}+U_{\mathrm{b}})\varphi_{\mathrm{c}}}{r}\frac{\partial h}{\partial\theta}+\frac{6\mu(U_{\mathrm{j}}-U_{\mathrm{b}})\sigma}{r}\frac{\partial\varphi_{\mathrm{s}}}{\partial\theta}+12\mu\varphi_{\mathrm{c}}\frac{\partial h}{\partial t} \quad (1)$$

Where, h is the nominal oil film thickness (smooth surface), r is the radius of the bearing, p is the oil film pressure, μ is the lubricant viscosity, U_{j} and U_{b} are the surface velocities of journal and bearing respectively, φ_θ and φ_y are the pressure flow factors in tangential and axial directions respectively, φ_{s} is the shear flow factor, φ_{c} is the contact factor, and σ is the standard deviation of combined roughness.

According to relevant literatures (pair and Cheng, 1978; pair and Cheng, 1979; Wu and Cheng, 1989), these factors are expressed as follows:

$$\varphi_\theta=\varphi_y=1-0.90e^{-0.56H} \quad (2)$$

$$\varphi_{\mathrm{s}}=\begin{cases}1.899H^{0.98}e^{-0.92H+0.05H^2} & H\le 5\\ 1.126e^{-0.25H} & H>5\end{cases} \quad (3)$$

$$\varphi_{\mathrm{c}}=\begin{cases}e^{-0.6912+0.782H-0.304H^2+0.0401H^3} & 0\le H<3\\ 1 & H\ge 3\end{cases} \quad (4)$$

Where, $H=h/\sigma$.

2.2 Metal contact pressure

According to the elastic contact model proposed by Greenwood and Tripp (1970), the formula for calculating the contact pressure is as follows:

$$\begin{cases}p_{\mathrm{t}}=\dfrac{16\sqrt{2}}{15}\pi(\eta\beta\sigma)^2\sqrt{\dfrac{\sigma}{\beta}}EF_{2.5}(H)\\ F_{2.5}(H)=\dfrac{1}{\sqrt{2\pi}}\displaystyle\int_H^{\infty}(s-H)^{2.5}e^{-\frac{s^2}{2}}ds\\ \dfrac{1}{E}=\dfrac{1}{2}\left(\dfrac{1-v_1^2}{E_1}+\dfrac{1-v_2^2}{E_2}\right)\end{cases} \quad (5)$$

Where, β is the radius of asperity, η is the areal density of asperities, E is the equivalent modulus of elasticity, $E_{1,2}$ are the Young's modulus of the bearing and the journal respectively, and $v_{1,2}$ are the Poisson's ratios.

2.3 Carrying capacity for mixed lubrication

The carrying capacity of the bearing in the mixed lubrication state includes two parts: the load carrying capacity of the oil film generated by the fluid dynamic pressure and the load carrying capacity of the metal contact. The calculation formula is:

$$W = \iint p ds + \iint p_t ds \tag{6}$$

2.4 Compressor and conditions

The research object is the main bearing and sub bearing of a single cylinder rotary compressor, and the three-dimensional model of the compressor is shown in Figure 1.

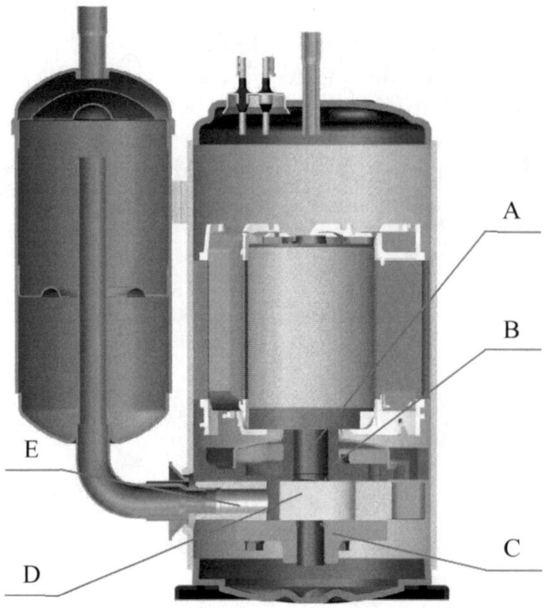

A – Crankshaft B – Main bearing C – Sub bearing D – Rolling piston E – Cylinder

Figure 1: 3D model of rotary compressor

The basic parameters and operation conditions of the compressor are given in Table 1 and Table 2 respectively. The bearing motion and friction characteristics of the compressor at three kinds of rotation speeds (1800rpm, 3600rpm and 5400rpm) are calculated. The effects of wide-diameter ratio, clearance and lubricant viscosity on improving the minimum oil film thickness are further analyzed.

Table 1: Main dimensions of compressor

Main dimensions	Displacement, V_c	Diameter of main journal, D_m	Width of main bearing, B_m	Diameter of sub journal, D_s	Width of sub bearing, B_s	clearance, C
Parameter	10.2 cm^3	13 mm	35 mm	11 mm	14 mm	20μm

Table 2: Operation conditions

Temperature Points	Evaporation	Condensation	Superheat	Subcooling
Parameter	7.2℃	46℃	11.1K	8.3K

3. RESULTS AND DISCUSSIONS

3.1 Variations of Axis locus and friction power with rotation speed

As a variable-speed compressor, it is necessary to evaluate the bearing lubrication within the allowable operating speed range. Fig. 2 and Fig. 3 respectively show the variations of the main bearing and the sub bearing's axis locus with the crankshaft speed. It can be seen from the figure that the relative eccentricity always decreases with the increase of speed, because the bearing capacity of oil film increases with the increase of frequency, but the load of bearing changes very little with the speed.

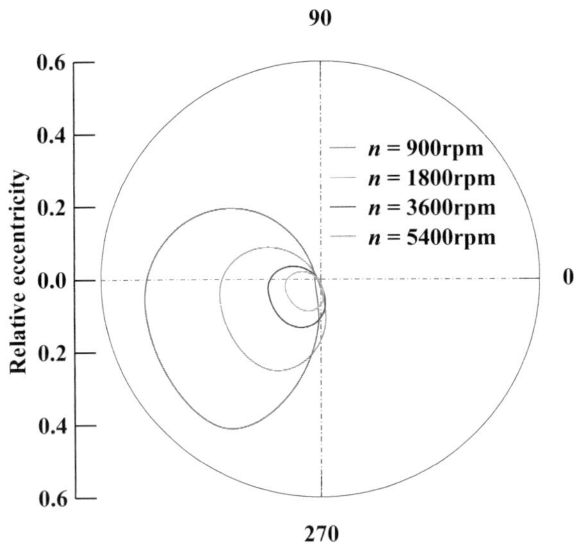

Figure 2: Axis locus of the main bearing

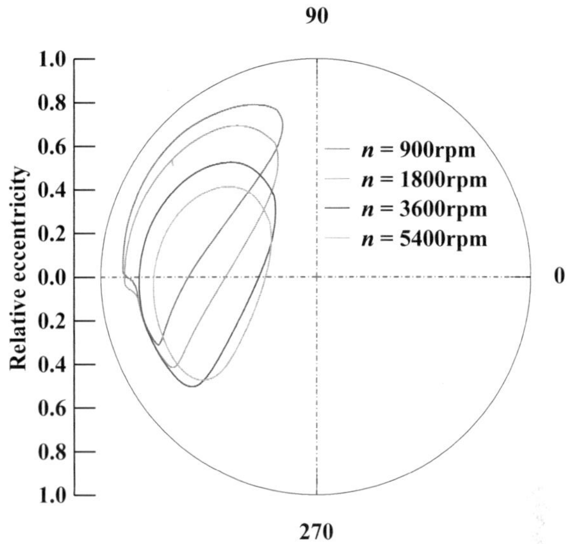

Figure 3: Axis locus of the sub bearing

Compared with Fig. 2 and Fig. 3, the axis loci of main bearing are mostly in the range of 180 ° ~ 270 °, but the axis loci of sub bearing are mostly in the range of 90 ° ~ 180 °. The reason for this difference is that the diameter and width of the sub bearing are smaller than that of the main bearing. The lubrication state of the sub bearing is mixed lubrication while that of the main bearing is fluid lubrication. In mixed lubrication, both metal contact and oil film play a role of bearing, but their directions are different, so the direction of resultant force is deflected compared with the direction of oil film force.

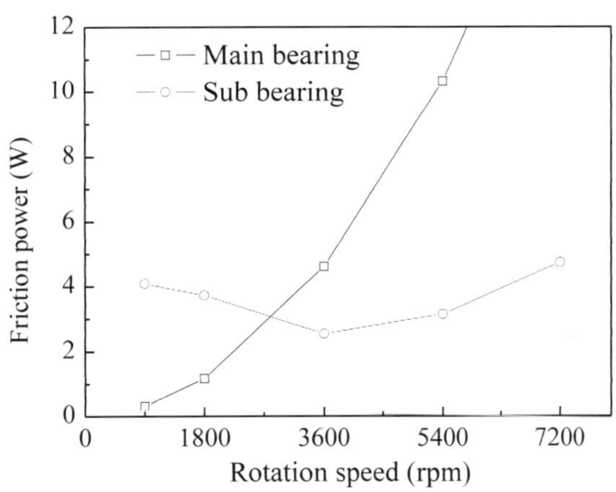

Figure 4: Friction power of the main and sub bearings

3.2 Influences of width-diameter ratio, oil viscosity and bearing clearance

In order to improve the lubrication performance of the sub bearing, the influences of the key parameters such as width-diameter ratio B/D, oil viscosity μ, clearance C on the minimum oil film thickness and carrying capacity are analyzed. Fig. 5 shows the variations of the minimum oil film thickness with the crankshaft angle under different widths, and Fig. 6 shows the variations of oil film carrying capacity C_O and metal contact carrying capacity C_C with the crankshaft angle under different widths. Fig. 7 shows the variations of minimum oil film thickness with crankshaft angle under different oil viscosities and Fig. 8 shows the variations of minimum oil film thickness with crankshaft angle under different clearance.

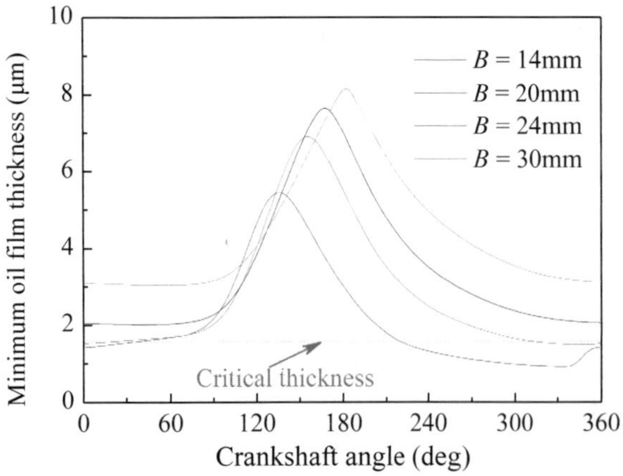

Figure 5: Effect of width on minimum oil film thickness

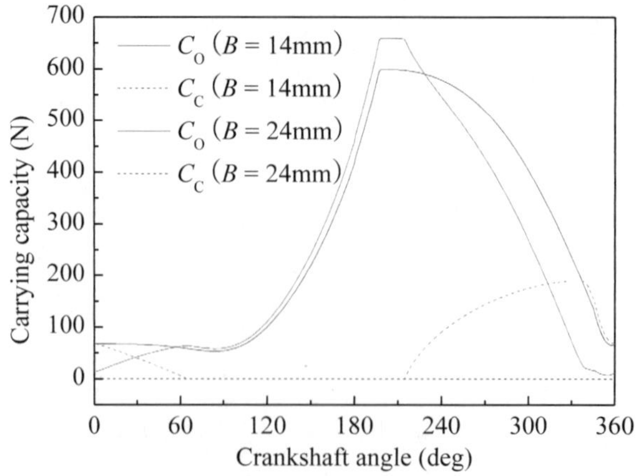

Figure 6: Effect of width on carrying capacity

It can be seen from Fig. 5 that the minimum oil film thickness increases as the bearing width B increases. When the width is 14 mm or 20 mm, the minimum oil film thickness is less than the critical oil film thickness ($h = 3\sigma$), so the bearing is in the mixed lubrication state. In a running cycle, the smaller the bearing width is, the greater the proportion of the time in the mixed lubrication state is. It can be seen from Fig. 6 that the oil film and metal contact both play a bearing role when the bearing width is 14mm (B/D= 1.27), and the angle range of metal contact bearing is the same as that of the minimum oil film thickness in Fig. 5 which is lower than the critical oil film thickness. When the bearing width is 24mm (B/D = 2.18), the bearing is in the fluid lubrication state, so the metal contact carrying capacity is zero.

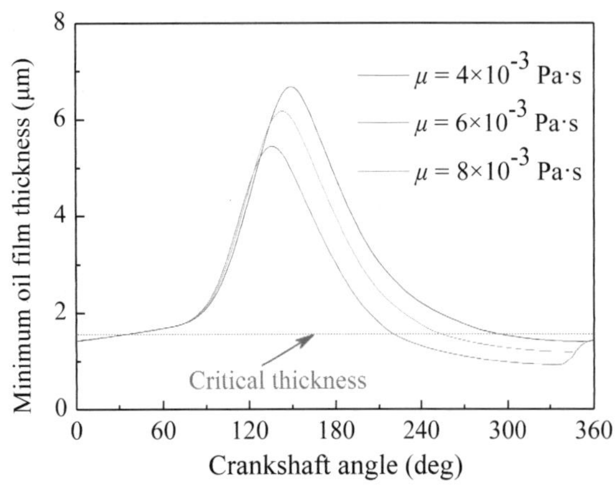

Figure 7: Effect of oil viscosity on minimum oil film thickness

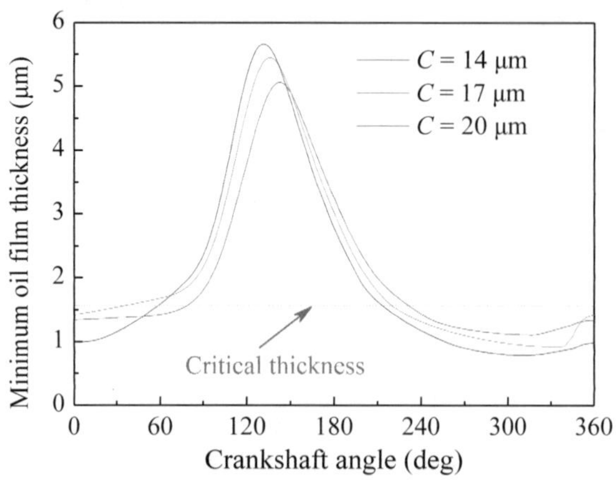

Figure 8: Effect of bearing clearance on minimum oil film thickness

As can be seen from Fig. 7, with the increase of the oil viscosity, the minimum oil film thickness increases, and the crankshaft angle range below the critical thickness decreases. Even if the viscosity of lubricating oil is doubled, metal contact cannot be avoided. In addition, considering the adverse effect of high viscosity oil in low temperature environment, we think that choosing higher viscosity lubricating oil is not a feasible measure.

It can be seen from Fig. 8 that there is an optimal clearance to make the minimum oil film thickness better. Based on the Reynolds' boundary condition, the hydrodynamic effect (the first item on the right side of formula (1)) will be weakened when the clearance increases, and the squeeze effect (the last item on the right side of formula (1)) will play a major role under the dynamic load conditions. If the clearance is too small, the crankshaft angle range where the minimum oil film thickness is below the critical oil film thickness will increase.

3.3 Optimization of sub bearing

According to the above analysis, the width-diameter ratio and clearance of the sub bearing in the rotary compressor are not good design. We adjusted these two parameters and confirmed the improvement effect through the compressor performance test. Fig. 9 shows a comparison of the physical photos of the two schemes.

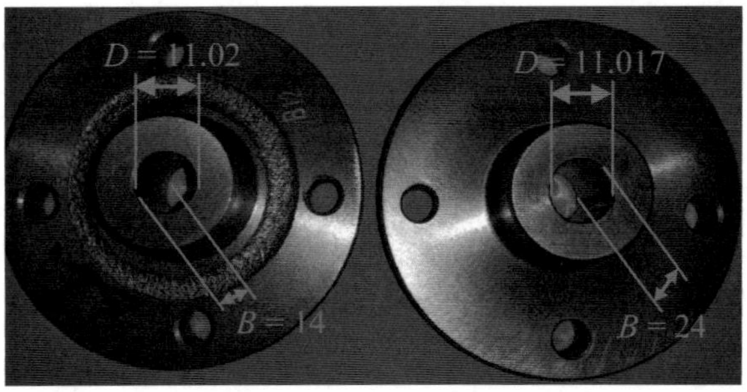

(a) Original scheme (b) Improved scheme

Figure 9: Comparison between the original scheme and the improved scheme

Fig. 10 shows the simulation results of the sub bearing friction power of the two schemes, and Fig. 11 shows the compressor COP of the two schemes at different rotation speeds. It can be seen from Figure 10 that the rotation speed of the minimum friction power is reduced from 3600 rpm to 1800 rpm. When n = 1800 rpm, the power of the improved sub bearing is 84.8% lower than that of the original scheme. It can be seen from Fig. 11 that the COP of the improved compressor are significantly improved at low speed. For example, the cop of the improved scheme is increased by 1.4% when n = 1800 rpm, but the cop of the improved scheme is reduced by 0.47% when n = 7200rpm. As we all know, the influence of low speed performance on APF is far greater than that of high speed, so the improved scheme in this paper will be beneficial to improve the comprehensive energy efficiency level of the air conditioner.

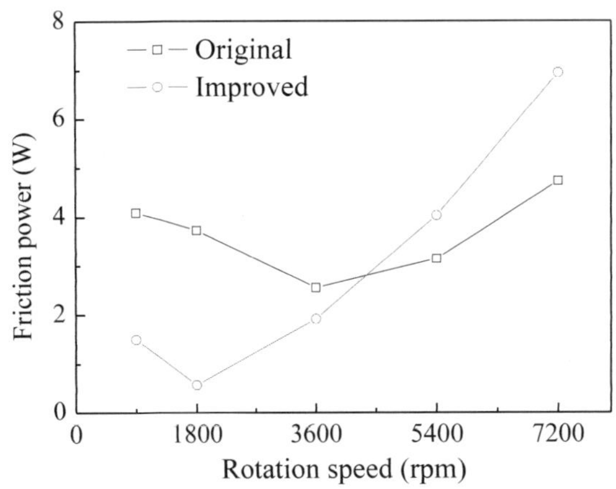

Figure 10: Comparison of the friction power

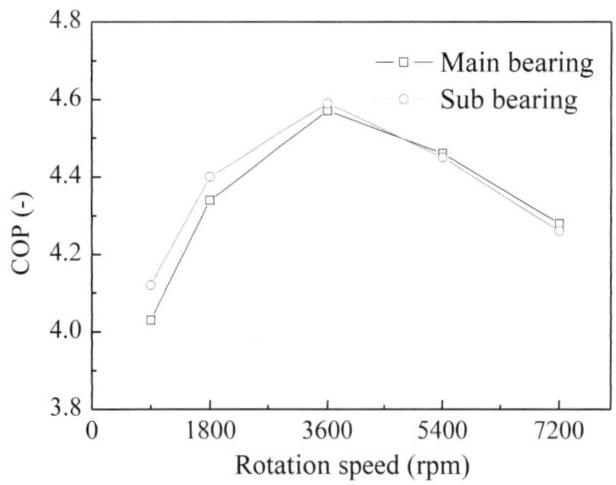

Figure 11: Comparison of the COP

4 CONCLUSIONS

Based on the average Reynolds' equation of hydrodynamic lubrication, the motion and load-carrying characteristics of the bearings of the rotary compressor are simulated. According to the simulation results, the sub bearing was optimized and the compressors were tested for performance. The following conclusions can be obtained from the analysis:

- At a lower rotation speed ($n \leq 1800$ rpm), the minimum oil film thickness of the sub bearing is smaller than the critical oil film thickness, and the friction power increases significantly with the decrease of rotating speed.

- The minimum film thickness increases with the increase of the width-diameter ratio and the oil viscosity, but increases first and then decreases with the increase of the clearance.

- By optimizing the width-diameter ratio and clearance of the bearing, the friction power of the sub bearing is reduced by more than 80%, and the experiment shows that the performance of the rotary compressor at low rotation speed is improved by more than 1%.

REFERENCES

Fu, Y., Zhou, X., Guo, H., & Mei, P. (2016). Lubrication Analysis of Journal Bearings in R410A Rotary Compressor. *International Compressor Engineering Conference*, Paper 2509.

Greenwood J. A., & Tripp J. H. (1970). The contact of two nominally flat rough surfaces. *Proceedings of the Institution of Mechanical Engineers*, *185*(46): 625-633.

Hirayama, T., Miura, K., Hattori, H., & Ito, Y. (2006). Numerical Analysis for Mixed Lubrication in Journal Bearings of Rotary Compressors. *International Compressor Engineering Conference*, Paper 1786.

Ito, Y., Hattori, H., & Miura, K. (2010). Numerical Analysis for Rotating Motion of a Rolling Piston in Rotary Compressors - Effective Factors for Characteristics of Rotating Motion of a Rolling Piston. *International Compressor Engineering Conference*, Paper 1963.

Kong, L., Zhao, Q., Ren, L., Xu, J., Cheng, X., & Hu, Y. (2016). The Influence of Main Bearing Parameters on The Bearing Wear in Rotary Compressor. *International Compressor Engineering Conference*, Paper 2445.

Matsui, M., Kitsunai, Y., & Inagaki, K. (2010). High Efficiency Development of a Reciprocating Compressor by Clarification of Loss Generation in Bearings. *International Compressor Engineering Conference*, Paper 2025.

Mi, J., & Meng, Y. (2016). THD analysis of rolling piston and journal bearings in rotary compressors. *Tribology Transactions*, *59*(2), 195–207.

Okada, K., & Kuyama, K. (1982). Motion of Rolling Piston in Rotary Compressor. *International Compressor Engineering Conference*, Paper 391.

Patir, N., & Cheng, H. S. (1978). Application of Average Flow Model for Determining Effects of Three-Dimensional Roughness on Partial Hydrodynamic Lubrication. *Trans. ASME*, *Journal of Lubrication Technology*, *100*(1), 12-17.

Patir, N., & Cheng, H. S. (1979). Application of Average Flow Model to Lubrication between Rough Sliding Surfaces. *Trans. ASME*, *Journal of Lubrication Technology*, *101*(4), 220-230.

Wu, C. W., & Zheng, L. Q. (1989). An Average Reynolds Equation for Partial Film Lubrication With a Contact Factor. *Journal of Tribology*, *111*(1), 188-191.

Theoretical Analysis of Noise and Vibration Generated by Gas Pressure Pulsation in Hermetic Compressors

Dazhuang HE[1]*, Davide ZIVIANI[1], Yangfan LIU[1]

[1]Purdue University, School of Mechanical Engineering,
West Lafayette, Indiana, USA
he385@purdue.edu; dziviani@purdue.edu; liu278@purdue.edu

* Corresponding Author

ABSTRACT

Due to the intermittent nature of discharge flow in hermetic compressors, gas pulsation is a major source of noise, vibration and harshness (NVH) effects. In the presented study on gas pulsation noise of hermetic compressors, a model was developed where the thermodynamics aspects of the compression process and the vibro-acoustics aspects are coupled together. The vibro-acoustics submodel is based on finite element method, and it describes the two-way coupled interaction of compressor cavity acoustics and shell vibration. The thermodynamics submodel is based on compressor model developed in previous studies, PDSim. It is used to compute the discharge gas pulsations that excite the cavity. Meanwhile, the acoustics response in the cavity is coupled with the thermodynamics submodel though proper modeling of discharge valves. Simulations were conducted to validate the models developed in the current work as well as the to demonstrate how these modeling tools can help compressor manufacturers to gain better understanding of the physical reasons behind NVH effects of compressors. These tools can also provide guidelines on NVH oriented design optimization of compressors.

1. INTRODUCTION

The operation of HVAC compressors inevitably yields NVH effects. The mitigation of NVH effects of compressors is an important factor not only to ensure customer satisfaction, but also to meet noise standards depending on the application. For a positive displacement compressor, pulsating flow and its interaction with the structure is a major NVH source. The existing studies on gas pulsation induced noise and vibration in positive displacement compressors are mostly based on the transfer matrix method (Soedel, 2007), in which gas pulsation is modelled as acoustics perturbation that propagates in a system consisting of lumped components. Transfer matrix method is capable to estimate transmission loss of gas pulsation across mufflers (Park et al., 2004). Experiment results have shown the validity of transfer matrix method in terms of transmission loss prediction (Chen and Huang, 2004). However, in the operation of compressor, the propagation of gas pulsation is coupled with valve motion, which makes it difficult to determine the actual cavity response through the muffler. Attempts were also made to coupled valve motion and gas pulsation. It has been shown that the propagation and reflection of pressure wave affects general performance by modulation valve motion (Yoshimura et al. 2002). Method of characteristic was used to calculate the propagation and reflection of pressure waves and agreement with experiment results was shown (Akashi et al., 2000). Finite element method is yet another tool that has been used to analyze mufflers. Besides transmission loss, which can be predicted by using transfer matrix method, additional information of muffler acoustic performance was obtained by doing finite element analysis (Svendsen, 2004). Researchers also noticed that the major source of air-borne noise of hermetic compressor are the cavity resonances. Cavity resonances of hermetic compressors were experimentally and numerically analyzed, and modifications of muffler shape were proposed to mitigate the noise induced by cavity resonances (Suh et al., 1998).

On the other hand, the noise induced by hermetic shell vibration has been extensively studied theoretically. The coupling between compressor assembly and hermetic shell was theoretically analyzed based on mathematical solution of shell vibration (Conrad and Soedel, 1992). In the theoretical analysis, the shell was assumed to be of cylindrical

shape without curved domes. Finite element method was used to analyze the effect of dome shape of hermetic shell on round radiation (Gillian and Di Flora, 1992).
The studies mentioned above are limited for a specific aspect of compressor. The sound and vibration of compressor shell, cavity, valve and suction/discharge line have been studied separately. However, a compression process consists of different phenomena which interact with each other in a short period of time. The gas pulsation noise and vibration are generated through acoustical response of the compressor cavity as well as vibration response of the hermetic shell, both of which require modeling of three-dimensional continuous systems. And the thermodynamic processes in the working chamber(s), the valve motion, gas pulsations in the gas paths, shell vibration and sound radiation are strongly coupled. Therefore, a comprehensive model that accounts for the mutual effects is necessary to identify viable solutions for noise reduction.

2. COMPRESSOR MECHANISTIC MODEL

The rotary rolling piston compressor is the reference compressor type under investigation. Rolling piston compressors usually features either a single or dual-cylinder configurations depending on the capacity range as well as if the compressor is a single or two-stage. A complete geometric model of rolling piston compressor has been developed and integrated into the existing PDSim platform (Bell et al., 2020).
The compressor mechanistic model is based on mass and energy conservation of compressor working chamber. Based on the conservation laws, two ordinary differential equations with respect to crank angle are derived. Temperature T and density ρ are used as independent thermodynamic variables. In general, the ordinary differential equations take the following form:

$$\frac{d\rho}{dt} = f(t, \rho, \dot{m}), \quad \frac{dT}{dt} = g\left(t, \rho, \dot{m}, \frac{d\rho}{dt}\right), \tag{1}$$

where f and g are linear functions with respect to the variables.
The equations are iteratively solved within a framework based on a core structure that can be readily applied to most types of positive displacement compressors by modifying the geometric parameters and other features/characteristics of each machine. The core structure includes working chamber volume calculations, evaluation of leakage flows, heat transfer within the working chamber, mechanical and frictional losses, suction and discharge valves (where applicable) and an overall energy balance to account for additional heat losses through the compressor shell.
The compressor mechanistic model has been coupled with vibro-acoustics model. It calculates discharge mass flow of refrigerant $\dot{m}(t)$. Then the mass flow of refrigerant is converted into the strength of acoustics sources that excite the cavity within compressor shell

3. VIBRO-ACOUSTICS MODEL

The vibration of hermetic compressors induced by discharge gas pulsation is a fluid-structure interaction phenomenon. The cavity is simultaneously excited by both the refrigerant gas discharge and the vibrating hermetic shell. Meanwhile, the hermetic shell is excited by the acoustic field in the cavity which, in turn, depends on the shell vibration itself. Specifically, a vibrating shell displaces the fluid in the cavity, so the motion of shell vibration can be modelled as a distribution of volume sources on the outer boundary of the cavity. Similarly, the cavity sound pressure on the inner boundary of the shell can be modelled as distributed load defined on boundary surface.

The mathematical formulations that govern thin shell vibration and cavity resonance are known as Love's equation and wave equation respectively (Fahy and Gardonio, 2007). Love's equation consists of five partial differential equations, among which three equations are independent and two equations are compatibility equations. A concise form of Love's equation is:

$$L[\boldsymbol{u}] + \rho h \frac{\partial^2 \boldsymbol{u}}{\partial t^2} = \boldsymbol{f}, \tag{2}$$

where $\boldsymbol{u}$ is shell displacement, $\boldsymbol{f}$ is fluid loading for current study, and L is a linear differential operator with respect to space. The governing equation for cavity resonance is simply wave equation:

$$\frac{1}{c^2}\frac{\partial^2\phi}{\partial t^2} - \nabla^2\phi = q, \tag{3}$$

where ϕ denotes particle velocity potential, q is the source term that represent the excitation of the cavity, which, in the current study, is related to the strength of the refrigerant flow rate or pressure pulsation at the discharge end of a compressor, as well as the equivalent source induced by shell vibration.

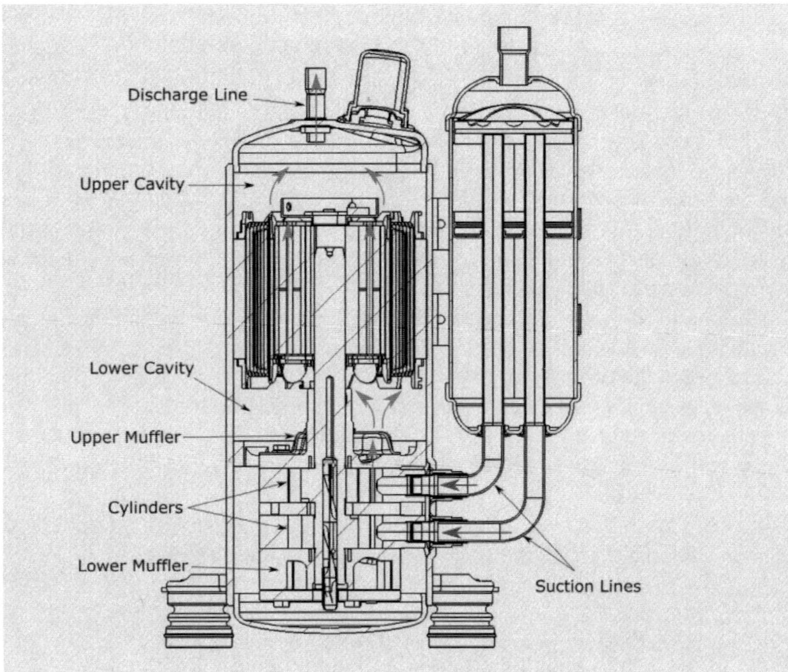

Figure 1: Typical structure of a rolling piston compressor

Apparently, the effects of fluid-structure interaction are contained in the right-hand side of equation (2) and equation (3). There are two strategies in handling fluid-structure interaction: one-way coupling and two-way coupling. In one-way coupling, the exterior shell is assumed to be rigid wall when computing the sound pressure in the cavity, then the pressure distribution on the interior wall of the shell is used as the excitation to the shell. In two-way coupling, the cavity is excited by the source inside it as well as the vibrating shell simultaneously, while the shell is excited by the cavity, whose response in turn affects the response of the cavity. The two-way coupling better reflects the actual physics, whereas one-way coupling computation is faster, and the theoretically exact solution of one-way coupling can be found. The assumption behind one-way coupling is that the cavity response induced by shell is negligible.
Assuming orthogonal modes of shell and cavity can be found, the projection of equation (2) and (3) onto orthogonal modes yields a set of ordinary differential equation of modal participation factors:

$$\begin{aligned} \ddot{\eta}_m + \omega_m^2\eta_m &= \frac{\iint_S (f_1U_{1m} + f_2U_{2m} + f_3U_{3m})dS}{\rho h \iint_S (U_{1m}^2 + U_{2m}^2 + U_{3m}^2)dS} - \frac{\rho_0 \sum_n \dot{\xi}_n \iint_S \Phi_n(U_{1m} + U_{2m} + U_{3m})dS}{\rho h \iint_S (U_{1m}^2 + U_{2m}^2 + U_{3m}^2)dS}, \\ \ddot{\xi}_n + \omega_n^2\xi_n &= \frac{-c_0^2 \iiint_V q\Phi_n dV}{\iiint_V \Phi_n^2 dV} - \frac{c_0^2 \sum_m \ddot{\eta}_m \iint_S \Phi_n(U_{1m} + U_{2m} + U_{3m})dS}{\iiint_V \Phi_n^2 dV}. \end{aligned} \tag{4}$$

In both equations, the integral

$$\iint\limits_S \Phi_n(U_{1m} + U_{2m} + U_{3m})dS$$

appears on the right-hand side. This is the integral of the product of a cavity mode and a shell mode on shell surface, which indicates the extent of coupling between a cavity mode and a shell mode. If the estimated magnitude of this coupling terms is small, the one-way coupling approach can be implemented. For one-way coupling, either the structure or the fluid is driven merely by external excitation, so one of equations (4) can be solved without considering coupling effect. After one of solution of equations (4) is obtained, it is substituted into the coupling term of the other equation as a known term. In our study, a numerical simulation of the cavity-shell coupled response were performed. The result indicated that the shell vibration displacement is significantly larger for one-way coupling than it is for two-way coupling. The difference in shell response suggests that the two-way coupling assumption is necessary for the study of compressor gas pulsation noise and vibration.

4. VALVE MODEL

Valve motion is a key factor that determines the gas pulsation pressure and flow rate at the ports and, in turn, determines the noise and vibration due to gas pulsations. In terms of thermodynamic process, the discharge valve model reacts to the compression process inside the working chamber and calculates the mass flow rate through the port. On the other hand, in terms of acoustics, the discharge valve motion is a key factor that determines the gas pulsation pressure and flow rate at the ports and, in turn, influences the noise and vibration due to discharge gas pulsations. The importance of discharge valve suggests that an accurate valve dynamics model is necessary to predict accurate discharge gas pulsation noise from compressors.
The original compressor mechanistic model integrated a simple valve model, in which the gas pulsation profile and the flow rate are calculated based on assumption that the pressure at the back side of the valve plate is mean discharge pressure (i.e., the mean pressure in the compressor cavity). However, it is only a simplified treatment, since two other fluctuating pressure components are not included: (1) the pressure difference introduced by gas flowing from one side of the valve plate to the other side and (2) the pressure applied at the back side in the form of acoustic wave in the cavity. The effects of these two addition pressure components will be considered in the improved valve model.

4.1 Forces Applied on the Valve

The force $F(t)$ is exerted on a certain point on the valve. The exact form of $F(t)$ is given by:

$$F(t) = \frac{1}{2}AC_D\rho u^2 + \rho A u^2 + (p_h - p_l) + p_a,$$

where the first term $\frac{1}{2}AC_D\rho u^2$ is the drag force induced by the gas flow, the second term $\rho A u^2$ stems from deceleration of the flow, the third term $(p_h - p_l)$ is the pressure difference between discharge gas and cavity gas and the last term p_a represent the additional pressure difference causing by cavity sound field and indicates the coupling between compression cycle and cavity sound field. p_a is induced by the pulsating mass flow going through the valve, meanwhile it affects the dynamics of the valve that modulate the gas flow.
The significance of term $\rho A u^2$ and $(p_h - p_l)$ varies as valve opening changes. When the opening is small, the pressure difference dominates, and the momentum decrease of gas is negligible. When the valve opening is large, on the contrary, the pressure difference tends to be equalized and the flow momentum decrease becomes significant.

4.2 Valve Dynamics

In our application, the reed valves of compressors are modelled as cantilever beams. The equation and boundary conditions that describe the forced vibration of a cantilever beam is:

$$EI\frac{\partial^4 u}{\partial x^4} + \rho A\frac{\partial^2 u}{\partial t^2} + \lambda\frac{\partial u}{\partial t} = F(t)\delta(x - x_0),$$

$$B.C.\begin{cases} u(0,t) = \dfrac{\partial u}{\partial x}(0,t) = 0 \\ \dfrac{\partial^2 u}{\partial x^2}(L,t) = \dfrac{\partial^3 u}{\partial x^3}(L,t) = 0 \end{cases}. \tag{5}$$

In the right-hand side of equation (5), F is the excitation that consists of gas flow momentum, pressure difference the valve port, drag force and the pressure difference induced by the acoustics wave within the cavity.

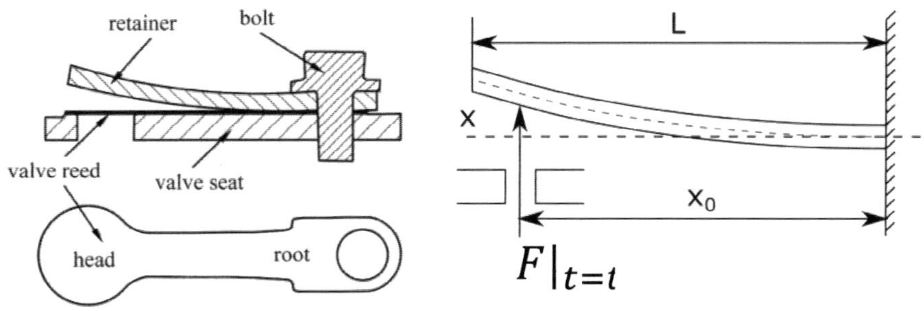

Figure 2: reed valve (left) and cantilever beam (right)

By using the method of separation of variables, the forced response valve reed can be expressed in terms of a combination of modes. If N_v valve modes are considered, modal expansion will result in a N_v 2nd order linear ODE's. In the following formulation, every 2nd order equation is formulated as a system of two 1st order ODE's for the convenience of model integration. The consistency of equation order allows us to solve the whole problem conveniently by means of numerical integration. The formulation for valve dynamic is:

$$\frac{d\eta_{v,k}}{dt} = \xi_{v,k},$$

$$\frac{d\xi_{v,k}}{dt} = F_{v,k}(\rho, p(\rho, T), \dot{m}, p_a) - 2\zeta_k \omega_{v,k} \xi_{v,k} - \omega_{v,k}^2 \eta_{v,k}, \tag{6}$$

$$u = \sum_{k}^{N_v} X_k \eta_{v,k}.$$

where $\eta_{v,k}$ is the modal participation factor, X_k is *k-th* valve mode, $\omega_{v,k}$ is the resonance frequency of mode X_k, $F_{v,k}$ and ζ_k are load and damping for mode X_k.

4.3 Refrigerant Gas Flow Model

The purpose of developing a valve model is to calculate the instantaneous mass flow going through the valve. The pulsating mass flow excites the cavity as a volume source. Moreover, in the control volume analysis used in the compressor model, the mass flow going across control volume boundaries is a key component in either mass or energy conservation equation.

The mass flow rate calculation is based on 1-D isentropic flow theory. The valve port is treated as a diverging nozzle and the flow is driven by the pressure difference and goes through the valve port. The nozzle throat is the location where the cross-section area along the flow path reaches minimum. The throat location also depends on the extent of valve opening. According to compressible flow theories, the mass flow rate through the valve is:

$$\dot{m} = C_{valve} A (2\rho_h p_h)^{\frac{1}{2}} \left\{ \frac{\gamma}{\gamma - 1} \left[\left(\frac{p_l}{p_h}\right)^{\frac{2}{\gamma}} - \left(\frac{p_l}{p_h}\right)^{\frac{\gamma+1}{\gamma}} \right] \right\}^{\frac{1}{2}}, \tag{7}$$

where C_{valve} is an empirical flow coefficient, γ is specific heat ratio, p_h and ρ_h are in-cylinder properties and p_l is the back pressure The nozzle is said to be choked as pressure ratio reaches

$$\frac{p_l}{p_h} = \left(1 + \frac{\gamma - 1}{2}\right)^{\frac{-\gamma}{\gamma - 1}}.$$

The mass flow rate reaches the maximum as the flow is choked, and it does not change anymore even if the back pressure p_l decreases further.

5. MODEL INTEGRATION

Model integration means the vibro-acoustic model and compressor mechanistic model working together. In the vibro-acoustic model, the mass flow rate of refrigerant given by compressor mechanistic model is used as the excitation to the model. The vibro-acoustic model then gives an acoustic load applied on the discharge valve and affects valve dynamics in return. A rigorous mathematical description helps us tackling the mutual interaction between vibro-acoustic model and compressor mechanistic model.

By using intermediate variable, the order of equation (4) can be lowered. Equation (4) can be formulated as a set of 1st order ODE's:

$$\frac{d\eta_{s,n}}{dt} = \xi_{s,n}, \qquad \frac{d\eta_{c,m}}{dt} = \xi_{c,m},$$
$$\frac{d\xi_{s,n}}{dt} = \sum_{n}^{N_s} C_{s,nm}\xi_{c,m} - \omega_{s,n}^2\eta_{s,n}, \qquad \frac{d\xi_{c,m}}{dt} = Q_m(\dot{m}) + \sum_{m}^{N_c} C_{c,mn}\xi_{s,n} - \omega_{c,m}^2\eta_{c,m}, \tag{8}$$
$$w = \sum_{n}^{N_s} W_n\eta_{s,n}, \qquad \phi = \sum_{m}^{N_c} \Phi_m\eta_{c,m}, \qquad p_a = -\rho_0\frac{\partial\phi}{\partial t},$$

where subscripts s and c correspond to shell and cavity respectively. w and ϕ are shell displacement and cavity velocity potential. W_n and Φ_m are *n-th* and *m-th* shell and cavity modes. $\eta_{s,n}$ and $\eta_{c,m}$ are modal participation factors of shell and cavity modes. N_s and N_c are the numbers of shell and cavity modes. The coupling between shell and cavity are reflected in the coupling terms $\sum_n^{N_s} C_{s,nm}\xi_{c,m}$ and $\sum_m^{N_c} C_{c,mn}\xi_{s,n}$.

By combining equation (1), (6), (7) and (8) together, a system of $2(N_v + N_s + N_c) + 7$ equations is obtained, with unknowns $\rho, T, \dot{m}, p_a, u, w, \phi, \eta_{v,k}, \xi_{v,k}, \eta_{s,n}, \xi_{s,n}, \eta_{c,m}, \xi_{c,m}$. The number of unknown variables is the same as the number of equations. The complete mathematical formulation allows the whole system to be solved simultaneously by means of numerical integration.

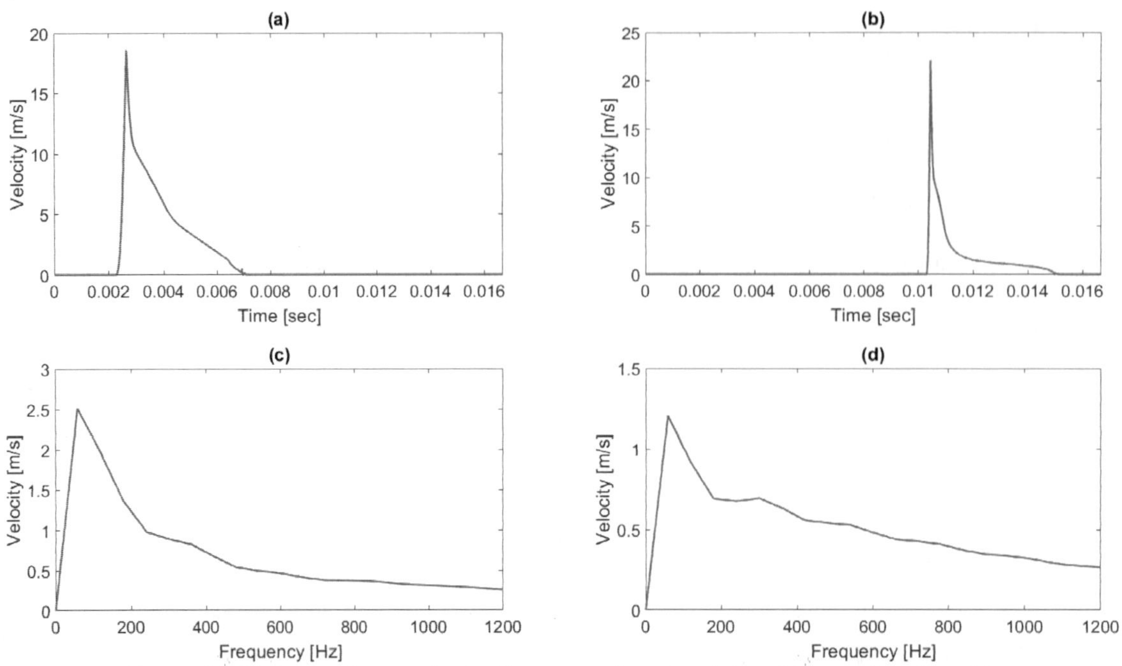

Figure 3: Discharge flows velocity for upper chamber (a) and lower chamber (b)
One-sided Fourier coefficients of discharge flows velocity for upper chamber (c) and lower chamber (d)

6. RESULTS AND DISCUSSION

6.1 Discharge Mass Flow Variation

The compressor mechanistic model generates the discharge mass flow variation over a cycle of operation. The periodic mass flow variation is used as the input of the vibro-acoustics model. The corresponding frequency response is computed. The operation of compressor is at 3600 rpm. The data obtained from compressor mechanistic model can be accordingly transformed to time histories sampled at 491520 Hz. For each set of data, 8192 samples are generated. The samples of instantaneous mass flow though discharge valve is shown in Figure 3. The DC components of signals shown in Figure 3(a) and 3(b) were neglected. The samples cover one exact cycle of compressor operation, so that Fourier coefficients can be estimated by doing DFT. Fourier coefficients in Figure 3(c) and 3(d) shows distinct peak at 60 Hz, which is the operation frequency of the compressor.

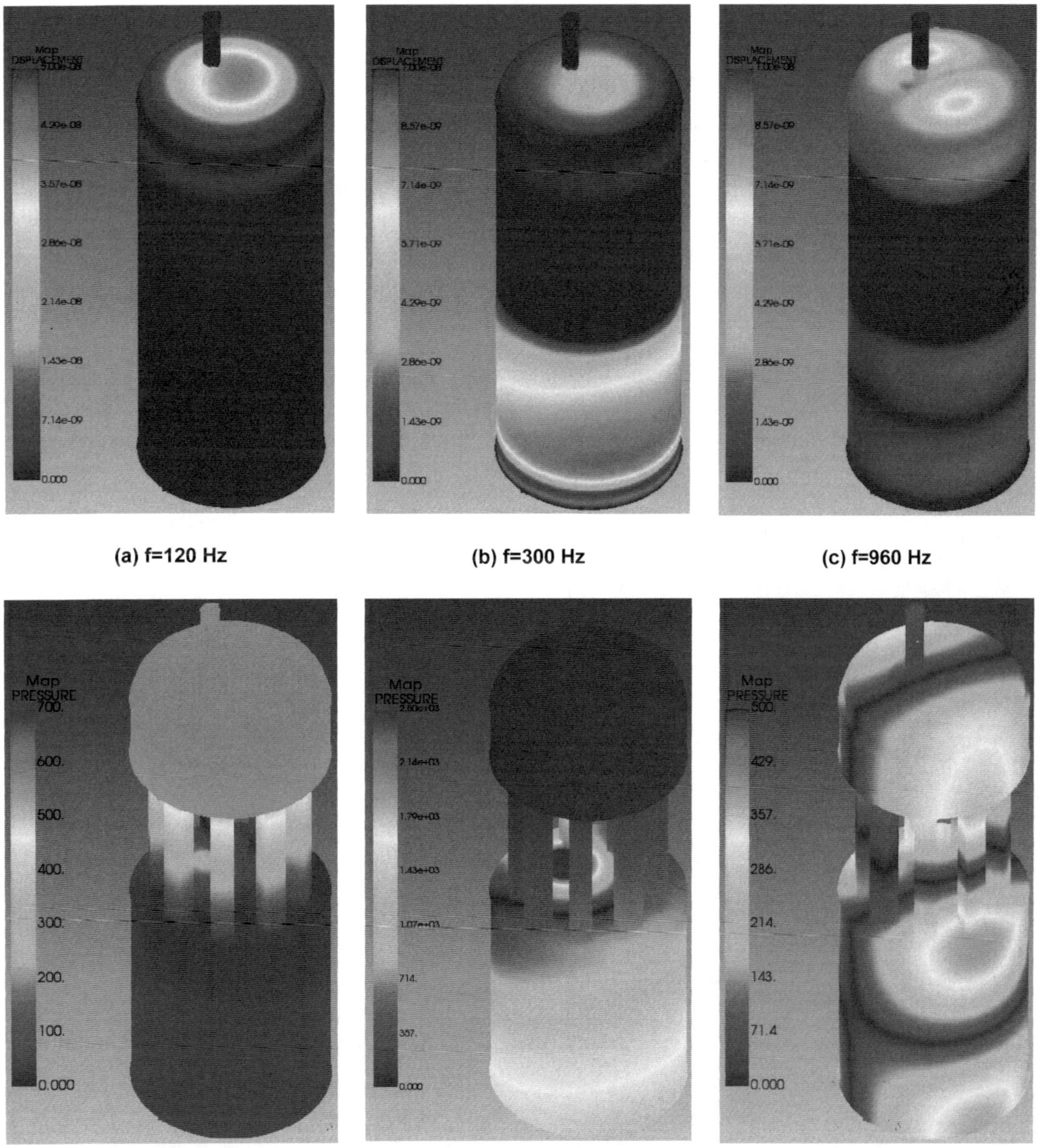

(a) f=120 Hz (b) f=300 Hz (c) f=960 Hz

(d) f=120 Hz (e) f=300 Hz (f) f=960 Hz

Figure 4: Frequency response of shell ((a), (b), (c)) and cavity ((d), (e), (f))

6.2 Finite Element Analysis of Cavity, Shell and Exterior Fluid

A numerical simulation of the cavity-shell coupled response were performed using a geometry that is simplified based on the actual geometry of a parallel dual-cylinder rotary compressor. The working fluid in the cavity is R-410a and the shell is made of steel with 3.5 mm thickness. The upper and lower cavities are connected by 9 channels between motor coils. The monopole excitation is located approximately at the discharge port of the upper muffler, from which high-pressure refrigerant is discharged into the lower cavity. The magnitude of monopole is computed based on discharge velocities. The frequency range of analysis is 60 Hz-1200 Hz. Results at 120 Hz, 300 Hz and 960 Hz are shown in Figure 4. Relatively large response occurs on the two caps of the shell, whereas the response on the shell side wall is less significant, except for the response at f = 300 Hz (Fig. 4(b)). The cap responses between 60 Hz and 900 Hz are similar to the pattern of (0,0) circular membrane mode like what is shown in Fig. 4(a). At frequencies higher than 900 Hz, the dominant mode of upper cap becomes the (1,0) circular membrane mode, while on the lower cap the (0,0) circular membrane mode is still the dominant mode. The pattern of frequency response of shell indicates that the two shell caps are major sound radiators. From the noise reduction point of view, the noise radiated by the shell can be mitigated by either changing mass or stiffness of caps to avoid cap resonance or adding damping mechanism.

The vibrating outer surface of the shell is an excitation to the exterior fluid. The technique of infinite element is used in the computation of exterior acoustics. Figure 5 shows the frequency response in finite element domain of exterior fluid. The hot spots also confirm that the shell caps are the major noise radiator.

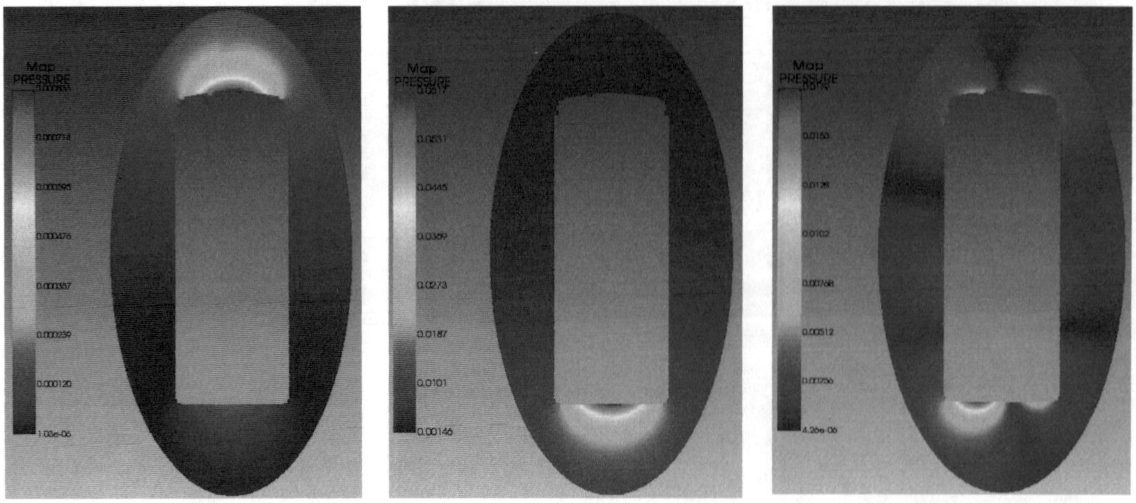

(a) f=120 Hz (b) f=300 Hz (c) f=960 Hz

Figure 5: Frequency response of exterior fluid

Outside the finite element domain, virtual microphones are set in X-Y, Y-Z and Z-X planes at $r = 0.5$ m, $r = 1$ m and $r = 5$ m. Figure 6 shows the virtual microphone set located at $r = 0.5$ m. The directivity pattern at 120 Hz, 300 Hz and 960 Hz are shown in Figure 7. At 120 Hz and 300 Hz, the exterior radiation is of uniform pattern, which is a result of (0,0) circular membrane mode shown in Fig. 5(a). The radiation at 960 Hz shows distinct directivity because of more complex shapes of higher order shell modes. The large response on shell side wall at 300 Hz also induces greater noise radiation in X-Y plane, which is about 20 times greater than side radiation at 120 Hz and 960 Hz in terms of sound pressure. The differences of exterior radiation pattern potentially suggest different structure modifications to mitigate NVH effects.

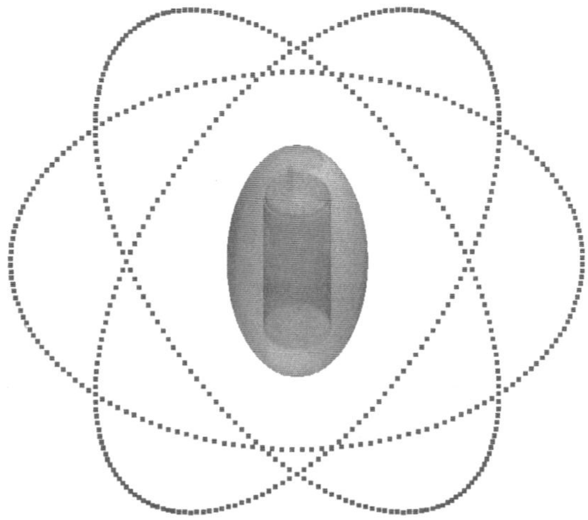

Figure 6: Virtual microphones at r = 0.5 m

7. CONCLUSION

A comprehensive theoretical model that couples the thermodynamic and mechanical aspects of the compression process with the structural and acoustic parts, with a focus on noise and vibration induced by gas pulsation is developed. Vibration of shell and cavity resonances of hermetic compressors are analyzed theoretically. The compressor modelling for rolling piston compressor in PDSim platform is done, and the discharge variations at the two discharge valves are obtained from the compressor mechanistic model. Coupled analysis between the compressor model and the vibro-acoustics model is done by feeding the mass flow variation obtained from PDSim into the two-way coupled vibro-acoustics simulation. The valve model has been developed to resolve the dynamics of the reed valves in compressors and serves as the component that couples the compression cycle and cavity acoustics field. The structure-fluid coupled vibration was studied and some simulation results were obtained using a simplified geometry of a hermetic compressor.

NOMENCLATURE

DFT	discrete Fourier transform
HVAC	heating, ventilation, and air conditioning
NVH	noise, vibration and harshness
ODE	ordinary differential equation

Subscript

c	cavity
s	shell
v	valve

REFERENCES

Soedel, W. (2007). *Sound and vibrations of positive displacement compressors*. Boca Raton, FL: CRC Press.

Park, J.I. and Adams, D.E. (2004). Modeling and Simulation of the Suction Process in a Multi-Cylinder Automotive Compressor. *Proceedings of the 2004 International Compressor Engineering Conference*, Paper C110. West Lafayette, IN, Purdue University.

Chen, L. and Huang, Z.S. (2004). Analysis of Acoustic Characteristics of the Muffler on a Rotary Compressor. *Proceedings of the 2004 International Compressor Engineering Conference*, Paper C015. West Lafayette, IN, Purdue University.

Yoshimura, T., Akashi, H., Yagi, A., and Tsuboi, K. (2002). The Estimation of Compressor Performance Using a Theoretical Analysis of the Gas Flow Through the Muffler Combined with Valve Motion. *Proceedings of the 2002 International Compressor Engineering Conference*, 507–514. West Lafayette, IN, Purdue University.

Akashi, H., Yagi, A., Sugimoto, S., and Yoshimura, T. (2000). Influence of Pressure Wave in a Suction Path on Performances in Reciprocating Compressor. *Proceedings of the 2000 International Compressor Engineering Conference,* 611–618. West Lafayette, IN, Purdue University.

Svendsen, C. (2004). Acoustics of Suction Mufflers in Reciprocating Hermetic Compressors. *Proceedings of the 2004 International Compressor Engineering Conference*, Paper C029. West Lafayette, IN, Purdue University.

Suh, K., Lee, H., Oh, W.S., and Jung, W.H. (1998). An Analysis of the Hermetic Reciprocating Compressor Acoustic System. *Proceedings of the 1998 International Compressor Engineering Conference*, 361–366. West Lafayette, IN, Purdue University.

Conrad, D.C. and Soedel, W. (1992). Modeling of Compressor Shell Vibrations Excited by a Rotor Imbalance. *Proceedings of the 1992 International Compressor Engineering Conference*, 759–768. West Lafayette, IN, Purdue University.

Gilliam, D.R. and Di Flora, M.A. (1992). The Effect of the Dome Shape of a Hermetic Compressor Housing on Sound Radiation. *Proceedings of the 1992 International Compressor Engineering Conference*, 955–965. West Lafayette, IN, Purdue University.

Bell, I.H., Ziviani, D., Lemort, V., Bradshaw, C.R., Mathison, M., Horton, W.T., Braun, J.E., Groll, E.A. (2020). PDSim: A general quasi-steady modeling approach for positive displacement compressors and expanders. *International Journal of Refrigeration*, *110*, 310–322.

Fahy, F., & Gardonio, P. (2007). *Sound and structural vibration: Radiation, transmission and response*. Amsterdam: Elsevier/Academic.

ACKNOWLEDGEMENT

The author would like to thank the Center for High Performance Buildings group for the support and encouragement of this project, and especially from those members on the Project Management Subcommittee.

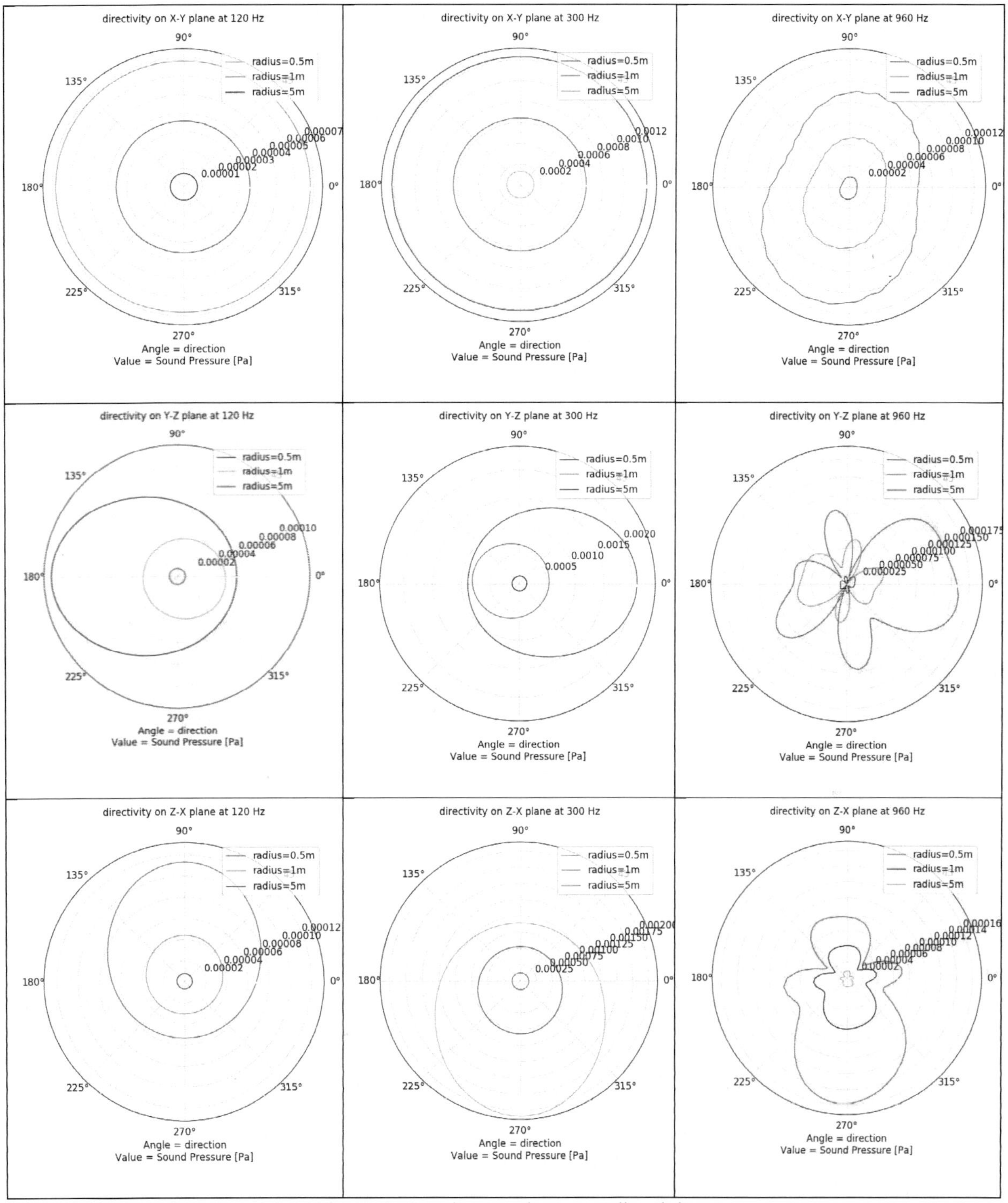

Figure 7: Exterior sound pressure directivity

Characterization of Oil Droplets in the Upper Cavity of a Rotary Compressor

Puyuan Wu[1], Jun Chen[1*], Paul E. Sojka[1], Yang Li[2], Hongjun Cao[2]

[1] School of Mechanical Engineering, Purdue University
West Lafayette, IN, USA

[2] Guangdong Meizhi Compressor Co., Ltd.,
Foshan, Guangdong, China

* Corresponding Author: junchen@purdue.edu

ABSTRACT

The lubricant oil is used to lubricate the bearings and seal the clearance between the sliding parts in the rotary compressor equipped in an air conditioning system. However, a portion of the oil is atomized and exhausted with the refrigerant flow, which reduces the system's efficiency and reliability. In this study, a high-resolution shadowgraph system is applied to a modified rotary compressor to visualize the oil droplets in its upper cavity. The oil droplet size distribution is studied statistically at different radial locations and different crankshaft's frequencies. It is observed that more oil droplets with larger D_{10} and D_{32} are generated as the rotating frequency increases. There are fewer droplets moving toward the center of the compressor. More oil droplets in the upper cavity at higher frequency might increase the Oil Discharge Rate (ODR) if these droplets are carried into the downstream discharge tube. On the other hand, larger droplets at a higher frequency may reduce ODR since the appropriate aerodynamic design may retain these droplets in the upper cavity without moving into the discharge tube. The result can assist designers in improving the CFD analysis of compressors and ultimately reducing ODR.

1. INTRODUCTION

Rotary compressors rely on high-speed revolving rolling pistons to compress low-temperature, low-pressure gas-phase refrigerant into superheated vapor that will be discharged into the internal space of the compressor through the discharge port on the cylinder. Lubricant oil is used to lubricate and seal the clearance of sliding parts in the compressor, such as the rolling piston and the cylinder, the vane and the cylinder. Part of the oil in the cylinder will be discharged during the discharge process of the cylinder and atomize into some droplets. Some droplets will follow the gas phase flow, pass through the rotor/stator, enter the compressor's upper cavity, and finally escape from the compressor. The ratio of the escaped oil to the total oil volume is characterized as the Oil Discharge Rate (ODE). The exhausting oil droplets can accumulate in the other parts of the AC system, e.g., the condenser and the evaporator, and significantly lower their heat transfer efficiency and reliability. Thus, lowering ODE is always a key issue for compressor optimization.

Since the atomization of the lubricant oil occurring at the discharge port is complicated and highly coupled to many factors, the oil droplets produced can have different sizes, velocities, and morphological features. Both computational and experimental works are reported about characterizing oil droplets' behavior in the rotary compressor. Wu et al. apply $k-\varepsilon$ model to compute the flow field of refrigerant vapor and use Discrete Phase Method (DPM) method to predict the trajectories of oil particles whose sizes are determined by Rosin-Rammler model (equation 5). Mass flux at the discharge groove and the pressure distribution at the discharge tube are prescribed as inlet and outlet boundary conditions, respectively. In the simulation, oil particles can be bounced off by moving walls or be trapped by stationary walls. Their results show that increasing the upper chamber's height and inserting the discharge tube into the upper chamber can reduce ODR as a result of the improved flow field, which is also verified by experiments (Wu et al., 2015). Noh et al. apply a similar method to analyze the multi-phase flow in a rotary compressor. Their results show that 53% of total oil droplets discharged from the compressor are from the stator core cut and 14% of total oil droplets are from the coil winding area close to the muffler's discharge hole. The percentage will change when the position of the discharge hole on the muffler changes. The number of oil droplets and ODR can be reduced by adding plate type vane in the compressor model used in CFD. They also investigate the oil droplet behavior in the lower cavity, higher cavity, stator core cut, and coil winding area in a rotary compressor experimentally. The mean diameter (D_{10}) is

130 μm in the lower chamber at a rotating frequency of 40 Hz, while over 93.5% of droplets are smaller than 150 μm in the upper chamber. The mean diameter decrease with increased rotating frequency. The speed of droplets in stator core cut and coil winding areas are also evaluated qualitatively.

However, to the best of our knowledge, oil droplets' size distribution in the surrounding area of the discharge tube has not been studied. In this study, the upper cavity refers to the space at the height of the discharge tube's inlet. The modification and optical setup are introduced in the next section, and the statistical analysis of droplets' size is presented in the result section. The conclusion is given in the last section.

2. METHODOLOGY

A twin-cylinder, variable frequency rotary compressor is used as the baseline model. The compressor uses R410a and lubricant of polyol ester, and its discharge pressure (P_D), suction pressure (P_S), and suction temperature (T_S) can be adjusted through a hot-gas bypass test rig. The working conditions are listed in Tab. 1.

Table 1: Experiment Conditions

Crankshaft's Rotating Frequency (Hz)	P_D (MPa)	P_S (MPa)	T_S (°C)
30	2.09	1.02	18.1
60	2.54	0.87	12.8
90	3.00	0.71	6.7

Three sapphire optical windows are mounted at the height above the rotor and stator to observe the oil droplets in the compressor's upper cavity, and a lens tube capable of length adjustment is applied to collect data at different radial locations.

A local Cartesian coordinate ($x_3y_3z_3$) is built for the upper cavity measurement, and its origin is at the center of the discharge tube's inlet.

A high-resolution shadowgraph system, as shown in Fig. 1, is set up with a LED light source placed at the 0° window and a 3.2 megapixel CMOS camera placed at the 180° window. The Field of View (FOV) and the Depth of Focus (DOF) of the shadowgraph system are determined by a scale target placed at the focal plane and rotated by 45° with respect to the focal plane. DOF is then $\cos 45°$ times the focused length.

The imaging processing method is used to extract oil droplets recorded by shadowgraph. The routine of the method includes exposure enhancement, de-noise with bilateral filter, calculating the grayscale gradients with Sobel operator and Non-maximum Suppression, Canny operator for edge detection and error checking with Convolutional Neural-Network (CNN).

The uncertainty of the droplet diameter determined from shadowgraph images is estimated to be 1 Pixel (Δl) of the digital imaging sensor. Here we define the characteristic relative uncertainty of droplet diameter as $\Delta l/D_{32}$.

Several characteristic diameters are applied in the current work to study the droplet's size distribution statistically:

Number mean diameter:

$$D_{10} = \frac{1}{N}\sum_{i=0}^{N} D_i. \tag{1}$$

where D_i is the diameter of the ith droplet.

Surface mean diameter (D_{20}):

$$D_{20} = (\frac{1}{N}\sum_{i=0}^{N}(D_i)^2)^{1/2}. \tag{2}$$

Volume mean diameter (D_{30}):

$$D_{30} = (\frac{1}{N}\sum_{i=0}^{N}(D_i)^3)^{1/3}. \tag{3}$$

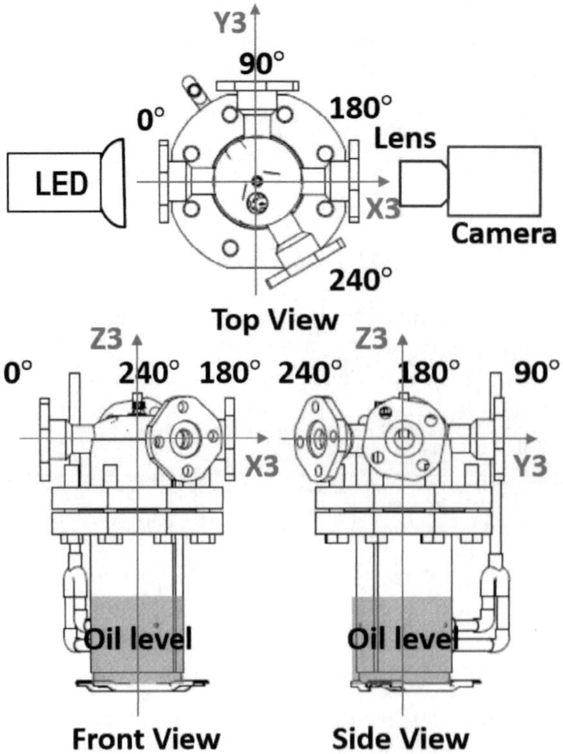

Figure 1: Modification of the discharge tube and optical setup.

Sauter Mean Diameter (SMD or D_{32}):

$$D_{32} = \frac{\frac{1}{N}\sum_{i=0}^{N}(D_i)^3}{\frac{1}{N}\sum_{i=0}^{N}(D_i)^2}. \tag{4}$$

Rosin-Rammler model is used to describe the distribution of droplet sizes:

$$1 - Q = \exp\left(-(\frac{D}{D_{0.632}})^q\right), \tag{5}$$

where Q is the total volume of droplets with a diameter less than D. $D_{0.632}$ denotes the characteristic diameter that 63.2% of the total volume is in droplets with a smaller diameter. q is a constant that describes the uniformity of the droplets' size, and its value increases as the droplets become more uniform.

3. RESULTS AND DISCUSSION

At the height of the discharge tube, the high-resolution droplet's images at different radial positions and frequencies ranging from 30 to 90 Hz are collected, as shown in Fig. 2. The dark region in the background is the discharge tube inserted into the compressor. It can be qualitatively described as that the droplet concentration at each radial location increases as the frequency increases. This is because that more oil is discharged from the cylinder at higher frequencies. Droplets' diameters are extracted and analyzed with the images and are listed in Table. 2-4. D_{10} ranges from 132 to 154 mm, 159 to 188 mm, and 204 to 216 mm at 30, 60, and 90 Hz, respectively. As illustrated in Fig. 3, the characteristic mean diameters have variations within 20% along the radial direction. When comparing droplet size at different frequencies, it is observed that the characteristic mean diameters increase as the frequency increases. The reason is that the refrigerant vapor can carry larger droplets to the height of the discharge tube due to larger velocities. Also, since the gas phase's density increases by 35% due to the higher pressure, while the liquid phase's density only increases by 4% due to the lower solubility of refrigerant as the frequency increases from 30 to 90 Hz, the ratio of the drag force applied to the droplet to droplet's mass is larger at a higher frequency.

On the other hand, the q values of Rosin-Rammler equation (Equ. 5) at the height of the discharge tube suggest that the droplet's size is close uniform there. It can be explained as that larger droplet existing in the lower area of the compressor is more difficult to be carried to the height of the discharge tube due to the gravity and larger mass, which is also confirmed by the PDF curves (Fig. 4).

Table 2: Characterized Diameter of Oil Droplets at the height of Discharge tube at 30 Hz.

X_3 (mm)	D_{10} (um)	D_{20} (um)	D_{30} (um)	D_{32} (um)	$D_{0.632}$ (um)	q
2.5	138	142	146	155	178	3.89
20.9	143	146	150	157	176	5.15
32.9	154	159	166	179	197	4.36
44	152	155	159	168	190	4.61
61.5	132	137	142	154	184	3.41

Table 3: Characterized Diameter of Oil Droplets at the height of Discharge tube at 60 Hz.

X_3 (mm)	D_{10} (um)	D_{20} (um)	D_{30} (um)	D_{32} (um)	$D_{0.632}$ (um)	q
2.5	184	189	193	203	224	4.96
20.9	188	193	200	213	243	4.30
32.9	180	184	188	197	221	5.24
44	182	186	190	198	220	5.32
61.5	159	163	167	176	200	4.42

Table 4: Characterized Diameter of Oil Droplets at the height of Discharge tube at 90 Hz.

X_3 (mm)	D_{10} (um)	D_{20} (um)	D_{30} (um)	D_{32} (um)	$D_{0.632}$ (um)	q
2.5	216	222	228	241	264	5.50
20.9	209	217	227	247	288	3.50
32.9	204	212	220	238	260	4.67
44	216	223	231	248	268	4.80
61.5	208	213	219	232	254	5.08

4. CONCLUSIONS

In the upper cavity of the rotary compressor, it is observed that, firstly, at 30 to 90 Hz, droplets' size has a close-uniform distribution along the radial direction. The droplets near the wall are slightly smaller than those in the inner region. Secondly, the characteristic mean diameters increase as the frequency increases. Since the flow rate increases at higher frequency, larger flow rates and larger vertical velocity are easier to carry larger droplets. Moreover, as the gas phase density becomes larger at higher frequency, the drag force applied to the droplet is larger. Thirdly, the q values of Rosin-Rammler fitting in all cases are larger than 3.4 and can be greater than 5 in some cases, suggesting that the span of droplet size distribution is small.

REFERENCES

Noh, K., Min, B., Song, S., Yang, J., Choi, G., & Kim, D. (2016, October). Oil droplet behavior with the position of a muffler discharge hole in the shell of a rotary compressor. *Journal of Mechanical Science and Technology*, *30*(10), 4579–4589.

Wu, J., Hu, J., Zhang, J., Sun, M., Wang, G., & Chen, A. (2015). A cfd model for the oil discharge ratio in rotary compressors. *Proceedings of the Institution of Mechanical Engineers, Part A: Journal of Power and Energy*, *229*(2), 151–159.

ACKNOWLEDGMENT

The project is sponsored by Guangdong Meizhi Compressor Co., Ltd..

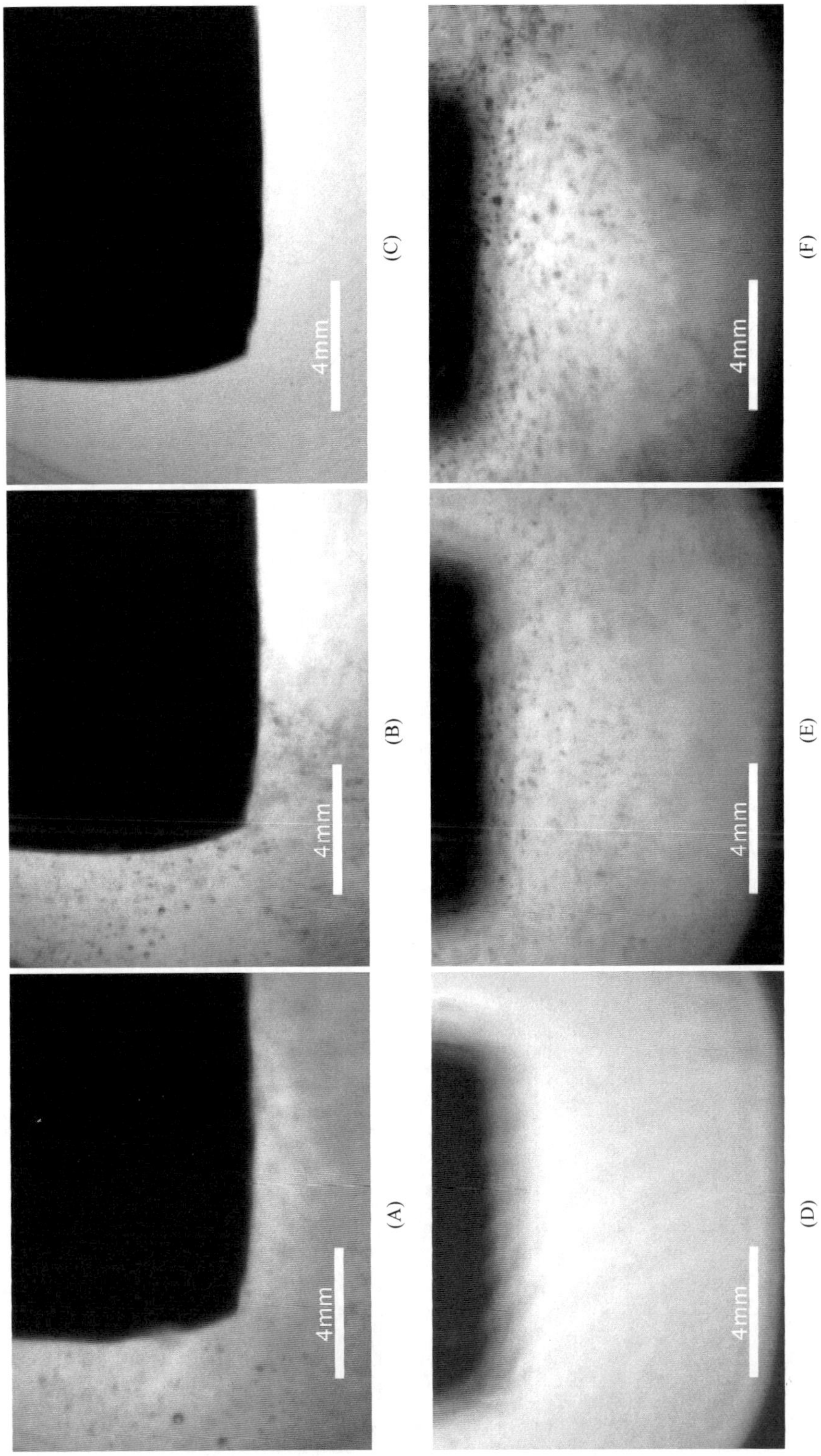
4mm
(A)
4mm
(B)
4mm
(C)
4mm
(D)
4mm
(E)
4mm
(F)

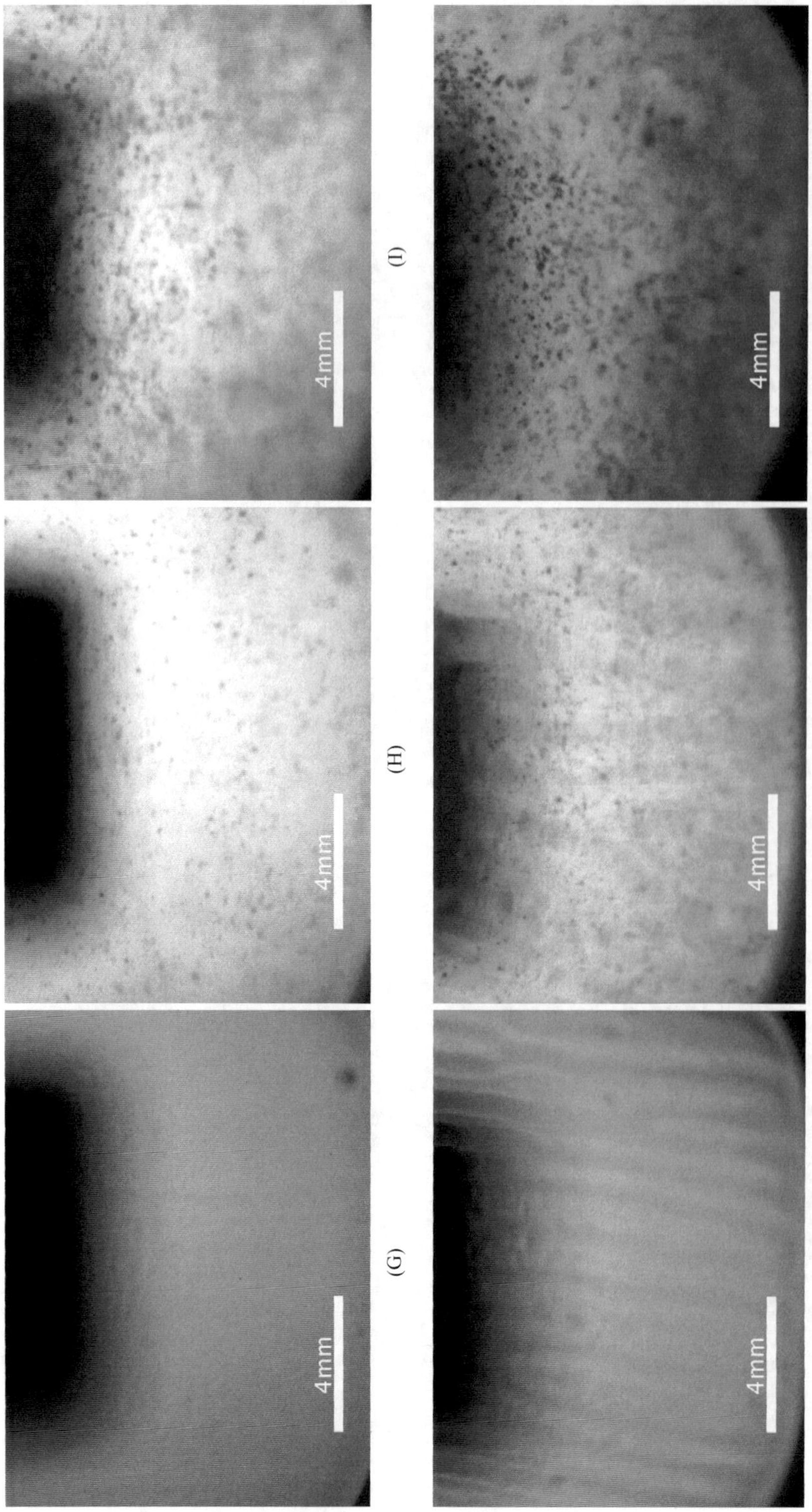
4mm
(G)
4mm
(H)
4mm
(I)
4mm
(J)
4mm
(K)
4mm
(L)

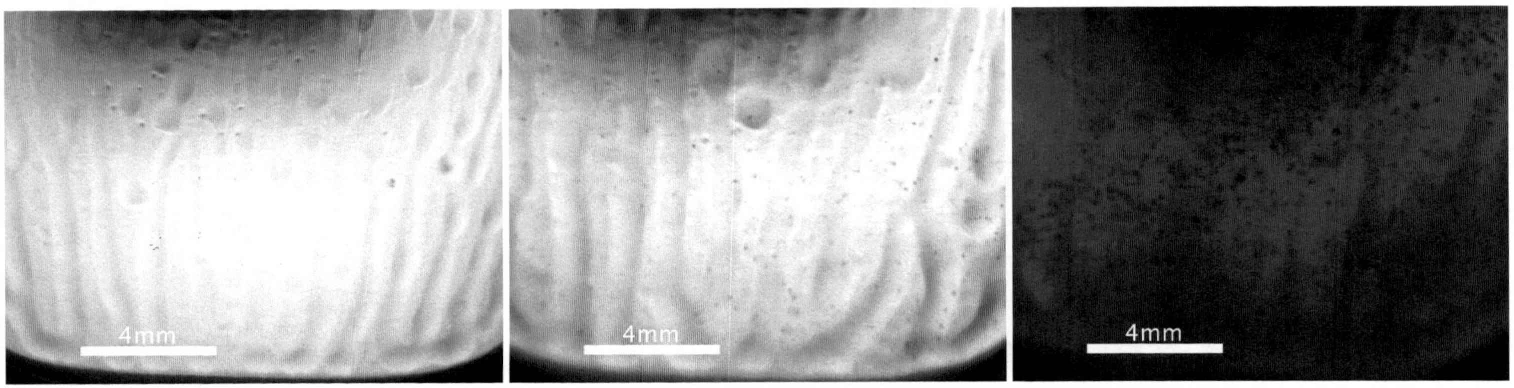

(M) (N) (O)

Figure 2: Snapshots of droplets at the height of the discharge tube. (a) 30 Hz, X_3=2.5 mm. (b) 60 Hz, X_3=2.5 mm. (c) 90 Hz, X_3=2.5 mm. (d) 30 Hz, X_3=20.9 mm. (e) 60 Hz, X_3=20.9 mm. (f) 90 Hz, X_3=20.9 mm. (g) 30 Hz, X_3=32.9 mm. (h) 60 Hz, X_3=32.9 mm. (i) 90 Hz, X_3=32.9 mm. (j) 30 Hz, X_3=44 mm. (k) 60 Hz, X_3=44 mm. (l) 90 Hz, X_3=44 mm. (m) 30 Hz, X_3=61.5 mm. (n) 60 Hz, X_3=61.5 mm. (o) 90 Hz, X_3=61.5.5 mm.

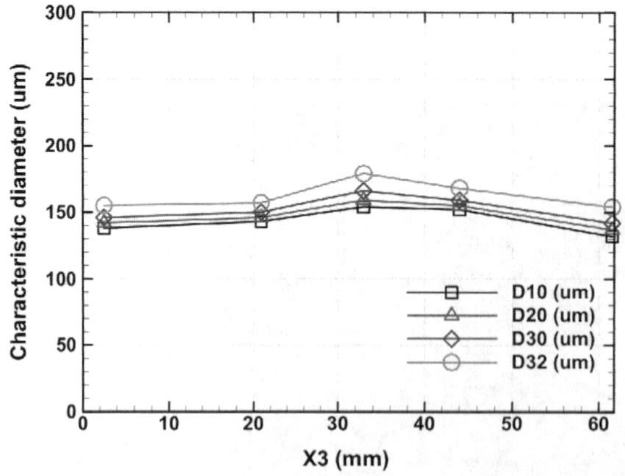

(A)

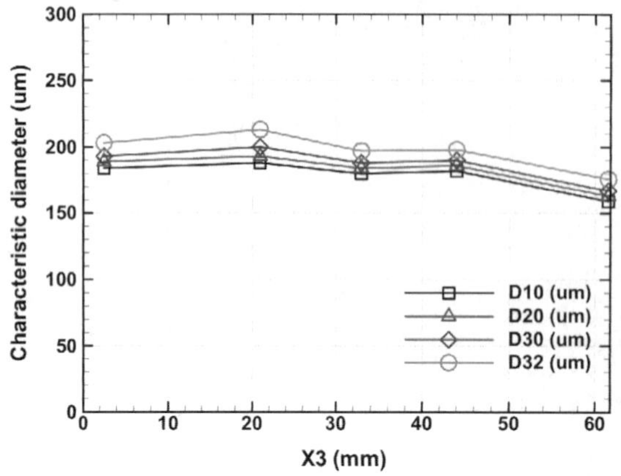

(B)

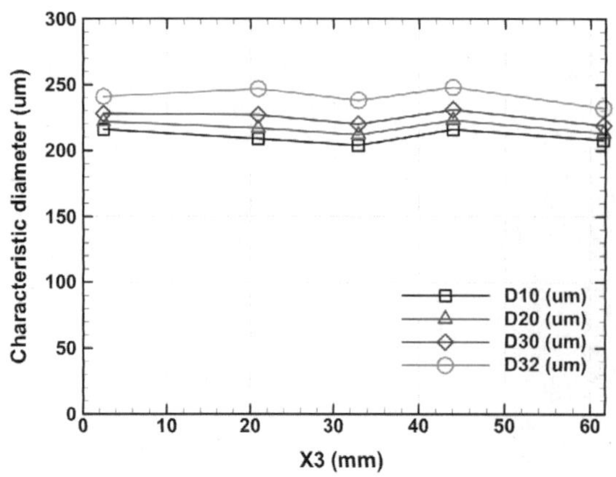

(C)

Figure 3: Characteristic mean diameters' distribution along the radial direction. (a) 30 Hz. (b) 60 Hz. (c) 90 Hz.

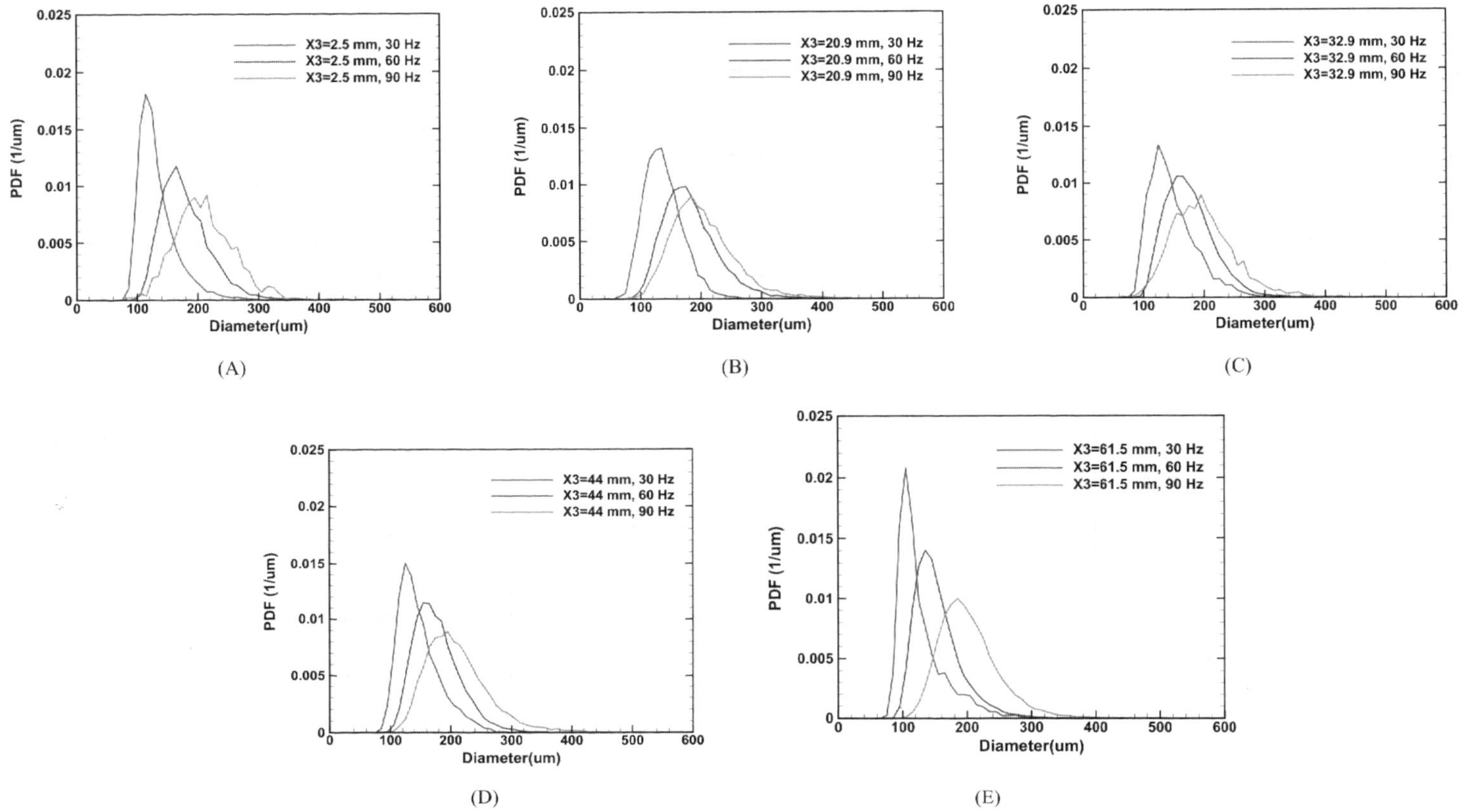

Figure 4: Oil droplets size distribution. (a) X_3=2.5 mm. (b) X_3=20.9 mm. (c) X_3=32.9 mm. (d) X_3=44 mm. (e) X_3=61.5 mm.

An Experimental Study on Bearing Characteristics of Compressor Considering Refrigerant Dissolution and Friction Heat

Tatsuya SASAKI[1,2]*, Yoshimi IKEDA[1], Shinya SASAKI[2]

[1] Mitsubishi Electric Corporation, Advanced Technology R&D Center,
Amagasaki, Hyogo, Japan
Sasaki.Tatsuya@ea.MitsubishiElectric.co.jp

[2]Tokyo University of Science, Department of Mechanical Engineering, Faculty of Engineering,
Katsushika-ku, Tokyo, Japan

*Corresponding Author

ABSTRACT

There is a growing need to expand the upper limit of the operating range of compressors for air-conditioning and refrigeration systems. There are two design approaches to expanding the upper limit. One is to expand the stroke volume, which increases the load of the journal bearing along with the increase of compressive load of the refrigerant. The other is to increase the rotation speed, where the load capacity of the journal bearing decreases due to lowering the viscosity of the refrigeration oil caused by shear heating of the oil film. Incidentally, the refrigerant is dissolved in the refrigeration oil that lubricates the journal bearing, and the refrigerant solubility and the dissolving viscosity change depending on the temperature and pressure conditions. It is impossible to ignore changes in viscosity of refrigeration oil due to friction heat and refrigerant dissolution at high rotation speeds and in the high load range. We have evaluated the influence of frictional heat generation, solubility and viscosity change on journal bearing characteristics by thermal fluid lubrication analysis. In this study, we developed an experimental apparatus to evaluate bearing characteristics in a refrigerant environment to verify the validity of the Thermo-Hydrodynamic Lubrication (THL) analysis and also the effects of frictional heat, and changes in refrigerant solubility on bearing characteristics were experimentally evaluated. As a result, we found that the bearing temperature decreased in the reverse wedge region of the journal bearing and we could obtain data to support the hypothesis that the heat of refrigeration oil was taken away while the refrigerant was separated from the refrigeration oil.

1. INTRODUCTION

Recently, to meet the demand for high-power and high-efficiency air-conditioning systems and refrigerators, expanding the operating range of compressors is required. The authors worked on securing the reliability of the journal bearing that supports the shaft rotation, which is one of the issues in expanding the operation range. There are two ways to expand the upper limit of compressors: increasing the stroke volume or increasing the maximum operating frequency. In the expansion of the stroke volume, the load of the journal bearing increases with the increase of the refrigerant compression load generated by the compression refrigerant. On the other hand, when the rotation speed is increased, the load capacity of the journal bearing decreases as refrigeration oil viscosity is reduced due to heat generation of the oil film of the journal bearing clearance. In the past, the main design method of journal bearings solved Reynolds's equation assuming the oil viscosity was constant, and then comprehended the bearing characteristics (minimum oil film thickness, oil film pressure, oil film temperature, friction coefficient, etc.). However, refrigerant is dissolved in the refrigeration oil that lubricates the sliding part of the compressor, and its solubility and viscosity actually change depending on temperature and pressure conditions. From past studies, it is known that the refrigerant dissolution concentration and oil viscosity change instantaneously depending on pressure and temperature in the micro-channel (Aritoshi, 2009). The same phenomenon is thought to occur at the bearing clearance. In particular, this change cannot be ignored in high load and high speed regions and it is assumed that the change in the refrigerant dissolution concentration has a great influence on the bearing characteristics. We have evaluated the influence of frictional heat generation, solubility and viscosity change on journal bearing characteristics by thermal fluid lubrication analysis (Sasaki, 2019). In this study, in order to verify the validity of THL analysis, we developed an experimental apparatus for bearings, measured bearing characteristics under the refrigerant environment and compared them with analysis results.

2. OUTLINE OF THE THERMOHYDRODYNAMIC LUBRICATION METHOD

Figure 1 shows a model of a journal bearing in THL analysis. The bearing load was a static load acting in the direction of $\theta = 180°$. It is assumed that in the steady operation of compressors, a certain amount of refrigerant is dissolved in the refrigeration oil supplied from the oil supply port. Oil film pressure and temperature are obtained under the conditions of flow rate continuity and energy conservation law of refrigeration oil flowing through minute elements of bearing clearance. Refrigerant solubility and dissolving viscosity obtained from oil film pressure and temperature are determined by functionalizing with appropriate approximate expressions. The distribution of oil film pressure is repeatedly calculated from the distribution of oil film temperature and dissolving viscosity, and convergence calculation is performed so that the integral value of oil film pressure balances with the bearing load. The prerequisites for the analysis model are as follows: (1) Provide adiabatic boundary conditions on the shaft and bearing surface. (2) Consider the separation heat associated with the change in refrigerant solubility of the refrigeration oil. (3) The temperature, refrigerant solubility and viscosity of the refrigeration oil supplied to the bearing are the assumed values in the environment in the compressor, and the refrigeration oil is supplied only from the oil supply port. (4) The boundary conditions at the bearing ends are given constant temperature and pressure according to the environment in the compressor.

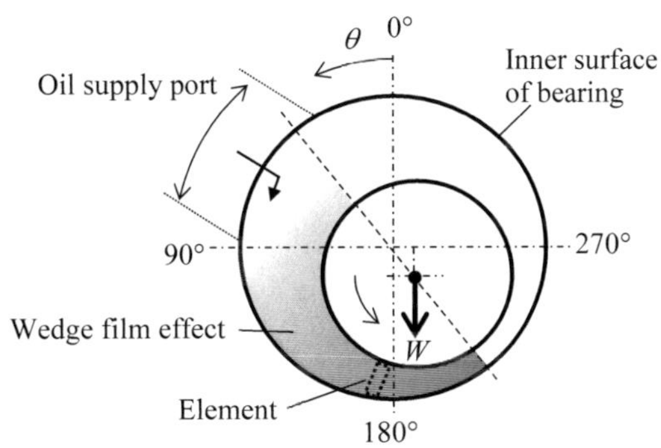

Figure 1: Model of journal bearing

3. EXPERIMENT

3.1 Apparatus

To verify the validity of the THL analysis, we have developed an experimental apparatus that can evaluate the characteristics of journal bearing in a refrigerant environment. Figure 2 shows experimental apparatus. The main motor and the test section were connected via a torque meter in the pressure vessel. The refrigeration oil in the oil tank is supplied to the test section via a flow meter by a pump motor installed at the bottom of the vessel. A constant-temperature circulator is installed in the oil tank to control the refrigeration oil to a constant temperature. The main motor and pump motor are controlled independently by the control board. The test bearing is placed in the middle of the support bearing that supports the test shaft, and the load is applied from the test bearing to the test shaft by pushing the bearing holder upward with the load mechanism via a load cell. An eddy current gap sensor is embedded inside the test shaft to measure the displacement of the bearing clearance (i.e., the oil film thickness). The measurement data is stored in the logger via the slip ring. The gap sensor was fixed near the shaft surface with bolts, filled around the sensor head with resin, and cured at high temperatures. Then, the shaft surface was polished so that the distance from the sensor head to the shaft surface becomes 0.2 mm, and a smooth cylindrical curved surface was formed. Figure 3 shows a bearing cross-sectional view. The port for oil film pressure measurement is represented by P, and that for bearing temperature measurement is represented by T. The bearing has 5 ports for each measurement. θ is defined from the upper part of the bearing to the rotational direction of the shaft. The angle of each measurement port is shown in parentheses. For the pressure measurement point, a through hole of ϕ0.8 mm was made in the bearing inner surface, and the pressure gauge was inserted from the back of the bearing to seal at the threaded part. The temperature measurement point was placed set back 1 mm from the bearing inner surface, and a ϕ1.6 mm K-type thermocouple was inserted.

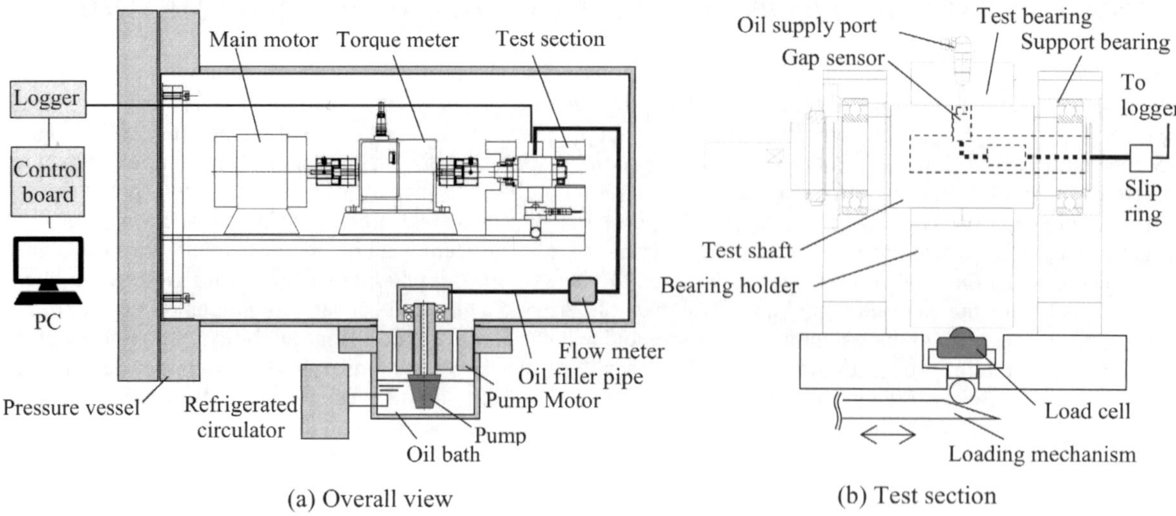

Figure 2: Experimental apparatus

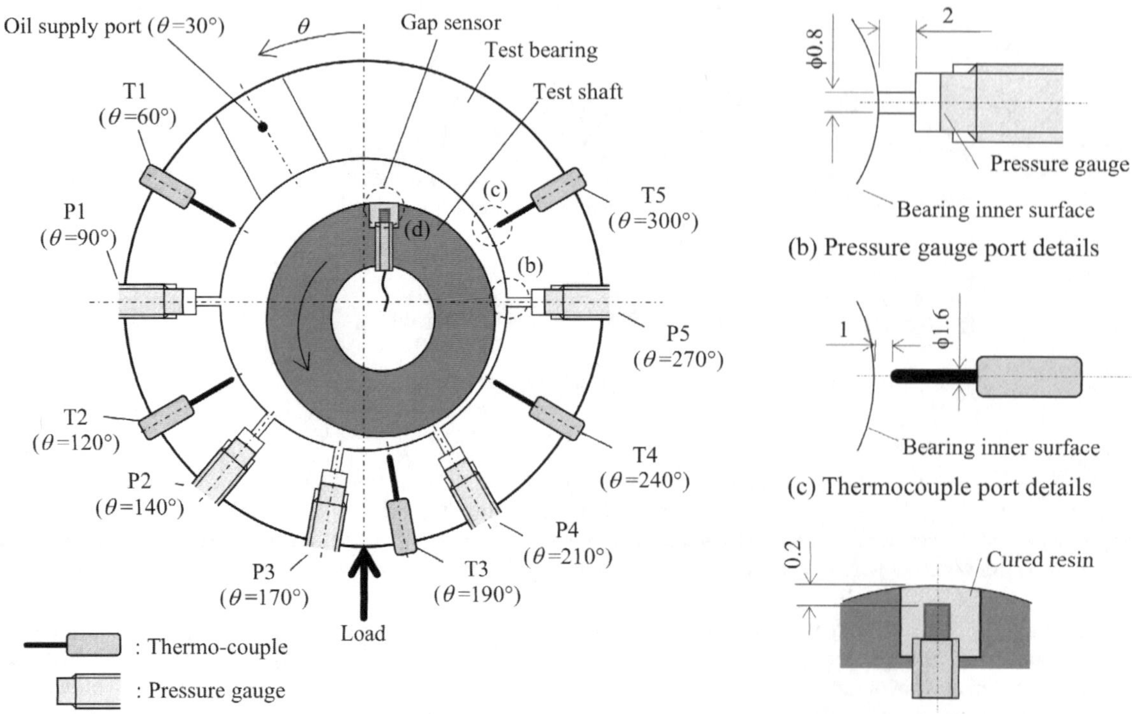

Figure 3: Bearing cross-sectional view

3.2 Test conditions

Tables 1 and 2 show the bearing specifications and experimental conditions, respectively. In the experiment, R410A is used as the refrigerant, and POE that is miscible with R410A is used as the refrigeration oil. The refrigerant was filled into the pressure vessel so that the pressure became 1 MPaG with the refrigerant sufficiently dissolved in the refrigeration oil. The inlet oil temperature was at 303 K. Bearing characteristics (oil film pressure, bearing temperature, friction coefficient, oil film thickness) are measured by changing the bearing pressure and circumferential velocity in the range shown in Table 2. The circumferential velocity are 2.5, 5.7, and 11.3 m / s, which are halfway values. The reason for this value is that the rotation speed [rps] was set to a nice round number and the circumferential velocity was converted from the rotation speed.

Table 1: Bearing specifications

Length to diameter ratio, L/D [-]		0.8
Radial clearance ratio, C/R [-]		1.2×10^{-3}
Oil supply port area [mm^2]		200
Material	Shaft	Steel
	Bearing	Cast iron
Surface roughness	Shaft	Ra0.2
	Bearing	Ra0.2

Table 2: Experimental conditions

Bearing pressure [MPa]	0.3–3.1
Circumferential velocity [m/s]	2.5, 5.7, 11.3
Refrigerant & refrigeration oil	R410A & POE
Ambient pressure [MPaG]	1.0
Inlet oil temperature [K]	303

4. CHARACTERISTICS OF EXPERIMENTAL APPARATUS

Preliminary tests were conducted in the atmosphere to understand the trends in the data obtained with the experimental apparatus. The test conditions are set to oil temperature 313 K, bearing pressure 2.4 MPa, circumferential velocity 11.3 m/s, and oil supply flow rate 4100 mm^3/s.

4.1 Output voltage of gap sensor

Figure 4 shows the output voltage waveform of the gap sensor. A periodic sine wave can be observed with the rotation of the shaft, which corresponds to the change of the oil film thickness. The cause for the voltage value overshooting every cycle is that the gap sensor passes through the oil supply port. The rotation angle of the shaft can be calculated from the overshoot occurrence timing and converted to the oil film thickness for each rotation angle. Under the constant bearing pressure and circumferential velocity, an arbitrary cycle data in the steady state is extracted, and the oil film thickness is calculated using the calibration curve of the output voltage and oil film thickness obtained in advance. Figure 5 shows the oil film thickness distribution in one rotation. The oil film thickness h was made dimensionless with a diameter clearance of $2C$. For the bearing angle range of 15-45°, the data was deleted because the gap sensor is in the phase of passing through the oil supply port. It was found that the experimental results agreed well with the analytical results and that appropriate results could be obtained.

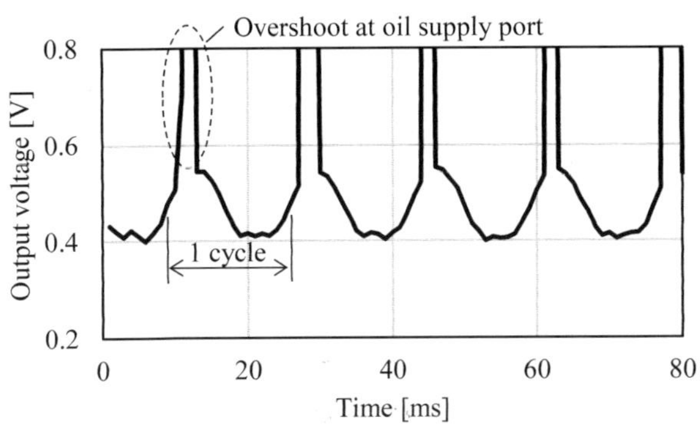

Figure 4: Output voltage of Gap sensor

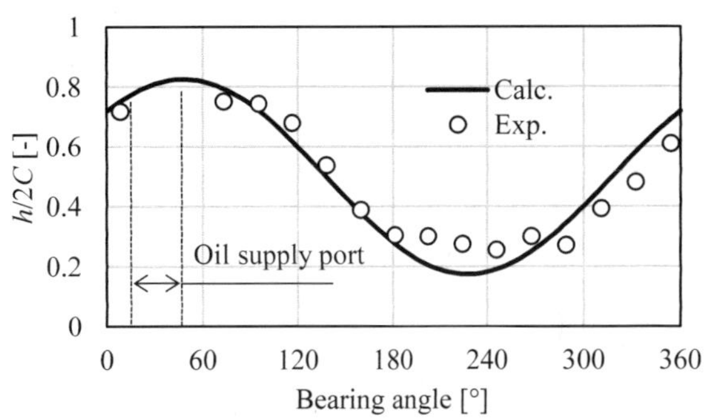

Figure 5: Oil film thickness

4.2 Oil film pressure and bearing temperature

Figure 6 shows oil film pressure distribution in one rotation. Two types of analysis results are also shown. One is the conventional lubrication analysis results of journal bearings assuming that the viscosity is constant, and the other is the THL analysis results considering the viscosity change caused by frictional heat. The experimental results agree well with the THL analysis results. Figure 7 shows the bearing temperature distribution. Here, as shown in Figure 3, the bearing temperature is the temperature at the point set back 1 mm in the radial direction from the bearing inner surface. On the other hand, the analysis result is the oil film temperature. The experimental results are closer to the THL analysis than the conventional analysis results assuming a constant viscosity. In the range of 0 to 240°, the experimental result becomes higher going downstream. This is because the oil film thickness gradually decreases and the shear stress increases. However, the error between the analysis and the experimental values is large in the reverse wedge region after 240°. In the experiment, it is considered that the temperature does not easily rise due to the effect of heat radiation to the bearing and bearing holder. Also, in the experiment, oil leaked from the end of the bearing when reaching the phase of minimum oil film thickness angle (around 240° in Figure 5) from the oil supply port at 30°. Therefore, after 240°, cavitation reduces the amount of shear heat. Since the analysis ignores the effect of cavitation, it is considered that the oil film temperature in the reverse wedge region is overestimated. It is necessary to make improvements in the future in the oil film temperature analysis method and the heat radiation from bearings in the test equipment. However, since the bearing temperature qualitatively agrees with the THL analysis results, in this study, the experiment was performed in a refrigerant environment using this experimental apparatus. In the following chapters, we will clarify the effect of viscosity change caused by frictional heat and refrigerant solubility on bearing characteristics.

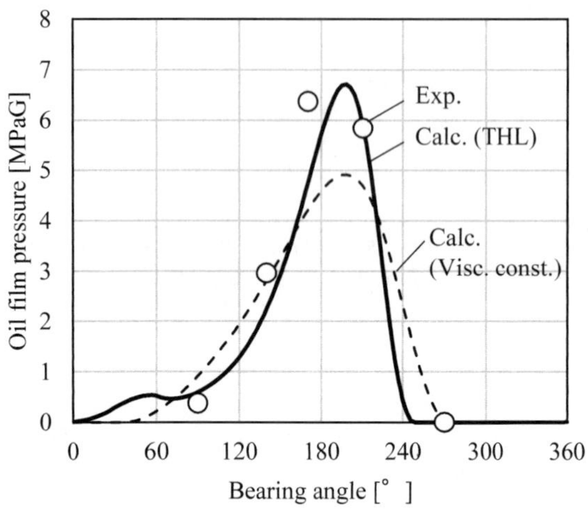

Figure 6: Oil film pressure

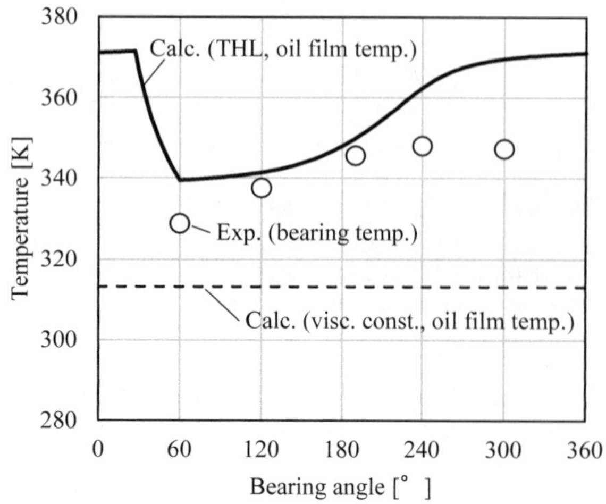

Figure 7: Bearing temperature

5. EXPERIMENTAL RESULTS

5.1 Effect of bearing load and circumferential velocity

The bearing characteristics (oil film pressure, oil film thickness, bearing temperature) were evaluated under the conditions shown in Table 2. Figure 8 shows the oil film pressure distribution in one rotation. The legend also shows the supply oil flow rate under each test condition. Here, the supply oil flow rate varies depending on the conditions, even though the rotational speed of the pump motor is constant. This is because the bearing clearance at the position of the oil supply port changes depending on the bearing pressure and the circumferential velocity. At any circumferential velocity, the maximum value of the oil film pressure increases with the increase of the bearing pressure. It can be seen that the experimental results agree well with the results of THL analysis, which considers the viscosity change caused by frictional heat and refrigerant solubility.

Bearing pressure [MPa]	Legend		Supply oil flow rate [mm³/s]		
	Exp.	Calc. (THL)	@2.5m/s	@5.7m/s	@11.3m/s
0.3	□	···········	1900	1900	2600
1.0	△	- - - -	2100		4000
1.4	○	———		3500	
1.7	■	▪▪▪▪▪	2100		3800
2.4	▲	— —	2100	3700	3100
3.1	●	━━━		3700	2600

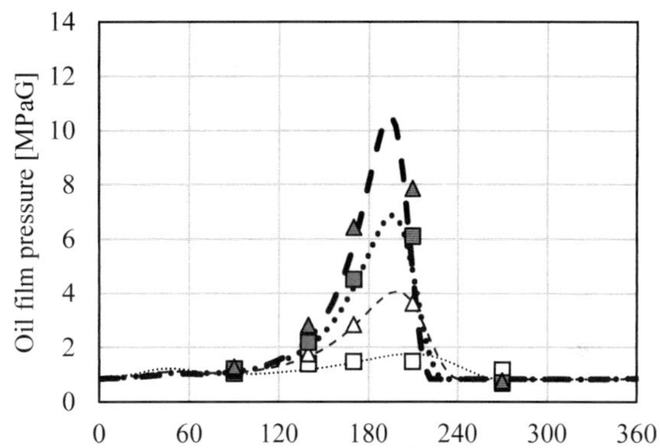

(a) 2.5 m/s

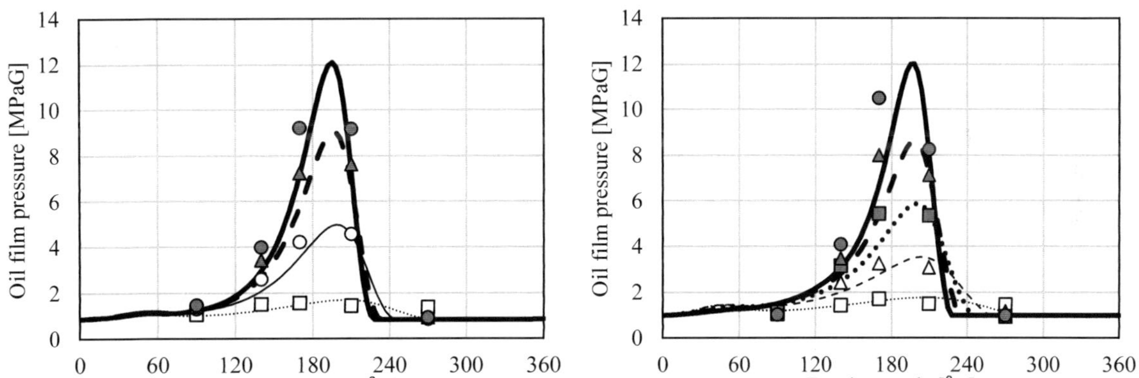

(b) 5.7 m/s

(c) 11.3 m/s

Figure 8: Oil film pressure

Figure 9 shows the oil film thickness distribution in one rotation. We could see experimentally how the minimum oil film thickness decreased with the increase of bearing pressure. The axis center coordinates were calculated from the oil film thickness for each bearing angle using least squares approximation, and converted to the journal center locus. Figure 10 shows the journal center locus. The eccentricity ratio increases as the circumferential velocity decreases. In addition, as the bearing pressure increases, the journal center captures an arc-shaped locus which is general bearing characteristics. Therefore, it is thought that a reasonable result was obtained as a verification of the analysis.

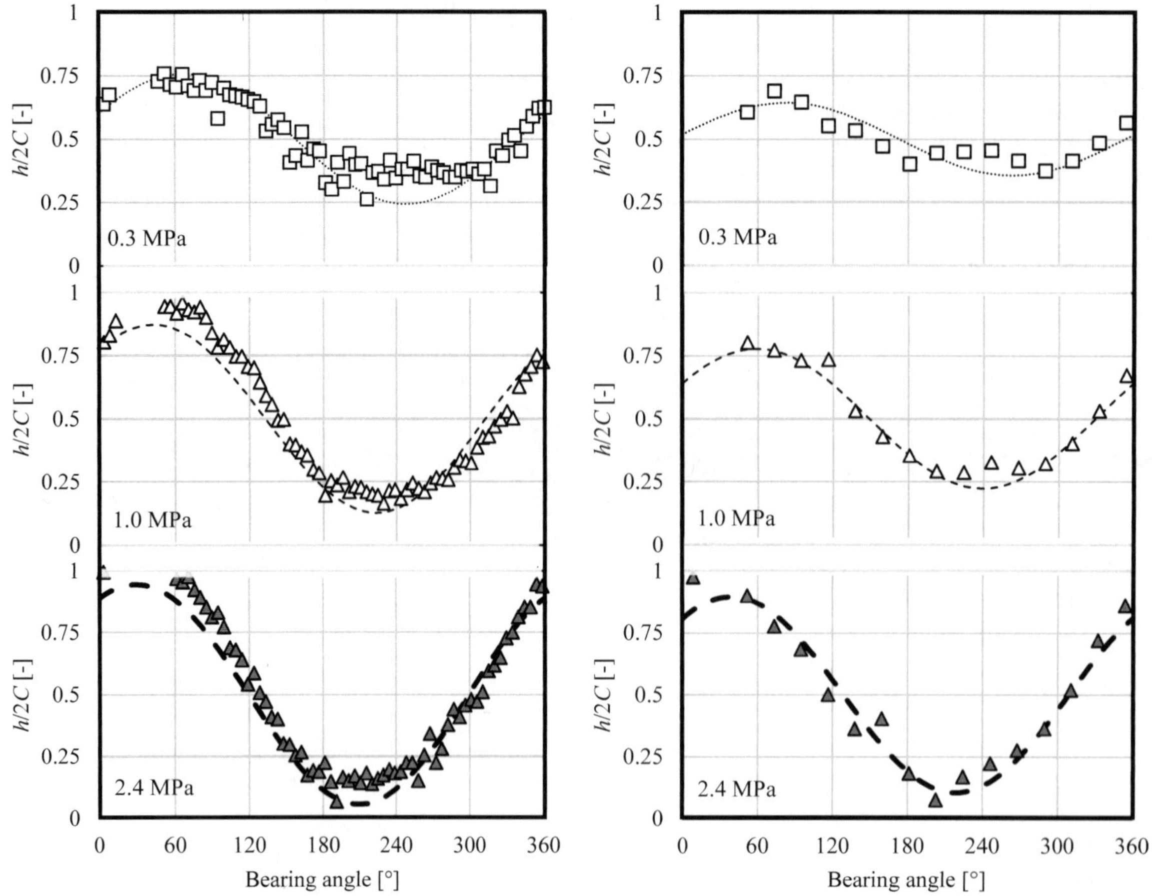

(a) 2.5 m/s

(b) 11.3 m/s

Figure 9: Oil film thickness

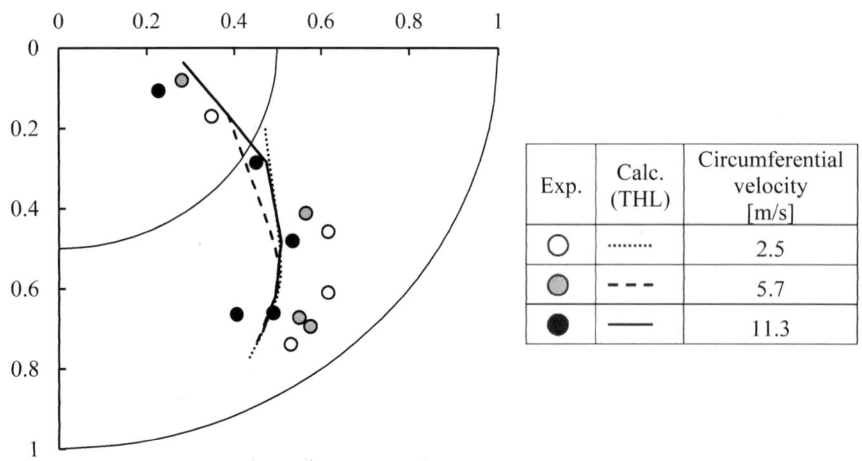

Figure 10: Journal center locus

Figure 11 shows bearing temperature distribution. As shown in Figure 7, the bearing temperature has an error from the analysis. Therefore, the effects of bearing pressure and circumferential velocity were evaluated using only the experimental results. The bearing temperature at a circumferential velocity of 2.5 m/s was almost the same regardless of the bearing pressure, and was constant in the circumferential direction of the bearing. When the circumferential velocity is 5.7 or 11.3 m/s, it can be confirmed that the bearing temperature increases from the oil supply port going downstream. This tendency becomes more prominent as the bearing pressure is larger. Also, as the bearing pressure increases, the bearing temperature decreases more downstream than the bearing angle of the minimum oil film thickness (see Figure 9). This has not been confirmed in atmospheric experiments (Figure 7) and is a phenomenon peculiar under the refrigerant environment. In the reverse wedge region, the pressure decreases rapidly and the refrigerant separates from the refrigeration oil, so it is considered that the separation heat generated at this time lowers the temperature of the refrigeration oil.

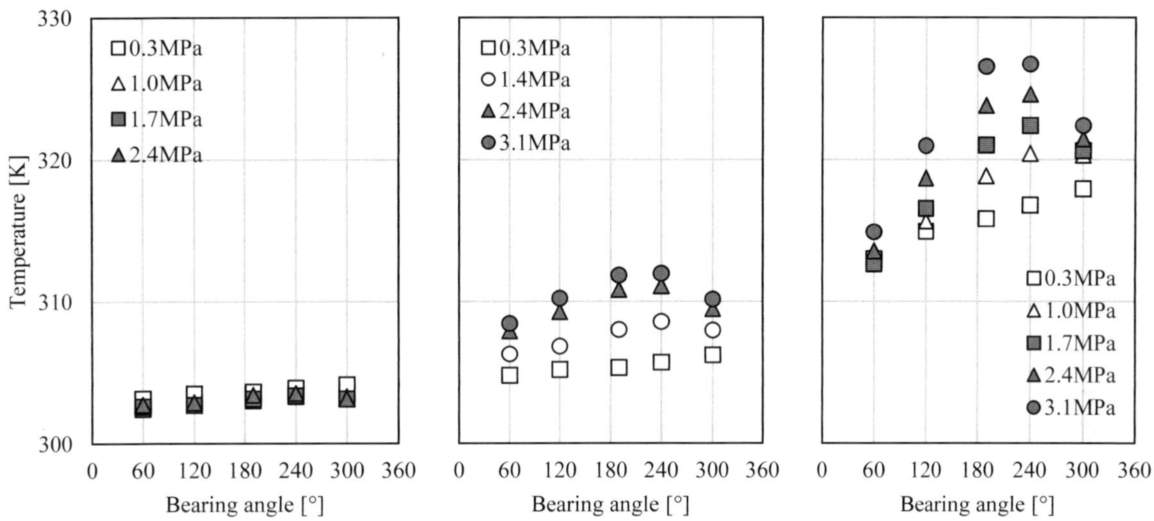

(a) 2.5 m/s (b) 5.7 m/s (c) 11.3 m/s

Figure 11: Bearing temperature

5.2 Relationship between Sommerfeld number and bearing characteristics

The friction coefficient and the eccentricity ratio are arranged relative to the Sommerfeld number. The friction coefficient is converted from the friction torque measured by a torque meter. As shown in Figure 2, the test section has two support bearings in addition to the test bearing, and the measured torque is the sum of these values. The friction torque is expressed by equation (1). Here, r, μ, and W are the journal radius, the friction coefficient, and the bearing load. The subscripts b and j indicate the support bearing and the test bearing, respectively. The load acting on the support bearing is expressed by equation (2). The friction coefficient is obtained by equation (3) from equation (1) and equation (2). Using a general value of 0.0015 for the friction coefficient of the support bearing (Hjertzen, 1973), the friction coefficient of the test bearing was calculated.

$$T = 2r_b\mu_b W_b + r_j\mu_j W \tag{1}$$

$$W_b = \frac{1}{2}W \tag{2}$$

$$\mu_j = \frac{T - r_b\mu_b W}{r_j W} \tag{3}$$

Figures 12 and 13 show the friction coefficient and the eccentricity ratio for the Sommerfeld number. For simplicity, the viscosity at the oil supply port was used to calculate the Sommerfeld number. As shown in Figure 12, the lower the circumferential velocity, the higher the friction coefficient when compared at the same Sommerfeld number. Considering that the oil film temperature is proportional to the bearing temperature in Figure 11, it is considered that the oil film temperature is lower and the viscosity is higher at lower circumferential velocity. As shown in Figure 13, the eccentricity ratio is high when the circumferential velocity is high for the same Sommerfeld number. It is considered that the viscosity decreases due to heat generation and dissolution of the refrigerant.

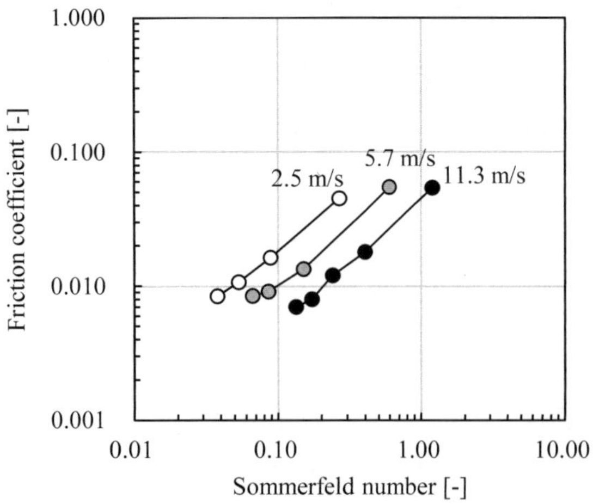

Figure 12: Friction coefficient

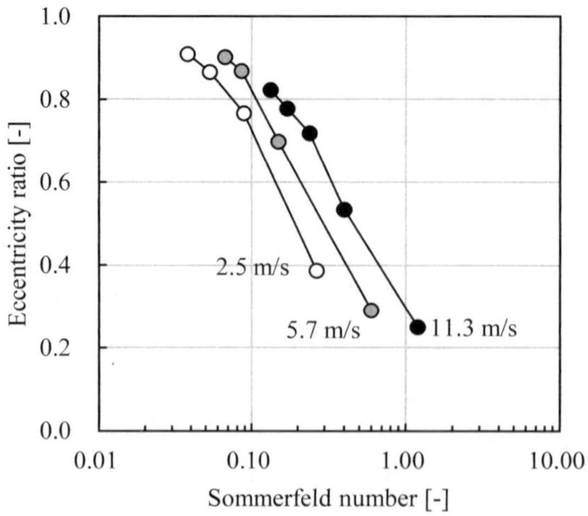

Figure 13: Eccentricity ratio

6. CONCLUSIONS

In order to verify the validity of the THL analysis in consideration of the viscosity change caused by the oil film temperature and refrigerant solubility changes in the bearing clearance for the journal bearing of compressors, we developed an experimental apparatus to evaluate the journal bearing characteristics (oil film thickness, oil film pressure, bearing temperature, friction coefficient) in a refrigerant environment. As a result of evaluating the basic characteristics of this experimental apparatus and the journal bearing characteristics in the combination of R 410A/POE, the following findings were obtained.

(1) The oil film pressure, bearing temperature, and oil film thickness of the journal bearing in the atmosphere agree better with the analysis results that take into account the viscosity change caused by frictional heat than the lubrication analysis results with a constant viscosity. It was clarified that viscosity change in the bearing clearance affected the bearing characteristics.
(2) The oil film pressure and oil film thickness in the refrigerant environment agreed well with the THL analysis, and it was clarified that the analysis method was valid.
(3) When the circumferential velocity is 5.7 m/s or more and the bearing pressure is 2.4 MPa or more in the refrigerant environment, the bearing temperature in the reverse wedge region decreases compared to the phase of the minimum oil film thickness. This is presumed to be due to the fact that the refrigerant dissolved in the refrigeration oil separates as a result of the decrease in the oil film pressure and takes heat from the refrigeration oil.
(4) The eccentricity ratio increases as the circumferential velocity increases under the same Sommerfeld number conditions in the refrigerant environment. It is considered that this is because, in the wedge region, the viscosity decreases due to dissolution of refrigerant in refrigeration oil and heat generation.

REFERENCES

Aritoshi, H., Tanaka, S., & Nakahara, T. (2009). Study on Variation of Refrigerant Concentration in Lubricating Film under High Pressure Atmosphere of CO2 as a Refrigerant. *Proceedings of World Tribology Congress 2009* (719). Kyoto, Japan: Japanese Society of Tribologists.

Hjertzen, D.G. (1973). Selection of rolling bearings. In M.J. Neale (Eds.), *Tribology Handbook* (A18). London, U.K.: Butterworths.

Sasaki, T., Ikeda, Y., Ishizono, F., Sato, K., & Sasaki, S. (2019). Study on Bearing Characteristics of Compressor Considering Refrigerant Dissolution and Friction Heat -1st Report: Analysis Modeling and Theory Construction-. *Transactions of the Japan Society of Refrigerating and Air Conditioning Engineers, 36(4)*, 199.

Evaluating Compressor Performance with Low-GWP Refrigerants by Simulation Model

Erika WADA[1]*, Tao LYU[1], Kenji TOJO[1], Yonezou IKUMI[1], Hiroo NAKAMURA[1], Kiyoshi SAITO[1]

[1]Waseda University, Department of Fundamental Science,
3-4-1 Okubo Shinjuku-ku, Tokyo, Japan
(Phone, Fax:81-03-5286-3259, E-mail:erwd@asagi.waseda.jp)

* Corresponding Author

ABSTRACT

Nowadays, as social awareness toward global warming grows, it is required to switch to lower global warming potential refrigerants in air-conditioning and refrigeration systems. Therefore, several new refrigerants are developed, and we must establish a performance evaluation method to select appropriate refrigerants. In this study, we developed a high precision scroll compressor simulation model while considering leakages and heat transfer between the compression chambers of gas and lubricant, evaluated compression characteristic, and compared R410A, R32, and R466A. We also conducted heat pump cycle simulation and considered the relationship between efficiency and refrigerant properties.

1. INTRODUCTION

In recent years, as worldwide interest toward global warming has grown and the Paris Agreement and the Kigali amendment regulate conventional refrigerants, it is required to develop next-generation refrigerants with limited environmental impact. Therefore, it is needed to develop a reliable heat pump cycle simulator for evaluating various refrigerants to select appropriate next-generation refrigerants quickly. Especially, compressor has a great influence for heat pump cycle performance, so appropriate modeling of the compressor is of critical importance.

In this study, we developed a compressor model considering leakages and heat transfer between the compression chambers of gas and lubricant on a scroll compressor widely used in the field of refrigeration and air conditioning. We conducted compressor simulation based on the model (Tojo *et al.*, 1986) (Suefuji *et al.*, 1993), evaluated compressor performance, and visualized leakage and heat transfer between adjacent compression chambers. Furthermore, we clarified heat pump cycle, performance. Three refrigerants are targeted; R410A, which was used in conventional air conditioning, R32, which is currently the mainstream, and R466A, which is attracting attention as a next-generation refrigerant. Table 1 shows main characteristics and theoretical COP of each refrigerant obtained under the conditions shown in Table 2.

Table 1: Properties of refrigerant

Refrigerant	R410A	R32	R466A
Composition (Mass %)	R32/R125=50/50wt%	R32=100wt%	R32/R125/CF3I =49/11.5/39.5wt%
GWP(AR4)	2090	675	733
Safety classification	A1	A2L	A1
Specific heat ratio	1.38	1.49	1.43
Density($kg \cdot m^{-3}$)	38.5	27.8	45.8
Cooling capacity($kJ \cdot kg^{-1}$)	165	250	149
Cooling capacity($kJ \cdot m^{-3}$)	6374	6953	6837
Theoretical COP	6.65	6.81	6.71

Table 2: Condition of temperature

Super heat	℃	5
Sub cool	℃	3
Evaporating temperature	℃	8.5
Condensing temperature	℃	42

2. COMPRESSOR MODEL

A positive displacement compressor for refrigerants draws refrigerant gas into the compression chamber, reduces its volume by mechanical power, and discharges it. Scroll compressor geometry is obtained from geometrical relation of the involute spiral (Tojo *et al.*, 1986). In the process of compressing gas, the state of gas in the compression chamber continuously changes in time. As shown in Fig. 1, there are work with volume change, leakages between adjacent chambers with different pressures, heat transfers between gas, wall surface, and lubricant in a compression chamber. In addition, gas and lubricant do not change in phase, and the specific volume of lubricant does not change with pressure. Also, lubricant behaves like a mist contained in the gas, and that there is no heat transfer between lubricant and wall surface.

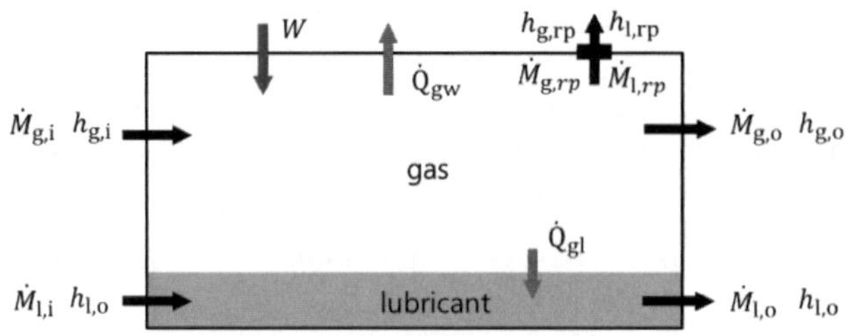

Figure 1: Compressor chamber model

2.1 Fundamental Equations

The energy and mass conservation of the gas and lubricant in the compression chamber are expressed below.

$$\frac{dM_g u_g}{dt} = \dot{M}_{g,i} h_{g,i} - \left(\dot{M}_{g,out} + \dot{M}_{g,rp}\right) h_{g,o} + W - \dot{Q}_{gw} - \dot{Q}_{gl} \tag{1}$$

$$\frac{dM_l u_l}{dt} = \dot{M}_{l,i} h_{l,i} - \left(\dot{M}_{l,o} + \dot{M}_{l,rp}\right) h_{l,o} + \dot{Q}_{gl} \tag{2}$$

$$\frac{dM_g}{dt} = \dot{M}_{g,i} - \dot{M}_{g,o} - \dot{M}_{g,rp} \tag{3}$$

$$\frac{dM_l}{dt} = \dot{M}_{l,i} - \dot{M}_{l,o} - \dot{M}_{l,rp} \tag{4}$$

2.2 Leakage

The area of the clearance is obtained from the model of Tojo *et al.* (1986). The amount of leakage is calculated by using the apparent specific heat ratio and gas temperature, assuming thermal equilibrium and the homogeneity of the two-phase flow of gas and lubricant, as in the model of Fujiwara *et al.* (1985). For the mixture of refrigerant and lubricant, the lubricant is always included in the mass fraction φ. The homogenous mass flow rate of leaked gas and lubricant mixture flowing through the gap of the high-pressure side (Chamber 1) to the low-pressure side (Chamber 2) is calculated by using Equation (5), which is the mass flow rate of a compressible fluid passing through an orifice. Apparent specific heat ratio of the mixture is obtained by Equation (7).

$$\dot{M} = \begin{cases} cA\sqrt{\dfrac{2\kappa_{\mathrm{gl}}}{\kappa_{\mathrm{gl}}-1}P_1\rho_1\left\{\left(\dfrac{P_2}{P_1}\right)^{\frac{2}{\kappa_{\mathrm{gl}}}} - \left(\dfrac{P_2}{P_1}\right)^{\frac{\kappa_{\mathrm{gl}}+1}{\kappa_{\mathrm{gl}}}}\right\}} & (P_2 > P_c) \\ cA\sqrt{P_1\rho_1\left(\dfrac{2}{\kappa_{\mathrm{gl}}+1}\right)^{\frac{\kappa_{\mathrm{gl}}+1}{\kappa_{\mathrm{gl}}-1}}} & (P_2 \le P_c) \end{cases} \tag{5}$$

$$P_c = P_1\left(\frac{2}{\kappa_{\mathrm{gl}}+1}\right)^{\frac{\kappa_{\mathrm{gl}}}{\kappa_{\mathrm{gl}}-1}} \tag{6}$$

$$\kappa_{\mathrm{gl}} = \frac{C_p M_g + C_l M_l}{C_v M_g + C_l M_l} \tag{7}$$

2.3 Heat Transfer

Heat transfer area between the gas and the wall is equivalent to the inner surface area of the compression chamber, which is the sum of top, bottom, and side surfaces can be determined by the geometric parameters. Heat transfer coefficients between the refrigerant and the wall surface are calculated by the Nusselt number using the equation for forced convection heat transfer in a circular tube (Kays, 1980). The temperature of lubricant assumed to be same as the temperature of gas exchanging heat instantly.

$$\dot{Q}_{\mathrm{gw}} = h_{\mathrm{gw}} S_{\mathrm{gw}} (T_g - T_w) \tag{8}$$

$$\mathrm{Nu} = C\,\mathrm{Re}^m\,\mathrm{Pr}^n (C = 0.022, m = 0.8, n = 0.5) \tag{9}$$

3. CALCULATING METHOD

3.1 Compressor Simulation

Based on the model in the previous section, compressor simulations are performed using R410A, R32, and R466A refrigerants, assuming a scroll compressor for a residential R410A air-conditioning system. The refrigerant state quantities from suction to discharge are calculated for each angle on MATLAB using the Runge-Kutta method along with the boundary conditions at the inlet and outlet of the compressor. Under the conditions shown in Table 3, volumetric efficiency, adiabatic efficiency, suction completed temperature, and discharge temperature are calculated by changing the condensing temperature. When the condensing temperature is 42°C, the change in the amount of leakages between adjacent chambers and heat transfer from compression chamber wall to refrigerant gas are visualized. Refrigerants properties are obtained from REFPROP10.

Table 3: Specifications of compressor simulation

Displacement volume	cc	8
Rotational speed	rps	50
Evaporation temperature (dew point)	℃	8.5
Super heat	℃	5
Lubricant circulating ratio	mass%	2.0

3.2 Heap Pump Cycle Simulation

We evaluated performance of heat pump cycle by assuming operation of air conditioner for room under the conditions shown in Table 4 based on the compressor efficiency obtained by the compressor simulation. The volumetric efficiency and adiabatic efficiency of the compressor are calculated by means of a regression analysis with the compression ratio as variable, and are calculated in the compressor module during the cycle calculation. Energy Flow + M (Ohno, 2012) simulation platform is used, which enables convenient system analysis even when the refrigerant or refrigerant circuit is changed, based on the modular analysis theory. The value of mass charge and displacement volume is shown in Table 5. Mass charge is determined to maximize COP, and the displacement volume is determined so that cooling capacity is consistent with the value shown in Table 4.

Table 4: Specifications of heat pump cycle simulation

Cooling capacity		kW	2.8
Compressor	Rotational speed	rps	50
Indoor fan	Power consumption	W	20
Outdoor fan		W	40
Indoor temperature		℃	27
Outdoor temperature		℃	35

Table 5: Condition of calculation

Refrigerant	Mass charge	Displacement volume
	kg	cc
R410A	0.87	8.00
R32	0.66	7.50
R466A	0.92	7.95

4. RESULTS AND CONSIDERARION

4.1 Compressor Simulation

Figure 2 shows the volumetric efficiency, adiabatic efficiency, suction completed temperature, and the discharge temperature when the condensing temperature is changed. The horizontal axis is compression ratio, which is the ratio of the discharge pressure to the suction pressure. Volumetric efficiency increases as the compression ratio decrease, and adiabatic efficiency reaches its maximum when the compression ratio is from 2.5 to 2.7. Suction completed temperature and discharge temperature increase as the compression ratio increases. When using R32, volumetric efficiency is 2-4% lower and adiabatic efficiency is 1-2% lower than that of R410A, and the suction completed temperature is 3-10℃ higher and the discharge temperature is 10-34℃ higher than that of R410A. When using R466A, adiabatic efficiency is 2% higher than that of R410A in high compression ratio region.

The reason why adiabatic efficiency has the peak is thought to be decrease in the indicated efficiency. Adiabatic efficiency is the product of volumetric efficiency and the indicated efficiency. When the compression ratio decreases, the indicated efficiency decreases due to over-compression. When the compression ratio increase, indicated work increases due to reverse flow from the discharge side and re-expansion because the pressure has not increased sufficiently when the compression ends and starts to discharge.

Figure 3 shows the effect of leakage in the axial gap when the condensing temperature is 42 ° C. The vertical axis is the ratio of the leakage mass flow rate to the suction completed mass in the chamber, and the horizontal axis is the rotation from the start of the suction process. It can be seen that R32 is greatly affected by such leakage. Figure 4 shows the effect of heat transfer from compression chamber at the same condition. The vertical axis is the ratio of the heat transfer rate to the heat capacity of the suction completed mass. It can be also seen that R32 is greatly affected by heat transfer.

As reported in Table 1, the reason why the compressor efficiency decreases when using R32 is considered to be the large specific heat ratio of R32. Since the specific heat ratio is large, rises of temperature and pressure are large on the same compression volume ratio and the temperature and pressure in the compression chamber are higher than those of other refrigerants. Therefore, mass flow rate of the leakage increase, the amount of heat transfer increase, and volumetric efficiency decrease.

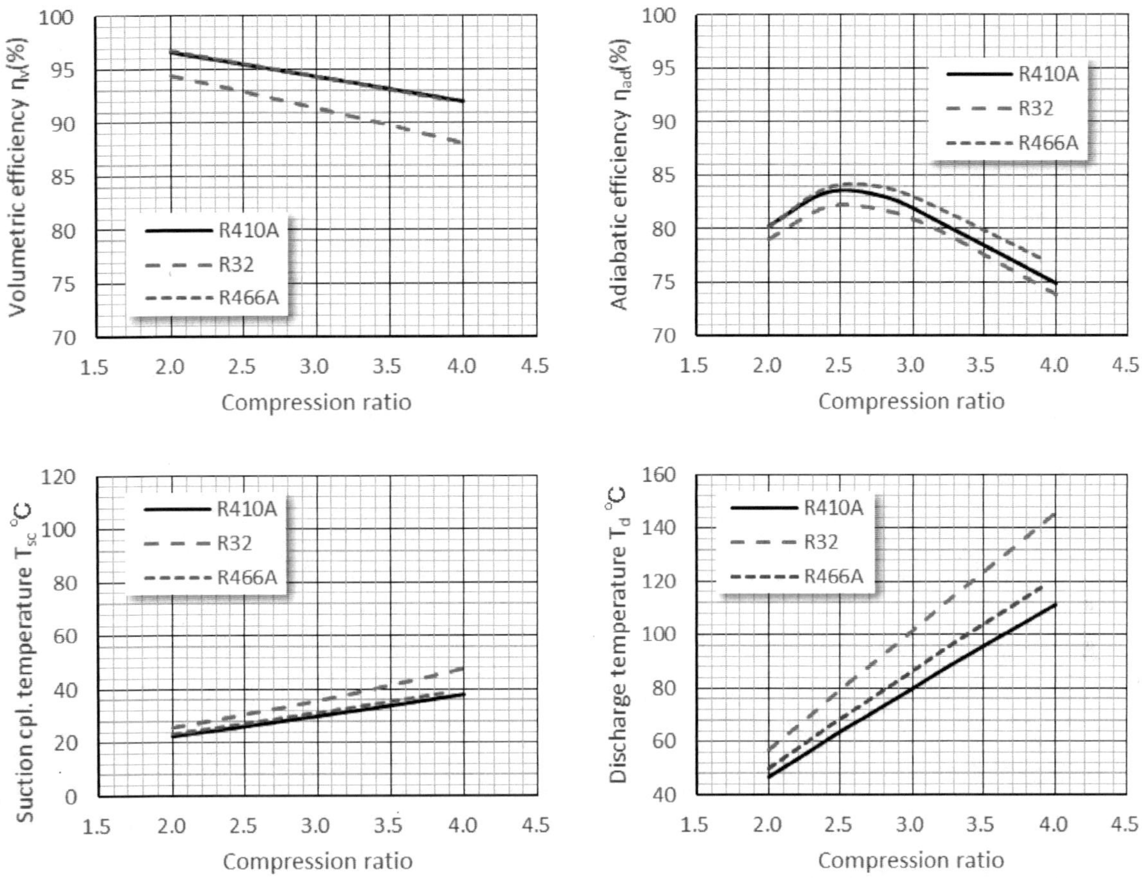

Figure 2: Results of compressor simulation

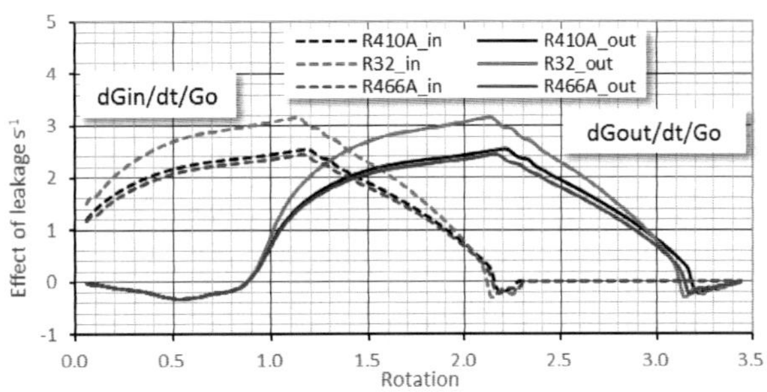

Figure 3: Leakage flow rate through axial wrap clearance

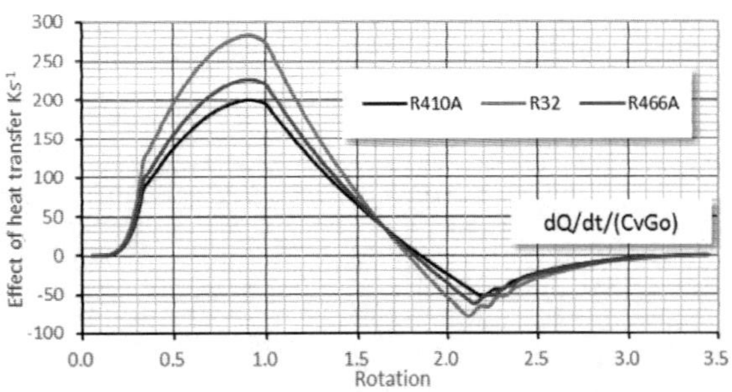

Figure 4: Heat transfer from compression chamber wall

4.2 Heap Pump Cycle Simulation

Figure 5 shows COP and cooling capacity when the outdoor temperature is changed. It can be seen that both COP and cooling capacity decrease as the outdoor temperature increases. COP of R32 is 4-5% higher and that of R466A is 2% lower than R410A.

Both volumetric efficiency and adiabatic efficiency of R32 are low in the compressor simulation, but COP is high in the cycle simulation. This can be explained by considering cooling effect and density of R32. Mass flow rate of refrigerant is small because cooling effect is large, and density is low, so the pressure loss in the evaporator and the suction connection pipe is small. Consumption power of the compressor is small, and finally COP is higher for the case of R32.

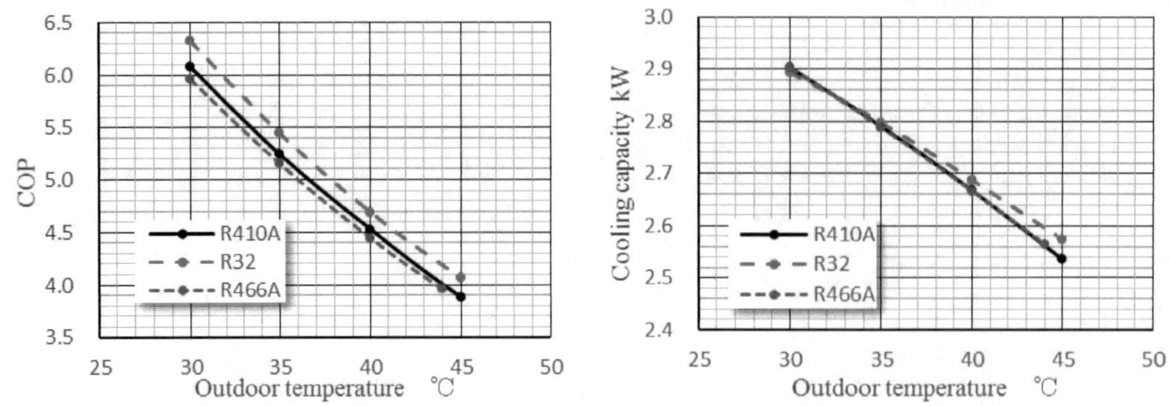

Figure 5: Results of heat pump cycle simulation

5. CONCLUSIONS

In this study, the following conclusions were obtained by the compressor simulation and the heat pump cycle simulation using three types of refrigerants (R410A, R32, R466A) for the case of room air conditioner.

1. As a result of the compressor simulation, volumetric efficiency of R32 is 2-4% lower and adiabatic efficiency is 1-2% lower than those of R410A. On the other hand, volume efficiency of R466A was almost the same as that of R410A, and adiabatic efficiency was 0-2% higher.

2. As a result of the compressor simulation, effect of leakage and heat transfer are increase when R32 is used.

3. As a result of the heat pump cycle simulation, COP of R32 is 4-5% higher than R410A, and that of R466A is 2% lower than R410A.

NOMENCLATURE

A	area of clearance	(m^2)
C	specific heat	($kJ \cdot kg^{-1} \cdot K^{-1}$)
c	flow coefficient	(-)
h	specific enthalpy	($kJ \cdot kg^{-1}$)
M	mass	(kg)
P	pressure	(Pa)
Q	heat transfer	(kJ)
S	surface	(m^2)
T	temperature	(K)
t	time	(s)
u	specific internal energy	($kJ \cdot kg^{-1}$)
W	work	(kJ)

Pr	Prandtl number	(-)
Nu	Nusselt number	(-)
Re	Reynolds number	(-)
η	efficiency	(-)
ρ	density	($kg \cdot m^{-3}$)
κ	specific heat ratio	(-)
COP	coefficient of performance	(-)

Subscript

g	gas
l	lubricant
w	wall
gl	mixture of gas and lubricant
p	pressure
v	volume
i	in
o	out
rp	relief port
c	choke
sc	suction completed
d	discharge

REFERENCES

[1] Tojo K., Ikegawa M., Maeda N., Machida S., Shiibayashi M., Uchikawa N., "Computer Modeling of Scroll Compressor with Self-Adjusting Back Pressure Mechanism", Proc of the 8th Int. Comp. Eng. Conf. at Purdue (1986).

[2] Fujiwara M., Kasuya K., Matsunaga T., Watanabe M., "Performance Analysis of Screw Compressor", Trans. Jpn. Soc. Mech. Eng., Vol.51, No.466, pp1816-1824 (1985). (in Japanese)

[3] W. M. Kays and M. E. Crawford, Convective Heat and Mass Transfer, McGraw-Hill, (1980), pp. 238-248.

[4] K. Ohno, K. Saito, Proceedings of the 46th Japanese Joint Conference on Air-conditioning and Refrigeration, 119 (2012).

[5] Compressors for Air Conditioning and Refrigeration, JSRAE, (2013), pp. 1-33, 83-98, 187-202.

[6] Suefuji K., Shiibayashi M., Tojo K., "Performance Analysis of Hermetic Scroll Compressor", Trans. JAR, Vol.10, No2, pp.319-326 (1993). (in Japanese)

[7] Tojo K., "Evolution and Future Prospects of Scroll Compressor Technology", Trans. JAR, Vol.37, No2, pp.123-144 (2020). (in Japanese)

[8] Chen Y., Groll E. A., Halm N. P., Braun J. E., "A Comprehensive Model of Scroll Compressor", Int. Journal of Refrigeration., Vol.25, PP.731-750 (2002).

[9] Itoh T., Fujitani M., Sakai Y., "Leakage Characteristics of Scroll Compressor by Two Phase Flow Model in Consideration of Wall Oil Film Thickness", Trans. Jpn. Soc. Mech. Eng., Vol.68, No.671 (2002). (in Japanese)

[10] Kawamura R., Iwatake W., Shimizu M., Tatsuwaki K., "Analysis of Thermal Fluid Behavior during Compression Process of Scroll Compressor", Mitsubishi Denki Gihou, 93(6), pp.27-30 (2019). (in Japanese)

ACKNOWLEDGEMENT

This result was obtained as a result of the commissioned work (JPNP18005) of the New Energy and Industrial Technology Development Organization (NEDO). We would like to express our gratitude here.

Fluid-structure Coupled Analysis on Clearance of Oil-free Twin Screw Blowers

Yiqin Zhang[1], Bingqi Wang[1], Shizhong Sun[1], Xi Pan[1], Zhilong He[1], Ziwen Xing [1]*

[1]School of Energy and Power Engineering, Xi'an Jiaotong University Xi'an, Shaanxi, China
(113709188502, zwxing@mail.xjtu.edu.cn)

* Corresponding Author

ABSTRACT

With the development of modern industry, the requirement of high-pressure air without oil is becoming even more strict in many fields, such as food industry, sewage disposal, electroplating, etc. Due to the wide range of pressure and airflow with high efficiency, oil-free twin screw blowers become more favored in industrial applications, compared with centrifugal blowers and lobe blowers. For oil-free twin screw air blowers, clearance is a key factor affecting the performance. In practical applications, due to the difference between suction and discharge temperature, inhomogeneous thermal expansion can cause the rotors to rub or even get stuck. How to accurately design the clearance is an urgent problem to be solved. This paper presents a three-dimensional CFD simulation model of an oil-free twin screw air blower, in which gas flow with turbulence effects, heat transfer and leakage are taken into account. Pressure ratio varied from 1.5 to 2.2 with the flow rate from 44.11 m3/min to 40.53 m3/min when the rotation speed of the male rotor kept 4000rpm. The performance of the blower, such as power, volumetric efficiency and isentropic efficiency and the deformation of rotors and casing under unsteady temperature and pressure field are calculated. The results of fluid-structure coupled analysis can provide guidance on the design and optimization of the twin screw blower.

1. INTRODUCTION

The positive displacement (PD) blowers have a lots of application in different fields. Specifically, the blowers with the high pressure oxygen supplying have an important position in fuel cell system[10]. Hsieh and Zhou[5] showed the comparison of the flow rates and pressure between cylindrical and screw type Roots blowers, then the result illustrated that the fluctuation of the screw curve pressure is lower. Arsie et al[1] presented a high efficiency screw blowers with a large range of pressure ratio. In addition, He and Wei[4,12] showed that twin screw blowers have well stability and high performance including all appropriate PD machines.

The basic structure of twin screw blowers is two screw-shaped rotors installed on parallel shafts, cylinder that enclose the rotors and end covers that close cylinder. The space made up of these three is divided into suction chamber, compression chamber and exhaust chamber. The fluid is sucked from inlet area and then compressed in compression area and discharged to outlet area finally when the rotors rotate in the opposite direction. In this process, there are leakage and heat transfer which affect the performance of oil-free screw blower. With the development of screw blower, it is found that the most important factor affecting the efficiency is leakage[2,3,11]。 To improve the performance of the twin screw blowers and explore the clearance flow in the internal flow field, many studies of twin screw blower focused on the leakage flows through clearance gaps both experimentally and numerically. Kovacevic et al[7]have presented two approaches to solve the simulation of an oil free air screw compressor with 'N' rotor profile of 3/5 lobe combination in Computational Fluid Dynamics (CFD) software. In addition, the calculations of two CFD models which include pressure in the compressor chamber, mass flow rate, indicated power and the volumetric efficiency are validated with the experimental measurement of the prototype compressor. Spille-Kohoff et al[9] have reported the simulation of a dry screw compressor whose roots mesh is meshed by an appropriative hexahedral meshing program ——Twinmesh and show time-resolved results for torques, forces, interlobe pressure, mass flow, and heat flow.

Sun et al[8] used SCORG to generate the dynamic mesh to simulate a roots blower and obtain the velocity field to demonstrate the leakage flow in tip gap and lobe gap. And the results were validated with the experimental investigation by using phase-locked particle image velocimetry. He et al[6] have presented a new method to replace

the dynamic mesh approach in the CFD calculation and analysis of twin screw compressor and through the use of new methods, the gap between the model can be reduced to millimeter level or even zero so that the simulation accuracy and reliability of the analysis results are effectively improved.
However, most researches aim at CFD simulation of the fluid domain, CFD simulation of fluid-structure coupling for screw blower is rarely studied. This paper presents a mathematical model of an oil-free twin screw air blower, which consists of fluid domain and rotor domain. After the coupling calculation of the fluid domain and the rotor domain, this paper shows the results, such as unsteady temperature field, unsteady pressure field, and unsteady velocity vector field of fluid domain. The study provides strategies to the design of twin screw blowers and can predict thermal deformation of the rotors.

2. THEORETICAL MODEL

An oil-free twin screw compressor named LG180 is analyzed in this study. In this section, the geometry model and mesh generation and numerical set-up are reported.

2.1 The Geometry of twin screw compressors

The 3D model which were analyzed by ANSYS CFX included a pair of "4*6" rotors, the cavity between the casing and the rotors, the suction and discharge ports. The gaps between the rotors and the casing is 0.15mm and between two rotors were 0.2mm. Besides, the axial suction gap was 1mm and the axial discharge gap was 0.15mm. Fig.1a shows a pair of "4*6" rotors whose length is 480mm. The suction and discharge ports only had the radial access, as seen in Fig.1b.

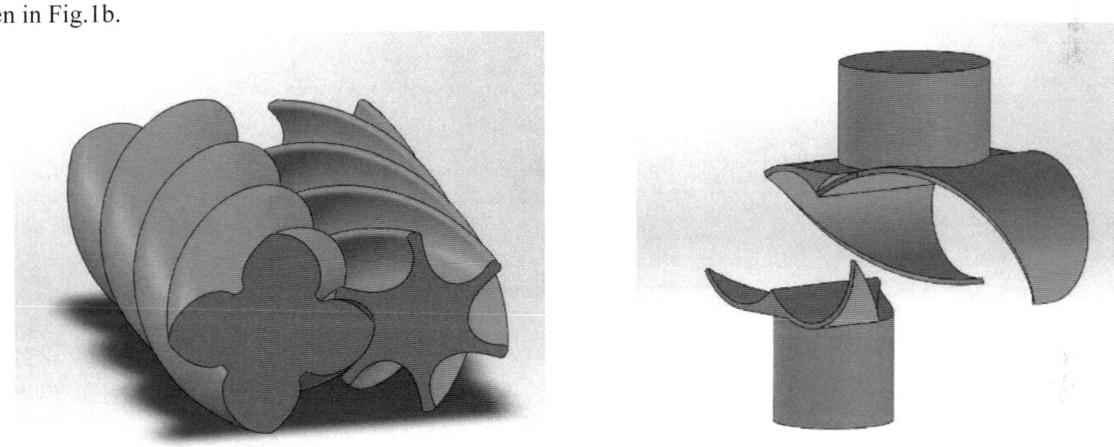

a) Geometry of rotors b) Geometry of suction and discharge ports

Figure1: Geometry of rotors and stationary ports

2.2 Mesh Generation

The significant problem in screw compressor simulation is that all simulation processes involved need to consider huge difference of length scales starting a fraction of a millimeter in the gaps up to hundreds of millimeter in the working chamber. The deform-and-remeshing method or dynamic method was used to address the issue. However, the mesh quality of both methods is usually bad because of the high rotation speed of rotors.
To prove the quality of meshes, customized grid generation was used. For each time step, the method generates a separate set of grids which improve the mesh quality widely. In this project, Twinmesh was used to generate the customized mesh where is the fluid between compression chamber and rotors, as Fig.2a showed. Fig.2b and Fig.2c shows the static solid rotors mesh and fluid suction and discharge ports regions, respectively. Meshes of the interface between radial ports and cavity was encrypted. The customized meshes consist of 5,000,000 elements. All meshes added up to about 6,300,000 elements.

a) Grids of deforming domain

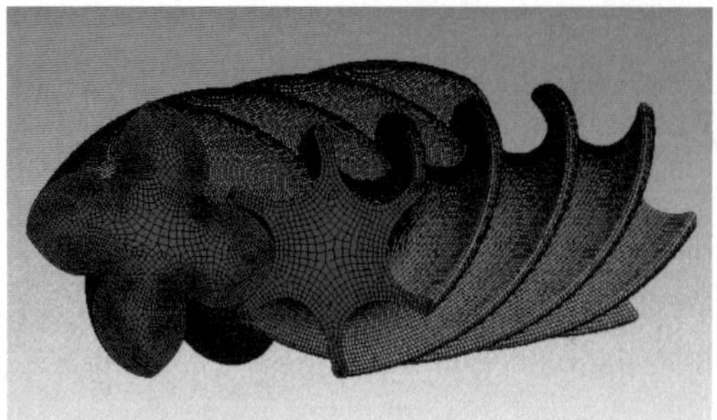

b) Grids of rotors

C) Grids of suction and discharge ports

Figure2: Meshes of simulation

2.3 Simulation Setup

The whole fluid meshes are imported into CFX by using Twinmesh, and then the solid rotor meshes are imported into CFX individually, and the solid rotor domains and fluid-solid interfaces should be manually set. In these simulations, in order to calculate the heat conduction of the rotor, the interface condition is set as conservative interface flux. The Shear Stress Transport (SST) model was used, which unites the advantages of k-w model and k-e model and converges well in the near wall and the main flow direction. Transient term discrete method of the governing equations was the implicit time step -- a second order backward Euler. The rotation speed of rotors is set as 4000 rpm. The ideal gas model is adopted for the gas and the rotor is adopted for the aluminum.
The boundary conditions used in the numerical simulation are reported in Table 1.

Table 1 Boundary conditions for the numerical simulation.

Quanlity	**Inlet**	**Outlet**
Pressure [bar]	1.	1.8
Temperature [K]	293	No Gradient
Turbulent kinetic energy	5%	No Gradient

3. Simulation results

3.1 Mass Flow

Fig. 3 shows the transformation of the inlet and outlet flow with the Angle. It can be found that the flow rate changes periodically with the turning Angle, and the fluctuation of the flow rate at the discharge ports is greater than that at the suction ports. At a 360Angle, the outlet flow fluctuates 4 times, because the male rotor has 4 lobes. Moreover, the simulated flow rate was 42.37 m3/min, which was consistent with the experimental results.

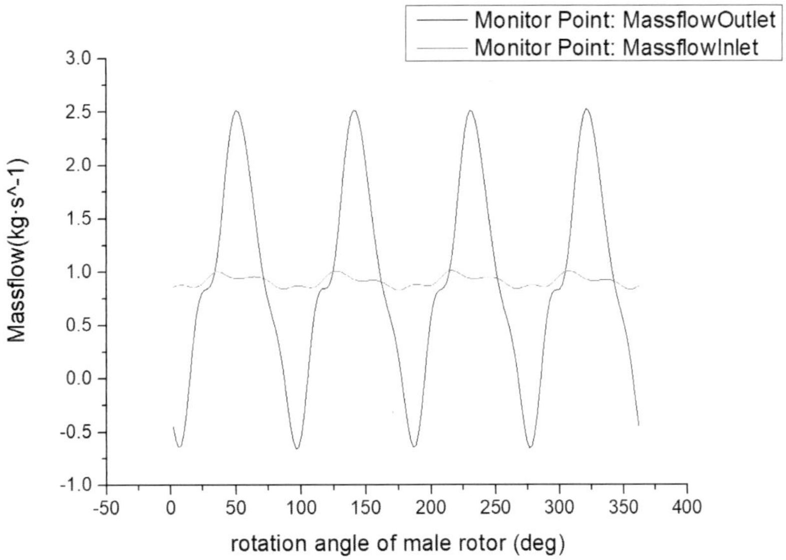

Figure3: Mass flow variation with rotation angle

3.2 Pressure field

Fig.4a shows the pressure cloud diagram when the pressure ratio is 1.8 and the angle of the male rotor is 180°. The entire chamber is divided into suction, compression, and discharge regions according to pressure. It can be found that once the cavity is disconnected from the suction end or male and female rotors begin to mesh, the pressure in the cavity increases gradually with the rotation of the rotor. And when the cavity moves to connect with the discharge port, the exhaust begins. Fig.4b shows the pressure cloud diagram when the angle of the male rotor changes to 110°, it can be found that the position on both sides of the radial exhaust orifice of the rotor which is far away from the central exhaust orifice exists a lot of over-compression. This will further increase the leakage at the exhaust. Fig.5 shows the pressure distribution on mid cross section at the180° of the male rotor. A small area of high pressure can be found in the middle of the rotor. This is due to the chamber has just been connected with the exhaust port, high pressure gas backflow into the chamber.

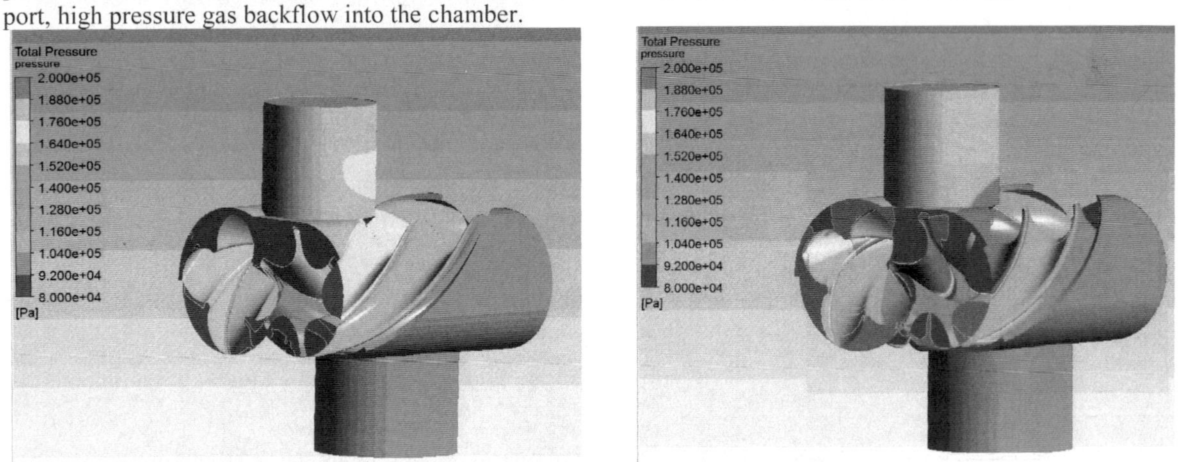

a) Φ =180° b) Φ =110°

Figure4: Pressure on casing wall and rotor surface

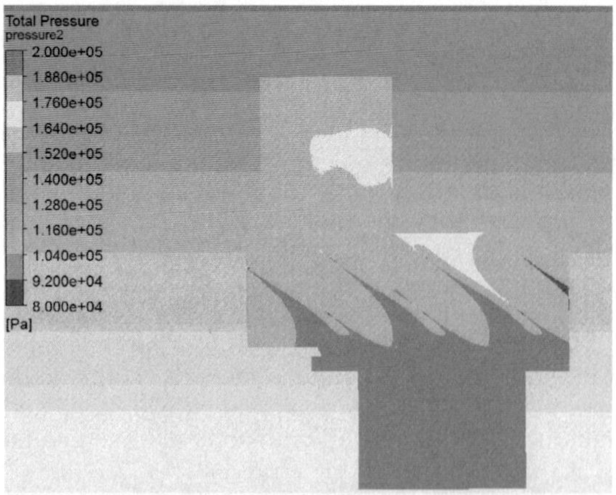

Figure5: Pressure on mid cross section

3.3 velocity field

Fig.6 shows the streamline diagram. It can be found that the velocity is very high at the tip clearance, the gear clearance and the discharge end clearance. Fig. 7 shows the velocity change of the tip clearance with different angles of the rotors at the cross section 0.1m from the exhaust end face. Fig.8 shows the vector diagram of the velocity on the central section. It can be found that the velocity vector of the tip clearance and the gear clearance is large. The maximum leakage velocity of tip clearance is 200m/s. The maximum leakage velocity of tooth tip clearance is 300m/s; The maximum leakage velocity of exhaust end clearance is 300m/s. Fig.7c shows the vortex when the right chamber is connected to the discharge ports. This is because the high-pressure chamber leaks to the low-pressure chamber through the clearance, causing the gas from the discharge ports to flow back into the high-pressure chamber. This is also the reason why the exhaust port in Fig. 4 has over-compression regions.

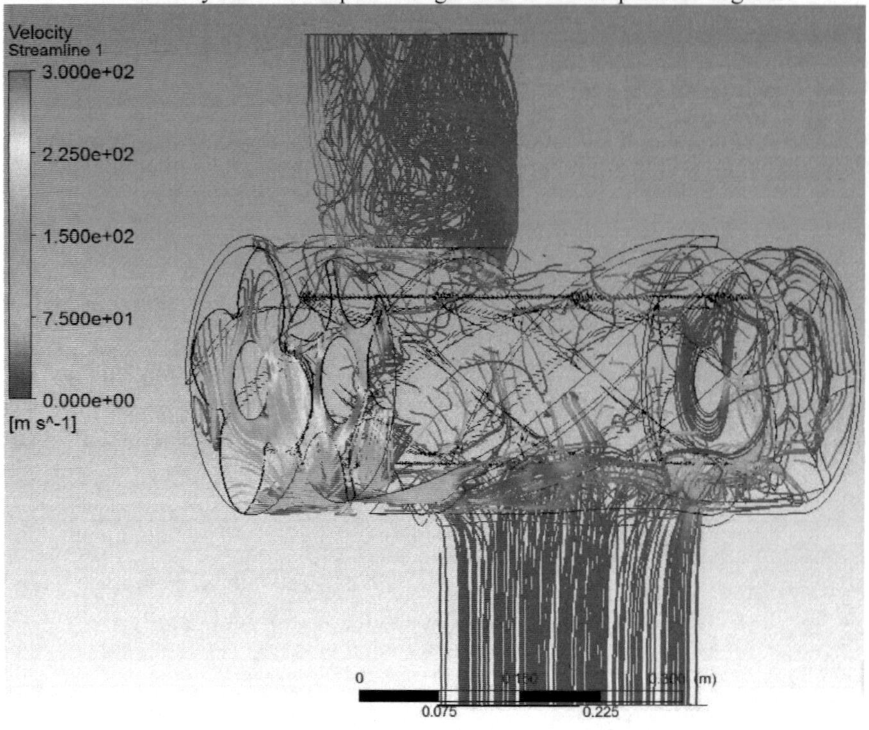

Figure6: VELOCITY flow diagram

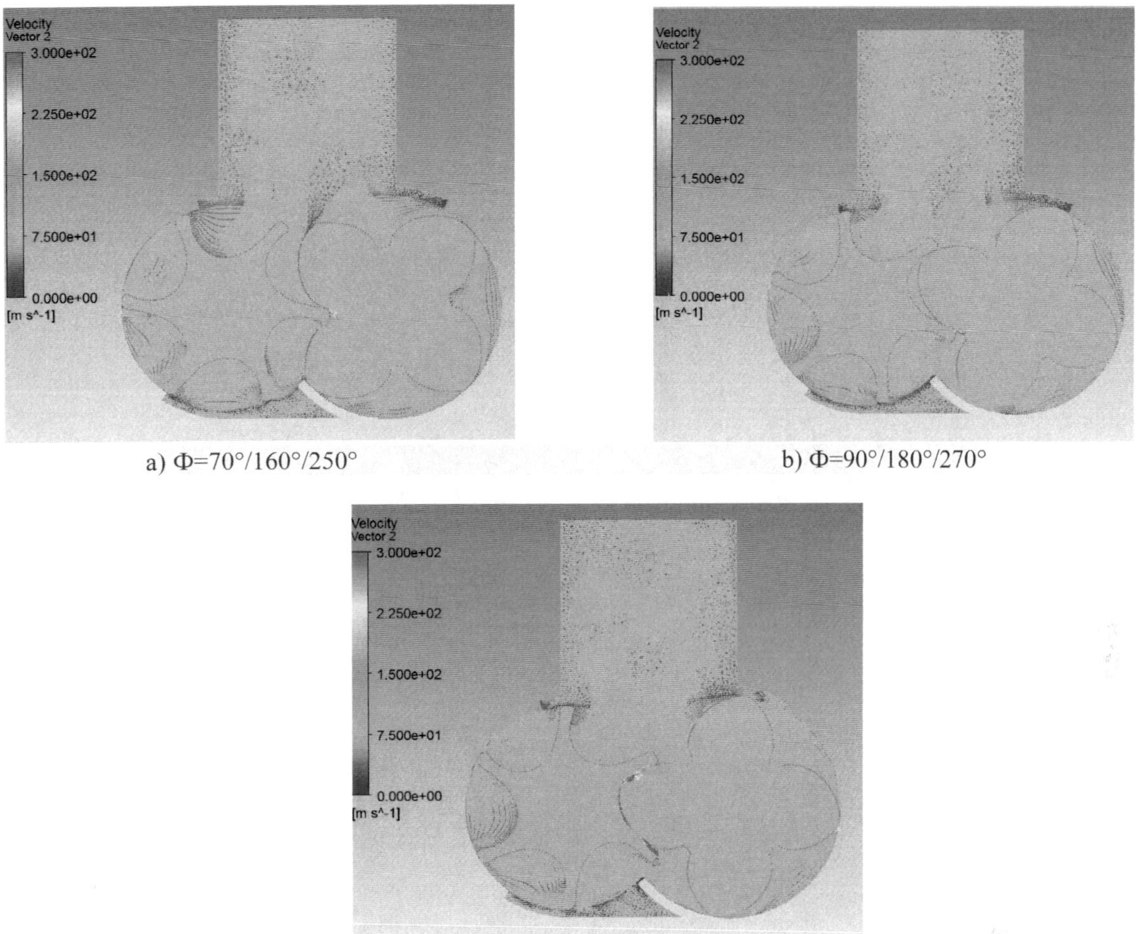

a) Φ=70°/160°/250°   b) Φ=90°/180°/270°

C) Φ=110°/200°/290°

Figure7: VELOCITY field on the cross section 0.1m from the exhaust end face with rotation angle

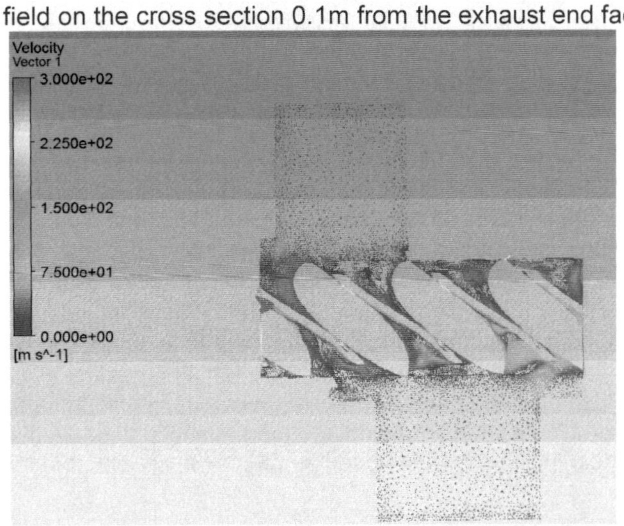

Figure8: VELOCITY field on mid cross section

3.4 rotor temperature field

The method of conjugate heat transfer was used to calculate the rotor force and heat flux on the walls of each time step. The actual time in CFD simulation is less than 1 second, but the direct heat transfer between solid and fluid in the rotor takes a long time, so the time scale of solid rotor is set as 1000 to accelerate the calculation of the actual temperature of the rotor. Finally, the temperature distribution and thermal deformation of the rotor surface are obtained. Fig.9 shows the rotor temperature distribution at the 180 angle of the male rotor. The rotor temperature at the exhaust end is326K, and that at the suction end is293K.

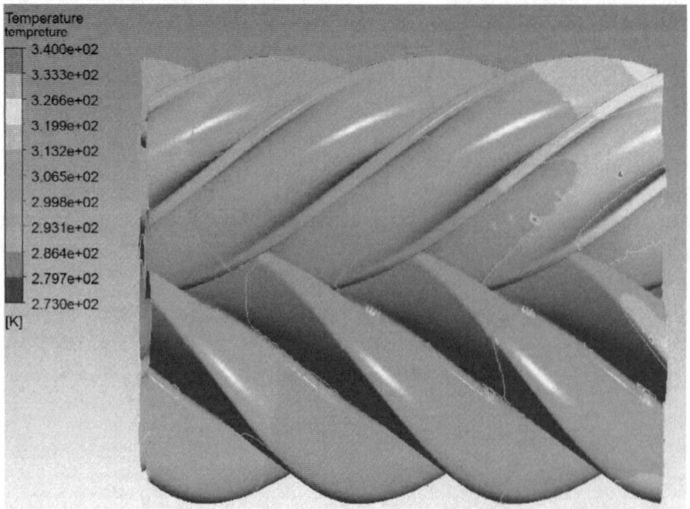

Figure9: Rotor temperature field at the 180 angle of the male rotor.

6. CONCLUSIONS

In this paper, CFD simulation was carried out for a working condition of the twin screw oil-free blower, in which the rotor has 4-6 teeth, the rotation speed is 6000rpm and the exhaust pressure is 1.8Bar. Twinmesh is used to generate a structured mesh that considers both small gaps and large chamber. The calculation method of CFX is SST model. The following results can be obtained from simulation results.

- Both the export flow and the import flow show a cyclical fluctuation of 90°, and the fluctuation amplitude of the export flow is much larger than that of the import flow. This is because the pressure of the exhaust orifice fluctuates greatly with the change of the angle. In addition, the simulated flow rate was 42.37m3/min, which was consistent with the experimental results and reflected the correctness of the simulation.
- The pressure field is stable on the whole, and the pressure is evenly distributed in the suction chamber and the compression chamber, but there are some over-compression areas at the exhaust end face and the engagement between the teeth. And when the compression chamber is just connected with the exhaust orifice, the pressure of the exhaust end face changes greatly. This is caused by the gas not being discharged in time. At the same time, these high pressure areas are also the main cause of pressure pulsation and noise.
- The velocity field should be affected by the pressure field. At the gap between the high pressure area and the low pressure area, there is a high speed backflow gas. The maximum leakage velocity of the exhaust clearance is 300m/s. The maximum leakage speed of the clearance between teeth is 300m/s. The maximum clearance leakage rate is 200m/s. When the compression chamber is just connected with the discharge ports, the maximum tooth tip leakage rate exists.
- The rotor temperature field is calculated with the fluid region by conjugate heat transfer. The highest temperature of the exhaust end face of the rotor is 326K, and the lowest temperature of the suction end face is 273K.

REFERENCES

[1] Arsie, I., Di Domenico, A., Pappalardo, L., Pianese, C., & Sorrentino, M. (2006, January). Steady-state analysis and energetic comparison of air compressors for PEM fuel cell systems. In ASME 2006 4th International Conference on Fuel Cell Science, Engineering and Technology (pp. 353-364). American Society of Mechanical Engineers Digital Collection..
[2] Fleming, J. S., & Tang, Y. (1995). The analysis of leakage in a twin screw compressor and its application to performance improvement. Proceedings of the Institution of Mechanical Engineers, Part E: Journal of Process Mechanical Engineering, 209(2), 125-136.
[3] Fong, Z. H., & Huang, F. C. (2006). Evaluating the interlobe clearance and determining the sizes and shapes of all the leakage paths for twin-screw vacuum pump. Proceedings of the Institution of Mechanical Engineers, Part C: Journal of Mechanical Engineering Science, 220(4), 499-506.
[4]He, Y., Xing, L., Zhang, Y., Zhang, J., Cao, F., & Xing, Z. (2018). Development and experimental investigation of an oil-free twin-screw air compressor for fuel cell systems. Applied Thermal Engineering, 145, 755-762.
[5] Hsieh, C. F., & Zhou, Q. J. (2015). Fluid analysis of cylindrical and screw type Roots vacuum pumps. Vacuum, 121, 274-282.
[6] He Xueming., Shi Guojiang., Wu Meiping., Zhang Rong1., & Ji Xiaogang. (2018). Exploring and Applying a New Method of Analyzing CFD of Twin Screw Compressor. Mechanical Science and Technology for Aerospace Engineering, 37(2), 211-219.
[7] Kovacevic, A., Rane, S., Stosic, N., Jiang, Y., Lowry, S., & Furmanczyk, M. (2014). Influence of approaches in CFD solvers on performance prediction in screw compressors.
[8] Sun, S. H., Kovacevic, A., Bruecker, C., Leto, A., Singh, G., & Ghavami, M. (2018, September). Numerical and experimental analysis of transient flow in Roots blower. In IOP Conference Series: Materials Science and Engineering (Vol. 425, No. 1, p. 012024). IOP Publishing.
[9] Spille-Kohoff, A., Hesse, J., & El Shorbagy, A. (2015). CFD simulation of a screw compressor including leakage flows and rotor heating. In IOP Conference Series: Materials Science and Engineering (Vol. 90, No. 1, p. 012009). IOP Publishing.
[10] Venturi, M., Sang, J., Knoop, A., & Hornburg, G. (2012). Air supply system for automotive fuel cell application (No. 2012-01-1225). SAE Technical Paper..
[11] Vimmr, J., & Ondrey, F. (2006). Numerical simulation of leakage flow between moving rotor and housing of screw compressor. In Proc. Conf. Modelowanyie Inzynierskie (Vol. 32, pp. 461-468).
[12] Yu, W., Sichuan, X., & Ni, H. (2015). Air compressors for fuel cell vehicles: An systematic review. SAE International Journal of Alternative Powertrains, 4(1), 115-122..

---

[1]

[2]

[3]

AUTHOR INDEX

AUTHOR INDEX

AUTHOR INDEX